REVIEWS in M
AND GEOCH

G000166202

Volume 83

Petrochronology:
Methods and Applications

EDITORS

Matthew J. Kohn
Boise State University, USA

Martin Engi & Pierre Lanari
University of Bern, Switzerland

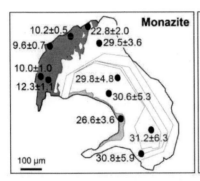
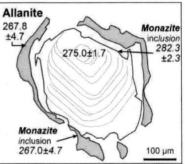

Cover image: Thorium compositional maps of monazite (X-ray counts, cps) and allanite (atoms per formula unit, apfu). *Left:* Monazite crystal from a Greater Himalayan Sequence orthogneiss, central Nepal. Ellipses show locations of ion probe Th–Pb analyses. Core shows oscillations, probably formed during igneous crystallization. An age of ca. 27 Ma likely reflects prograde metamorphic overprinting (23 Ma age straddles two chemical domains). An age of 10–11 Ma reflects hydrothermal replacement; based on Corrie and Kohn (2011). *Right:* Allanite from the Cima d'Asta pluton, southern Alps (NE Italy). Magmatic allanite core with oscillatory zoning (dated at 275.0 ± 1.7 Ma (2s) by quadrupole LA-ICP-MS) formed by breakdown of early-magmatic monazite, preserved as relics (282.3 ± 2.3 Ma). During hydrothermal alteration Th-rich allanite rim (267.8 ± 4.7 Ma) formed, and its partial breakdown produced a new generation of monazite (267.0 ± 4.7 Ma); based on Burn (2016), compare Chapter 12, this volume.

Series Editor: Ian Swainson

MINERALOGICAL SOCIETY of AMERICA
GEOCHEMICAL SOCIETY

DE GRUYTER

Reviews in Mineralogy and Geochemistry, Volume 83

Petrochronology: Methods and Applications

ISSN 1529-6466 (print)
ISSN 1943-2666 (online)
ISBN 978-0-939950-05-8

COPYRIGHT 2017

THE MINERALOGICAL SOCIETY OF AMERICA
3635 CONCORDE PARKWAY, SUITE 500
CHANTILLY, VIRGINIA, 20151-1125, U.S.A.
WWW.MINSOCAM.ORG
www.degruyter.com

Petrochonology:
Methods and Applications

83   *Reviews in Mineralogy and Geochemistry*   **83**

FROM THE SERIES EDITOR

It has been a pleasure working with the volume editors and authors on this 83rd volume of *Reviews in Mineralogy and Geochemistry*. Several chapters have associated supplemental figures and or tables that can be found at the MSA website. Any future errata will also be posted there.

Ian P. Swainson, Series Editor
Vienna, Austria

PREFACE

"Thy friendship makes us fresh"
Charles, King of France, Act III, Scene III
(Henry VI, Part 1, by William Shakespeare)

Friendship does indeed make us fresh—fresh in our enthusiasm, fresh in our creativity, and fresh in our collaborative potential. Indeed, it is the growing friendship between petrology and geochronology that has given rise to the new field of petrochronology. This, in turn, has opened a new array of methods to investigate the history of the geologic processes that are encoded (oh, so tantalizingly close!) in rocks, and to develop a broad new array of questions about those processes.

All friendships have their initiations and growth periods, and the origins and evolution of petrochronology are discussed in some detail in the Introduction. In brief, petrochronology has been practiced for many decades, but was first labeled in 1997. The seeds for this specific volume were planted in 2013, watered in 2015, and (we hope) will thrive as a resource through the coming decades. Indeed, it is hard to envision any future work involving the geochronology of igneous or metamorphic rocks in the context of tectonics and petrogenesis that is not somehow petrochronologic.

Our overall goal in this volume is to capture a high-resolution image of the state of petrochronology during its ascendance, not simply for historical purposes, but rather to provide a solid foundation for the future. We have striven to corral the very best practitioners in the field in hopes that their wisdom can help train new generations of petrochronologists, and inspire them to greater enthusiasm and more diverse research directions. The high quality of each chapter suggests that we might just have succeeded in this endeavor!

We thank all the authors for their immense investment of time and resources to pull off the writing of this book. Similarly, the reviewers worked overtime to temper the sometimes soft metal of each chapter (often on regrettably short notice from the editors…). Ian Swainson rapidly turned around our manuscripts, hardly giving us rest between submission of final versions and editing proofs. We appreciate his remarkable attention to detail and unflagging patience. We also thank our governmental, corporate, society, and university sponsors who helped support the accompanying short courses: the US National Science Foundation, Cameca & Nu Instruments, ESI, Selfrag, the European Association of Geochemistry, The European Geosciences Union, the Geochemical Society, Société Française de Minéralogie et Cristallographie, Boise State University, and the University of Bern. Last, but not least, we thank our families and close friends for somehow managing to put up with us over the long two years that it took to bring about this book.

Matthew J. Kohn, *Boise, Idaho, USA*
Martin Engi, *Bern, Switzerland*
Pierre Lanari, *Bern, Switzerland*
March 2017

1529-6466/17/0083-0000$00.00 (print)
1943-2666/17/0083-0000$00.00 (online)
http://dx.doi.org/10.2138/rmg.2017.83.0

Petrochronology

83   *Reviews in Mineralogy and Geochemistry*   **83**

TABLE OF CONTENTS

1   Significant Ages—An Introduction to Petrochronology

Martin Engi, Pierre Lanari, Matthew J. Kohn

2   Phase Relations, Reaction Sequences and Petrochronology

Chris Yakymchuk, Chris Clark, Richard W. White

3 Local Bulk Composition Effects on Metamorphic Mineral Assemblages
Pierre Lanari, Martin Engi

4 Diffusion: Obstacles and Opportunities in Petrochronology

Matthew J. Kohn, Sarah C. Penniston-Dorland

5 Electron Microprobe Petrochronology

Michael L. Williams, Michael J. Jercinovic, Kevin H. Mahan, Gregory Dumond

6   Petrochronology by Laser-Ablation Inductively Coupled Plasma Mass Spectrometry

Andrew R. C. Kylander-Clark

7   Secondary Ionization Mass Spectrometry Analysis in Petrochronology

Axel K. Schmitt, Jorge A. Vazquez

8   Petrochronology and TIMS

Blair Schoene, Ethan F. Baxter

9   Zircon: The Metamorphic Mineral

Daniela Rubatto

10 Petrochronology of Zircon and Baddeleyite in Igneous Rocks: Reconstructing Magmatic Processes at High Temporal Resolution

Urs Schaltegger, Joshua H.F.L. Davies

11 Hadean Zircon Petrochronology

T. Mark Harrison, Elizabeth A. Bell, Patrick Boehnke

12   Petrochronology Based on REE-Minerals: Monazite, Allanite, Xenotime, Apatite

Martin Engi

13 Titanite Petrochronology

Matthew J. Kohn

14   Petrology and Geochronology of Rutile

Thomas Zack, Ellen Kooijman

15   Garnet: A Rock-Forming Mineral Petrochronometer

E.F. Baxter, M.J. Caddick, B. Dragovic

16   Chronometry and Speedometry of Magmatic Processes using Chemical Diffusion in Olivine, Plagioclase and Pyroxenes

Ralf Dohmen, Kathrin Faak, Jon D. Blundy

RiMG Series

Reviews in Mineralogy & Geochemistry
Vol. 83 pp. 1-12, 2017
Copyright © Mineralogical Society of America

Significant Ages—An Introduction to Petrochronology

Martin Engi and Pierre Lanari

Institute of Geological Sciences
University of Bern
Baltzerstrasse 3
3012 Bern
Switzerland

engi@geo.unibe.ch

pierre.lanari@geo.unibe.ch

Matthew J. Kohn

Department of Geosciences
Boise State University
Boise, Idaho, 83725
U.S.A.

mattkohn@boisestate.edu

INTRODUCTION AND SCOPE

Question: Why "Petrochronology"? Why add another term to an already cluttered scientific lexicon?

Answer: Because petrologists and geochronologists need a term that describes the unique, distinctive way in which they apply geochronology to the study of igneous and metamorphic processes. Other terms just won't do.

Such evolution of language is natural and well-established. For instance, "Geochronology" was originally coined during the waning stages of the great Age-of-the-Earth debate as a means of distinguishing timescales relevant to Earth processes from timescales relevant to humans (Williams 1893). Eighty-eight years later, Berger and York (1981) coined the term "Thermochronology," which has evolved as a branch of geochronology aimed at constraining thermal histories of rocks, where (typically) the thermally activated diffusive loss of a radiogenic daughter governs the ages we measure. Thermochronology may now be distinguished from "plain vanilla" geochronology, whose limited purpose, in the words of Reiners et al. (2005), is "…exclusively to determine a singular absolute stratigraphic or magmatic [or metamorphic] formation age, with little concern for durations or rates of processes" that give rise to these rocks.

Neither of these terms describes what petrologists do with chronologic data. A single date is virtually useless in understanding the protracted history of magma crystallization or metamorphic pressure–temperature evolution. And we are not simply interested in thermal histories, but in chemical and baric evolution as well. Rather, we petrologists and geochronologists strive to understand rock-forming processes, and the rates at which they occur, by integrating numerous ages into the petrologic evolution of a rock. It is within this context that a new discipline, termed "Petrochronology", has emerged[1]. In some sense petrochronology may be considered the sister of thermochronology: petrochronology typically focuses on the processes leading up

---

[1] Several parts of this introduction are taken from a discussion that took place in the forum GEO-METAMORPHISM in June, 2013.

1529-6466/17/0083-0001$05.00 (print)
1943-2666/17/0083-0001$05.00 (online)   http://dx.doi.org/10.2138/rmg.2017.83.1

to the formation of igneous and metamorphic rocks—the minerals and textures we observe and the processes that formed them—whereas thermochronology emphasizes cooling processes in the wake of igneous, metamorphic, and tectonic events. Typically petrochronology is "hot", thermochronology is "cold". While each field has its unique features, and while their disciplinary boundaries overlap, each complements the other.

Any rock sample we study, whether igneous, sedimentary, or metamorphic, results from the transformation of one or more previous rocks. Petrologists and geochemists have found that such transformation rarely erases a rock's memory completely, instead most samples contain relics from more than one stage of their evolution. Whether and how these affect an age determination is essentially a question of resolution—both spatial and chronometric—i.e., of isotopic and chemical analysis. Analytical efforts in petrochronology typically find that several stages or generations of mineral formation are evident in any single rock sample, in which case we conclude that such a rock does not have, *sensu stricto*, one age.

In fact, one is led to wonder what the term "age" may signify in everyday geologic usage. It might seem clear what is meant, for example, by the age of a basaltic lava flow: the time of deposition or solidification. But what is the age of a meta-basalt? Does it refer to the point on the prograde path when its mineralogy and texture would define it as "metamorphic", and no longer igneous? The pressure peak? The maximum temperature? The point on the retrograde path where mineralogy and chemistry no longer change measurably? And by what methods can that singular metamorphic age be measured? Actually, defining "an" age of a volcanic rock presents its own problems. How do we choose among the ages of initial melting, magma movement or rejuvenation, crystallization of antecrysts and duration of residence in a magma chamber, eruption, or solidification? And what does an age mean for a clastic rock, where each grain may have a slightly different parent, and materials may be reworked. The concept of "an age" really makes sense only within a defined petrogenetic context.

This recognition leads us to a practical definition: Petrochronology is the branch of Earth science that is based on the study of rock samples and that links time (i.e., ages or duration) with specific rock-forming processes and their physical conditions. Petrochronology is founded in petrology and geochemistry, which define a petrogenetic context or delimit a specific process, to which chronometric data are then linked.

SIGNIFICANCE OF AGE DATA

Chronometers essentially rely on the behavior of specific elements, more specifically of ions or isotopes, in a mineral (or group of minerals) chosen for dating. To assess chronometric data, two criteria need to be combined: transport properties (diffusivity), and analytical quality. Diffusion of the elements relevant to the chronometric system in the chosen mineral, say Pb in zircon or Ar in feldspar, sets basic limits on how a chronometer can be applied and what meaning the age data have: Where diffusion is relatively fast, the age of a mineral refers to the time when radiogenic daughter material started to accumulate in that mineral. Such systems are used for *thermochronology*, and the resulting *cooling age* specifies the amount of time that mineral has remained below a particular temperature, called the *closure temperature* (T_c). This age will correspond to the time of formation only if that mineral crystallized (and remained) at temperatures below T_c or was somehow shielded from diffusive loss of the daughter isotope. By contrast, in systems where diffusion is extremely slow, commonly for the U–Pb, Th–Pb, Lu–Hf, or Sm–Nd decay systems, minerals remain closed up to high temperature, and radio-isotopic dating commonly returns *formation ages*, i.e., the time of crystallization. For igneous rocks, this may reflect a pulse of crystallization, catalyzed by transport or degassing events or possibly related to magma recharge. For metamorphic rocks, it may represent a particular mineral reaction, an event of fluid infiltration, or deformation-induced recrystallization.

Typically, spatially resolved age data are combined with compositional data of accessory and/or major phases to recover specific *P–T–t* or *T–t* points and paths. In many cases, data for individual growth zones of minerals are required and/or several chronometers are combined to secure the interpretation. Although we commonly focus on collecting ages with high spatial and chronologic resolution, it is not actually the age that matters most to geologic investigations but rather the process itself and our ability to characterize rates.

Chronologic quality is a delicate subject in petrochronology because analytical accuracy and the behavior of elements involved in the system used are both important, and element behavior is difficult to quantify. Traditionally the quality of an age determination, say for garnet using the Sm–Nd system, focused on the MSWD or statistical likelihood that all data conform to a single crystallization event. Today, in part through vastly improved analytical capabilities, we recognize that protracted growth of garnet should *rarely* produce a good MSWD (Kohn 2009). Our ability to measure age differences among different growth zones means that data scatter *should* nearly always exceed analytical uncertainties. While this failure of traditional tests of chronologic significance might undermine some past investigations (there is no single age that we can assign to such a garnet), it opens a vast new spectrum of scientific inquiry related to the growth dynamics of minerals, and consequently the processes that drive mineral growth. So, instead of asking what the age of a pluton is, we might instead ask how rapid is magma transport, what catalyzes degassing and consequent melt crystallization, and how these are related to magma recharge. Similarly, in metamorphic petrology and tectonics, we can now ask how rapid was heating/cooling or burial/exhumation, how are these related to thrusting or extensional processes, and what does this history imply about large-scale deformation of the crust? Addressing these types of questions means that a mineral age must be linked to some other petrogenetic indicator, whether it be other associated minerals of petrogenetic or textural significance, major element chemistry, trace element patterns, or independently estimated temperature or pressure.

PETRO-CHRONO-LOGICAL APPROACH AND AMBITION

A basic and generic petrochronological approach may be formulated in five steps:

1. Identify one or more specific stage(s) of the metamorphic, magmatic and/or structural evolution in a given sample based on textural criteria such as overgrowths, inclusions, fabric alignments, etc. In deformed rocks, it is helpful and often necessary to work with oriented samples and sections, such that fabric characteristics of local assemblages can be related to observed fabrics in hand specimens and to meso- to megascopic observations.

2. Document the phase relations among minerals based on images and compositions as determined using electron beam methods (scanning electron microscopy: SEM; electron probe micro-analysis: EPMA), laser ablation–inductively coupled plasma–mass spectrometry (LA-ICP-MS) and/or secondary ion mass spectrometry (SIMS) analysis.

3. Attempt to relate one or more specific growth zones of a suitable robust chronometer (e.g., U–Pb, Th–Pb, Lu–Hf, Sm–Nd) to each stage. This step often requires support from trace element data to verify coexistence.

4. Use thermobarometric techniques such as multi-equilibrium, isochemical phase diagrams[2], Thermoba-Raman-try, and empirical thermobarometry to constrain the Pressure–temperature (*P–T*) conditions of local equilibria.

5. If steps 1–4 were successful, use a microdating technique to analyze the isotopic ratios in each suitably large growth zone. This can be done by *in situ* analysis or microdrilling for ID-TIMS.

---

[2] Also known as "pseudosections", but this term may be misunderstood. Spear FS, Pattison DRM, Cheney JT (2016) The metamorphosis of metamorphic petrology. Geological Society of America Special Papers 523 doi:10.1130/2016.2523(02)

Several research groups have applied this type of approach. Ambitious applications have attempted to reconstruct pressure–temperature–deformation-time (*P–T–D–t*) paths in units that experienced a complex tectono-metamorphic evolution. As an example of the potential of petrochronology, it is instructive to summarize results obtained for a specific area, the Dora Maira terrane in the Western Italian Alps. A variety of techniques have helped elucidate its evolution, with implications for other fields, from tectonic modeling to experimental petrology.

An example: *P–T–t* path for geodynamic and tectonic modeling

Prescient researchers realized long ago that *P–T–t* paths of rocks from different structural positions in an orogenic belt can constrain models of large-scale tectonic processes (Thompson and England 1984; Shi and Wang 1987). Theoretical *P–T–t* paths from numerical models started appearing before they could be reconstructed from rock samples. Numerical models were first restricted to one-dimension problems (Oxburgh and Turcotte 1974; England and Thompson 1984), but they rapidly evolved to complex geometry in two-dimensions (Peacock 1990; Ruppel and Hodges 1994). At the dawn of the 21st century, dynamic thermo-mechanical models were developed and intensively used to predict theoretical *P–T–D–t* trajectories of rock units involved in various scenarios of subduction and collision zones (e.g., Beaumont et al. 2001; Gerya et al. 2002; van Keken et al. 2002, and many others inpired by these). While much progress has been made since then in investigating geodynamic processes by means of numerical models (Gerya 2011 and references therein), stark differences occur between many model *P–T–t* trajectories and the natural record, especially for continental collisions (Kohn 2008) and for prograde subduction paths at pressures below 2 GPa (Penniston-Dorland et al. 2015).

An excellent example of how *P–T–D–t* paths may be determined from the study of specific samples comes for the ultra-high pressure (UHP) slice of Dora-Maira located in the Western Alps of Italy. This massif is renowned as one of two localities where coesite inclusions were first discovered in continental rocks, specifically in garnet porphyroblasts (Chopin 1984). As shown in Figure 1, this terrain now sports one of the best constrained *P–T–D–t* paths for a UHP continental fragment in the world. Detailed field studies of the Dora-Maira UHP slice document it as a coherent piece of continental crust (1 km thick, exposed over 10 by 5 km), composed of granitic gneiss with intercalated (K- or Na-bearing) whiteschists, metabasites (calcsilicate eclogites and eclogite boudins), and marbles (Vialon 1966; Chopin et al. 1991). Several studies have demonstrated that this tectonic slice was subducted to depths greater than ~100 km during the early stages of Alpine collision (e.g., Chopin et al. 1991; Chopin and Schertl 1999; Rubatto and Hermann 2001; Hermann 2003). *P–T* conditions based on several prograde and retrograde reactions and local assemblages were derived using thermodynamic modeling of phase equilibria for various rock types (Fig. 1). In addition, Hermann (2003) compared the observed phase assemblages and compositions in K-bearing whiteschists with experimentally determined petrogenetic grids and calibrated fluid-absent equilibria, which reduced uncertainties in *P–T* conditions. Dating of accessory minerals included zircon, monazite, and rutile, with results delimiting a restricted time span of 3.2 Ma for the UHP event (e.g., Gebauer et al. 1997; Rubatto and Hermann 2001; Gauthiez-Putallaz et al. 2016). The vertical burial and exhumation rates for Dora Maira are >3.5 to 4 cm/yr, much faster than the subduction rate expected from the horizontal velocity of 1.5 cm/yr inferred by plate motion paths (Handy et al. 2010). The tightly constrained petrochronologic datasets for this terrane have triggered the curiosity and imagination of modelers who are keen on understanding geodynamic processes. So far, most of the available thermo-mechanical models predict both lower peak temperatures and much lower burial and exhumation rates (e.g., Fig. 7 in Yamato et al. 2008). The lower temperatures may reflect the neglect of frictional heating or heat advection from fluids in the numerical experiments (Yamato et al. 2008; Penniston-Dorland et al. 2015).

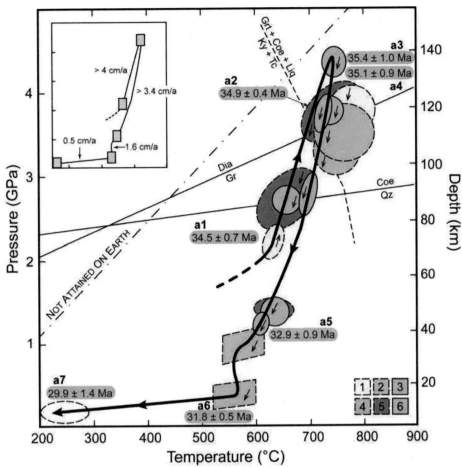

Figure 1. Synthetic *P–T–t* path for the UHP unit of Dora Maira. Mineral reactions among quartz, coesite, graphite and diamond were calculated using the thermodynamic database of Berman (1988). The reaction Ky + Tc → Grt + Coe + Liq is from Gauthiez-Putallaz et al. (2016). *P–T* estimates for natural mineral assemblages refer to the following lithologies: [1] K-bearing whiteschist (Schertl et al. 1991; Gauthiez-Putallaz et al. 2016); [2] Na-bearing whiteschist (Compagnoni et al. 1995); [3] marbles (Castelli et al. 2007); [4] calcsilicate eclogites (Rubatto and Hermann 2001); [5] eclogite boudins (Groppo et al. 2007). Experimental data [6] are from samples of K-bearing whiteschist (Hermann 2003). Select ages **a1** and **a2** are from Gauthiez-Putallaz et al. (2016), **a3** from Gebauer et al. (1997) and **a4, a5, a6, a7** from Rubatto and Hermann (2001). Note that prograde ages (**a1, a2**), while nominally younger than the pressure peak (**a3, a4**), are not distinguishable within their 2σ-uncertainties. These ages delimit a maximal time span of 0.7 Ma for burial of the Dora Maira UHP massif from 2.2–2.5 GPa to 3.5 GPa. **Inset:** burial and exhumation rates from Rubatto and Hermann (2001) and Gauthiez-Putallaz et al. (2016). Abbreviations: Coe: coesite; Dia: diamond; Gr: graphite; Grt: garnet; Ky: kyanite; Liq: silicate melt; Qz: quartz; Tc: talc.

Some models can reproduce the Alpine subduction *P–T* trajectories (Gerya et al. 2008; Butler et al. 2014), but only for such slow subduction (which induces higher temperatures) that the duration of the HP–UHP stages is overestimated by a factor of up to 2.5 (Fig. 2). This large discrepancy has promoted alternative proposals that the pressure recorded by the assemblages may exceed lithostatic pressure, and the observed (U)HP parageneses may have formed at shallower depths (Gerya et al. 2008; Reuber et al. 2016). Such proposals might reconcile durations of metamorphism, but would imply that slab-top geotherms in

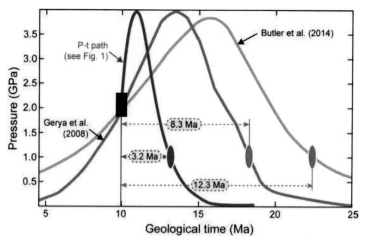

Figure 2. Comparison between the synthetic pressure–time path of Figure 1 (dark) and the representative *P–T* paths (light) from model V1.5 (Butler et al. 2014) and model 1 (Gerya et al. 2008) for the Dora Maira UHP massif. These two thermal–mechanical models were selected because they show the highest burial and exhumation rates. The duration of the HP–UHP stages suggested by the petrochronological data is shown between the first recorded prograde stage ($P \geq 2\,GPa$) to the exhumation to crustal levels ($P < 1\,GPa$). The mismatch between durations as recorded by rocks (brief, ~3 Ma) vs. models (long, >8 Ma) implies that natural subduction occurred faster than models are capable of explaining, i.e., rocks can be faster than models.

subduction zones are even hotter than rocks indicate, because the metamorphic temperature recorded by a rock would occur at a shallower depth than the apparent pressure implies. If tectonic overpressures were high, this would exacerbate the already stark discrepancy in *P–T* conditions between samples (thermobarometry) and models: rocks record generally hotter conditions, models predict generally colder conditions (Penniston-Dorland et al. 2015). So far, the models that favor overpressure by introducing a mechanically heterogeneous crust have not succeeded in reproducing the *P–T–t* path of the Dora Maira unit. Nor has thermobarometry in this terrane discovered significant pressure-discrepancies, such as one would expect from rock types with such different rheologies as those investigated. Indeed, the range of pressures documented from weak and strong lithologies indicates no significant over- and underpressures, thus thermobarometry does not appear to have registered appreciable deviations from lithostatic pressure. Clearly, alternative geodynamic processes are needed to understand continental subduction (see Guillot et al. 2009 for a review).

Methods of choice, choice of methods

The three sections of this volume present the state of the art and progress regarding

• BASICS: conceptual approaches used in petrochronology (3 chapters);

• METHODS: developments of analytical techniques (4 chapters);

• MINERALS: specific potential and use of mineral groups (8 chapters).

Numerous approaches may be used, so there is no best and certainly no unique approach one could recommend for any specific research ambition or focus. Consequently, current and future studies must make choices. Ideally, petrochronologists will be able to select tools in combination, and in practice much will depend on the specific expertise of the investigators and the available hardware in analytical labs, etc.

However, we must emphasize that the likelihood of obtaining convincing results depends on nothing so much as finding the "right" samples. All of the approaches presented in this book are time-intensive, and they promise success only when applied to suitable samples. It is seldom evident in the field whether a sample will have the right combination of promising compositions and textures. Experience shows that while many samples may interview for the job of petrochronologic investigation, few are worth hiring, so extensive sampling is recommended. The best machines in the world cannot replace sample selection and detailed documentation. Critical requirements include good structural control in the field, careful microscopy and then accurate chemical / isotopic analysis.

A brief walk through the present volume will expose some of the developments in petrochronology and the main principles and purposes behind them.

Thermodynamic modeling of mineral assemblages and compositions (Chapters 2–3). Quantifying conditions of rock formation is an essential requirement of petrochronology. This task has become quite manageable in metamorphic rocks because thermodynamic models and software based on these are readily available. It is now standard procedure to compute phase diagrams for the approximate bulk composition of a sample under investigation, which allows multi-equilibrium thermobarometry, supports the interpretation of microscopically visible phase relations and reaction textures, all useful to design strategies for *in situ* age dating (Yakymchuk et al. 2017). This approach has proven very helpful, but phase diagrams depict equilibrium relations, and many rock samples contain disequilibrium features (e.g., zoned minerals). Lanari and Engi (2017) emphasize the importance of choosing the appropriate bulk composition if the approach used is based on equilibrium phase relations. That chapter presents successful methods, based on X-ray maps, to accommodate compositional heterogeneity, such as arise where porphyroblasts grow or partial melting occurs. However, the interpretation of ages measured on minerals that grew out of equilibrium, e.g., due to reaction overstepping (Spear et al. 2014), remains a concern.

The fundamentals of modeling igneous rocks are not specifically covered in this volume, but the most prevalent models of melting and solidification are MELTS (Ghiorso and Sack 1995), PETROLOG3 (Danyushevsky and Plechov 2011), and MELTS derivatives (Ghiorso et al. 2002; Gualda et al. 2012). While these models appear to work reasonably well for mafic and some rhyolitic compositions, thermodynamic models for compositionally intermediate rocks are lacking, in part because adequate mixing models for amphibole have not yet been assembled. Peraluminous rocks also pose modeling challenges. Still, MELTS and PETROLOG3 provide good starting points, at least for mafic rocks. When combined with accessory mineral solubilities (e.g., Watson and Harrison 1983; Montel 1993; Stepanov et al. 2012; Boehnke et al. 2013), the growth of monazite and zircon can also be modeled.

Diffusion can set limits and open new doors (Chapter 4) As discussed in Kohn and Penniston-Dorland (2017), diffusion can homogenize pre-existing chemical heterogeneities, biasing certain types of petrologic calculations, but can also induce chemical zoning that may be useful for determining rates of geologic processes. One aspect is that diffusivity limits the temperature range in which specific thermometers and chronometers can be trusted, a major worry in some types of chronology (e.g., $^{40}Ar/^{39}Ar$ dating). However, quite a different aspect is the door that diffusion opens for delimiting the duration of thermal events by modeling the time-dependent flux of components in response to composition gradients. Diffusion chronometry also is a central topic in Chapter 16 (Dohmen et al. 2017).

Analytical methods for mineral chronometry (Chapter 5–8). Petrochronological work has continually stretched and exploited the range of analytical techniques available. Spatial resolution is a concern to itself, but linking chronometric data with geochemical analyses remains a key challenge. Consequently, much effort has focused on obtaining both types of data using a single instrument, and many chapters emphasize not only the advantages but also the disadvantages of each technique. Nearly all petrologic studies rely on the electron probe microanalyzer (EPMA) to characterize major and minor element chemistry of minerals, which is required for modeling (Chapters 1–2; Yakymchuk et al. 2017; Lanari and Engi 2017).

From the perspective of combined chronologic and compositional analysis:

•*EPMA* (Chapter 5, Williams and Jercinovic 2017) offers unparalleled spatial resolution for polished sections and grain mounts (ca. 1 μm), but has relatively poor compositional resolution (typically a few tens of ppm) and age resolution, in part because it cannot distinguish isotopes.

•*Laser-ablation inductively-coupled plasma mass spectrometry* (LA-ICP-MS; Chapter 6, Kylander-Clark 2017) and *secondary ion mass spectrometry* (SIMS or ion microprobe; Chapter 7, Schmitt and Vazquez 2017) offer superior compositional resolution (sub-ppm), but with only moderate resolution of ages (a few percent, absolute) and position (typically > 10 μm). However, depth profiling into crystal faces can permit sub-μm spatial resolution (e.g., Grove and Harrison 1999; Carson et al. 2002; Cottle et al. 2009; Kohn and Corrie 2011; Smye and Stöckli 2014).

•*Thermal ionization mass spectrometry* (TIMS; Chapter 8, Schoene and Baxter 2017) offers the highest chronologic precision and accuracy, but the poorest spatial resolution (whole grains or partial grains). Nonetheless, zoning in large crystals, such as garnet, can provide direct estimates of mineral growth rates that can be linked directly to tectonic processes.

Preferred clocks and their role in petrochronology (Chapters 9–16). Accessory minerals have dominated much of classical chronometry, whereas major rock-forming minerals (and reactions among these) have played key roles in determining petrogenetic conditions. Over the past decade, the boundaries between these two camps have gradually crumbled, and combined thermometry, trace element fingerprinting, and chronometry are based increasingly on one and the same mineral grain (or subgrain). From the perspective of published chronologic utility:

•*Zircon* (Chapter 9, Rubatto 2017; Chapter 10, Schaltegger and Davies 2017; Chapter 11, Harrison et al. 2017) remains the most popular chronometer because of its high Pb retentivity (Cherniak and Watson 2001), preservation of multiple growth zones (Corfu et al. 2003), resistance to abrasion during transport as a clastic grain, and use of trace elements for thermometry (Watson et al. 2006) as well as chemical fingerprinting (Rubatto 2002; Whitehouse and Platt 2003). Zircon petrochronology from all three basic rock types is discussed in this volume. Metamorphic zircon requires *in situ* analysis by LA-ICP-MS or SIMS because of its complex zoning but sub-wt% U contents. It provides a key monitor of *P–T–t* paths and fluid processes (Rubatto 2017), even if it remains difficult to pinpoint precisely the conditions, especially of pressure, at which zircon formed. Igneous zircon records magmatic processes (Schaltegger and Davies 2017), where short timescales commonly necessitate high-precision TIMS analysis (Schoene and Baxter 2017). Detrital zircon preserves chemical, isotopic, mineralogical (inclusion) and chronologic records of past Earth processes; Harrison et al. (2017) illustrate the power and impact of this information most comprehensively through their review of Hadean zircons, which are commonly analyzed using SIMS or LA-ICP-MS.

•*REE minerals* (Chapter 12, Engi 2017), especially *monazite and allanite*, are increasingly popular targets of petrochronology, owing to their high Th and U contents, low cation diffusion rates (high Pb retentivity; e.g., Cherniak et al. 2004), and, for monazite, low initial Pb (Parrish 1990). Multiple chemically distinct domains in each grain require *in situ* analysis, and can in favorable cases be tied to specific mineral reactions and hence to *P–T* conditions. While monazite can be dated using EPMA, LA-ICP-MS, and SIMS, allanite's high common Pb content requires isotopic methods of analysis, typically LA-ICP-MS and SIMS because of intra-crystalline zoning. *Xenotime* and *apatite* have received less scrutiny, and Engi (2017) discusses how more research could explore better their petrochronologic value.

•*Titanite* (Chapter 13, Kohn 2017) has long been analyzed for U–Pb ages, but until recently its high closure temperature for Pb and other cations was underappreciated, and ages were assumed to reflect cooling (see Frost et al. 2000). That is, geochronologists linked titanite's utility to thermochronology rather than petrochronology. Like allanite, multiple domains with relatively high common Pb require *in situ* isotopic analysis via LA-ICP-MS or SIMS, and, like zircon, a direct thermometer—the Zr-in-titanite thermometer (Hayden et al. 2008)—links temperature to age. Examples of applications commonly focus on *P–T–t* paths, but new research is developing its use in understanding igneous petrogenesis.

•*Rutile* (Chapter 14, Zack and Kooijman 2017) is not so commonly analyzed as other accessory minerals because it contains virtually no Th, and U contents are typically < 10 ppm. These disadvantages are offset by extremely low common Pb contents, such that high-U rutile can be analyzed *in situ* using LA-ICP-MS or SIMS. Zack and Kooijman (2017) emphasize rutile's utility in understanding crystallization conditions for low-temperature metamorphic rocks, early stage cooling for high-temperature metamorphic rocks, and rutile's ability to recover temperatures directly through Zr-in-rutile thermometry. Textural context can be of paramount importance, as it is for most petrochronology involving accessory minerals.

•*Garnet* (Chapter 15, Baxter et al. 2017) is, of course, a major mineral in many rocks, and it is presented in all its glory for its ability to constrain petrogenetic processes through combined major and minor element zoning plus Lu–Hf and Nd–Sm isotopic dating. Generally low parent-daughter ratios and extremely low decay rates require TIMS analysis for accurate dating. While inclusions in garnet commonly can be very helpful in relating garnet growth to specific *P–T* stages or deformation fabrics, certain inclusions jeopardize garnet dating. These challenges are reviewed by Baxter et al (2017), as are the techniques used to minimize the effects on ages.

•*Major igneous minerals, including olivine, plagioclase and pyroxenes* (Chapter 16, Dohmen et al. 2017), round out this volume and provide useful petrochronologic constraints. Although U-series dating can be applied (with care) to igneous systems, much emphasis is placed on diffusion profiles. These allow the duration and rates of various processes to be estimated, e.g., cooling rates for volcanic bodies and magma chamber time scales, such as crystal growth rates and magma residence time.

EVOLUTION OR REVOLUTION?

The concept of petrochronology was originally proposed (Fraser et al. 1997) to unify the fields of metamorphic petrology and geochronology. These authors demonstrated that the breakdown of a major phase such as garnet or amphibole can trigger zircon growth, and thus geochronological data can be linked to petrologically derived pressure–temperature conditions, helping piece together a *P–T*–time path.

Previously, we outlined linkages among different branches of science that contribute to the field of petrochronology. As a snapshot of the evolution thus far, we compare how the evolution in each field or technique is reflected in citations over the past years (Fig. 3). For the fields of "Petrology" and "Geochronology + Petrology" the citation frequencies increased linearly from 1970 to ~2002, then growth rates doubled; "Petrochronology" was a term barely cited ten years ago, taking off only after 2010—but look at the rate! Citations of "Thermobarometry" seem to have been growing linearly since the early 80s, as have references to petrological software, i.e., Thermocalc (since the late 80s), "PerpleX", and "Theriak-Domino" (both since ~2005). However, exponential growth is evident for the chronometric techniques. Citations of SIMS (including SHRIMP) lead the pack, but LA-ICP-MS is catching up.

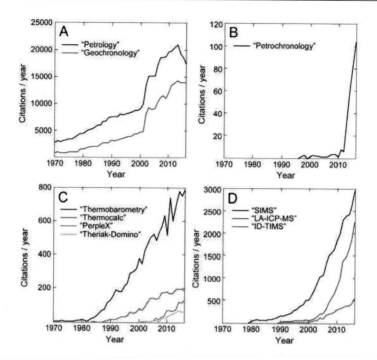

Figure 3. Citations per year, found in Google Scholar (February 2017), using the following criteria, **A:** "Petrology", "Geochronology + Geochemistry"; **B:** "Petrochronology"; **C:** "Thermobarometry", "Thermocalc", "Perple_X", "Theriak-Domino"; **D:** "SIMS or SHRIMP + Geochronology + Petrology", "LA-ICPMS or LA-ICP-MS + Geochronology + Petrology", "ID-TIMS + Geochronology + Petrology".

The field of petrochronology is evidently young, yet spectacular progress has been made already (Kohn 2016), as evident in the literature and reviewed in this volume. Depending on the scope of applications, studies cover a wide range of scales, from microns to mountain belts. Our understanding has thus deepened, from small-scale processes in the evolution of magmas and rocks, their duration and tempo, all the way to the scale of orogeny, lithospheric tectonics, the nature of Early Earth, and the dynamics of continental recycling.

For these reasons, the time to dedicate a RiMG volume to petrochronology is ripe, for the field is burgeoning, combining many diverse directions, aiming to elucidate time and tempo of processes that have shaped the Earth (and sister planetary bodies) at all scales. We trust that this RiMG volume, compiling current petrochronological methods and reviewing applications, will stimulate and nourish future work in this field. It is with this intent that the authors of each chapter have outlined current limits and identify potentially useful directions for study.

ACKNOWLEDGMENTS

It is a pleasure to thank Ian Swainson, the Editor of the RiMG Series. He has provided expedient and unfailing support in producing this volume. We also thank Frank Spear for his perceptive comments that helped us improve this Introduction. This work was funded by US-NSF grants EAR-1321897, -1419865, and -1545903 to MJK and by the Swiss Nationalfonds grants 200021-117996/1, 200020-126946, and -146175 to ME.

REFERENCES

Baxter EF, Caddick MJ, Dragovic B (2017) Garnet: A rock-forming mineral petrochronometer. Rev Mineral Geochem 83:469–533

Beaumont C, Jamieson RA, Nguyen M, Lee B (2001) Himalayan tectonics explained by extrusion of a low-viscosity crustal channel coupled to focused surface denudation. Nature 414:738–742

Berger GW, York D (1981) $^{40}Ar/^{39}Ar$ dating of the Thanet gabbro, Ontario: looking through the Grenvillian metamorphic veil and implications for paleomagnetism. Can J Earth Sci 18:266–273

Berman RG (1988) Internally consistent thermodynamic data for minerals in the system $Na_2O–K_2O–CaO–MgO–FeO–Fe_2O_3–Al_2O_3–SiO_2–TiO_2–H_2O–CO_2$. J Petrol 29:445–522

Butler JP, Beaumont C, Jamieson RA (2014) The Alps 2: Controls on crustal subduction and (ultra)high-pressure rock exhumation in Alpine-type orogens. J Geophys Res: Solid Earth 119:5987–6022 doi:10.1002/2013JB010799

Castelli D, Rolfo F, Groppo C, Compagnoni R (2007) Impure marbles from the UHP Brossasco–Isasca Unit (Dora–Maira Massif, western Alps): evidence for Alpine equilibration in the diamond stability field and evaluation of the $X(CO_2)$ fluid evolution. J Metamorph Geol 25:587–603

Chopin C (1984) Coesite and pure pyrope in high-grade blueschists of the Western Alps: a first record and some consequences. Contrib Mineral Petrol 86:107–118 doi:10.1007/bf00381838

Chopin C, Schertl H-P (1999) The UHP unit in the Dora-Maira massif, Western Alps. Int Geol Rev 41:765–780

Chopin C, Henry C, Michard A (1991) Geology and petrology of the coesite-bearing terrain, Dora-Maira massif, Western Alps. Eur J Mineral 3:263–291

Compagnoni R, Hirajima T, Chopin C (1995) Ultra-high-pressure metamorphic rocks in the Western Alps. *In:* Ultrahigh Pressure Metamorphism. Coleman RG, Wang X (eds) Cambridge University Press, Cambridge, p 206–243

Dohmen R, Faak K, Blundy JD (2017) Chronometry and speedometry of magmatic processes using chemical diffusion in olivine, plagioclase and pyroxenes. Rev Mineral Geochem 83:535–576

Engi M (2017) Petrochronology based on REE-minerals: monazite, allanite, xenotime, apatite. Rev Mineral Geochem 83: 365–418

England PC, Thompson AB (1984) Pressure–temperature–time paths of regional metamorphism I. Heat transfer during the evolution of regions of thickened continental crust. J Petrol 25:894–928

Fraser G, Ellis DJ, Eggins S (1997) Zirconium abundance in granulite-facies minerals, with implications for zircon geochronology in high-grade rocks. Geology 25:607–610 doi:10.1130/0091–7613(1997)025<0607:ZAIGFM>2.3.CO;2

Gauthiez-Putallaz L, Rubatto D, Hermann J (2016) Dating prograde fluid pulses during subduction by in situ U–Pb and oxygen isotope analysis. Contrib Mineral Petrol 171:15 doi:10.1007/s00410-015-1226-4

Gebauer D, Schertl HP, Brix M, Schreyer W (1997) 35 Ma old ultrahigh-pressure metamorphism and evidence for very rapid exhumation in the Dora Maira Massif, Western Alps. Lithos 41:5–24

Gerya T (2011) Future directions in subduction modeling. J Geodynam 52:344–378 doi:10.1016/j.jog.2011.06.005

Gerya TV, Stöckhert B, Perchuk AL (2002) Exhumation of high-pressure metamorphic rocks in a subduction channel: A numerical simulation. Tectonics 21:1056, doi:1010.1029/2002TC001406

Gerya TV, Perchuk LL, Burg J-P (2008) Transient hot channels: perpetrating and regurgitating ultrahigh-pressure, high temperature crust–mantle associations in collision belts. Lithos 103:236–256

Groppo C, Lombardo B, Castelli D, Compagnoni R (2007) Exhumation history of the UHPM Brossasco-Isasca Unit, Dora-Maira Massif, as inferred from a phengite–amphibole eclogite. Int Geol Rev 49:142–168 doi:10.2747/0020–6814.49.2.142

Guillot S, Hattori K, Agard P, Schwartz S, Vidal O (2009) Exhumation processes in oceanic and continental subduction contexts: A review. *In:* Subduction Zone Geodynamics. Lallemand S, Funiciello F (eds) Springer Berlin Heidelberg, Berlin, Heidelberg, p 175–205

Handy MR, Schmid SM, Bousquet R, Kissling E, Bernoulli D (2010) Reconciling plate-tectonic reconstructions of Alpine Tethys with the geological–geophysical record of spreading and subduction in the Alps. Earth-Sci Rev 102:121–158 doi:10.1016/j.earscirev.2010.06.002

Harrison TM, Bell EA, Boehnke P (2017) Hadean zircon petrochronology. Rev Mineral Geochem 83:329–363

Hermann J (2003) Experimental evidence for diamond-facies metamorphism in the Dora–Maira massif. Lithos 70:163–182 doi:10.1016/S0024-4937(03)00097-5

Kohn MJ (2016) Metamorphic chronology—a tool for all ages: Past achievements and future prospects. Am Mineral 101:25–42

Kohn MJ (2017) Titanite petrochronology. Rev Mineral Geochem 83:419–441

Kohn MJ, Penniston-Dorland SC (2017) Diffusion: Obstacles and opportunities in petrochronology. Rev Mineral Geochem 83:103–152

Kylander-Clark ARC (2017) Petrochronology by laser-ablation inductively coupled plasma mass spectrometry. Rev Mineral Geochem 83:183–198

Lanari P, Engi M (2017) Local bulk composition effects on metamorphic mineral assemblages. Rev Mineral Geochem 83:55–102

Oxburgh ER, Turcotte DL (1974) Thermal gradients and regional metamorphism in overthrust terrains with special reference to the Eastern Alps. Schweiz Mineral Petrogr Mitt 54:642–662

Peacock SM (1990) Numerical simulation of metamorphic pressure–temperature–time paths and fluid production in subducting slabs. Tectonics 9:1197–1211

Penniston-Dorland SC, Kohn MJ, Manning CE (2015) The global range of subduction zone thermal structures from exhumed blueschists and eclogites: Rocks are hotter than models. Earth Planet Sci Lett 428:243–254 doi:10.1016/j.epsl.2015.07.031

Reiners PW, Ehlers TA, Zeitler PK (2005) Past, present, and future of thermochronology. Rev Mineral Geochem 58:1–18

Reuber G, Kaus BJP, Schmalholz SM, White RW (2016) Nonlithostatic pressure during subduction and collision and the formation of (ultra)high-pressure rocks. Geology 44:343–346 doi:10.1130/g37595.1

Rubatto D, Hermann J (2001) Exhumation as fast as subduction? Geology 29:3–6

Rubatto D (2017) Zircon: The metamorphic mineral. Rev Mineral Geochem 83:261–295

Ruppel C, Hodges KV (1994) Pressure–temperature–time paths from two-dimensional thermal models: Prograde, retrograde, and inverted metamorphism. Tectonics 13:17–44 doi:10.1029/93TC01824

Schaltegger U, Davies JHFL (2017) Petrochronology of zircon and baddeleyite in igneous rocks: Reconstructing magmatic processes at high temporal resolution. Rev Mineral Geochem 83:297–328

Schertl H-P, Schreyer W, Chopin C (1991) The pyrope–coesite rocks and their country rocks at Parigi, Dora Maira Massif, Western Alps: detailed petrography, mineral chemistry and *P–T* path. Contrib Mineral Petrol 108:1–21

Schmitt AK, Vazquez JA (2017) Secondary ionization mass spectrometry analysis in petrochronology. Rev Mineral Geochem 83:199–230

Schoene B, Baxter EF (2017) Petrochronology and TIMS. Rev Mineral Geochem 83:231–260

Shi Y, Wang C-Y (1987) Two-dimensional modeling of the *P–T–t* paths of regional metamorphism in simple overthrust terrains. Geology 15:1048–1051 doi:10.1130/0091–7613(1987)15<1048:tmotpp>2.0.co;2

Spear FS, Thomas JB, Hallett BW (2014) Overstepping the garnet isograd: a comparison of QuiG barometry and thermodynamic modeling. Contrib Mineral Petrol 168:1–15

Spear FS, Pattison DRM, Cheney JT (2016) The metamorphosis of metamorphic petrology. Geol Soc Am Spec Papers 523 doi:10.1130/2016.2523(02)

Thompson AB, England PC (1984) Pressure–temperature–time paths of regional metamorphism II. Their inference and interpretation using mineral assemblages in metamorphic rocks. J Petrol 25:929–955

van Keken PE, Kiefer B, Peacock SM (2002) High-resolution models of subduction zones: Implications for mineral dehydration reactions and the transport of water into the deep mantle. Geochem Geophys Geosystem 3:1–20 doi:10.1029/2001GC000256

Vialon P (1966) Etude géologique du massif cristallin Dora–Maira, Alpes cottiennes internes, Italie. Thèse d'état, Université de Grenoble

Williams HS (1893) The making of the geological time-scale. J Geol 1:180–197

Williams ML, Jercinovic MJ, Mahan KH, Dumond G (2017) Electron microprobe petrochronology. Rev Mineral Geochem 83:153–182

Yamato P, Burov E, Agard P, Le Pourhiet L, Jolivet L (2008) HP–UHP exhumation during slow continental subduction: Self-consistent thermodynamically and thermomechanically coupled model with application to the Western Alps. Earth Planet Sci Lett 271:63–74 doi:10.1016/j.epsl.2008.03.049

Yakymchuk C, Clark C, White RW (2017) Phase relations, reaction sequences and petrochronology. Rev Mineral Geochem 83:13–53

Zack T, Kooijman E (2017) Petrology and geochronology of rutile. Rev Mineral Geochem 83:443–468

Reviews in Mineralogy & Geochemistry
Vol. 83 pp. 13-53, 2017
Copyright © Mineralogical Society of America

2

Phase Relations, Reaction Sequences and Petrochronology

Chris Yakymchuk

Department of Earth and Environmental Sciences
University of Waterloo
Waterloo, Ontario
Canada
N2 L 3G1

cyakymchuk@uwaterloo.ca

Chris Clark

Department of Applied Geology
Curtin University
Perth
Western Australia
Australia
6102

C.Clark@curtin.edu.au

Richard W. White

Institute of Geoscience
University of Mainz
Mainz
Germany
D-55099

rwhite@uni-mainz.de

INTRODUCTION

At the core of petrochronology is the relationship between geochronology and the petrological evolution of major mineral assemblages. The focus of this chapter is on outlining some of the available strategies to link inferred reaction sequences and microstructures in metamorphic rocks to the ages obtained from geochronology of accessory minerals and datable major minerals. Reaction sequences and mineral assemblages in metamorphic rocks are primarily a function of pressure (P), temperature (T) and bulk composition (X). Several of the major rock-forming minerals are particularly sensitive to changes in P–T (e.g., garnet, staurolite, biotite, plagioclase), but their direct geochronology is challenging and in many cases not currently possible. One exception is garnet, which can be dated using Sm–Nd and Lu–Hf geochronology (e.g., Baxter et al. 2013). Accessory mineral chronometers such as zircon, monazite, xenotime, titanite and rutile are stable over a relatively wide range of P–T conditions and can incorporate enough U and/or Th to be dated using U–Th–Pb geochronology. Therefore, linking the growth of P–T sensitive major minerals to accessory and/or major mineral chronometers is essential for determining a metamorphic P–T–t history, which is itself critical for understanding metamorphic rocks and the geodynamic processes that produce them (e.g., England and Thompson 1984; McClelland and Lapen 2013; Brown 2014).

1529-6466/17/0083-0002$05.00 (print)
1943-2666/17/0083-0002$05.00 (online)

http://dx.doi.org/10.2138/rmg.2017.83.2

Linking the ages obtained from accessory and major minerals with the growth and breakdown of the important *P–T* sensitive minerals requires an understanding of the metamorphic reaction sequences for a particular bulk rock composition along a well-constrained *P–T* evolution. Fortunately, the phase relations and reaction sequences for the most widely studied metamorphic protoliths (e.g., pelites, greywackes, basalts) can be determined using quantitative phase equilibria forward modelling (e.g., Powell and Holland 2008). Comprehensive activity–composition models of the major metamorphic minerals in large chemical systems (e.g., White et al. 2014a) allow the calculation of phase proportions and compositions for a given rock composition along a metamorphic *P–T* path. For accessory minerals, subsolidus growth and breakdown can be modelled in some cases using phase equilibria modelling (e.g., Spear 2010; Spear and Pyle 2010). Suprasolidus accessory mineral behaviour can be investigated by coupling phase equilibria modelling with the experimental results of accessory mineral solubility in melt (Kelsey et al. 2008; Yakymchuk and Brown 2014b). This technique provides a basic framework for interpreting the geological significance of accessory mineral ages in suprasolidus metamorphic rocks.

In this chapter, we use phase equilibria modelling techniques to investigate the reaction sequences for three common rock types (pelite, greywacke, MORB) along several *P–T* paths and explore how these sequences relate to accessory mineral growth and dissolution with a particular focus on zircon and monazite and to a lesser extent apatite. First, we review the important major rock-forming minerals that are used to link metamorphic reaction sequences to trace element chemistries in accessory minerals. Second, we summarize the current understanding of the controls of accessory mineral growth and breakdown in metamorphic rocks during a *P–T* evolution. Third, we use phase equilibria modelling to examine the reaction sequences for common rock types along several schematic *P–T* paths and discuss the implications for petrochronology. Finally, we examine some of the complicating factors for reconciling the behaviour of accessory minerals in natural systems with the predictions from phase equilibria modelling.

MAJOR MINERALS

Understanding the growth and consumption of the major rock-forming minerals is important in accessory mineral petrochronology for four reasons. First, major minerals may contain significant quantities of the essential structural constituents of accessory minerals commonly used as chronometers. In some cases, accessory minerals can grow directly from the breakdown of major minerals. Examples of this are zircon growth from the release of Zr during the breakdown of garnet (Fraser et al. 1997; Degeling et al. 2001) and ilmenite (Bingen et al. 2001). Second, these minerals may represent important repositories for trace elements, and thus the growth and breakdown of major minerals will influence the availability of these elements for incorporation by growing accessory minerals. Third, the major minerals are important hosts for inclusions of accessory minerals. The breakdown of the major minerals may liberate included accessory minerals into the reaction volume of the rock or alternatively may sequester these minerals away allowing their preservation when they would otherwise be consumed in a reaction sequence (e.g., Montel et al. 2000). Fourth, the microstructural relationships between accessory and major minerals provide context for delineating the *P–T* history of a metamorphic rock.

Garnet and plagioclase are the major minerals most commonly used in petrochronology because of their distinctive trace element behaviour. Linking their growth and breakdown to ages obtained from accessory mineral chronometers requires an understanding of the bulk composition controls on their stabilities. Below, we outline the controls on the growth and breakdown of garnet and plagioclase, which are of particular importance for petrochronology studies. Our focus is on linking mineral growth to metamorphic reactions and we do not discuss the minerals that are important for thermochronology studies (e.g., amphibole, biotite, muscovite, K-feldspar).

Garnet

Garnet is one of the most useful minerals for constraining metamorphic grade and its high density and strong partitioning of cations forms the basis of many useful thermobarometers (e.g., Spear 1995; Caddick and Kohn 2013). It is a common metamorphic mineral for many different protoliths (pelites, mafic rocks, ultramafic rocks, calc-silicates) and has been extensively used in petrochronology studies (e.g., Vance and O'Nions 1990; Vance and Mahar 1998; Harris et al. 2004). Coupling *P–T* estimates from garnet with monazite U–Pb geochronology can also be used to constrain *P–T–t* paths (e.g., Foster et al. 2000; Gibson et al. 2004; Dragovic et al. 2016). Garnet can also be directly dated using Sm–Nd and Lu–Hf geochronology (e.g., Baxter et al. 2013, 2017, this volume). Although linking *P–T* information from garnet-bearing assemblages with garnet geochronology is very powerful (e.g., Mulcahy et al. 2014; Dragovic et al. 2012, 2015), direct dating of garnet is challenging and is not yet extensively used.

An important control on the stability of garnet during metamorphism is the bulk rock MnO content (e.g., Symmes and Ferry 1992; Mahar et al. 1997; White et al. 2014b). For pelites, higher bulk rock Mn concentrations stabilize garnet at lower temperatures in the greenschist facies and at lower pressure in the amphibolite facies (White et al. 2014b). The ratio of Fe/Mg is also an important control on garnet growth. For example, relatively magnesium-rich pelites have restricted garnet stability fields (e.g., White et al. 2014a) and in very Mg-rich bulk rock compositions garnet may not even grow along common *P–T* paths (e.g., Fitzsimons et al. 2005). Therefore, depending on the bulk composition of the rock, garnet can be first stabilized in the greenschist facies, the amphibolite facies or not at all.

Garnet is the most important major mineral sink for the heavy rare earth elements (HREE) and yttrium in metamorphic rocks (Bea et al. 1994). Accessory minerals such as zircon, monazite, xenotime, apatite, epidote and allanite are also important repositories for the HREE and Y. Thus, the growth and breakdown of garnet plays an important role in the HREE and Y budgets in rocks and links the trace element chemistry of accessory mineral chronometers to the *P–T* information obtained from garnet-bearing metamorphic assemblages (Pyle et al. 2001; Foster et al. 2002; Rubatto 2002; Kohn and Malloy 2004; Rubatto and Hermann 2007; Taylor et al. 2015). In general, the equilibrium distribution coefficients of the HREE between garnet and zircon are close to unity for Gd to Lu at high temperature (Taylor et al. 2015). For Yttrium, concentrations in monazite are ~1.5 orders of magnitude higher than in garnet (Bea et al. 1994). Xenotime has Y concentrations that are ~2 orders of magnitude higher than in monazite (Pyle et al. 2001). However, in most cases, garnet is substantially more modally abundant than monazite and xenotime and thus the breakdown or growth of garnet will exert first-order controls on the Y and HREE budget of the rock. Zircon and monazite crystallization during garnet growth will result in relatively low concentrations of HREE and Y in these minerals because these elements are partitioned into garnet. By contrast, garnet breakdown during accessory mineral growth can result in HREE- and Y-enriched trace element concentrations of the new growth zones of accessory minerals (e.g., Foster et al. 2002, 2004; Rubatto 2002).

Plagioclase

Plagioclase is extremely common in metamorphic rocks and is stable over a wide range of *P–T* conditions. Plagioclase breakdown with increasing pressure is an important part of the reaction sequences for high-pressure metamorphic rocks and eclogites. Similar to garnet, plagioclase is commonly a key mineral in thermobarometers that provides important *P–T* information in many different rock types (Ghent 1976; Molina et al. 2015; Wu 2015). For petrochronology, plagioclase (and to a lesser extent K-feldspar) strongly partitions Eu and Sr over the other major rock-forming minerals (e.g., Gromet and Silver 1983). The growth and breakdown of plagioclase can be tied to Eu anomalies in rare earth element patterns of accessory minerals (Rubatto et al. 2013; Holder et al. 2015; Regis et al. 2016). Strontium concentrations have been used to link accessory mineral growth to the timing of plagioclase breakdown; this is a particularity important relationship in ultrahigh pressure metamorphic rocks (Finger and Krenn 2007; Kylander-Clark et al. 2013; Holder et al. 2015).

ACCESSORY MINERALS

Like the major minerals, accessory minerals are involved in the reaction sequence experienced by metamorphic rocks. The behaviour of some accessory minerals used in petrochronology such as epidote, titanite and rutile can be quantified along a *P–T* evolution using phase equilibria modelling. However, the behaviour of zircon and monazite—the most commonly used mineral chronometers—is more difficult to quantify with current phase equilibria modelling techniques (e.g., Spear and Pyle 2010; Kelsey and Powell 2011) as they contain key elements commonly not considered in model chemical systems.

Epidote

Epidote is common in low–medium pressure metabasites, metamorphosed intermediate rocks and calc-silicates (e.g., Grapes and Hoskin 2004) as well as in high- to ultrahigh-pressure metamorphic rocks (e.g., Enami et al. 2004). The chemical controls on epidote stability are mainly bulk rock concentrations of Ca and Al as well as f_{O_2} (e.g., Enami et al. 2004). The epidote-group minerals can be important repositories for Zr (e.g., Frei et al. 2004; Kohn et al. 2015) and LREE in metamorphic rocks (Frei et al. 2004; Janots et al. 2008). Therefore, the breakdown of Zr-rich epidote may have the potential to generate metamorphic zircon. Coupled allanite and epidote breakdown may produce new monazite (e.g., Janots et al. 2008) if there is also a source of phosphorus such as xenotime or apatite.

Titanite

Titanite is common in mafic rocks and in some low-grade metasedimentary rocks and it can be used as a barometer in metabasites (e.g., Kapp et al. 2009). An important factor in the stability of titanite is the relative activities of Ca and Al as influenced by the bulk rock ratio of Ca to Al and the accompanying mineral assemblage. High Ca activities favour titanite over ilmenite and high Al activities favour anorthite (and ilmenite) over titanite (Frost et al. 2001). Consequently, titanite is a common mineral in low-Al compositions such as calc-silicate and mafic rocks whereas aluminous metasedimentary rocks generally have ilmenite as the main Ti-bearing mineral above the greenschist facies.

Titanite can be directly dated using U–Pb geochronology, but there is some uncertainty about the closure temperature for Pb, which is generally considered to be around 600 °C (Warren et al. 2012; Spencer et al. 2013; Stearns et al. 2015; Kirkland et al. 2016), depending on cooling rate. Diffusion profiles of trace elements in titanite can be used to determine the timing and duration of cooling (e.g., geospeedometry) and this is covered in more detail in Kohn (2017, this volume). Titanite thermometry uses the Zr concentration of titanite to estimate temperature (Hayden et al. 2008). However, application of this thermometer requires an estimate of titania activity (a_{TiO_2}), which, in the absence of rutile, is difficult to constrain and can vary along a metamorphic evolution depending on the reaction sequences and mineral assemblages as discussed below.

Rutile

Rutile is commonly stable in relatively reduced bulk rock compositions (e.g., Diener and Powell 2010) and is particularity useful in high-pressure metamorphic rocks (Zack and Kooijman 2017, this volume). Similar to titanite, rutile can be directly dated with U–Pb geochronology (e.g., Mezger et al. 1989; Clark et al. 2000; Zack et al. 2011; Ewing et al. 2015) and it can be used as a geospeedometer (Smye and Stockli 2014). One of the main uses of rutile is as a thermometer that uses the Zr concentration of rutile to estimate temperature (Zack et al. 2004; Watson et al. 2006; Ferry and Watson 2007; Tomkins et al. 2007; Hofmann et al. 2013; Taylor-Jones and Powell 2015). These applications are discussed in more detail by Zack and Kooijman (2017, this volume).

Metamorphic reaction sequences with or without rutile have important implications for applying the Ti-in-quartz thermometer (e.g., Chambers and Kohn 2012; Ashley and Law 2015) and the Ti-in-zircon thermometer (e.g., Ferry and Watson 2007), both of which use a_{TiO_2} as a variable. In rutile-bearing systems a_{TiO_2} is buffered at 1.0. In rutile-absent systems, application of the thermometer requires an estimate of a_{TiO_2} that is less that 1.0. Figure 1 shows the difference of calculated temperatures using the Ti-in-quartz thermometer (Wark and Watson 2006) and the Ti-in-zircon thermometer (Ferry and Watson 2007) as a function of temperature for a range of a_{TiO_2} values.

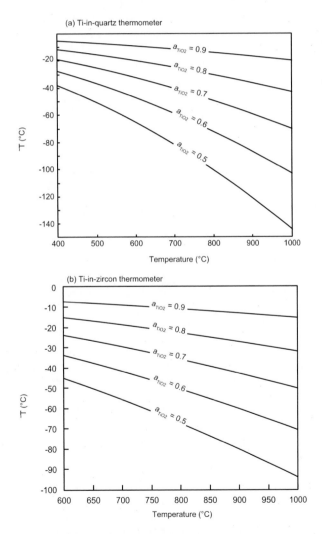

Figure 1. Underestimation of metamorphic temperature when the bulk rock value of aTiO$_2$ is less than 1.0 (the true value is shown by the contours), but a value of 1.0 is assumed in: (a) the Ti-in-quartz thermometer of Wark and Watson (2006), and (b) the Ti-in-zircon thermometer of Ferry and Watson (2007).

For example, consider an amphibolite-facies rock at 650 °C with a true a_{TiO_2} value 0.6. If an a_{TiO_2} value of 1.0 is assumed for Ti-in-quartz thermometry, the result would underestimate the true temperature by ~50 °C (Fig. 1a). Similarly, using an a_{TiO_2} value of 1.0 when the true value is 0.6, the Ti-in-zircon thermometer would underestimate the true temperature by ~40 °C (Fig. 1b). Phase equilibria modelling provides one method of determining a_{TiO_2} in metamorphic rocks at different stages of a *P–T* evolution in rutile-absent rocks (e.g., Ashley and Law 2015). We explore this in more detail during our discussion of particular reaction sequences.

Zircon

Zircon is the main repository of Zr in most igneous and metamorphic rocks and is the most widely used accessory mineral for U–Pb geochronology (Hoskin and Schaltegger 2003; Rubatto 2017, this volume). Many metamorphic rocks contain relict zircon that may be detrital or igneous in origin and may also be inherited from a previous metamorphic event. For petrochronology, metamorphic zircon can be used to date different portions of a *P–T* evolution depending on the growth mechanisms. The main processes that increase the mode of zircon in metamorphic rocks include solid-state growth and the crystallization from anatectic melt. In some cases, zircon may precipitate from hydrothermal fluids (e.g., Schaltegger 2007).

Solid-state zircon growth can occur from the breakdown of other Zr-rich minerals such as garnet (Fraser et al. 1997; Degeling et al. 2001), ilmenite (Bingen et al. 2001), rutile (Ewing et al. 2014) and possibly amphibole (Sláma et al. 2007). This zircon can be used to date major metamorphic assemblage changes during prograde (e.g., Fraser et al. 1997) or retrograde metamorphism (e.g., Degeling et al. 2001) depending on when the Zr-rich mineral breaks down during the reaction sequence. However, major minerals such as garnet, rutile and hornblende can accommodate more Zr as temperature increases and, in general, zircon will be consumed and prograde zircon growth is expected to be limited (Kohn et al. 2015).

The crystallization of anatectic melt typically drives new zircon growth in suprasolidus metamorphic rocks (Watson 1996; Roberts and Finger 1997; Schaltegger et al. 1999; Vavra et al. 1999; Hermann et al. 2001; Kelsey et al. 2008; Yakymchuk and Brown 2014b). Experimental studies show that the concentration of Zr in melt needed to maintain equilibrium with zircon increases with temperature (e.g., Harrison and Watson 1983; Gervasoni et al. 2016) and the compositional parameter *M*, which is the cation ratio of $[Na + K + 2Ca]/[Al \times Si]$ (Watson and Harrison 1983; Boehnke et al. 2013). Generally, the concentration of Zr needed for saturation of intermediate melts (high *M* values) is higher than for more felsic melts (low *M* values). In suprasolidus metamorphic rocks, the most important factor is the amount of anatectic melt present in the system (Kelsey et al. 2008, 2011; Yakymchuk and Brown 2014b). As the fraction of melt increases during prograde metamorphism, zircon is expected to break down to maintain Zr saturation of the melt in an equilibrated system. Consequently, prograde zircon growth is expected to be limited above the solidus. In general, melt crystallization during cooling from peak *T* is expected to be the main mechanism for zircon growth in suprasolidus metamorphic rocks. This is supported by ranges of concordant ages that reflect protracted zircon growth during melt crystallization from peak *T* in migmatites (e.g., Korhonen et al. 2013b, 2014).

New zircon growth can also occur at the expense of pre-existing zircon with no change in the mode of zircon in the rock. Recrystallization of metamict zircon in the presence of a fluid may be an important factor for some prograde zircon (e.g., Rubatto and Hermann 2003; Hay and Dempster 2009). Ostwald ripening has been proposed as a mechanism that could produce prograde growth of zircon in suprasolidus metamorphic rocks (Vavra et al. 1999; Nemchin et al. 2001; Kawakami et al. 2013). Ostwald ripening (or second phase coarsening) is a process whereby small solids are preferentially dissolved and precipitate on existing larger solids to reduce the total surface free energy (e.g., Tikare and Cawley 1998). While this process has been studied and debated for the major minerals in metamorphic rocks (Miyazaki 1991, 1996;

Carlson 1999, 2000) it has not been as extensively studied for accessory minerals—an exception being Nemchin et al. (2001). Nonetheless, it is a possible mechanism for prograde growth of zircon in suprasolidus metamorphic rocks, though involving no net modal increase in zircon.

Zircon can also be used as a thermometer in igneous and metamorphic rocks. The Ti-in-zircon thermometer has been increasingly applied to high-temperature metamorphic rocks over the last decade. Kelsey and Hand (2015) compiled Ti-in-zircon temperatures from UHT rocks and noted that 62% of the results fall below the UHT threshold of 900 °C. One of the possible reasons for this discrepancy is that some UHT rocks do not contain rutile and are therefore undersaturated in TiO_2 ($a_{TiO_2} < 1.0$). Values of a_{TiO_2} can range from 1.0 when rutile is in equilibrium with growing zircon to values as low as 0.6 (e.g., Hiess et al. 2008). Using a value of 1.0 when rutile is absent will underestimate the true temperature. Again, the presence and absence of rutile during a reaction sequence has important consequences for applying mineral thermometers in metamorphic rocks.

Monazite

Monazite is generally more reactive than zircon in subsolidus metamorphic rocks and has been extensively used for U–Pb geochronology in aluminous bulk compositions (e.g., Parrish 1990; Engi 2017, this volume). Monazite can be detrital in origin (Smith and Barreiro 1990; Kingsbury et al. 1993; Suzuki et al. 1994; Rubatto et al. 2001), produced through solid-state reactions (e.g., Rubatto et al. 2001; Wing et al. 2003), precipitated from a fluid (e.g., Ayers et al. 1999) or can crystallize from anatectic melt (e.g., Stepanov et al. 2012).

Solid-state monazite growth during prograde metamorphism occurs from the breakdown of LREE-rich precursors, which can include: allanite for high-Ca bulk compositions (Spear and Pyle 2010; Wing et al. 2003; Finger et al. 2016), LREE-rich clays (Copeland et al. 1971), Th or LREE oxides and hydrous phosphates (Spear and Pyle 2002). The most studied solid-state reaction is the growth of monazite at the expense of allanite in bulk compositions with sufficient Ca and LREE to grow allanite at lower grade (e.g., Janots et al. 2008). In natural examples, this reaction has been spatially correlated with the garnet-in isograd (Catlos et al. 2001; Foster et al. 2004), the staurolite-in isograd (Smith and Barriero 1990; Kohn and Malloy 2004; Corrie and Kohn 2008) and the kyanite or sillimanite-in isograds (Wing et al. 2003; Štípská et al. 2015).

Spear (2010) used phase equilibria modelling to examine the *P–T* conditions of the allanite to monazite transition in several bulk rock compositions relative to the average pelite of Shaw (1956). The reaction boundaries for monazite-in at the expense of allanite for these various compositions are summarized in Figure 2. A higher bulk rock CaO concentration allows allanite to persist to higher temperatures whereas a higher Al_2O_3 concentration stabilizes monazite at lower temperatures. However, the results of this modelling of monazite growth in rocks with variable bulk rock Al_2O_3 concentrations by Spear (2010) are different to those observed in some natural examples. For example, Wing et al. (2003) found that pelites from New England with relatively elevated Ca and/or Al concentrations can preserve allanite to higher temperatures. Gasser et al. (2012) found no link between bulk rock concentrations of CaO and Al_2O_3 and the timing of monazite growth. Therefore, the bulk compositional controls of monazite growth at the expense of allanite are not always clear, but it appears that monazite can grow at a range of temperatures and pressures along a prograde *P–T* path.

The relative amounts of iron and magnesium in a bulk rock composition may also play an important role in the reactions that produce subsolidus monazite. For example, Fitzsimons et al. (2005) showed that monazite in pelitic schists from Western Australia was generated during garnet-breakdown to staurolite and was interpreted to record the timing of peak metamorphism. However, samples that were too magnesium-rich to grow garnet or too iron-rich to grow staurolite at the expense of garnet yielded older monazite ages. Fitzsimons et al. (2005) interpreted these older ages to represent greenschist-facies monazite growth.

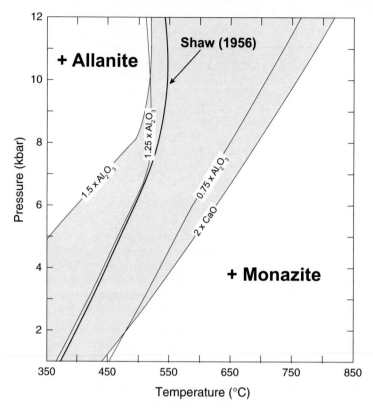

Figure 2. Compilation of the range of *P–T* conditions for the allanite to monazite reaction for variable bulk rock pelite compositions. Curves are from Spear (2010).

In suprasolidus metamorphic rocks, monazite breakdown and growth is controlled mainly by dissolution into and crystallization from anatectic melt. Experimental studies demonstrate that monazite dissolution into melt is a function of temperature, pressure, and the bulk composition of the melt (Montel 1986; Rapp and Watson 1986; Rapp et al. 1987; Skora and Blundy 2012; Stepanov et al. 2012; Duc-Tin and Keppler 2015). The solubility of monazite increases with higher H_2O concentrations in melt (Stepanov et al. 2012) and with decreasing phosphorus concentrations in the melt (Duc-Tin and Keppler 2015). Studies that couple experimentally-determined solubility equations of monazite with phase equilibria modelling show that monazite dissolution increases during prograde metamorphism above the solidus and that for UHT metamorphism most monazite will be completely consumed along the prograde path (Kelsey et al. 2008; Yakymchuk and Brown 2014b). Based on these theoretical models of monazite behaviour, no prograde monazite growth is expected above the solidus. This contrasts with studies of natural rocks that record prograde suprasolidus monazite growth (e.g., Hermann and Rubatto 2003; Hacker et al. 2015). Johnson et al. (2015) suggest that apatite dissolution during prograde metamorphism may have contributed to LREE saturation of the melt and resulted in prograde suprasolidus monazite crystallization. However, the role of apatite breakdown and monazite growth in suprasolidus metamorphic rocks is still poorly understood.

Finally, similar to zircon, Ostwald ripening could produce prograde monazite growth in suprasolidus metamorphic rocks (e.g., Nemchin and Bodorkos 2000). However, this has not been extensively studied and may only apply to monazite growth just above the solidus where the modal proportion of anatectic melt is relatively low.

Xenotime

Xenotime is a common phosphate in low- to high-grade metamorphic rocks (e.g., Franz et al. 1996; Bea and Montero 1999; Pyle and Spear 1999; Spear and Pyle 2002). It can be directly dated using U–Pb geochronology (e.g., Parrish 1990; Rasmussen 2005; Sheppard et al. 2007; Janots et al. 2009; Crowley et al. 2009) and can be used as a thermometer when in equilibrium with monazite and garnet (Pyle et al. 2001). Some of the proposed mechanisms of metamorphic xenotime growth are: (1) growth due to dissolution/reprecipitation of detrital or diagenetic zircon (Dawson et al. 2003; Rasmussen et al. 2011), (2) growth during to the breakdown of detrital zircon (Franz et al. 2015), (3) growth during the breakdown of allanite and/or monazite (Janots et al. 2008), (4) crystallization from anatectic melt (Pyle and Spear 1999; Crowley et al. 2009), and (5) growth during the breakdown of Y-rich garnet (e.g., Pyle and Spear 1999).

There are three important controls on the stability of xenotime in metamorphic rocks. First, higher bulk rock Yttrium concentrations allow xenotime to persist to higher pressures and temperatures (Spear and Pyle 2010). Second, similar to monazite and zircon, partial melting will result in xenotime consumption in order to contribute to REE and P saturation of the melt (e.g., Wolf and London 1995; Duc-Tin and Keppler 2015). Third, xenotime will be consumed during the growth of garnet, which partitions Y and HREE (Pyle and Spear 1999, 2000; Spear and Pyle 2002). Consequently, xenotime in the matrix of high-grade garnet-bearing rocks is relatively rare although xenotime inclusions may be present in garnet (e.g., Pyle and Spear 1999). By contrast, xenotime may be present in the matrix at various metamorphic grades in garnet-absent rocks.

PHASE EQUILIBRIA MODELLING

Phase assemblages in metamorphic rocks generally change through continuous or multivariant reactions rather than discontinuous or univariant reactions (e.g., Stüwe and Powell 1995; Kelsey and Hand 2015). Petrogenetic grids display discontinuous reactions and may yield the false impression that reactions occur over narrow intervals along a P–T path (e.g., Vernon 1996). In reality, mineral (and melt) modes and compositions are continuously changing along a P–T evolution and these variations represent the reaction sequence of the rock. The particular reaction sequence experienced by a metamorphic rock is dependent on bulk composition, the P–T path and whether the system is open or closed (e.g., White and Powell 2002; Brown and Korhonen 2009; Yakymchuk and Brown 2014a). The phase assemblages expected for a particular bulk rock composition over a range of P–T conditions are commonly depicted using P–T pseudosections, phase diagrams constructed for a fixed bulk composition (e.g., Hensen 1971; Spear et al. 2016). More detailed information about the reaction sequences can be determined by using the intersection of a P–T path with isopleths (or contours) of different variables (e.g., mineral modes and compositions) on a pseudosection or by constructing mode–temperature or mode–pressure diagrams (e.g., White et al. 2011).

Here, we present pseudosections and mode-box diagrams (P/T–mode plots) for different bulk chemical compositions to investigate the reaction sequences along common P–T paths and we discuss the consequences for interpreting the ages obtained from accessory mineral geochronology. The effects of fractionation of growing porphyroblasts and changes in bulk composition are discussed later. Because we use phase equilibria modelling, it is assumed that there are no kinetic controls (e.g., nucleation barriers or sluggish diffusion) that impact the growth or dissolution of major and accessory minerals. Although this assumption may not be strictly valid in some circumstances (Watt and Harley 1993; Pattison and Tinkham 2009; Gaides et al. 2011; Pattison and Debuhr 2015), the modelling here provides a general framework for investigating reaction sequences.

Bulk compositions

Three bulk chemical compositions are modelled that represent the most common protoliths investigated during most metamorphic studies. For the subsolidus and suprasolidus *P–T* paths, this includes an average amphibolite-facies metapelite (Ague 1991) and an average passive margin greywacke (Yakymchuk and Brown 2014a). These two compositions were chosen because they are expected to dominate passive margin turbidite sequences that are involved in orogenesis at convergent plate boundaries. A mid-ocean ridge basalt (MORB) from Sun and McDonough (1989) was also used to investigate a suprasolidus reaction sequence for a typical mafic protolith. The modelled bulk compositions are summarized in Table 1. Note that we do not consider melt loss nor fractionation of minerals away from the reacting volume in the modelling here, both of which can modify the composition of the equilibrium volume during a metamorphic evolution (e.g., White and Powell 2002; Evans 2004; Guevara and Caddick 2016; Mayne et al. 2016).

Computational Methods

Forward phase equilibria modelling is used to evaluate the changes to metamorphic mineral assemblages along several schematic *P–T* paths. There are two main approaches to phase equilibria modelling. The first approach is to use Gibbs free energy minimization to determine the stable mineral assemblage at a given *P–T* condition and there are two commonly used software packages available for this approach, including Perple_X (Connolly and Petrini 2002) and Theriak–Domino (de Capitani and Brown 1987; de Capitani and Petrakakis 2010). A second approach is to determine the solution of simultaneous non-linear equations to build up an array of points and lines that make up a metamorphic phase diagram (THERMOCALC: Powell and Holland 1988, 2008; Powell et al. 1998). Both of these approaches require an internally consistent thermodynamic database derived from the results of experimental studies. There are several commonly-used databases available: Berman (1988), Holland and Powell (1998) and most recently Holland and Powell (2011). Finally, the calculations require activity–composition models that relate end-member proportions to end-member activities for the solid-solution minerals as well as for complex fluids and melt.

Here, calculations were performed using THERMOCALC v.3.40 (Powell and Holland 1988) using the internally consistent dataset of Holland and Powell (2011). For metapelite and greywacke compositions, modelling was undertaken in the $MnO-Na_2O-CaO-K_2O-FeO-MgO-Al_2O_3-SiO_2-H_2O-TiO_2-Fe_2O_3$ (MnNCKFMASHTO) chemical system using the activity–composition relations in White et al. (2014a,b). An average MORB composition was investigated in the NCKFMASHTO chemical system using the activity-composition models from Green et al. (2016). Phases modelled as pure end-members are quartz, rutile, titanite, aqueous fluid (H_2O), kyanite and sillimanite. Mineral abbreviations are from Holland and Powell (2011) with the exception of titanite (ttn).

Table 1. Bulk compositions used for phase equilibria modelling (mol%)

	Figures	H_2O	SiO_2	Al_2O_3	CaO	MgO	FeO	K_2O	Na_2O	TiO_2	MnO	O
					subsolidus							
metapelite	3,4	+	64.58	13.65	1.59	5.53	8.03	2.94	2.00	0.91	0.17	0.60
greywacke	5,6	+	77.62	8.20	1.15	3.85	4.21	1.26	2.85	0.52	0.35	0.32
					suprasolidus							
metapelite	7,8,13	6.24	60.55	12.80	1.49	5.18	7.52	2.76	1.88	0.85	0.16	0.60
greywacke	9,10,13	2.61	75.35	7.96	1.12	3.74	4.09	1.22	2.77	0.51	0.34	0.31
MORB	11,12	6.19	42.58	10.45	12.27	14.76	9.68	0.22	2.62	0.66	–	0.58

Note: + H_2O in excess, – not considered

Phase equilibria modelling of subsolidus and suprasolidus systems requires different approaches to approximate the bulk H_2O content of the system. For subsolidus phase equilibria modelling, the amount of H_2O in each bulk composition was set to be in excess, such that $aH_2O = 1$. By contrast, suprasolidus rocks are generally not considered to have excess H_2O as the very small amount of free H_2O at the solidus is partitioned into anatectic melt (Huang and Wyllie 1973; Thompson 1982; Clemens and Vielzeuf 1987; White and Powell 2002; White et al. 2005). Along the solidus, the solubility of H_2O in anatectic melt increases with pressure. Therefore, for phase equilibria modelling of suprasolidus rocks, the amount of H_2O in the bulk composition must be set, and was adjusted so that the melt is just saturated with H_2O at solidus at 8 kbar. If the modelled prograde path crossed the solidus at lower or higher pressures, the quantity of melt produced will be slightly overestimated and underestimated, respectively.

Modelled *P–T* paths

Several simplified typical *P–T* paths for subsolidus and suprasolidus systems are modelled to investigate the reaction sequences, changes in mineral modes, amount of melt generated and/or consumed, variations in bulk rock a_{TiO_2} and the consequences for interpreting the results of accessory mineral geochronology. Two subsolidus and two suprasolidus *P–T* paths were investigated for the metapelite and greywacke compositions that represent different tectonic evolutions. We also model one suprasolidus *P–T* path for a MORB composition applicable to high-pressure mafic granulites.

For the subsolidus systems, the first *P–T* path has a clockwise trajectory and is representative of collisional orogenesis (e.g., England and Thompson 1984; Thompson and England 1984). An important feature of this *P–T* path is that peak *P* occurs before peak *T*, which is consistent with a relatively long residence time in the core of an orogenic belt. The model path ('clockwise' in Figs 3, 5) contains four segments: (1) heating and burial up to 10 kbar and 550 °C, (2) isobaric heating up to 650 °C, (3) decompression and minor heating to 8 kbar and 660 °C, and (4) decompression and cooling to 6.2 kbar and 610 °C.

The second subsolidus *P–T* path has a 'hairpin' trajectory where peak *P* and peak *T* nearly coincide (e.g., 'hairpin' in Figs 3, 5). This style of *P–T* path is also common in collisional orogenesis (e.g., Brown 1998; Kohn 2008) and reflects relatively short residence times at depth and can represent crustal thickening in an accretionary zone of a propagating orogen (Jamieson et al. 2004). The model hairpin *P–T* path contains three segments: (1) heating and burial up to 7 kbar and 610 °C, (2) heating and decompression to 6 kbar and 650 °C, and (3) decompression and cooling to 4 kbar and 600 °C. Peak metamorphism is at the same temperature for both subsolidus *P–T* paths, but the pressures at peak *T* are different.

Two *P–T* paths for suprasolidus rocks that are typical of granulite-facies metamorphism are investigated. First, a clockwise *P–T* path is modelled (e.g., 'clockwise' in Figs 7, 9) that is associated with crustal thickening and heating and is typical of many granulite terranes (e.g., Clark et al. 2011). This *P–T* path contains four segments: (1) increase in pressure and heating across the solidus up to 9 kbar at 850 °C, (2) isobaric heating up to 900 °C, (3) isothermal decompression to 7 kbar, and (4) cooling and decompression to 5 kbar at 750 °C.

Second, a suprasolidus counterclockwise (or anticlockwise) *P–T* path is considered (e.g., 'counterclockwise' in Figs 7, 9) and is common in high temperature–low pressure terranes (e.g., Clarke et al. 1987; Collins and Vernon 1991) and some ultrahigh temperature terrains (e.g., Korhonen et al. 2013a, 2014). This *P–T* path reflects heating prior to thickening (usually at peak *T*) followed by near-isobaric cooling. In some cases, this has been attributed to the inversion of a hot back-arc basin during collisional orogenesis (Clark et al. 2014). The modelled counterclockwise *P–T* path contains four segments: (1) isobaric heating up to 830 °C and 5 kbar, (2) heating and an increase in pressure up to 860 °C and 6 kbar, (3) cooling and an increase in pressure to 850 °C and 7 kbar, and (4) isobaric cooling to 750 °C.

A high-pressure clockwise path was also chosen to model the reaction sequences for high-pressure mafic granulites (e.g., O'Brien and Rötzler 2003) and to investigate the growth and breakdown of garnet for these rocks. This *P–T* path contains four segments: (1) heating and an increase in pressure up to 12 kbar and 850 °C, (2) isobaric heating up to 950 °C, (3) isothermal decompression to 8 kbar, and (4) cooling and decompression to 800 °C and 6 kbar. The decompression segments of this *P–T* path are also compatible with the decompression segments of some ultrahigh-pressure metamorphic rocks (Hacker et al. 2010; Chen et al 2013; Xu et al 2013).

Modelling suprasolidus zircon and monazite dissolution

For the suprasolidus *P–T* paths, the growth and dissolution of zircon and monazite are modelled using the method of Kelsey et al. (2008). Note that the bulk composition used in this modelling is fixed without fractionation of elements into growing porphyroblasts (e.g., zirconium into garnet) or loss of melt. These factors are discussed later. First, the saturation concentrations of the melt in ppm are calculated as follows. The major element composition of the anatectic melt is calculated at a specified *P–T*. This composition of the melt is combined with the solubility equations of Boehnke et al. (2013) for zircon and Stepanov et al. (2012) for monazite, and the stoichiometric concentrations of Zr in zircon (497,664 ppm Zr) and LREE in monazite (566,794 ppm LREE) to determine the saturation concentrations of Zr and LREE in ppm (Kelsey et al. 2008). These initial calculations do not account for the proportion of anatectic melt or the bulk-rock concentrations of Zr and LREE. Second, the saturation concentrations of Zr or LREE (ppm) are multiplied by the proportion of anatectic melt in the system (in oxide mole proportion where each phase is normalized to a one oxide basis—this is approximately equivalent to vol%) at *P–T* to arrive at concentrations in ppm that are required to saturate the melt in the equilibration volume of the rock. Finally, these values are divided by the assumed bulk-rock chemical concentrations of Zr and LREE. The Zr and LREE in pelites and the greywackes are generally similar to each other and to the values of metasedimentary migmatites; a value of 150 ppm was chosen here (e.g., Yakymchuk and Brown 2014b). For MORB, the Zr concentration was set at 103 ppm, which is an average of global MORB compositions (White and Klein 2014). The sensitivity of these calculations to bulk rock concentrations of Zr and LREE and the assumptions and limitations of this methodology are discussed by Kelsey et al. (2008) and Yakymchuk and Brown (2014b). The result of the calculations is the proportion of zircon or monazite dissolution required to saturate the anatectic melt in Zr and LREE in the equilibration volume of the rock. This value is subtracted from 100% and the results are reported as the percent of zircon and monazite remaining relative to the amounts existing at the fluid-present solidus for each *P–T* path.

One important limitation of the monazite saturation equation used here is that it does not account for phosphorus in anatectic melt, which can affect monazite stability (e.g., Duc-Tin and Keppler 2015). The main repository of phosphorus in most metamorphic rocks is apatite, which is expected to break down during partial melting (Wolf and London 1994). Phosphorus saturation in melt is a function of temperature, SiO_2 concentration (Harrison and Watson 1984) and the aluminum saturation index of the melt (ASI = molar $[Al_2O_3]$ / $[K_2O + Na_2O + CaO]$, Wolf and London 1994). The solubility of apatite in melt increases with rising ASI, which is common during prograde partial melting in migmatites (e.g., Johnson et al. 2015). For example, an increase in the ASI from 1.1 to 1.2 increases the solubility of apatite by an order of magnitude (Wolf and London 1994). The breakdown of LREE-rich apatite during prograde metamorphism may contribute to LREE saturation of anatectic melt and monazite crystallization (e.g., Johnson et al. 2015; Rocha et al. 2016). Here, we use melt ASI as a qualitative tool to investigate the behaviour of apatite during partial melting and the consequences for prograde monazite growth in suprasolidus metamorphic rocks.

SUBSOLIDUS PHASE RELATIONS AND REACTION SEQUENCES

Phase relations and reaction sequences for two subsolidus $P–T$ paths for two bulk compositions are discussed below for a closed system where bulk composition does not change. Fractionation of porphyroblasts away from the reacting volume is not considered in the calculations here.

Metapelite

Phase relations. The pseudosection for the amphibolite-facies metapelite is shown in Figure 3. Biotite is stable at $T>450–460\,°C$ with increasing pressure. Garnet is stable across most of the diagram except at low P and low T and at $P<5\,kbar$ just below the solidus. The staurolite-in

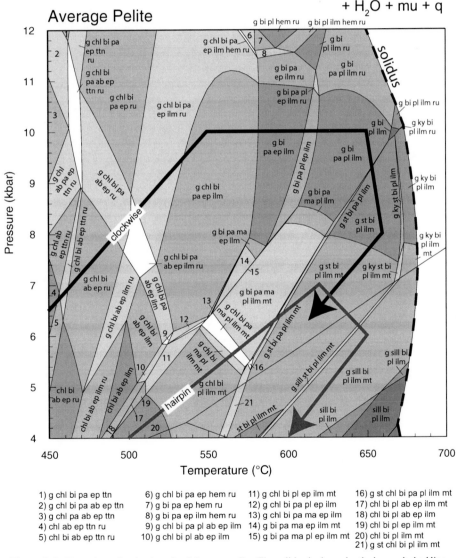

Figure 3. $P–T$ pseudosection for the subsolidus metapelite. The solidus is shown by the heavy dashed line.

1) g chl bi pa ep ttn
2) g chl bi pa ab ep ttn
3) g chl pa ab ep ttn
4) chl ab ep ttn ru
5) chl bi ab ep ttn ru

6) g chl bi pa ep hem ru
7) g bi pa ep hem ru
8) g bi pa ep ilm hem ru
9) g chl bi pa pl ab ep ilm
10) g chl bi pl ab ep ilm

11) g chl bi pl ep ilm mt
12) g chl bi pa pl ep ilm
13) g chl bi pa ma ep ilm
14) g bi pa ma ep ilm mt
15) g bi pa ma pl ep ilm mt

16) g st chl bi pa pl ilm mt
17) chl bi pl ab ep ilm mt
18) chl bi pl ab ep ilm
19) chl bi pl ep ilm mt
20) chl bi pl ilm mt
21) g st chl bi pl ilm mt

field boundary extends from 550 °C at 4 kbar up to 650 °C at 10 kbar. The stability field of the aluminosilicate minerals (kyanite and sillimanite) ranges from $T > 580$ °C at 4 kbar to $T > 660$ °C at 10 kbar. Rutile is stable at low P at low temperatures and at $P > 10.5$ at $T > 550$ °C.

Reaction sequence for the clockwise P–T path. A mode-box for the reaction sequence for the clockwise P–T path is shown in Figure 4a. During the prograde segment of the clockwise P–T path, chlorite is progressively consumed to produce biotite (starting at 455 °C) and then garnet (starting at 460 °C). Chlorite is exhausted by 580 °C. Rutile is completely consumed by ~530 °C. Epidote breaks down along the prograde path and the epidote-out field boundary is encountered at 620 °C. The complete breakdown of epidote at 620 °C may result in minor zircon crystallization or more likely the growth of new rims on pre-existing zircon.

The staurolite stability field is encountered during the heating and decompression segment at 9 kbar and 650 °C. In natural examples, prograde monazite growth has been associated with garnet, staurolite and kyanite/sillimanite growth in metapelites (Catlos et al. 2001; Smith and Barriero 1990; Wing et al. 2003; Kohn and Malloy 2004; Fitzsimons et al. 2005; Corrie and Kohn 2008), although kyanite/sillimanite is not encountered along the P–T path modelled here.

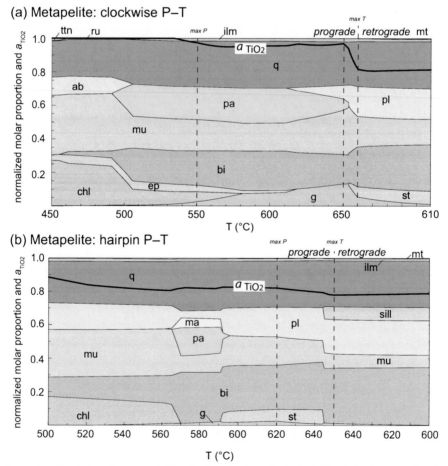

Figure 4. (a) Mode-box diagram and $a\mathrm{TiO_2}$ for the subsolidus clockwise P–T path for the metapelite. (b) Mode-box diagram and $a\mathrm{TiO_2}$ for the subsolidus hairpin P–T path.

In general, monazite could grow at different points along the prograde *P–T* path. Because garnet growth is predicted to be continuous up to the staurolite-in field boundary, new monazite grown during this segment of the *P–T* path is expected to have low concentrations of HREE and Y; these elements are expected to partition into garnet. Staurolite growth is at the expense of garnet and occurs during the decompression portion of the prograde path (e.g., Florence and Spear 1993). Monazite formed during this segment is expected to have elevated HREE and Y concentrations due to the breakdown of garnet and the age would record the initial stages of decompression from peak *P*. By contrast, zircon is expected to undergo only minor changes to its mode and will be slightly consumed during subsolidus prograde metamorphism (e.g., Kohn et al. 2015). Two possible exceptions are minor zircon growth during the breakdown of Zr-rich epidote and/or rutile. However, at low temperatures, rutile is not expected to be zirconium rich.

The stable mineral assemblage at the metamorphic peak of 660 °C and 8 kbar includes garnet, staurolite, biotite, plagioclase, ilmenite, muscovite and quartz. In general, the mode of monazite in metapelites is expected to reach a maximum at peak *P–T* and monazite is not expected to grow during cooling (e.g., Spear and Pyle 2010). An additional factor is that monazite is susceptible to dissolution/reprecipitation in the presence of fluids (e.g., Williams et al. 2011). While fluid-mediated dissolution/reprecipitation should not increase the net mode of monazite, it can occur at almost any point along the *P–T* evolution. Therefore, it is possible that monazite in a subsolidus metapelite can record a range of ages that vary from monazite-in up to the metamorphic peak and dissolution/precipitation may result in ages that record retrogression (Harlov et al. 2005; Williams et al. 2011). On the other hand, zircon is expected to be relatively unreactive and may not record prograde to peak metamorphism in subsolidus metamorphic rocks.

After the exhaustion of rutile during the heating and burial segment the value of a_{TiO_2} decreases to 0.95 initially and remains nearly constant until the decompression and heating segment, where a_{TiO_2} drops to 0.8. During the subsequent decompression and cooling segment the a_{TiO_2} value remains stable at 0.8. After rutile is exhausted, any growth of new zircon or modification of existing zircon will occur when $a_{TiO_2} < 1.0$; this needs to be considered when applying the Ti-in-zircon thermometer.

Reaction sequence for the hairpin P–T path. A mode-box for the reaction sequence for the hairpin *P–T* path is shown in Figure 4b. The prograde reaction sequence includes garnet-in at 520 °C and 4.5 kbar, staurolite-in at 590 °C and 6.5 kbar and sillimanite-in at 640 °C and 6 kbar. Garnet both grows and is consumed during the prograde portion of the reaction sequence. The value of a_{TiO_2} decreases from 0.88 at 500 °C to 0.79 at peak *T*. For the decompression and cooling segment, a small amount of biotite grows at the expense of garnet; the modes of the other minerals are little changed and the a_{TiO_2} value remains close to 0.79.

As with the clockwise *P–T* path, monazite growth may occur at various times during the prograde segment of the hairpin *P–T* path. Because the mode of garnet increases and decreases multiple times during the prograde reaction sequence, linking Y and HREE in monazite would be challenging in the absence of microstructural context, such as monazite in coronae surrounding resorbed garnet. For example, HREE and Y-rich monazite could be generated during: (1) the heating and burial segment during garnet breakdown to staurolite, (2) the heating and decompression segment during garnet breakdown to biotite near the metamorphic peak, or (3) garnet breakdown during the retrograde segment of the *P–T* path. Furthermore, xenotime breakdown during prograde metamorphism would also contribute to the growth of Y and HREE-rich monazite with or without a contribution from garnet (e.g., Spear and Pyle 2010). Therefore, for the hairpin *P–T* path modelled here, HREE- or Y-rich monazite can record burial and/or exhumation.

Greywacke

Phase relations. The pseudosection for the greywacke is shown in Figure 5. Garnet is stable across the entire diagram. Titanite is stable at $T < 460–470\,°C$ with increasing pressure. Biotite is stable across the diagram except at the high-*P*-low-*T* corner. Chlorite is unstable above $600\,°C$ at 7 kbar and unstable above $530\,°C$ at 12 kbar. Albite is stable below $565\,°C$ at 4 kbar and below $620\,°C$ at 11 kbar. White mica (muscovite and/or paragonite) is stable up to the solidus at $P > 10$ kbar and at $T < 570\,°C$ at 4 kbar. There are two important sets of quasi-linear field boundaries with positive slopes in *P*–*T* space. First is the set of boundaries that range from $480\,°C$ at 4 kbar up to $640\,°C$ at 12 kbar represent the breakdown of epidote to produce plagioclase. A second set of fields extends from $570\,°C$ at 4 kbar to the solidus with increasing *P* that represent the consumption of paragonite to produce sillimanite or kyanite.

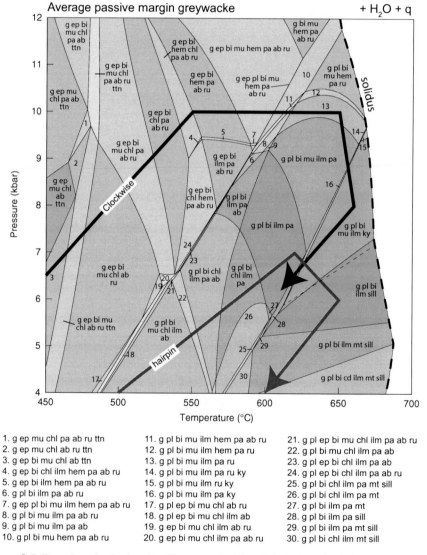

1. g ep mu chl pa ab ru ttn
2. g ep mu chl ab ru ttn
3. g ep bi mu chl ab ttn
4. g ep bi chl ilm hem pa ab ru
5. g ep bi ilm hem pa ab ru
6. g pl bi ilm pa ab ru
7. g ep pl bi mu ilm hem pa ab ru
8. g pl bi mu ilm pa ab ru
9. g pl bi mu ilm pa ab
10. g pl bi mu hem pa ab ru
11. g pl bi mu ilm hem pa ab ru
12. g pl bi mu ilm hem pa ru
13. g pl bi mu ilm pa ru
14. g pl bi mu ilm pa ru ky
15. g pl bi mu ilm ru ky
16. g pl bi mu ilm pa ky
17. g pl ep bi mu chl ab ru
18. g pl ep bi mu chl ilm ab
19. g ep bi mu chl ilm ab ru
20. g ep bi mu chl ilm pa ab ru
21. g pl ep bi mu chl ilm pa ab ru
22. g pl bi mu chl ilm pa ab
23. g pl ep bi chl ilm pa ab
24. g pl ep bi chl ilm pa ab ru
25. g pl bi chl ilm pa mt sill
26. g pl bi chl ilm pa mt
27. g pl bi ilm pa mt
28. g pl bi ilm pa sill
29. g pl bi ilm pa mt sill
30. g pl bi chl ilm mt sill

Figure 5. *P–T* pseudosection for the subsolidus greywacke. The solidus is shown by the heavy dashed line.

Reaction sequence for the clockwise P–T path. A mode-box for the assemblage sequence for the clockwise *P–T* path is shown in Figure 6a. For the clockwise *P–T* path the important changes along the prograde reaction sequence include: (1) the growth of rutile commencing at 460 °C resulting in a_{TiO_2} increasing from 0.9 at 450 °C to 1.0 at 460 °C, (2) the complete consumption of titanite by 475 °C, (3) the growth of biotite and garnet at the expense of chlorite and muscovite from 450 °C to 550 °C, which results in the complete consumption of chlorite by 550 °C, (4) the breakdown of epidote to produce 2 mol% plagioclase at ~605 °C, (5) the consumption of ~0.5 mol% garnet to produce plagioclase and hematite from 605 °C to 620 °C, and (6) the growth of 0.5 mol% garnet from 620 °C to 650 °C. Similar to the metapelite, the breakdown of Zr-rich epidote may result in minor zircon growth.

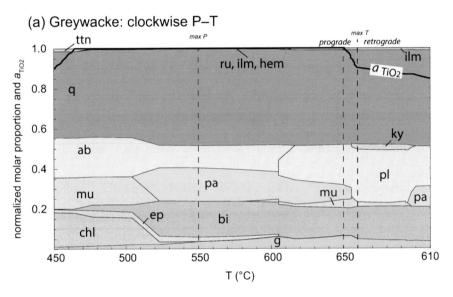

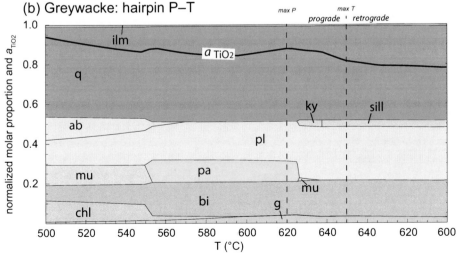

Figure 6. (a) Mode-box diagram and aTiO$_2$ for the subsolidus clockwise *P–T* path for the greywacke composition. (b) Mode-box diagram and aTiO$_2$ for the subsolidus hairpin *P–T* path.

Decompression and heating from 10 kbar at 650 °C to 8 kbar at 660 °C results in: (1) the disappearance of rutile at 9.8 kbar and a subsequent drop in a_{TiO_2}, and (2) the breakdown of paragonite and garnet to produce kyanite, biotite and plagioclase at 8.8 kbar. The peak metamorphic assemblage at 660 °C and 8 kbar includes garnet, kyanite, muscovite, biotite, plagioclase and ilmenite. Further decompression and cooling results in the growth of biotite at the expense of garnet and a gradual decrease in a_{TiO_2}.

Like the metapelite, garnet breakdown occurs during a portion of the heating and burial segment of the *P–T* path (at ~600 °C) and during the heating and decompression segment immediately before the metamorphic peak. Therefore, Y- and HREE-rich monazite associated with garnet breakdown could record the timing of burial and/or exhumation.

Reaction sequence for the hairpin P–T path. A mode-box for the reaction sequence for the hairpin *P–T* path is shown in Figure 6b. The hairpin *P–T* path has three notable differences in its reaction sequence compared with the clockwise path for the greywacke. First, the *P–T* path does not intersect the epidote stability field. Second, the value of a_{TiO_2} is always less than one and decreases from 0.95 to 0.80 along the *P–T* path. And third, white mica is completely consumed by the end of the *P–T* path. In contrast to the multiple garnet growth/consumption segments along the clockwise *P–T* path, the mode of garnet increases during burial and decreases during decompression for the hairpin path. Therefore, Y- and HREE-depleted monazite is predicted to record burial and monazite enriched in these elements would document garnet-breakdown during decompression and/or cooling and may also record xenotime breakdown.

SUPRASOLIDUS PHASE RELATIONS AND REACTION SEQUENCES

Phase relations and reaction sequences for three suprasolidus scenarios are discussed below for a closed system where bulk composition does not change and the water content is fixed to just saturate the rock in H_2O at 8 kbar at the solidus. This approach provides important first-order constraints on the reaction sequences for the modelled compositions; it does not take into account melt loss and the associated effects on rock fertility, solidus temperature and zircon and monazite stability. The consequences of open system behaviour on the reaction sequence and accessory mineral stability are discussed later.

Metapelite

Phase Relations. The pseudosection for the metapelite is shown in Figure 7. Ilmenite and plagioclase are stable across the entire diagram. Garnet is stable at high pressures across the diagram except at $T < 700$ °C at $P < 5.5$ kbar. Cordierite is stable at high-T–low-P conditions. Orthopyroxene is not stable in this diagram. Rutile is restricted to $P > 10$–11.5 kbar across the modelled temperature range. The three important partial melting reactions for the metapelite include: (1) the consumption of any free aqueous fluid to produce melt at the wet solidus, which ranges from 660 °C at low pressures to 710 °C at high pressure, (2) the incongruent breakdown of muscovite to produce K-feldspar, which is represented by a narrow low-variance field that extends from the wet solidus at low *P* to $T > 780$ °C at $P > 12$ kbar, and (3) the progressive breakdown of biotite at temperatures above the muscovite stability field to produce either garnet at higher pressure or cordierite ± garnet at lower pressure. Biotite is exhausted by 850 °C at $P > 7$ kbar and melting progresses via the consumption of quartz and feldspar.

Reaction sequence for the clockwise P–T path. The mode-box and titania activity for the clockwise assemblage sequence are shown in Figure 8a. The predicted amount of monazite and zircon remaining as well as melt ASI are shown in Figure 8b. During the

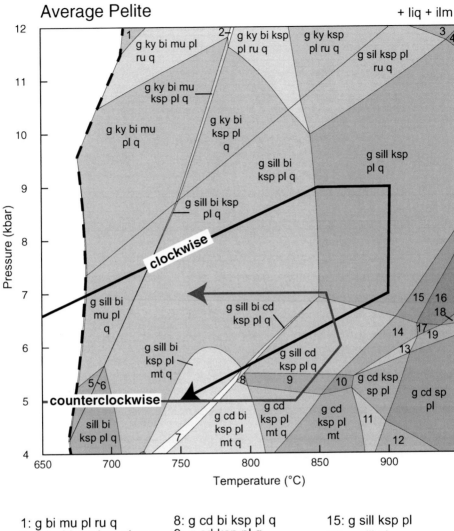

Figure 7. *P–T* pseudosection for suprasolidus metapelite. The solidus is shown by the heavy dashed line.

1: g bi mu pl ru q
2: g ky bi mu ksp pl ru q
3: g sill ksp ru q
4: g sill ksp q
5: sill bi mu pl q
6: sill ksp bi mu pl q
7: g sill cd bi ksp pl mt q

8: g cd bi ksp pl q
9: g cd ksp pl q
10: g cd ksp pl
11: g cd ksp sp pl mt
12: g cd sp pl mt
13: g sill cd ksp sp pl
14: g sill cd ksp pl

15: g sill ksp pl
16: g sill pl
17: g sill cd pl
18: g sill sp pl
19: g sill cd sp pl

prograde segment of the clockwise *P–T* path, the metapelite begins to melt at the wet solidus at ~680 °C through the consumption of any free H_2O as well as quartz and plagioclase. The amount of aqueous fluid at the solidus is expected to be small given the limited porosity of high-grade metamorphic rocks (e.g., Yardley and Valley 1997). Therefore, the amount of melt produced at the wet solidus is expected to be limited. For the modelled metapelite, the amount of melt produced is 4 mol%, although this is mostly an artefact of the modelling which assumes H_2O saturation of the solidus at 8 kbar.

Metapelite: clockwise P–T

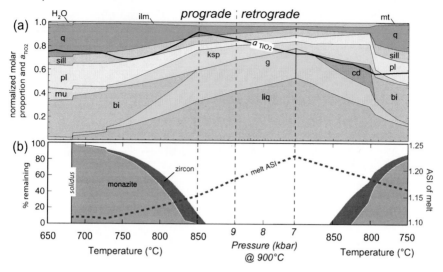

Metapelite: counterclockwise P–T

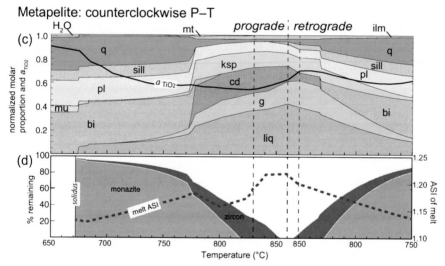

Figure 8. (a) Mode-box diagram and $a\mathrm{TiO_2}$ for the suprasolidus clockwise *P–T* path for the metapelite. (b) Amount of zircon and monazite remaining relative to the amount at the solidus and the ASI value of melt. (c) Mode-box diagram and $a\mathrm{TiO_2}$ for the counterclockwise *P–T* path for the metapelite. (d) Amount of zircon and monazite remaining relative to the amount at the solidus for the counterclockwise *P–T* path.

After any free H_2O is consumed, partial melting continues with increasing temperature through the consumption of muscovite up until 728 °C. This produces an additional 3 mol% melt with a generally constant melt ASI. At 728 °C and 7.5 kbar, melting proceeds via the incongruent breakdown of muscovite to generate K-feldspar over a narrow (~2 °C) temperature range. This narrow field produces an additional 3 mol% melt. After muscovite is exhausted, the rock contains ~9 mol% melt and approximately 8% of the zircon and 12% of the monazite that was present at the solidus is predicted to have been consumed. A minor amount of apatite breakdown is expected in order to saturate the anatectic melt in phosphorus.

After muscovite is completely consumed, partial melting continues through the consumption of biotite, plagioclase and quartz to produce melt, garnet, K-feldspar and ilmenite. This produces an additional 32 mol% melt. Because biotite is an important host of accessory mineral inclusions (e.g., Watson et al. 1989), biotite breakdown may release inclusions of monazite and zircon into the reaction volume of the rock. The liberation of these accessory minerals may contribute to LREE and Zr saturation of the anatectic melt. On the other hand, some zircon and monazite may be included in growing garnet and will be sequestered away from the reaction volume. These minerals will be shielded from dissolution and are more likely to preserve inherited (or detrital) ages as well as any subsolidus to early suprasolidus prograde metamorphic ages. The inclusion of accessory minerals in major minerals effectively reduces the Zr and LREE available to the system and proportionally more zircon and monazite dissolution will be required to maintain melt saturation in these elements (e.g., Yakymchuk and Brown 2014b). Therefore, the sequestration of zircon and monazite in stable peritectic minerals such as garnet will promote the dissolution of accessory minerals along grain boundaries in the matrix of the rock with increasing T.

Titania activity is predicted to reach its highest value of 0.92 at 850 °C. At this point, all of the monazite and 85% of the zircon are expected to be consumed. However, during this interval, the ASI of the melt increases from 1.10 to 1.15, which may result in enhanced apatite dissolution. For example, Pichavant et al. (1992) estimate that a similar increase in ASI at 800 °C and 5 kbar would change the P_2O_5 concentration of melt from 0.50 to 0.75 wt%. If apatite is LREE rich, the dissolution of apatite may delay the complete dissolution of monazite to higher temperatures or, in extreme cases, may even promote prograde monazite crystallization (e.g., Johnson et al. 2015).

After the complete consumption of biotite, the hydrous minerals have been exhausted and the residuum is essentially composed of anhydrous minerals. Melting continues through the continued consumption of quartz and K-feldspar up to the modelled peak temperature of 900 °C. The ASI of the melt reaches 1.18 at peak T and significant apatite dissolution is likely. At this point, the metapelite contains 43 mol% melt and both monazite and zircon are absent. After zircon and monazite are completely consumed, further anatexis is expected to generate melt that is undersaturated in Zr and LREE. Therefore, even with the breakdown of LREE-rich apatite, prograde monazite crystallization is not expected above ~850 °C.

Isothermal decompression from 9 to 7 kbar produces an additional 11 mol% melt at the expense of quartz and K-feldspar. At this point the rock contains the maximum amount of melt of 54 mol%. During this decompression segment, melt ASI increases from 1.18 to 1.23 and a_{TiO_2} decreases from 0.86 to 0.74. Approximately 1 mol% garnet is consumed during decompression, which would liberate some HREE and Y into the reaction volume. If significant apatite dissolution occurs and promotes monazite crystallization during decompression, this may be reflected as elevated HREE concentrations in monazite. However, the melt is undersaturated in LREE given the complete exhaustion of monazite at 850 °C so new monazite growth is unlikely unless the apatite is very enriched in LREE or the modal proportion of apatite is high.

Decompression and cooling from peak T results in: (1) melt crystallization, (2) the growth of K-feldspar and quartz until the cordierite stability field is reached at ~880 °C and 6.7 kbar, (3) garnet and cordierite consumption to produce biotite, (4) new zircon and monazite growth, (5) a decrease in melt ASI and a_{TiO_2}. Melt crystallization and cordierite growth are concomitant with the consumption of garnet and sillimanite; this is a common reaction sequence for high-temperature decompression in migmatites. Zircon or monazite produced over this reaction interval is expected to be enriched in HREE and Y (e.g., Yakymchuk et al. 2015). Protracted monazite and zircon growth will occur from peak T to the solidus. U–Pb zircon and monazite ages that spread down Concordia have been interpreted to record protracted growth during cooling and melt crystallization in various migmatite terranes with clockwise P–T evolutions (Korhonen et al. 2012; Reno et al. 2012; Morrisey et al. 2014; Walsh et al. 2015).

Reaction sequence for the counterclockwise P–T path. The mode-box and titania activity for the counterclockwise assemblage sequence are shown in Figure 8c. The amount of monazite and zircon remaining as well as melt ASI are shown in Figure 8d. During isobaric heating at 5 kbar, partial melting begins at the wet solidus at ~670 °C and continues through the breakdown of muscovite to produce a total 8 mol% melt by 685 °C. Melting continues via the progressive breakdown of biotite to produce peritectic garnet and K-feldspar. By 765 °C, 17 mol% melt is present and roughly 80% and 60% of the initial amount of zircon and monazite, respectively, remains. From the solidus up to 765 °C, the a_{TiO_2} value of the system has decreased from 0.9 to 0.6 and the ASI of the melt has increased from 1.13 to 1.18. An increase in melt ASI increases the solubility of apatite; this may liberate some LREE and P that could contribute to minor monazite crystallization because the melt is predicted to be saturated with respect to the LREE. Similar to the clockwise *P–T* path, the breakdown of biotite may liberate zircon and/or monazite that was sequestered away from the reacting volume.

Cordierite enters the phase assemblage at ~770 °C and melting continues through the breakdown of biotite, sillimanite and plagioclase to produce K-feldspar, garnet and cordierite. Sillimanite is completely consumed by 780 °C and biotite is exhausted by 800 °C. During the interval from 770 °C to 800 °C, approximately 15 mol% melt is generated and melt ASI decreases slightly from 1.18 to 1.16. Although the change in ASI would decrease the solubility of apatite, this may be counteracted by the additional melt generation during biotite breakdown and apatite growth is not expected.

In the absence of biotite, melting continues through the breakdown of quartz, plagioclase and K-feldspar, which results in a progressively drier melt. Monazite and zircon are predicted to be completely consumed by 820 °C and 850 °C, respectively. The increase in pressure near the metamorphic peak results in minor cordierite consumption to produce garnet. At the metamorphic peak, the metapelite has generated ~44 mol% melt.

Isobaric cooling at 6.5 kbar from 860 °C to 750 °C results in: (1) melt crystallization; (2) new zircon and monazite growth commencing at 855 °C and 825 °C, respectively; (3) a decrease in melt ASI, which would decrease the solubility of apatite and contribute to apatite crystallization; (4) the retrogression of cordierite and garnet to biotite; (5) the consumption of K-feldspar and the growth of plagioclase; and (6) a decrease in a_{TiO_2}. The concurrent breakdown of garnet and growth of zircon, monazite and plagioclase may result in elevated HREE and Y concentrations and more pronounced negative Eu anomalies in newly crystallized accessory minerals. New zircon growth occurs when the a_{TiO_2} value of the system ranges from 0.6 to 0.7. Applying the Ti-in-zircon thermometer assuming an a_{TiO_2} value of 1.0 would underestimate the true temperature by ~50 °C (Fig. 1b). Protracted retrograde monazite and zircon growth is expected during melt crystallization along the isobaric cooling segment of the counterclockwise *P–T* path (e.g., Korhonen et al. 2013b).

Greywacke

Phase relations. The pseudosection for the greywacke is shown in Figure 9. When compared with the metapelite, the greywacke composition is less fertile (e.g., Clemens and Vielzeuf 1987; Thompson 1996; Vielzeuf and Schmidt 2001; Johnson et al. 2008; Yakymchuk and Brown 2014a), contains different mineral assemblages, and yields different reaction sequences. For the greywacke, quartz, ilmenite, plagioclase and garnet are stable across the entire diagram. Cordierite becomes stable at pressures of 5–7 kbar with increasing temperature. Orthopyroxene is stable at *P* < 6 kbar and *T* > 800 °C. Rutile is stable at *P* > 9–11 kbar. Muscovite is stable at *P* > 7–12 kbar with increasing temperature. Similar to the metapelite, partial melting of the greywacke begins at the wet solidus at temperatures of ~670–700 °C. A narrow low-variance field representing the breakdown of muscovite to produce K-feldspar is restricted to *P* > 10.8 kbar; this contrasts with a similar field in the metapelite pseudosection that extends from < 4 to > 12 kbar (Fig. 7). For the greywacke

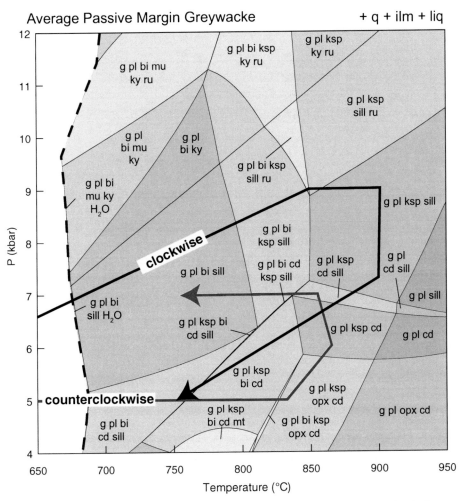

Figure 9. *P–T* pseudosection for the suprasolidus greywacke. The solidus is shown by the heavy dashed line.

composition, biotite breakdown generates garnet at high pressure, cordierite at low pressure, and orthopyroxene at low pressure and high temperature.

Reaction sequence for the clockwise P–T path. The mode-box and titania activity for the clockwise assemblage sequence are shown in Figure 10a. The predicted amount of monazite and zircon remaining along with the ASI of melt are shown in Figure 10b. Along the clockwise *P–T* path, partial melting begins at ~675 °C at the wet solidus and muscovite is not stable. Therefore, further melting proceeds through the breakdown of biotite and sillimanite to produce garnet. Similar to the metapelite, growing garnet has the potential to capture inherited or prograde monazite and zircon allowing their preservation to higher temperatures. K-feldspar becomes stable at 800 °C and 8.2 kbar and biotite is completely consumed by 850 °C. Up to this point, the greywacke has produced 14 mol% melt, which is less than the 34 mol% melt generated by the pelite for the same *P–T* path. Up to 850 °C, melt is produced gradually for the greywacke composition whereas melting of the metapelite occurs as a pulse in the narrow muscovite–K-feldspar field followed by more gradual melting due to biotite breakdown.

Greywacke: clockwise P–T

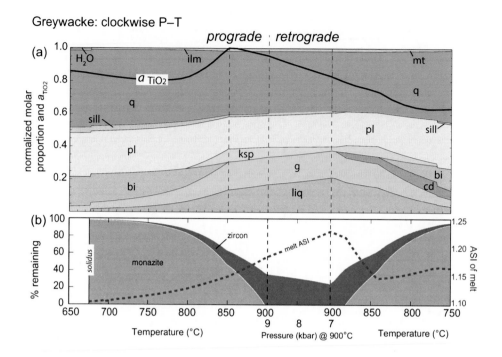

Greywacke: counterclockwise P–T

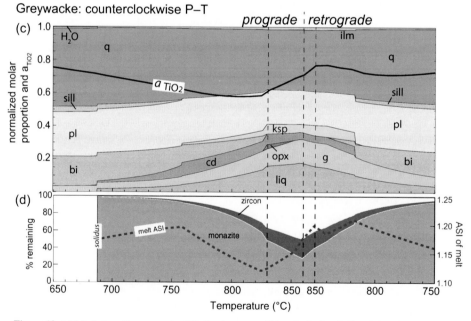

Figure 10. (a) Mode-box diagram and $a\mathrm{TiO_2}$ for the suprasolidus clockwise *P–T* path for the greywacke. (b) Amount of zircon and monazite remaining relative to the amount at the solidus and the ASI value of melt. (c) Mode-box diagram and $a\mathrm{TiO_2}$ for the counterclockwise *P–T* path for the greywacke. (d) Amount of zircon and monazite remaining relative to the amount at the solidus for the counterclockwise *P–T* path.

Beyond the loss of biotite from the stable assemblage, melting continues through the breakdown of K-feldspar, plagioclase and quartz and an additional 4 mol% melt is produced by 900 °C. Although monazite dissolution is modelled to continue from 850 to 900 °C, some prograde monazite crystallization may be possible in this temperature range for three reasons. First, the melt is saturated with respect to LREE due to progressive monazite dissolution. Second, after the exhaustion of biotite the melt becomes progressively drier, which decreases the solubility of monazite. Third, melt ASI increases, which increases the dissolution of apatite and may liberate enough LREE to support new monazite growth. However, unless this monazite is sequestered away in a growing peritectic mineral (such as garnet) it is expected to be consumed during further heating and partial melting.

All of the monazite and ~65% of the zircon are consumed by 900 °C. Isothermal decompression from 9 kbar to 7 kbar produces an additional 4 mol% melt and results in minor (<1 mol%) garnet consumption. Melt ASI increases from 1.18 to 1.23 during decompression, which enhances the solubility of apatite. However, the melt is undersaturated with respect to LREE and new monazite growth is not expected during decompression. Approximately 10% of the initial amount of zircon present at the solidus is consumed during decompression and titania activity decreases from 0.95 to 0.85.

Cooling and decompression from peak T to 750 °C results in: (1) melt crystallization, (2) the consumption of garnet and sillimanite to produce cordierite commencing at 890 °C and biotite starting at 840 °C, (3) a decrease in a_{TiO_2} from 0.85 to 0.65, (4) a decrease in melt ASI from 1.23 to 1.16, (5) the growth of new zircon (likely as overgrowths on existing zircon), and (6) the crystallization of monazite starting at ~885 °C. Zircon and monazite growth occurs during garnet breakdown and these accessory minerals are expected to have elevated Y and HREE concentrations. The application of Ti-in-zircon thermometry should use an a_{TiO_2} value ranging from 0.85 to 0.65. Assuming an a_{TiO_2} value of 1.0 would underestimate temperatures by up to 50 °C (Fig. 1b).

Reaction sequence for the counterclockwise P–T path. The mode-box and titania activity for the counterclockwise assemblage sequence are shown in Figure 10c. The amount of monazite and zircon remaining as well as melt ASI are shown in Figure 10d. Similar to the clockwise *P–T* path for the greywacke, the prograde segment of the counterclockwise *P–T* path generates melt gradually in contrast to the more pulsed melting in the metapelite. Melting commences at the wet solidus and proceeds via the breakdown of biotite to produce cordierite and garnet followed by K-feldspar at 760 °C. Orthopyroxene enters the phase assemblage at 825 °C and biotite is completely consumed by 830 °C. The value of a_{TiO_2} decreases from ~0.80 to 0.60 up to the orthopyroxene-in field boundary and then increases for the remainder of the prograde path. The increase in pressure near peak T results in the breakdown of orthopyroxene to produce garnet (e.g., White et al. 2008). By the end of the heating segment, ~50% of the zircon and ~70% of the monazite that was present at the solidus has been consumed. Melt ASI varies between 1.16 and 1.11 during the prograde path. As with the clockwise *P–T* path, melting above the biotite-stability field (in this case from 830–860 °C) has the potential to generate some prograde monazite if LREE-rich apatite is consumed. A total of 18 mol% melt is predicted to be generated during heating, which is significantly less than the 44 mol% produced along the same *P–T* path for the metapelite.

Consistent with the metapelite for the same *P–T* path, isobaric cooling results in melt crystallization and monazite and zircon growth. The mode of garnet decreases by 7 mol% and K-feldspar breaks down to sillimanite in the presence of melt at 815 °C. Newly crystallized monazite and zircon are predicted to have elevated Y and HREE due to the breakdown of garnet. Zircon growth occurs when the a_{TiO_2} value of the system is ~0.7.

Average mid ocean ridge basalt

Phase relations. The *P–T* pseudosection for an average MORB composition is shown in Figure 11. The wet solidus has a negative slope from 4 to 11.5 kbar and a positive slope at $T > 11.5$ kbar. The temperature of the wet solidus ranges from 620 to 700 °C over the modelled *P–T* range. Garnet is stable from 10–14 kbar with decreasing temperature. Orthopyroxene enters the assemblage at temperatures of 800 °C to 900 °C with increasing *P* and is not stable above 10 kbar at high *T*. Garnet and orthopyroxene are only stable together at pressures of 9.5–10 kbar and at $T > 900$ °C. Rutile is stable at $T > 790$ °C at $P > 7$ kbar.

Reaction sequence for the clockwise P–T path. The mode-box and titania activity for the reaction sequence as well as the amount of zircon remaining and melt ASI are shown in Figure 12. Melting starts at the wet solidus at 630 °C and 9.7 kbar and generates ~ 2 mol% melt and ~ 2 mol% clinopyroxene at the expense of epidote and biotite. A minor amount (~2 mol%) of zircon is expected to be consumed in order to saturate the melt in Zr. Epidote can be an important source of Zr (e.g., Kohn et al. 2015) and the breakdown of Zr-rich epidote may result in Zr saturation and minor zircon crystallization at this stage. After the exhaustion of biotite at ~650 °C, melting continues via the breakdown of hornblende, titanite and quartz to produce an additional 4 mol% melt as well as peritectic clinopyroxene by 800 °C and 11.4 kbar. During this portion of the prograde path, a_{TiO_2} increases from 0.6 at the wet solidus up to 0.9 at 800 °C. The amount of zircon remaining is ~80 mol% of the amount present at the solidus. Melt ASI increases slightly from 0.99 to 1.00.

Garnet becomes stable at ~800 °C and melting proceeds through the breakdown of hornblende and titanite. Garnet growth may include zircon grains that would be isolated from the reacting volume of the rock and zircon could be preserved to higher *T*. Rutile becomes stable at 818 °C and titanite is completely consumed by 821 °C. The peak pressure of 12 kbar is reached at 850 °C and at this point the system contains ~13 mol% melt, 10 mol% garnet and 65% of the zircon has been consumed. Although zircon is expected to be consumed during partial melting, the breakdown of Zr-rich amphibole (e.g., Sláma et al. 2007) may yield enough Zr to oversaturate the melt and grow new zircon; this zircon may be relatively depleted in HREE in response to the presence of garnet in the rock.

The isobaric heating segment of the *P–T* path at peak *P* produces garnet at the expense of hornblende and quartz. Quartz is exhausted at 905 °C. This has implications for applying the Ti-in-zircon thermometer, which uses a_{SiO_2} as a variable. Zircon is completely consumed by 880 °C and the melt produced at higher *T* is expected to be undersaturated in Zr. A consequence of this is that prograde zircon growth at $T > 880$ °C is unlikely because any excess Zr due to the breakdown of other minerals (e.g., amphibole) will be incorporated into the Zr-undersaturated melt. At the metamorphic peak of 950 °C, the system contains ~28 mol% melt, ~20 mol% garnet and ~14% hornblende.

Isothermal decompression from 12 to 10 kbar results in significant garnet consumption (from 20 to 9 mol%), hornblende growth (from 14 to 24 mol%) and minor melt consumption (from 28 to 27 mol%). Zirconium liberated from garnet breakdown is expected to be partitioned between the Zr-undersaturated melt and hornblende. Consequently, zircon crystallization is unlikely. Orthopyroxene enters the assemblage at 10.1 kbar and grows at the expense of garnet, which is exhausted by 9.7 kbar. If garnet is completely consumed, any liberated zircon may become available to the reacting volume and will likely be consumed into the Zr-undersaturated melt. Rutile is completely consumed by 9.1 kbar and a_{TiO_2} decreases with further decompression. At the end of the isothermal decompression segment of the *P–T* path the amount of orthopyroxene reaches 5 mol%. No new zircon growth is expected during the isothermal decompression segment because: (1) there is a <1 mol% melt change during isothermal decompression and the melt is already significantly undersaturated in Zr, (2) the M value of the melt (cation ratio of [Na + K + 2Ca]/ [Al × Si]) increases from 1.6 to 1.7,

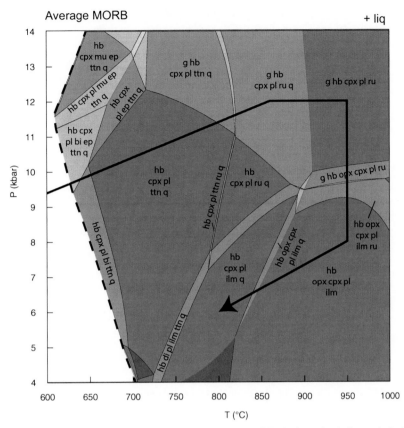

Figure 11. *P–T* pseudosection for the suprasolidus MORB. The solidus is shown by the heavy dashed line.

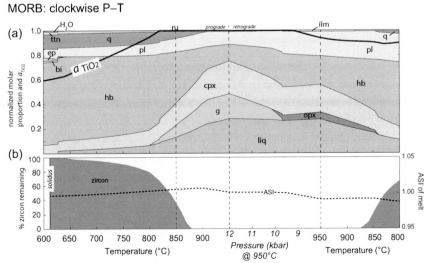

Figure 12. (a) Mode-box diagram and *a*TiO₂ for the suprasolidus clockwise *P–T* path for the MORB. (b) Amount of zircon remaining relative to the amount at the solidus and the ASI value of melt.

which results in an increases the Zr required to saturate the melt (Harrison and Watson 1983; Boehnke et al. 2013), (3) the mode of hornblende increases from 14 to 23 mol%, likely accommodating significant amounts of Zr at high temperature (e.g., Kohn et al. 2015).

Cooling and decompression from 950 °C at 8 kbar to 800 °C at 6 kbar results in the consumption of 17 mol% melt and the complete breakdown of orthopyroxene by ~830 °C. Zirconium saturation of the melt is reached at ~870 °C and zircon begins to crystallize. At ~850 °C, approximately 20% of the amount of zircon originally present at the solidus has grown back and is expected to be enriched in HREE because there is no garnet present in the system. Zircon growth at this stage occurs in a system with an a_{TiO_2} value of 0.9 and in the absence of quartz, which is important for the application of Ti-in-zircon thermometry.

Summary of reaction sequence modelling

Linking the trace element concentrations of accessory minerals to the key major minerals (e.g., garnet and plagioclase) requires an understanding of the reaction sequences for a particular bulk composition. Key minerals like garnet and plagioclase can grow or be consumed multiple times along a *P–T* path during heating and cooling or burial and exhumation (Figs 3–10). For example, garnet is predicted to grow in the subsolidus greywacke composition during heating and burial and garnet breakdown occurs during cooling and exhumation for the hairpin *P–T* path (Fig. 6b). By contrast, garnet growth and consumption occurs multiple times along the heating and increasing pressure segment of the *P–T* paths for the subsolidus metapelite composition (Fig. 4). Therefore, high-Y and high-HREE zones in monazite that can be linked to garnet breakdown may reflect the burial and/or exhumation portions of a *P–T* path.

Titania activity also varies along the modelled *P–T* paths and this needs to be taken into account when applying Ti-in-zircon or Ti-in-quartz thermometers. In general, new zircon growth in subsolidus rocks is expected to be limited because major minerals such as hornblende and garnet can accommodate more Zr with increasing temperature. One exception to this may be minor zircon growth through the breakdown of Zr-rich epidote. For the suprasolidus *P–T* paths for the metapelite and greywacke, new zircon growth is expected during cooling and melt crystallization; this growth is predicted to occur when bulk rock a_{TiO_2} is less than one for all of the modelled *P–T* paths. If the Ti-in-zircon thermometer is applied with an a_{TiO_2} value of one, then the result will be an underestimation in peak metamorphic temperatures by up to ~40 °C. For the MORB composition, new zircon growth occurs at $a_{TiO_2} < 1.0$ as cooling and melt crystallization occurs outside the stability field of rutile (Fig. 12).

In suprasolidus metamorphic rocks, zircon and monazite are expected to be consumed along the prograde path and new growth is generally predicted to occur along the cooling path. Zircon and monazite dissolution is non-linear and the rate increases with increasing temperature. For the clockwise *P–T* path for the metapelite and greywacke, an increase in the melt ASI leads to more apatite dissolution. Because the melt is saturated in LREE with respect to monazite up to 820 °C, apatite breakdown may liberate enough LREE to promote new monazite growth. However, after the exhaustion of monazite, the melt is expected to be undersaturated in LREE and prograde monazite growth at the expense of LREE-rich apatite is unlikely. For the counterclockwise *P–T* path for the metapelite and greywacke, melt ASI increases and decreases during the *P–T* evolution and monazite growth from apatite breakdown will be more complex to interpret.

The metapelite is expected to lose most of the inherited or subsolidus prograde monazite and zircon during heating above the solidus (unless these mineral are sequestered away from the reaction volume) whereas a larger proportion of these minerals remains in the greywacke composition except for the clockwise *P–T* path where monazite is completely consumed for the greywacke. The difference reflects the fertility of the two rocks; pelites generate more melt than the greywacke and require more Zr and LREE to saturate this melt. Therefore, less fertile compositions, such as greywackes, are more likely to preserve subsolidus zircon and monazite.

COMPLICATING FACTORS

Changes in effective bulk composition

Phase equilibria modelling requires an assessment of the effective bulk composition of a system that is used to model the P–T phase relations for a rock (e.g., Stüwe 1997). The effective bulk composition approximates the composition 'available' to the rock in which equilibrated mineral assemblages and reaction sequence develop. In natural systems, the effective bulk composition may change along a P–T path (e.g., Guevara and Caddick 2016), which has subsequent implications for modelling reaction sequences in both subsolidus and suprasolidus metamorphic rocks. There are two main mechanisms that can change the effective bulk composition of a metamorphic system: fractionation of elements into growing porphyroblasts and melt loss.

Growing porphyroblasts can fractionate certain elements into their cores that become unavailable to the reacting system in the remainder of the rock. For example, chemical zoning in garnet is commonly preserved in metamorphic rocks because cation diffusion in garnet is relatively slow (e.g., Chakraborty and Ganguly 1992). The preferential partitioning of elements into garnet cores reduces their effective composition in the reactive volume of the rock (e.g., Lanari and Engi 2017, this volume). Some of the consequences for using phase diagrams to infer metamorphic conditions considering garnet fractionation include reduced stability fields for garnet (Gaides et al. 2008) and other minerals (Zuluaga et al. 2005; Moynihan and Pattison 2013) as well as changes in mineral compositional isopleths for garnet (Evans 2004; Gaides et al. 2006) and plagioclase (Moynihan and Pattison 2013). While crystal fractionation needs to be assessed on a case-by-case basis, it is generally most important to consider for greenschist- and amphibolite-facies metamorphic assemblages where mineral chemistries are particularity useful for determining P–T conditions. At higher grades, the use of mineral composition isopleths is generally less effective due to retrograde exchange reactions (e.g., Spear and Florence 1992; Kohn and Spear 2000; Pattison et al. 2003). For zircon, the fractionation of Zr into growing garnet (e.g., Kohn et al. 2015) may reduce the Zr budget available for zircon growth.

In high-grade metamorphic rocks that underwent anatexis the preservation of peritectic minerals and lack of extensive retrogression supports melt drainage during prograde metamorphism (Fyfe 1973; Powell 1983; White and Powell 2002; Guernina and Sawyer 2003; Reno et al. 2012). Melt loss produces progressively more refractory bulk compositions, which results in elevated solidus temperatures in the residual rocks. In migmatites that have undergone melt loss, suprasolidus zircon and monazite growth is generally expected to occur during cooling from peak T to the elevated solidus (e.g., Kelsey et al. 2008; Spear and Pyle 2010; Yakymchuk and Brown 2014b). Therefore, rocks that have experienced identical P–T histories but variable amounts of melt loss and have different solidus temperatures should record a range of ages (e.g., Korhonen et al. 2013b).

Bulk composition and the suprasolidus behaviour of zircon and monazite

For the modelled suprasolidus reaction sequences, the preservation of subsolidus (e.g., prograde or inherited) zircon and monazite is mainly related to the fertility of the rocks. The metapelite generates more melt along the same P–T path than the greywacke. Consequently, zircon and monazite are completely consumed for the metapelite composition along both P–T paths (Figs. 8a, b) whereas some zircon and/or monazite can survive in the greywacke. An additional factor is the bulk rock content of Zr for zircon and LREE and phosphorus for monazite (e.g., Kelsey et al. 2008; Yakymchuk and Brown 2014b). Both of these factors are explored together in Figure 13.

The temperature–composition diagrams in Figure 13 illustrate the variation in melt mode and the stability of zircon and monazite for compositions ranging linearly from the metapelite (left side) to that of the greywacke (right side). The diagrams are isobaric and were calculated at 7 kbar because this pressure intersects the main melt producing reactions for both bulk compositions. The common reaction to both compositions is the breakdown of biotite to produce cordierite at ~840–845 °C (Fig. 13a). The amount of melt in the metapelite

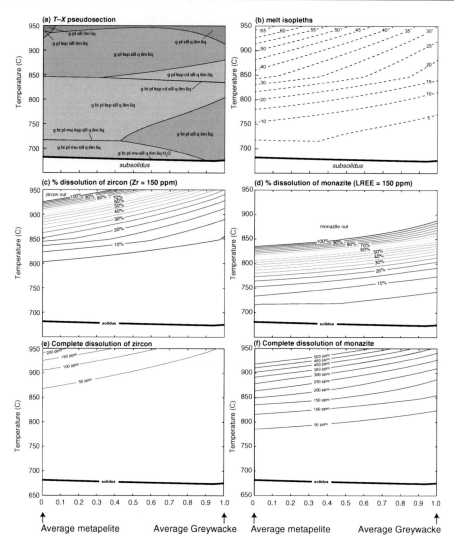

Figure 13. (a) *T–X* pseudosection calculated at 7 kbar showing the change in mineral assemblages for compositions ranging from the metapelite (left side) to the greywacke (right side). (b) *T–X* diagram with melt isopleths (mol%). (c) *T–X* diagram with the calculated amount of zircon dissolution assuming a bulk Zr concentration of 150 ppm. (d) Calculated amount of monazite dissolution assuming a bulk LREE concentration of 150 ppm. (e) *T–X* diagram with contours for the complete dissolution of zircon for a range of bulk rock Zr concentrations. (f) *T–X* diagram with contours for the complete dissolution of monazite for a range of bulk rock LREE concentrations.

composition is roughly twice the amount in the greywacke composition over the range of modelled temperatures (Fig. 13b). For a bulk composition of 150 ppm Zr, zircon is completely consumed in the metapelite composition by 925 °C whereas only 30% of the zircon has been consumed in the greywacke composition at the same temperature (Fig. 13c). For a bulk composition of 150 ppm LREE, monazite is completely consumed by 830 °C for the metapelite and 880 °C for the greywacke (Fig. 13d). The slopes of the dissolution contours are steeper for zircon than for monazite in Figures 13c and 13d. This indicates that zircon dissolution is more sensitive to bulk composition (metapelite vs. greywacke) than monazite.

The sensitivity of zircon and monazite dissolution to bulk composition of Zr and LREE is explored in Figures 13e and 13f. The contours represent the complete dissolution of zircon and monazite for various bulk rock Zr and LREE contents. For low-Zr pelites with concentrations of 50 ppm, zircon is completely consumed by 860 °C, which is 65 °C lower than for a closer-to-average Zr concentration of 150 ppm. Greywackes with low concentrations of Zr (e.g., 50 ppm) can still preserve zircon up to UHT conditions due to their lower fertility. Monazite dissolution contours in Figure 13f have shallower slopes than those for zircon (Fig. 13e), which again suggests that monazite dissolution is less sensitive to bulk composition. A metapelite with low concentrations of LREE (e.g., 50 ppm) is predicted to lose monazite by 780 °C and monazite is expected to be completely consumed in the greywacke composition by 820 °C. Preserving subsolidus monazite to UHT conditions requires very LREE-rich compositions of >400 ppm for the metapelite and >180 ppm for the greywacke. Therefore, a strategy for finding subsolidus prograde zircon and monazite in migmatites is to choose samples with the highest concentrations of Zr and LREE.

The breakdown of LREE-rich apatite is a potential mechanism to promote prograde suprasolidus monazite growth (e.g., Johnson et al. 2015). This is most likely to occur when the anatectic melt is saturated with respect to monazite in LREE. After the exhaustion of monazite, the melt is expected to be undersaturated and LREE liberated from the breakdown of apatite is not predicted to generate new prograde monazite growth. In principle, the higher the bulk rock LREE the longer monazite will persist during a suprasolidus heating path and the longer the melt will remain saturated with respect to LREE (Fig. 13f). A further consideration is that bulk compositions with low phosphorus concentrations may lose apatite during the prograde path and would result in melt that is undersaturated in P. This may promote further monazite dissolution instead of monazite growth. Therefore, LREE and P-rich bulk compositions should be targeted for accessory mineral geochronology to constrain the timing of suprasolidus prograde monazite growth.

Effects of open system behaviour on accessory minerals

Dissolution/re-precipitation of accessory minerals in metamorphic rocks due to the infiltration of an externally derived fluid has been documented in experiments and studies of natural samples (e.g., Tomaschek et al. 2003; Crowley et al. 2008; Harlov and Hetherington 2010; Blereau et al. 2016). The careful integration of petrography with the chemistries of these minerals can be used to provide information on the timing of fluid infiltration (e.g., Williams et al. 2011) and on fluid chemistry (e.g., Harlov et al. 2011; Taylor et al. 2014; Shazia et al. 2015). However, the timing of fluid ingress relative to the metamorphic history and the chemistry of these fluids are highly variable and should to be assessed on a case-by-case basis.

For migmatites that have lost melt, the extraction of melt saturated in Zr or LREE will also change the effective composition of the residuum (e.g., Rapp et al. 1987), which has subsequent consequences for zircon and monazite dissolution. For example, consider the muscovite to K-feldspar melting reaction for the metapelite along the clockwise *P–T* path at ~730 °C and 7.5 kbar and consider bulk rock values of 150 ppm for LREE and Zr. This reaction involves a large positive volume change (e.g., Powell et al. 2005), which may be accommodated by melt extraction from the rock. When muscovite is exhausted the rock contains ~9 mol% melt. The saturation concentrations of LREE and Zr in the melt at this point are 195 ppm and 137 ppm, respectively. Assuming 8 mol% melt (approximately equivalent to vol%) is extracted (leaving 1 mol% in the rock along grain boundaries), mass balance can be used to determine the amount of Zr and LREE left in the system. For this example, the effective concentration of LREE in the rock decreases to 146 ppm and the effective concentration of Zr increases to 151 ppm. Although the changes to the bulk composition are minor in this example, monazite is predicted to be completely consumed at lower T and zircon may persist to higher T than for closed system (Fig. 13). For rocks with very low concentrations of Zr and LREE, melt extraction can have a more significant impact on the effective bulk concentrations of these elements and the stability of zircon and monazite (e.g., Yakymchuk and Brown 2014b).

Inclusion/host relationships

An important consideration for accessory mineral reactivity is their inclusion in the major rock-forming minerals (Watson et al. 1989; Bea et al. 2006). Watson et al. (1989) showed that for a Himalayan migmatite sample roughly 78% of the zircon mass is located along grain boundaries and the remaining 28% is included in major minerals (predominately biotite and garnet), though how representative this is of typical migmatitic gneiss is unknown. Consequently, the breakdown of major minerals may liberate accessory minerals into the reacting volume of the rock that would otherwise be sequestered away.

Inclusions of zircon and monazite that are isolated from the reaction volume may also reduce the effective bulk rock concentration of Zr and LREE (Yakymchuk and Brown 2014b). For example, consider a bulk rock composition of LREE with half of the monazite sequestered away from the reaction volume as inclusions. For a metapelite with a bulk rock LREE composition of 300 ppm and considering that half of this is unavailable, the effective concentration of LREE is 150 ppm. The complete dissolution of monazite is modelled to occur at 830 °C in contrast to 870 °C for the scenario where all monazite is available for reaction (Fig. 13f). Therefore, applying the models in Figure 13 to natural examples requires an estimate of the amount of zircon or monazite sequestered away from the reaction volume as inclusions as well as an estimate of the amount of Zr and LREE locked away in the major minerals.

The heterogeneous distribution of melt and minerals in high-grade metamorphic rocks also has implications for the dissolution and preservation of accessory minerals. Even for an initially homogenous protolith, in situ melt may be spatially associated with peritectic minerals in isolated patches; this produces a heterogeneous melt framework (e.g., White et al. 2004). Zircon and monazite proximal to the zones of incipient melting and in chemical communication with this melt are more likely to contribute to Zr and LREE saturation of the melt whereas more distal grains may not. Consequently, detrital, inherited or prograde (subsolidus or early suprasolidus) zircon and monazite are more likely to be preserved in domains away from incipient melting whereas post-peak and retrograde zircon and monazite may be spatially associated with in situ leucosome.

CONCLUDING REMARKS

One important facet of petrochronology is to link the ages of accessory mineral chronometers to the P–T information obtained from major rock-forming minerals in metamorphic rocks. The growth and consumption of major minerals is important because these minerals: (1) may contain the necessary essential structural constituents to promote accessory mineral growth directly from their breakdown, (2) are repositories of the trace elements used to link accessory mineral chronometers to P–T conditions (e.g., Sr and Eu related to stability of plagioclase as well as Y and HREE reflecting the growth/consumption of garnet), (3) are important hosts for accessory mineral inclusions, and (4) play a role in controlling the component activities (e.g., a_{TiO_2}) along a P–T evolution. The main controls on accessory mineral behaviour differ between subsolidus and suprasolidus metamorphic conditions.

For subsolidus metamorphism, zircon is generally unreactive and monazite can grow during the prograde and retrograde segments. Linking ages from these accessory mineral to a metamorphic history requires an understanding of the major mineral reaction sequence as well as the behaviour of accessory minerals like xenotime, apatite and allanite. Major minerals such as garnet or plagioclase may experience growth and breakdown stages at any point along a *P–T* path: linking their behaviour to the trace element chemistries of accessory minerals requires an assessment of the reaction sequence for a particular rock along a well-constrained *P–T* path.

For suprasolidus metamorphism, phase equilibria modelling predicts that both zircon and monazite will be consumed during prograde metamorphism and grow during cooling and melt crystallization. However, this contrasts with some studies that have convincingly showed evidence for suprasolidus prograde zircon and monazite growth. For monazite, apatite dissolution may have contributed to minor prograde monazite growth if the anatectic melt is saturated in LREE. For zircon, solid-state breakdown of major minerals that contain appreciable quantities of Zr may facilitate prograde zircon growth. Ostwald ripening may also play a role in the prograde growth of both zircon and monazite, but this mechanism is still incompletely understood.

ACKNOWLEDGMENTS

We thank Mark Caddick and Dave Waters for thorough and perceptive reviews and Pierre Lanari for his patient editorial work. Nonetheless, the authors are responsible for any misinterpretations or omissions that persist. CY was partially funded by a National Sciences and Engineering Research Council of Canada Discovery Grant.

REFERENCES

Ague JJ (1991) Evidence for major mass transfer and volume strain during regional metamorphism of pelites. Geology 19:855–858

Ashley KT, Law RD (2015) Modeling prograde TiO_2 activity and its significance for Ti-in-quartz thermobarometry of pelitic metamorphic rocks. Contrib Mineral Petrol 169:1–7

Ayers JC, Miller C, Gorisch B, Milleman J (1999) Textural development of monazite during high-grade metamorphism: Hydrothermal growth kinetics, with implications for U, Th–Pb geochronology. Am Mineral 84:1766–1780

Baxter EF, Caddick MJ, Dragovic B (2017) Garnet: A rock–forming mineral petrochronometer. Rev Mineral Geochem 83:469–533

Baxter EF, Scherer EE (2013) Garnet geochronology: timekeeper of tectonometamorphic processes. Elements 9:433–438

Bea F, Montero P (1999) Behavior of accessory phases and redistribution of Zr, REE, Y, Th, and U during metamorphism and partial melting of metapelites in the lower crust: an example from the Kinzigite Formation of Ivrea-Verbano, NW Italy. Geochim Cosmochim Acta 63:1133–1153

Bea F, Pereira MD, Stroh A (1994) Mineral/leucosome trace-element partitioning in a peraluminous migmatite (a laser ablation-ICP-MS study). Chem Geol 117:291–312

Bea F, Montero P, Ortega M (2006) A LA–ICP–MS evaluation of Zr reservoirs in common crustal rocks: implications for Zr and Hf geochemistry, and zircon-forming processes. Can Mineral 44:693–714

Berman RG (1988) Internally-consistent thermodynamic data for minerals in the system $Na_2O–K_2O–CaO–MgO–FeO–Fe_2O_3–Al_2O_3–SiO_2–TiO_2–H_2O–CO_2$. J Petrol 29:445–522

Bingen B, Austrheim H, Whitehouse M (2001) Ilmenite as a source for zirconium during high-grade metamorphism? Textural evidence from the Caledonides of Western Norway and implications for zircon geochronology. J Petrol 42:355–375

Blereau E, Clark C, Taylor RJM, Johnson TJ, Fitzsimons I, Santosh M (2016) Constraints on the timing and conditions of high-grade metamorphism, charnockite formation and fluid–rock interaction in the Trivandrum Block, southern India. J Metamorph Geol 34:527–549.

Boehnke P, Watson EB, Trail D, Harrison TM, Schmitt AK (2013) Zircon saturation re-revisited. Chem Geol 351:324–334

Brown M (1998) Unpairing metamorphic belts: *P–T* paths and a tectonic model for the Ryoke Belt, southwest Japan. J Metamorph Geol 16:3–22

Brown M (2014) The contribution of metamorphic petrology to understanding lithosphere evolution and geodynamics. Geosci Front 5:553–569

Brown M, Kothonen FJ (2009) Some remarks on melting and extreme metamorphism of crustal rocks. *In*: Physics and Chemistry of the Earth's Interior. Gupta AK, Dasgupta S (eds) Springer, p 67–87

Caddick MJ, Kohn MJ (2013) Garnet: Witness to the evolution of destructive plate boundaries. Elements 9:427–432

Carlson WD (1999) The case against Ostwald ripening of porphyroblasts. Can Mineral 37:403–414

Carlson WD (2000) The case against Ostwald ripening of porphyroblasts: Reply. Can Mineral 38:1029–1031

Catlos EJ, Harrison TM, Kohn MJ, Grove M, Ryerson FJ, Manning CE, Upreti BN (2001) Geochronologic and thermobarometric constraints on the evolution of the Main Central Thrust, central Nepal Himalaya. J Geophys Res B: Solid Earth 106:16177–16204

Chakraborty S, Ganguly J (1992) Cation diffusion in aluminosilicate garnets: experimental determination in spessartine-almandine diffusion couples, evaluation of effective binary diffusion coefficients, and applications. Contrib Mineral Petrol 111:74–86

Chambers JA, Kohn MJ (2012) Titanium in muscovite, biotite, and hornblende: Modeling, thermometry, and rutile activities of metapelites and amphibolites. Am Mineral 97:543–555

Chen Y-X, Zheng Y-F, Hu Z (2013) Synexhumation anatexis of ultrahigh-pressure metamorphic rocks: petrological evidence from granitic gneiss in the Sulu orogen. Lithos 156:69–96

Clark DJ, Hensen BJ, Kinny PD (2000) Geochronological constraints for a two-stage history of the Albany–Fraser Orogen, Western Australia. Precambrian Res 102:155–183

Clark C, Fitzsimons ICW, Healy D, Harley SL (2011) How does the continental crust get really hot? Elements 7:235–240

Clark C, Kirkland CL, Spaggiari CV, Oorschot C, Wingate MTD, Taylor RJ (2014) Proterozoic granulite formation driven by mafic magmatism: An example from the Fraser Range Metamorphics, Western Australia. Precambrian Res 240:1–21

Clarke GL, Guiraud M, Powell R, Burg JP (1987) Metamorphism in the Olary Block, South Australia: compression with cooling in a Proterozoic fold belt. J Metamorph Geol 5:291–306

Clemens JD, Vielzeuf D (1987) Constraints on melting and magma production in the crust. Earth Planet Sci Lett 86:287–306

Collins WJ, Vernon RH (1991) Orogeny associated with anticlockwise *PTt* paths: Evidence from low-*P*, high-*T* metamorphic terranes in the Arunta inlier, central Australia. Geology 19:835–838

Connolly JAD, Petrini K (2002) An automated strategy for calculation of phase diagram sections and retrieval of rock properties as a function of physical conditions. J Metamorph Geol 20:697–708

Copeland RA, Frey FA, Wones DR (1971) Origin of clay minerals in a Mid-Atlantic Ridge sediment. Earth Planet Sci Lett 10:186–192

Corrie SL, Kohn MJ (2008) Trace-element distributions in silicates during prograde metamorphic reactions: Implications for monazite formation. J Metamorph Geol 26:451–464

Crowley JL, Brown RL, Gervais F, Gibson HD (2008) Assessing inheritance of zircon and monazite in granitic rocks from the Monashee Complex, Canadian Cordillera. J Petrol 49:1915–1929

Crowley JL, Waters DJ, Searle MP, Bowring SA (2009) Pleistocene melting and rapid exhumation of the Nanga Parbat massif, Pakistan: Age and *P–T* conditions of accessory mineral growth in migmatite and leucogranite. Earth Planet Sci Lett 288:408–420

Dawson GC, Krapež B, Fletcher IR, McNaughton NJ, Rasmussen B (2003) 1.2 Ga thermal metamorphism in the Albany–Fraser Orogen of Western Australia: consequence of collision or regional heating by dyke swarms? J Geol Soc London 160:29–37

de Capitani C, Brown TH (1987) The computation of chemical equilibrium in complex systems containing non-ideal solutions. Geochim Cosmochim Acta 51:2639–2652

de Capitani C, Petrakakis K (2010) The computation of equilibrium assemblage diagrams with Theriak/Domino software. Am Mineral 95:1006–1016

Degeling H, Eggins S, Ellis DJ (2001) Zr budgets for metamorphic reactions, and the formation of zircon from garnet breakdown. Mineral Mag 65:749–758

Diener JFA, Powell R (2010) Influence of ferric iron on the stability of mineral assemblages. J Metamorph Geol 28:599–613

Dragovic B, Samanta LM, Baxter EF, Selverstone J (2012) Using garnet to constrain the duration and rate of water-releasing metamorphic reactions during subduction: an example from Sifnos, Greece. Chem Geol 314:9–22

Dragovic B, Baxter EF, Caddick MJ (2015) Pulsed dehydration and garnet growth during subduction revealed by zoned garnet geochronology and thermodynamic modeling, Sifnos, Greece. Earth Planet Sci Lett 413:111–122

Dragovic B, Guevara VE, Caddick MJ, Baxter EF, Kylander-Clark ARC (2016) A pulse of cryptic granulite-facies metamorphism in the Archean Wyoming Craton revealed by Sm–Nd garnet and U–Pb monazite geochronology. Precambrian Res 283:24–49

Duc-Tin Q, Keppler H (2015) Monazite and xenotime solubility in granitic melts and the origin of the lanthanide tetrad effect. Contrib Mineral Petrol 169:1–26

Enami M, Liou JG, Mattinson CG (2004) Epidote minerals in high *P/T* metamorphic terranes: Subduction zone and high-to ultrahigh-pressure metamorphism. Rev Mineral Geochem 56:347–398

Engi M (2017) Petrochronology based on REE–minerals: monazite, allanite, xenotime, apatite. Rev Mineral Geochem 83:365–418

England PC, Thompson AB (1984) Pressure–temperature–time paths of regional metamorphism I. Heat transfer during the evolution of regions of thickened continental crust. J Petrol 25:894–928

Evans TP (2004) A method for calculating effective bulk composition modification due to crystal fractionation in garnet-bearing schist: implications for isopleth thermobarometry. J Metamorph Geol 22:547–557

Ewing TA, Rubatto D, Hermann J (2014) Hafnium isotopes and Zr/Hf of rutile and zircon from lower crustal metapelites (Ivrea–Verbano Zone, Italy): implications for chemical differentiation of the crust. Earth Planet Sci Lett 389:106–118

Ewing TA, Rubatto D, Beltrando M, Hermann J (2015) Constraints on the thermal evolution of the Adriatic margin during Jurassic continental break-up: U–Pb dating of rutile from the Ivrea–Verbano Zone, Italy. Contrib Mineral Petrol 169:1–22

Ferry JM, Watson EB (2007) New thermodynamic models and revised calibrations for the Ti-in-zircon and Zr-in-rutile thermometers. Contrib Mineral Petrol 154:429–437

Finger F, Krenn E (2007) Three metamorphic monazite generations in a high-pressure rock from the Bohemian Massif and the potentially important role of apatite in stimulating polyphase monazite growth along a *PT* loop. Lithos 95:103–115

Finger F, Krenn E, Schulz B, Harlov D, Schiller D (2016) "Satellite monazites" in polymetamorphic basement rocks of the Alps: Their origin and petrological significance. Am Mineral 101:1094–1103

Fitzsimons IC, Kinny PD, Wetherley S, Hollingsworth DA (2005) Bulk chemical control on metamorphic monazite growth in pelitic schists and implications for U–Pb age data. J Metamorph Geol 23:261–277

Florence FP, Spear FS (1993) Influences of reaction history and chemical diffusion on PT calculations for staurolite schists from the Littleton Formation, northwestern New Hampshire. Am Mineral 78:345–359

Foster G, Kinny P, Vance D, Prince C, Harris N (2000) The significance of monazite U–Th–Pb age data in metamorphic assemblages; a combined study of monazite and garnet chronometry. Earth Planet Sci Lett 181:327–340

Foster G, Gibson H, Parrish RR, Horstwood MSA, Fraser J, Tindle A (2002) Textural, chemical and isotopic insights into the nature and behaviour of metamorphic monazite. Chem Geol 191:183–207

Foster G, Parrish RR, Horstwood MS, Chenery S, Pyle J, Gibson HD (2004) The generation of prograde *P–T–t* points and paths; a textural, compositional, and chronological study of metamorphic monazite. Earth Planet Sci Lett 228:125–142

Franz G, Andrehs G, Rhede D (1996) Crystal chemistry of monazite and xenotime from Saxothuringian-Moldanubian metapelites, NE Bavaria, Germany. Euro J Mineral:1097–1118

Franz G, Morteani G, Rhede D (2015) Xenotime-(Y) formation from zircon dissolution–precipitation and HREE fractionation: an example from a metamorphosed phosphatic sandstone, Espinhaço fold belt (Brazil). Contrib Mineral Petrol 170:1–22

Fraser G, Ellis D, Eggins S (1997) Zirconium abundance in granulite-facies minerals, with implications for zircon geochronology in high-grade rocks. Geology 25:607–610

Frei D, Liebscher A, Franz G, Dulski P (2004) Trace element geochemistry of epidote minerals. Rev Mineral Geochem 56:553–605

Frost BR, Chamberlain KR, Schumacher JC (2001) Sphene (titanite): phase relations and role as a geochronometer. Chem Geol 172:131–148

Fyfe WS (1973) The granulite facies, partial melting and the Archaean crust. Philos Trans R Soc London, Ser A 273:457–461

Gaidies F, Abart R, De Capitani C, Schuster R, Connolly JAD, Reusser E (2006) Characterization of polymetamorphism in the Austroalpine basement east of the Tauern Window using garnet isopleth thermobarometry. J Metamorph Geol 24:451–475

Gaidies F, De Capitani C, Abart R (2008) THERIA_G: a software program to numerically model prograde garnet growth. Contrib Mineral Petrol 155:657–671

Gaidies F, Pattison DRM, De Capitani C (2011) Toward a quantitative model of metamorphic nucleation and growth. Contrib Mineral Petrol 162:975–993

Gasser D, Bruand E, Rubatto D, Stüwe K (2012) The behaviour of monazite from greenschist facies phyllites to anatectic gneisses: an example from the Chugach Metamorphic Complex, southern Alaska. Lithos 134:108–122

Gervasoni F, Klemme S, Rocha-Júnior ERV, Berndt J (2016) Zircon saturation in silicate melts: a new and improved model for aluminous and alkaline melts. Contrib Mineral Petrol 171:1–12

Ghent ED (1976) Plagioclase–garnet–Al_2SiO_5–quartz: a potential geobarometer–geothermometer. Am Mineral 6:710–714

Gibson HD, Carr SD, Brown RL, Hamilton MA (2004) Correlations between chemical and age domains in monazite, and metamorphic reactions involving major pelitic phases: an integration of ID-TIMS and SHRIMP geochronology with Y–Th–U X-ray mapping. Chem Geol 211:237–260

Grapes RH, Hoskin PWO (2004) Epidote group minerals in low–medium pressure metamorphic terranes. Rev Mineral Geochem 56:301–345

Green ECR, White RW, Diener JFA, Powell R, Palin RM (2016) Activity–composition relations for the calculation of partial melting equilibria in metabasic rocks. J Metamorph Geol, doi: 10.1111/jmg.12211

Gromet LP, Silver LT (1983) Rare earth element distributions among minerals in a granodiorite and their petrogenetic implications. Geochim Cosmochim Acta 47:925–939

Guernina S, Sawyer EW (2003) Large-scale melt-depletion in granulite terranes: An example from the Archean Ashuanipi Subprovince of Quebec. J Metamorph Geol 21:181–201

Guevara VE, Caddick MJ (2016) Shooting at a moving target: phase equilibria modelling of high-temperature metamorphism. J Metamorph Geol 34:209–235.

Hacker BR, Andersen TB, Johnston S, Kylander-Clark ARC, Peterman EM, Walsh EO, Young D (2010) High-temperature deformation during continental-margin subduction & exhumation: The ultrahigh-pressure Western Gneiss Region of Norway. Tectonophysics 480:149–171

Hacker BR, Kylander-Clark ARC, Holder R, Andersen TB, Peterman EM, Walsh EO, Munnikhuis JK (2015) Monazite response to ultrahigh-pressure subduction from U–Pb dating by laser ablation split stream. Chem Geol 409:28–41

Harlov DE, Hetherington CJ (2010) Partial high-grade alteration of monazite using alkali-bearing fluids: Experiment and nature. Am Mineral 95:1105–1108

Harlov DE, Wirth R, Förster H-J (2005) An experimental study of dissolution–reprecipitation in fluorapatite: fluid infiltration and the formation of monazite. Contrib Mineral Petrol 150:268–286

Harlov DE, Wirth R, Hetherington CJ (2011) Fluid-mediated partial alteration in monazite: the role of coupled dissolution–reprecipitation in element redistribution and mass transfer. Contrib Mineral Petrol 162:329–348

Harris NBW, Caddick M, Kosler J, Goswami S, Vance D, Tindle AG (2004) The pressure–temperature–time path of migmatites from the Sikkim Himalaya. J Metamorph Geol 22:249–264

Harrison TM, Watson EB (1983) Kinetics of zircon dissolution and zirconium diffusion in granitic melts of variable water content. Contrib Mineral Petrol 84:66–72

Harrison TM, Watson EB (1984) The behavior of apatite during crustal anatexis: equilibrium and kinetic considerations. Geochim Cosmochim Acta 48:1467–1477

Hay DC, Dempster TJ (2009) Zircon behaviour during low-temperature metamorphism. J Petrol 50: 571–589.

Hayden LA, Watson EB, Wark DA (2008) A thermobarometer for sphene (titanite). Contrib Mineral Petrol 155:529–540

Hensen BJ (1971) Theoretical phase relations involving cordierite and garnet in the system $MgO–FeO–Al_2O_3–SiO_2$. Contrib Mineral Petrol 33:191–214

Hermann J, Rubatto D (2003) Relating zircon and monazite domains to garnet growth zones: age and duration of granulite facies metamorphism in the Val Malenco lower crust. J Metamorph Geol 21:833–852

Hermann J, Rubatto D, Korsakov A, Shatsky VS (2001) Multiple zircon growth during fast exhumation of diamondiferous, deeply subducted continental crust (Kokchetav Massif, Kazakhstan). Contrib Mineral Petrol 141:66–82

Hiess J, Nutman AP, Bennett VC, Holden P (2008) Ti-in-zircon thermometry applied to contrasting Archean metamorphic and igneous systems. Chem Geol 247:323–338

Hofmann AE, Baker MB, Eiler JM (2013) An experimental study of Ti and Zr partitioning among zircon, rutile, and granitic melt. Contrib Mineral Petrol 166:235–253

Holder RM, Hacker BR, Kylander-Clark ARC, Cottle JM (2015) Monazite trace-element and isotopic signatures of (ultra) high-pressure metamorphism: Examples from the Western Gneiss Region, Norway. Chem Geol 409:99–111

Holland TJB, Powell R (1998) An internally consistent thermodynamic data set for phases of petrological interest. J Metamorph Geol 16:309–343

Holland TJB, Powell R (2011) An improved and extended internally consistent thermodynamic dataset for phases of petrological interest, involving a new equation of state for solids. J Metamorph Geol 29:333–383

Hoskin PWO, Schaltegger U (2003) The composition of zircon and igneous and metamorphic petrogenesis. Rev Mineral Geochem 53:27–62

Huang WL, Wyllie PJ (1973) Melting relations of muscovite-granite to 35 kbar as a model for fusion of metamorphosed subducted oceanic sediments. Contrib Mineral Petrol 42:1–14

Jamieson RA, Beaumont C, Medvedev S, Nguyen MH (2004) Crustal channel flows: 2. Numerical models with implications for metamorphism in the Himalayan–Tibetan orogen. J Geophys Res B: Solid Earth 109:2156–2202

Janots E, Engi M, Berger A, Allaz J, Schwarz JO, Spandler C (2008) Prograde metamorphic sequence of REE minerals in pelitic rocks of the Central Alps: implications for allanite–monazite–xenotime phase relations from 250 to 610 C. J Metamorph Geol 26:509–526

Janots E, Engi M, Rubatto D, Berger A, Gregory C, Rahn M (2009) Metamorphic rates in collisional orogeny from in situ allanite and monazite dating. Geology 37:11–14

Johnson TE, White RW, Powell R (2008) Partial melting of metagreywacke: a calculated mineral equilibria study. J Metamorph Geol 26:837–853

Johnson TE, Clark C, Taylor RJM, Santosh M, Collins AS (2015) Prograde and retrograde growth of monazite in migmatites: An example from the Nagercoil Block, southern India. Geosci Front 6:373–387

Kapp P, Manning CE, Tropper P (2009) Phase-equilibrium constraints on titanite and rutile activities in mafic epidote amphibolites and geobarometry using titanite-rutile equilibria. J Metamorph Geol 27:509–521

Kawakami T, Yamaguchi I, Miyake A, Shibata T, Maki K, Yokoyama TD, Hirata T (2013) Behavior of zircon in the upper-amphibolite to granulite facies schist/migmatite transition, Ryoke metamorphic belt, SW Japan: constraints from the melt inclusions in zircon. Contrib Mineral Petrol 165:575–591

Kelsey DE, Powell R (2011) Progress in linking accessory mineral growth and breakdown to major mineral evolution in metamorphic rocks: a thermodynamic approach in the Na_2O–CaO–K_2O–FeO–MgO–Al_2O_3–SiO_2–H_2O–TiO_2–ZrO_2 system. J Metamorph Geol 29:151–166

Kelsey DE, Hand M (2015) On ultrahigh temperature crustal metamorphism: phase equilibria, trace element thermometry, bulk composition, heat sources, timescales and tectonic settings. Geosci Front 6:311–356

Kelsey DE, Clark C, Hand M (2008) Thermobarometric modelling of zircon and monazite growth in melt-bearing systems: examples using model metapelitic and metapsammitic granulites. J Metamorph Geol 26:199–212

Kingsbury JA, Miller CF, Wooden JL, Harrison TM (1993) Monazite paragenesis and U–Pb systematics in rocks of the eastern Mojave Desert, California, USA: implications for thermochronometry. Chem Geol 110:147–167

Kirkland CL, Spaggiari CV, Johnson TE, Smithies RH, Danišík M, Evans N, Wingate MTD, Clark C, Spencer C, Mikucki E, McDonald BJ (2016) Grain size matters: Implications for element and isotopic mobility in titanite. Precambrian Res 278:283–302

Kohn MJ (2008) *PTt* data from central Nepal support critical taper and repudiate large-scale channel flow of the Greater Himalayan Sequence. Geol Soc Am Bull 120:259–273

Kohn MJ (2017) Titanite petrochronology. Rev Mineral Geochem 83:419–441

Kohn MJ, Malloy MA (2004) Formation of monazite via prograde metamorphic reactions among common silicates: implications for age determinations. Geochim Cosmochim Acta 68:101–113

Kohn MJ, Spear F (2000) Retrograde net transfer reaction insurance for pressure–temperature estimates. Geology 28:1127–1130

Kohn MJ, Corrie SL, Markley C (2015) The fall and rise of metamorphic zircon. Am Mineral 100:897–908

Korhonen FJ, Brown M, Grove M, Siddoway CS, Baxter EF, Inglis JD (2012) Separating metamorphic events in the Fosdick migmatite–granite complex, West Antarctica. J Metamorph Geol 30:165–192

Korhonen FJ Brown M, Clark C, Bhattacharya S (2013a) Osumilite–melt interactions in ultrahigh temperature granulites: phase equilibria modelling and implications for the *P–T–t* evolution of the Eastern Ghats Province, India. J Metamorph Geol 31:881–907

Korhonen FJ, Clark C, Brown M, Bhattacharya S, Taylor R (2013b) How long-lived is ultrahigh temperature (UHT) metamorphism? Constraints from zircon and monazite geochronology in the Eastern Ghats orogenic belt, India. Precambrian Res 234:322–350

Korhonen FJ, Clark C, Brown M, Taylor RJM (2014) Taking the temperature of Earth's hottest crust. Earth Planet Sci Lett 408:341–354

Kylander-Clark ARC, Hacker BR, Cottle JM (2013) Laser-ablation split-stream ICP petrochronology. Chem Geol 345:99–112

Lanari P, Engi M (2017) Local bulk composition effects on metamorphic mineral assemblages. Rev Mineral Geochem 83:55–102

Mahar EM, Baker JM, Powell R, Holland TJB, Howell N (1997) The effect of Mn on mineral stability in metapelites. J Metamorph Geol 15:223–238

Mayne MJ, Moyen JF, Stevens G, Kaisl Aniemi L (2016) Rcrust: a tool for calculating path-dependent open system processes and application to melt loss. J Metamorph Geol 34: 663–682.

McClelland WC, Lapen TJ (2013) Linking time to the pressure–temperature path for ultrahigh-pressure rocks. Elements 9:273–279

Mezger K, Hanson GN, Bohlen SR (1989) High-precision U–Pb ages of metamorphic rutile: application to the cooling history of high-grade terranes. Earth Planet Sci Lett 96:106–118

Miyazaki K (1991) Ostwald ripening of garnet in high P/T metamorphic rocks. Contrib Mineral Petrol 108:118–128

Miyazaki K (1996) A numerical simulation of textural evolution due to Ostwald ripening in metamorphic rocks: A case for small amount of volume of dispersed crystals. Geochim Cosmochim Acta 60:277–290

Molina JF, Moreno JA, Castro A, Rodríguez C, Fershtater GB (2015) Calcic amphibole thermobarometry in metamorphic and igneous rocks: New calibrations based on plagioclase/amphibole Al–Si partitioning and amphibole/liquid Mg partitioning. Lithos 232:286–305

Montel, J-M Kornprobst, J Vielzeuf, D (2000) Preservation of old U–Th–Pb ages in shielded monazite: example from the Beni Bousera Hercynian kinzigites (Morocco). J Metamorph Geol 18:335–342

Montel J-M (1986) Experimental determination of the solubility of Ce-monazite in SiO_2–Al_2O_3–K_2O–Na_2O melts at 800 C, 2 kbar, under H_2O-saturated conditions. Geology 14:659–662

Morrissey LJ, Hand M, Raimondo T, Kelsey DE (2014) Long-lived high-*T*, low-*P* granulite facies metamorphism in the Arunta Region, central Australia. J Metamorph Geol 32:25–47

Moynihan DP, Pattison DRM (2013) An automated method for the calculation of *P–T* paths from garnet zoning, with application to metapelitic schist from the Kootenay Arc, British Columbia, Canada. J Metamorph Geol 31:525–548

Mulcahy SR, Vervoort JD, Renne PR (2014) Dating subduction-zone metamorphism with combined garnet and lawsonite Lu–Hf geochronology. J Metamorph Geol 32:515–533

Nemchin AA, Bodorkos S (2000) Zr and LREE concentrations in anatectic melt as a function of crystal size distributions of zircon and monazite in the source region. Geol Soc Am, Abstracts and Programs: 52286

Nemchin AA, Giannini LM, Bodorkos S, Oliver NHS (2001) Ostwald ripening as a possible mechanism for zircon overgrowth formation during anatexis: theoretical constraints, a numerical model, and its application to pelitic migmatites of the Tickalara Metamorphics, northwestern Australia. Geochim Cosmochim Acta 65:2771–2788

O'Brien PJ, Rötzler J (2003) High-pressure granulites: formation, recovery of peak conditions and implications for tectonics. J Metamorph Geol 21:3–20

Parrish RR (1990) U–Pb dating of monazite and its application to geological problems. Can J Earth Sci 27:1431–1450

Pattison DRM, Tinkham DK (2009) Interplay between equilibrium and kinetics in prograde metamorphism of pelites: an example from the Nelson aureole, British Columbia. J Metamorph Geol 27:249–279

Pattison DRM, Chacko T, Farquhar J, McFarlane CRM (2003) Temperatures of granulite-facies metamorphism: constraints from experimental phase equilibria and thermobarometry corrected for retrograde exchange. J Petrol 44:867–900

Pattison DRM, DeBuhr CL (2015) Petrology of metapelites in the Bugaboo aureole, British Columbia, Canada. J Metamorph Geol 33:437–462

Pichavant M, Montel J-M, Richard LR (1992) Apatite solubility in peraluminous liquids: Experimental data and an extension of the Harrison-Watson model. Geochim Cosmochim Acta 56:3855–3861

Powell R (1983) Processes in granulite-facies metamorphism. *In:* Proceedings of the Geochemical Group of the Mineralogical Society: Migmatites, melting and metamorphism. Atherton, MP, Gribble, CD (eds). Shiva, Nantwich, p 127–139

Powell R, Holland TJB (1988) An internally consistent dataset with uncertainties and correlations: 3. Applications to geobarometry, worked examples and a computer program. J Metamorph Geol 6:173–204

Powell R, Holland TJB (2008) On thermobarometry. J Metamorph Geol 26:155–179

Powell R, Holland T, Worley B (1998) Calculating phase diagrams involving solid solutions via non-linear equations, with examples using THERMOCALC. J Metamorph Geol 16:577–588

Powell R, Guiraud M, White RW (2005) Truth and beauty in metamorphic phase-equilibria: conjugate variables and phase diagrams. Can Mineral 43:21–33

Pyle JM, Spear FS (1999) Yttrium zoning in garnet: coupling of major and accessory phases during metamorphic reactions. Geol Mat Res 1:1–49

Pyle JM, Spear FS (2000) An empirical garnet (YAG)–xenotime thermometer. Contrib Mineral Petrol 138:51–58

Pyle JM, Spear FS, Rudnick RL, McDonough WF (2001) Monazite–xenotime–garnet equilibrium in metapelites and a new monazite–garnet thermometer. J Petrol 42:2083–2107

Rapp RP, Watson EB (1986) Monazite solubility and dissolution kinetics: implications for the thorium and light rare earth chemistry of felsic magmas. Contrib Mineral Petrol 94:304–316

Rapp RP, Ryerson FJ, Miller CF (1987) Experimental evidence bearing on the stability of monazite during crustal anaatexis. Geophys Res Lett 14:307–310

Rasmussen B (2005) Radiometric dating of sedimentary rocks: the application of diagenetic xenotime geochronology. Earth Sci Rev 68:197–243

Rasmussen B, Fletcher IR, Muhling JR (2011) Response of xenotime to prograde metamorphism. Contrib Mineral Petrol 162:1259–1277

Regis D, Warren CJ, Mottram CM, Roberts NMW (2016) Using monazite and zircon petrochronology to constrain the P–T–t evolution of the middle crust in the Bhutan Himalaya. J Metamorph Geol

Reno BL, Piccoli PM, Brown M, Trouw RAJ (2012) In situ monazite (U–Th)–Pb ages from the Southern Brasília Belt, Brazil: constraints on the high-temperature retrograde evolution of HP granulites. J Metamorph Geol 30:81–112

Roberts MP, Finger F (1997) Do U–Pb zircon ages from granulites reflect peak metamorphic conditions? Geology 25:319–322

Rocha BC, Moraes R, Möller A, Cioffi CR, Jercinovic MJ (2016) Timing of anatexis and melt crystallization in the Socorro–Guaxupé Nappe, SE Brazil: Insights from trace element composition of zircon, monazite and garnet coupled to U–Pb geochronology. Lithos

Rubatto D (2017) Zircon: The metamorphic mineral. Rev Mineral Geochem 83:261–295

Rubatto D (2002) Zircon trace element geochemistry: partitioning with garnet and the link between U–Pb ages and metamorphism. Chem Geol 184:123–138

Rubatto D, Hermann J (2003) Zircon formation during fluid circulation in eclogites (Monviso, Western Alps): implications for Zr and Hf budget in subduction zones. Geochim Cosmochim Acta 67:2173–2187

Rubatto D, Hermann J (2007) Experimental zircon/melt and zircon/garnet trace element partitioning and implications for the geochronology of crustal rocks. Chem Geol 241:38–61

Rubatto D, Williams IS, Buick IS (2001) Zircon and monazite response to prograde metamorphism in the Reynolds Range, central Australia. Contrib Mineral Petrol 140:458–468

Rubatto D, Chakraborty S, Dasgupta S (2013) Timescales of crustal melting in the higher Himalayan crystallines (Sikkim, Eastern Himalaya) inferred from trace element-constrained monazite and zircon chronology. Contrib Mineral Petrol 165:349–372

Schaltegger U (2007) Hydrothermal zircon. Elements 3:51–79

Schaltegger U, Fanning CM, Günther D, Maurin JC, Schulmann K, Gebauer D (1999) Growth, annealing and recrystallization of zircon and preservation of monazite in high-grade metamorphism: conventional and in-situ U–Pb isotope, cathodoluminescence and microchemical evidence. Contrib Mineral Petrol 134:186–201

Sheppard S, Rasmussen B, Muhling JR, Farrell TR, Fletcher IR (2007) Grenvillian-aged orogenesis in the Palaeoproterozoic Gascoyne Complex, Western Australia: 1030–950 Ma reworking of the Proterozoic Capricorn Orogen. J Metamorph Geol 25:477–494

Shaw DM (1956) Geochemistry of pelitic rocks. Part III: Major elements and general geochemistry. Geol Soc Am Bull 67:919–934

Shazia JR, Harlov DE, Suzuki K, Kim SW, Girish-Kumar M, Hayasaka Y, Ishwar-Kumar C, Windley BF, Sajeev K (2015) Linking monazite geochronology with fluid infiltration and metamorphic histories: Nature and experiment. Lithos 236:1–15

Skora S, Blundy J (2012) Monazite solubility in hydrous silicic melts at high pressure conditions relevant to subduction zone metamorphism. Earth Planet Sci Lett 321:104–114

Sláma J, Košler J, Pedersen RB (2007) Behaviour of zircon in high-grade metamorphic rocks: evidence from Hf isotopes, trace elements and textural studies. Contrib Mineral Petrol 154:335–356

Smith HA, Barreiro B (1990) Monazite U–Pb dating of staurolite grade metamorphism in pelitic schists. Contrib Mineral Petrol 105:602–615

Smye AJ, Stockli DF (2014) Rutile U–Pb age depth profiling: A continuous record of lithospheric thermal evolution. Earth Planet Sci Lett 408:171–182

Spear FS (1995) Metamorphic phase equilibria and pressure–temperature–time paths. Mineral Soc Am. Washington

Spear FS (2010) Monazite–allanite phase relations in metapelites. Chem Geol 279:55–62

Spear FS, Florence FP (1992) Thermobarometry in granulites: pitfalls and new approaches. Precambrian Res 55:209–241

Spear FS, Pyle JM (2002) Apatite, monazite, and xenotime in metamorphic rocks. Rev Mineral Geochem 48:293–335

Spear FS, Pyle JM (2010) Theoretical modeling of monazite growth in a low-Ca metapelite. Chem Geol 273:111–119

Spear FS, Pattison DRM, Cheney JT (2016) The metamorphosis of metamorphic petrology. Geol Soc Am Sp Pap 523:SPE523-502

Spencer KJ, Hacker BR, Kylander-Clark ARC, Andersen TB, Cottle JM, Stearns MA, Poletti JE, Seward GGE (2013) Campaign-style titanite U–Pb dating by laser-ablation ICP: implications for crustal flow, phase transformations and titanite closure. Chem Geol 341:84–101

Stearns MA, Hacker BR, Ratschbacher L, Rutte D, Kylander-Clark ARC (2015) Titanite petrochronology of the Pamir gneiss domes: Implications for middle to deep crust exhumation and titanite closure to Pb and Zr diffusion. Tectonics 34:784–802

Stepanov AS, Hermann J, Rubatto D, Rapp RP (2012) Experimental study of monazite/melt partitioning with implications for the REE, Th and U geochemistry of crustal rocks. Chem Geol 300:200–220

Štípská P, Hacker BR, Racek M, Holder R, Kylander-Clark ARC, Schulmann K, Hasalová P (2015) Monazite Dating of Prograde and Retrograde *P–T–d* paths in the Barrovian terrane of the Thaya window, Bohemian Massif. J Petrol 56:1007–1035

Stüwe K (1997) Effective bulk composition changes due to cooling: a model predicting complexities in retrograde reaction textures. Contrib Mineral Petrol 129:43–52

Stüwe K, Powell R (1995) *PT* paths from modal proportions: application to the Koralm Complex, Eastern Alps. Contrib Mineral Petrol 119:83–93

Sun S-S, McDonough W (1989) Chemical and isotopic systematics of oceanic basalts: implications for mantle composition and processes. Geol Soc London, Sp Pub 42:313–345

Suzuki K, Adachi M (1994) Middle Precambrian detrital monazite and zircon from the Hida gneiss on Oki-Dogo Island, Japan: their origin and implications for the correlation of basement gneiss of Southwest Japan and Korea. Tectonophysics 235:277–292

Symmes GH, Ferry JM (1992) The effect of whole-rock MnO content on the stability of garnet in pelitic schists during metamorphism. J Metamorph Geol 10:221–237

Taylor RJM, Clark C, Fitzsimons ICW, Santosh M, Hand M, Evans N, McDonald B (2014) Post-peak, fluid-mediated modification of granulite facies zircon and monazite in the Trivandrum Block, southern India. Contrib Mineral Petrol 168:1–17

Taylor RJM, Harley SL, Hinton RW, Elphick S, Clark C, Kelly NM (2015) Experimental determination of REE partition coefficients between zircon, garnet and melt: a key to understanding high-*T* crustal processes. J Metamorph Geol 33:231–248

Taylor-Jones K, Powell R (2015) Interpreting zirconium-in-rutile thermometric results. J Metamorph Geol 33:115–122

Thompson AB (1982) Dehydration melting of pelitic rocks and the generation of H$_2$O-undersaturated granitic liquids. Am J Sci 282:1567–1595

Thompson AB (1996) Fertility of crustal rocks during anatexis. Geol Soc Am Spec Pap 315:1–10

Thompson AB, England PC (1984) Pressure—temperature—time paths of regional metamorphism II. Their inference and interpretation using mineral assemblages in metamorphic rocks. J Petrol 25:929–955

Tikare V, Cawley JD (1998) Application of the Potts model to simulation of Ostwald ripening. J Am Ceram Soc 81:485–491

Tomaschek F, Kennedy AK, Villa IM, Lagos M, Ballhaus C (2003) Zircons from Syros, Cyclades, Greece—recrystallization and mobilization of zircon during high-pressure metamorphism. J Petrol 44:1977–2002

Tomkins HS, Powell R, Ellis DJ (2007) The pressure dependence of the zirconium-in-rutile thermometer. J Metamorph Geol 25:703–713

Vance D, Mahar E (1998) Pressure–temperature paths from *PT* pseudosections and zoned garnets: potential, limitations and examples from the Zanskar Himalaya, NW India. Contrib Mineral Petrol 132:225–245

Vance D, O'Nions RK (1990) Isotopic chronometry of zoned garnets: growth kinetics and metamorphic histories. Earth Planet Sci Lett 97:227–240

Vavra G, Schmid R, Gebauer D (1999) Internal morphology, habit and U–Th–Pb microanalysis of amphibolite-to-granulite facies zircons: geochronology of the Ivrea Zone (Southern Alps). Contrib Mineral Petrol 134:380–404

Vernon RH (1996) Problems with inferring *P–T–t* paths in low-P granulite facies rocks. J Metamorph Geol 14:143–153

Vielzeuf D, Schmidt MW (2001) Melting relations in hydrous systems revisited: application to metapelites, metagreywackes and metabasalts. Contrib Mineral Petrol 141:251–267

Walsh AK, Kelsey DE, Kirkland CL, Hand M, Smithies RH, Clark C, Howard HM (2015) *P–T–t* evolution of a large, long-lived, ultrahigh-temperature Grenvillian belt in central Australia. Gondwana Res 28:531–564

Wark DA, Watson EB (2006) TitaniQ: a titanium-in-quartz geothermometer. Contrib Mineral Petrol 152:743–754

Warren CJ, Grujic D, Cottle JM, Rogers NW (2012) Constraining cooling histories: rutile and titanite chronology and diffusion modelling in NW Bhutan. J Metamorph Geol 30:113–130

Watson EB (1996) Dissolution, growth and survival of zircons during crustal fusion: kinetic principles, geological models and implications for isotopic inheritance. Geol Soc Am Sp Pap 315:43–56

Watson EB, Harrison TM (1983) Zircon saturation revisited: temperature and composition effects in a variety of crustal magma types. Earth Planet Sci Lett 64:295–304

Watson EB, Vicenzi EP, Rapp RP (1989) Inclusion/host relations involving accessory minerals in high-grade metamorphic and anatectic rocks. Contrib Mineral Petrol 101:220–231

Watson EB, Wark DA, Thomas JB (2006) Crystallization thermometers for zircon and rutile. Contrib Mineral Petrol 151:413–433

Watt GR, Harley SL (1993) Accessory phase controls on the geochemistry of crustal melts and restites produced during water-undersaturated partial melting. Contrib Mineral Petrol 114:550–566

White RW, Powell R (2002) Melt loss and the preservation of granulite facies mineral assemblages. J Metamorph Geol 20:621–632

White RW, Powell R, Halpin JA (2004) Spatially-focussed melt formation in aluminous metapelites from Broken Hill, Australia. J Metamorph Geol 22:825–845

White RW, Pomroy NE, Powell R (2005) An in situ metatexite–diatexite transition in upper amphibolite facies rocks from Broken Hill, Australia. J Metamorph Geol 23:579–602

White RW, Powell R, Baldwin JA (2008) Calculated phase equilibria involving chemical potentials to investigate the textural evolution of metamorphic rocks. J Metamorph Geol 26:181–198

White RW, Stevens G, Johnson TE (2011) Is the crucible reproducible? Reconciling melting experiments with thermodynamic calculations. Elements 7:241–246

White RW, Powell R, Johnson TE (2014a) The effect of Mn on mineral stability in metapelites revisited: New *a–x* relations for manganese-bearing minerals. J Metamorph Geol 32:809–828

White RW, Powell R, Holland TJB, Johnson TE, Green ECR (2014b) New mineral activity–composition relations for thermodynamic calculations in metapelitic systems. J Metamorph Geol 32:261–286

White WM, Klein EM (2014) 4.13—Composition of the Oceanic Crust. *In:* Treatise on Geochemsitry (Second Edition). Holland, HD, Turekian, K (ed). Elsevier, Oxford, p 457–496

Williams ML, Jercinovic MJ, Harlov DE, Budzyń B, Hetherington CJ (2011) Resetting monazite ages during fluid-related alteration. Chem Geol 283:218–225

Wing BA, Ferry JM, Harrison TM (2003) Prograde destruction and formation of monazite and allanite during contact and regional metamorphism of pelites: petrology and geochronology. Contrib Mineral Petrol 145:228–250

Wolf MB, London D (1994) Apatite dissolution into peraluminous haplogranitic melts: an experimental study of solubilities and mechanisms. Geochim Cosmochim Acta 58:4127–4145

Wolf MB, London D (1995) Incongruent dissolution of REE-and Sr-rich apatite in peraluminous granitic liquids: Differential apatite, monazite, and xenotime solubilities during anatexis. Am Mineral 80:765–775

Wu CM (2015) Revised empirical garnet–biotite–muscovite–plagioclase geobarometer in metapelites. J Metamorph Geol 33:167–176

Xu H, Ye K, Song Y, Chen Y, Zhang J, Liu Q, Guo S (2013) Prograde metamorphism, decompressional partial melting and subsequent melt fractional crystallization in the Weihai migmatitic gneisses, Sulu UHP terrane, eastern China. Chem Geol 341:16–37

Yakymchuk C, Brown M (2014a) Consequences of open-system melting in tectonics. J Geol Soc London 171:21–40

Yakymchuk C, Brown M (2014b) Behaviour of zircon and monazite during crustal melting. J Geol Soc London 171:465–479

Yakymchuk C, Brown M, Clark C, Korhonen FJ, Piccoli PM, Siddoway CS, Taylor RJM, Vervoort JD (2015) Decoding polyphase migmatites using geochronology and phase equilibria modelling. J Metamorph Geol 33:203–230

Yardley BWD, Valley JW (1997) The petrologic case for a dry lower crust. J Geophys Res B: Solid Earth 102:12173–12185

Zack T, Kooijman E (2017) Petrology and geochronology of rutile. Rev Mineral Geochem 83:443–467

Zack T, Moraes R, Kronz A (2004) Temperature dependence of Zr in rutile: empirical calibration of a rutile thermometer. Contrib Mineral Petrol 148:471–488

Zack T, Stockli DF, Luvizotto GL, Barth MG, Belousova E, Wolfe MR, Hinton RW (2011) In situ U–Pb rutile dating by LA-ICP-MS: [208]Pb correction and prospects for geological applications. Contrib Mineral Petrol 162:515–530

Zuluaga CA, Stowell HH, Tinkham DK (2005) The effect of zoned garnet on metapelite pseudosection topology and calculated metamorphic *PT* paths. Am Mineral 90:1619–1628

Reviews in Mineralogy & Geochemistry
Vol. 83 pp. 55–102, 2017
Copyright © Mineralogical Society of America

3

Local Bulk Composition Effects on Metamorphic Mineral Assemblages

Pierre Lanari and Martin Engi

Institute of Geological Sciences
University of Bern
Baltzerstrasse 3
CH-3012 Bern
Switzerland

pierre.lanari@geo.unibe.ch

martin.engi@geo.unibe.ch

INTRODUCTION AND SCOPE

Plate tectonic forcing leads to changes in the physical conditions that affect the lithosphere. In response to such changes, notably the local temperature (T) and pressure (P), rocks evolve dynamically. Processes mostly involve mineral transformations, i.e., solid-state reactions, but (hydrous) fluids are often involved, and partial melting may occur in the Earth's middle and lower crust. While these chemical reactions reflect the tendency of natural systems to reduce their Gibbs free energy, metamorphic rocks commonly preserve textural and mineralogical relics, such as compositionally zoned minerals. Where relics are present, thermodynamic equilibrium clearly was not attained during the evolution of the rock.

Petrochronology seeks to establish a temporal framework of petrologic evolution, and for this purpose it is essential to determine the P–T conditions prevailing at several stages. When analyzing a rock sample it is thus critical:

(a) to recognize whether several stages of its evolution can be discerned,

(b) to document the minerals that formed or were coexisting at each stage, and

(c) to estimate at what physical conditions this happened.

If (and only if) a chronometer then can be associated to one of these stages—or better yet several chronometers to different stages—then the power of petrochronology becomes realizable.

This chapter is concerned with a basic dilemma that results directly from steps (b) and (c) above: P–T conditions are determined on the basis of mineral barometers and thermometers, which mostly rest on the assumption of chemical (or isotopic) equilibrium, yet the presence of relics is proof that thermodynamic equilibrium was not attained. One way out of the dilemma is to analyze reaction mechanisms and formulate a model based on non-equilibrium thermodynamics and kinetics (Lasaga 1998). While this can be fruitful for understanding fundamental aspects of metamorphic petrogenesis, there are more direct ways to address the limited scope needed for petrochronology. The alternative pursued here seeks to define the scale(s) of equilibration, i.e., to determine spatial limits within which chemical equilibration can reasonably be assumed.

The past fifty years have seen rapid developments in forward thermodynamic models aimed at retrieving the conditions of equilibration from local assemblages. These models are rooted in chemical equilibrium theory—the concept that rocks successively re-equilibrate along

1529-6466/17/0083-0003$05.00 (print)
1943-2666/17/0083-0003$05.00 (online)

http://dx.doi.org/10.2138/rmg.2017.83.3

at least part of their evolution or remain at least close to chemical equilibrium. Since the advent of electron probe micro-analysis (EPMA), mineral chemical data have been combined with textural observations in thin section, and in many rocks different local mineral assemblages have been documented. The question is what caused these differences: Do they represent temporally distinct stages of a single chemical system or coeval domains of locally different chemical composition? This question is known as the *N*-dimensional tie-line problem (Greenwood 1967): Can two mineral assemblages be related to one another by a balanced chemical reaction or must the observed differences in their mineralogy be attributed to differing bulk compositions? A rigorous answer is possible,[1] provided the frozen-in assemblages reflect chemical equilibrium at the time of formation—a hypothesis to be tested. In any case, equilibrium assemblages reflect the chemical composition of rocks, and for forward chemical modeling of assemblages that composition must be known. Textural evidence in polymineralic rocks typically suggests some chemical heterogeneity, as the modal distributions of the minerals (i.e., their volumetric abundance) commonly varies at centimeter- to millimeter-scale, or less. Chemical heterogeneity can be inherited, e.g., from interlayered strata, or it may form during metamorphism, by metamorphic differentiation (Orville 1969). Either way, the effects of chemical heterogeneity should be considered in forward modeling, especially in clearly domainal rocks. As petrochronological studies focus on dating minerals in their local context, petrogenetic conditions ought to be reliably determined by modeling at similarly small scale.

This papers examines some of the major causes for and consequences of chemical heterogeneity in rocks, which can affect the scale of equilibration and thus the modeled mineral assemblages. The central question addressed is: how can a realistic bulk composition be found, such that model results can be compared in detail to a documented sample? A completely systematic treatment is beyond the scope of this paper, but as some fundamental aspects are involved in applying equilibrium thermodynamics to metamorphic rocks, the necessary basics are presented in the first section. Some of the case studies discussed in the second half of this paper also invoke kinetic aspects. These are cited where necessary but are not reviewed here; several resources treat the kinetics of metamorphic processes (see for example the book of Lasaga 1998). In the following sections we analyze specific cases such as the porphyroblastic growth of garnet that fractionates the reactive bulk composition and the link with mineral inclusion chronology. We also develop in a more general way the formation and evolution of textural domains in rocks, including the chemical subsystems and potential gradients. In the last section, we review analytical methods to estimate local bulk compositions using standardized X-ray maps (and the program XMapTools, Lanari et al. 2014b). We then show some applications to petrochronological analysis, and we outline perspectives for future research.

THEORETICAL BASIS AND LIMITS
OF FORWARD THERMODYNAMIC ANALYSIS

Based the pioneering work of Bowen (1913) and Goldschmidt (1911), equilibrium thermodynamics has proven a powerful conceptual framework to develop forward chemical modeling. Recent progress in the accuracy and efficiency of such techniques has had a major impact on the evolution of metamorphic petrology (Spear et al. 2016). It now is straightforward to model at least some of the main transformational stages experienced by crystalline rocks (Yakymchuk et al. 2017, this volume and references therein). Many of the previously daunting technical obstacles, such as the need for sophisticated solution models in complex systems involving solid and fluid phases, now appear much less evident, as the available software is making use of thermodynamic databases. However, despite periodic updates of such databases,

---

[1] Greenwood showed that linear algebraic examination is sufficient to determine whether two samples (or assemblages) overlap in composition space. If they do, yet they show different assemblages, this necessarily reflects a difference in physical conditions (e.g., of pressure or temperature).

models for some mineral systems remain imperfectly calibrated. For many bulk compositions predicted phase relations are reliably known, but for others the models may be partly flawed, and this is easily overlooked. Furthermore, predictions based on classical thermodynamics are limited to systems in equilibrium, and the application to metamorphic rocks is not self-evident where they contain non-equilibrium phenomena. In particular, the definition of an appropriate equilibration volume suitable for modeling can be a challenging task, as natural rocks typically record several stages of the metamorphic history through compositional zoning, mineral relics and textural relationships. Nevertheless, our current understanding of the metamorphic processes in the Earth's crust has much improved due to the comparison of model predictions with the observed mineral assemblages and compositions. Such predictions rely on the application of forward chemical modeling based on the bulk rock composition of natural samples. In a recent review, Powell and Holland (2010) concluded that "the success of using forward equilibrium model to study metamorphic rocks provides, at least in part, an *a posteriori* justification of the assumption" [of equilibrium]. On the other hand, some authors have warned the community about the limits of equilibrium models and have demonstrated, for instance, that kinetic impediments to reaction may prevent metamorphic rocks from attaining rock-wide chemical equilibrium along their prograde crystallization paths (Carlson et al. 2015 and references therein). It is also increasingly recognized that nucleation and growth can be kinetically inhibited (Gaidies et al. 2011; Spear et al. 2014), and so models based on reaction affinity rather than equilibrium thermodynamics may be required.

Gibbs free energy minimization

The equilibrium phase assemblage in a chemically closed system for any specified set of conditions is classically determined using the principle of Gibbs free energy minimization. The chemical equilibrium condition of any system of fixed composition is:

$$\text{minimize}\, G_{\text{system}} = \sum_k n_k g_k \tag{1}$$

where n_k is the number of moles and g_k the molar Gibbs free energy of phase k. The interested reader would note that this chemical equilibrium condition is subject to two additional constraints: (1) all $n_k \geq 0$ with $\sum n_k = 1$ and (2) the mass balance equations (e.g., $\sum x_k^i n_k = x_{\text{system}}^i$ for component i with x_k^i the composition in phase k) are satisfied. The computation of the equilibrium phase assemblage requires values for the molar Gibbs free energy of pure phases and a formulation of the relation between composition and activity for each solution phase. Consequently Gibbs free energy minimization relies on thermodynamic data for the standard state thermodynamic properties, *PVT*-equations of state, and solution models for solids and fluids. Successful application of forward equilibrium models requires an internally consistent thermodynamic database with accurate solution models and a robust estimate for the bulk composition of the system.

The development and successive improvements of internally-consistent thermodynamic databases for solid and fluids (Berman 1988; Holland and Powell 1988, 1998, 2011; Gottschalk 1996; Miron et al. 2016) are one of the greatest advances in the field of metamorphic petrology in the last three decades. Using the Gibbs free energy minimization principle, these databases facilitate the creation of phase diagrams that describe which mineral phases are stable as a function of temperature (*T*), pressure (*P*), H_2O and CO_2 activity (aH_2O, aCO_2), oxygen fugacity (fO_2) or chemical potential of a component i (μ_i). A significant move occurred in the late 80's with the development of modeling programs including Thermocalc (Powell et al. 1998; Powell and Holland 2008), Perple_X (Connolly 1990, 2005) and Theriak-Domino (De Capitani and Brown 1987; de Capitani and Petrakakis 2010). Such programs can be used to generate isochemical equilibrium phase diagrams (elsewhere called 'pseudosection', but this term may engender confusion as noted by Spear et al. 2016), which map assemblages predicted at minimum free energy for the chemical system of interest (Eqn. 1). It is owing to the remarkable power and availability of these programs that isochemical equilibrium phase diagrams have become so widely used.

The concept of chemical equilibrium and its application to metamorphic rocks

A volume of rock is said to be in chemical equilibrium when the quantities and compositions of the phases involved do not change in time without an external influence. Equilibrium is a macroscopic concept that is defined in terms of macroscopic variables such as P, T and chemical potential (or activity or composition) in the considered system, which is commonly known as the equilibration volume. It is important to note that the macroscopic variables are large-scale average quantities, which are subject to fluctuations at a microscopic scale. Equilibrium thermodynamics requires that, in response to any change in P–T conditions, the equilibration volume (with a given composition vector $\mathbf{x}$) will adjust its phase assemblage, i.e., the modes and compositions of all minerals, in an attempt to reach or maintain chemical equilibrium. By definition, the principles of chemical equilibrium predict the final state of a system, independent of the path by which the system arrived at its present state (see below).

A schematic thin section of metapelite consisting of $Grt + Chl + Ms + Bt + Pl + Qz + H_2O$ (mineral abbreviations are from Whitney and Evans 2010) at ~525 °C and 6 kbar is shown in Figure 1a. The mineral phases are assumed to be homogeneous in composition and coexist at chemical equilibrium for the given P–T conditions. Any increase in P or T would require

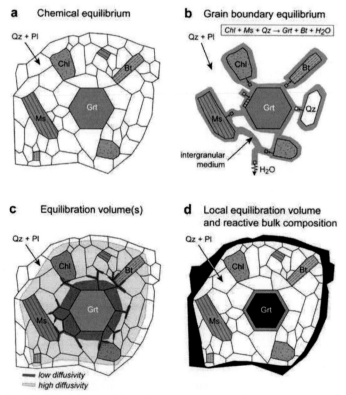

Figure 1. Schematic thin section consisting of garnet, chlorite, muscovite, biotite in a quartz–plagioclase matrix. (a) Chemical equilibrium model: all mineral phases are homogeneous and coexist in equilibrium at the P–T conditions of interest. (b) Grain boundary equilibrium model and assemblage re-equilibration under higher temperature conditions through the reaction $Chl + Ms + Qz \rightarrow Grt + Bt + H_2O$. (c) Two equilibrium volumes are shown, one for an element (dark gray) with low diffusivity, a second one (light gray) assuming fast transport. (d) The local equilibrium volume involves only the rim of the zoned minerals (here garnet) and a homogenous domain (matrix) of a section. The reactive bulk composition is the composition of the local equilibrium volume, which excludes the domains shown in black.

adjustments through the reaction $Chl + Ms + Qz \rightarrow Grt + Bt + H_2O$ (Fig. 1b) to attain a new equilibrium state. As the solid-state transformations are achieved by dissolution, precipitation and transport of the elemental species through the intergranular medium, the problem can be reduced to a grain boundary equilibrium model. This concept is a step forward in the application of chemical equilibrium theory to analyze metamorphic rocks but it requires an approximation of the bulk composition of the intergranular medium. However some fundamental concerns appear as soon as we abandon this idealized scenario. Most of them are related to the size and the composition of the equilibration volume. From a theoretical point of view, transport and reaction kinetics can establish and maintain chemical potential gradients in the system and avoid the achievement of chemical equilibrium, i.e., an equilibration volume cannot be defined. An additional concern comes from the observation that a metamorphic rock is seldom truly homogeneous and often contains evidence of disequilibrium, even at grain-scale, such as chemical zoning or mineral relics. In this case it can be challenging to define an appropriate equilibration volume to be modeled. Those cases are separately addressed below.

Reaction kinetics. Equilibrium thermodynamic models in closed systems are based on the equilibrium condition (Eqn. 1), and they neglect the fact that solid-state reactions cannot strictly proceed to equilibrium (Carlson et al. 2015 and references therein) simply because the driving force—chemical potential gradients between reactants and products—would be eliminated. Consequently, equilibrium thermodynamics ignores (1) the time-courses of mineral assemblage transformations, (2) the specific mechanism by which rocks crystallize, and (3) the rate at which that crystallization occurs. Any evolution of texture from metastable states to a stable state is a transient feature, and intermediate states are completely obliterated if the rock reaches equilibrium. From a microscopic point of view, the successive transformations are achieved via re-equilibration, e.g., dissolution and precipitation or pseudomorphic replacement of mineral phases and by inter- and intragranular diffusion of the chemical components (Fig. 1b). The driving forces for diffusion are the chemical potential gradients that are established between minerals such as the reactants and products of a metamorphic reaction or between different groups of minerals, such as between domains (Fisher 1973). In this context, equilibration is attained when the chemical potentials of each the chemical component in the system of interest is equalized across a rock. Several studies on equilibrium have demonstrated that deformation and fluid influx can help to achieve equilibration (Brodie and Rutter 1985; Wintsch 1985; Foster 1991; Erambert and Austrheim 1993).

Frozen chemical potential gradients. The modeling of rock volumes that attained chemical equilibrium, such as shown in the idealized scenario of Figure 1a, is relatively straightforward using Equation (1) and an appropriate thermodynamic database. However, nature provides abundant examples of diffusion-controlled structures such as coronas, which typically result from incomplete metamorphic transformation due to low rates of elemental movement by intragranular diffusion. Indeed chemical zoning in coronas may also reflect chemical potential gradients, which must be distinguished from zoning due to changes in *P–T* conditions (Indares and Rivers 1995). Diffusion-controlled structures can be identified in thin section because they typically have a strong spatial organization with textural mineral zones showing sharp changes in compositions at zone boundaries. Reaction fronts are arranged in an orderly sequence of increasing or decreasing chemical potential (Fisher 1973). Non-equilibrium thermodynamics (also known as non-classical or irreversible thermodynamics) has been used to analyze such inhomogeneous systems; their study requires the knowledge of the rates of reactions (Fisher 1973; Foster 1977, 1986, 1999; Fisher and Lasaga 1981; Johnson and Carlson 1990; Carlson and Johnson 1991; Lasaga 1998).

Mineral compositional zoning and mineral relics. Growth related compositional zoning and mineral relics are clear evidence of disequilibrium in metamorphic rocks. It was recognized early after the introduction of the EPMA to the geosciences that garnet porphyroblasts can freeze compositional zoning (Hollister 1966). Such textures reflect changes in $P-T$ conditions, indicating that the crystal as a whole did not equilibrate with the matrix during growth (Atherton 1968; Tracy et al. 1976). In such cases some part of the minerals that formed earlier (e.g., garnet) is effectively removed from the reactive part of the rock (the effects of metamorphic fractionation are intensely discussed below, in the section related to garnet porphyroblast growth). A few years later, Tracy (1982) reported a list of 18 metamorphic minerals showing similar compositional zoning. As most metamorphic rocks exhibit compositionally zoned minerals, equilibrium at best was established at a local scale only (Fig. 1c). Hence, to apply equilibrium thermodynamics, more sophisticated models are required.

The principle of local equilibrium. The principle of local equilibrium (or mosaic equilibrium) restricts the investigation to subsystems that are small enough that the attainment of an equilibrium state can be assumed (Korzhinskii 1936, 1959; Thompson 1955, 1959, 1970). Local equilibrium is considered only at a scale over which variations in chemical potential— and all physical variables—are negligible. Since chemical heterogeneity in a sample is primarily evident in solid solutions, local equilibria address only those situations where the minerals (or individual zones thereof) are chemically uniform. The principle of local equilibrium significantly extends thermodynamic modeling to cases that would otherwise require a kinetic description (White et al. 2008). This includes, as we shall see, the domain in which chemical potential gradients were frozen in and the mineral relics (e.g., cores) were isolated from the reactive part of the rock. The concept of local equilibrium is thus useful for forward equilibrium models, but it demands that we define the equilibration volume to be investigated.

Bulk rock composition *vs* reactive bulk composition

In the literature of metamorphic petrology, the term bulk rock composition generally refers to the average chemical composition of a whole-rock sample analyzed, for example by X-ray fluorescence spectrometry (XRF). In the framework of local equilibrium, the analysis can be restricted to a specific region or domain—its size often is less than thin-section scale— in a rock, chosen for its uniform chemical composition (or mineral modes). In this case we refer to the concept of local bulk composition. Note that bulk rock compositions and local bulk compositions are measured quantities and may or may not be relevant for modeling, i.e., representative of the equilibrium volume to be investigated.

Various studies over the past 15 years struggled to define a relevant bulk composition for a given sample situation, and this question typically was raised when models using the bulk rock composition rendered unrealistic results (e.g., Warren and Waters 2006). As discussed above, the relevant bulk composition of a rock for equilibrium models is the composition of the (presumed) equilibration volume at a specific stage of the evolution, and it may exclude certain refractory or inert minerals that are observed in a domain, but may be shielded from reactions. The composition of the equilibration volume is known as the effective bulk composition or reactive bulk composition (Fig. 1d). The term effective bulk composition was originally introduced by Tracy (1982) and reused in several studies (Hickmott et al. 1987; Spear 1988b; Stüwe 1997; Marmo et al. 2002; Evans 2004; Tinkham and Ghent 2005a), but in the following we decided to use the term reactive bulk composition as it is most descriptive in a modeling framework. As will be demonstrated below in the section on garnet porphyroblast growth, the reactive bulk composition changes along a $P-T$ trajectory because of compositional fractionation (Tracy 1982; Spear 1988b; Spear et al. 1990).

Open questions on bulk composition effects and the size of equilibration volume

Forward equilibrium models based on the assumption of local equilibrium require the use of reactive bulk compositions, which in some favorable cases can be approximated by local bulk compositions or calculated using fractional crystallization models. The choice of an appropriate or representative rock or domain composition can be critical, as it may significantly affect the results of modeling. Aware of the perils, Kelsey and Hand (2015) warned that "this topic is in some ways like the elephant in the room". The question is: How sensitive are predictions made by equilibrium models to variations in the bulk rock composition? Detailed investigations have been conduced at the regional scale (Tinkham et al. 2001; White et al. 2003), but it appears that only a few studies have explicitly addressed this question at the sample scale. The paucity of knowledge is due in part to the difficulty of estimating uncertainties in the bulk rock composition and then propagating their effect.

Are realistic equilibration volumes impossible to identify exactly, as claimed by Stüwe (1997)? Despite many studies over more than 25 years (e.g., Powell and Downes 1990; Vance and Mahar 1998; Brown 2002; Marmo et al. 2002; Powell et al. 2005), it remains intriguing which parameters control the size of the equilibration volume. Some progress has been made in accounting for phases that should be included or excluded from the equilibration volume (Lanari et al. 2017). If an intergranular medium (specifically a fluid) is present in a rock, transport distances of dissolved species in this medium are critical (Carlson 2002); but what if mobility was limited to volume diffusion? Orders of magnitude difference in transport result from this uncertainty, especially in highly deformed rocks, since strain can increase the dislocation density, reactivity, and transport rates. In any case, diffusion rates are temperature-dependent, so equilibration volume is expected to increase during heating (and drastically decrease upon cooling, as rocks tend to "dry up"), and as shown in Figure 1c, it may differ substantially for different chemical species (Carlson 1989; Spear and Daniel 2001). This is particularly critical for some of the trace elements relevant to petrochronology (e.g., Gatewood et al. 2015). An essential parameter controlling the extent of local (dis)equilibrium is the rate at which local reactions proceed towards equilibrium compared to the rate of change in physical parameters (P–T–X, e.g., the heating rate). If kinetics are favorable, the system can respond rapidly by adjusting the mode (i.e., the volumetric abundance) and composition of each phase in the assemblage to the P–T–X conditions of the equilibration volume; if so, all the minerals in that domain will be stable at the same reactive bulk composition. If not, either a smaller domain size (or component space) must be chosen, or a kinetic approach is required. More work is needed to quantify the size of the equilibration volume, and future studies should take into account the rate at which rocks "travelled through P–T space", i.e., to link petrogenetic analysis to chronology.

PORPHYROBLAST GROWTH

Porphyroblastic growth implies a continuous supply of mineral constituents and a relatively high activation energy barrier for nucleation hence growth of existing grains is favored over formation of new nuclei. The most common example is garnet, whose unique chemical and mechanical properties can record evidence of a potentially complex path that the host rock experienced during a petrogenetic cycle (Spear et al. 1991; Vance and Mahar 1998; Caddick and Kohn 2013; Ague and Axler 2016). Though diffusion can bias the record at high temperatures (Anderson and Olimpio 1977; Tracy 1982; Caddick et al. 2010; Ganguly 2010; Kohn and Penniston-Dorland 2017, this volume), chemical zoning is commonly preserved. In metapelites for instance, garnet commonly exhibits clear compositional zoning with a systematic, bell-shaped decrease of Mn from core to rim (spessartine fraction, X_{sps} in Fig. 2a). But garnet may also provide clues about polycyclic metamorphism, where core and rim relate to two distinct metamorphic cycles (Fig. 2b). Garnet is quite resilient to mechanical and chemical weathering. In fact, metamorphic overgrowths on detrital garnet (Manzotti and Ballèvre 2013) have been recognized (Fig. 2c). Strong compositional zoning in both mono- and polycyclic garnet demonstrates its refractory nature; it does not, in general, completely re-equilibrate with the matrix during growth (Atherton 1968), though outside its stability field garnet readily interacts with hydrous fluid or melt to form secondary phases, for example,

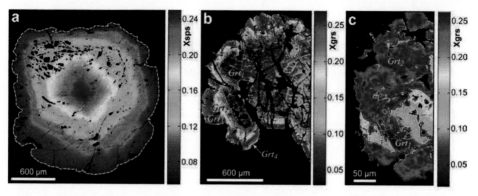

Figure 2. High-resolution end-member compositional maps of garnet grains from the Western Alps. (a) Typical bell-shaped zoning of spessartine of garnet from a metasediment of the Mont Emilius Klippe (Burn 2016); $X_{sps} = Mn/(Mn + Ca + Fe + Mg)$. (b) Grossular zoning in a garnet crystal from an eclogitic micaschist of the Sesia Zone (Giuntoli 2016; Lanari et al. 2016); $X_{grs} = Ca/(Mn + Ca + Fe + Mg)$. This garnet porphyroblast recorded two distinct cycles of metamorphism: a Permian LP–HT core (Grt_1) and three Upper Cretaceous (Alpine) HP–LT rims (Grt_2, Grt_3, Grt_4). (c) Grossular zoning in polycyclic garnet from the Zone Houllière with a LT overgrowth of spessartine-rich hydrothermal garnet (Grt_2); the core (Grt_1) is a detrital fragment that preserves pre-depositional internal zoning (Dupuis 2012).

retrograde chlorite. But where preserved, garnet serves as an archive of the *P–T–X* conditions of formation, as long as intracrystalline diffusion remained negligible (Spear et al. 1984; Florence and Spear 1991; Caddick et al. 2010; Kohn 2014). This fortunate ability to record and preserve chemical zoning is the main reason why garnet compositions have been so extensively used as an indicator of *P–T* conditions throughout its wide stability range. A method referred to as isopleth thermobarometry for garnet is based on Duhem's theorem, which simply states that *P–T* conditions determine the composition and modal abundance of all phases in a closed system at equilibrium; hence the intersection of garnet isopleths for 3–4 endmember components in garnet (Fe, Mg, Ca, Mn) in *P–T* space indicate whether equilibrium was closely approached and, if so, under what *P–T* conditions. Various modeling techniques have been specially designed to be applied to garnet thermobarometry (Spear and Selverstone 1983; Spear et al. 1984, 1991; Evans 2004; Zeh 2006; Gaidies et al. 2008, 2011; Konrad-Schmolke et al. 2008; Schwarz et al. 2011; Moynihan and Pattison 2013; Vrijmoed and Hacker 2014; Lanari et al. 2017). However, determining accurate *P–T* conditions by modeling garnet zoning profiles demands some understanding of the interplay between chemical equilibrium (e.g., Spear and Daniel 2001; Gaidies et al. 2008), reaction kinetics (e.g., Gaidies et al. 2011; Pattison et al. 2011; Schwarz et al. 2011), and post crystallization intragranular diffusion (e.g., Ganguly et al. 1996; Caddick et al. 2010).

Several issues are separately addressed in the next paragraphs: (1) We first recall some basics of nucleation and growth. (2) Porphyroblast growth is then addressed, i.e., the successive alienation of core parts, as they typically become isolated from the reactive part of the rock, thus segregating components. In terms of the reaction volume in the matrix, this process causes and progressively enhances chemical fractionation of the reactive bulk composition. (3) The effects of fractionation on isopleth thermobarometry are then quantitatively illustrated for two examples: A pelitic schist that experienced Barrovian metamorphism (Moynihan and Pattison 2013), and a typical MORB that underwent eclogite facies transformation (Konrad-Schmolke et al. 2008). (4) Different automated strategies are currently in use to retrieve *P–T* information from garnet zoning based on equilibrium thermodynamics. These are presented and briefly discussed, then garnet resorption is addressed for those models. (5) Finally, the concept of distinct growth stages is reviewed based on some studies that link such metamorphic stages to age data.

Crystal nucleation and growth

Four main processes are involved in the nucleation and growth of a porphyroblast crystal: (1) Dissolution of source material provides potential nutrient material; (2) Nutrients generated at various locations and/or from different sources are transported through the intergranular medium to nucleation sites. (3) Nucleation occurs at the atomic scale where chemical species from the transport medium and possibly local reactant phases rearrange into a cluster of the product phase(s) of sufficient size to be thermodynamically stable. (4) Further precipitation onto an existing surface is termed crystal growth. The reaction interface encompasses both the detachment/dissolution at the reactant site and the attachment/precipitation of nutrients at the product site. In this conceptual model, kinetics may affect a metamorphic reaction during nucleation, during intergranular transport, and at the reaction interface. The slowest one of these three processes usually dominates the overall reaction kinetics (Pattison et al. 2011). Detachment and attachment are surface-processes whereas element transfer is a transport-process. Nucleation and growth can be affected by microstructures (favorable site model) and by reaction overstepping.

Favorable site model. Microstructural features may catalyze nucleation by increasing the local free energy, which amounts to reducing the critical energy barrier. Such is the case, for example, in crenulation hinges (Bell et al. 1986) or at grain boundaries, dislocations and cracks (Gaidies et al. 2011). Several lines of evidence support the interpretation that garnet growth is more rapid along triple-grain intersections than along boundaries separating only two grains (Spear and Daniel 2001). The heterogeneous distribution of porphyroblasts in a rock may reflect effects of the local bulk composition and/or deformation. For example, dissolution may occur along shear-dominated zones such as the crenulation limbs, whereas nucleation occurs at the crenulation hinges (Bell et al. 1986; Vernon 1989; Williams 1994). Several local processes may thus be involved in the overall transformation process leading to porphyroblast growth (Carmichael 1969).

Reaction overstepping. Even if a product phase is nominally part of the thermodynamically stable assemblage of the reaction volume, it may fail to nucleate (Waters and Lovegrove 2002). A metamorphic reaction proceeds only if and when the mechanical or chemical energy overcomes the kinetic barriers for nucleation. This delay can be viewed as a disequilibrium state required to gain the amount of energy required to initiate the interface between reactants and products that will initiate the reaction. The so-called interfacial energy (Gaidies et al. 2011 and references therein) controls the departure from equilibrium required before a phase (in our case garnet) nucleates, and the free energy of the system can attain a lower energy state. Reaction overstepping can be estimated as the difference in Gibbs free energy between the thermodynamically stable, but not yet crystallized products and the metastable reactants. This quantity is known as the reaction affinity (e.g., Fisher and Lasaga 1981; Lasaga 1986; White 2013). P–T–reaction affinity maps computed with Theriak-Domino have been used to predict metamorphic reaction overstepping (Pattison et al. 2011). Such models demonstrate, for example, that thermal overstepping may have some dependence on the heating rate (Gaidies et al. 2011; Pattison et al. 2011).

Models of equilibrium and transport control

In the framework of modeling garnet growth we distinguish two end-member models.

Equilibrium control. Chemical equilibrium is maintained at the rim of all growing phases at all times if transport rates are faster than the rates of surface processes. Equilibrium thermodynamics can be used to predict the distribution of all components among the various phases of the system. This grain boundary equilibrium model (Fig. 1b) implies that the rim composition of all the crystallizing phases will be identical and dictated by P–T conditions and other intensive variables such as fluid composition and f_{O_2}.

Transport control. If transport is the limiting factor, gradients in chemical potential are established (at least for some elements) and may evolve via the intergranular medium. Consequently, the composition of the growing phase will be controlled by the flux of these elements to the reaction interface. As thermodynamics predicts the macroscopic equilibrium state only, a kinetic description must be used to model how and how fast equilibrium will be approached at a specific reaction site.

In rocks, the attainment of chemical equilibrium and departures from it depend on scale, in both space and time. Several considerations are necessary to decide what modeling approach is most appropriate for a given situation; strong cases can be made for combining the equilibrium control and transport control approaches (e.g., Carlson et al. 2015). When the goal is to analyze and quantify petrogenesis of rock samples as a basis for detailed chronology, we regard the following considerations as essential:

- Equilibrium control may occur on different scales for different elements (Spear and Daniel 2001). Specifically, highly charged or large ions often are far less mobile than smaller or less highly charged ions (Fig. 1c).

- The concept of "chemical homogeneity" is useless beyond upper and lower spatial limits; these limits essentially depend on rock texture (e.g., layering, grain size), which evolves during petrogenesis. The spatial scale of chemical uniformity is bound to change by rock-forming processes, e.g., as a few components are sequestered by porphyroblasts or simply as static Ostwald ripening or dynamic recrystallization proceeds, by mechanical grain size reduction (grinding), or differentiation induced by plastic deformation, partial melting or metasomatic processes involving fluids.

- Trace elements contribute little in terms of the overall free energy of a rock. Equally minor is the contribution to the driving force towards equilibrium that chemical potential gradients in the trace elements may add.

- A few accessory minerals (e.g., monazite, zircon, titanite, allanite, apatite) largely sequester certain trace elements of particular interest to petrochronology. Their transport (by diffusion) is typically slow because they include large, highly charged ions, such as U^{4+}, Th^{4+}, and REE^{3+}, and they are involved in heterovalent coupled exchange mechanisms, but their mobility may be enhanced by deformation, radiation damage, and reactive fluids (e.g., Tropper et al. 2011). Modelers should be aware that accessory minerals commonly occur as relics (often armored) that appear to have metastably survived high temperatures, even partial melting.

- In general, some parts of a rock's reaction path may be kinetically sensitive while others are not.

Taken together, these considerations indicate that sober modeling should recognize the potentials and limits of equilibrium and transport models, and possible combinations (Gaidies et al. 2011; Schwarz et al. 2011). A conceptual division of samples may be most appropriate in practice. For instance, it remains a largely open question to what extent trace element mobility will allow chemical potential gradients to be retained (or to disappear) at the temporal scale of rock-forming processes.[2] Reaction-driven exchanges of these trace elements between major rock-forming minerals and accessory phases, possibly mediated by fluids, may or may not be sufficient to even out concentration gradients. Current understanding of transport properties is quite insufficient for the case of highly charged minor and trace elements. The extent to which small and sparsely present grains are able to communicate deserves careful study in the future, as this promises to tighten the links between accessory mineral chronometry and petrogenetic interpretations. For example, relations between zircon equilibration and Zr-in-rutile thermometry are well documented (Meyer et al. 2011; Taylor-Jones and Powell 2015; Kohn et al. 2016).

---

[2]Note that these time scales could well be much shorter than could be resolved by mineral chronology (say $< 10^3$–10^5 years).

With our present goal of quantifying petrogenetic conditions, we concentrate on equilibrium control in this chapter with the assumption of local equilibrium that reduces the need to consider transport. We show that the power of thermodynamics can be enhanced and several limitations reduced by restricting the analysis to local spatial domains.

Fractionation of the reactive bulk composition during porphyroblast growth

Some component of growth zoning in garnet such as Mn are largely controlled by reactive bulk composition change generated by crystal fractionation. Mn distribution in garnet often shows a progressive decrease with a typical "bell-shaped" zoning profile from the center toward the rim (Fig. 2a, 3a). In this case the segregation of Mn during garnet growth is directly controlled by the progressive depletion of Mn in the surrounding matrix. Many studies explained such observation by a fractionation process following a Rayleigh's (or Pfann's) model (Hollister 1966; Atherton 1968; Cygan and Lasaga 1982; Evans 2004). Based on those results, a simple method was proposed by Evans (2004) to generate composition versus modal proportion curves for garnet. This method models the effects of the crystal fraction on the MnO content of the reactive bulk rock using the Rayleigh fractionation equation:

$$C_{grt}^{MnO} = C_{bulk}^{MnO} K_d \left(1 - w_{grt}\right)^{K_d - 1} \tag{2}$$

where C_{grt}^{MnO} and C_{bulk}^{MnO} are the concentrations of MnO (in oxide wt%); K_d is the bulk distribution coefficient (by mass) for the element between garnet and matrix ($C_{grt}^{MnO} / C_{mtx}^{MnO}$) and w_{grt} is the mass fraction of garnet in the rock. For $K_d > 1$, the species partitions into the porphyroblast; a characteristic profile across a grain should be "bell shaped" (Fig. 3a), whereas for $K_d < 1$ the species partitions into the matrix, i.e., growing porphyroblasts are depleted relative to the matrix and show a U-shaped profile. If neither *P–T* conditions nor equilibrium phase relations change drastically during early garnet growth, K_d can be approximated by the ratio between C_{grt}^{MnO} in the earliest garnet nucleus and C_{bulk}^{MnO} (Gaidies et al. 2006). As partitioning among minerals depends on temperature, and because changes in *T* drive mineral growth, K_d is expected to change as the mineral grows (Hollister 1966; Kohn 2014).

To quantify the effects of fractionation of the reactive bulk composition, a case study is selected for which the prograde *P–T* conditions of garnet growth are well constrained. Moynihan and Pattison (2013) analyzed garnet porphyroblasts in a garnet- and staurolite-bearing schist (sample DM-06-128) from the southern Omineca belt of the Canadian Cordillera that underwent Barrovian metamorphism peaking at middle amphibolite facies during the Early Cretaceous. The evolution involved heating and burial along a linear *P–T* trend, followed by a heating-dominated stage accompanied first by exhumation, then renewed burial. Garnet porphyroblasts grew along a *P–T* path from 500 °C, 5 kbar to 570 °C, 7 kbar. From the zoning profile reported in Moynihan and Pattison (2013), we obtained a K_d value of 62 using the approximation of Gaidies et al. (2006). The corresponding MnO composition of garnet was calculated using Equation (2) and plotted against the volume percentage of garnet produced (Fig. 3c). Assuming a single population of garnet with ~5 mm diameter for a total amount of 6 vol% of garnet in the rock (as predicted by the thermodynamic model) the zoning profile matches Rayleigh fractionation with $K_d = 62 \pm 10$, reproducing to a first order the chemical zoning in MnO for this rock. In the matrix Mn is distributed between chlorite and ilmenite (Moynihan and Pattison 2013).

This technique of Evans (2004) has been successfully used in various cases to model garnet fractionation based on MnO zoning profiles (e.g., Gaidies et al. 2006; Sayab 2006; Groppo and Castelli 2010; Vitale Brovarone et al. 2011; Cheng and Cao 2015). Zoning profiles of other elements may reflect more complicated fractionation, with non-Rayleigh behavior and variable K_d during garnet growth. This is expected to occur along a *P–T* path that has successively involved more than one garnet-producing reactions as for most amphibolite- to upper amphibolite-facies

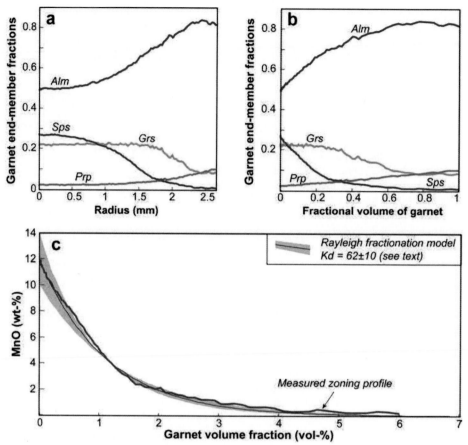

Figure 3. Garnet compositional zoning from a schist that experienced Barrovian metamorphism (modified from Moynihan and Pattison 2013) against (a) the radius, (b) the fractional volume of a single garnet porphyroblast, (c) the garnet volume fraction assuming a mode of 6 vol% garnet and a single garnet size population. The Rayleigh fractionation model for $K_d = 62 \pm 10$ is shown by the black curve in the gray domain.

rocks. For both simple and complex growth histories, quantitative models based on Gibbs free energy minimization must be favored to approximate the changes in the reactive bulk composition.

Equilibrium crystallization models *vs* fractional crystallization models

Two models incorporating or ignoring chemical fractionation effects during porphyroblast growth are presented and compared here. For garnet we show that (1) fractionation has a strong effect on the amount of garnet predicted, and (2) isopleth thermobarometry systematically produces erroneous *P–T* estimates if fractionation is ignored.

Equilibrium crystallization models (ECM). These classic models rely on isochemical *P–T* equilibrium phase diagrams for the given bulk rock composition and isopleth thermobarometry (Spear 1988a). Such a model cannot 'predict' chemical zoning because all the phases are assumed to re-equilibrate at any *P–T* conditions. In theory, the use of ECM models must be restricted to well-equilibrated mineral assemblages showing no relics, no chemical zoning in the mineral phases, and no partial local re-equilibration. In practice, such diagrams are often used to reveal first-order compositional and modal trends that help to interpret metamorphic mineral growth and composition changes (Kohn 2014), but the results may be inaccurate.

Fractional crystallization model (FCM). These models assume chemical fractionation of some mineral phases (usually garnet) that are produced at each step of a *P–T* path, where the grains produced are immediately isolated from the reactive part of the rock (Spear 1988b; Spear et al. 1991; Gaidies et al. 2008; Konrad-Schmolke et al. 2008). This is equivalent to removing the fractionating phase from the system. Step-wise growth modeling is applied, with the composition and volume fraction of the fractionating phases from the previous step(s) being isolated or removed from the reactive system.

The first major difference between FCM and ECM is the modal amount of the phase produced along a given *P–T* path. In their study, Konrad-Schmolke et al. (2008) documented this effect of garnet fractionation under eclogite facies conditions using a typical MORB composition. Garnet is predicted to be stable along a prograde trajectory from 525 °C, 15 kbar to 750 °C, 35 kbar using ECM (Fig. 4 in Konrad-Schmolke et al. 2008). By contrast, garnet growth ends at 650 °C and 25 kbar using FCM because the matrix is strongly depleted due to garnet fractionation (Fig. 5 in Konrad-Schmolke et al. 2008). Above that limit, garnet is no longer stable in the new reactive bulk composition. There will be a significant difference between two tectono-metamorphic scenarios based on the results from FCM and ECM models. In the present example, the FCM shows that garnet does not 'record' the pressure peak reached by the rock. It is important to note that the volume of garnet predicted by ECM is systematically higher than the volume predicted by FCM (Fig. 4). Accurate modeling of mineral modes along any *P–T* trajectory requires the use of FCM as soon as compositional zoning is observed in a sample.

To evaluate the consequences of garnet crystallization on the reactive bulk composition of typical metamorphic rocks, we again analyze the two examples of garnet growth in (1) a metapelitic schist along a typical LP–HT trajectory (Moynihan and Pattison 2013) and (2) a metabasite with a typical MORB composition along a HP-LT trajectory (Konrad-Schmolke et al. 2008). The changes in the reactive bulk rock compositions predicted using a FCM are plotted in Figure 5 against the volume percentage of garnet produced. Note that in the metapelitic schist only 6 vol% of garnet is produced along the *P–T* trajectory. For the metabasite, ~13 vol% is produced along the selected *P–T* trajectory. For the metapelitic schist, all the components

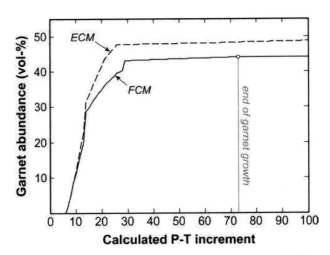

Figure 4. Volume of garnet (in vol%) predicted by ECM and FCM for a typical MORB composition along the *P–T* path used by Konrad-Schmolke et al. (2008). Model calculated for the system SiO_2–Al_2O_3–FeO–Fe_2O_3–MgO–CaO–Na_2O using the thermodynamic dataset tc55.txt (Holland and Powell 1998), oxygen fugacity controlled by hematite-magnetite buffer.

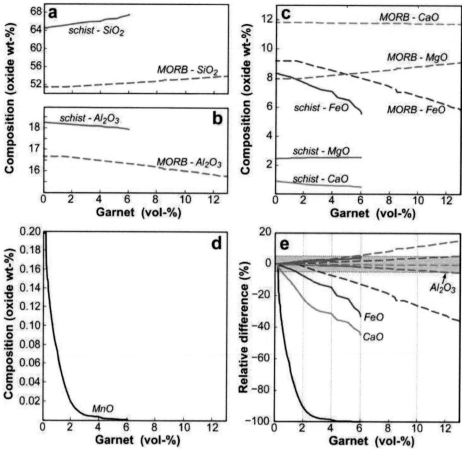

Figure 5. Evolution of the reactive bulk rock composition using a fractional crystallization model (FCM) for garnet growth in pelitic schist (Moynihan and Pattison 2013) and MORB (Konrad-Schmolke et al. 2008) along a LP–HT and a HP–LT path, respectively. Note that along the respective *P–T* paths only 6 vol% of garnet is produced in the schist, whereas in meta-MORB the garnet mode reaches 40 vol% garnet (shown here to 12 vol% only, but the same trend continues). (e) Relative differences of the reactive bulk compositions to the original bulk rock composition for the two samples; note that oxides are color-coded. The gray band outlines relative differences of <5%.

(except Al_2O_3 and MgO) show significant variation during garnet growth (Fig. 5e). As previously discussed, the matrix quickly becomes depleted in MnO, i.e., the matrix loses >90% of the initial MnO once the garnet fraction reaches 2 vol%. To a lesser extent CaO and FeO decrease and finally reach 60 and 70% of their initial value once 6 vol% garnet formed. Fractionation is less pronounced for the metabasite (Fig. 5). Still, FeO and MgO are significantly affected once the garnet proportion reaches 6 vol%. From these two examples we can conclude that garnet growth systematically affects the reactive bulk composition.

What is the effect of FCM on equilibrium phase diagrams? For small fraction of garnet produced, FCM has a minor effect on the major phase relations (Zuluaga et al. 2005; Groppo et al. 2007; Moynihan and Pattison 2013). However, the compositional differences of garnet predicted stable by ECM and FCM at given *P–T* conditions are systematic and large enough to affect isopleth thermobarometry (Spear 1988b; Evans 2004; Gaidies et al. 2006; Chapman et al. 2011). To evaluate this, the program GrtMod is useful because it searches the *P–T* conditions for which the model composition of garnet best matches the measured composition (Lanari et al. 2017), and it can be used to check differences in garnet isopleth thermobarometry between FCM and ECM model. For the schist DM-06-128 (Moynihan and Pattison 2013), this comparison was made along the original *P–T* path (500–650 °C, 3–12 kbar), taking three steps (at 2, 4, and 6 vol% garnet produced). Results are reported in Table 1. It is evident that ECM does not find satisfactory isopleth intersection in the *P–T* space for the three selected cases. In the first case (2 vol%) a 'best match' is found, but with a relatively high residual (Co value in Lanari et al. 2017), and it is predicted at higher pressure. Deviations are significant for almandine and grossular components (modeled: 0.74 and 0.16; measured: 0.78 and 0.14). Such deviations are due to the overestimation of CaO and FeO in the bulk rock composition, and these affect the composition of phases coexisting with garnet (Fig. 5). For the second and third cases (4 and 6 vol% of garnet), the deviations between modeled and observed compositions become even larger, i.e., they show much higher residuals. In both cases GrtMod found two distinct local minima but the lowest Co values always occur at higher pressure. Deviations of 0.10 and 0.04 are systematically observed in almandine and grossular components.

From these sensitivity tests we can conclude that FCM must always be used to model porphyroblasts growth if compositional growth zoning is observed. Note that the compositional changes become significant as soon as porphyroblasts represent >2 vol% in pelitic systems and >4 vol% in mafic systems (Fig. 5). In the following FCM are used to model growth and resorption of garnet porphyroblasts.

Table 1. Differences in *P–T* results between fractional crystallization models (FCM) and equilibrium crystallization model (ECM) for three stages of garnet growth (2, 4 and 6 vol% of garnet growth) along the prograde *P–T* path of Moynihan and Pattison (2013). Co is the residuum calculated by GrtMod. A value of Co < 0.3 generally indicates a good fit between model and measured compositions (Lanari et al. 2017).

	T (°C)	P (bar)	Co	X_{Prp}	X_{Grs}	X_{Alm}	X_{Sps}
Case 1: 2 vol% of garnet produced							
Reference	554	6916		0.055	0.141	0.779	0.024
FCM	554	6916	0.0001	0.055	0.141	0.779	0.024
ECM - S1	555	11996	0.0489	0.073	0.156	0.737	0.034
Case 2: 4 vol% of garnet produced							
Reference	571	6427		0.081	0.086	0.831	0.002
FCM	571	6427	0.0002	0.081	0.086	0.831	0.002
ECM - S1	572	11994	0.1017	0.103	0.126	0.744	0.027
ECM - S2	584	7690	0.1092	0.110	0.122	0.737	0.031
Case 3: 6 vol% of garnet produced							
Reference	583	6797		0.104	0.082	0.814	0.000
FCM	583	6797	0.0001	0.104	0.082	0.814	0.000
ECM - S1	581	11996	0.0869	0.121	0.115	0.739	0.025
ECM - S2	591	7731	0.0920	0.126	0.112	0.734	0.029

Automated fractional crystallization models designed to retrieve $P–T$ paths from chemical zoning

Examples of simplified fractionation modeling based on BSE or X-ray images and electron microprobe analyses can be found in many studies (e.g., Marmo et al. 2002; Tinkham and Ghent 2005a; Zuluaga et al. 2005; Zeh 2006; Caddick et al. 2007). Four methods presented here are based on forward equilibrium models and require no manual alteration of the reactive bulk composition to account for material sequestered in garnet crystal cores.

- The first automated method models garnet zoning and the composition of coexisting phases along an arbitrarily selected $P–T$ path using Gibbs free energy minimization. The generated compositional profiles are then compared against data along high-resolution zoning profiles analyzed by electron microprobe (Konrad-Schmolke et al. 2008; Hoschek 2013; Robyr et al. 2014). These models merely test if the chosen $P–T$ trajectory is in accordance with the observed zoning (or vice versa). In fact, the three studies cited found some mismatch between model compositions and observed zoning profiles, in some cases with systematic and large discrepancies (>0.08 in the end-member proportions). Is the $P–T$ trajectory at fault or the assumption of grain boundary equilibrium? The two following approaches aim to derive the optimal $P–T$ trajectory and proposed numerical methods to improve its selection.

- Moynihan and Pattison (2013) provide a MATLAB©-based program linked to Theriak (de Capitani and Brown 1987; de Capitani and Petrakakis 2010) that uses an inverse modeling strategy to derive the "best" $P–T$ trajectory by minimizing a misfit parameter, i.e., the weighted differences between measured and model compositions. For any garnet spot analysis, the best $P–T$ conditions are found by matching the model against the measured composition, and the procedure is successively applied to all analyzed points from core to rim. Using this approach, Moynihan and Pattison (2013) successfully modeled the observed zoning profile (see their Fig. 6a). The match is excellent and demonstrates that in favorable cases, equilibrium thermodynamics in the context of fractional crystallization can be successfully used to model porphyroblast growth, provided the $P–T$ path is part of the model optimization.

- Similarly, Vrijmoed and Hacker (2014) provided a MATLAB©-script linked to Perple_X (Connolly 1990, 2005). It uses a brute-force inverse computational method to determine the best $P–T$ trajectory by minimizing the differences between predicted and the entire measured garnet compositional profiles along different trajectories. This routine examines a multitude of paths from an unspecified starting $P–T$ to a predetermined maximum $P–T$ point. The best match is selected such that all endmember mole fractions of the fractionated garnet show least discrepancy between data and model. Several tests based on published zoning profiles data show that brute force can pay off.

- Finally, a fourth approach is of interest: Program Theria_G (Gaidies et al. 2008) allows the simulation of porphyroblast nucleation and growth using a FCM for any given $P–T$ trajectory, and it takes into account further possible modifications driven by intragranular multi-component diffusion. Theria_G uses the Gibbs free energy minimization routine of Theriak and simulates the formation of a garnet population with variable grain size that can be compared with observations. The $P–T$ trajectory can also be part of the model optimization as shown by Moynihan and Pattison (2013).

Crystal resorption and implications on fractional crystallization models

The models presented above take into account the fractionation due to garnet growth and how this affects the reactive bulk composition. However, garnet may also be affected by local resorption (de Béthune et al. 1975; Kohn and Spear 2000; Ague and Axler 2016), in some extreme cases leading to atoll or mushroom garnets (Cheng et al. 2007; Faryad et al. 2010;

Robyr et al. 2014). More commonly, the production of staurolite and biotite at the expense of garnet and chlorite causes the resorption of garnet and a step in the zoning profile (Spear 1991). Garnet resorption has been invoked to explain the peripheral increase in MnO (dubbed near-rim kick-up) observed in some grains. Typically this results from partial consumption or dissolution of garnet crystals during cooling and exhumation, which returns Mn to the reactive matrix (Kohn and Spear 2000), and Mn can show retrodiffusion features in garnet (de Béthune et al. 1975). The amount of Mn in the near-rim kick-up has been used as a semi-quantitative measure of the volume of garnet dissolved. Kohn and Spear (2000) estimated that 45% of garnet was dissolved in an amphibolite facies metapelite from the central Himalaya of Nepal. Based on forward equilibrium models (see below) and different types of evidence Lanari et al. (2017) estimated that 50% of garnet in eclogitic micaschist was dissolved during HP metamorphism before a new episode of garnet growth. As shown above garnet chemical fractionation has a strong impact on the reactive bulk composition. Fractional crystallization drives the reactive bulk composition away from the garnet composition. Thus, resorption of garnet drives the reactive bulk composition back towards the garnet composition.

The effects of garnet resorption on forward equilibrium models were investigated by Lanari et al. (2017), who proposed a numerical strategy and a MATLAB©-based program (GrtMod) linked to Theriak. It allows numerical simulation of the evolution of garnet based on compositions of successive growth zones, optimization of successive local reactive bulk compositions, accommodating resorption and/or fractionation of previously crystallized garnet. Each growth stage here is defined as an interval during which garnet grows at fixed *P–T–X* conditions while in equilibrium with the same stable matrix assemblage. This approach uses compositional maps (instead of profiles) to define the growth zones and to estimate average compositions.

Timing of porphyroblast growth

In porphyroblast petrochronology, two different ways are generally used to link metamorphic age (*t*) with temperature and pressure: (1) in-situ dating of the porphyroblast or (2) textural correlation between the porphyroblast and dateable accessory minerals either trapped as inclusions or chemically correlated to distinct zones within a porphyroblast (e.g., Pyle and Spear 1999; Regis et al. 2014).

In situ dating of garnet porphyroblasts is the most attractive method to constrain the time interval of growth because porphyroblasts commonly contain inclusions that help constrain *P–T*, as can the chemical zones that recorded successive growth stages. Several potentially useful isotopic systems (Sm–Nd, Rb–Sr, Lu–Hf) are extensively discussed in Baxter et al. (2017, this volume). Rb–Sr dating of garnet is restricted to a few cases because it is not always realistic to assume that the whole-rock or matrix compositions adequately reflect the reactive bulk composition at the time garnet grew (Sousa et al. 2013). Sm–Nd is most commonly used because Sm and Nd generally are uniformly distributed in garnet (Kohn 2009), whereas Lu and Hf may be restricted towards early garnet growth (because Lu strongly fractionates into the core, depleting the matrix in Lu). In addition, much more material is required per analysis to obtain ages of specific zones using Lu/Hf. Sm/Nd is most powerful for dating specific garnet zones (e.g., Pollington and Baxter 2010), since improved analytical techniques now permit analysis of very small volumes (Harvey and Baxter 2009; Dragovic et al. 2012), such as single growth zones mapped by electron microprobe (Gatewood et al. 2015). Ages are calculated based on isochrons for each garnet zone and the surrounding matrix, and such ages rest on assumptions about the extent of trace element homogenization.

Instead of applying direct garnet dating many studies have used textural correlations between the porphyroblast and dateable accessory minerals trapped as inclusions. An example of monazite is presented in the following paragraphs. Other examples are given in Williams et al. (2017, this volume).

Textural correlation is conceptually simple (e.g., Kohn 2016) and obviously requires in situ dating of accessory phases to preserve textural relations. In particular, U–Th–Pb dating of monazite has become one of the primary tools for constraining the timing of moderate to high-grade metamorphism (Harrison et al. 2002). Of course, the youngest inclusions only set an upper age limit to the enclosing garnet growth zone. If an accessory formed just before it got trapped in a growing porphyroblast, it presumably was isolated from the matrix and thus protected from later re-equilibration, recrystallization or overgrowth. Such an (idealized) scenario should generate an age gradient among inclusions from the core of the porphyroblast to its rim.

Examples. Mottram et al. (2015) dated monazite included in garnet from pelitic schist of the Lesser Himalayan Sequence. Monazite trapped in garnet cores gave (common-Pb and Th-corrected) $^{238}U/^{206}Pb$ ages of 20.7 ± 2.2 Ma (from 4 single-spot analyses), inclusions in the mantle surrounding the core show ages of 17.9 ± 0.5 Ma (3 analyses), and inclusions in the rim 15.8 ± 1 Ma (4 analyses). Three distinct stages of garnet growth with different inclusion patterns are visible in chemical maps. As xenotime was absent in this sample, one might expect monazite dissolution during garnet growth (Spear and Pyle 2010). The process invoked by Mottram et al. (2015) assumes that monazite grains present in the matrix underwent continuous and complete re-equilibration during garnet growth. However, it is not clear by what process such re-equilibration happened nor whether there merely has been overgrowth. Inclusions trapped by the growing garnet were shielded, since diffusion of U, Th and Pb in garnet is slow. Many other studies have reported evidence that garnet porphyroblasts provided such shielding and found monazite inclusions trapped in garnet porphyroblasts to be older than in the matrix (Foster et al. 2000, 2004; Martin et al. 2007; Hoisch et al. 2008). Yet, to derive robust estimates on the time and tempo of garnet growth, careful documentation of the inclusion location is required, e.g., reporting the distance of the inclusion from the core (in equatorial sections). In addition, inclusions that seem connected to the matrix by hairline fractures or microcracks must be avoided, as these may have been subject to interaction with the matrix or re-equilibration (Montel et al. 2000; Martin et al. 2007). For instance, Martin et al. (2007) compiled 196 in situ Th–Pb dates of monazite inclusions obtained by LA-ICP-MS and found that microcracks allowed communication between the interior of the garnet and the matrix, facilitating dissolution, recrystallization, and intergrowth formation.

Assuming that continuous re-equilibration of matrix monazite does happen, as suggested by Mottram et al. (2015), is it the rule or the exception? Some studies have indicated incomplete re-equilibration to interpret the observed age distribution. For instance, monazite ages in a garnet–biotite schist from Bhagirathi valley in Garhwal Himalaya (Foster et al. 2000) range between 40 and 25 Ma. The age of monazite *m1* included in garnet cores is ~39.5 Ma, whereas monazite *m2* in garnet rims spread in age between 36 and 41 Ma (Fig. 6). However, instead of invoking partial re-equilibration, we contend that the data indicate partial dissolution of *m1*-monazite, and discrete overgrowth by *m2*. In the matrix, no *m1* and only one survivor of *m2* was found (Fig. 6), suggesting strong dissolution of monazite *m1* and *m2* before the formation of a new generation *m3* at 30–27 Ma. So, we suggest partial replacement rather than "re-equilibration". To ascertain whether this (or some alternative) process was responsible for monazite growth, geochemical tracers for monazite should be analyzed, and readily interpretable ones are available: Th and Y contents and their zoning (Kohn et al. 2005; Williams et al. 2017, this volume) are most helpful.

A model of partial replacement is also supported by the comprehensive dataset of spatially distributed ages reported by Hoisch et al. (2008). Monazite inclusions were dated (SIMS, Th–Pb) in several garnet grains from upper amphibolite facies pelitic schist in the northern Grouse Creek Mountains, Utah. Compositional maps exhibit three successive growth zones with distinct

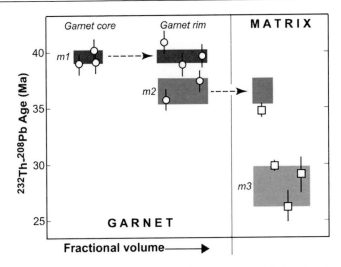

Figure 6. Monazite ages from grains trapped as inclusions in garnet (circles) and grains in the matrix (squares) of a biotite schist from Bhagirathi valley, Garhwal Himalaya (modified after Foster et al. 2000).

compositions from a core to a mantle and a rim (inset in Fig. 7a). The age of inclusions from different grains (*gm1b-e* and *gm3h*) correlated with chemical zoning in the three zones, as shown in plots against the fractional garnet volume (Fig. 7). Ages generally decrease toward garnet rims, and Hoisch et al. (2008) interpreted the monazite ages as decreasing linearly with increasing mode (volume) of garnet. Kohn (2016) pointed out that this interpretation implies that analytical errors were underestimated. Certainly the monazite inclusion ages in the three growth zones show scatter beyond the analytical error (Fig. 7a). An alternative interpretation is to regard each inclusion age as a maximum age estimate of a single growth stage (Fig. 7b). Thus partial preservation of older monazite ages in discrete, successive growth zones is visible and suggestive of a process of incomplete and continuous replacement of monazite, along the lines previously discussed.

Is partial or complete replacement the only process to explain the age trend recorded by monazite inclusions? Monazite textures give some indications: Grains from a sillimanite-bearing metapelite from the Hunza Valley in Pakistan studied by Foster et al. (2004) exhibit complex zoning patterns with four distinct growth zones (Fig. 8), suggesting a succession of partial resorption and precipitation stages. Monazite *m1* grew slightly prior to garnet and was then partially dissolved during growth of the garnet core (Fig. 8). Monazite *m2* grew after the garnet core and was then captured as inclusions in garnet rims. The enrichment in Y found in rim garnet was interpreted as the prograde breakdown or consumption of xenotime (Fig. 8). Because monazite *m3* is not observed as inclusions in garnet, it is likely that it grew coevally with xenotime during the garnet resorption stage. Finally the last monazite *m4* grew locally in the matrix, in equilibrium with xenotime. Some monazite grains in the matrix exhibit the four growth zones, but with clear resorption features separating them. The preservation of several growth zones of monazite in this sample is well supported by both textures and chemical zoning (Fig. 8). Resorption of monazite appears to be correlated to garnet growth and provides some constraints on timing. Based on the textural relationships (Fig. 8), we can conclude that the garnet core grew between 80 and 68 Ma and garnet rims between 60 and 58 Ma. A break of 8 Ma occurred between the two stages of growth of garnet (core and rim). This dataset with its extensive documentation of the textural and chemical relationships also provided a fairly precise estimate (2.4 ± 1.2 °C/Ma) of the heating rate experienced by the sample (Foster et al. 2004).

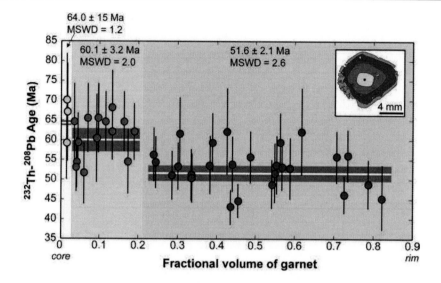

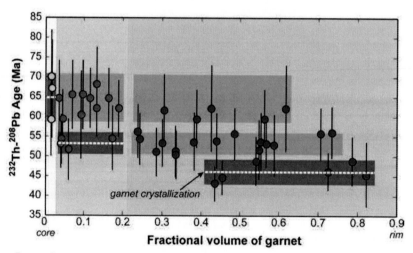

Figure 7. Ages from monazite inclusions (2σ) trapped in three garnet crystals (*gm1b*, *gm3h*, *gm1e*) from upper amphibolite facies pelitic schist in the northern Grouse Creek Mountains, Utah plotted against the fractional volume of garnet (modified after Hoisch et al. 2008). The garnet crystals were divided into three successive growth zones (see inset); ages are plotted in corresponding colors. (a) Average age and weighted uncertainty of inclusion ages from the three growth zones. (b) Alternative interpretation of the age distribution, assuming partial re-equilibration of monazite grains during growth or overgrowth.

Any petrochronological interpretation demands good documentation of both inclusion texture and chemical zoning. From the selected examples of textural correlation, several main conclusions can be drawn: (1) Age patterns can be complex because of partial replacement (for example, by dissolution–reprecipitation, Putnis 2009; Grand'Homme et al. 2016). (2) *P–T* modeling based on the concept of FCM is required, incorporating an inverse numerical strategy to obtain the most likely *P–T* trajectory (Moynihan and Pattison 2013; Vrijmoed and Hacker 2014; Lanari et al. 2017). This is the price to be paid if we want to obtain robust *P–T–t* estimates for successive growth zones. If, in addition, inclusion ages are combined with host (here garnet) dating, this adds certainty, especially to the significance of the earliest inclusion age.

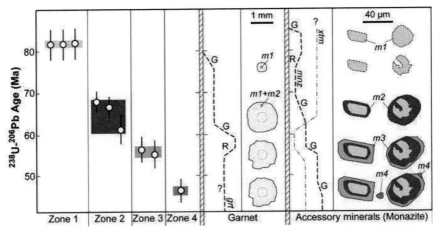

Figure 8. Monazite petrochronology on inclusions and matrix grains from a sillimanite bearing metapelite from the Hunza Valley in Pakistan (modified after Foster et al. 2004). The ages are reported together with schematic volume fractions of garnet, monazite, and xenotime through time and the corresponding textures. Abbreviations: G, growth; R, resorption, mnz, monazite; xtm, xenotime; grt, garnet. The four monazite zones (*m1...m4*) correspond to the different generations discussed by Foster et al. (2004).

BULK ROCK COMPOSITION EFFECTS ON MELT PRODUCTION
AND ACCESSORY MINERALS

Many studies have pointed out the strong effect bulk rock composition has on major and accessory mineral stability and on melt production (e.g., Spear 1993; Stevens et al. 1997; Pickering-Witter and Johnston 2000; Tinkham et al. 2001; Evans and Bickle 2005; Janots et al. 2008; Kelsey and Hand 2015; Yakymchuk et al. 2017, this volume; Zack and Kooijman 2017, this volume). To illustrate this effect in a petrochronological framework, we present two case studies of the regional metamorphic aureole in the Mt Stafford area, central Australia. A metamorphic field gradient from low-pressure greenschist- to granulite-facies is exposed, and Greenfield et al. (1996) divided the terrain into five metamorphic zones (Fig. 9a). The Proterozoic metasedimentary rocks of the Mt Stafford area comprise aluminous metapelites interbedded with metapsammite layers and cordierite or amphibolite granofels on a centimeter to meter scale. In the migmatite zones the interbedded metapelite and metapsammite form bedded migmatites preserving sedimentary features (Greenfield et al. 1996) that suggest in situ melting (caused by a series of biotite breakdown reactions) without substantial migration of melt (Vernon and Collins 1988). Bulk rock compositions are constant across the various zones (Greenfield et al. 1996), confirming the absence of significant melt mobilization. This result was essential to use the bulk rock composition as representative of the equilibration volume. Based on this assumption, White et al. (2003) reconstructed the melt production history across the successive zones (Fig. 9b) using equilibrium phase diagrams computed for the bulk rock compositions of Greenfield (1997). Metapelite produced more melt at lower temperatures, whereas metapsammite experienced additional major melt-production at higher temperatures (zone 5 in Fig. 9). This example shows the control of the bulk rock composition on the melt production history. It is important to note that this approach would not work in case of large-scale melt migration that would change the reactive bulk composition.[3]

---

[3] In such cases a melt extraction step could be added in modeling along a *P–T* loop. For example, 6 vol% of melt could be extracted once the proportion reaches the melt connectivity transition at 7 vol%, and a new reactive bulk composition would be calculated for use in the next steps.

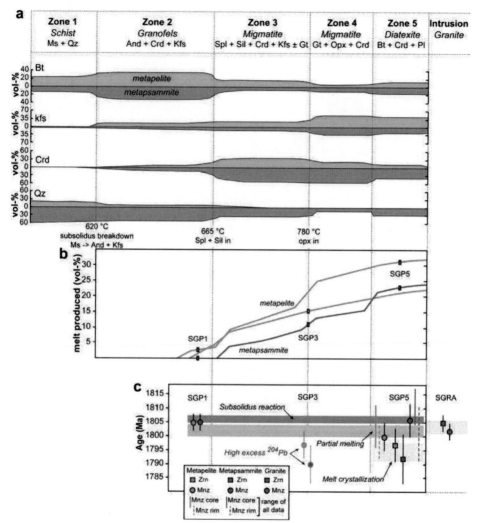

Figure 9. Petrochronological investigation of a metamorphic aureole from the Mt Stafford area in central Australia. (a) Metamorphic zones and mineral modes from Greenfield et al. (1996). (b) Melt fractions for two metapelite and one metapsammite samples from White et al. (2003). Sample numbers refer to those with age data for accessory phases (Rubatto et al. 2006). (c) Ages of monazite and zircon. Dates for SGP3 are not considered here because of possible excess ^{204}Pb, as discussed by Rubatto et al. (2006).

Taking the scenario and quantification of the melt history from White et al. (2003), age data (Rubatto et al. 2006) for these same rocks can be put into context. The metamorphic behavior of monazite and zircon depends on host-rock composition (Wing et al. 2003; Fitzsimons et al. 2005; Rubatto 2017, this volume), and metapelites and metapsammites affect inherited zircon and monazite in different ways. In the case of the Mt Stafford aureole, monazite cores show a less pronounced Eu anomaly than the rims, and both are interpreted as due to prograde growth, with the amount of coexisting potassic feldspar increasing (Rubatto et al. 2006). In the higher-grade samples, monazite rims may have formed later, as they show an increase in Gd/Lu, which reflects strong fractionation of Lu into garnet cores. Therefore, the monazite rims most likely grew during a limited period of garnet resorption close to peak metamorphic conditions. This scenario is supported by the compositional zoning of garnet that exhibits clear evidence of core resorption (Rubatto et al. 2006). Trace element maps (Th and Y) of the last monazite growth zone would help in interpreting how the last monazite formed. However, the solubility of monazite as well as the melt volume increase with temperature (Montel 1993; Stepanov et al. 2012), one would expect resorption not growth (Kelsey et al. 2008). Chemical characterization of monazite—ideally based on compositional maps (Mahan et al. 2006; Williams et al. 2007, 2017, this volume)—is best done prior to chronological microanalysis (Kohn et al. 2005). Other studies have reported that monazite rims may also form during cooling from crystallizing melt (Pyle and Spear 2003; Kohn et al. 2004; Johnson et al. 2015).

In migmatites and residual granulites new metamorphic growth of zircon occurs because the concentration of Zr and light rare earth elements (LREE) increases in the melt during cooling (Kohn et al. 2015). The corresponding age data thus reflect high-temperature retrogression (Roberts and Finger 1997; Whitehouse and Platt 2003; Kelsey and Powell 2011; Yakymchuk and Brown 2014; Kohn et al. 2015), and indeed zircon ages from the Mt Stafford aureole are slightly younger than monazite ages (Fig. 9c), supporting growth upon cooling and melt crystallization.

In this set of samples, we observe that (1) the ages between monazite (core and rim) and zircon are slightly different, and (2) a systematic shift occurs between the weighted mean ages of accessory minerals from higher-grade samples of metapsammitic versus metapelitic compositions. For instance, monazite ages in the former are slightly older than those in the latter (Fig. 9), though this shift in age is almost within analytical error. The apparent age differences are most likely due to migmatites reaching the solidus at different temperatures for the different local bulk rock compositions (Yakymchuk and Brown 2014). This example can be used to demonstrate that if chemical and age differences do exist between two metasedimentary layers with different bulk rock compositions, successive *P–T–t* investigations should be made for the distinct layers. The same applies to distinct domains occurring within a single layer, but in the case of Mt Stafford these have not been investigated. Implications for domainal rocks are discussed in the next section.

LOCAL REACTIONS AND FORMATION OF DOMAINAL ROCKS

Chemical differentiation processes are essential to the formation of many metamorphic rocks (e.g., schist, hornfels, banded gneiss, migmatite). While differentiation occurs at various spatial scales, we focus here on evidence in single rock specimens and on processes that lead to chemical segregation in these, thus producing compositional domains. Once more, our goal is essentially to "read rocks", i.e., to understand samples petrogenetically and quantify the conditions at which they formed. We showed above that models rooted in chemical thermodynamics remain powerful, if they are cleverly applied to analyze rocks that have not experienced strong chemical differentiation (e.g., melt loss) or chemical segregation (zoned porphyroblasts). In general, complexity levels increase if we consider rocks with textural and chemical heterogeneity—both commonly evident in interesting samples. Clearly such rocks did not reach anything like rock-wide chemical equilibrium, but documenting local domains allows us to investigate them and paves the way to understand their petrogenesis.

In this section we examine further evidence for and consequences of domain formation from various rock types. Touching briefly on isolated segregations, such as corona structures or pseudomorphs after porphyroblasts, we mostly concentrate on analyzing spatially more extensive and organized domains, such as typically occur in migmatites.

Evidence of local reactions in discrete textural domains

Domain formation requires an initial step of physical segregation of material (DeVore 1955) to create spatial heterogeneity in composition. Such initial heterogeneity may form, for example, by mass transfer during a period of interaction with a reactive fluid (e.g., Beinlich et al. 2010) or by metamorphic differentiation (Fletcher 1977; Foster 1981, 1999), enhanced by solid-state transformations (e.g., Brouwer and Engi 2005; Lanari et al. 2013), deformation (e.g., Mahan et al. 2006; Goncalves et al. 2012; López-Carmona et al. 2014), or partial melting (Milord et al. 2001; Kriegsman and Nyström 2003).

As shown above, prograde growth of garnet porphyroblasts generates strong local heterogeneity in the rock composition, and this effect is not limited to garnet; indeed it may be even more pronounced for porphyroblastic lawsonite or kyanite if they are later involved as reactants in replacement reactions. Such reactions may respond to these unusually Ca- and Al-rich domains, which are commonly identified because they preserve the original shape of the porphyroblast (as pseudomorphs), and they generate new mineral assemblages very distinct from the rest of the rock matrix (Carmichael 1969; Selverstone and Spear 1985; Foster 1986; Carlson and Johnson 1991; Elvevold and Gilotti 2000; Ballevre et al. 2003; Brouwer and Engi 2005; Zhang et al. 2009; Verdecchia et al. 2013; López-Carmona et al. 2014). In some cases, such domains can be modeled as a closed system. For example, Brouwer and Engi (2005) combined backscatter electron images with electron microprobe spot analyses to obtain local bulk compositions of four domains in retrogressed kyanite-bearing eclogite from the Central Swiss Alps. The models account for the development of plagioclase symplectites involving very Al-rich phases like corundum, hercynite, and staurolite, which are not normally expected in rocks of basaltic bulk composition. This technique can be generalized to any textural domain for which the local bulk compositions are calculated from mineral compositions and estimated mineral modes (Tóth et al. 2000; Korhonen and Stout 2005; Mahan et al. 2006; Riel and Lanari 2015; Cenki-Tok et al. 2016; Guevara and Caddick 2016) assuming limited chemical interaction among the domains (see below). In delimiting domains, a certain amount of judgment is needed, and some assumptions are implied (Elvevold and Gilotti 2000; Kelsey and Hand 2015).

Fluid influx can cause metamorphic differentiation as well, and it may lead to significant changes in the local reactive composition (Putnis and John 2010). Using an example from the subduction complex of the Tianshan mountains (China), Beinlich et al. (2010) reported the transitional conversion from blueschist to eclogite accompanied by a change in composition (from Ca-poor to Ca-rich) along a profile sampled perpendicular to a vein (10–15 cm thick). Equilibrium phase diagrams computed for the two distinct bulk rock compositions indicate that both mineral assemblages were stable at the same peak $P–T$ conditions (21 ± 1.5 kbar and $510 \pm 30\,°C$). Here the fluid affected the local composition and probably played a significant role to achieve chemical equilibrium. However, it has long been debated (e.g., Ridley 1984) why blueschists and (low-temperature) eclogites coexist in some areas; differences in bulk composition (Brovarone et al. 2011) or in $P–T$ conditions (Davis and Whitney 2006) may be responsible.

The presence of textural domains and compositional differences among these may or may not imply chemical potential gradients. If chemical exchange between domains was very inefficient, each domain may be regarded as essentially a closed system, and thermodynamic modeling poses no problem. If transport was very limited for only a few chemical components (e.g., of trace elements), but efficient for the others, then chemical potential gradients between domains may have been essentially zero for most components, i.e., (partial) equilibrium was maintained between the domains. However, wherever local mineral assemblages within textural domains indicate substantial chemical potential gradients between them, transport (of these components) between the domains evidently happened, but at rates too low to attain overall equilibrium.

Chemical potential gradients and element transfer between domains

If two systems A and B are in thermal equilibrium and are open with regard to a mobile component i, then μ_i must be the same in both systems, otherwise there is a tendency for transfer of i between the two systems. When such a distribution has been attained, the free energy of the total system is at minimum with regard to the distribution of component i (Thompson 1955). This model of mobile component has been termed selective chemical interaction. An example of this principle concerns migmatites, which are domainal rocks with a leucosome and a melt-depleted melanosome (here considered a residue). The segregation of solids and melt and the subsequent evolution of their mineral assemblages can be modeled using $T-X$ residue–melt equilibrium phase diagrams (White et al. 2001), these are binary diagrams displaying the evolution of the phase assemblage for reactive bulk composition lying between a residue (melanosome or restite) and a melt-rich domain (leucosome). Of interest here is the role that diffusion of H_2O can play between leucosome and residue in the crystallization history of (segregated) melt in contact with the residue. If the physical separation of melt from its residue is purely mechanical and happens at chemical equilibrium, there are no gradients in chemical potential. However, subsequent cooling results in chemical re-equilibration of now separate domains of different reactive bulk compositions. Now chemical potential gradients will be established, in particular $\mu(H_2O)$ will rise upon crystallization of the melt-rich domain. Where in contact with restite, equilibration between the two domains requires diffusion of H_2O from the leucosome to the melanosome. White and Powell (2010) used this model of selective chemical interaction to show that the diffusive interaction of H_2O between residue and segregated melt—aiming to equalize $\mu(H_2O)$—promotes the crystallization of anhydrous quartzofeldspathic products in the leucosome and hydration of the residue. Further studies are warranted to identify evidence of interaction for other elements between the two domains, possibly with different length scales. It is crucial to understand such scales if equilibrium thermodynamics (based on local bulk composition, see below) may be applied or not.

Chemical potential gradient within domains

Frozen-in chemical potential gradients can also be observed within single domains provided that diffusion-controlled structures such as coronae are preserved. An interesting approach based on equilibrium thermodynamics and quantitative chemical potential diagrams has been proposed by White et al. (2008) who reconstruct the chemical potential gradients preserved in the final corona structure. The same approach was applied by Schorn and Diener (2016) to investigate coronae developed at the expense of magmatic plagioclase and orthopyroxene during the gabbro-to-eclogite transformation. So far, such models are restricted to simple systems with some general assumptions on components in excess or assumed to be immobile.

Size of the equilibrium volume versus scale of the model

As discussed earlier, the size of the equilibrium volume is controlled by the diffusion rate of the elements in the intergranular medium. Knowing diffusion rates is essential as they may set a spatial limit within which rocks can maintain global chemical equilibrium as they evolve. Because experimental data are sparse or may not apply, few reliable estimates on rates of intergranular diffusion are as yet available; they have instead been extracted from natural examples (Carlson 2002 and references therein). Because it is difficult to constrain the duration of intergranular diffusion tightly, these rates are not very well known, but Carlson (2010) succeeded in estimating the rate of intergranular diffusion of Al and its dependence on temperature for H_2O-saturated and fluid-undersaturated media (Fig. 10). Data for Al diffusion are essential because many studies found the diffusive transport of Al to exercise the dominant control on overall rates of reaction for aluminosilicates (Carmichael 1969; Foster 1977, 1981, 1983; Carlson 1989, 2002, 2010; White et al. 2008). To the extent that this may apply, the diagram in Figure 10 sets limits on the length scales over which reactions and

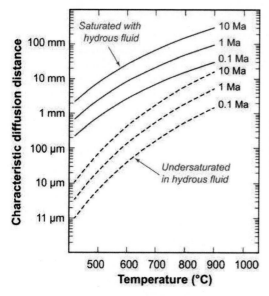

Figure 10. Characteristic intergranular diffusion length scale for Al in a medium saturated in hydrous fluid (blue curves) and undersaturated in fluid (black dashed curves); modified from Carlson (2010).

chemical equilibration can be expected for a range of temperature and typical duration of metamorphic transformations. For example, the data indicate that at upper amphibolite facies conditions (say 600–700 °C), domains 0.5 mm apart will require 1 Ma to equilibrate if grain boundaries are not fluid-saturated, whereas domains 5 mm apart should equilibrate in 100 ka when saturated with hydrous fluid. But do fluids ever persist that long?

In any case, this sort of analysis may serve to indicate whether we may rely on chemical equilibrium models to infer petrogenetic conditions from domainal rocks. Where valid, it is useful to analyze *P–X* or *T–X* phase diagrams, calculated for a range in local bulk rock compositions (*X*1…*X*2), corresponding to different domains. Individual textural domains can be analyzed to determine the local bulk compositions (Tóth et al. 2000; Brouwer and Engi 2005; Riel and Lanari 2015; Cenki-Tok et al. 2016) and then the equilibrium assumption can be tested by comparing predictions from forward thermodynamic models against observed phase relations. To support such comparisons, we propose an analytical method to quantify variations in the local bulk composition based on standardized X-ray maps.

QUANTITATIVE MAPPING OF THE LOCAL BULK COMPOSITION AS A BASIS FOR MODELING

It is possible to correct for the effects of chemical fractionation and chemical differentiation described in the previous sections using quantitative compositional maps (expressed in oxide wt%) and to extract local compositions from such a map. The first application (to our knowledge) that used quantitative compositional maps to obtain the local bulk composition was published by Marmo et al. (2002) who focused on two eclogite samples showing zoned garnet porphyroblasts. The authors examined the effect of garnet core and mantle fractionation on the bulk rock composition using compositional maps with a standardization based on the semi-empirical Bence and Albee (1968) matrix correction algorithm (Clarke et al. 2001). Major progress since then involved a density correction, clarifications on the choice of domain boundaries, and a correct extrapolation from 2D to 3D. This last point is tricky; for example it is easy to overestimate the contribution of garnet cores, as surface fractions need to be properly converted to volume fractions, but equally easy to underestimate cores—simply by mapping a grain sectioned off-center.

The influence of various assumptions on the local bulk composition estimates is shown below, but an adequate mapping strategy is definitely required to produce high-quality standardized maps.

Quantitative X-ray mapping

Since the first X-ray "spot maps" measured using an electron probe micro-analyzer (Cosslett and Duncumb 1956), both the instruments and techniques have been greatly improved. Although quantitative methods (Kohn and Spear 2000; Clarke et al. 2001; De Andrade et al. 2006) and computer programs (Tinkham and Ghent 2005b; Lanari et al. 2014c) are available to obtain oxide wt% maps, many studies still combine uncorrected 'semi-quantitative' X-ray maps with profiles of high-precision point analyses.

A quantitative method requires a correction called 'analytical standardization' that transforms the number of collected photons (i.e., X-ray intensity) using either a semi-empirical Bence and Albee (1968) matrix correction (Clarke et al. 2001) or high-resolution spot analyses as internal standard (De Andrade et al. 2006). The technique of internal standardization provides accurate compositional maps and has recently been integrated into the software XMapTools (Lanari et al. 2014c). Quantitative compositional mapping allows measuring the natural variability of the mineral phases at the scale of a thin section and thus is useful in petrogenetic analysis (e.g., Kohn and Spear 2000). Compositional mapping has been successfully combined with multi-equilibrium thermobarometry (Vidal et al. 2006; Yamato et al. 2007; Fiannacca et al. 2012; Lanari et al. 2012, 2013, 2014a,b; Pourteau et al. 2013; Grosch et al. 2014; Loury et al. 2015, 2016; Trincal et al. 2015; Scheffer et al. 2016) and forward thermodynamic modeling (Abu-Alam et al. 2014; Lanari et al. 2017) or may be used to extract mineral modes (Cossio et al. 2002; Martin et al. 2013) and local compositions (Centrella et al. 2015; Riel and Lanari 2015; Mészaros et al. 2016).

Strategy to derive local bulk composition from X-ray maps using XMapTools

Local bulk composition can be easily estimated based on quantitative maps, for which the composition at each pixel is expressed in oxide wt%. The analytical procedure is detailed in Appendix 1. The local bulk composition of any desired part of the section mapped is spatially integrated, using the (oxide wt%) map and applying a density correction for each mineral phase identified. Density values are estimated using Theriak. This procedure is coded as a function available in XMapTools 2.3.1 and allows extracting the composition (in oxide wt%) of any domain directly from the compositional maps by polygonal boundaries selected by the user.

To illustrate the simplicity of this procedure, we selected an orthogneiss from the Glacier–Rafray klippe in the Western Italian Alps (Burn 2016). X-ray maps were acquired using a beam current of 100 nA and dwell time of 70 ms. These maps were standardized using XMapTools 2.3.1 and the method described in Lanari et al. (2014c). The map size is 10.24×10.24 mm (corresponding to 1024×1024 pixels with a pixel size of 10 μm; the total measurement time is ~42h). This orthogneiss contains 38 vol% quartz, 33% albite, 21% phengite, 4% K-feldspar, 2% epidote, and 2% actinolite. The local bulk composition LBC1 was extracted from a square domain of ~100 mm^2 (dashed line in Fig. 11). To estimate the uncertainty in the composition resulting from the domain selection, a sensitivity test was performed using a Monte-Carlo simulation that randomly changed the position of the corners and thus the shape of the selected domain (Fig. 11). In this example 100 permutations were used with a displacement of each corner of ± 10 pixels (1σ assuming a Gaussian distribution). The composition of LBC1 is shown in Table 2 with the associated standard deviations and its relative uncertainty. The sensitivity test shows that there is very little variation (<1% for all the elements) in the composition due to this local domain selection.

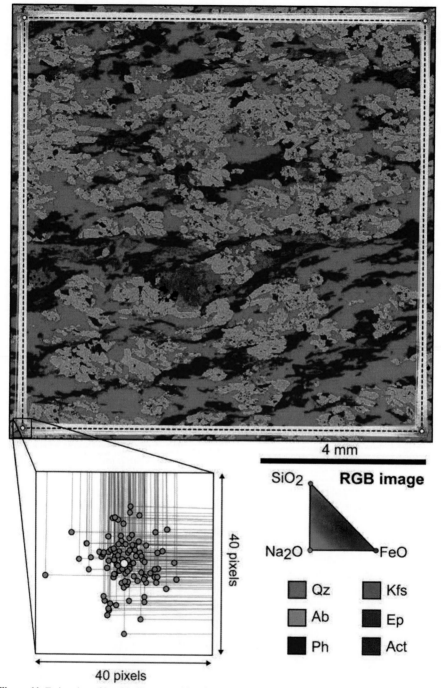

Figure 11. Estimation of local bulk composition for an orthogneiss from the Western Alps. Variably sized areas were considered, one delimitation shown by dashed black outline, gray lines in inset show variations used in sensitivity test performed by Monte-Carlo simulation (see text). Analysis based on XMapTools; the composite compositional map shown is color-coded in an RGB image displaying variations in SiO_2–Na_2O–FeO, but any combination of oxides or element ratios could be chosen.

Table 2. Local bulk composition LBC1 of an orthogneiss from the Glacier-Rafray Klipee in the Western Alps (see text) calculated using XMapTools (see Fig. 11). Abbreviations: Stdev. Standard deviation; Unc. Relative uncertainty.

	Mean	**Stdev. (2σ)**	**Unc. (%)**
SiO_2	75.420	0.057	0.076
Al_2O_3	13.230	0.033	0.249
FeO	1.230	0.006	0.488
MgO	0.940	0.004	0.426
CaO	0.810	0.008	0.942
Na_2O	3.360	0.018	0.522
K_2O	3.090	0.017	0.551
Total	98.08		

Gibbs free energy minimization for the local bulk composition

A similar approach is used to propagate this relative uncertainty through the forward thermodynamic models. Computations were made using program Bingo-Antidote designed to estimate the optimal P–T conditions using equilibrium models (Lanari and Duesterhoeft 2016). The thermodynamic database JUN92.bs (Berman 1988 and subsequent updates) was selected for this test in the system SiO_2–Al_2O_3–FeO–Fe_2O_3–MgO–CaO–Na_2O–K_2O–H_2O. Excess oxygen (0.09 mol%) in the bulk composition was added to stabilize the observed amount of epidote. Negligible amounts of hematite (0.3 vol%) are predicted by the model (and ignored in the following). Mineral modes and compositions extracted from the domain LBC1 (Fig. 11) were modeled at 475 °C and 12.6 kbar (Fig. 12a), the estimated conditions of equilibration of the assemblage quartz + albite + phengite + K-feldspar + epidote. The dispersion (±2σ) of the mineral modes predicted by the equilibrium model was estimated from 2000 permutations and is less than 1.6% (relative) for quartz, albite, phengite, and epidote, and 2.4% for K-feldspar (Fig. 12b). The corresponding information can also be extracted for mineral compositions. For example the model phengite composition is $Si^{4+} = 3.481 \pm 0.002$ atoms per formula unit (2σ) and $XMg = 0.874 \pm 0.008$. This technique allows propagating the relative uncertainty in composition through the equilibrium models. As mentioned previously, few such sensitivity analyses have been reported (Kelsey and Hand 2015), and thus modeled modes have rarely been compared to observed mineral phases that contribute to the bulk composition (Warren and Waters 2006).

In the present example, an arbitrary ~10 × 10 mm² domain was mapped and used to estimate the local bulk composition that was then taken as a reactive bulk composition for modeling. The overall quality of the model was excellent, which would seem to support the choice of this domain as an equilibrium domain. But is such a comparison sufficient to conclude that equilibrium was established at this spatial scale for the major elements? The answer is probably yes—at least for the elements of interest. However, for petrochronological studies the scale of equilibration for trace elements would be particularly relevant to verify whether slowly diffusing species attained uniform concentrations via an intergranular medium. To pursue this topic, the approach should be extended to use LA-ICP-MS quantitative maps. In this case a FCM would be needed as the accessory phases almost certainly act like porphyroblasts and trace constituents would not be part of the equilibration volume or reactive bulk composition.

Advantages of the micro-mapping approach

The approach described above can be used to test several textural domains from the same map dataset. Using Bingo-Antidote, it is thus possible to select a polygon and directly calculate the corresponding model assemblage, modes, and compositions for any P–T conditions.

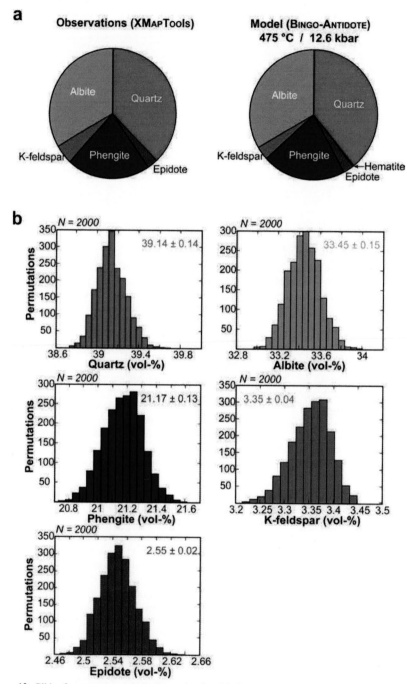

Figure 12. Gibbs free energy minimization using local bulk compositions (based on Fig. 11) and the program Bingo-Antidote. (a) Comparison between observed (left) and modeled (right) mineral modes. (b) Results of a sensitivity test (2000 permutations) performed to evaluate the effect of the domain selection on predicted mineral modes. Relative uncertainty in mode reported at 2σ level.

This strategy provides appropriate remedies to refine the estimation of the composition of the reactive part of a rock. For instance, most accessory phases are generally ignored in equilibrium phase diagrams, and this can lead to an erroneous bulk rock composition. Apatite is a good example because it contains > 50 wt% CaO, and if a simplified system does not incorporate phosphorous, the amount of CaO from the bulk rock composition available for the other phases should be corrected for the CaO stored in apatite. However, correcting CaO based on P_2O_5 contents would introduce error if much monazite is present in the sample. This problem is easily solved using the micro-mapping approach: Apatite pixels can simply be ignored (a software option), and thus the CaO value of the local bulk composition is directly suitable for modeling. A similar approach can be used if any mineral present is considered non-reactive.

Potential artifacts affecting the local bulk composition estimates

To illustrate potential effects of some artifacts on the local bulk composition estimate and the equilibrium models, we use compositional maps of a mafic eclogite boudin from the Atbashi Range in the Kyrgyz South Tien Shan (Loury et al. 2015). The sample contains garnet porphyroblasts (with quartz and omphacite inclusions) in a matrix of omphacite, rutile, and ilmenite.

Geometric effects. Geometric effects arise as a consequence of sectioning a 3D texture and are most consequential if the surface fractions of successive growth zones of crystals cannot be linearly correlated with their volume fractions. For example, garnet typically has a dodecahedral crystal habit that can be approximated by a spherical geometry (Fig. 13b). If such a garnet grain is cut near its crystal center (see for example Fig. 1), any bulk crystal composition extracted from 2D compositional maps will overestimate the contribution of the core at the expense of the rim. If the cut is not equatorial, the section visible may not be representative of the zoning at all. The effect of such geometric bias is explored by comparing the composition of a single garnet grain extracted by first averaging from the 2D-map (Fig. 13a) or by converting the composition of each 2D-annulus to a 3D-hollow sphere (Fig. 13b) and then integrating these. This effect is generally neglected for local bulk composition estimates (Marmo et al. 2002) but for spherical domains it is judicious to use a 3D extrapolation (Mészaros et al. 2016). In the case of garnet from the mafic eclogite of the Atbashi Range, MnO is enriched in garnet core, and the average MnO composition of garnet is overestimated by 36% if integration is based on the 2D surface of a single grain (Fig. 13a). Similar deviations are observed for other elements (MgO −29%; and FeO +7%). For a map containing many grains, such geometric effects are partially compensated by stochastic sampling of different sections (Fig. 13c). In our example, integrating all of the garnet composition in 2D cuts the discrepancy in half, but it remains + 16% in MnO, −17% in MgO, and + 4% in FeO. It is also possible to evaluate the average garnet composition using a single crystal that is cut near the center (to be established with a diagnostic element such as Mn) and the spherical correction. Then this average garnet composition can be mixed with a complementary fraction of matrix for which an average composition is extracted from the maps.

Chemical equilibrium and the arbitrary choice of domains. Choosing or delimiting an appropriate local domain is obviously problem-dependent and usually non-trivial. For equilibrium thermodynamics to be applicable, the spatial domain selected for modelling should be no larger than the equilibration volume. (Of course any smaller domain might be selected). The problem is that the size limit is not known. Modellers often just hope that the domains are uniform in composition, i.e., that the mobility of the (slowest) components considered in their model was sufficient to equilibrate compositions over the chosen domain size. Generally, decisions should be based on (1) textural criteria such as the apparent homogeneity and representativeness of the domain compared to the entire specimen, or (2) independent transport criteria such as the maximum length scale of chemical equilibration predicted by the diffusion rate of Al (see Fig. 10).

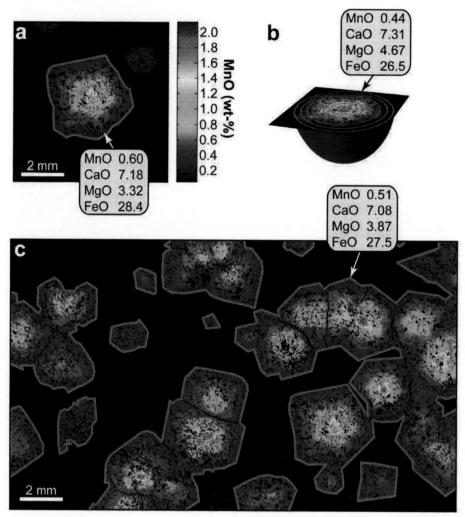

Figure 13. Average composition of garnet single grain estimated using XMapTools. (a) Method 1: 2D averaging of all the garnet pixels of a single crystal. (b) Method 2: 3D extrapolation using seven ellipsoids (details of procedure available in user guide of program, http://www.xmaptools.com). The 2D technique overestimates MnO (absolute: 0.16 wt%; relative difference (Δ) = 36%) and FeO (1.9 wt%; Δ 7%), underestimation of Cao (0.13 wt%; Δ 2%) and MgO (1.35 wt%; Δ 29%). (c) Method 3: 2D averaging of garnet pixels in all grains results in overestimation of MnO (0.07 wt%; Δ 16%) and FeO (1.0 wt%; Δ 4%) and underestimation of Cao (0.23 wt%; Δ 3%) and MgO (0.8 wt%; Δ 17%).

For a Kyrgyz mafic eclogite from Loury et al. (2015), the predicted length scale of intergranular diffusion of Al in 1 Ma, assuming a hydrous fluid present, is about 2 mm (Fig. 10). As garnet in this case crystallized at the expense of chlorite and lawsonite, it is likely that the intergranular medium was saturated in hydrous fluid, though perhaps only episodically, not for the entire 1 Ma. It is interesting to note that this spatial scale corresponds to the minimum distance between the centers of neighboring porphyroblasts observed in the section (Fig. 14). However the grains show remarkably regular chemical zoning in Ca and X_{Mg} ($Mg^{2+}/(Mg^{2+}+Fe^{2+})$) at the thin section scale (Fig. 14), suggesting that chemical equilibrium was maintained at the periphery of growing phases at all times (equilibrium control growth model assuming grain boundary equilibrium), which implies a much larger scale for chemical equilibration of the other major elements.

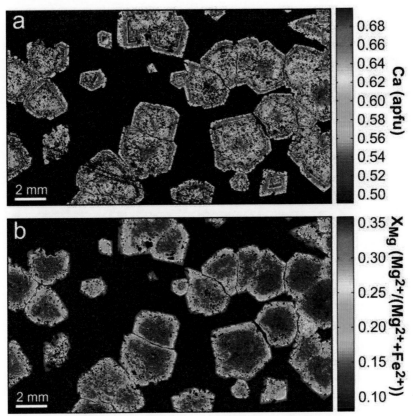

Figure 14. Compositional maps of garnet grains of a mafic eclogite boudin from the Atbashi Range in Kyrgyz South Tien Shan (Loury et al. 2015). Color code is for structural formula (apfu: atoms per formula unit); (a) Ca and (b) X_{Mg}.

Another strategy to ensure that the local bulk composition estimate is robust is to check the sensitivity of the composition and model predictions on the polygon selection (Fig. 15). The compositional map of the Kyrgyz mafic eclogite was sampled using a floating rectangular window, the size of which was increased from $6.8 \times 6.8\,mm^2$ to $20.5 \times 12\,mm^2$. The local bulk composition was successively calculated after each increment, and results are shown in Figure 15b. This experiment shows that the local bulk composition can be very sensitive to the area selected. The fluctuations are visible in the variable proportion of garnet (from 32 to 45 vol%, Fig. 15). Once some 80% of the area is integrated, the composition no longer changes significantly (<5% for all elements), as shown in the sensitivity tests.

Seeing such variations in the local bulk compositions (Fig. 15b), the critical question is: how sensitive are model predictions and *P–T* estimates? To appreciate the effect of this, the compositions of the four steps used in the above test (i.e., 18%, 50%, 80% and 100% of the area, Fig. 15) were used at 550 °C and 25 kbar, with Bingo-Antidote and the thermodynamic dataset tc55.txt (Holland and Powell 1998) in the system SiO_2–TiO_2–Al_2O_3–FeO–Fe_2O_3–MnO–MgO–CaO–Na_2O; the oxygen fugacity was controlled by the QFM buffer. The modeled mineral modes are in line with the mineral modes observed in the four areas (Fig. 15c), indicating that the predicted modes are correct. By contrast, the modeled garnet compositions are slightly different for the four models, with deviations of 0.03 in almandine

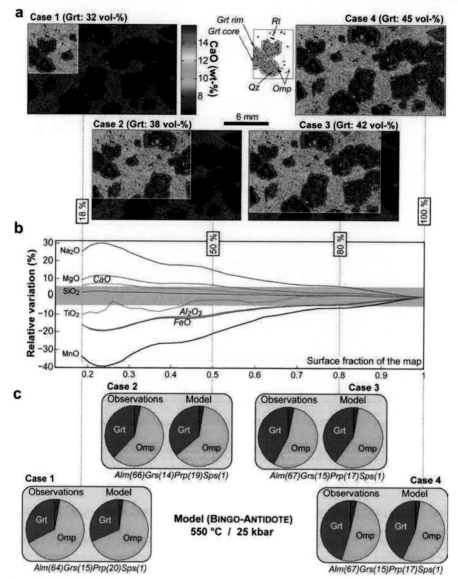

Figure 15. Estimated local bulk composition of a mafic eclogite boudin from the Atbashi Range in the Kyrgyz South Tien Shan (Loury et al. 2015). (a,b) The sensitivity test is based on different areas of a thin section (from 18% to 100% of the map surface). Note variation in predicted garnet mode. Compositional map shown for CaO (wt-%); small inset: Grt: garnet, Omp: omphacite, Qz: quartz, Rt: rutile. (c) Modeled and observed mineral modes for the four cases shown in (a): 18%, 50%, 80% and 100% modeled at 550 °C and 25 kbar (see text).

and pyrope contents (Fig. 15c). To explore the bias of such differences to *P–T* estimates, the best isopleth intersection of a reference garnet core composition (Grt$_{ref}$ corresponding to garnet predicted stable for the entire map composition; Alm$_{67}$Grs$_{15}$Prp$_{17}$Sps$_1$; case 4 in Fig. 15) was calculated for successive local bulk rock compositions using GrtMod (Fig. 16). Note that this example is used essentially to illustrate the concept: only one composition corresponding to a hypothetical garnet core is modeled, without expanding the case to FCM.

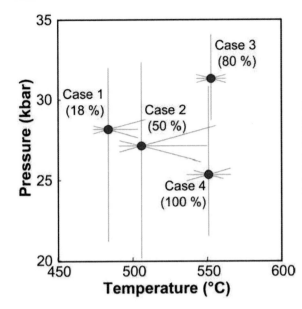

Figure 16. *P–T* conditions of Grt$_{ref}$ (see text) for the four bulk compositions extracted from the areas shown in Figure 14 (Cases 1, 2, 3 and 4) computed with GrtMod. The errors bars show the relative uncertainty of the equilibrium model (i.e., isopleth position) resulting from typical uncertainties in garnet composition.

Grt$_{ref}$ is predicted stable at lower temperature using the local composition of a small area (480 ± 25 °C) compared to larger areas (550 ± 18 °C). The errors bars reported in Figure 14 indicate the relative uncertainty of the equilibrium model (i.e., the isopleth position) resulting from typical uncertainties in garnet composition. The results presented here show that the differences in local bulk composition, depending on the domain size integrated, definitely can affect the garnet isopleths and hence *P–T* estimates. While somewhat technical and problem-dependent, this selection has significant consequences on *P–T* estimation.

The illustration example of the Kyrgyz mafic eclogite was selected because it contains several limits that have been intensively discussed in this review. First, it involves garnet porphyroblasts that are compositionally zoned, and an optimized strategy would require a FCM to obtain the *P–T* segment along which garnet grew. Second, the shape of garnet porphyroblasts requires the use of 3D extrapolation to avoid geometric effects on the local bulk composition estimates.

If a domain is assumed to be in chemical equilibrium at given *P–T* conditions, any smaller subdomain should yield a similar *P–T* estimate. Figure 16 shows substantial *P–T* differences for the domains shown in Figure 15, because garnet compositional zoning (disequilibrium) has a strong effect on the local bulk compositions. Apart from this, it is important to note that differences among thermodynamic data sets (standard state properties and solution models; not tested here) may also affect the *P–T* predictions.

Toward systematic quantitative trace element micro-mapping to address petrochronological problems

Quantitative mapping of diagnostic trace elements is crucial for many petrochronological applications. Over the past two decades, EMPA has been quite effectively used to produce such maps (e.g., Spear and Kohn 1996; Chernoff and Carlson 1999), but recent improvements in LA-ICP-MS analysis and software support (Paul et al. 2012; Rittner and Müller 2012) have fostered applications using quantitative maps of trace elements and isotopes (Becker et al. 2007; Woodhead et al. 2007; Stowell et al. 2010; Duval et al. 2011; Netting et al. 2011; Šelih and van Elteren 2011; Peng et al. 2012; Paul et al. 2014; Gatewood et al. 2015; Ubide et al. 2015).

For instance a garnet from the Peaked Hill shear zone, Reynolds Range, central Australia, has been mapped by this technique (Raimondo et al. 2017), with successive raster images of the focused laser beam stitched together to form a 2D representation of the trace element distribution. Quantification was initially performed using Iolite (Woodhead et al. 2007; Paton et al. 2011), and subsequent image processing and manipulation by XMapTools (Lanari et al. 2014c). The maps reveal significant decoupling between the major and trace element zoning patterns, with smooth radial zoning in Fe, Mg, Ca and Mn juxtaposed against a discrete annular structure and successive satellite peaks for REEs and Cr (Fig. 17). Importantly, the growth and dissolution of accessory phases can be directly linked to garnet evolution through the superposition of sharp satellite peaks in Zr and HREE (zircon) and P, Th and LREEs (monazite). Such micron-scale features are clearly resolved by trace element mapping, but easy to miss with more coarsely spaced spot analyses. Thus the increased spatial resolution and mass range of LA-ICP-MS mapping offer a powerful means to place garnet growth in a specific paragenetic context and integrate it with temporal constraints provided by accessory phase geochronometers or direct Lu–Hf/Sm–Nd dating.

CONCLUSIONS AND PERSPECTIVES

The examples presented in this review indicate that care is needed when modeling a rock with evident compositional heterogeneity or textural domains. Textural evidence in rocks—notably mineral zoning—shows that rocks adapt to changes in conditions during their evolution, but that equilibration is limited by kinetics of transport. Nonetheless, equilibrium models—and especially thermodynamic forward models—remain powerful for petrogenetic analysis and allow the retrieval of $P–T$ conditions, provided that partial and local assemblages are analyzed.

Zoned porphyroblasts can significantly fractionate the reactive bulk composition, and this effect must be included in setting up FCM's. In the case of garnet, it has been shown that ECM's significantly overestimate the predicted modes and affect the $P–T$ estimates made by isopleths thermobarometry. Dating porphyroblast growth remains attractive, based on textural correlation and in situ dating techniques. Correlation efforts to link ages with porphyroblast growth (or resorption) conditions can profit from quantitative element mapping of both major and trace elements. Documenting how concentrations in REE, Y, Th, and U evolve in successive growth zones of the host and in the included accessories should be useful for testing whether or not host and inclusions are in exchange equilibrium (i.e., may or may not be regarded as coeval). More fundamentally, such combined analysis promises much needed insight into the mobility of trace elements relative to major elements.

A bulk rock composition determined by XRF analysis of an entire sample can be very far from the composition of a local assemblage, hence thermobarometry based on isopleths will be flawed, even observed phase relations may compare poorly with predictions. Instead, the reactive bulk composition must be approximated by taking account the textural evidence and preserved local mineral compositions. X-ray maps yield a solid basis to estimate the local bulk composition, if the spatial domain to be modeled is carefully delimited in thin section. Predictions from local equilibrium models must be critically evaluated, checking the textural and mineralogical record preserved in a sample against the computed phase relations for it. Using a program such as Bingo-Antidote also allows the sensitivity of predictions to the chosen domain boundaries to be tested. In our experience the local composition needs to be iteratively refined, or it may turn out that different chemical subsystems should be chosen to analyze different parts of a rock's evolution. The goal must be to gain a fairly detailed understanding of the petrogenetic stages visible in a sample, which ensures that $P–T$ conditions estimated using isopleths are robust. Such a basis allows a meaningful strategy to select datable (sub)grains for in situ analysis, which we consider critical to trustworthy petrochronology.

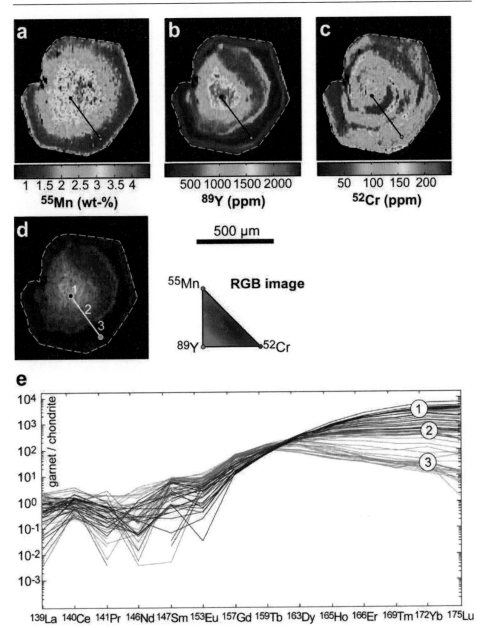

Figure 17. Minor and trace element compositional maps of a garnet grain from the Peaked Hill shear zone, Reynolds Range (Raimondo et al. 2017), central Australia. (a) Mn in wt%; (b) Y in ppm; (c) Cr in ppm; (d) Composite RGB image; (d) Chondrite-normalized REE patterns of garnet along the profile drawn in a, b, c and d (one curve in (e) per pixel). Concentrations given are for the element of interest, isotope numbers show which isotope was used to monitor element abundance.

Some elements are far less mobile than others, and gradients in concentration (and thus in chemical potential) may be preserved that render equilibrium modeling inappropriate, i.e., relatively immobile species (including some trace elements) may have to be excluded.

At present, there is limited knowledge regarding the mobility of many trace elements, notably for actinides, Pb and REE that are critical to chronology. It remains a challenge to develop approaches that integrate local equilibrium models for domains in which these elements show evidence of limited transport. To improve our understanding, concentration gradients, both within and among domains, should be documented for elements with very different transport properties (ionic charge and size of species dominant for transport). Such maps provide integrals over the relative mobility and may allow us to distinguish chemical subsystems for which local equilibrium modeling is reliable vs. those that are sensitive to kinetics. To reconstruct local chemical potential gradients, it would be intriguing to invert the model, i.e., the compositional gradient that can be observed between two textural domains. Extending earlier studies (Carlson 2002), the characterization of the frozen-in gradients will help quantify extents of disequilibrium.

ACKNOWLEDGMENTS

We thank M. Burn, F. Giuntoli, F. Guillot, C. Loury and T. Raimondo for providing sample information and compositional maps used in Figures 2 (MB, FGi, FGu), 11 (MB), 13–14–15 (CL), 17 (TR). We acknowledge stimulating discussions with R. Berman, C. de Capitani, E. Duesterhoeft, J. Hermann and D. Rubatto. Reviews from H. Stowell and G. Clarke, as well as helpful comments from M. Kohn that led us to improve this paper are gratefully acknowledged, as is Matt's editorial handling. The Swiss National Science Foundation (Project 200020-146175) and the Faculty of Science of the University of Bern have supported work presented here.

APPENDIX 1—LOCAL BULK COMPOSITIONS FROM OXIDE WEIGHT PERCENTAGE MAPS

Let us consider a domain of rock composed of three mineral phases Min_1, Min_2 and Min_3, each homogeneous in composition C_1^i, C_2^i and C_3^i of the oxides of the element i. C is expressed in oxide wt%. This is convenient here because chemical analyses of silicate minerals are commonly reported in wt% of the oxides determined. The local bulk composition of this domain C_{LB} can be calculated as:

$$C_{LB} = w_1 C_1^i + w_2 C_2^i + w_3 C_3^i \tag{A1}$$

With w_1, w_2 and w_3 the mass fractions of the mineral phases Min_1, Min_2 and Min_3. This relation can be generalized for a map of a given domain containing n pixels:

$$C_{LB} = \sum_{j=1}^{n} w_j C_j^i \tag{A2}$$

w_j and C_j^i are the mass fraction and composition in oxide weight percentage of pixel j. The use of Relation (A2) is not straightforward, as it requires the knowledge of the mass fraction of every pixel that may belong to different phases, each with a different molar mass.

On the other hand, the pixel fraction of a phase k is a good approximation of the surface covered by this phase and can be extrapolated to a volume fraction. To a first approximation, it is often assumed that:

$$v_k = s_k \tag{A3}$$

In metamorphic petrology, this relation may be reasonable if (i) the sample was sectioned perpendicular to the foliation or schistosity, (ii) the compositional map is acquired on an unaltered rock surface devoid of local compositional heterogeneities, (iii) the size of the map is sufficient to ensure good sampling, (iv) the resolution of the map is high enough to avoid issues with the smaller grain size population, and (v) 3D effects are negligible. Possible pitfalls and issues are discussed under 'Potential artifacts affecting the local bulk composition estimates'. The relationship between the mass fraction and the volume fraction is:

$$w_k = \frac{\rho_k}{\rho_{\text{mixture}}} v_k \qquad (A4)$$

with ρ_k being the density of the phase k and the average density ρ_{mixture} of the domain. Integrating the density correction in Equation (A2) leads to a more convenient expression of the local bulk composition of the domain:

$$C_{\text{LB}} = \sum_{j=1}^{n} \frac{\rho_k}{\rho_{\text{mixture}}} v_j C_j^i \qquad (A5)$$

From this relationship it is possible to extract the local bulk composition of a domain using the average density of every phase involved.

In case of multi-phase assemblages the density correction is required to predict accurate local bulk compositions. For example, most of the studies discussed in this review did not correct for density differences between the considered minerals (M. Tóth et al. 2000; Marmo et al. 2002; Brouwer and Engi 2005; Cenki-Tok et al. 2016), and this can lead to large discrepancies for elements sequestered by dense mineral phases (e.g., Fe in magnetite or garnet).

Bulk rock compositions can also be extracted from oxide weight percentage maps. In this case, a domain with as many grains as practical should be selected. Based on the two examples discussed in this paper, we suggest that the width of the selected surface must be at least 8–10 times the 'average' size of the largest grains. This criterion (i.e., >65 times the 'average' grain surface) ensures that the composition is fairly representative of the entire composition of the domain, as long as the section is uniform in mineral assemblage.

REFERENCES

Abu-Alam TS, Hassan M, Stüwe K, Meyer SE, Passchier CW (2014) Multistage tectonism and metamorphism during Gondwana collision: Baladiyah Complex, Saudi Arabia. J Petrol 55:1941–1964, doi:10.1093/petrology/egu046

Ague JJ, Axler JA (2016) Interface coupled dissolution–reprecipitation in garnet from subducted granulites and ultrahigh-pressure rocks revealed by phosphorous, sodium, and titanium zonation. Am Mineral 101:1696–1699, doi:10.2138/am-2016-5707

Anderson DE, Olimpio JC (1977) Progressive homogenization of metamorphic garnets, South Morar, Scotland; evidence for volume diffusion. Can Mineral 15:205–216

Atherton MP (1968) The variation in garnet, biotite and chlorite composition in medium grade pelitic rocks from the Dalradian, Scotland, with particular reference to the zonation in garnet. Contrib Mineral Petrol 18:347–371, doi:10.1007/bf00399696

Ballevre M, Pitra P, Bohn M (2003) Lawsonite growth in the epidote blueschists from the Ile de Groix (Armorican Massif, France): a potential geobarometer. J Metamorph Geol 21:723–735, doi:10.1046/j.1525-1314.2003.00474.x

Baxter EF, Caddick MJ, Dragovic B (2017) Garnet: A rock–forming mineral petrochronometer. Rev Mineral Geochem 83:469–533

Becker JS, Zoriy M, Becker JS, Dobrowolska J, Matusch A (2007) Laser ablation inductively coupled plasma mass spectrometry (LA-ICP-MS) in elemental imaging of biological tissues and in proteomics. J Anal At Spectrom 22:736–744, doi:10.1039/B701558E

Beinlich A, Klemd R, John T, Gao J (2010) Trace-element mobilization during Ca-metasomatism along a major fluid conduit: Eclogitization of blueschist as a consequence of fluid–rock interaction. Geochim Cosmochim Acta 74:1892–1922, doi:10.1016/j.gca.2009.12.011

Bell TH, Rubenach MJ, Fleming PD (1986) Porphyroblast nucleation, growth and dissolution in regional metamorphic rocks as a function of deformation partitioning during foliation development. J Metamorph Geol 4:37–67, doi:10.1111/j.1525–1314.1986.tb00337.x

Bence AE, Albee AL (1968) Empirical correction factors for the electron microanalysis of silicates and oxides. J Geol 76:382–403

Berman RG (1988) Internally consistent thermodynamic data for minerals in the system $Na_2O–K_2O–CaO–MgO–FeO–Fe_2O_3–Al_2O_3–SiO_2–TiO_2–H_2O–CO_2$. J Petrol 29:445–522

Bowen NL (1913) The melting phenomena of the plagioclase feldspars. Am J Sci Ser 4 35:577–599, doi:10.2475/ajs. s4-35.210.577

Brodie KH, Rutter EH (1985) On the relationship between deformation and metamorphism, with special reference to the behavior of basic rocks. *In:* Metamorphic Reactions: Kinetics, Textures, and Deformation. Thompson AB, Rubie DC (eds). Springer New York, p.138–179

Brouwer FM, Engi M (2005) Staurolite and other high-alumina phases in Alpine eclogite: Analysis of domain evolution. Can Mineral 43:105–128

Brovarone AV, Groppo C, Hetenyi G, Compagnoni R, Malavieille J (2011) Coexistence of lawsonite-bearing eclogite and blueschist: phase equilibria modelling of Alpine Corsica metabasalts and petrological evolution of subducting slabs. J Metamorph Geol 29:583–600, doi:10.1111/j.1525–1314.2011.00931.x

Brown M (2002) Retrograde processes in migmatites and granulites revisited. J Metam. Geol 20:25–40

Burn M (2016) LA-ICP-QMS Th–U/Pb allanite dating: methods and applications. PhD thesis University of Bern

Caddick MJ, Kohn MJ (2013) Garnet: Witness to the evolution of destructive plate boundaries. Elements 9:427–432, doi:10.2113/gselements.9.6.427

Caddick MJ, Bickle MJ, Harris NBW, Holland TJB, Horstwood MSA, Parrish RR, Ahmad T (2007) Burial and exhumation history of a Lesser Himalayan schist: Recording the formation of an inverted metamorphic sequence in NW India. Earth Planet Sci Lett 264:375–390

Caddick MJ, Konopásek J, Thompson AB (2010) Preservation of garnet growth zoning and the duration of prograde metamorphism. J Petrol 51:2327–2347, doi:10.1093/petrology/egq059

Carlson WD (1989) The significance of intergranular diffusion to the mechanisms and kinetics of porphyroblast crystallization. Contrib Mineral Petrol 103:1–24

Carlson WD (2002) Scales of disequilibrium and rates of equilibration during metamorphism. Am Mineral 87:185–204, doi:10.2138/am-2002-2-301

Carlson WD (2010) Dependence of reaction kinetics on H_2O activity as inferred from rates of intergranular diffusion of aluminium. J Metamorph Geol 28:735–752, doi:10.1111/j.1525–1314.2010.00886.x

Carlson WD, Johnson CD (1991) Coronal reaction textures in garnet amphibolites of the Llano Uplift. Am Mineral 76:756–772

Carlson WD, Pattison DRM, Caddick MJ (2015) Beyond the equilibrium paradigm: How consideration of kinetics enhances metamorphic interpretation. Am Mineral 100:1659–1667, doi:10.2138/am-2015-5097

Carmichael DM (1969) On the mechanism of prograde metamorphic reactions in quartz-bearing pelitic rocks. Contrib Mineral Petrol 20:244–267

Cenki-Tok B, Berger A, Gueydan F (2016) Formation and preservation of biotite-rich microdomains in high-temperature rocks from the Antananarivo Block, Madagascar. Int J Earth Sci 105:1471–1483, doi:10.1007/s00531-015-1265-0

Centrella S, Austrheim H, Putnis A (2015) Coupled mass transfer through a fluid phase and volume preservation during the hydration of granulite: An example from the Bergen Arcs, Norway. Lithos 236–237:245–255, doi:10.1016/j.lithos.2015.09.010

Chapman AD, Luffi PI, Saleeby JB, Petersen S (2011) Metamorphic evolution, partial melting and rapid exhumation above an ancient flat slab: insights from the San Emigdio Schist, southern California. J Metamorph Geol 29:601–626, doi:10.1111/j.1525–1314.2011.00932.x

Cheng H, Cao D (2015) Protracted garnet growth in high-P eclogite: constraints from multiple geochronology and *P–T* pseudosection. J Metamorph Geol 33:613–632, doi:10.1111/jmg.12136

Cheng H, Nakamura E, Kobayashi K, Zhou Z (2007) Origin of atoll garnets in eclogites and implications for the redistribution of trace elements during slab exhumation in a continental subduction zone. Am Mineral 92:1119–1129, doi:10.2138/am.2007.2343

Chernoff CB, Carlson WD (1999) Trace element zoning as a record of chemical disequilibrium during garnet growth. Geology 27:555–558

Clarke GL, Daczko NR, Nockolds C (2001) A method for applying matrix corrections to X-ray intensity maps using the Bence–Albee algorithm and Matlab. J Metamorph Geol 19:635–644, doi:10.1046/j.0263–4929.2001.00336.x

Connolly JAD (1990) Multivariate phase diagrams: An algorithm based on generalized thermodynamics. Am J Sci 290:666–718

Connolly JAD (2005) Computation of phase equilibria by linear programming: a tool for geodynamic modeling and its application to subduction zone decarbonation. Earth Planet Sci Lett 236:524–541

Cossio R, Borghi A, Ruffini R (2002) Quantitative modal determination of geological samples based on X-ray multielemental map acquisition. Microsc Microanal 8:139–149, doi:10.1017/S1431927601020062

Cosslett VE, Duncumb P (1956) Micro-analysis by a flying-spot X-ray method. Nature 177:1172–1173

Cygan RT, Lasaga AC (1982) Crystal growth and the formation of chemical zoning in garnets. Contrib Mineral Petrol 79:187–200, doi:10.1007/bf01132887

Davis PB, Whitney DL (2006) Petrogenesis of lawsonite and epidote eclogite and blueschist, Sivrihisar Massif, Turkey. J Metamorph Geol 24:823–849

de Andrade V, Vidal O, Lewin E, O'Brien P, Agard P (2006) Quantification of electron microprobe compositional maps of rock thin sections: an optimized method and examples. J Metamorph Geol 24:655–668, doi:10.1111/j.1525-1314.2006.00660.x

de Béthune P, Laduron D, Bocquet J (1975) Diffusion processes in resorbed garnets. Contrib Mineral Petrol 50:197–204, doi:10.1007/bf00371039

de Capitani C, Brown TH (1987) The computation of chemical equilibrium in complex systems containing non-ideal solutions. Geochim Cosmochim Acta 51:2639–2652

de Capitani C, Petrakakis K (2010) The computation of equilibrium assemblage diagrams with Theriak/Domino software. Am Mineral 95:1006–1016, doi:10.2138/am.2010.3354

DeVore GW (1955) The role of adsorption in the fractionation and distribution of elements. J Geol 63:159–190, doi::10.1086/626242

Dragovic B, Samanta LM, Baxter EF, Selverstone J (2012) Using garnet to constrain the duration and rate of water-releasing metamorphic reactions during subduction: An example from Sifnos, Greece. Chem Geol 314–317:9–22, doi:10.1016/j.chemgeo.2012.04.016

Dupuis M (2012) Macro- et microstructure de l'éventail briançonnais. MSc thesis Université de Lille 1

Duval M, Aubert M, Hellstrom J, Grün R (2011) High resolution LA-ICP-MS mapping of U and Th isotopes in an early Pleistocene equid tooth from Fuente Nueva-3 (Orce, Andalusia, Spain). Quat Geochronol 6:458–467, doi:10.1016/j.quageo.2011.04.002

Elvevold S, Gilotti JA (2000) Pressure–temperature evolution of retrogressed kyanite eclogites, Weinschenk Island, North–East Greenland Caledonides. Lithos 53:127–147, doi:10.1016/S0024-4937(00)00014-1

Erambert M, Austrheim H (1993) The effect of fluid and deformation on zoning and inclusion patterns in poly-metamorphic garnets. Contrib Mineral Petrol 115:204–214, doi:10.1007/bf00321220

Evans TP (2004) A method for calculating effective bulk composition modification due to crystal fractionation in garnet-bearing schist; implications for isopleth thermobarometry J Metamorph Geol 22:547–557

Evans KA, Bickle MJ (2005) An investigation of the relationship between bulk composition, inferred reaction progress and fluid flow parameters for layered micaceous carbonates from Maine, U.S.A. J Metamorph Geol 23:181–197

Faryad SW, Klápová H, Nosál L (2010) Mechanism of formation of atoll garnet during high-pressure metamorphism. Mineral Mag 74:111–126, doi:10.1180/minmag.2010.074.1.111

Fiannacca P, Lo Pò D, Ortolano G, Cirrincione R, Pezzino A (2012) Thermodynamic modeling assisted by multivariate statistical image analysis as a tool for unraveling metamorphic P–T-d evolution: an example from ilmenite–garnet-bearing metapelite of the Peloritani Mountains, Southern Italy. Mineral Petrol 106:151–171, doi:10.1007/s00710-012-0228-4

Fisher GW (1973) Nonequilibrium thermodynamics as a model for diffusion-controlled metamorphic processes. Am J Sci 273:897–924, doi:10.2475/ajs.273.10.897

Fisher GW, Lasaga AC (1981) Irreversible thermodynamics in petrology. Rev Mineral Geochem 8:171–207

Fitzsimons ICW, Kinny PD, Wetherley S, Hollingsworth DA (2005) Bulk chemical control on metamorphic monazite growth in pelitic schists and implications for U–Pb age data. J Metamorph Geol 23:261–277, doi:10.1111/j.1525-1314.2005.00575.x

Fletcher RC (1977) Quantitative theory for metamorphic differentiation in development of crenulation cleavage. Geology 5:185–187, doi:10.1130/0091-7613

Florence FP, Spear FS (1991) Effects of diffusional modification of garnet growth zoning on *PT* path calculations. Contrib Mineral Petrol 107:487–500

Foster CT (1977) Mass transfer in sillimanite-bearing pelitic schists near Rangeley, Maine. Am Mineral 62:727–746

Foster CT (1981) A thermodynamic model of mineral segregations in the lower sillimanite zone near Rangeley, Maine. Am Mineral 66:260–277

Foster CT (1983) Thermodynamic models of biotite pseudomorphs after staurolite. Am Mineral 68:389–397

Foster CT (1986) Thermodynamic models of reactions involving garnet in a sillimanite/staurolite schist. Mineral Mag 50:427–439

Foster CT (1991) The role of biotite as a catalyst in reaction mechanisms that form sillimanite. Can Mineral 29:943–963

Foster CT (1999) Forward modeling of metamorphic textures. Can Mineral 37:415–429

Foster G, Kinny P, Vance D, Prince C, Harris N (2000) The significance of monazite U–Th–Pb age data in metamorphic assemblages; a combined study of monazite and garnet chronometry. Earth Planet Sci Lett 181:327–340

Foster G, Parrish RR, Horstwood MSA, Chenery S, Pyle J, Gibson HD (2004) The generation of prograde *P–T–t* points and paths; a textural, compositional, and chronological study of metamorphic monazite. Earth Planet Sci Lett 228:125–142, doi:10.1016/j.epsl.2004.09.024

Gaidies F, Abart R, de Capitani C, Schuster R, Connolly JAD, Reusser E (2006) Characterization of polymetamorphism in the Austroalpine basement east of the Tauern Window using garnet isopleth thermobarometry. J Metamorph Geol 24:451–475, doi:10.1111/j.1525–1314.2006.00648.x

Gaidies F, de Capitani C, Abart R (2008) THERIA_G: a software program to numerically model prograde garnet growth. Contrib Mineral Petrol 155:657–671, doi:10.1007/s00410-007-0263-z

Gaidies F, Pattison DRM, de Capitani C (2011) Toward a quantitative model of metamorphic nucleation and growth. Contrib Mineral Petrol 162:975–993, doi:10.1007/s00410-011-0635-2

Ganguly J (2010) Cation diffusion kinetics in aluminosilicate garnets and geological applications. Rev Mineral Geochem 72:559–601

Ganguly J, Chakraborty S, Sharp TG, Rumble D (1996) Constraint on the time scale of biotite-grade metamorphism during Acadian orogeny from a natural garnet–garnet diffusion couple. Am Mineral 81:1208–1216

Gatewood MP, Dragovic B, Stowell HH, Baxter EF, Hirsch DM, Bloom R (2015) Evaluating chemical equilibrium in metamorphic rocks using major element and Sm–Nd isotopic age zoning in garnet, Townshend Dam, Vermont, USA. Chemical Geology 401:151–168, doi:10.1016/j.chemgeo.2015.02.017

Giuntoli F (2016) Assembly of continental fragments during subduction at HP: Metamorphic history of the central Sesia Zone (NW Alps). PhD thesis University of Bern

Goldschmidt VM (1911) Die Kontaktmetamorphose im Kristianiagebiet. In Kommission bei J. Dybwad, Kristiania

Goncalves P, Oliot E, Marquer D, Connolly JAD (2012) Role of chemical processes on shear zone formation: an example from the Grimsel metagranodiorite (Aar Massif, Central Alps). J Metamorph Geol 30:703–722, doi:10.1111/j.1525–1314.2012.00991.x

Gottschalk M (1996) Internally consistent thermodynamic data for rock-forming minerals in the system SiO_2–TiO_2–Al_2O_3–Fe_2O_3–CaO–MgO–FeO–K_2O–Na_2O–H_2O–CO_2. EurJMineral9:175–223,doi:10.1127/ejm/9/1/0175

Grand'Homme A, Janots E, Seydoux-Guillaume A-M, Guillaume D, Bosse V, Magnin V (2016) Partial resetting of the U–Th–Pb systems in experimentally altered monazite: Nanoscale evidence of incomplete replacement. Geology 44:431–434

Greenfield JE (1997) Migmatite formation at Mt Stafford, central Australia. PhD thesis, University of Sidney

Greenfield JE, Clarke GL, Bland M, Clark DJ (1996) In-situ migmatite and hybrid diatexite at Mt Stafford, central Australia. J Metamorph Geol 14:413–426, doi:10.1046/j.1525–1314.1996.06002.x

Greenwood HJ (1967) The N-dimensional tie-line problem. Geochimi Cosmochim Acta 31:465–490

Groppo C, Castelli D (2010) Prograde *P–T* evolution of a lawsonite eclogite from the Monviso meta-ophiolite (Western Alps): Dehydration and redox reactions during subduction of oceanic FeTi-oxide gabbro. J Petrol 51:2489–2514, doi:10.1093/petrology/egq065

Groppo C, Lombardo B, Rolfo F, Pertusati P (2007) Clockwise exhumation path of granulitized eclogites from the Ama Drime range (Eastern Himalayas). J Metamorph Geol 25:51–75, doi:10.1111/j.1525–1314.2006.00678.x

Grosch EG, McLoughlin N, Lanari P, Erambert M, Vidal O (2014) Microscale mapping of alteration conditions and potential biosignatures in basaltic–ultramafic rocks on early earth and beyond. Astrobiology 14:216–228, doi:10.1089/ast.2013.1116

Guevara VE, Caddick MJ (2016) Shooting at a moving target: phase equilibria modelling of high-temperature metamorphism. J Metamorph Geol 34:209–235, doi:10.1111/jmg.12179

Harrison TM, Catlos EJ, Montel J-M (2002) U–Th–Pb dating of phosphate minerals. Rev Mineral Geochem 48:524–558, doi:10.2138/rmg.2002.48.14

Harvey J, Baxter EF (2009) An improved method for TIMS high precision neodymium isotope analysis of very small aliquots (1–10 ng). Chem Geol 258:251–257, doi:10.1016/j.chemgeo.2008.10.024

Hickmott DD, Shimizu N, Spear FS, Selverstone J (1987) Trace-element zoning in a metamorphic garnet. Geology 15:573–576, doi:10.1130/0091–7613(1987)15<573:tziamg>2.0.co;2

Hoisch TD, Wells ML, Grove M (2008) Age trends in garnet-hosted monazite inclusions from upper amphibolite facies schist in the northern Grouse Creek Mountains, Utah. Geochimica et Cosmochimica Acta 72:5505–5520, doi:10.1016/j.gca.2008.08.012

Holland TJB, Powell R (1988) An internally consistent thermodynamic data set for phases of petrological interest. J Metamorph Geol 8:89–124

Holland TJB, Powell R (1998) An internally consistent thermodynamic data set for phases of petrological interest. J Metamorph Geol 16:309–343

Holland TJB, Powell R (2011) An improved and extended internally consistent thermodynamic dataset for phases of petrological interest, involving a new equation of state for solids. J Metamorph Geol 29:333–383, doi:10.1111/j.1525–1314.2010.00923.x

Hollister LS (1966) Garnet zoning: an interpretation based on the Rayleigh fractionation model. Science 154:1647–1651

Hoschek G (2013) Garnet zonation in metapelitic schists from the Eclogite Zone, Tauern Window, Austria: comparison of observed and calculated profiles. Eur J Mineral 25:615–629, doi:10.1127/0935–1221/2013/0025–2310

Indares A, Rivers T (1995) Textures, metamorphic reactions and thermobarometry of eclogitized metagabbros; a Proterozoic example. Eur J Mineral 7:43–56

Janots E, Engi M, Berger A, Allaz J, Schwarz J-O, Spandler C (2008) Prograde metamorphic sequence of REE-minerals in pelitic rocks of the Central Alps: implications for allanite–monazite–xenotime phase relations from 250 to 610 °C. J Metamorph Geol 26:509–526, doi:10.1111/j.1525–1314.2008.00774.x

Johnson CD, Carlson WD (1990) The origin of olivine–plagioclase coronas in metagabbros from the Adirondack Mountains, New York. J Metamorph Geol 8:697–717, doi:10.1111/j.1525–1314.1990.tb00496.x

Johnson TE, Clark C, Taylor RJM, Santosh M, Collins AS (2015) Prograde and retrograde growth of monazite in migmatites: An example from the Nagercoil Block, southern India. Geosci Front 6:373–387, doi:10.1016/j.gsf.2014.12.003

Kelsey DE, Hand M (2015) On ultrahigh temperature crustal metamorphism: Phase equilibria, trace element thermometry, bulk composition, heat sources, timescales and tectonic settings. Geosci Front 6:311–356, doi:10.1016/j.gsf.2014.09.006

Kelsey DE, Powell R (2011) Progress in linking accessory mineral growth and breakdown to major mineral evolution in metamorphic rocks: a thermodynamic approach in the $Na_2O–CaO–K_2O–FeO–MgO–Al_2O_3–SiO_2–H_2O–TiO_2–ZrO_2$ system. J Metamorph Geol 29:151–166, doi:10.1111/j.1525–1314.2010.00910.x

Kelsey DE, Clark C, Hand M (2008) Thermobarometric modeling of zircon and monazite growth in melt-bearing systems: examples using model metapelitic and metapsammitic granulites. J Metamorph Geol 26:199–212

Kohn MJ (2009) Models of garnet differential geochronology. Geochim Cosmochim Acta 73:170–182, doi:10.1016/j.gca.2008.10.004

Kohn MJ (2014) 4.7 - Geochemical zoning in metamorphic minerals A2. *In:* Treatise on Geochemistry (Second Edition). Turekian KK, (ed) Elsevier, Oxford, p 249–280

Kohn MJ (2016) Metamorphic chronology — a tool for all ages: Past achievements and future prospects. Am Mineral 101:25–42, doi:10.2138/am-2016-5146

Kohn MJ, Spear FS (2000) Retrograde net transfer reaction insurance for pressure–temperature estimates. Geology 28:1127–1130

Kohn MJ, Wieland MS, Parkinson CD, Upreti BN (2004) Miocene faulting at plate tectonic velocity in the Himalaya of central Nepal. Earth Planet Sci Lett 228:299–310

Kohn MJ, Wieland MS, Parkinson CD, Upreti BN (2005) Five generations of monazite in Langtang gneisses: implications for chronology of the Himalayan metamorphic core. J Metamorph Geol 23:399–406, doi:10.1111/j.1525–1314.2005.00584.x

Kohn MJ, Corrie SL, Markley C (2015) The fall and rise of metamorphic zircon. Am Mineral 100:897–908

Kohn MJ, Penniston-Dorland SC, Ferreira JC (2016) Implications of near-rim compositional zoning in rutile for geothermometry, geospeedometry, and trace element equilibration. Contrib Mineral Petrol 171:78

Kohn MJ, Penniston–Dorland SC (2017) Diffusion: Obstacles and opportunities in petrochronology. Rev Mineral Geochem 83:103–152

Konrad-Schmolke M, O'Brien PJ, De Capitani C, Carswell DA (2008) Garnet growth at high- and ultra-high pressure conditions and the effect of element fractionation on mineral modes and composition. Lithos 103:309–332

Korhonen FJ, Stout JH (2005) Borosilicate- and phengite-bearing veins from the Grenville Province of Labrador: evidence for rapid uplift. J Metamorph Geol 23:297–311, doi:10.1111/j.1525–1314.2005.00577.x

Korzhinskii DS (1936) Mobility and inertness of components in metasomatosis. Izv Akad Nauk SSSRm Ser Geol 1:58–60

Korzhinskii DS (1959) Physicochemical Basis of the Analysis of the Paragenesis of Minerals. Consultants Bureau, New York

Kriegsman LM, Nyström AI (2003) Melt segregation rates in migmatites: review and critique of common approaches. Geological Society, London, Special Publications 220:203–212

Lanari P, Duesterhoeft E (2016) Thermodynamic modeling using BINGO-ANTIDOTE: A new strategy to investigate metamorphic rocks. Geophys Res Abstr 18:EGU2016-11363

Lanari P, Guillot S, Schwartz S, Vidal O, Tricart P, Riel N, Beyssac O (2012) Diachronous evolution of the alpine continental subduction wedge: evidence from *P–T* estimates in the Briançonnais Zone houillere (France—Western Alps). J Geodyn 56–57:39–54

Lanari P, Riel N, Guillot S, Vidal O, Schwartz S, Pêcher A, Hattori KH (2013) Deciphering high-pressure metamorphism in collisional context using microprobe mapping methods: Application to the Stak eclogitic massif (northwest Himalaya). Geology 41:111–114, doi:10.1130/g33523.1

Lanari P, Wagner T, Vidal O (2014a) A thermodynamic model for di-trioctahedral chlorite from experimental and natural data in the system $MgO–FeO–Al_2O_3–SiO_2–H_2O$: applications to *P–T* sections and geothermometry. Contr Mineral Petrol 167–968, doi:10.1007/s00410-014-0968-8

Lanari P, Rolland Y, Schwartz S, Vidal O, Guillot S, Tricart P, Dumont T (2014b) *P–T–t* estimation of deformation in low-grade quartz–feldspar-bearing rocks using thermodynamic modelling and $^{40}Ar/^{39}Ar$ dating techniques: example of the Plan-de-Phasy shear zone unit (Briançonnais Zone, Western Alps). Terra Nova 26:130–138, doi:10.1111/ter.12079

Lanari P, Vidal O, De Andrade V, Dubacq B, Lewin E, Grosch EG, Schwartz S (2014c) XMapTools: A MATLAB©-based program for electron microprobe X-ray image processing and geothermobarometry. Comp Geosci 62:227–240, doi:10.1016/j.cageo.2013.08.010

Lanari P, Giuntoli F, Burn M, Engi M (2017) An inverse modeling approach to obtain *P–T* conditions of metamorphic stages involving garnet growth and resorption. Eur J Mineral in press, doi:10.1127/ejm/2017/0029-2597

Lasaga AC (1986) Metamorphic reaction rate laws and development of isograds. Mineral Mag 50:359–373

Lasaga AC (1998) Kinetic Theory in the Earth Sciences. Princeton University Press, Princeton, New Jersey, 822 p.

López-Carmona A, Abati J, Pitra P, Lee JKW (2014) Retrogressed lawsonite blueschists from the NW Iberian Massif: *P–T–t* constraints from thermodynamic modelling and $^{40}Ar/^{39}Ar$ geochronology. Contrib Mineral Petrol 167:1–20, doi:10.1007/s00410-014-0987-5

Loury C, Rolland Y, Guillot S, Mikolaichuk AV, Lanari P, Bruguier O, Bosch D (2015) Crustal-scale structure of South Tien Shan: implications for subduction polarity and Cenozoic reactivation. Geol Soc, London, Spec Publ 427, doi:10.1144/sp427.4

Loury C, Rolland Y, Cenki-Tok B, Lanari P, Guillot S (2016) Late Paleozoic evolution of the South Tien Shan: Insights from *P–T* estimates and allanite geochronology on retrogressed eclogites (Chatkal range, Kyrgyzstan). J Geodyn 96:62–80, doi:10.1016/j.jog.2015.06.005

Mahan KH, Goncalves P, Williams ML, Jercinovic MJ (2006) Dating metamorphic reactions and fluid flow: application to exhumation of high-*P* granulites in a crustal-scale shear zone, western Canadian Shield. J Metamorph Geol 24:193–217, doi:10.1111/j.1525–1314.2006.00633.x

Manzotti P, Ballèvre M (2013) Multistage garnet in high-pressure metasediments: Alpine overgrowths on Variscan detrital grains. Geology 41:1151–1154

Marmo BA, Clarke GL, Powell R (2002) Fractionation of bulk rock composition due to porphyroblast growth: effects on eclogite facies mineral equilibria, Pam Peninsula, New Caledonia. J Metamorph Geol 20:151–165, doi:10.1046/j.0263–4929.2001.00346.x

Martin AJ, Gehrels GE, DeCelles PG (2007) The tectonic significance of (U,Th)/Pb ages of monazite inclusions in garnet from the Himalaya of central Nepal. Chem Geol 244:1–24, doi:10.1016/j.chemgeo.2007.05.003

Martin C, Debaille V, Lanari P, Goderis S, Vandendael I, Vanhaecke F, Vidal O, Claeys P (2013) REE and Hf distribution among mineral phases in the CV–CK clan: A way to explain present-day Hf isotopic variations in chondrites. Geochim Cosmochim Acta 120:496–513, doi:10.1016/j.gca.2013.07.006

Mészaros M, Hofmann BA, Lanari P, et al. (2016) Petrology and geochemistry of feldspathic impact-melt breccia Abar al' Uj 012, the first lunar meteorite from Saudi Arabia. Meteorit Planet Sci 51:1830–1848

Meyer M, John T, Brandt S, Klemd R (2011) Trace element composition of rutile and the application of Zr-in-rutile thermometry to UHT metamorphism (Epupa Complex, NW Namibia). Lithos 126:388–401

Milord I, Sawyer EW, Brown M (2001) Formation of diatexite migmatite and granite magma during anatexis of semi-pelitic metasedimentary rocks: an Example from St. Malo, France. J Petrol 42:487–505, doi:10.1093/petrology/42.3.487

Miron GD, Wagner T, Kulik DA, Heinrich CA (2016) Internally consistent thermodynamic data for aqueous species in the system Na–K–Al–Si–O–H–Cl. Geochim Cosmochim Acta 187:41–78, doi:10.1016/j.gca.2016.04.026

Montel J-M (1993) A model for monazite/melt equilibrium and application to the generation of granitic magmas. Chemical Geology 110:127–146

Montel J-M, Kornprobst J, Vielzeuf D (2000) Preservation of old U–Th–Pb ages in shielded monazite: example from the Beni Bousera Hercynian kinzigites (Marocco). J Metamorph Geol 18:335–342

Mottram CM, Parrish RR, Regis D, Warren CJ, Argles TW, Harris NBW, Roberts NMW (2015) Using U–Th–Pb petrochronology to determine rates of ductile thrusting: Time windows into the Main Central Thrust, Sikkim Himalaya. Tectonics 34:2014TC003743, doi:10.1002/2014TC003743

Moynihan DP, Pattison DRM (2013) An automated method for the calculation of *P–T* paths from garnet zoning, with application to metapelitic schist from the Kootenay Arc, British Columbia, Canada. J Metamorph Geol 31:525–548, doi:10.1111/jmg.12032

Netting A, Payne J, Wade B, Raimondo T (2011) Trace element micro-analytical imaging via laser ablation inductively coupled plasma mass spectrometry (LA-ICP-MS). Microsc Microanal 17:566–567, doi:10.1017/S1431927611003709

Orville PM (1969) A model for metamorphic origin of thin layered amphibloites. Am J Sci 267:64–86

Paton C, Hellstrom J, Paul B, Woodhead J, Hergt J (2011) Iolite: Freeware for the visualisation and processing of mass spectrometric data. J Anal At Spectrom 26:2508–2518, doi:10.1039/C1JA10172B

Pattison DRM, De Capitani C, Gaidies F (2011) Petrological consequences of variations in metamorphic reaction affinity. J Metamorph Geol 29:953–977, doi:10.1111/j.1525–1314.2011.00950.x

Paul B, Paton C, Norris A, Woodhead J, Hellstrom J, Hergt J, Greig A (2012) CellSpace: A module for creating spatially registered laser ablation images within the Iolite freeware environment. J Anal At Spectrom 27:700–706, doi:10.1039/C2JA10383D

Paul B, Woodhead JD, Paton C, Hergt JM, Hellstrom J, Norris CA (2014) Towards a method for quantitative LA-ICP-MS imaging of multi-phase assemblages: mineral identification and analysis correction procedures. Geostand Geoanal Res 38:253–263, doi:10.1111/j.1751-908X.2014.00270.x

Peng S, Hu Q, Ewing RP, Liu C, Zachara JM (2012) Quantitative 3-D Elemental Mapping by LA-ICP-MS of a Basaltic Clast from the Hanford 300 Area, Washington, USA. Environ Sci Technol 46:2025–2032, doi:10.1021/es2023785

Pickering-Witter J, Johnston AD (2000) The effects of variable bulk composition on the melting systematics of fertile peridotitic assemblages. Contrib Mineral Petrol 140:190–211, doi:10.1007/s004100000183

Pollington AD, Baxter EF (2010) High resolution Sm–Nd garnet geochronology reveals the uneven pace of tectonometamorphic processes. Earth Planet Sci Lett 293:63–71, doi:10.1016/j.epsl.2010.02.019

Pourteau A, Sudo M, Candan O, Lanari P, Vidal O, Oberhänsli R (2013) Neotethys closure history of Anatolia: insights from ^{40}Ar–^{39}Ar geochronology and P–T estimation in high-pressure metasedimentary rocks. J Metamorph Geol 31:585–606, doi:10.1111/jmg.12034

Powell R, Downes J (1990) Garnet porphyroblast-bearing leucosomes in metapelites: mechanisms, phase diagrams, and an example from Broken Hill, Australia. *In:* High-temperature metamorphism and crustal anatexis. Springer, p 105–123

Powell R, Holland TJB (2008) On thermobarometry. J Metamorph Geol 26:155–179, doi:10.1111/j.1525-1314.2007.00756.x

Powell R, Holland T (2010) Using equilibrium thermodynamics to understand metamorphism and metamorphic rocks. Elements 6:309–314, doi:10.2113/gselements.6.5.309

Powell R, Holland T, Worley B (1998) Calculating phase diagrams involving solid solutions via non-linear equations, with examples using THERMOCALC. J Metamorph Geol 16:577–588, doi:10.1111/j.1525-1314.1998.00157.x

Powell R, Guiraud M, White RW (2005) Truth and beauty in metamorphic phase-equilibria: conjugate variables and phase diagrams. Can Mineral 43:21–33

Putnis A (2009) Mineral replacement reactions. Rev Mineral Geochem 70:87–124

Putnis A, John T (2010) Replacement processes in the earth's crust. Elements 6:159–164, doi:10.2113/gselements.6.3.159

Pyle JM, Spear FS (1999) Yttrium zoning in garnet: coupling of major and accessory phases during metamorphic reactions. Geol Mater Res 1:1–49

Pyle JM, Spear FS (2003) Four generations of accessory-phase growth in low-pressure migmatites from SW New Hampshire. Am Mineral 88:338–351

Raimondo T, Payne J, Wade B, Lanari P, Clark C, Hand M (2017) Trace element mapping by LA-ICP-MS: assessing geochemical mobility in garnet. Contrib Mineral Petrol, in press.

Regis D, Rubatto D, Darling J, Cenki-Tok B, Zucali M, Engi M (2014) Multiple metamorphic stages within an eclogite-facies terrane (Sesia Zone, Western Alps) revealed by Th–U–Pb petrochronology. J Petrol 55:1429–1456, doi:10.1093/petrology/egu029

Ridley JR (1984) The significance of deformation associated with blueschist facies metamorphism on the Aegean island of Syros. *In:* The Geological Evolution of the Eastern Mediterranean. Vol 17. Dixon JE, Robertson AHF (eds) Geological Society, 545–550

Riel N, Lanari P (2015) Techniques, méthodes et outils pour la quantification du métamorphisme. Géochroniques 136:53–60

Rittner M, Müller W (2012) 2D mapping of LA-ICPMS trace element distributions using R. Comput Geosci 42:152–161, doi:10.1016/j.cageo.2011.07.016

Roberts MP, Finger F (1997) Do U–Pb zircon ages from granulites reflect peak metamorphic conditions? Geology 25:319–322, doi:10.1130/0091-7613(1997)025<0319:dupzaf>2.3.co;2

Robyr M, Darbellay B, Baumgartner LP (2014) Matrix-dependent garnet growth in polymetamorphic rocks of the Sesia zone, Italian Alps. J Metamorph Geol 32:3–24, doi:10.1111/jmg.12055

Rubatto D (2017) Zircon: The metamorphic mineral. Rev Mineral Geochem 83:261–295

Rubatto D, Hermann J, Buick IS (2006) Temperature and bulk composition control on the growth of monazite and zircon during low-pressure anatexis (Mount Stafford, Central Australia). J Petrol 47:1973–1996, doi:10.1093/petrology/egl033

Sayab M (2006) Decompression through clockwise P–T path: implications for early N–S shortening orogenesis in the Mesoproterozoic Mt Isa Inlier (NE Australia). J Metamorph Geol 24:89–105, doi:10.1111/j.1525-1314.2005.00626.x

Scheffer C, Vanderhaeghe O, Lanari P, Tarantola A, Ponthus L, Photiades A, France L (2016) Syn- to post-orogenic exhumation of metamorphic nappes: Structure and thermobarometry of the western Attic-Cycladic metamorphic complex (Lavrion, Greece). J Geodyn 96:174–193, doi:10.1016/j.jog.2015.08.005

Schorn S, Diener JFA (2016) Details of the gabbro-to-eclogite transition determined from microtextures and calculated chemical potential relationships. J Metamorph Geol, doi:10.1111/jmg.12220

Schwarz J-O, Engi M, Berger A (2011) Porphyroblast crystallization kinetics: The role of the nutrient production rate. J Metamorph Geol 29, doi:10.1111/j.1525-1314.2011.00927.x

Šelih VS, van Elteren JT (2011) Quantitative multi-element mapping of ancient glass using a simple and robust LA-ICP-MS rastering procedure in combination with image analysis. Anal Bioanal Chem 401:745–755, doi:10.1007/s00216-011-5119-8

Selverstone J, Spear FS (1985) Metamorphic *P–T* Paths from pelitic schists and greenstones from the south-west Tauern Window, Eastern Alps. J Metamorph Geol 3:439–465, doi:10.1111/j.1525-1314.1985.tb00329.x

Sousa J, Kohn MJ, Schmitz MD, Northrup CJ, Spear FS (2013) Strontium isotope zoning in garnet: implications for metamorphic matrix equilibration, geochronology and phase equilibrium modelling. J Metamorph Geol 31:437–452, doi:10.1111/jmg.12028

Spear FS (1988a) The Gibbs method and Duhem's theorem: The quantitative relationships among *P*, *T*, chemical potential, phase composition and reaction progress in igneous and metamorphic systems. Contrib Mineral Petrol 99:249–256, doi:10.1007/bf00371465

Spear FS (1988b) Metamorphic fractional crystallization and internal metasomatism by diffusional homogenization of zoned garnets. Contrib Mineral Petrol 99:507–517, doi:10.1007/bf00371941

Spear FS (1991) On the interpretation of peak metamorphic temperatures in light of garnet diffusion during cooling. J Metamorph Geol 9:379–388, doi:10.1111/j.1525-1314.1991.tb00533.x

Spear FS (1993) Metamorphic Phase Equilibria and Pressure–temperature–Time Paths. Monograph Mineral Soc Am, Washington, D.C.

Spear FS, Daniel CG (2001) Diffusion control of garnet growth, Harpswell Neck, Maine, USA. J Metamorph Geol 19:179–195

Spear FS, Kohn MJ (1996) Trace element zoning in garnet as a monitor of crustal melting. Geology 24:1099–1102

Spear FS, Pyle JM (2010) Theoretical modeling of monazite growth in a low-Ca metapelite. Chem Geol 273:111–119

Spear FS, Selverstone J (1983) Quantitative *P–T* paths from zoned minerals: Theory and applications. Contrib Mineral Petrol 83:348–357

Spear FS, Selverstone J, Hickmott D, Crowley P, Hodges KV (1984) *P–T* paths from garnet zoning: A new technique for deciphering tectonic processes in crystalline terranes. Geology 12:87–90

Spear FS, Kohn MJ, Florence FP, Menard T (1990) A model for garnet and plagioclase growth in pelitic schists: implications for thermobarometry and *P–T* path determinations. J Metamorph Geol 8:683–696, doi:10.1111/j.1525-1314.1990.tb00495.x

Spear FS, Kohn MJ, Florence FP, Menard T (1991) A model for garnet and plagioclase growth in pelitic schists: Implications for thermobarometry and *P–T* path determinations. J Metamorph Geol 8:683–696

Spear FS, Thomas JB, Hallett BW (2014) Overstepping the garnet isograd: a comparison of QuiG barometry and thermodynamic modeling. Contrib Mineral Petrol 168:1–15

Spear FS, Pattison DRM, Cheney JT (2016) The metamorphosis of metamorphic petrology. Geol Soc Am Spec Papers 523, doi:10.1130/2016.2523(02)

Stepanov AS, Hermann J, Rubatto D, Rapp RP (2012) Experimental study of monazite/melt partitioning with implications for the REE, Th and U geochemistry of crustal rocks. Chem Geol 300–301:200–220, doi:10.1016/j.chemgeo.2012.01.007

Stevens G, Clemens JD, Droop GTR (1997) Melt production during granulite-facies anatexis: experimental data from "primitive" metasedimentary protoliths. Contrib Mineral Petrol 128:352–370, doi:10.1007/s004100050314

Stowell H, Tulloch A, Zuluaga C, Koenig A (2010) Timing and duration of garnet granulite metamorphism in magmatic arc crust, Fiordland, New Zealand. Chem Geol 273:91–110, doi:10.1016/j.chemgeo.2010.02.015

Stüwe K (1997) Effective bulk composition changes due to cooling: a model predicting complexities in retrograde reaction textures. Contrib Mineral Petrol 129:43–52, doi:10.1007/s004100050322

Taylor-Jones K, Powell R (2015) Interpreting zirconium-in-rutile thermometric results. J Metamorph Geol 33:115–122

Thompson J (1959) Local equilibrium in metasomatic processes. *In:* Researches in Geochemistry 1, Abelson PH (Ed) Wiley, New York, p. 427–457

Thompson JB (1955) The thermodynamic basis for the mineral facies concept. Am J Sci 253:65–103, doi:10.2475/ajs.253.2.65

Thompson JB (1970) Geochemical reaction and open systems. Geochim Cosmochim Acta 34:529–551, doi:10.1016/0016-7037(70)90015-3

Tinkham DK, Ghent ED (2005a) Estimating *P–T* conditions of garnet growth with isochemical phase-diagrams sections and the problem of effective bulk-composition. Can Mineral 43:35–50, doi:10.2113/gscanmin.43.1.35

Tinkham DK, Ghent ED (2005b) XRMapAnal: A program for analysis of quantitative X-ray maps. Am Mineral 90:737–744, doi:10.2138/am.2005.1483

Tinkham DK, Zuluaga CA, Stowell HH (2001) Metapelite phase equilibria modeling in MnNCKFMASH: The effect of variable Al_2O_3 and MgO/(MgO+FeO) on mineral stability. Geol Mater Res 3:1–42

Tóth M, Grandjean V, Engi M (2000) Polyphase evolution and reaction sequence of compositional domains in metabasalt: A model based on local chemical equilibrium and metamorphic differentiation. Geol J 35:163–183

Tracy RJ (1982) Compositional zoning and inclusions in metamorphic minerals. Rev Mineral Geochem 10:355–397

Tracy RJ, Robinson P, Thompson AB (1976) Garnet composition and zoning in the determination of temperature and pressure of metamorphism, central Massachusetts. Am Mineral 61:762–775

Trincal V, Lanari P, Buatier M, Lacroix B, Charpentier D, Labaume P, Muñoz M (2015) Temperature micromapping in oscillatory-zoned chlorite: Application to study of a green-schist facies fault zone in the Pyrenean Axial Zone (Spain). Am Mineral 100:2468–2483, doi:10.2138/am-2015-5217

Tropper P, Manning CE, Harlov DE (2011) Solubility of $CePO_4$ monazite and YPO_4 xenotime in H_2O and H_2O–NaCl at 800 °C and 1 GPa: Implications for REE and Y transport during high-grade metamorphism.Chem Geol 282:58–66, doi:10.1016/j.chemgeo.2011.01.009

Ubide T, McKenna CA, Chew DM, Kamber BS (2015) High-resolution LA-ICP-MS trace element mapping of igneous minerals: In search of magma histories. Chem Geol 409:157–168, doi:10.1016/j.chemgeo.2015.05.020

Vance D, Mahar E (1998) Pressure–temperature paths from PT pseudosections and zoned garnets: potential, limitations and examples from the Zanskar Himalaya, NW India. Contrib Mineral Petrol 132:225–245

Verdecchia SO, Reche J, Baldo EG, Segovia-Diaz E, Martinez FJ (2013) Staurolite porphyroblast controls on local bulk compositional and microstructural changes during decompression of a St–Bt–Grt–Crd–And schist (Ancasti metamorphic complex, Sierras Pampeanas, W Argentina). J Metamorph Geol 31:131–146, doi:10.1111/jmg.12003

Vernon RH (1989) Porphyroblast-matrix microstructural relationships: recent approaches and problems. Geol Soc, London, Spec Publ 43:83–102, doi:10.1144/gsl.sp.1989.043.01.05

Vernon RH, Collins WJ (1988) Igneous microstructures in migmatites. Geology 16:1126–1129, doi:10.1130/0091–7613(1988)016<1126:imim>2.3.co;2

Vidal O, De Andrade V, Lewin E, Munoz M, Parra T, Pascarelli S (2006) P–T–deformation–Fe^{3+}/Fe^{2+} mapping at the thin section scale and comparison with XANES mapping: application to a garnet-bearing metapelite from the Sambagawa metamorphic belt (Japan). J Metamorph Geol 24:669–683, doi:10.1111/j.1525–1314.2006.00661.x

Vitale Brovarone A, Beltrando M, Malavieille J, Giuntoli F, Tondella E, Groppo C, Beyssac O, Compagnoni R (2011) Inherited Ocean–Continent Transition zones in deeply subducted terranes: Insights from Alpine Corsica. Lithos 124:273–290, doi:10.1016/j.lithos.2011.02.013

Vrijmoed JC, Hacker BR (2014) Determining P–T paths from garnet zoning using a brute-force computational method. Contrib Mineral Petrol 167:1–13, doi:10.1007/s00410-014-0997-3

Warren CJ, Waters DJ (2006) Oxidized eclogites and garnet-blueschists from Oman: P–T path modelling in the NCFMASHO system. J Metamorph Geol 24:783–802, doi:10.1111/j.1525–1314.2006.00668.x

Waters DJ, Lovegrove DP (2002) Assessing the extent of disequilibrium and overstepping of prograde metamorphic reactions in metapelites from the Bushveld Complex aureole, South Africa. J Metamorph Geol 20:135–149, doi:10.1046/j.0263–4929.2001.00350.x

White RW, Powell R (2010) Retrograde melt–residue interaction and the formation of near-anhydrous leucosomes in migmatites. J Metamorph Geol 28:579–597, doi:10.1111/j.1525–1314.2010.00881.x

White RW, Powell R, Holland TJB (2001) Calculation of partial melting equilibria in the system Na_2O–CaO–K_2O–FeO–MgO–Al_2O_3–SiO_2–H_2O (NCKFMASH). J Metamorph Geol 19:139–153, doi:10.1046/j.0263–4929.2000.00303.x

White RW, Powell R, Clarke GL (2003) Prograde Metamorphic Assemblage evolution during partial melting of metasedimentary rocks at low pressures: Migmatites from Mt Stafford, Central Australia. J Petrol 44:1937–1960, doi:10.1093/petrology/egg065

White RW, Powell R, Baldwin JA (2008) Calculated phase equilibria involving chemical potentials to investigate the textural evolution of metamorphic rocks. J Metamorph Geol 26:181–198, doi:10.1111/j.1525–1314.2008.00764.x

White WM (2013) Geochemistry. Wiley-Blackwell

Whitehouse MJ, Platt JP (2003) Dating high-grade metamorphism—constraints from rare-earth elements in zircon and garnet. Contrib Mineral Petrol 145:61–74, doi:10.1007/s00410-002-0432-z

Whitney DL, Evans BW (2010) Abbreviations for names of rock-forming minerals. Am Mineral 95:185–187, doi:10.2138/am.2010.3371

Williams ML (1994) Sigmoidal inclusion trails, punctuated fabric development, and interactions between metamorphism and deformation. J Metamorph Geol 12:1–21, doi:10.1111/j.1525–1314.1994.tb00001.x

Williams ML, Jercinovic MJ, Hetherington CJ (2007) Microprobe monazite geochronology: Understanding geologic processes by integrating composition and chronology. Ann Rev Earth Planet Sci 35:137–175, doi:10.1146/annurev.earth.35.031306.140228

Williams ML, Jercinovic MJ, Mahan KH, Dumond G (2017) Electron microprobe petrochronology. Rev Mineral Geochem 83:153–182

Wing BA, Ferry JM, Harrison TM (2003) Prograde destruction and formation of monazite and allanite during contact and regional metamorphism of pelites: petrology and geochronology. Contrib Mineral Petrol 145:228–250

Wintsch RP (1985) The possible effects of deformation on chemical processes in metamorphic fault zones. *In:* Metamorphic Reactions: Kinetics, Textures, and Deformation. Thompson AB, Rubie DC (eds). Springer New York, New York, NY, p 251–268

Woodhead JD, Hellstrom J, Hergt JM, Greig A, Maas R (2007) Isotopic and elemental imaging of geological materials by laser ablation inductively coupled plasma-mass spectrometry. Geostand Geoanal Res 31:331–343, doi:10.1111/j.1751-908X.2007.00104.x

Yakymchuk C, Brown M (2014) Behaviour of zircon and monazite during crustal melting. J Geol Soc 171:465–479, doi:10.1144/jgs2013-115

Yakymchuk C, Clark C, White RW (2017) Phase relations, reaction sequences and petrochronology. Rev Mineral Geochem 83:13–53

Yamato P, Agard P, Burov E, Le Pourhiet L, Jolivet L, Tiberi C (2007) Burial and exhumation in a subduction wedge: Mutual constraints from thermomechanical modeling and natural *P–T–t* data (Schistes Lustrés, western Alps). J Geophys Res: Solid Earth 112:B07410, doi:10.1029/2006JB004441

Zack T, Kooijman E (2017) Petrology and geochronology of rutile. Rev Mineral Geochem 83:443–467

Zeh A (2006) Calculation of garnet fractionation in metamorphic rocks, with application to a flat-top, Y-rich garnet population from the Ruhla Crystalline Complex, Central Germany. J Petrol 47:2335–2356, doi:10.1093/petrology/egl046

Zhang L, Wang Q, Song S (2009) Lawsonite blueschist in Northern Qilian, NW China: *P–T* pseudosections and petrologic implications. J Asian Earth Sci 35:354–366, doi:10.1016/j.jseaes.2008.11.007

Zuluaga CA, Stowell HH, Tinkham DK (2005) The effect of zoned garnet on metapelite pseudosection topology and calculated metamorphic *P–T* paths. Am Mineral 90:1619–1628, doi:10.2138/am.2005.1741

Reviews in Mineralogy & Geochemistry
Vol. 83 pp. XXX-XXX, 2017
Copyright © Mineralogical Society of America

4

Diffusion: Obstacles and Opportunities in Petrochronology

Matthew J. Kohn

Department of Geosciences
Boise State University
Boise, ID 83725
USA

mattkohn@boisestate.edu

Sarah C. Penniston-Dorland

Department of Geology
University of Maryland
College Park, MD 20742
USA

sarahpd@umd.edu

INTRODUCTION AND SCOPE

Many of the approaches in petrochronology are rooted in the assumption of equilibrium. Diffusion is an expression of disequilibrium: the movement of mass in response to chemical potential gradients, and isotopes in response to isotopic gradients. It is extremely important that we be aware of how the effects of diffusion can place obstacles across our path towards petrochronologic enlightenment. Conversely the effects of diffusion also provide opportunities for understanding rates, processes, and conditions experienced by rocks. The enormity of the field does not permit us to provide a comprehensive review of either the mathematics of diffusion or quantitative data that have been obtained relevant to the interpretation of diffusive processes in rocks and minerals. Many resources cover these topics, including RiMG volume 72 (*Diffusion in Minerals and Melts*; Zhang and Cherniak 2010) and several textbooks (Crank 1975; Glicksman 2000). Particularly relevant to the discussion of petrochronology are summaries of the theory and controls on diffusion (Brady and Cherniak 2010; Zhang 2010), as well as diffusion rates in feldspar (Cherniak 2010a), accessory minerals (Cherniak 2010b), garnet (Ganguly 2010), mica, pyroxene, and amphibole (Cherniak and Dimanov 2010), and melts (Zhang and Ni 2010; Zhang et al. 2010). Rather than duplicate that material, our goal is to explore the obstacles and opportunities presented by the effects of diffusion as they inform the rates of petrologic processes. To achieve this goal, we emphasize key principles and illustrative examples.

Quantitative interpretation of the effects of diffusion assumes predictability of numerous factors that may affect chemical or isotopic transport, including temperature, initial and boundary conditions, water and oxygen fugacities, activities of other components, multiple mechanisms of diffusion, and crystal chemistry ('coupling' of the substitution of elements into different crystallographic sites). Additionally, the extraction of meaningful ages, durations of events, and temperatures requires considerable petrologic and contextual information to place a diffusive interval correctly into the petrologic history of a rock. Bulk-rock scale interpretations of mass transport occurring through an intergranular medium require additional assumptions about metamorphic porosity and tortuosity, the nature of the intergranular

1529-6466/17/0083-0004$05.00 (print)
1943-2666/17/0083-0004$05.00 (online)
http://dx.doi.org/10.2138/rmg.2017.83.4

medium, the diffusivity of the element or isotope through that medium, the host minerals of the element/isotope, and whether host minerals precipitate, dissolve, recrystallize, or remain inert except for diffusion during mass transfer.

In this chapter we highlight mineralogical and petrologic factors influencing diffusion and the interpretation of diffusion profiles. The chapter is divided into 3 sections:

- Part 1. Diffusion theory and controls on diffusivity, including the effects of crystal chemistry on inhibiting or enhancing the mobility of atoms within minerals.

- Part 2. Models of diffusive processes relevant to petrochronology to predict how boundary conditions and mechanisms of diffusion may affect composition profiles.

- Part 3. Examples of natural systems, emphasizing how assumptions about diffusive processes, particularly boundary conditions, affect interpretations for a variety of applications.

We explore examples where the effects of diffusion impede our understanding of the rock record as well as examples where we can exploit the process of diffusion to illuminate the duration and rates of geological processes. We emphasize concepts that have guided our own research, in hopes that these principles in turn guide future petrochronologic studies. One theme that runs through the examples is that understanding how chemical or isotopic compositions have changed on a mineral surface (the boundary condition) proves singularly challenging, and that differences in assumptions about boundary conditions can give rise to radically different interpretations of petrogenetic and tectonic processes. Getting the boundary condition right is central to accurate petrochronologic interpretations. Of similar importance is understanding the initial conditions, especially temperature. Because diffusion is thermally activated, an error in temperature of only 25 °C can easily change calculated timescales by a factor of 2.

PART 1: DIFFUSION THEORY AND CONTROLS ON DIFFUSIVITY

This section reviews principles of diffusion as well as atomic-scale processes that influence diffusion rates: Fick's laws, multicomponent diffusion, and the dependencies of diffusion rates on crystal-chemical controls and defects (including the influence of oxygen fugacity, water fugacity, and strain).

Fick's laws

Fick's first and second laws (written here in simple one-dimensional form) underlie the quantitative interpretation of the effects of diffusion in minerals and rocks:

$$J_i = -D_i \frac{\partial C_i}{\partial x} \tag{1}$$

$$\frac{\partial C_i}{\partial t} = D_i \frac{\partial^2 C_i}{\partial x^2} \tag{2}$$

These equations relate the flux and chemical concentration of species i (J_i, C_i), time (t), diffusivity (D_i) and distance (x). Diffusion of elements or isotopes within mineral grains can result in loss (or gain) of these constituents, creating smoothly varying compositions across mineral grains that represent 'frozen-in' diffusion profiles. The spatial scale of these features increases over time as constituents diffuse. Measuring compositions along profiles across such grains can provide constraints on the duration and rates of metamorphic and igneous processes (e.g., Ganguly 2002; Costa et al. 2008; Chakraborty 2008).

Ultimately, estimates of diffusivity, whether experimental or empirical, rely on solutions to Equations (1) and (2). Strictly speaking, chemical fluxes respond to a chemical potential gradient ($d\mu/dx$), which can differ from concentration gradient (dC/dx) due to non-ideal thermodynamic interactions (activity coefficients), and to cross-coupling among multiple interdiffusing species. Ganguly (2002) develops requisite theory more comprehensively, and his work serves as an example of how Equations (1) and (2) may be modified for chemically complex, thermodynamically non-ideal mineral systems. Indeed, chemical potential can be important in designing experiments and understanding experimental results (Watson 1982), especially if it controls defect densities (Jollands et al. 2014). Fortunately, many questions in petrochronology involve diffusion of trace elements, which should follow Henry's law (constant activity coefficient). Thus, commonly the chemical potential and concentration gradients are proportional to each other (Watson and Baxter 2007; however see also Costa et al. 2003). How chemical potential influences defect densities in natural crystals presents a new direction for future research (Jollands et al. 2014).

The Arrhenius relationship describes the diffusivity (D) as a function of temperature (T, in K):

$$D = D_0 \cdot \exp\left\{\frac{-E}{RT}\right\} \tag{3}$$

where D_0 is a pre-exponential constant, E is activation energy (which is commonly taken to be constant, but can be slightly pressure dependent), and R is the gas constant (see Zhang (2010) for a review of the empirical and theoretical foundations for this relationship). Although different temperature intervals may be characterized by different D_0–E values for a specific ion in a specific mineral, reflecting the influence of intrinsic vs. extrinsic defects, most data are interpreted using a single Arrhenius expression, i.e., single D_0 and E values. Brady and Cherniak (2010) tabulate experimental estimates of D_0 and E for a wide range of minerals and diffusing species. The exponential dependence of D on T means that accurate estimates of temperature are needed for accurate interpretations.

Multicomponent diffusion

Diffusion of an element within a mineral can be affected by diffusive properties of other elements within the same mineral. When interdiffusion occurs among at least 3 elements, a substantially different D for one element can cause so-called "uphill" diffusion, or diffusion against a concentration gradient. This counterintuitive process can occur even when components mix ideally, as a consequence of cross-linkages among diffusing species—decreases in concentration of one species must be balanced by increases in another (e.g., Lasaga 1979; Ganguly 2002, 2010; Krishna 2015). In three-component systems, Equation (1) expands to a series of two coupled equations:

$$\begin{pmatrix} J_1 \\ J_2 \end{pmatrix} = -\begin{bmatrix} D_{11} & D_{12} \\ D_{21} & D_{22} \end{bmatrix} \begin{pmatrix} \diagup \\ C_2 \diagup \end{pmatrix} \tag{4}$$

where the D's are diffusion coefficients that take into account all chemical components in the system of interest, including ideal and non-ideal interactions. A flux equation is not needed for component 3 because the fluxes must sum to zero, so J_3 is a dependent variable. Taking Fe–Mg–Ca interdiffusion in garnet, with Fe as component 1, Mg as component 2, and Ca as the dependent component, $-D_{\text{FeFe}} \cdot dC_{\text{Fe}}/dx$ represents the diffusive response of Fe to its own concentration gradient while $-D_{\text{FeMg}} \cdot dC_{\text{Mg}}/dx$ represents the diffusive response of Fe to the concentration gradient of Mg. Different expressions have been proposed for the D's in Equation (4), as referenced to either self-diffusion or interdiffusion coefficients. Again taking Fe–Mg-Ca interdiffusion in garnet, Equation (4) can be expressed in terms of Maxwell–Stefan interdiffusion coefficients[1] ($\mathcal{D}$'s, Krishna 2015) as:

---

[1] Maxwell–Stefan interdiffusion coefficients in Equations (5) and (6) differ from the Fick matrix D's in Equation (4) and should not be equated. For example, Maxwell–Stefan $\mathcal{D}_{12} = \mathcal{D}_{21}$ whereas in general Fick matrix $D_{12} \neq D_{21}$ (see example in Eqn. 7).

$$\begin{pmatrix} J_{Fe} \\ J_{Mg} \end{pmatrix} = -\frac{1}{d} \begin{bmatrix} \mathcal{D}_{FeCa}\left[X_{Fe}\mathcal{D}_{MgCa} + \left(1 - X_{Fe}\right)\mathcal{D}_{FeMg}\right] & X_{Fe}\mathcal{D}_{MgCa}\left[\mathcal{D}_{FeCa} - \mathcal{D}_{FeMg}\right] \\ X_{Mg}\mathcal{D}_{FeCa}\left[\mathcal{D}_{MgCa} - \mathcal{D}_{FeMg}\right] & \mathcal{D}_{MgCa}\left[X_{Mg}\mathcal{D}_{FeCa} + \left(1 - X_{Mg}\right)\mathcal{D}_{FeMg}\right] \end{bmatrix} \begin{pmatrix} \dfrac{dX_{Fe}}{dx} \\ \dfrac{dX_{Mg}}{dx} \end{pmatrix} \quad (5)$$

where X is mole fraction and:

$$d = X_{Mg}\mathcal{D}_{FeCa} + X_{Fe}\mathcal{D}_{MgCa} + X_{Ca}\mathcal{D}_{FeMg} \quad (6)$$

Although the diagonal terms, D_{11} and D_{22} in Equation (4), are always positive (a component tries to diffuse down its own concentration gradient), the off-diagonal terms, D_{12} and D_{21}, can be negative, as they are in garnet because D_{FeCa} and D_{MgCa} are smaller than D_{FeMg} (e.g., Ganguly et al. 1998; Vielzeuf et al. 2007; Vielzeuf and Saúl 2011). This means that for garnets that contain Ca, a steep gradient in Fe would tend to drive backwards diffusion of Mg, while a steep gradient in Mg would tend to drive backwards diffusion of Fe. Some garnet interdiffusion experiments do show clear uphill diffusion (Vielzeuf et al. 2007), and for one of these (Fig. 1A), Vielzeuf and Saúl (2011) inferred coefficients for Equation (4) of:

$$\begin{pmatrix} J_{Fe} \\ J_{Mg} \end{pmatrix} = -\begin{bmatrix} 5.86x10^{-19} & -1.02x10^{-19} \\ -5.50x10^{-19} & 1.18x10^{-19} \end{bmatrix} \begin{pmatrix} \dfrac{dX_{Fe}}{dx} \\ \dfrac{dX_{Mg}}{dx} \end{pmatrix} \quad (7)$$

Uphill diffusion of Mg in this experiment resulted from a large initial gradient in Fe [abs($dX_{Fe}/dx) \gg 0$] and a small initial gradient in Mg [abs($dX_{Mg}/dx) \sim 0$], such that the off-diagonal term, D_{MgFe} ($=-5.50 \times 10^{-19}$ m²/s), dominated Mg diffusion. Diffusion of Mg also affected Fe ($D_{FeMg} < 0$), but the large initial Fe gradient and small initial Mg gradient mean that the effect is not obvious (Fig. 1A). More generally, uphill diffusion is apparent when strong gradients occur in two components with different D's (in this case Ca and Fe), and a third component has a very small initial composition gradient (Mg). Carlson (2006) discusses a natural example where Ca in garnet may have diffused against its concentration gradient.

Gradients in major element compositions can also influence diffusion rates of other elements, including trace elements, because major elements dominate chemical potentials (Costa et al. 2003). In plagioclase, for example, isothermal models of the diffusive interaction across a step function in anorthite content (inset, Fig. 1B) and a corresponding step function in trace Mg predict uphill diffusion of Mg (Fig. 1B). This unusual diffusive behavior occurs for Mg even though, under dry conditions, the anorthite content is essentially inert to diffusive modification (e.g., Grove et al. 1984), and Mg^{2+} simply exchanges for Ca^{2+}. The anorthite content influences the activity of the Mg-component in plagioclase, which in turn affects the off-diagonal terms in Equation (4) (Costa et al 2003). Modeling that does not account for the zoning in anorthite content (lower inset, Fig. 1B) shows quite different profiles. Consequently, any interpretation of Mg zoning in plagioclase would have to account for these non-ideal terms.

Crystal-chemical controls on diffusion

Fully understanding cation diffusivities requires close consideration of crystal chemistry and cation substitution mechanisms. While some elements of chronologic or petrologic significance can diffuse via simple cation exchange (e.g., $Pb^{2+} \Leftrightarrow Ca^{2+}$ in titanite, $U^{4+} \Leftrightarrow Zr^{4+}$ in zircon, $Fe^{2+} \Leftrightarrow Mg^{2+}$ in garnet, etc.), others require coupled exchanges (e.g., $^{VIII}Sm^{3+} + {}^{VI}Mg^{2+} \Leftrightarrow {}^{VIII}Ca^{2+} + {}^{VI}Al^{3+}$ in garnet, $U^{4+} + Ca^{2+} \Leftrightarrow 2Ce^{3+}$ in monazite, $Yb^{3+} + P^{5+} \Leftrightarrow Zr^{4+} + Si^{4+}$ in zircon, etc.). As a very general rule of thumb, low valence cations in larger crystallographic sites diffuse faster than high valence cations in small sites (e.g., Brady and Cherniak 2010). For a

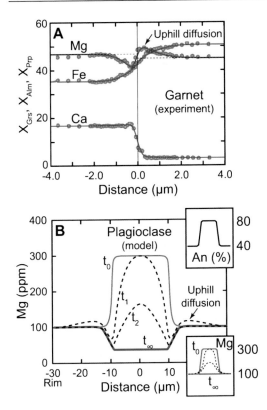

Figure 1. Examples of multicomponent diffusion phenomena. (A) Experimental garnet diffusion couple, showing uphill diffusion of Mg due to strong initial gradients in Ca and Fe. Original Mg composition shown by dashed lines. Modified from Vielzeuf and Saúl (2011). (B) Models of Mg diffusion in plagioclase. An initial step function in anorthite (upper inset) is correlated with an initial step function in Mg (t_0). Relaxation of the Mg profile (t_1, t_2) towards equilibrium (t_∞) develops oscillations and shows uphill diffusion because of thermodynamically non-ideal interaction between the anorthite component and Mg. Lower inset shows how diffusion would proceed without effect of zoning in anorthite component. Modified from Costa et al. (2003).

fixed valence, smaller cations generally diffuse faster than larger cations (Brady and Cherniak 2010), although such a simple relationship does not always hold for melts (Zhang 2010). Chronologically useful radioactive isotopes and their daughters tend to have different ionic radii and different ionic charges than the dominant cation for the site in which they substitute. So, diffusion of these elements tends to be different than expected from ionic radius and site characteristics alone. For example, REE^{3+} diffuse more slowly in garnet than divalent cations, not necessarily because ionic radii are so different ($Ca^{2+} = 1.12$Å, $Fe^{2+} = 0.92$Å, $Lu^{3+} = 0.98$Å, $Nd^{3+} = 1.11$Å), but because trivalent cations in the cubic site must be charge coupled to other cations, e.g., $^{VIII}REE^{3+} + {}^{VIII}Na^+ \Leftrightarrow {}^{VIII}Mg^{2+} + {}^{VIII}Mg^{2+}$, $^{VIII}REE^{3+} + {}^{VI}Mg^{2+} \Leftrightarrow {}^{VIII}Mg^{2+} + {}^{VI}Al^{3+}$, or $^{VIII}REE^{3+} + {}^{IV}Al^{3+} \Leftrightarrow {}^{VIII}Mg^{2+} + {}^{IV}Si^{4+}$ (Fig. 2A; Carlson 2012). Charge coupling involving vacancies may also influence diffusivity (Jollands et al. 2014).

Charge coupling readily explains the slow diffusion of Li in garnet at a rate comparable to Y and Yb (Cahalan et al. 2014). Ordinarily, one might expect a monovalent cation to diffuse faster than divalent or trivalent cations in the same site (Brady and Cherniak 2010). Cahalan et al. (2014), however, show that natural diffusion profiles of Li and REE in garnet are indistinguishable in form and ~6 times shorter (so D is lower) than for Fe, Mg, Mn and Ca (Fig. 2B). Coupled substitution in the dodecahedral site of Li^+ and REE^{3+} for $2(Mg, Fe, Mn, Ca)^{2+}$ may limit its diffusivity, so that while Li^+ diffusion might be fast when exchanging with other monovalent cations, charge coupling to the slow-diffusing REE may constrain D_{Li} to equal DREE (Cahalan et al. 2014). In contrast, charge coupling should not impede diffusion of Li isotopes because ^7Li can exchange by a simple mechanism with ^6Li. Consequently, Li isotopic gradients may be erased at high temperature, even while Li concentration gradients are preserved. Slow diffusion due to

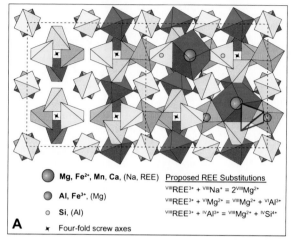

Figure 2. (A) Crystal structure and chemistry of garnet projected along *a*-axis showing tetrahedral, octahedral, and distorted cubic sites. Detail increases from left (tetrahedral sites only, showing 4-fold screw axes) to right (all 3 sites). Cation sizes illustrated around two cubic sites on right-hand side. Dashed lines delineate unit cell. Geochemically important cation substitutions and REE substitution mechanisms are listed. Crystal structure image based on http://www.uwgb.edu/dutchs/petrology/Garnet Structure.HTM. (B) Evidence for slow diffusion of Li and REE (short compositional gradients; coupled substitutions) vs. fast diffusion of divalent Fe (long compositional gradient; simple substitution) in natural garnet from the Makhavinekh Lake Pluton contact aureole, Labroador. Dashed lines show assumed initial compositions. Modified from Carlson (2006) and Cahalan et al. (2014).

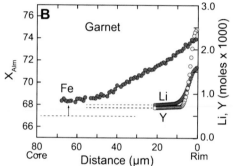

charge coupling is also well known from studies of feldspars. At 800 °C interdiffusion of alkali-site Na^+ for Ca^{2+} in plagioclase (charge coupled to tetrahedral site Si^{4+} and Al^{3+}) is ~10 orders of magnitude slower than simple Na^+ and K^+ interdiffusion in alkali feldspar (Brady 1995). Thus, in plutonic plagioclase, Ca isotopes might reset to low-temperature (through diffusional self-exchange with other Ca atoms), while the anorthite content is virtually invariant and retains information about high-temperature crystallization conditions.

Some species may be distributed among different sites, which may influence their diffusive behavior. For example, Li in olivine appears to occupy both an octahedral site and an interstitial position (Dohmen et al. 2010). Interpretation of Li diffusion profiles in olivine therefore requires more complex models of diffusive transfer. Diffusion of interstitial Li is faster than octahedral Li (Dohmen et al. 2010), and published data on Li distributions in olivine (e.g., Jeffcoate et al. 2007; Parkinson et al. 2007; Rudnick and Ionov 2007) suggest that the slower Li diffusion mechanism dominates in nature. This diffusion rate is, on average, an order of magnitude higher than diffusion of Fe and Mg in olivine (Dohmen et al. 2010), so the diffusion of Li provides information about shorter duration events and possibly processes occurring at lower temperature than diffusion of Fe and Mg. Similarly, a two-species model of lithium diffusion appears to operate in clinopyroxene (Richter et al. 2014), such that Li isotopic compositions take longer to homogenize than Li concentrations. Other elements also have multiple substitution mechanisms (e.g., Si in rutile, Golden et al. 2015) that presumably convey complex diffusive behavior.

Dependence of D on defects

Ultimately, diffusion depends on defects in the crystal structure, including thermal defects, defects arising from chemical substitutions, and strain defects or dislocations. To illustrate the importance of defects, Chakraborty (2008) draws an analogy to a parking lot: a fully packed parking lot allows no movement of automobiles, whereas at least one vacancy (defect) allows movement. Thermal defects are point defects including vacancies and interstitialcies (atoms that occupy positions interstitial to the ideal crystallographic sites). Point defects might involve multiple neighboring sites or interstitial positions, but do not extend into lines or planes. Thermal defects form spontaneously and increase in abundance at higher temperature, causing an increase in diffusion rates. Referring to the parking lot analogy, a greater number of open vacancies, coupled with more vigorously active drivers, allows greater mobility. Thermal defects are referred to as intrinsic defects because their presence is intrinsic to any compound, no matter how chemically pure. Substitutional defects are associated with substitution of aliovalent ions[2], for example Cr^{3+} substitution for Mg^{2+} in olivine, Fe^{3+} substitution for Ti^{4+} in rutile, etc. Such defects are also point defects and can depend on parameters such as f_{H_2O} or f_{O_2}, which affect interstitial(cy) concentrations (e.g., H^+ or OH^-) and valence state, especially Fe^{3+}. Substitutional defects are referred to as extrinsic defects because their concentrations are dependent on the chemical potentials of the substituting atoms or compounds. Strain defects, more commonly called dislocations, are distortions to the crystal lattice caused by deformation independent of crystal chemistry. Dislocations occupy lines or planes within a crystal and can close upon themselves, forming dislocation loops.

f_{O_2} *dependence.* Minerals whose cation diffusion rates appear to depend on f_{O_2} include olivine (see summaries of Dohmen et al. 2007; Dohmen and Chakraborty 2007; Chakraborty 2010), rutile (Sasaki et al. 1985), pyroxenes (see Cherniak and Dimanov 2010; Dohmen et al. 2016), garnet (Chakraborty and Ganguly 1992) and sodic feldspar (see Cherniak 2010a). These studies reveal two fundamental principles. First, oxidation state can influence defect densities by altering the valence of transition metals. The oxidation of Fe^{2+} to Fe^{3+} with increasing f_{O_2} can create octahedral site vacancies, whereas the reduction of Ti^{4+} to Ti^{3+} with decreasing f_{O_2} can create oxygen vacancies (although interstitial cations appear to dominate diffusion rates in rutile for several elements; Sasaki et al. 1985; Nowotny et al. 2008). Thus, high f_{O_2} for olivine, pyroxene, garnet, etc. and low f_{O_2} for rutile are expected to induce higher defect densities and faster diffusion. This concept helps reconcile otherwise disparate cation diffusion rates determined experimentally in garnet (Chakraborty and Ganguly 1992), olivine (Dohmen et al. 2007), and pyroxene (Dohmen et al. 2016). Second, the importance of f_{O_2}-induced defects diminishes whenever the substitution of aliovalent cations produces a greater proportion of total vacancies. Substitutional defects are expected to dominate diffusion in ferro-magnesian silicates at low f_{O_2} (low Fe^{3+} concentration), or in rutile at high f_{O_2} (low Ti^{3+} concentration). A change in diffusion rates with f_{O_2} was proposed (Chakraborty 1997) and observed experimentally (Dohmen et al. 2007), with faster diffusion dominated by defects resulting from f_{O_2}-dependent transition metals and slower diffusion dominated by defects resulting from f_{O_2}-independent aliovalent cations.

Exceptions to these principles include diffusion rates of Fe in labradorite (Behrens et al. 1990) and Eu in orthopyroxene (Cherniak and Liang 2007), which are both lower at higher f_{O_2}. This phenomenon probably reflects slower diffusion for more highly charged ions (Fe^{3+} and Eu^{3+} rather than Fe^{2+} and Eu^{2+}), where oxidation of a minor or trace constituent does

---

[2] An aliovalent (a.k.a. heterovalent) ion has a different charge than the ion for which it substitutes in the endmember compound. Substitution of aliovalent ions forms defects only when charge balance is achieved through vacancies rather than coupled substitution of cations on other sites. For example, substitution of aliovalent Ca^{2+} for Na^+ in albite does not normally induce vacancies because it is charge balanced by substitution of Al^{3+} for Si^{4+} to form the plagioclase solid solution.

not change defect densities appreciably. No f_{O_2} dependence was observed for Zr and Hf diffusion in natural rutile (Cherniak et al. 2007a), for O and Si in forsterite (Jaoul et al. 1981, 1983), or for Fe–Mg interdiffusion in pyroxene (Müller et al. 2013) likely because substitutional defects dominate. In this respect, just because a ferro-magnesian mineral crystallizes at high f_{O_2} does not mean it will necessarily have a different diffusion rate than the same mineral that crystallizes at low f_{O_2}. This is because, for crystallization at high f_{O_2}, charge balance likely occurs due to coupled substitution involving cations rather than vacancies. For example, in pyroxenes, substitution of M1-site Fe^{3+} for Mg^{2+} at higher f_{O_2} would likely be charge balanced by M2-site substitution of Na^+ for Ca^{2+} or Mg^{2+}, or by T-site substitution of Al^{3+} for Si^{4+}, not by vacancies. Nonetheless, as f_{O_2} changes during cooling along an f_{O_2} buffer, *in situ* oxidation-reduction of transition metals should change defect densities and diffusion rates. Thus, diffusion parameters for minerals should be reported and applied along an f_{O_2} buffer (e.g., Chakraborty and Ganguly 1992).

f_{H_2O} ***dependence.*** Oxygen diffusion in minerals commonly shows much larger D's and smaller E's in water- or mixed-fluid-present experiments at 0.05–1 GPa than when $P_{H_2O} < 1$ bar (e.g., Zhang 2010; Brady and Cherniak 2010; Farver 2010). In general, these different D's are referred to as "wet" vs. "dry". Enhanced D at high P_{H_2O} or f_{H_2O} is usually ascribed to rapid diffusion of trace molecular H_2O in the mineral as interstitial defects combined with exchange between the H_2O diffusant and lattice oxygen (Doremus 1969; Zhang et al. 1991; Zhang 2010; Farver 2010). Materials with higher molecular H_2O contents and H_2O diffusivities show higher overall O diffusivities (Zhang et al. 1991). Peck et al. (2003) and Bowman et al. (2011) proposed that the physical presence of an aqueous phase is necessary to enhance D. However, if an equilibrium reaction can be written between molecular H_2O in a fluid phase and in a mineral, e.g., $H_2O_{(aqueous)} = H_2O_{(mineral)}$, Henry's Law would require that a mineral's trace molecular H_2O content should be proportional to f_{H_2O}. Even if a fluid is not present, most mineral assemblages contain hydrous minerals that should buffer f_{H_2O} to high values and impart relatively high H_2O contents in minerals (Kohn 1999). By analogy, f_{H_2O} controls mineral "water" content (actually H^+ and OH^-) in nominally anhydrous minerals through thermodynamic equilibria (e.g., Keppler and Bolfan-Casanova 2006), even when water is not present, for example in experiments investigating partitioning between nominally anhydrous minerals and basaltic melt (Aubaud et al. 2004).

Cation diffusion rates can also increase slightly with increasing f_{H_2O}. Silicon diffusion in olivine and wadsleyite appears to occur via Si-vacancy complexes, and the abundance of these vacancies increases with increasing H^+ and f_{H_2O} (Costa and Chakraborty 2008; Shimojuku et al. 2010; Fei et al. 2013). Increasing H^+ concentration at higher f_{H_2O} also increases Fe–Mg interdiffusion rates in olivine (Wang et al. 2004; Hier-Majumder et al. 2005) and in Fe-Mg oxide (Demouchy et al. 2007) because interstitial protons cause an increase in octahedral site vacancies.

Strain dependence. Logically, if higher defect concentrations increase diffusion rates, and deformation increases defect density, deformation should increase diffusion rates. If so, two different states may be considered (Cohen 1970): dislocation-assisted diffusion occurs when prior deformation has created dislocations, whereas strain-enhanced diffusion occurs when dislocations form simultaneously with diffusion. Whether deformation significantly enhances diffusion rates has proved surprisingly difficult to demonstrate. Oxygen diffusion in albite is ~5 times faster in pre-deformed vs. non-deformed crystals (Yund et al. 1981) whereas ongoing deformation does not measurably affect Na–K interdiffusion in albite-adularia diffusion couples (Yund et al. 1989). However, disordering rates of tetrahedral Al–Si in feldspar do increase with increasing strain and strain rate (Yund and Tullis 1980; Kramer and Seifert 1991).

The effect of strain on trace elements in zircon has recently received close scrutiny. Commonly, zircons are first imaged using cathodoluminescence and electron back-scatter diffraction to map out chemical patterns and to identify sub-grain misorientations, which result from deformation. Ion probe or atom probe analysis then allows comparison of the chemistry of deformed vs. non-deformed regions or of dislocations themselves. Most studies show increased concentrations of REE, U, and Th, and decreased concentrations of Ti and Pb in regions containing dislocations (Reddy et al. 2006, 2016; Timms et al. 2006, 2011; Piazolo et al. 2012, 2016; MacDonald et al. 2013). Such compositional effects will bias trace element patterns, U–Th–Pb ages, and trace element thermometers, so understanding mechanisms of chemical change is potentially important to petrochronologic research. Atom probe studies now demonstrate that anomalously high and low concentrations spatially correlate with dislocations, especially along sub-grain boundaries (Reddy et al. 2016; Piazolo et al. 2016).

Why would trace elements accumulate at dislocations? For reasons discussed in detail elsewhere (see chapters 5 and 7 of Sutton and Balluffi 2003), diffusion of solutes from the crystal interior to grain surfaces typically decreases free energies. This behavior can be understood either from the macroscopic thermodynamics of grain boundaries (Gibbs 1874–1878; Cahn 1979) or from the perspective that grain boundaries can readily accommodate a wider range of cations than a crystal lattice, which has only a few specific sites. Consequently, a trace constituent that fits poorly in the crystal lattice (higher free energy) might find a more compatible bonding position along a grain boundary (lower free energy). Thus, creation of new sub-grain boundaries during deformation should scavenge trace elements (just as grain coarsening might liberate trace elements to form new accessory mineral grains; Corrie and Kohn 2008).

How do trace elements accumulate at dislocations? Accumulation could occur either as dislocations sweep through a crystal and coalesce into sub-grain boundaries, or via diffusion through the crystal lattice after dislocations become stationary (Piazolo et al. 2016; Reddy et al. 2016). For interstitial cations, elastic modeling predicts that they will diffuse towards dislocations (Cottrell and Bilby 1949), forming a "Cottrell atmosphere" of enhanced trace element concentration. Examples in zircon may include interstitial Be, Mg, and Al (Reddy et al. 2016). This process tends to pin dislocations, increasing the resistance of the solid to further deformation. However, a Cottrell atmosphere does not explain high concentrations of substitutional cations at dislocations, such as Y, so other factors have been invoked, including diffusion of combined oxygen vacancies plus aliovalent cations (Reddy et al. 2016).

After migration of a linear or planar dislocation has ceased, fast diffusion along the dislocation may continue, resetting trace element thermometers and U–Th–Pb ages in the immediately adjacent host crystal (Piazolo et al. 2016). Fast diffusion along grain boundaries also implies that some trace elements derive from the matrix, not solely from the host crystal. Conversely, matrix minerals must serve as the host for trace elements whose concentrations decrease in the host crystal, such as Ti in zircon.

PART 2: PETROCHRONOLOGIC CONCEPTS

This section reviews principles of diffusion that are directly relevant to petrochronology: closure temperature, geospeedometry, diffusion through porous media (fluid–rock interactions), major and trace element thermometry, reaction rates, and resetting of isochrons.

Closure temperature

Closure temperature (T_c) is usually defined in reference to thermochronologic data—there is a transition temperature, T_c, above which virtually all radiogenic daughter (d^*) is lost through diffusion, and below which virtually all radiogenic daughter is retained; the age of a mineral reflects the time (t) at which it passed through that transition temperature (Dodson 1973; Fig. 3A). Following more closely the derivation of Dodson (1973; Fig. 3A), closure temperature reflects the fact that, due to an exponential decrease in D with decreasing T (Eqn. 3), a mineral transitions from essentially maintaining equilibrium with an imposed boundary condition throughout the entire crystal, to adopting a composition that is independent of changes to the boundary

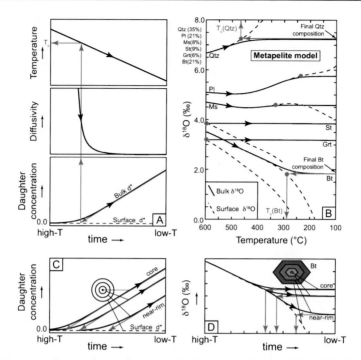

Figure 3. Closure temperature concepts and models. Black arrows show forward (normal time) progression of temperature, diffusivity and composition. Red arrows show backwards projections. (A) Radiogenic isotopes. At high temperature, D is large, and no radiogenic daughter accumulates—the bulk crystal maintains a concentration of 0, equivalent to the fixed boundary condition. At low temperature, D is small, and radiogenic daughter accumulates according to the age equation (approximately linearly for short times), independent of the boundary condition. T_c is defined as the projection of the trace of accumulated d^* vs. time to the boundary condition (red arrow). Modified from Dodson (1973). (B) Stable isotope model of a typical metapelite (modified from Kohn and Valley 1998). Model cooling rate was 5 °C/Ma, grain sizes were 1 mm diameter, and diffusion rates under hydrothermal conditions were assumed. In this example, staurolite (St) and garnet (Grt) crystallize below their closure temperatures, whereas quartz (Qtz), plagioclase (Pl), muscovite (Ms), and biotite (Bt) crystallize above their closure temperatures and their oxygen isotope compositions evolve during cooling. At high temperature the bulk $\delta^{18}O$ values of these minerals (excepting staurolite and garnet) track their respective boundary conditions. At low temperature, isotope compositions become invariant. T_c for each mineral is defined as the point on the boundary condition at which surface composition matches final composition [i.e., the projection of the final composition (dots on each mineral trajectory) to the surface trace; red arrows illustrate the concept for quartz and biotite]. These points are mathematically equivalent to the analytical expression of closure temperature defined by Dodson (1973), and do not depend on mineral assemblage. (C, D) Position-dependent closure temperature for radiogenic and stable isotope systems. Cores of crystals close earlier and reflect higher T_c's than positions closer to rims.

condition. Geometrically, T_c is then defined as the projection of the bulk mineral composition vs. t onto the trace of its boundary composition vs. t (Dodson 1973; Kohn and Valley 1998), with a conversion from time to temperature according to the cooling history (Fig. 3A). From this broader perspective, the thermochronologic T_c represents the projection of the accumulation of daughter (d^*) vs. t onto the boundary condition, $d^*=0$ (Fig. 3A). This projection corresponds with the measured age, t, of the mineral. The advantage of the broader definition of T_c is that it applies to all diffusion-controlled systems, irrespective of diffusing species, mineral assemblage, etc. (Fig. 3B; Kohn and Valley 1998). This definition also more meaningfully reveals the significance of boundary conditions in defining closure for minerals. Closure temperature has also been used to describe the apparent stable isotope temperature recorded between two minerals, but normally this is referred to as an apparent temperature, or T_{app}.

Mathematically, T_c for an entire crystal is (Dodson 1973):

$$T_c = \frac{E/R}{\ln\left(\dfrac{ART_c^2 D_0 / a^2}{E \cdot dT/dt}\right)} \tag{8}$$

where R is the gas constant, A is a constant reflecting the geometry of diffusion (8.7 for a plane sheet, 27 for a cylinder, and 55 for a sphere), a is the radius or half-width of the crystal, and dT/dt is the cooling rate. The equation must be solved iteratively, and T_c is fixed for a given geometry, crystal size, cooling rate, E and D_0. Although Equation (8) was developed for radiogenic systems, it applies even when boundary conditions exhibit complex behavior, including stable isotopic systems (Kohn and Valley 1998; Fig. 3B)—closure depends on the intrinsic ability of a mineral to respond to changes in its boundary condition, not on how that boundary condition changes. For example, although muscovite and plagioclase boundaries can exhibit compositional inflections and even oscillations, the closure temperature as evaluated numerically is indistinguishable from the closure temperature as calculated using Equation (8) (Fig. 3B; Kohn and Valley 1998).

In principle, the composition of a mineral can be used to constrain cooling rate because T_c and hence composition depend on cooling rate (Eqn. 8). For example, $\delta^{18}O$ of quartz in a schist will be higher in a more slowly cooled rock (lower T_c) and lower in a more rapidly cooled rock (higher T_c; Giletti 1986). Because cooling rate appears as a logarithmic term in Equation (8), however, cooling rates estimated this way carry extremely large uncertainties—a large change in cooling rate normally does not change T_c or composition much.

Geospeedometry

Closure can also be considered as a condition that initiates in the core of a crystal and sweeps outward as temperature decreases with time. Because the core is farthest from the boundary condition, it closes earliest and at the highest temperature whereas points closer to the rim of a crystal close later and at lower temperatures (Fig. 3C–D). Dodson (1986) modeled the position-dependent closure temperature for slow cooling:

$$T_c(x) = \frac{E/R}{\ln\left(\dfrac{\gamma R T_c^2 D_0 / a^2}{E \cdot dT/dt}\right) + 4S_2(x)} \tag{9}$$

where x is fractional position relative to the crystal center (0 at the center, 1 at the rim), γ is the exponential of Euler's constant, and $4S_2(x)$ is a position-dependent term (smaller towards the core, larger towards the rim). The bulk T_c for a crystal (Eqn. 8) is simply Equation (9) integrated over the volume of the crystal. Equation (9) implies that age or compositional zoning in a mineral can provide useful information about the cooling history, and forms a conceptual foundation for geospeedometry, or the determination of cooling rates.

One assumption in deriving Equation (9) is that initial temperatures are sufficiently high and cooling rates sufficiently low that the core of a crystal is able to equilibrate with the boundary condition imposed at the crystal surface. The parameter M describes this characteristic:

$$M = D_{T_0} \tau / a^2 \tag{10}$$

where D_{T_0} is the diffusivity at maximum temperature, τ is the time required for D to drop by a factor of E (which depends on the cooling rate), and a is the characteristic length scale of the crystal. Thus $M^{1/2}$ is the characteristic diffusion distance at T_0 divided by the characteristic length scale of the crystal. Large values of M (>1) lead to Equation (9). Ganguly and Tirone (1999) developed an analytical expression, evaluated numerically, to describe systems with $M \le 1$, where the core of the crystal remains relatively inert to diffusive resetting:

$$T_c(x) = \frac{E/R}{\ln\left(\dfrac{\gamma R T_c^2 D_o / a^2}{E \cdot dT/dt}\right) + 4S_2(x) + g(x)} \tag{11}$$

and $g(x)$ is a correction term for small M.

Although analytical expressions have been developed to invert zoning profiles in minerals so as to calculate cooling histories and relate them to tectonic processes (e.g., Lasaga 1983; Ganguly et al. 2000), many researchers employ numerical methods. Most use a Markov-chain Monte Carlo (MCMC) approach, which differs from classic Monte Carlo (totally random) approaches in requiring that temperature continually decrease for each time increment. That is, for each new time step (t_{i+1}), a new temperature (T_{i+1}) is chosen randomly between the temperature of the previous time step (T_i at t_i) and zero, i.e., $T_i > T_{i+1} > 0$. A model diffusion profile is calculated for each random cooling history, compared with the observed profile, and goodness of fit criteria calculated. An assemblage of T–t paths is then contoured for quality of fit (e.g., Ketcham 2005; Gallagher 2012).

A key outcome of modeling and sensitivity tests is that inversion of chronologic zoning is inherently more precise for constraining cooling rates than inversion of temperature-sensitive chemical zoning (Kohn 2013, 2016; see also Lindström et al. 1991). Here we show this using Equation (9) to model chemical and chronologic diffusion profiles at constant 25 °C/Ma cooling, then re-calculate temperatures and times resulting from perturbations in E commensurate with typical diffusion parameter uncertainties (Fig. 4). We assume $E = 250$ kJ and $D_0 = 1 \times 10^{-9}$ m^2/s, yielding a T_c of ~700 °C in the core of a 100 µm-radius spherical crystal. These values are similar to Pb diffusion in rutile (Cherniak 2000). An uncertainty in both E and $\log_{10}(D_0)$ of $\pm10\%$ (2σ) is assumed, well within typical 2σ variations between ~$\pm4\%$ to ~$\pm40\%$ (e.g., see Brady and Cherniak 2010). Because D_0 correlates almost perfectly with E, a 10% increase or decrease in E is assumed to increase or decrease $\log_{10}(D_0)$ by 10%.

For the chronologic zoning model, each point in a crystal has a measured age that corresponds to the position-dependent closure temperature of the radiogenic daughter. That is, we know Δt from direct chronologic measurements on at least 2 different positions in the crystal, and we use Equation (9) with different values of E (and D_0) to solve for ΔT_c at the same positions. Even accounting for correlated changes to D_0, calculated T_c's shift upward for higher E, and downward for lower E. Thus, variations in E (and D_0) shift the entire cooling history to higher or lower temperatures (vertical double-arrow in Fig. 4), but with virtually no change to either ΔT_c or the cooling rate (24–26 °C/Ma). That is, chronologic zoning retains information about cooling rate, but with some ambiguity about the specific temperature interval.

For chemical zoning, each point in a crystal has a measured composition that directly defines T_c. For example, if diffusion zoning of Zr in a rutile crystal reflects resetting of the Zr-in-rutile thermometer, each position-dependent Zr concentration represents a position-dependent T_c. But T_c depends on cooling rate, too (Eqn. 9). So, if we change E (and D_0), we must change the cooling rate to obtain the same (measured) T_c within a crystal: cooling must be slower with a higher E to obtain the same T_c and faster with a lower E (horizontal double-arrow, Fig. 4). Even modest variations in E and D_0 require very large changes to cooling rate to obtain the same T_c (between ~2 and ~300 °C/Ma). This sensitivity occurs because dT/dt appears in the logarithm of Equation (9). Whereas chemical zoning retains precise information about temperature, it poorly constrains cooling rate (Lindström et al. 1991; Kohn 2016). A similar result was obtained from numerical models of major element zoning in

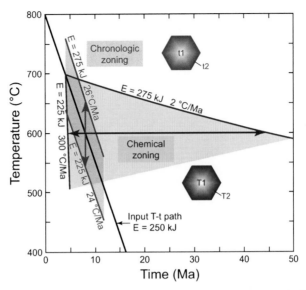

Figure 4. Sensitivity of cooling history inversions (geospeedometry) to uncertainty in diffusivities based on chronologic vs. chemical zoning. Hexagons show schematic crystals with diffusion zoning in either ages (t_1, t_2) or temperatures (T_1, T_2). Input T–t path is 25 °C/Ma, with an activation energy for diffusion of 250 kJ. Perturbations of E by ±10% (±25 kJ, with commensurate changes to D_0) shift cooling history for chronologic zoning to higher or lower temperatures (vertical double-arrow), but retain basic T–t shape with cooling rates between 24 and 26 °C/Ma. Same perturbations to E for chemical zoning shift temperatures to younger or older ages (horizontal double-arrow), and cause radical changes to T–t shape with cooling rates between ~2 and ~300 °C/Ma.

garnet, specifically an uncertainty in E of ~±15% (with corresponding error in D_0) propagates to an uncertainty in dT/dt of ~±1 order of magnitude (Spear 2014).

Diffusion in porous media: fluid–rock interaction

Mass transport through porous media, including diffusion, presents a large literature which we can only briefly discuss here. For more comprehensive development of these concepts, see Bear (1988). In geological applications, smoothly varying stable isotope profiles at the hand sample to outcrop scale across lithologic discontinuities may reflect chemical transport through rocks via both advection and diffusion, and provide constraints on the duration of metamorphic fluid–rock interactions (e.g., Bickle and McKenzie 1987; Bickle and Baker 1990; Baumgartner and Ferry 1991). Chemical transport between rocks likely occurs through an intergranular medium because volume diffusion through the solid in many cases is too slow at moderate metamorphic temperatures (Bickle and McKenzie 1987; Cartwright and Valley 1991). The type of intergranular medium (dry grain boundaries, fluid, or melt) will affect the diffusivity of the element or isotope, so it is important to consider petrologic evidence for the type of medium likely involved in such transport.

Chemical and isotopic profiles collected across compositional boundaries can be modeled using a one-dimensional solution to the mass continuity equation (Bickle and McKenzie 1987):

$$\frac{\partial C}{\partial t} K_e + v\phi \frac{\partial C}{\partial x} = D_e \frac{\partial^2 C}{\partial x^2} \tag{12}$$

where C is concentration, x is distance from the contact, t is time, v is the fluid velocity, ϕ is the porosity, D_e is the effective diffusivity, and K_e is the effective partition coefficient. Equation (12) balances the time variation of concentration in the solid (1st term) and the effects of advection (2nd term) against diffusion (3rd term). The effective partition coefficient takes into account the porosity (ϕ) of the rocks as well as the fractionation of the elements or isotopes into the solid or fluid (Bickle and McKenzie 1987; Cartwright and Valley 1991, Baumgartner and Valley 2001):

$$K_e = \frac{\rho_s K_c}{\rho_f}(1-\phi)+\phi \tag{13}$$

where ρ_s and ρ_f are the density of the solid and fluid respectively and K_c is the partition coefficient between the mineral and fluid (isotopic exchange is assumed to occur between solid and an intergranular medium). In multi-mineralic rocks, a bulk fluid–rock partition coefficient is determined by weighting each fluid–mineral partition coefficient according to the abundance of that mineral and summing over all minerals. For most stable isotopes (e.g., oxygen, carbon, sulfur) the delta notation (e.g., $\delta^{18}O$) may be used instead of the concentration over relatively small variations (see Baumgartner and Rumble 1988; Baumgartner and Valley 2001). Equation (13) links composition profiles to the solid–fluid partition coefficient and porosity (Fig. 5A, B). The effective diffusivity D_e is defined as (Bickle and Baker 1990):

$$D_e = \phi D \tau \tag{14}$$

where τ is the tortuosity of the system and D is the diffusivity of the element or isotope in the medium (aqueous fluid or melt). Equation (14) links composition profiles to porosity, diffusivity of solute in the fluid, and tortuosity (Fig. 5B–D). Modeled profiles are matched to the data to solve for the diffusive distance, $\sqrt{D_e \Delta t K_e^{-1}}$. The duration ($\Delta t$) of a metamorphic event is then determined for given values of the mineral and fluid densities, mineral–fluid partition coefficient, porosity, tortuosity and intergranular diffusivity.

Major element thermometry

Diffusive resetting and other retrograde processes occurring in a rock affect temperatures recorded in mineral compositions. Here we discuss the differential impact of exchange reactions, which simply exchange elements between two minerals without changing modes (e.g., Fe–Mg exchange), as contrasted with net transfer reactions, which drive significant changes to the mode of one or more minerals. In general, net transfer reactions cause changes in concentration for numerous elements (e.g., Fe, Mg, Ca and Mn in garnet), whereas exchange reactions do not (e.g., Fe and Mg only).

Major element thermometry commonly emphasizes Fe–Mg exchange between garnet and other silicates, for example biotite. Models of diffusional resetting of apparent garnet–biotite temperatures illuminate the impact of Retrograde Exchange Reactions (ReER's) vs. Retrograde Net Transfer Reactions (ReNTR's; Spear 1991; Spear and Florence 1992; Kohn and Spear 2000) on calculated temperatures. In nearly all models, matrix biotite is assumed to homogenize its composition, whereas the rim of the garnet is allowed to equilibrate with the matrix, inducing diffusive fluxes of Fe and Mg into and out of the garnet, and producing a diffusion profile on the rim of the garnet. If only ReER's occur, matrix biotite becomes more Mg-rich, and garnet–biotite apparent temperatures are always equal to or less than the peak metamorphic temperature (Fig. 6A). The garnet core paired with matrix biotite most closely approximates the peak metamorphic temperature, but always underestimates it, at least slightly, even if the mode of biotite is large (Fig. 6A). These models implicitly underlie many interpretations of metamorphic temperatures and diffusion profiles: if exchange reactions alone occur, calculated temperatures are always minima (lower T's for smaller garnets), and boundary conditions are modeled using simple Fe–Mg exchange equilibria.

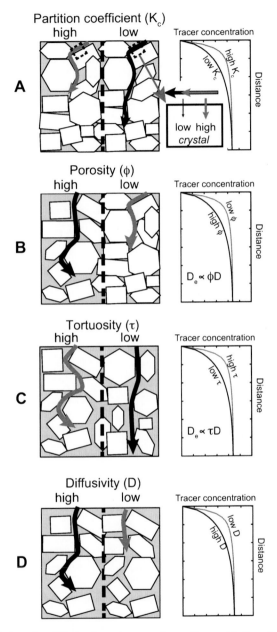

Figure 5. Factors influencing diffusion in an intergranular medium and bulk rock tracer concentration profiles. Schematic diagram for each factor illustrates mineral grains surrounded by intergranular medium (shaded region). Arrows indicate distance of diffusion for each scenario. Graphs to the right show the expected differences in diffusion profiles. (A) Partition coefficient between minerals and fluid. Inset shows flux from intergranular medium to crystal. A higher partition coefficient results in a shorter diffusion distance. (B) Porosity. A higher porosity increases effective D and results in a greater diffusion distance ($D_e \propto \phi D$). (C) Tortuosity. A higher tortuosity results in a shorter diffusion depth. (D) Diffusion coefficient in the intergranular medium. A higher diffusion coefficient results in a greater diffusion distance.

Unlike ReER's, if ReNTR's occur (specifically growth of biotite at the expense of garnet, e.g., in the reaction garnet + K-feldspar + melt = biotite + sillimanite + plagioclase + quartz), matrix biotite becomes more Fe-rich (Fig. 6B). Pairing a garnet core with matrix biotite then yields a temperature that exceeds the peak metamorphic temperature (Fig. 6B). In fact, smaller garnet crystals, which give lower apparent temperatures with ReER models, give higher apparent temperatures in their cores with ReNTR models (500 μm vs. 5 mm radius garnets, Fig. 6A, B). Nearly all high-grade garnets show Fe–Mg diffusion profiles near their rims, but this fact alone does not justify a particular model. Rather, identifying whether ReNTR's occurred during cooling, for example by considering textures or zoning in other chemical constituents or mineralogical evidence for ReNTR's, is crucial to correct thermometric interpretation (see Spear and Florence 1992; Kohn and Spear 2000; Spear 2004). In detail, the shapes of the diffusion profiles also differ, which impacts geospeedometry calculations based on Fe–Mg zoning in garnet.

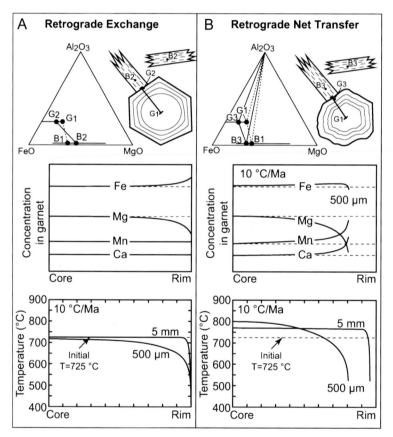

Figure 6. Impact of diffusion and reaction type on major element thermometry, showing differences between retrograde exchange reactions (ReER's) and retrograde net transfer reactions (ReNTR's). Models assume a cooling rate of 10 °C/Ma, an initial temperature of 725 °C, and a homogeneous original garnet composition. ReER model assumes a volume ratio of garnet/biotite of 0 (infinite biotite, so biotite composition does not change); ReNTR model assumes a volume ratio of garnet/biotite of 0.2. Dashed lines indicate original compositions. Based on Spear (1991) and Spear and Florence (1992). (A) Exchange reactions cause divergence of garnet ("G") and biotite ("B") compositions, such that matrix biotite becomes more Mg-rich. Zoning in garnet occurs only in Fe and Mg, and apparent temperature is always at or below peak temperature. (B) Net transfer reactions drive zoning in all garnet components, while biotite shifts towards more Fe-rich compositions. Apparent temperature is low on the rim of the garnet, but exceeds peak temperature in garnet core.

Trace element thermometry

The diffusion of trace elements within minerals also responds to matrix–mineral equilibria. Here we focus on Zr-in-rutile thermometry because it has been investigated extensively and exhibits curious features that have spawned increasing debate about rutile's diffusional resistance to resetting and about crystal-chemical controls on trace element mobility more generally. These issues may have broad applicability to many accessory minerals.

The Zr content of rutile in equilibrium with quartz and zircon is temperature-dependent (Zack et al. 2004; Watson et al. 2006), and experiments indicate that the diffusivity of Zr in rutile should be sufficiently fast that diffusive resetting is expected at temperatures above 700–750 °C (Cherniak et al. 2007a). For example, Zr-in-rutile thermometry from matrix grains yields temperatures of ~700 °C in rocks of the Greater Himalayan Sequence in central Nepal that likely reached temperatures of 750–800 °C (Kohn 2008) and ~735 °C from granulites from East Antarctica that likely reached temperatures > 900 °C (Pauly et al. 2016). However, two perplexing questions have been raised recently about Zr-in-rutile thermometry: why does zircon or baddeleyite sometimes precipitate in rutile interiors (Kooijman et al. 2012; Ewing et al. 2013; Pape et al. 2016), and why are very high (up to 900 °C) Zr-in-rutile temperatures sometimes recorded if Zr diffusion is so fast? One hypothesis is that slow Si diffusion could explain high Zr-in-rutile temperatures through a resulting elevation of μ_{SiO_2} (Taylor-Jones and Powell 2015). This hypothesis assumes that silicon concentrations increase with increasing temperature, and that Si diffuses more slowly than Zr (Taylor-Jones and Powell 2015). While neither assumption is as yet verified, the implications can be considered qualitatively. Here, we discuss the importance of local equilibrium on diffusive reequilibration and the formation of Zr and Si diffusion profiles (see also Kohn et al. 2016).

To understand variations in the chemical potential of Zr and Si, we first assume that both Si and Zr homogenize if initial temperatures are sufficiently high (Fig. 7A). During cooling, Zr and Si concentrations are predicted to decrease at rutile rims to maintain equilibrium with matrix quartz and zircon, inducing diffusion loss profiles (Fig. 7B). Within rutile, Si and Zr each diffuse in response to their respective chemical potential gradients (μ_{SiO_2} and μ_{ZrO_2} respectively), which for trace elements in a nearly pure mineral are proportional to the natural logarithm of their concentration gradients via Henry's Law. Because substitution of Si and Zr into rutile is not crystal-chemically coupled, and because the interior of the grain contains neither quartz nor zircon that can buffer chemical potentials, each element diffuses independently of the other. Therefore, slow diffusion of Si does not impede Zr diffusion, each trace element simply develops a different profile reflecting different diffusivities and chemical potential gradients.

As Si and Zr contents in the rutile interior progressively deviate from rim concentrations, the interior of the rutile becomes supersaturated with quartz and zircon. If zircon does not precipitate, decreasing temperature and ZrO_2 content at the rutile rim increases the diffusion gradient and the magnitude of core ZrO_2 supersaturation (Fig. 7C), possibly crossing the baddeleyite saturation threshold and catalyzing baddeleyite growth. If zircon does nucleate as an inclusion in the rutile interior, both Zr and Si contents decrease adjacent to the inclusion, inducing new diffusion loss profiles (Fig. 7D). Outward fluxes of Zr and Si produce matrix zircon, whereas inward fluxes produce zircon inclusions. Because Zr solubility in rutile likely exceeds Si solubility (Taylor-Jones and Powell 2015), Si concentrations at the interface with a zircon inclusion drop below quartz saturation (so $a_{SiO_2} < 1.0$), and Zr concentrations are buffered at higher values than on the rutile rim to maintain equilibrium with zircon (Fig. 7D). The Zr concentration at the zircon inclusion and matrix interfaces, however, is always lower than the maximum Zr elsewhere in the rutile interior (Fig. 7D).

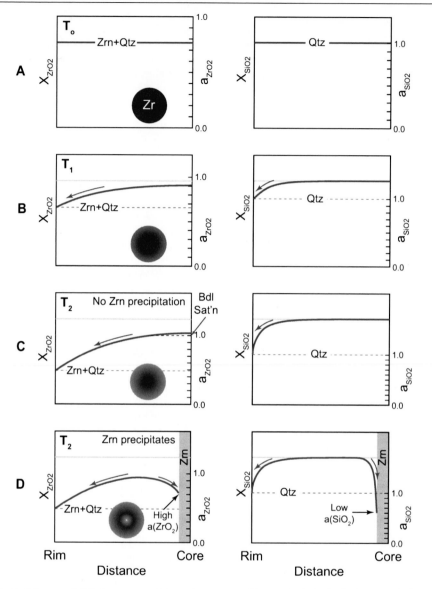

Figure 7. Impact of diffusion on trace element thermometry, illustrated schematically for the evolution of Zr and Si concentrations and activities in rutile (red lines and curves) during cooling from high temperature (based on Kohn et al. 2016). Dashed lines are concentrations of Zr and Si in equilibrium with zircon (Zrn) and quartz (Qtz) in the matrix, darker vs. lighter shading within circle indicates higher vs. lower concentration of Zr within rutile. Activities are relative to quartz and baddeleyite. (A) At high initial temperature, T_0, rutile has homogeneous Zr and Si concentrations in equilibrium with quartz and zircon; $a_{ZrO_2} < 1$. (B) With decreasing temperature (T_1), rim concentrations decrease to maintain equilibrium with matrix quartz and zircon, inducing diffusive loss of Zr and Si; in the rutile interior, $a_{SiO_2} > 1$ and $a_{ZrSiO_4} > 1$ (i.e., quartz and zircon are supersaturated). (C) With further decreasing temperature (T_2), quartz and zircon become increasingly supersaturated, and the rutile core may become supersaturated in baddeleyite (bdl sat'n). (D) If zircon precipitates in the rutile interior (gray region on right), Zr and Si will diffuse toward the zircon to reach equilibrium with zircon (without quartz). If Si concentrations are low, Zr at the zircon inclusion boundary will be higher than at the rutile rim, but lower than elsewhere in the rutile grain. Si concentrations will be lower at the zircon inclusion boundary because, although zircon is a saturating phase, quartz is not.

Although the concepts illustrated in Figure 7 help explain formation of zircon and baddeleyite inclusions inside high-T rutile, as well as 700–750 °C Zr-in-rutile temperatures in rocks that were once hotter, they do not explain how unusually high Zr-in-rutile temperatures can be preserved. Such high temperatures require that boundary concentrations did not maintain equilibrium with matrix quartz and zircon (Kohn et al. 2016). That is, high Zr-in-rutile temperatures do not necessitate slow Zr diffusion, as some have argued (e.g., Kooijman et al. 2012) but instead require shutdown of mineral equilibration via chemical transport along grain boundaries. This topic is explored further in the discussion of natural examples of trace element diffusion profiles.

Reaction rates

Commonly, diffusion profiles are modeled and interpreted assuming that the physical position of a crystal surface is fixed. Reacting minerals, however, also form and preserve diffusion profiles. These situations are generally referred to as moving boundary problems, and have different solutions than fixed boundary problems. If specific mass balance and reaction rate criteria are met, diffusion profiles can be used to infer mineral reaction rates (Jackson 2004; Lucassen et al. 2010; Cruz-Uribe et al. 2014). Here we discuss more generally how reacting minerals might develop diffusion profiles and how those profiles may inform rates of petrogenetic processes.

Understanding the time evolution of reaction (Fig. 8A) helps predict compositional trends for different scenarios (Figs. 8B–E). In each case, material α transforms to material β at a constant rate. The boundary moves at a constant velocity, v, which is proportional to the reaction rate. Thus, to obtain the reaction rate, we must estimate v. At equilibrium, α and β exhibit constant partitioning, $k = C_\beta / C_\alpha$, where C_α is the concentration of the diffusant in α and C_β is the concentration in β. For example, α and β might be olivine melt and crystalline olivine, and the diffusant could be any major, minor or trace element with $k \neq 1$, e.g., Fe/Mg, Ni, Li, etc. Our illustrations assume $k < 1$ (the element prefers reactant phase α), but for $k > 1$, our figures would simply be inverted vertically. Diffusion in β is assumed to be slow, so that it preserves composition profiles, but infinitely fast diffusion in β does not change the main conclusions regarding profiles in α. Compositional evolution of α and β depends on whether the α–β interface is wholly isolated from equilibrating with the matrix or melt ("closed system"), or whether the matrix participates as a source or sink of the diffusant ("open system"), as well as whether partitioning is preserved at the α–β interface.

The closed system model has a unique analytical solution for the diffusion profile in phase α, which is derived by transforming Equation (2) from a fixed coordinate system (x) to a moving coordinate system (z) where $z = x - vt$:

$$C(z) = C_0 + C_0 \left(\frac{1}{k} - 1 \right) e^{(-vz/D)} \qquad (15)$$

where D and C_0 are the diffusion coefficient and original concentration of the diffusant in phase α (Fig. 8B; Jackson 2004). At steady state, α transforms to β at a constant rate, and both the diffusion profile defined by Equation (15) and the boundary between α and β move at constant velocity, v. To conserve mass, the steady state concentration of any new β must be C_0, fixing the steady-state boundary condition on the surface of α at C_0/k where k is the partition coefficient as described above (Fig. 8B). Even if α and β are not compositionally comparable, the same principles apply but require a scaling factor to conserve the mass of the diffusant (Lucassen et al. 2010). For example, if titanite ($CaTiSiO_5$) grows at the expense of rutile (TiO_2), rutile contributes only ~40% of the total mass of titanite. So, on a mass basis, the concentration of a trace element (e.g., Zr, Nb) in the steady-state titanite should be ~40% of the concentration in the original rutile, assuming the matrix does not serve as a source or sink of the trace element.

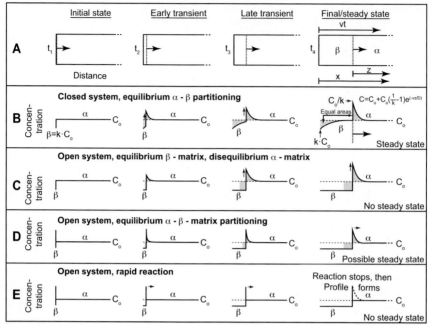

Figure 8. Models of reaction rates and induced diffusion profiles. (A) Time series, showing movement of boundary between α and β (the same progression of the boundary is shown in Figs. 8B–E). Phase β replaces phase α at a constant rate, and coordinate system can be transformed from fixed (x) to moving (z). The velocity (v) is proportional to the reaction rate. (B) Closed system with equilibrium partitioning of major, minor or trace elements between α and β (Jackson 2004). During initial growth of phase β, a deficit in diffusant occurs in β that is balanced by an excess of diffusant at the rim of α (shaded regions). During this transient phase the excess diffusant in α increases, while the concentration in β increases to maintain equilibrium with α. This excess forms a diffusion profile in α, which at steady state moves at the same rate as the boundary between α and β. (C–E) Open system in which some diffusant enters matrix. Mass excess in α does not match mass deficit in β: shaded area in β (equivalent to shaded area in α) is smaller than the total area in β. (C) Matrix maintains equilibrium with β, but not with α, so boundary composition of α and α–β partitioning change with time. Diffusion profile in α never reaches steady state. (D) Matrix maintains equilibrium with both α and β. Steady state is possible with same solution as Equation (15). (E) Matrix maintains equilibrium with both α and β, but rapid reaction prevents diffusion profile from developing in α until after reaction ceases.

The closed-system steady-state model further predicts that the first-formed β has concentration $k·C_0$ (a consequence of the α–β equilibrium condition), and that the mass deficit that develops in β balances the mass excess present in the diffusion profile in α (a consequence of mass conservation; shaded regions, Fig. 8B). Thus, if D is low in phase β, the process by which the diffusion profile develops in phase α simultaneously produces a complementary profile in β (Fig. 8B; Jackson 2004). In addition, after the system achieves steady state, phase α should exhibit concentration C_0/k at the interface between α and β, while the concentration ratio of β to α should be k (Fig. 8B).

What happens if mass is not conserved? For example, perhaps phase equilibria involving matrix minerals control the composition of phase β, rather than partitioning with phase α. If so, the composition of phase β is fixed, and the mass deficit in β will not match the mass excess in α (Fig. 8C). Such a scenario might occur if some of the diffusant escapes the local α–β reaction region, traversing phase β to form or enter some other phase in the matrix (e.g., Lucassen et al. 2010). Depending on the rate of diffusant loss to the matrix, the formation of a steady-state diffusion profile in α is not assured (Fig. 8C). Indeed, as phase β progressively armors phase a, loss of diffusant from α probably decreases, and the mass of diffusant in the diffusion profile in α gradually increases. No steady state exists, so no reaction rate can be calculated.

Alternatively (Figs. 8D, E), perhaps equilibrium with matrix minerals establishes not only a constant composition for product phase β, but also a constant boundary condition for phase α, independent of the mass balance that Equation (15) requires. A steady state diffusion profile can form in α, mathematically comparable to Equation (15), but with a different maximum concentration (Fig. 8D). It is unclear, however, whether steady state can ever be assumed for such a system. Relaxing mass balance constraints opens alternate interpretations for how diffusion profiles might form. For example, reaction might be too fast to form a diffusion profile in α, but the fixed boundary condition forms a diffusion profile after the reaction has ceased (Fig. 8E). The form of the resulting diffusion profile (an error function), is not easily distinguishable from the exponential form of Equation (15) (Lucassen et al. 2010).

In sum, steady-state closed-system reaction (Fig. 8B) is the only scenario that permits unequivocal reaction rate estimates and makes several testable predictions: an increasing concentration in the product phase (β) towards the α–β boundary, a concentration in β that is either the same as the interior of α (for 1:1 replacement; Jackson 2004) or in proportion to the amount of reactant α that is present in β (Lucassen et al. 2010), and equilibrium partitioning between α–β at their interface. Conversely, a constant concentration profile in β, mass deficits between phases α and β, or disequilibrium at the α–β interface would challenge a steady-state reaction model and its analytical solution. Such data might place limits on the extraction of information about other kinetic processes, such as cooling rate from the shape of the diffusion profile (Fig. 8E), but not on the extraction of reaction rates.

Diffusive resetting of isochrons

Differential diffusivity of parent and daughter isotopes in a mineral can change chronologically useful isotopic ratios after a mineral has crystallized and produce ages that no longer correspond to the crystallization age. Commonly, geochronologists assume that resetting responds solely to diffusion of the daughter isotope, and that the parent isotope is inert, so that the measured age is younger than the crystallization age. A large and well-trodden literature is devoted to this phenomenon, especially [40]Ar loss (see Dodson 1973; McDougall and Harrison 1999). While preferential resetting of the daughter isotope generally holds true for U–Pb and K–Ar because U has a higher charge than Pb, and Ar is not electrostatically bonded in a crystal, preferential diffusion of the parent can also occur. This phenomenon has been recognized and quantitatively characterized for the Lu–Hf system in garnet (e.g., Kohn 2009; Bloch et al. 2015; Bloch and Ganguly 2015), and leads to unusual chronologic behavior. In this system, the parent, [176]Lu^{3+}, diffuses about an order of magnitude faster than the daughter, [176]Hf^{4+} (Bloch et al. 2015) and whereas Hf is generally immune to diffusive resetting except for ultrahigh-temperature rocks or sub-mm diameter garnets, strong zoning in Lu is subject to diffusional relaxation, even at moderate temperatures (~750 °C; Kohn 2009; Bloch and Ganguly 2015).

Differential uptake of Lu but not Hf during garnet growth forms a bell-shaped Lu concentration profile, while Hf concentrations are uniformly low (Fig. 9A; Lapen et al. 2003, Kohn 2009). Strong Lu concentration gradients drive outward diffusion from garnet cores, reducing [176]Lu/[177]Hf ratios; if resorption occurs, Lu can diffuse inward at garnet rims, increasing [176]Lu/[177]Hf (Fig. 9B; Kohn 2009). Where metamorphic temperatures are high enough (>800 °C), grain sizes small enough (<1 mm), and/or the duration of metamorphism long enough (tens of Ma), Lu may be completely homogenized, however in most metamorphic environments these criteria are not met (Bloch and Ganguly 2015). Consequently, intracrystalline diffusion of Lu, if it occurs after significant ingrowth of [176]Hf, should produce a counterclockwise rotation of isochrons, potentially yielding ages that are older than the crystallization age of the garnet (Fig. 9B; Kohn 2009).

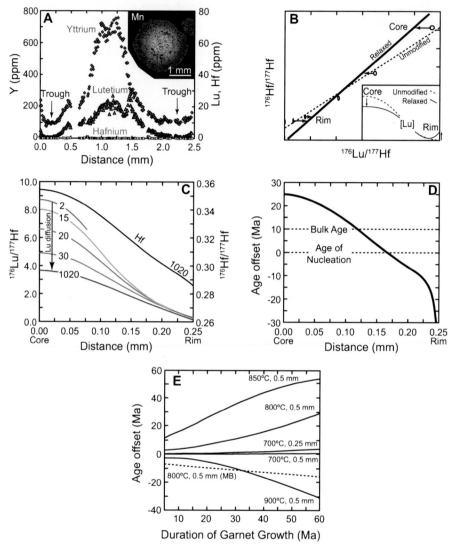

Figure 9. Diffusive resetting of isochrons, modified from Kohn (2013) and Bloch and Ganguly (2015). (A) Trace element zoning across garnet from a ~750 °C rock in the Himalaya, showing bell-shaped distributions of Y and Lu, but increases towards rim (forming troughs), indicating resorption and back-diffusion during cooling. Inset shows Mn X-ray map of garnet from same rock. (B) Schematic illustration of how Lu diffusion causes $^{176}Lu/^{177}Hf$ to decrease in garnet core and increase at garnet rim, rotating isochrons to older ages. Inset illustrates Lu zoning. (C) Fully quantitative model of Lu zoning in garnet through a metamorphic cycle. Each curve represents the Lu profile at specific times (in Ma) since garnet nucleation. Garnet grows from 500 °C to 750 °C over a 20 Ma interval, so its radius progressively lengthens up to 20 Ma, then cools at a rate of 2 °C/Ma, ultimately achieving an age of 1020 Ma. Diffusion of Lu during and after growth drives down Lu concentration in garnet core. Resorption and Lu increase at rim were not modeled. (D) Age offset (difference between modeled age for garnet growth with and without diffusion) generously provided by E Bloch. Loss of Lu from garnet cores leads to unsupported ^{176}Hf and spuriously older ages. Younger ages towards rim reflect position-dependent closure. (E) Age offset as a function of garnet growth rate, peak temperature, and final garnet radius. Solid curves are for metapelitic compositions, dashed curve (MB) is for metabasaltic composition. The curve for 900 °C develops opposite trend because Hf diffusion becomes significant.

Fully quantitative numerical simulations of Lu and Hf uptake and diffusion (Bloch and Ganguly 2015; Fig. 9C–E) demonstrate that garnet cores can lose Lu during both prograde and retrograde metamorphism (Fig. 9C), leading to garnet core ages that exceed the nucleation age by over 20 Ma (Fig. 9D). Even the bulk apparent age can exceed the nucleation age by ~10 Ma (Bloch and Ganguly 2015; Fig. 9D). Garnets from metabasaltic rocks may not show the same age bias, however, because Lu is not so strongly fractionated in garnet cores (Lapen et al. 2003) so the extent of diffusional bias is reduced (Bloch and Ganguly 2015). The Sm–Nd system behaves differently, because parent and daughter have almost identical diffusivities (Tirone et al. 2005; Carlson 2012), and because neither becomes strongly zoned during garnet growth (Lapen et al. 2003; Kohn 2009; Bloch and Ganguly 2015). Clearly care must be taken in interpreting Lu–Hf garnet ages, especially for relatively small (<~1 mm), zoned garnets that grew between 700 and 900 °C (Fig. 9E; Bloch and Ganguly 2015). Despite these problems, Lu–Hf ages can be used to constrain the duration of prograde metamorphism if factors such as peak temperature and cooling rate are determined independently, and if Lu–Hf dating is combined with Sm–Nd dating, which experiences different but predictable age bias (Lapen et al. 2003; Kohn 2009; Bloch and Ganguly 2015).

PART 3: EXAMPLES

This section focuses on examples of the effects of diffusion that are relevant to petrochronology: natural contraints on diffusion rates, geospeedometry (from major element, trace element, and radiogenic isotope zoning), reaction rates, timescales of magmatic processes and magma ascent rates, duration of metamorphism, and fluid–rock interaction.

Natural constraints on D

Natural data are commonly interpreted using experimentally determined diffusion parameters to characterize geological processes. But the inverse exercise—using natural data from well-characterized geological systems to constrain diffusivities—can help refine E and D_0 by extending the ranges of intensive parameters that control D, especially temperature, beyond what is possible in experiments. Such comparisons are needed to validate diffusion-based petrochronologic applications in other settings. The vast majority of natural data are consistent with experimental data, but in a few instances natural and experimental D's are not reconcilable. These instances may provide important insight into mechanisms of diffusion, or recommend alternative experimental methods more relevant to natural systems. Here we discuss examples illustrating consistency between natural data and high-T experiments, but also a few exceptional cases where natural and experimental data appear in conflict—in these cases natural data typically imply much lower D's than experiments.

Before discussing these examples, we emphasize that the most difficult problem in using natural rocks to estimate D's is that solutions to the diffusion equation all involve the combined parameter $D \cdot \Delta t$, where Δt is a characteristic time interval; because D depends exponentially on temperature (Eqn. 3), Δt is generally approximated with the duration (time-interval) the rock spends near maximum T. Although maximum T may also be subject to considerable error (Spear 2014), robustly constraining Δt often proves even more difficult. In metamorphic rocks, there are few a priori constraints on how long peak metamorphism "should" last. For example, some studies have argued for relatively short metamorphic events (e.g., $\Delta t \ll 1$ Ma in the Barrovian type locality; Ague and Baxter 2007), including rapid metamorphism of rocks via movement through thermal gradients at plate tectonic velocities (e.g., European Alps: Rubatto and Hermann 2001; Himalaya: Kohn et al. 2004; Kokchetav massif: Stepanov et al. 2016) or via thin-skinned duplexing at mid- to deep-crustal depths (northern Appalachians: Spear et al. 2012; Spear 2014). Subduction models of metamorphism similarly imply peak metamorphic $\Delta t \sim 1$–2 Ma (e.g., Gerya et al. 2002; Warren et al. 2008). Yet other studies argue for more protracted high temperatures ($\Delta t \sim 20$ Ma; e.g., Himalaya: Kohn and Corrie 2011; Dabie-Sulu: Liou et al. 2012), as do other types of thermal-mechanical models (e.g., Beaumont et al. 2001, 2004). Commonly, we know the maximum duration of metamorphism better than the minimum. For example, Δt is always smaller than the difference between depositional and late-stage cooling ages. Because D is determined inversely to Δt, a minimum D (maximum Δt) may be readily constrained, whereas a maximum D (minimum Δt) is more difficult to estimate. In several studies, the duration of metamorphism is modeled using thermal models of cooling plutons (e.g., garnet: Carlson 2006, 2012; monazite: McFarlane and Harrison 2006). Otherwise Δt is determined from regional chronologies. With typical analytical uncertainties of 1–2% for ages, errors in Δt for Archean and Proterozoic orogens (e.g., Grenville) can easily exceed 10 or 20 Ma. In contrast, for younger orogens (Himalaya, Alps), even large chronologic errors (e.g., 5–10%) result in much smaller errors in Δt.

As an illustration of potential ambiguities in determining D, Peck et al. (2003) argued for slow oxygen diffusion in zircons from the Adirondack Mountains, in part based on the preservation of isotopic differences on length scales of $\sim 50\,\mu m$ across zircon overgrowths. Many other data also suggest slow oxygen diffusion in zircon, so these interpretations are not unusual. However, Peck et al. (2003) assumed a peak temperature of $\sim 600\,°C$ based on regional thermobarometry (Bohlen et al. 1985) and estimated Δt to be 25–50 Ma based on a broad regional chronology for the Ottawan orogenic event (dated at ~ 1000 Ma). The chronology, in turn, was determined from conventional U–Pb analyses of zircon, monazite, titanite, and rutile separates from rocks distributed over large distances exceeding 100 km (Mezger et al. 1991).

Subsequent work on some of the same metamorphic rocks demonstrates that the zircon overgrowths actually formed ~ 1150 Ma during an altogether different orogeny (the Shawinigan event; Peck et al. 2010) and that temperatures locally reached at least $800\,°C$ (Storm and Spear 2005, 2009). The duration of the Shawinigan event is not well constrained, but is associated with anorthosite intrusions (e.g., McLelland et al. 2010). Arguably, if a short-lived thermal pulse during the Shawinigan catalyzed formation of metamorphic zircon overgrowths, and Ottawan overprinting was either low-temperature or brief (Δt as low as ~ 1 Ma, Page et al. 2010), D's for oxygen diffusion could be higher, perhaps by as much as 1–3 orders of magnitude. Conversely, higher peak metamorphic temperatures would require even lower D's than implied by the original assumed temperature of $600\,°C$, perhaps by as much as 5 orders of magnitude. In the absence of robust estimates of temperature and petrogenetically based chronologic data with small absolute errors, preferably from the same outcrops, calculations of diffusivities based on natural samples and comparison to experimental data can carry large uncertainties.

Garnet. Divalent cation diffusion in garnet has received close scrutiny, both experimentally and in natural systems (see summary of Ganguly 2010), and illustrates internal consistency for several elements. Early experiments yielded somewhat discrepant results until the effects of composition and oxygen fugacity were fully understood. Now a diverse set of experiments can be assembled into a self-consistent set of diffusion coefficients (Chakraborty and Ganguly 1992; Ganguly et al. 1998; Ganguly 2010). Among empirical estimates, Carlson (2006) is most comprehensive in attempting to reconcile experiments with empirical analysis at two localities: the Llano uplift in central Texas and the Makhavinekh Lake Pluton contact aureole in Labrador. Some questions have been raised about Carlson's estimates of pressure corrections and compositional dependencies (Ganguly 2010), and also temperatures assumed for the Makhavinekh contact aureole are higher than reported from mineral equilibria (McFarlane and Harrison 2006), which might introduce a systematic error of as much as 4% in $1/T$. Although it is difficult to account quantitatively for all sources of error when using natural data, such that corresponding errors in T and D may be large, there is still remarkable agreement among experimental and empirical data for Mg, Fe and Mn (Fig. 10A). Internally consistent constraints have also been derived for diffusion of trivalent cations (Carlson 2012). Experimental and empirical estimates of Ca diffusivities remain unusually discrepant, however (~4 orders of magnitude at 1000 °C; Fig. 10B). Data scatter for Ca diffusivity overlaps data for Mg, Fe, and Mn, but natural data have long demanded much lower D's for Ca than for Fe, Mg and Mn, in fact likely more similar to trivalent cations and oxygen (e.g., Spear and Kohn 1996; Kohn 2004; Vielzeuf et al. 2005). But exactly how much lower remains unknown. Regardless, even where absolute values of D cannot be constrained, relative D's can sometimes be inferred (Qian et al. 2010; but see also Till et al. 2015).

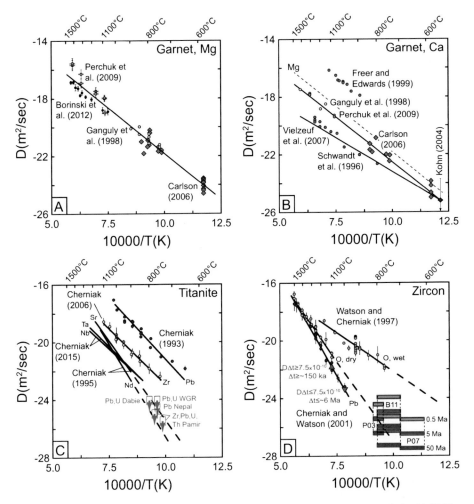

Figure 10. Constraints on diffusivities from experiments (circles) and natural data (other symbols). (A–B) Mg and Ca diffusion in garnet normalized to 1 GPa and f_{O_2} at the C–CO buffer, showing good internal consistency for Mg and large discrepancies for Ca. Borinski et al. (2012) data include reinterpretation of high-T experimental data from Ganguly et al. (1998), and Ganguly et al. (1998) correct low-T data of Cygan and Lasaga (1985) and Chakraborty and Rubie (1996) for f_{O_2} and pressure. Solid lines show general trends of all data (two illustrative lines for Ca). Maximum D for Ca must be lower than Mg D (Kohn 2004). (C) Diffusion data for titanite, highlighting discrepancies between experimental vs. natural data for Zr and Pb (Kohn 2017). Lines show trends of experimental data, dashed where extrapolated. (D) Diffusion data for Pb and O in zircon. Lines show trends of experimental data, dashed where extrapolated. Large triangles for Pb show constraints on $D\Delta t$ values and implications for minimum and maximum durations of peak metamorphism for Kokchetav and Bohemian massifs. Boxes show constraints on oxygen D for different durations of metamorphism (50, 5 and 0.5 Ma). P03 = Peck et al. (2003; temperatures adjusted for data in Storm and Spear 2005, 2009), P07 = Page et al. (2007; temperature adjusted for Bohlen et al. 1985), B11 = Bowman et al. (2011).

Titanite. In many rocks, titanite U–Pb ages are younger than zircon U–Pb ages, prompting most early researchers to conclude that Pb diffusion is sufficiently fast that titanite records only a cooling age, not a crystallization age (e.g., Mattinson 1978; Mezger et al. 1991; Heaman and Parrish 1991). Experimental data also suggest relatively high D's, implying typical closure temperatures of ~600°C (Cherniak 1993). Nearly every subsequent chronologic study has assumed Pb diffusion is fast, and that ages are reset. Curiously, the high diffusivity for Pb contrasts starkly with the low diffusivity of Sr (Cherniak 1995), which is also divalent, substitutes into the same crystallographic site, and has a nearly identical ionic radius (1.21Å for Sr and 1.23Å for Pb; Shannon 1976). An increasing body of data from diverse orogens now suggests that Pb diffuses much more slowly, likely as slowly as Sr (Fig. 10C; see summary in this volume of Kohn 2017). In fact, diffusivities of Pb, Zr, U and Th all appear comparable to Sr and Nd, and 2–4 orders of magnitude lower than experimental estimates for Pb and Zr (Fig. 10C; Kohn 2017). The consistency of empirical Pb and experimental Sr diffusivities accords with their respective charges and radii, whereas low empirical diffusivities for Zr, U and Th are consistent with their high ionic charges and substitution either into a small crystallographic site (Zr) or via coupled chemical exchanges (U and Th; Brady and Cherniak 2010; Cherniak 2010b).

Explanations for the discrepancies between experimental and natural data are lacking. Many of the Pb diffusion data were collected using a Pb implantation method, which damages the crystal lattice, but the damage quickly anneals and experiments using a PbS exchange medium yielded indistinguishable D's (Cherniak 1993). The Zr diffusion experiments show no dependence of D on Zr source or f_{O_2}. Still it would appear the experiments tap a fast-diffusion pathway not seen (yet) in nature. New experiments with different starting materials or methods seem warranted.

Zircon. Zircon is commonly assumed to be extremely resistant to diffusional modification, but is this rule of thumb always true? Whereas experimental data for Pb and O diffusion (dry) indeed suggest extremely low D's, wet experiments suggest higher D's for O (Watson and Cherniak 1997; Cherniak and Watson 2001), and modeling of H_2O fugacity in metamorphic rocks during cooling suggests that oxygen should diffuse in most minerals at rates commensurate with wet diffusion experiments (Kohn 1999). Do natural data show the expected widespread resetting of O, but retention of Pb?

Although few rocks retain such high temperatures for sufficiently long times that significant Pb diffusion is expected, resetting of Pb is possible for zircons from the Kokchetav and Bohemian massifs, depending on assumed durations at near-peak temperatures. In evaluating these data, a useful benchmark is that 90% diffusive resetting of the center of a sphere will occur at $D\Delta t / r^2 \geq 0.3$ (r=radius, Δt=duration at peak metamorphic T; Crank 1975), or $D\Delta t \geq 7.5 \times 10^{-10}$ m^2 for the core of a 50 μm radius crystal. For both massifs, assuming D from experiments, we can identify the timescales over which U–Pb ages will be preserved vs. erased. In the Bohemian massif, core, mantle and rim domains in zircon, interpreted as prograde, peak, and high-T retrograde growth, show no age differences (Kotková et al. 2016), possibly indicating resetting. Temperatures reached ~1100°C at UHP conditions, and retrograde overprinting occurred at ≥1050°C in the granulite facies (Haifler and Kotková 2016). At 1100°C, using D's from Cherniak and Watson (2001; "dry" experiments), 90% resetting would occur for durations of only ~150ka. Models of subduction and UHP metamorphism suggest that maximum temperatures may be retained for 1–2 Ma (Gerya et al. 2002; Warren et al. 2008), so resetting of zircon U–Pb ages in the Bohemian massif is therefore expected. In the Kokchetav massif, peak temperatures of 1000°C were insufficient to completely reset prograde and peak metamorphic zircon domains (Hermann et al. 2001; Katayama et al. 2001; Stepanov et al. 2016), implying ≤90% Pb loss. If experimental D's are correct, this requires $\Delta t \leq$ ~6 Ma, which can be reconciled with the 1–2 Ma peak metamorphic duration that some subduction models imply. Thus, retention of U–Pb ages at Kokchetav is possible. Whereas a shorter duration at peak temperature would not constrain diffusivities, a longer duration would imply slower diffusion ($D < \sim 10^{-23}$ m^2/s at 1000 °C, Fig. 10D). In sum, we know of no natural data that directly constrain Pb diffusivities, but additional petrochronologic analysis in a few exceptional complexes such as the Kokchetav and Bohemian massifs might provide limits. Diffusion profiles in other elements might be compared with U–Pb ages to determine diffusion rates relative to Pb.

Attempts to reconcile experimental vs. empirical oxygen diffusion rates in zircon have proved difficult. Microanalytical data reveal diffusion gradients in isotopic composition on scales as small as a few μm (Adirondacks: Peck et al. 2003; Page et al. 2007; Superior Province: Bowman et al. 2011), suggesting very slow diffusion. Unfortunately, durations at peak metamorphic conditions are known poorly for the areas studied. An example of ambiguity for the Adirondacks is discussed above. For the Superior Province, the duration near peak granulite-facies conditions is constrained only to $\leq 27 \pm 3$ Ma, based on bracketing ages of deposition (2667 ± 2 Ma) and of dikes that exhibit amphibolite-facies margins (2640 ± 2 Ma; Krogh 1993). Regardless of these details, Bowman et al. (2011) report data that seem irreconcilable with wet diffusion experiments: the duration of regional metamorphism would have to be unrealistically brief, 500–5000 years (Fig. 10D).

Some studies have proposed that a hydrous fluid must be present to produce oxygen isotope resetting, and that oxygen diffusion may be slow in fluid-free but high-f_{H_2O} rocks (Peck et al. 2003; Bowman et al. 2011). However, the compositional systematics of retrograde resetting of oxygen isotope compositions in quartz, feldspar, micas and oxides suggests that high "wet" diffusion rates apply, even in granulites that likely lacked a free fluid phase (Kohn 1999). That is, rocks may be physically dry during cooling but are thermodynamically buffered to relatively high f_{H_2O} and experience resetting as if they were wet. In addition, dehydration during prograde metamorphism should produce a fluid during the heating portion of each rock's *P–T* path. Unless heating rates exceeded ~100 °C/Ma, zircon δ^{18}O should be reset. A simpler, albeit more radical, interpretation is that the wet diffusion experiments are somehow biased. Growth of thin rims on zircon, which would lengthen apparent profiles and increase apparent *D*, is not supported (E. B. Watson, pers. comm. 2016). As suggested for titanite, perhaps the experiments tap a fast-diffusion pathway that is less effective in nature. That is, although f_{H_2O} does control oxygen resetting in cooling rocks (Zhang et al. 1991; Kohn 1999), natural wet diffusivities for oxygen in zircon are simply lower than measured experimentally, much as was inferred for Zr and Pb in titanite. One possible resolution to this quandary would be to re-investigate oxygen diffusion rates in zircon using fluid-absent, but high-f_{H_2O} experiments.

Dolomite–ankerite. Dolomite cores with ankerite overgrowths provide a natural diffusion couple for investigating the duration of biotite- to garnet-grade metamorphism in the northern Appalachiansm, USA (Ferry et al. 2015). Regional petrology suggests that ankerite rims grew before peak metamorphism, so the concentration profiles of Fe/Mg between carbonate cores and rims provide a measure of the maximum duration near peak metamorphic conditions. Experimentally determined diffusion rates for Fe and Mg in carbonates (Müller et al. 2012) imply durations at peak metamorphic conditions of ~1 year, which is clearly implausible. One complicating factor is f_{O_2}, which was up to 20 orders of magnitude lower during metamorphism than in experiments (Ferry et al. 2015). Most diffusion rate f_{O_2}-dependencies are fairly small: $D \propto f_{O_2}^a$ where the exponent *a* is typically ~1/6 (Chakraborty and Ganguly 1992; Dohmen et al. 2007, 2016), but can be as high as 1/5 to 1/3.2 (see Chakraborty 2010). Thus, corrections would increase calculated metamorphic durations by 3–6 orders of magnitude or between ~1 ka and ~1 Ma, within the range of other estimates in the same region ($\leq \sim 1$ Ma; Spear et al. 2012; Spear 2014). More accurate estimates of the duration of metamorphic events will require further investigation of Fe–Mg diffusion rates and their f_{O_2} dependence in ankerite and dolomite.

Geospeedometry from major element cation zoning

Major element zoning within garnet rims from the Valhalla Metamorphic Complex, British Columbia, Canada, has been used to estimate cooling rates in this high-grade metamorphic core complex (Spear and Parrish 1996; Ducea et al. 2003; Spear 2004). In these rocks, Fe/Mg increases towards the rims of garnets due to retrograde diffusional reequilibration with the matrix, and this zoning has been used to interpret cooling rates between ~825 and ~550 °C (Figs. 11A–B; Ducea et al. 2003; Spear 2004). These data exemplify petrology's central role in constraining models that are used to estimate cooling rates: differences in the retrograde reactions that are assumed to produce the zoning (ReER's vs. ReNTR's) radically change estimated rates and consequent tectonic interpretations of these rocks. In all models, the change in composition (mole fraction) with respect to time (dX/dT) was determined by varying cooling rate (dT/dt), and by using a temperature–composition path (dX/dt) to establish the equilibrium boundary condition on the rim of the garnet. The thermodynamics of the assumed governing reaction—whether ReER or ReNTR—define dX/dT. In both studies, the diffusivities of Chakraborty and Ganguly (1992) were used in a finite difference diffusion model to extract the change in composition over distance within the garnet.

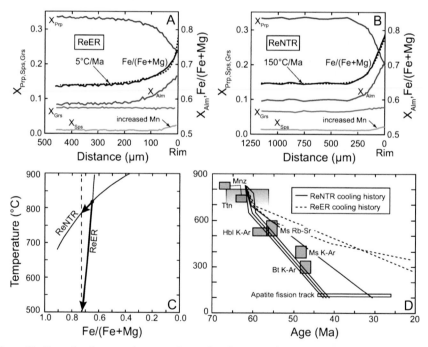

Figure 11. Example of geospeedometry using major element zoning. (A–B) Data and models for two garnets from the same sample from the Valhalla metamorphic complex, British Columbia, analyzed and interpreted in two different studies: Ducea et al. (2003) and Spear (2004). Towards garnet rims, X_{Alm} and Fe/(Fe+Mg) increase dramatically, X_{Prp} decreases dramatically, X_{Sps} increases, and X_{Grs} either remains flat or increases slightly. Prp = pyrope, Sps = spessartine, Grs = grossular, Alm = almandine. (A) A ReER model with a cooling rate of 5 °C/Ma (dashed black line) fits data (Ducea et al. 2003). (B) A ReNTR model with a cooling rate of 150 °C/Ma fits a similar set of data (Spear 2004). (C) Temperature vs. composition curves for different equilibria: Fe/(Fe+Mg) changes rapidly with temperature when governed by ReNTR equilibrium and slowly when governed by ReER equilibrium. Modified from Spear (2004). (D) Temperature vs. time paths determined from garnet zoning models and from independent petrochronologic and thermochronologic data. Modified from Spear (2004); most chronologic data from Spear and Parrish (1996). Mnz = monazite, Ttn = titanite, Hbl = hornblende, Ms = muscovite, Bt = biotite.

Ducea et al. (2003) calculated relatively slow cooling of ~5 °C/Ma, assuming a simple Fe–Mg ReER between matrix biotite and garnet rims (Fig. 11A). However, close examination of minerals found at garnet rims, chemical zoning within garnets, and textural examination of garnets demonstrate operation of a ReNTR (Spear 2004). For example, the mineral assemblage sillimanite + biotite + plagioclase + quartz occurs along garnet rims, and is likely a retrograde product after the reactants garnet + melt + K-feldspar (Spear 2004). Increases in Mn towards the rims of garnets (Fig. 11A–B; Ducea et al. 2003; Spear 2004) further indicate operation of a ReNTR because the garnet–biotite ReER affects only Fe–Mg, not Mn (Spear and Florence 1992; Spear 2004). Assuming that the retrograde reaction garnet + melt + K-feldspar = sillimanite + biotite + plagioclase + quartz defined the boundary condition at the garnet rim, an initial cooling rate of ~150 °C/Ma for < 1 Ma was calculated (Fig. 11B; Spear 2004; the reported range of possible cooling rates was 100–500 °C/Ma).

The main reason calculated cooling rates differ so substantially between studies is that the ReNTR yields a radically different equilibrium temperature-composition path (much larger dX/dT in garnet) than the ReER (Fig. 11C). This $T–X$ sensitivity requires that the measured change in garnet rim compositions occurred over a much smaller temperature interval and at higher temperature: $\Delta T \sim 75\,°C$ between 825 and 750 °C for the ReNTR vs. $\Delta T \sim 275\,°C$ between 825 and 550 °C for the ReER (Fig. 11C; Spear 2004). Recalling that the length of a diffusion profile generally scales as $(D \cdot \Delta t)^{1/2}$, and that D increases exponentially with increasing temperature (Eqn. 3), average D for a high-temperature ReNTR must be far larger than for a more protracted ReER (Fig. 11C). A larger D implies a smaller Δt and higher cooling rate. This result does not reflect any complications resulting from a moving boundary such as resorption of the garnet rim. Although resorption must have occurred (different profiles are not identical), dX/dT for the ReNTR is simply larger (Fig. 11C), driving a higher calculated cooling rate.

Modeling of Fe–Mg exchange between biotite inclusions and garnets refines the later-stage cooling history. Although faster initial cooling during operation of the ReNTR produced most of the Fe–Mg–Mn profile observed at garnet rims, the ReER continued at lower temperature. Compositions of biotite inclusions within garnet are sensitive to this later, lower temperature cooling and Fe–Mg exchange. These data imply lower cooling rates of ~25 °C/Ma for several Ma between ~600 and ~500 °C (Fig. 11D; Spear and Parrish 1996; Spear 2004).

A wealth of chronologic data permits comparison of different cooling rate models to an independently determined $T–t$ history (Fig. 11D). Spear and Parrish (1996) summarize most data, which we updated for muscovite and titanite closure temperatures (Jenkin 1997; Harrison et al. 2009; Kohn 2017); Spear (2004) also reevaluated monazite ages in a petrochronologic context. Altogether, these independent constraints on cooling history agree better with the ReNTR model for garnet rim zoning than with the ReER model (Fig. 11D; Spear 2004). As discussed in Spear (2004), rapid cooling is the expected consequence of rapid thrusting along a ramp. Thrust rates of cm/yr along ramps of ~20° slope should cool rocks at rates of ~100 °C/Ma or greater. Such rapid cooling may be more common in orogens than ReER-based geospeedometry would imply.

Geospeedometry from trace element zoning

A key conceptual outcome of Spear (2004) is that defining the temperature dependence of reactions and other boundary conditions governing rim concentrations plays a crucial role in interpreting data—different petrogenetic models can lead to radically different cooling rates. The same cautionary principles apply to trace element zoning—new data from amphibolite-facies rutile crystals from Catalina Schist amphibolites, California, suggest that different crystals of the same mineral in a single rock experience different boundary conditions during cooling (Kohn et al. 2016). Because rock-wide equilibrium is not maintained, a cooling rate cannot be calculated. Rather, diffusion profiles reflect highly localized equilibria and transport phenomena.

Kohn et al. measured near-rim trace element zoning in LA-ICP-MS depth profiles from separated rutile grains (Fig. 12A). Although profiles for a specific element can be similar, many profiles show boundary concentrations and trends that deviate strongly from theoretical expectations. For example, nearly all models of diffusive resetting follow Dodson (1973, 1986) in assuming the crystal surface equilibrates with the matrix throughout cooling. If so, Zr concentrations and Zr-in-rutile temperatures should trend towards zero. Instead only one profile conforms with this expectation (profile *a* in sample A14-57b, Fig. 12A), and all other profiles within the same sample and in other samples show markedly flatter trends (Fig. 12A, B). Moreover, crystals that have essentially indistinguishable Zr profiles (Fig. 12B) can have completely different Nb profiles (Fig. 12C), implying that Zr and Nb boundary conditions vary on small scales. Clearly if a rock is fully equilibrated during cooling, profiles should be consistent among grains.

A single model of diffusive resetting cannot explain all data. For example, crystal orientation (parallel vs. perpendicular to the *c*-axis) or spot proximity to crystal tips might explain some disparities (e.g., profiles *a* and *d*, Fig. 12A), but some profiles imply completely different cooling rates, e.g., ~15 °C/Ma for profile *a* vs. ≫100 °C/Ma for profile *e* (Fig. 12D). The rock experienced only one cooling history, so if different profiles imply different cooling rates, cooling rate is not reliably retrievable from these data. Kohn et al. (2016) proposed that flux limitations could explain many of the disparities among crystals. They modeled two scenarios— that the matrix supports only a maximum flux throughout its cooling history (defined in terms of the percent of the total flux experienced for an equilibrium model; Fig. 12E), and that the matrix not only undergoes high-*T* flux limitations, but also shuts down all equilibration at some cutoff temperature, arbitrarily taken as 550 °C (Fig. 12F). These models explain relatively flat profiles better (profiles *c* and *e*, Fig. 12D–F), as well as the trend towards a non-zero Zr concentration and Zr-in-rutile temperatures ≥600 °C on crystal rims (nearly all profiles: Fig. 12A–B).

Two results from this work warrant particular emphasis. First, the average profile says little about equilibrium behavior or cooling rates. Arguably, only profile *a* in sample A14-57b could be modeled to infer cooling history—it is the only one that undoubtedly trends towards zero Zr concentration, so can be modeled theoretically with the fewest assumptions. None of the profiles in sample A15-10 conform to simple theory, and instead require (as yet arbitrary) assumptions about proximal mass fluxes (Fig. 12E). Thus, the improved statistics of averaging more data would not improve confidence in cooling rate estimates. Second, with one exception (profile *a*) Zr data consistently indicate that there is a lower temperature at which equilibration among matrix minerals simply shuts down. Carlson (2012) concluded similarly—flattening of trace element zoning in the near-rim region of garnet crystals from the Makhavinekh Lake Pluton contact aureole is best explained if rim reequilibration ceased (trace element flux dropped to zero) at some elevated temperature. The two temperatures inferred in these studies differ considerably (~550 vs. ~770 °C), possibly reflecting physical differences in f_{H_2O} ("wetter" for Catalina, "drier" for Makhavinekh Lake) or rates of cooling (~25 °C/Ma for Catalina, up to 50 °C/Ma for Makhavinekh Lake). Clearly mineral rims do not always maintain equilibrium, and the circumstances under which they depart from equilibrium deserve further study.

Geospeedometry from chronologic zoning

Grove and Harrison (1999) were the first to attempt to quantify cooling histories using chronologic zoning. Their analysis of Th–Pb age zoning in Himalayan monazite using ion probe depth profiling implied cooling of ~100 °C/Ma, assuming diffusion rates from then-current experimental data (Smith and Giletti 1997). Subsequent experimental (Cherniak et al. 2004; Gardés et al. 2006, 2007) and empirical (McFarlane and Harrison 2006) investigations of Pb diffusion rates in monazite, however, suggest much lower diffusivities. The monazite age profile in these rocks more likely reflects late-stage growth rather than diffusional resetting during cooling.

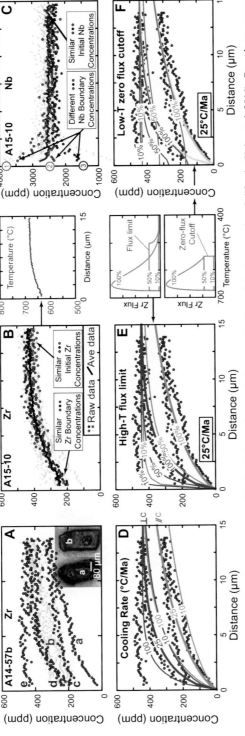

Figure 12. Trace element zoning and models for amphibolite-facies rutile crystals from the Catalina subduction complex, California (Kohn et al. 2016). Crystals were separated using conventional crushing, and only clean crystal faces were analyzed. Data were collected using LA-ICP-MS depth profiling (photomicrograph in (A)) Illustrative profiles are shown with darker symbols. (A–C) Some profiles show good internal consistency, but disparities in other profiles from the same sample require different boundary conditions during cooling. Even crystals with essentially equivalent profiles in one element (Zr) can show radically different profiles in another element (Nb). Inset [between (B) and (C)] shows temperature profile for averaged data; temperature does not approach zero at rim. (D–F) Models of Zr concentration vs. depth for sample A14-57b showing that no one model fits all data [profiles *a c, d* and *e* in (A)]. Diffusion parallel to the c-axis [lower curves in (D)] is about 10 times faster than diffusion perpendicular to c [upper curves in (D)]. Independent chronologic data suggest a cooling rate of ~25 °C/Ma for ca. 10Ma, but high-*T* cooling could have been faster or slower. Insets [between (E) and (F)] illustrate flux-limitation models. A minority of rutile grains were not flux limited, and their rim concentrations tend towards 0 ppm Zr (e.g., profile *a*) – these profiles correspond to most models of diffusional resetting. A majority of grains were flux limited and their concentration profiles are flatter (e.g., especially profile *e*).

A more recent example of diffusion-induced age zoning shows decreasing U–Pb ages towards the rims of rutile crystals separated from a high-grade gneiss from the Ivrea Zone, southern European Alps (Fig. 13; Smye and Stockli 2014). Multiple depth profiles were collected using LA-ICP-MS. All depth profiles show ages that increase from ~150 Ma near the rim to 170–200 Ma towards cores (Fig. 13A). Averaged data were inverted using a (Markov chain) Monte Carlo method with goodness of fit assigned using metrics of Ketcham (2005). Pooled inversions yielded a best-fit *T–t* history and confidence intervals, suggesting rapid cooling between ~185 and ~175 Ma, with slower cooling thereafter (Fig. 13B).

Although not emphasized in the original study, depth profiles from different rutile grains actually yield statistically distinct age trends, much as was observed for trace element zoning in Catalina Schist rutile. For example, although all profiles converge towards rim ages of ~150 Ma, some profiles between ~10 and ~30 μm display ages of ~200 Ma vs. ~170 Ma without any age overlap (Fig. 13A). Taken individually, these profiles imply quite different cooling histories, well outside the statistical bounds inferred for the averaged data (Fig. 13B). Because rutile crystals were collected from the same rock, they should all yield indistinguishable age profiles and *T–t* histories, whereas clearly they do not. Either some crystals were growing during cooling (so the profiles do not reflect diffusion alone), or the boundary conditions for Pb diffusion were not identical among different grains. Collecting trace element data, especially *T*-sensitive Zr, simultaneously with U–Pb ages might help resolve some of these ambiguities. Although Zr and Pb do not diffuse at the same rate (Cherniak 2000, Cherniak et al. 2007a), differences in boundary conditions or possible rim growth during cooling might be identifiable from the shapes of profiles. For example, a broader and younger profile in U–Pb age might correspond with lower Zr concentrations, suggesting late-stage, lower-*T* growth.

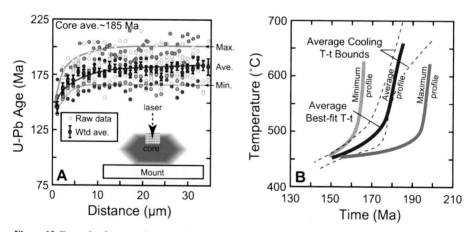

Figure 13. Example of geospeedometry using chronologic zoning. Age profiles and *T–t* histories determined via depth profiling of rutile crystals (each grain represented by different colored symbols) from a single rock collected from the Ivrea Zone, southern European Alps. Raw data from Smye and Stockli (2014). Modified from Kohn (2016) and Kohn et al. (2016). (A) U–Pb age profiles show decreasing ages towards rim, but several profiles are analytically distinct. Although rim ages appear to converge, boundary conditions could not have been equivalent among grains during initial cooling. Inset schematically illustrates how depth profiling data are collected using LA-ICP-MS. Min = minimum, Max = maximum, Wtd ave = weighted average. (B) Average cooling history (dark curve) and 95% confidence limits (dashed curve) calculated using a Monte Carlo approach and diffusion rates of Cherniak (2000). From Smye and Stockli (2014). Maximum and minimum *T–t* histories are more schematic and are based on bounding profiles from Figure 13A.

Reaction rates

Lucassen et al. (2010) introduced the geosciences to the mathematics of steady-state reaction as a basis for determining reaction rates (Jackson 2004), and provided data from an amphibolitized eclogite that can be used to test model predictions. Rutile porphyroblasts grew under eclogite-facies conditions in these rocks (~700 °C, 2 GPa), then titanite overgrew rutile in the amphibolite facies (~700 °C, 1 GPa; Ravna and Roux 2006; Fig. 14A, B). Determination of reaction rates assumes equilibrium between titanite and rutile (Fig. 14C, see also Fig. 8B, D), but the titanite and rutile may not have been in equilibrium with the matrix (forming chemically zoned titanite: Jackson 2004; Fig. 8B), or they could both have equilibrated with the matrix (forming homogeneous titanite; Fig. 8D). We refer to these different models as "closed system" when Ti and trace elements are wholly conserved between rutile and titanite and do not exchange with the matrix (Fig. 8B), and "open system" when titanite and the rutile grain boundaries completely equilibrate with the matrix (Fig. 8D). The open system model implies that the matrix can serve as a source or sink of major and trace elements. The molar volume of rutile is approximately 1/3 that of titanite, and the mass ratio is ~40%.

The closed system model predicts that a Zr compositional minimum should form in titanite 1/3 of the distance from the rutile–titanite interface to the edge of the titanite. This composition reflects that of the first titanite to form in equilibrium with rutile at the original position of the rutile crystal edge. If rutile has an initial concentration C_0, first-formed titanite should have a concentration $k \cdot C_0$ (Fig. 8B), where k is the trace element partition coefficient between titanite and rutile (at $T \sim 700$ °C, $k \sim 0.08$ for Zr as measured in ppm by weight; Watson et al. 2006, Hayden et al. 2008). To maintain mass balance, steady-state titanite should have a trace element concentration $0.4 \cdot C_0$ (Fig. 14C), and the rutile crystal edge should have concentration $0.4 \cdot C_0 / k$ (or, for Zr, $\sim 0.4 \cdot C_0 / 0.08 = 5 \cdot C_0$). A compositional well develops as titanite progressively grows inward and outward from the original titanite–rutile interface, with a composition that evolves from $k \cdot C_0$ (~$0.08 \cdot C_0$ for Zr) to $0.4 \cdot C_0$.

Under open system conditions with whole-rock equilibration, titanite trace element concentrations should be homogeneous and consistently reflect the temperature of reaction, whereas the rim of the rutile should adopt a commensurate composition (Fig. 14C). Because Zr contents of rutile increase with decreasing pressure (Tomkins et al. 2007), a rutile rim in equilibrium with the matrix should adopt a higher Zr concentration than in the core, inducing a diffusion profile.

Data from rutile crystals indeed show that trace element concentrations increase towards rutile–titanite interfaces (Fig. 14D, E), as expected from either closed- or open-system equilibrium models (Fig. 14C). Such increases are not limited to the more abundant trace elements (Zr and Nb), and include low-abundance geochemically similar elements (Hf and Ta). In detail, however, several observations contradict the closed-system model: titanite does not exhibit a compositional well, Zr concentrations are 2–3 times too high, and Nb concentrations are slightly too low (Fig. 14D, E). Lucassen et al. (2010) recognized the problems with mass conservation, noting that there was insufficient titanite developed adjacent to rutile to balance Ti, and that Zr concentrations in titanite were far too high to conform with Equation (15). Rather, Ti must have been exported to the matrix to produce additional titanite, while Zr may have been imported from zircon breakdown or possibly pyroxene to support the high Zr content of the titanite and produce the Zr diffusion profile (Lucassen et al. 2010). Major and trace element import–export is unsurprising because the reaction zone must import Ca and Si to form titanite from rutile. Significantly, the data also do not meet expectations of open system equilibration among rutile, titanite and matrix minerals. Zr-in-titanite and Zr-in-rutile temperatures (at the edge of the rutile) cannot be reconciled, even accounting for potentially lower a_{SiO_2} and a_{ZrSiO_4} at the reaction interface.

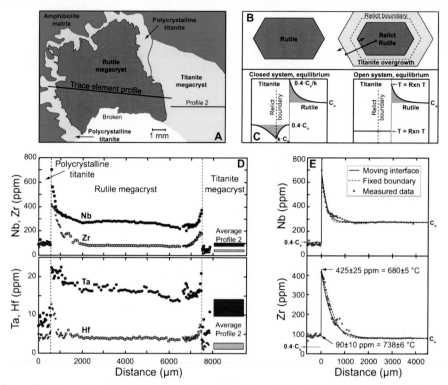

Figure 14. Data and models relevant to steady-state reaction and induced diffusion profiles. All data and model fits from Lucassen et al. (2010). (A) Sketch of coarse rutile with titanite overgrowths, amphibolitized eclogite, Tromsø Nappe, Norway. (B) Sketch illustrating growth of titanite at the expense of rutile, with associated volume change due to the difference in molar volumes. Small arrows show direction of titanite overgrowth. (C) Schematic models of reaction-induced diffusion profiles for rutile with titanite overgrowths. Equilibrium is maintained at the rutile–titanite interface, but may be closed system (does not equilibrate with the matrix; Jackson 2004; Fig. 8B) or open system (equilibrates with the matrix; Fig. 8D). For open system equilibration, trace element concentrations at the titanite–rutile boundary should correspond to equilibrium concentrations for the temperature of reaction. (D) Data from trace element profiles in Figure 14A, showing steep gradients near rutile rims, and relatively flat profiles in titanite. The spatial distribution of analyses for Profile 2 was not reported so an average and 2 s.e. is shown for comparison with model predictions. Different size data points were collected with different spot sizes and spatial resolutions. (E) Detail of left-hand sides of profiles in Figure 14D, fitted to a moving interface solution (Eqn. 15) or a fixed boundary condition (error function). Concentrations of Nb and Zr in titanite do not conform with expectations of a closed-system model, implying that the matrix served as a source or sink of trace elements. Interface concentrations of Zr do not yield the same temperature, implying disequilibrium partitioning.

Overall, no steady-state reaction model appears completely reconcilable with these data. While a steady-state solution fits the rutile diffusion profile reasonably well, so, too, does a fixed boundary (error function) solution (Fig. 14E). Physically, this means that reaction rates could have been so fast that no diffusion occurred in rutile during reaction, and only afterwards did a fixed boundary condition induce a diffusion profile (Fig. 8E). Thus it is impossible to know whether the diffusion profiles developed during the reaction of rutile to form titanite, or were superimposed subsequently (Lucassen et al. 2010). Additional profiles collected for several other examples of titanite overgrowths on rutile, including rocks of the Tromsø nappe, Norway (Cruz-Uribe et al. 2014), show similar trends and incompatibilities with steady-state models. The consistent shape of diffusion profiles documented by Lucassen et al. (2010) and Cruz-Uribe et al. (2014) does suggest some consistent petrologic process, but despite good fits to diffusion profiles using Equation (15), calculations of reaction rates should be viewed with caution.

Timescales of magmatic processes from zoning in crystals

Major and trace element profiles across igneous phenocrysts, including olivine, feldspar, and quartz, constrain timescales of magmatic processes including phenocryst residence times, duration of magma mixing events, and the time between a magma rejuvenation event and eruption. Costa et al. (2008) summarize theory, applications, and potential pitfalls in using diffusion profiles to constrain timescales of magmatic processes. Recent investigations have taken advantage of *in situ* methods, especially NanoSIMS. This increasingly high chemical and spatial resolution affords new insight into very short-duration events.

Minerals often develop oscillatory zoning as they grow from a melt, sometimes because of extrinsic factors like magma convection, mixing, or rejuvenation, but also because of localized undercooling and intrinsic mineral growth kinetics (see summary of Shore and Fowler 1996). Diffusional smoothing of potentially sharp oscillatory boundaries allows calculation of crystal residence times. The scale of zoning can be quite small, however, so quantifying the original compositional step and measuring compositional gradients accurately can pose a major analytical challenge. Oscillatory-zoned quartz phenocrysts from a Jurassic rhyolite of the El Quemado Complex (Chon Aike large igneous province), Patagonia, provide an opportunity to constrain crystal residence times in a silicic system (Fig. 15A, B). Although Ti diffusion in quartz is not strongly anisotropic (Cherniak et al. 2007b), for consistency Seitz et al. (2016) first imaged crystals using microtomography, then sectioned all crystals perpendicular to their c-axes. NanoSIMS measurements of Ti concentrations along traverses that cross boundaries between oscillatory bands reveal smoothly varying changes in Ti over scales of less than ten microns (Fig. 15B; Seitz et al. 2016). Diffusion modeling of several profiles consistently suggests residence times of four to six years for these quartz crystals in the magma.

Overgrowths on igneous phenocrysts can also be used to calculate timescales of magmatic processes, especially the time between magma rejuvenation and eruption, but with potential complications. Some common crystals, such as olivine, exhibit diffusional anisotropy (diffusion of divalent cations parallel to the *c*-axis is ~6 times faster than parallel to the *a*- and *b*-axes; Chakraborty 2010) while the 3-dimensional morphology of other crystals, such as plagioclase, can influence diffusion profiles (Costa et al. 2003). When diffusion profiles develop on scales approaching a significant proportion of a crystal, 2- and 3-dimensional models of diffusion may become necessary as errors in applying 1-dimensional models can range from factors of 0.1–25 (Shea et al. 2015a). Examples of 2- and 3-dimensional modeling of Fe–Mg zoning in olivine indicate timescales of months to years between magma rejuvenation and eruption (Costa and Dungan 2005; Shea et al. 2015b). The study of Shea et al. (2015b) is novel in using P zoning (P is virtually inert to diffusion) to identify the original core–rim boundary, which is needed for modeling, but difficult to identify from the distribution of fast-diffusing Fe and Mg alone.

At a finer scale, zoning in Ba, Sr, and Mg was measured by NanoSIMS across core-rim overgrowths from sanidine crystals of the Scaup Lake rhyolite, Yellowstone caldera (Fig. 15C–D; Till et al. 2015). Like the study of Shea et al. (2015b), analysis of multiple elements with different diffusivities confers significant modeling and interpretational advantages. Specifically, Ba diffuses about 1000 times more slowly than Mg, and about 70 times more slowly than Sr (Cherniak 1996, 2002; LaTourette and Wasserburg 1998). Thus, if overgrowths formed with sharp compositional steps in all elements, the steepness of the compositional gradient should increase from Mg to Sr to Ba, and diffusion modeling of the gradient should yield the same apparent duration. Not so, however! Assuming an initial compositional step, gradients imply quite different timescales, ranging from ~1000 years (Ba) and ~15 years (Sr) to ~0.8 years (Mg; Fig. 15D). If diffusion occurred for 1000 years (to match the Ba profile), Mg and Sr profiles should be considerably flatter, whereas if diffusion occurred for 0.8 years (to match the Mg profile), Ba and Sr profiles should be considerably steeper (Fig. 15D). Evidently, concentrations of these elements changed continuously, not discontinuously, as overgrowths formed and the melt evolved. Crystal growth produced compositional gradients,

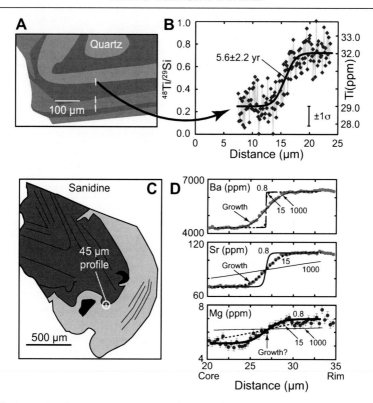

Figure 15. Timescales of magma chamber processes from zoning in crystals as measured using NanoSIMS. (A) Cathodoluminescence image of quartz phenocryst, El Quemado Complex rhyolite, Patagonia, showing oscillatory zoning. Crystal is sectioned perpendicular to its c-axis. Small vertical lines show locations of NanoSIMS traverses where Ti/Si was measured. Modified from Seitz et al. (2016). (B) NanoSIMS data showing calculated error function fit assuming an initial ~2 ppm step in Ti content. Width of symbols is equal to the spatial resolution of analyses. Modified from Seitz et al. (2016). (C) Sketch of backscattered electron image of sanidine phenocryst from Yellowstone caldera rhyolite, showing distinct rim that has overgrown core during magma rejuvenation. Tiny line within circle shows location of NanoSIMS traverse where Ba, Sr and Mg concentrations were measured. Thin lines illustrate compositional banding. Based on Till et al. (2015). (D) Trace element profiles along core–rim interface and diffusion models for 1000, 15, and 0.8 yr durations (numbers next to models). Although each profile can be fit assuming diffusional relaxation of a compositional step, similar compositional gradients when D differs by as much as 3 orders of magnitude imply irreconcilable timescales. The core–rim compositional change for Sr and Ba reflects growth, whereas the Mg profile probably reflects a combination of growth and diffusional relaxation (i.e., the model for Mg provides a maximum estimate of 0.8 Myr for the duration of diffusion). Data from Till et al. (2015).

not steps, and in fact the profiles for Sr and Ba must be virtually unmodified by diffusion. Each element provides a maximum time limit for diffusion—< 1000 years for Ba, < 15 years for Sr, and < 0.8 years for Mg—but the limit from Mg is most strict. Clearly if only Ba or Sr had been measured, a significantly different timescale would have been inferred. The resulting interpretation reveals much shorter timescales between the magma rejuvenation events and eruption (< 0.8 years; Till et al. 2015) and implies that geophysical evidence for increasing melt fraction beneath volcanoes may necessitate remarkably fast response to mitigate hazards.

Magma ascent rates from zoning in glass

Several studies have used diffusion profiles of volatile species in glasses to constrain rates of magmatic processes (e.g., Castro et al. 2005; Liu et al. 2007; Humphreys et al. 2008; Lloyd et al. 2014). Melt embayments in igneous phenocrysts provide an unusual opportunity to constrain magma ascent rates because they can trap melt, which will then attempt to expel volatile components such as H_2O and CO_2 during magma ascent and decompression. Bubble nucleation and dispersion in the melt outside the phenocryst should promote melt–bubble equilibrium in the magma body. For melt that is trapped in an embayment, however, volatiles must diffuse through the embayed melt to the nearest bubble to achieve equilibrium. Slow magma ascent allows the interior of embayed melt to equilibrate its volatile content, whereas rapid ascent does not. Thus, the H_2O and CO_2 content of the interior of embayed glass and the shape of the H_2O and CO_2 composition profiles within it can potentially constrain ascent rates (Liu et al. 2007; Humphreys et al. 2008; Lloyd et al. 2014).

Lloyd et al. (2014) measured composition profiles in major elements and volatiles (F, Cl, S, H_2O and CO_2) in basaltic glasses that were trapped in embayments in olivine recovered from basaltic tephra, Volcán de Fuego, Guatemala (Fig. 16A–B). During ascent, a bubble formed at the outlet of each embayment, establishing a boundary condition of steadily decreasing concentrations of H_2O and CO_2 on the margin of the embayed melt as the magma ascended and degassed. Decreasing trends of H_2O and CO_2 in the glass (Fig. 16B) were interpreted to reflect diffusion of volatiles towards the melt–bubble interface. Major element compositions also change systematically toward the embayment outlet, likely because of concurrent crystallization of olivine and pyroxene proximal to the bubble.

Modeling the H_2O and CO_2 profiles requires establishing initial H_2O and CO_2 contents of the original melt, modeling the pressure dependence of the volatile content of the melt in equilibrium with a gas bubble, and establishing the diffusivity of H_2O and CO_2. Contents of S, H_2O and CO_2 broadly correlate in melt inclusions from the same eruptive rocks (Lloyd et al. 2013), so the S content of the interior of the embayed glass, where it appears unaffected by diffusion, and the melt inclusion correlations for S–H_2O and for S–CO_2 were used to constrain the original H_2O and CO_2 contents. The boundary condition was modeled isothermally (1030 °C) as a function of pressure. Diffusivities were modeled at a fixed temperature but varied systematically with composition because H_2O diffusivity depends somewhat on major element composition (the glass is chemically zoned), and the diffusivity of CO_2 depends on H_2O content (Zhang et al. 2007).

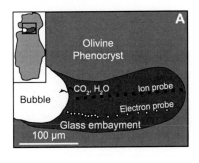
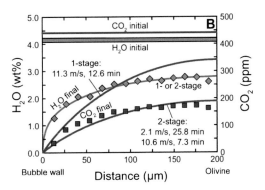

Figure 16. Timescales of magma ascent from zoning in glass (modified from Lloyd et al. 2014) (A) Sketch of backscattered electron image of olivine phenocryst with basaltic glass trapped in an embayment that opens to a gas bubble. Squares and dots show locations of ion probe and electron probe compositional analyses. Inset shows larger view of olivine crystal and its embayment. (B) Composition profiles of H_2O and CO_2 in embayment glass, showing gradual decreases towards embayment edge (bubble). Initial H_2O and CO_2 concentrations are based on melt S content in the context of melt inclusion compositions. H_2O profile can be fit with either a 1-stage or 2-stage ascent history, but CO_2 profile implies a more complex history with accelerating ascent.

In general, H_2O and S profiles can be explained with either a single-stage or multi-stage ascent history, whereas the low CO_2 content in the embayment interior but steep gradient near the bubble implies accelerating ascent. Some second-order effects that could be considered in future studies include loss of H_2O to the olivine crystal (although this should be much slower than diffusion through the melt; Gaetani et al. 2012) and concurrent changes in melt composition during ascent. Use of a fixed, chemically evolved profile, as is observed in the major element profile in the glass, probably underestimates ascent rates slightly because diffusivities decrease in more silicic melts (Zhang and Ni 2010). Because of the tradeoff between D and Δt in solutions to the diffusion equation, a smaller D assumed for the fully evolved melt leads to a slightly larger Δt and lower calculated ascent rate.

Duration of metamorphism

Once diffusion coefficients are known for minerals, the duration of metamorphic processes can be constrained by evaluating the variations in the elemental or isotopic composition along profiles across mineral grains. Ague and Baxter (2007) modeled variations in Sr concentrations in apatite from the classic Barrovian zones of northeast Scotland (Fig. 17A–B), as well as major-element zoning in garnet (Fig. 17C–D), to constrain the duration of thermal events during the regional metamorphism that accompanied the ~465 Ma Grampian orogeny. For both minerals, a compositional step was assumed, establishing an initial diffusion couple, and diffusion profiles were fitted to measured compositional gradients to estimate the duration of peak metamorphism. For apatite, inclusions in garnet show the same type of zoning as matrix grains, implying that rims formed during prograde metamorphism, and that compositional steps must have experienced the full extent of peak metamorphism. For garnets, a simple compositional step cannot be demonstrated, but gradual compositional changes during growth would lead to shorter estimates of the duration of peak metamorphism. The best fits to these diverse data range from < 100 ka to ~650 ka (Fig. 17A, 17D), regardless of metamorphic grade, suggesting that a very rapid thermal pulse (or pulses) was responsible for producing the observed regional metamorphism (Ague and Baxter 2007).

As Ague and Baxter (2007) emphasize, the data do not uniquely constrain the shape of the T–t path, rather they reflect the integral of the D–t curve. For example, T–t histories involving a single square pulse (model a, Fig. 17B), multiple instantaneous heat pulses with thermal relaxation (model b, Fig. 17B), or a more gradual heating-cooling history (model c, Fig. 17B) provide identical fits to the data (Fig. 17A), because their integrals of $D \cdot \Delta t$ are identical. Traditionally, crustal thickening has been thought to produce much longer timescales of metamorphism ranging from several Ma to several tens of Ma (e.g., England and Richardson 1977; England and Thompson 1984). Thus, many workers infer a contact metamorphic origin based on petrochronologic and thermochronologic data (Ague and Baxter 2007; Viete et al. 2013). This conclusion would be highly ironic insofar as "Barrovian metamorphism" is virtually synonymized with crustal thickening, not contact metamorphism, but would also accord with Barrow's original view that igneous intrusions, not crustal thickening, provided the heat to metamorphose these rocks (Barrow 1893; see also Harte and Hudson 1979). As an alternative to the view that brief thermal pulses require plutons, inversion of diffusion profiles from quartz inclusions (Ti) and garnet rims (Fe–Mg–Mn) suggests similarly brief durations of regional metamorphism in the northern Appalachians where intrusions are wholly lacking (Spear et al. 2012, Spear 2014). These results imply that brief regional metamorphic heating might result from discrete, tectonically driven events, e.g., from thin thrust slices or duplexing. Disentangling thin-skinned thrusting from other types of thermal pulses will require further petrochronologic investigations.

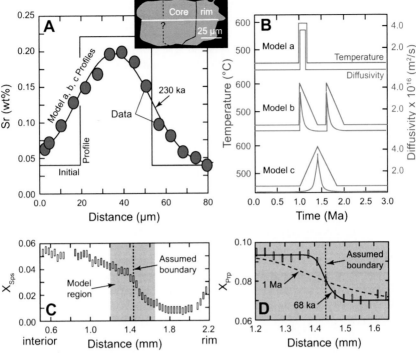

Figure 17. Diffusion profile-based estimates of the duration of metamorphism determined from zoning in apatite and garnet from Barrovian type region, Scotland. Modified from Ague and Baxter (2007). (A) Strontium zoning in apatite with 3 diffusionally equivalent models that correspond with different *T–t* histories. Backscattered electron image shows the core–rim boundary (dashed where inferred) and location of line traverse. (B) Models of duration of thermal pulses. Several different *T–t* histories yield equivalent extents of diffusional resetting. (C) Manganese zoning in garnet showing typical decrease outward from core, and increase at rim. Dashed line is location of assumed original compositional step. Gray = modeled region; Sps = spessartine. (D) Magnesium zoning in garnet from the modeled region, showing good fit for a duration at peak metamorphic conditions of ~68 ka, and poor fit for 1 Ma duration. Prp = pyrope.

Fluid–rock interactions

The duration and physical scale of fluid–rock interactions during metamorphism can be assessed by investigating whole-rock variations of fluid-mobile elements and isotopes. Most studies emphasize geochemical traverses that cross features such as veins or reaction zones that are associated with fluid–rock interaction. Such investigations focus on the record left behind by diffusion through a porous rock medium. Recent research emphasizes Li because it is highly soluble in metamorphic fluids, and its low valence and small ionic radius allow it to diffuse readily in a variety of materials. Lithium also enjoys the unusual characteristic that the large relative mass difference between its isotopes causes significant differences in ^6Li and ^7Li diffusivities. This leads to kinetic fractionation of Li isotopes, such that incomplete or unidirectional processes (e.g., diffusion along a chemical gradient, evaporation, etc.) transport ^6Li more rapidly than ^7Li and produce isotopic anomalies. The difference in diffusion coefficients between ^6Li and ^7Li is characterized by an empirical term β, defined as:

$$\frac{D_{6_{\text{Li}}}}{D_{7_{\text{Li}}}} = \left(\frac{m_{7_{\text{Li}}}}{m_{6_{\text{Li}}}} \right)^{\beta} \tag{16}$$

(Richter et al. 1999), where m is the mass of each isotope. In experiments, kinetic isotopic fractionations occur where ^{6}Li diffuses up to 3% faster than ^{7}Li in both silicate melt and water, causing fractionations of up to tens of permil (Richter et al. 2003, 2006). This difference in diffusion rates would not be detected in a concentration profile, but can be detected in profiles of measured isotopic ratios (δ^7Li). Thus, unlike more traditionally studied elements whose concentration gradients in rocks may be small (e.g., O), or whose isotopes diffuse at virtually the same rate (e.g., Sr), Li develops two different profiles—Li concentration and δ^7Li that permit simultaneous quantification of transport rates or durations.

Whole-rock Li concentrations and δ^7Li across fluid-related features such as veins (e.g., John et al. 2012), contacts with igneous rocks (Fig. 18A; e.g., Teng et al. 2006, Marks et al. 2007, Liu et al. 2010, Ireland and Penniston-Dorland 2015), or rock features diagnostic of fluid infiltration (e.g., relatively "dry" eclogite altered to "wet" blueschist, Penniston-Dorland et al. 2010; Fig. 18B–E) show isotopic fractionations up to ~30‰ and indicate transport distances up to tens of meters. An unusual, key feature of some data is that preferential transport of ^{6}Li means that δ^7Li values can shift more than the initial range between two rocks. For example, δ^7Li values of amphibolites next to the Tin Mountain pegmatite (Black Hills, South Dakota) approach −20‰, even though the initial amphibolite and pegmatite values were 0 to +10‰ respectively (Fig. 18A; Teng et al. 2006). Diffusion through an interconnected intergranular fluid, rather than through minerals, dominates transport of Li concentrations and isotopic compositions, which depends on time, effective diffusivity, permeability, temperature, and tortuosity (Fig. 5). Thus, if physical factors can be estimated, Li can be used as a geospeedometer to record the duration of fluid–rock interactions. Correlation of the variations in isotopic composition with petrologic indicators of fluid–rock interaction aids in the interpretation of such profiles. For example, in the alteration of eclogite to blueschist (Penniston-Dorland et al. 2010), the replacement of anhydrous omphacite and garnet in the eclogite by hydrous pumpellyite and chlorite in the blueschist (Fig. 18C–E) was associated with variations in Li concentrations and isotopic compositions.

Durations of fluid infiltration events inferred from Li concentration and isotope measurements may be quite short, for example only 10s to 100s of years during subduction metamorphism (Fig. 18B; Penniston-Dorland et al. 2010; John et al. 2012). Such pulsed flow may have implications for formation of melts in arcs, where melt transport may be pulsed on comparable timescales (Turner et al. 2001). A direct comparison of variations in oxygen and lithium isotopic compositions across a contact between the Bushveld Complex and metasedimentary rocks of the Phepane Dome demonstrates much faster diffusion of Li relative to O (Ireland and Penniston-Dorland 2015). Along this traverse, O isotopes vary over a distance < 1 m, while Li isotopes vary over a distance of ~100 m. These distances translate into estimates of a duration for the emplacement and cooling of the Bushveld Complex of 5 kyr to 5 Myr, depending on choices of diffusivities and porosities.

FUTURE DIRECTIONS

Ever improving *in situ* technologies have dramatically refined our ability to investigate diffusive processes in nature and constrain rates of petrogenetic processes. For example, NanoSIMS has increased spatial resolution (Fig. 15; Schmitt and Vazquez 2017) while LA-ICP-MS now allows composition profiles to be measured in ~1 minute (data in Fig. 12; Kylander-Clark 2017). Undoubtedly such analytical improvements will continue to advance petrochronologic research in the near future. Progress on the following topics would also help promote more accurate petrochronologic interpretations that invoke diffusive processes:

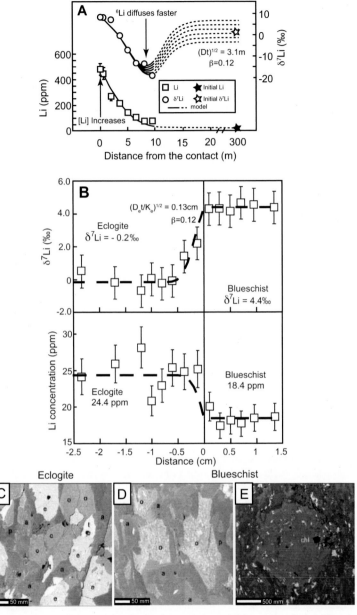

Figure 18. Constraints on fluid–rock interaction from whole-rock Li concentration and isotope profiles. (A) Data from amphibolites adjacent to Tin Mountain (South Dakota, USA) pegmatite along with diffusion model curves (modified from Teng et al. 2006). The different curves illustrate how the initial isotopic composition of the amphibolite does not strongly affect the model outcome. (B) Data across blueschist-eclogite contact in subduction-related tectonic mélange of the Franciscan Complex (California; modified from Penniston-Dorland et al. 2010). Constant Li and δ^7Li within blueschist layer indicate fluid infiltration parallel to the contact that altered the original eclogite to blueschist. Dashed lines show best-fit pinned boundary diffusion model. (C–E) Mineral textures support fluid-assisted alteration of eclogite. Eclogite retains relatively pristine minerals, including idioblastic garnet (not shown), but the blueschist contains altered omphacite (o), and chlorite replaces garnet. a = amphibole, chl = chlorite, e = epidote, p = phengite, t = titanite.

Boundary conditions

Accurately constraining boundary conditions is the single most important task if we are to obtain useful petrochronologic information from diffusion profiles. This problem can be framed in terms of the reactive reservoir of a rock: if mineral rims equilibrate with a reservoir, we must accurately identify its bounds (whole rock vs. proximal minerals; Lanari and Engi 2017) and composition (all minerals vs. some minerals, entire grains vs. partial grains, etc.; Baxter and DePaolo 2002; Sousa et al. 2013). In some instances (e.g., Nb in rutile), we do not know the reservoir and cannot even predict whether a trace element concentration should increase or decrease at a mineral rim during cooling. Yet many profiles are quite systematic, lending hope that we can usefully invert profiles and quantify petrologic processes. Developing methods for identifying the reactive bulk composition and for evaluating how it changes as a rock evolves will be key for interpreting such systematic changes in concentration. Although some theoretical models have been developed for specific problems during prograde metamorphism, for example Sr isotope evolution (Baxter and DePaolo 2002; Sousa et al. 2013), specific tests of assumed boundary conditions must be identified, such as the use of Mn profiles in garnet to assess ReER's vs. ReNTR's (Spear 1991; Spear and Florence 1992; Kohn and Spear 2000).

Empirical diffusivities

Diffusivities as measured experimentally and as determined from natural samples do not always correspond, e.g., Pb and Zr in titanite, O in zircon, and Fe–Mg in dolomite and ankerite. There is a continuing need to improve comparisons of natural and experimental systems to identify whether any experiments should be revisited. Errors in natural systems will always be large (only the most obvious discrepancies will be identifiable), but can be reduced by determining accurate timescales of metamorphism that are carefully evaluated in an appropriate petrologic context.

Controls on diffusivities

Applying diffusion models to constrain petrogenetic processes requires close understanding of the numerous factors that control D. The advent of trace element microanalysis makes this effort imperative because many trace elements have different valences than the major cations for which they substitute (e.g., REE^{3+} for divalent cations). Diverse mechanisms can accommodate the resulting charge imbalance, and D must depend on which other crystallographic sites are involved, and whether vacancies play a role. For example, D_{Li} in olivine depends on whether Li is sited octahedrally or interstitially (Dohmen et al. 2010). The diffusivity of trivalent REE in garnet probably depends on whether charge imbalance is accommodated through substitution into the cubic site (probably highest D), octahedral site (moderately low D), or tetrahedral site (lowest D). The proportionation of these substitutions probably depends on P–T conditions and bulk composition (mineral assemblage), so a full model of REE diffusion may well have to consider the full P–T–t and mineralogical evolution. How other intensive variables (e.g., f_{O_2} and chemical potential) affect defect densities and D is also poorly known.

Multiple element/isotope comparisons

Once we develop tight constraints on diffusivities for numerous elements and isotopes, measurement of poly-elemental profiles should constrain petrogenetic processes and their rates with increasing accuracy. Studies of igneous phenocrysts (Shea et al. 2015a,b; Till et al. 2015) illustrate the power of this approach, in that some elements closely reflect changes to magma chemistry while others more closely reflect rates of cooling and ascent. Analogously, if a cooling rate is proposed based on the diffusion zoning of one element or isotope in a mineral, diffusion profiles of all elements in that mineral should give the same cooling rate (within uncertainty), regardless of their diffusivities.

Tectonics

A major goal of diffusion-based petrochronology is to determine cooling histories and link them to tectonic processes. Several recent studies have inferred unexpectedly high cooling rates (Spear 2004, 2014; Ague and Baxter 2007; Spear et al. 2012), not all of them plausible (Ferry et al. 2015). Some cases implicate inadequate understanding of diffusion coefficients in natural systems (Ferry et al. 2015), but others are interpreted as resulting from brief thermal pulses, either magmatic (Ague and Baxter 2007) or tectonic (Spear 2004, 2014; Spear et al. 2012). A key goal of future research will be to determine whether the interpretations of brief pulses are real, rather than artifacts of poorly understood diffusive processes, and to examine their tectonic implications. Rapid thermal pulses imply either that magmas play a much greater role in heating the crust than petrologists have assumed, or that thin-skinned thrust processes are more common than structural geologists have inferred. The mounting conflicts between diffusion-based vs. thermal model-based rates of metamorphism suggest we may be on the verge of a scientific revolution (Kuhn 1962) that, when resolved, will provide a much better understanding of how crust is assembled.

Thermometry

Because diffusion is thermally activated (Eqn. 3) with an exponential dependence on temperature, tightly constraining maximum temperatures is crucial for accurate interpretations. The advent of trace element thermometry has improved the precision with which temperatures may be determined, but there are still lingering concerns about scales of equilibration and how much a thermometer may be reset during cooling. Further efforts are needed to hone the methods by which temperatures are determined.

ACKNOWLEDGMENTS

This research was funded by NSF grants EAR-1321897 and EAR-1419865 to MJK, NSF grant EAR-1419871 to SPD, and NSF grant EAR-1545903 to MJK and SPD. We thank Elias Bloch for sending model output for Lu and Hf diffusive resetting in garnet, Bruce Watson for helpful discussions, and Frank Spear, Rick Ryerson, and Martin Engi for especially insightful and helpful reviews. Lukas Baumgartner and the University of Lausanne are thanked for hosting SPD during much of the writing. We thank Frank for reminding us that we should never be in aReERs on our ReNTR's insurance.

REFERENCES

Ague JJ, Baxter EF (2007) Brief thermal pulses during mountain building recorded by Sr diffusion in apatite and multicomponent diffusion in garnet. Earth Planet Sci Lett 261:500–516

Aubaud C, Hauri EH, Hirschmann MM (2004) Hydrogen partition coefficients between nominally anhydrous minerals and basaltic melts. Geophys Res Lett 31, doi:10.1029/2004gl021341

Barrow G (1893) On an intrusion of muscovite–biotite gneiss in the southeastern Highlands of Scotland and its accompanying metamorphism. Quart J Geol Soc London 49:330–358

Baumgartner LP, Ferry JM (1991) A model for coupled fluid-flow and mixed-volatile mineral reactions with applications to regional metamorphism. Contrib Mineral Petrol 106:273–285

Baumgartner LP, Rumble D, III (1988) Transport of stable isotopes: I: Development of a kinetic continuum theory for stable isotope transport. Contrib Mineral Petrol 98:417–430

Baumgartner LP, Valley JW (2001) Stable isotope transport and contact metamorphism. Rev Mineral Geochem 43:415–467

Baxter EF, DePaolo DJ (2002) Field measurement of high temperature bulk reaction rates I: Theory and technique. Am J Sci 302:442–464

Behrens H, Johannes W, Schmalzried H (1990) On the mechanisms of cation diffusion processes in ternary feldspars. Phys Chem Miner 17:62–78

Beaumont C, Jamieson RA, Nguyen MH, Lee B (2001) Himalayan tectonics explained by extrusion of a low-viscosity crustal channel coupled to focused surface denudation. Nature 414:738–742

Beaumont C, Jamieson RA, Nguyen MH, Medvedev S (2004) Crustal channel flows: 1. Numerical models with applications to the tectonics of the Himalayan–Tibetan orogen. J Geophys Res 109, doi:B06406 10.1029/2003jb002809

Bear J (1988) Dynamics of Fluids in Porous Media. Dover Publications, New York

Bickle MJ, Baker J (1990) Advective–diffusive transport of isotopic fronts: An example from Naxos Greece. Earth Planet Sci Lett 97:78–93

Bickle MJ, McKenzie D (1987) The transport of heat and matter by fluids during metamorphism. Contributions to Mineralogy and Petrology 95:384–392

Bloch E, Ganguly J (2015) $^{176}Lu–^{176}Hf$ geochronology of garnet II: numerical simulations of the development of garnet–whole-rock $^{176}Lu–^{176}Hf$ isochrons and a new method for constraining the thermal history of metamorphic rocks. Contrib Mineral Petrol 169:1–16, doi:10.1007/s00410-015-1115-x

Bloch E, Ganguly J, Hervig R, Cheng W (2015) $^{176}Lu–^{176}Hf$ geochronology of garnet I: experimental determination of the diffusion kinetics of Lu^{3+} and Hf^{4+} in garnet, closure temperatures and geochronological implications. Contrib Mineral Petrol 169, doi:10.1007/s00410-015-1109-8

Bohlen SR, Valley JW, Essene EJ (1985) Metamorphism in the Adirondacks: I. Petrology, pressure and temperature. J Petrol 26:971–992

Borinski SA, Hoppe U, Chakraborty S, Ganguly J, Bhowmik SK (2012) Multicomponent diffusion in garnets I: general theoretical considerations and experimental data for Fe–Mg systems. Contrib Mineral Petrol 164:571–586, doi:10.1007/s00410-012-0758-0

Bowman JR, Moser DE, Valley JW, Wooden JL, Kita NT, Mazdab FK (2011) Zircon U–Pb isotope, $d^{18}O$ and trace element response to 80 m.y. of high temperature metamorphism in the lower crust: Sluggish diffusion and new records of Archean craton formation. Am J Sci 311:719–772, doi:10.2475/09.2011.01

Brady J (1995) Diffusion data for silicate minerals, glasses and liquids. *In:* Mineral Physics and Crystallography: A Handbook of Physical Constants, TJ Ahrens (ed) Am Geophys Union, p 269–290

Brady JB, Cherniak DJ (2010) Diffusion in minerals: an overview of published experimental diffusion data. Rev Mineral Geochem 72:899–920, doi:10.2138/rmg.2010.72.20

Cahalan RC, Kelly ED, Carlson WD (2014) Rates of Li diffusion in garnet: Coupled transport of Li and Y^+ REEs. Am Mineral 99:1676–1682, doi:10.2138/am.2014.4676

Cahn JW (1979) Thermodynamics of solid and fluid surfaces. *In:* Interfacial segregation. Johnson WC, Blakely J, (eds). American Society for Metals, Metals Park, OH, p 3–23

Carlson WD (2006) Rates of Fe, Mg, Mn, and Ca diffusion in garnet. Am Mineral 91:1–11

Carlson W (2012) Rates and mechanism of Y, REE, and Cr diffusion in garnet. Am Mineral 97:1598–1618

Cartwright I, Valley JW (1991) Steep oxygen-isotope gradients at marble–metagranite contacts in the Northwest Adirondack Mountains, New York, USA; products of fluid-hosted diffusion. Earth Planet Sci Lett 107:148–163

Castro JM, Manga M, Martin MC (2005) Vesiculation rates of obsidian domes inferred from H_2O concentration profiles. Geophys Res Lett 32, doi:10.1029/2005gl024029

Chakraborty S (1997) Rates and mechanisms of Fe–Mg interdiffusion in olivine at 980°–1300°C. J Geophys Res 102:12317–12331

Chakraborty S (2008) Diffusion in solid silicates: a tool to track timescales of processes comes of age. Ann Rev Earth Planet Sci 36:153–190

Chakraborty S (2010) Diffusion coefficients in olivine, wadsleyite and ringwoodite. Rev Mineral Geochem 72:603–639, doi:10.2138/rmg.2010.72.13

Chakraborty S, Ganguly J (1992) Cation diffusion in aluminosilicate garnets; experimental determination in spessartine-almandine diffusion couples, evaluation of effective binary, diffusion coefficients, and applications. Contrib Mineral Petrol 111:74–86

Chakraborty S, Rubie DC (1996) Mg tracer diffusion in aluminosilicate garnets at 750–850 degrees C, 1 atm. and 1300 degrees C, 8.5 GPa. Contrib Mineral Petrol 122:406–414

Cherniak DJ (1993) Lead diffusion in titanite and preliminary results on the effects of radiation damage on Pb transport. Chem Geol 110:177–194

Cherniak DJ (1995) Sr and Nd diffusion in titanite. Chem Geol 125:219–232

Cherniak DJ (1996) Strontium diffusion in sanidine and albite, and general comments on strontium diffusion in alkali feldspars. Geochim Cosmochim Acta 60:5037–5043

Cherniak DJ (2000) Pb diffusion in rutile. Contrib Mineral Petrol 139:198–207

Cherniak DJ (2002) Ba diffusion in feldspar. Geochim Cosmochim Acta 66:1641–1650

Cherniak DJ (2006) Zr diffusion in titanite. Contrib Mineral Petrol 152:639–647

Cherniak DJ (2010a) Cation diffusion in feldspars. Rev Mineral Geochem 72:691–733, doi:10.2138/rmg.2010.72.15

Cherniak DJ (2010b) Diffusion in accessory minerals: zircon, titanite, apatite, monazite and xenotime. Rev Mineral Geochem 72:827–869

Cherniak DJ (2015) Nb and Ta diffusion in titanite. Chem Geol 413:44–50, doi:10.1016/j.chemgeo.2015.08.010

Cherniak DJ, Dimanov A (2010) Diffusion in pyroxene, mica and amphibole. Rev Mineral Geochem 72:641–690, doi:10.2138/rmg.2010.72.14

Cherniak DJ, Liang Y (2007) Rare earth element diffusion in natural enstatite. Geochim Cosmochim Acta 71:1324–1340, doi:10.1016/j.gca.2006.12.001

Cherniak DJ, Watson EB (2001) Pb diffusion in zircon. Chem Geol 172:5–24

Cherniak DJ, Watson EB, Grove M, Harrison TM (2004) Pb diffusion in monazite: a combined RBS/SIMS study. Geochim Cosmochim Acta 68:207–226

Cherniak DJ, Manchester J, Watson EB (2007a) Zr and Hf diffusion in rutile. Earth Planet Sci Lett 261:267–279

Cherniak DJ, Watson EB, Wark DA (2007b) Ti diffusion in quartz. Chem Geol 236:65–74

Cohen M (1970) Self-diffusion during plastic deformation. Trans Japan Inst Metals 11:145–151

Corrie SL, Kohn MJ (2008) Trace-element distributions in silicates during prograde metamorphic reactions: implications for monazite formation. J Metamorph Geol 26:451–464, doi:10.1111/j.1525–1314.2008.00769.x

Costa F, Chakraborty S (2008) The effect of water on Si and O diffusion rates in olivine and implications for transport properties and processes in the upper mantle. Phys Earth Planet Int 166:11–29, doi:10.1016/j.pepi.2007.10.006

Costa F, Dungan M (2005) Short time scales of magmatic assimilation from diffusion modeling of multiple elements in olivine. Geology 33:837, doi:10.1130/g21675.1

Costa F, Chakraborty S, Dohmen R (2003) Diffusion coupling between trace and major elements and a model for calculation of magma residence times using plagioclase. Geochim Cosmochim Acta 67:2189–2200, doi:10.1016/s0016-7037(02)01345–5

Costa F, Dohmen R, Chakraborty S (2008) Time scales of magmatic processes from modeling the zoning patterns of crystals. Rev Mineral Geochem 69:545–594, doi:10.2138/rmg.2008.69.14

Cottrell AH, Bilby BA (1949) Dislocation theory of yielding and strain ageing of iron. Proc Phys Soc A 62:49–62

Crank J (1975) The Mathematics of Diffusion. Oxford University Press, London

Cruz-Uribe AM, Feineman MD, Zack T, Barth M (2014) Metamorphic reaction rates at ~650–800°C from diffusion of niobium in rutile. Geochim Cosmochim Acta 130:63–77, doi:10.1016/j.gca.2013.12.015

Cygan RT, Lasaga AC (1985) Self-diffusion of magnesium in garnet at 750° to 900°C. Am J Sci 285:328–350

Demouchy S, Mackwell SJ, Kohlstedt DL (2007) Influence of hydrogen on Fe–Mg interdiffusion in (Mg,Fe)O and implications for Earth's lower mantle. Contrib Mineral Petrol 154:279–289, doi:10.1007/s00410-007-0193-9

Dodson MH (1973) Closure temperature in cooling geochronological and petrological systems. Contrib Mineral Petrol 40: 259–274

Dodson MH (1986) Closure profiles in cooling systems. Mater Sci Forum 7:145–154

Dohmen R, Chakraborty S (2007) Fe–Mg diffusion in olivine II: point defect chemistry, change of diffusion mechanisms and a model for calculation of diffusion coefficients in natural olivine. Phys Chem Mineral 34:409–430, doi:10.1007/s00269-007-0158-6

Dohmen R, Becker H-W, Chakraborty S (2007) Fe–Mg diffusion in olivine I: experimental determination between 700 and 1,200°C as a function of composition, crystal orientation and oxygen fugacity. Phys Chem Mineral 34:389–407, doi:10.1007/s00269-007-0157-7

Dohmen R, Kasemann SA, Coogan L, Chakraborty S (2010) Diffusion of Li in olivine. Part I: Experimental observations and a multi species diffusion model. Geochim Cosmochim Acta 74:274–292, doi:10.1016/j.gca.2009.10.016

Dohmen R, Ter Heege JH, Becker H-W, Chakraborty S (2016) Fe–Mg interdiffusion in orthopyroxene. Am Mineral 101:2210–2221, doi:10.2138/am-2016-5815

Doremus R (1969) The diffusion of water in fused silica. *In:* Reactivity of Solids. Mitchell J, Devries R, Robers R, Cannon P (eds) Wiley, New York, p 667–673

Ducea MN, Ganguly J, Rosenberg EJ, Patchett PJ, Cheng W, Isachsen C (2003) Sm–Nd dating of spatially controlled domains of garnet single crystals: a new method of high-temperature thermochronology. Earth Planet Sci Lett 213:31–42

England PC, Richardson SW (1977) The influence of erosion upon the mineral facies of rocks from different metamorphic environments. J Geol Soc London 134:201–213

England PC, Thompson AB (1984) Pressure–temperature–time paths of regional metamorphism, Part I: Heat transfer during the evolution of regions of thickened continental crust. J Petrol 25:894–928

Ewing TA, Hermann J, Rubatto D (2013) The robustness of the Zr-in-rutile and Ti-in-zircon thermometers during high-temperature metamorphism (Ivrea-Verbano Zone, northern Italy). Contrib Mineral Petrol 165:757–779

Farver JR (2010) Oxygen and hydrogen diffusion in minerals. Rev Mineral Geochem 72:447–507, doi:10.2138/rmg.2010.72.10

Fei H, Wiedenbeck M, Yamazaki D, Katsura T (2013) Small effect of water on upper-mantle rheology based on silicon self-diffusion coefficients. Nature 498:213–215, doi:10.1038/nature12193

Ferry JM, Stubbs JE, Xu H, Guan Y, Eiler JM (2015) Ankerite grains with dolomite cores: A diffusion chronometer for low- to medium-grade regionally metamorphosed clastic sediments. Am Mineral 100:2443–2457, doi:10.2138/am-2015-5209

Freer R, Edwards A (1999) An experimental study of Ca-(Fe,Mg) interdiffusion in silicate garnets. Contrib Mineral Petrol 134:370–379

Gaetani GA, O'Leary JA, Shimizu N, Bucholz CE, Newville M (2012) Rapid reequilibration of H_2O and oxygen fugacity in olivine-hosted melt inclusions. Geology 40:915–918

Gallagher K (2012) Transdimensional inverse thermal history modeling for quantitative thermochronology. J Geophys Res: Solid Earth 117:B02408, doi:10.1029/2011jb008825

Ganguly J (2002) Diffusion kinetics in minerals: principles and applications to tectono-metamorphic processes. EMU Notes in Mineralogy 4:271–309

Ganguly J (2010) Cation diffusion kinetics in aluminosilicate garnets and geological applications. Rev Mineral Geochem 72:559–601

Ganguly J, Tirone M (1999) Diffusion closure temperature and age of a mineral with arbitrary extent of diffusion; theoretical formulation and applications. Earth Planet Sci Lett 170:131–140

Ganguly J, Cheng W, Chakraborty S (1998) Cation diffusion in aluminosilicate garnets; experimental determination in pyrope–almandine diffusion couples. Contrib Mineral Petrol 131:171–180

Ganguly J, Dasgupta S, Cheng W, Neogi S (2000) Exhumation history of a section of the Sikkim Himalayas, India: records in the metamorphic mineral equilibria and compositional zoning of garnet. Earth Planet Sci Lett 183:471–486

Gardés E, Jaoul O, Montel JM, Seydoux-Guillaume A-M, Wirth R (2006) Pb diffusion in monazite: an experimental study of $Pb^{2+} + Th^{4+} \Leftrightarrow 2Nd^{3+}$ interdiffusion. Geochim Cosmochim Acta 70:2325–2336

Gardés E, Montel J-M, Seydoux-Guillaume A-M, Wirth R (2007) Pb diffusion in monazite: New constraints from the experimental study of $Pb^{2+} \Leftrightarrow Ca^{2+}$ interdiffusion. Geochim Cosmochim Acta 71:4036–4043

Gerya TV, Stöckhert B, Perchuk AL (2002) Exhumation of high-pressure metamorphic rocks in a subduction channel: A numerical simulation. Tectonics 21:doi:10.1029/2002TC001406

Gibbs JW (1874–1878) On the equilibrium of heterogeneous substances. Trans Connecticut Acad Arts Sci 3:108–248, and 343–524

Giletti BJ (1986) Diffusion effects on oxygen isotope temperatures of slowly cooled igneous and metamorphic rocks. Earth Planet Sci Lett 77:218–228

Glicksman ME (2000) Diffusion in Solids. Field Theory, Solid-State Principles, and Applications. John Wiley and Sons, Inc., New York

Golden EM, Giles NC, Yang S, Halliburton LE (2015) Interstitial silicon ions in rutile TiO_2 crystals. Phys Rev B 91:134110

Grove M, Harrison TM (1999) Monazite Th–Pb age depth profiling. Geology 27:487–490

Grove TL, Baker MB, Kinzler RJ (1984) Coupled CaAl–NaSi diffusion in plagioclase feldspar: experiments and applications to cooling rate speedometry. Geochim Cosmochim Acta 48:2113–2121

Haifler J, Kotková J (2016) UHP–UHT peak conditions and near-adiabatic exhumation path of diamond-bearing garnet–clinopyroxene rocks from the Eger Crystalline Complex, North Bohemian Massif. Lithos 248–251:366–381, doi:10.1016/j.lithos.2016.02.001

Harrison TM, Célérier J, Aikman AB, Hermann J, Heizler MT (2009) Diffusion of ^{40}Ar in muscovite. Geochim Cosmochim Acta 73:1039–1051

Harte B, Hudson NFC (1979) Pelite facies series and temperatures and pressures of Dalradian metamorphism in E Scotland. Geol Soc Spec Publ 8:323–337

Hayden LA, Watson EB, Wark DA (2008) A thermobarometer for sphene (titanite). Contrib Mineral Petrol 155:529–540

Heaman L, Parrish R (1991) U–Pb geochronology of accessory minerals. *In*: Applications of Radiogenic Isotope Systems to Problems in Geology. Short Course Handbook. Vol 19. Heaman L, Ludden JN (eds) Mineral Assoc Canada, Toronto, Canada, p 59–102

Hermann J, Rubatto D, Korsakov A, Shatsky VS (2001) Multiple zircon growth during fast exhumation of diamondiferous, deeply subducted continental crust (Kokchetav Massif, Kazakhstan). Contrib Mineral Petrol 141:66–82

Hier-Majumder S, Anderson IM, Kohlstedt DL (2005) Influence of protons on Fe–Mg interdiffusion in olivine. J Geophys Res 110, doi:10.1029/2004jb003292

Humphreys MCS, Menand T, Blundy JD, Klimm K (2008) Magma ascent rates in explosive eruptions: Constraints from H_2O diffusion in melt inclusions. Earth Planet Sci Lett 270:25–40, doi:10.1016/j.epsl.2008.02.041

Ireland RHP, Penniston-Dorland SC (2015) Chemical interactions between a sedimentary diapir and surrounding magma: Evidence from the Phepane Dome and Bushveld Complex, South Africa. Am Mineral 100:1985–2000, doi:10.2138/am-2015-5196

Jackson K (2004) Kinetic Processes: Crystal Growth, Diffusion, and Phase Transitions in Materials. Wiley-VHC, Weinheim

Jaoul O, Poumellec M, Froidevaux C, Havette A (1981) Silicon diffusion in forsterite: a new constraint for understanding mantle deformation. *In*: Anelasticity in the Earth. Stacey FD, Paterson MS, Nicholas A, (eds). Am Geophys Union, Washington, D.C., p 95–100

Jaoul O, Houlier B, Abel F (1983) Study of ^{18}O diffusion in magnesium orthosilicate by nuclear microanalysis. J Geophys Res 88:613–624

Jeffcoate AB, Elliott T, Kasemann SA, Ionov D, Cooper K, Brooker R (2007) Li isotope fractionation in peridotites and mafic melts. Geochim Cosmochim Acta 71:202–218, doi:10.1016/j.gca.2006.06.1611

Jenkin GRT (1997) Do cooling paths derived from mica Rb–Sr data reflect true cooling paths? Geology 25:907–910

John T, Gussone N, Podladchikov YY, Bebout GE, Dohmen R, Halama R, Klemd R, Magna T, Seitz H-M (2012) Volcanic arcs fed by rapid pulsed fluid flow through subducting slabs. Nat Geosci 5:489–492, doi:10.1038/ngeo1482

Jollands MC, O'Neill HSC, Hermann J (2014) The importance of defining chemical potentials, substitution mechanisms and solubility in trace element diffusion studies: the case of Zr and Hf in olivine. Contrib Mineral Petrol 168, doi:10.1007/s00410-014-1055-x

Katayama I, Maruyama S, Parkinson CD, Terada K, Sano Y (2001) Ion micro-probe U–Pb zircon geochronology of peak and retrograde stages of ultrahigh-pressure metamorphic rocks from the Kokchetav Massif, northern Kazakhstan. Earth Planet Sci Lett 188:185–198

Keppler H, Bolfan-Casanova N (2006) Thermodynamics of water solubility and partitioning. Rev Mineral Geochem 62:193–230, doi:10.2138/rmg.2006.62.9

Ketcham RA (2005) Forward and inverse modeling of low-temperature thermochronometry data. Rev Mineral Geochem 58:275–314

Kohn MJ (1999) Why most "dry" rocks should cool "wet". Am Mineral 84:570–580

Kohn MJ (2004) Oscillatory- and sector-zoned garnets record cyclic (?) rapid thrusting in central Nepal. Geochem Geophys Geosyst 5:10.1029/2004gc000737, doi:Q12014 10.1029/2004gc000737

Kohn MJ (2008) *P–T–t* data from central Nepal support critical taper and repudiate large-scale channel flow of the Greater Himalayan Sequence. Geol Soc Am Bull 120:259–273, doi:10.1130/b26252.1

Kohn MJ (2009) Models of garnet differential geochronology. Geochim Cosmochim Acta 73:170–182

Kohn MJ (2013) Geochemical zoning in metamorphic minerals. *In*: Treatise on Geochemistry, volume 3, The Crust. Rudnick R (ed) Elsevier, p 229–261

Kohn MJ (2016) Metamorphic chronology—a tool for all ages: Past achievements and future prospects. Am Mineral 101:25–42

Kohn MJ (2017) Titanite petrochronology. Rev Mineral Geochem 83:419–441

Kohn MJ, Corrie SL (2011) Preserved Zr-temperatures and U–Pb ages in high-grade metamorphic titanite: evidence for a static hot channel in the Himalayan orogen. Earth Planet Sci Lett 311:136–143

Kohn MJ, Spear F (2000) Retrograde net transfer reaction insurance for pressure–temperature estimates. Geology 28:1127–1130, doi:10.1130/0091–7613(2000)028<1127:rntrif>2.3.co;2

Kohn MJ, Valley JW (1998) Obtaining equilibrium oxygen isotope fractionations from rocks: theory and examples. Contrib Mineral Petrol 132:209–224

Kohn MJ, Wieland MS, Parkinson CD, Upreti BN (2004) Miocene faulting at plate tectonic velocity in the Himalaya of central Nepal. Earth Planet Sci Lett 228:299–310, doi:10.1016/j.epsl.2004.10.007

Kohn MJ, Penniston-Dorland SC, Ferreira JSC (2016) Implications of near-rim compositional zoning in rutile for geothermometry, geospeedometry, and trace element equilibration. Contrib Mineral Petrol 171:78 DOI 10.1007/s00410-016-1285-1

Kooijman E, Smit MA, Mezger K, Berndt, J (2012) Trace element systematics in granulite facies rutile: implications for Zr geothermometry and provenance studies. J Metamorph Geol 30:397–412

Kotková J, Whitehouse M, Schaltegger U, D'Abzac F-XD (2016) The fate of zircon during UHT–UHP metamorphism: isotopic (U/Pb, $\delta^{18}O$, Hf) and trace element constraints. J Metamorph Geol 34:719–739 doi: 10.1111/jmg.12206

Kramer MJ, Seifert KE (1991) Strain enhanced diffusion in feldspars. *In*: Diffusion, Atomic Ordering, and Mass Transport. Ganguly J, (ed) Springer-Verlag, New York, p 286–303

Krishna R (2015) Uphill diffusion in multicomponent mixtures. Chem Soc Rev 44:2812–2836, doi:10.1039/c4cs00440j

Krogh TE (1993) High precision U–Pb ages for granulite metamorphism and deformation in the Archean Kapuskasing structural zone, Ontario: implications for structure and development of the lower crust. Earth Planet Sci Lett 119:1–18, doi: 10.1016/0012-821X(93)90002-Q

Kuhn TS (1962) The structure of scientific revolutions. University of Chicago Press, Chicago

Kylander–Clark ARC (2017) Petrochronology by laser–ablation inductively coupled plasma mass spectrometry. Rev Mineral Geochem 83:183–198

Lanari P, Engi M (2017) Local bulk composition effects on metamorphic mineral assemblages. Rev Mineral Geochem 83:55–102

Lapen TJ, Johnson CM, Baumgartner LP, Mahlen NJ, Beard BL, Amato JM (2003) Burial rates during prograde metamorphism of an ultra-high-pressure terrane: an example from Lago di Cignana, western Alps, Italy. Earth Planet Sci Lett 215:57–72

Lasaga AC (1979) Multicomponent exchange and diffusion in silicates. Geochim Cosmochim Acta 43:455–469

Lasaga AC (1983) Geospeedometry: An extension of geothermometry. *In*: Kinetics and Equilibrium in Mineral Reactions. Saxena SK (ed) Springer-Verlag, New York, p 81–114

LaTourrette T, Wasserburg GJ (1998) Mg diffusion in anorthite: implications for the formation of early solar system planetesimals. Earth Planet Sci Lett 158:91–108

Lindström R, Viitanen M, Juhanoja J, Holtta P (1991) Geospeedometry of metamorphic rocks; examples in the Rantasalmi-Sulkava and Kiuruvesi areas, eastern Finland; biotite–garnet diffusion couples. J Metamorph Geol 9:181–190

Liou JG, Zhang R, Liu F, Zhang Z, Ernst WG (2012) Mineralogy, petrology, U–Pb geochronology, and geologic evolution of the Dabie-Sulu classic ultrahigh-pressure metamorphic terrane, East-Central China. Am Mineral 97:1533–1543, doi:10.2138/am.2012.4169

Liu Y, Anderson AT, Wilson CJN (2007) Melt pockets in phenocrysts and decompression rates of silicic magmas before fragmentation. J Geophys Res 112, doi:10.1029/2006jb004500

Liu X-M, Rudnick RL, Hier-Majumder S, Sirbescu M-LC (2010) Processes controlling lithium isotopic distribution in contact aureoles: A case study of the Florence County pegmatites, Wisconsin. Geochem Geophys Geosystem 11:Q08014, doi:10.1029/2010gc003063

Lloyd AS, Ruprecht P, Hauri EH, Rose W, Gonnermann HM, Plank T (2014) NanoSIMS results from olivine-hosted melt embayments: Magma ascent rate during explosive basaltic eruptions. J Volcanol Geothermal Res 283:1–18

Lucassen F, Dulski P, Abart R, Franz G, Rhede D, Romer RL (2010) Redistribution of HFSE elements during rutile replacement by titanite. Contrib Mineral Petrol 160:279–295, doi:10.1007/s00410-009-0477-3

MacDonald JM, Wheeler J, Harley SL, Mariani E, Goodenough KM, Crowley Q, Tatham D (2013) Lattice distortion in a zircon population and its effects on trace element mobility and U–Th–Pb isotope systematics: examples from the Lewisian Gneiss Complex, northwest Scotland. Contrib Mineral Petrol 166:21–41, doi:10.1007/s00410-013-0863-8

Marks MAW, Rudnick RL, McCammon C, Vennemann T, Markl G (2007) Arrested kinetic Li isotope fractionation at the margin of the Ilímaussaq complex, South Greenland: Evidence for open-system processes during final cooling of peralkaline igneous rocks. Chem Geol 246:207–230, doi:10.1016/j.chemgeo.2007.10.001

Mattinson JM (1978) Age, origin, and thermal histories of some plutonic rocks from the Salinian Block of California. Contrib Mineral Petrol 67:233–245, doi:10.1007/bf00381451

McDougall I, Harrison TM (1999) Geochronology and thermochronology by the $^{40}Ar/^{39}Ar$ method. Oxford University Press, Oxford

McFarlane C, Harrison TM (2006) Pb-diffusion in monazite: Constraints from a high-T contact aureole setting. Earth Planet Sci Lett 250:376–384, doi:10.1016/j.epsl.2006.06.050

McLelland JM, Selleck BW, Bickford ME (2010) Review of the Proterozoic evolution of the Grenville Province, its Adirondack outlier, and the Mesoproterozoic inliers of the Appalachians. Geol Soc Am Mem 206:21–49

Mezger K, Rawnsley C, Bohlen S, Hanson G (1991) U–Pb garnet, sphene, monazite and rutile ages: Implications for the duration of high grade metamorphism and cooling histories, Adirondack Mts., New York. J Geol 99:415–428

Müller T, Cherniak D, Bruce Watson E (2012) Interdiffusion of divalent cations in carbonates: Experimental measurements and implications for timescales of equilibration and retention of compositional signatures. Geochim Cosmochim Acta 84:90–103, doi:10.1016/j.gca.2012.01.011

Müller T, Dohmen R, Becker HW, ter Heege JH, Chakraborty S (2013) Fe–Mg interdiffusion rates in clinopyroxene: experimental data and implications for Fe–Mg exchange geothermometers. Contrib Mineral Petrol 166:1563–1576, doi:10.1007/s00410-013-0941-y

Nowotny MK, Sheppard LR, Bak T, Nowotny J (2008) Defect chemistry of titanium dioxide. Application of defect engineering in processing TiO_2-based photocatalysts. J Phys Chem C 112:5275–5300

Page FZ, Ushikubo T, Kita NT, Riciputi LR, Valley JW (2007) High-precision oxygen isotope analysis of picogram samples reveals 2 μm gradients and slow diffusion in zircon. Am Mineral 92:1772–1775

Page FZ, Kita NT, Valley JW (2010) Ion microprobe analysis of oxygen isotopes in garnets of complex chemistry. Chem Geol 270:9–19

Pape J, Mezger K, Robyr M (2016) A systematic evaluation of the Zr-in-rutile thermometer in ultra-high temperature (UHT) rocks. Contrib Mineral Petrol 171, doi:10.1007/s00410-016-1254-8

Parkinson I, Hammond S, James R, Rogers N (2007) High-temperature lithium isotope fractionation: Insights from lithium isotope diffusion in magmatic systems. Earth Planet Sci Lett 257:609–621, doi:10.1016/j.epsl.2007.03.023

Pauly J, Marschall HR, Meyer H-P, Chatterjee N, Monteleone B (2016) Prolonged Ediacaran–Cambrian metamorphic history and short-lived high-pressure granulite-facies metamorphism in the H.U. Sverdrupfjella, Dronning Maud Land (East Antarctica): evidence for continental collision during Gondwana assembly. J Petrol 57:185–228, doi:10.1093/petrology/egw005

Peck WH, Valley JW, Graham CM (2003) Slow oxygen diffusion rates in igneous zircons from metamorphic rocks. Am Mineral 88:1003–1014

Peck WH, Bickford ME, McLelland JM, Nagle AN, Swarr GJ (2010) Mechanism of metamorphic zircon growth in a granulite-facies quartzite, Adirondack Highlands, Grenville Province, New York. Am Mineral 95:1796–1806, doi:10.2138/am.2010.3547

Penniston-Dorland SC, Sorensen SS, Ash RD, Khadke SV (2010) Lithium isotopes as a tracer of fluids in a subduction zone melange: Franciscan Complex, CA. Earth Planet Sci Lett 292:181–190

Perchuk AL, Burchard M, Schertl HP, Maresch WV, Gerya TV, Bernhardt HJ, Vidal O (2009) Diffusion of divalent cations in garnet: multi-couple experiments. Contrib Mineral Petrol 157:573–592, doi:10.1007/s00410-008-0353-6

Piazolo S, Austrheim H, Whitehouse M (2012) Brittle–ductile microfabrics in naturally deformed zircon: Deformation mechanisms and consequences for U–Pb dating. Am Mineral 97:1544–1563, doi:10.2138/am.2012.3966

Piazolo S, La Fontaine A, Trimby P, Harley S, Yang L, Armstrong R, Cairney JM (2016) Deformation-induced trace element redistribution in zircon revealed using atom probe tomography. Nat Commun 7:10490, doi:10.1038/ncomms10490

Qian Q, O'Neill HSC, Hermann J (2010) Comparative diffusion coefficients of major and trace elements in olivine at 950 °C from a xenocryst included in dioritic magma. Geology 38:331–334, doi:10.1130/g30788.1

Ravna EJK, Roux MRM (2006) Metamorphic evolution of the Tonsvika eclogite, Tromso Nappe—Evidence for a new UHPM Province in the Scandinavian Caledonides. Int Geol Rev 48:861–881

Reddy SM, Timms NE, Trimby P, Kinny PD, Buchan C, Blake K (2006) Crystal-plastic deformation of zircon: A defect in the assumption of chemical robustness. Geology 34:257, doi:10.1130/g22110.1

Reddy SM, van Riessen A, Saxey DW, Johnson TE, Rickard WD, Fougerouse D, Fischer S, Prosa TJ, Rice KP, Reinhard DA, Chen Y (2016) Mechanisms of deformation-induced trace element migration in zircon resolved by atom probe and correlative microscopy. Geochim Cosmochim Acta 195:158–170, doi:10.1016/j.gca.2016.09.019

Richter FM, Liang Y, Davis AM (1999) Isotope fractionation by diffusion in molten oxides. Geochim Cosmochim Acta 63:2853–2861

Richter FM, Davis AM, DePaolo DJ, Watson EB (2003) Isotope fractionation by chemical diffusion between molten basalt and rhyolite. Geochim Cosmochim Acta 67:3905–3923, doi:10.1016/s0016-7037(03)00174-1

Richter FM, Mendybaev RA, Christensen JN, Hutcheon ID, Williams RW, Sturchio NC, Beloso AD (2006) Kinetic isotopic fractionation during diffusion of ionic species in water. Geochim Cosmochim Acta 70:277–289, doi:10.1016/j.gca.2005.09.016

Richter F, Watson B, Chaussidon M, Mendybaev R, Ruscitto D (2014) Lithium isotope fractionation by diffusion in minerals. Part 1: Pyroxenes. Geochim Cosmochim Acta 126:352–370, doi:10.1016/j.gca.2013.11.008

Rubatto D, Hermann J (2001) Exhumation as fast as subduction? Geology 29:3–6

Rudnick R, Ionov D (2007) Lithium elemental and isotopic disequilibrium in minerals from peridotite xenoliths from far-east Russia: Product of recent melt/fluid–rock reaction. Earth Planet Sci Lett 256:278–293, doi:10.1016/j.epsl.2007.01.035

Sasaki J, Peterson NL, Hoshino K (1985) Tracer impurity diffusion in single-crystal rutile (TiO$_{2-x}$). J Phys Chem Solids 46:1267–1283

Schmitt AK, Vazquez JA (2017) Secondary ionization mass spectrometry analysis in petrochronology. Rev Mineral Geochem 83:199–230

Schwandt CS, Cygan RT, Westrich HR (1996) Ca self-diffusion in grossular garnet. Am Mineral 81:448–451

Seitz S, Putlitz B, Baumgartner LP, Escrig S, Meibom A, Bouvier A-S (2016) Short magmatic residence times of quartz phenocrysts in Patagonian rhyolites associated with Gondwana breakup. Geology 44:67–70, doi:10.1130/g37232.1

Shannon RD (1976) Revised effective ionic radii and systematic studies of interatomic distances in halides and chalcogenides. Acta Crystallogr Sect A A32:751–767

Shea T, Costa F, Krimer D, Hammer JE (2015a) Accuracy of timescales retrieved from diffusion modeling in olivine: A 3D perspective. Am Mineral 100:2026–2042, doi:10.2138/am-2015-5163

Shea T, Lynn KJ, Garcia MO (2015b) Cracking the olivine zoning code: Distinguishing between crystal growth and diffusion. Geology 43:935–938, doi:10.1130/g37082.1

Shimojuku A, Kubo T, Ohtani E, Nakamura T, Okazaki R (2010) Effects of hydrogen and iron on the silicon diffusivity of wadsleyite. Phys Earth Planet Int 183:175–182, doi:10.1016/j.pepi.2010.09.011

Shore M, Fowler AD (1996) Oscillatory zoning in minerals: a common phenomenon. Can Mineral 34:1111–1126

Smith HA, Giletti BJ (1997) Lead diffusion in monazite. Geochim Cosmochim Acta 61:1047–1055

Smye AJ, Stockli DF (2014) Rutile U–Pb age depth profiling: a continuous record of lithospheric thermal evolution. Earth Planet Sci Lett 408:171–182

Sousa JL, Kohn MJ, Schmitz MD, Northrup CJ, Spear FS (2013) Strontium isotope zoning in garnets: Implications for metamorphic matrix equilibration, geochronology, and phase equilibrium modeling. J Metamorph Geol 31:437–452

Spear FS (1991) On the interpretation of peak metamorphic temperatures in light of garnet diffusion during cooling. J Metamorph Geol 9:379–388

Spear FS (2004) Fast cooling and exhumation of the Valhalla metamorphic core complex, southeastern British Columbia. Int Geol Rev 46:193–209

Spear FS (2014) The duration of near-peak metamorphism from diffusion modelling of garnet zoning. J Metamorph Geol 32:903–914, doi:10.1111/jmg.12099

Spear FS, Florence FP (1992) Thermobarometry in granulites: Pitfalls and new approaches. J Precambrian Res 55:209–241

Spear FS, Kohn MJ (1996) Trace element zoning in garnet as a monitor of crustal melting. Geology 24:1099–1102

Spear FS, Parrish RR (1996) Petrology and cooling rates of the Valhalla complex, British Columbia, Canada. J Petrol 37:733–765

Spear FS, Ashley KT, Webb LE, Thomas JB (2012) Ti diffusion in quartz inclusions: implications for metamorphic time scales. Contrib Mineral Petrol, doi:10.1007/s00410-012-0783-z

Stepanov AS, Rubatto D, Hermann J, Korsakov AV (2016) Contrasting *P–T* paths within the Barchi-Kol UHP terrain (Kokchetav Complex): Implications for subduction and exhumation of continental crust. Am Mineral 101:788–807, doi:10.2138/am-2016-5454

Storm LC, Spear FS (2005) Pressure, temperature and cooling rates of granulite facies migmatitic pelites from the southern Adirondack Highlands, New York. J Metamorph Geol 23:107–130, doi:10.1111/j.1525-1314.2005.00565.x

Storm LC, Spear FS (2009) Application of the titanium-in-quartz thermometer to pelitic migmatites from the Adirondack Highlands, New York. J Metamorph Geol 27:479–494, doi:10.1111/j.1525-1314.2009.00829.x

Sutton AP, Balluffi RW (2003) Interfaces in Crystalline Materials. Clarendon Press

Taylor-Jones K, Powell R (2015) Interpreting zirconium-in-rutile thermometric results. J Metamorph Geol 33:115–122, doi:10.1111/jmg.12109

Teng F-Z, McDonough WF, Rudnick RL, Walker RJ (2006) Diffusion-driven extreme lithium isotopic fractionation in country rocks of the Tin Mountain pegmatite. Earth Planet Sci Lett 243:701–710, doi:10.1016/j.epsl.2006.01.036

Till CB, Vazquez JA, Boyce JW (2015) Months between rejuvenation and volcanic eruption at Yellowstone caldera, Wyoming. Geology 43:695–698, doi:10.1130/g36862.1

Timms NE, Kinny PD, Reddy SM (2006) Enhanced diffusion of uranium and thorium linked to crystal plasticity in zircon. Geochem Trans 7:10, doi:10.1186/1467-4866-7-10

Timms NE, Kinny PD, Reddy SM, Evans K, Clark C, Healy D (2011) Relationship among titanium, rare earth elements, U–Pb ages and deformation microstructures in zircon: Implications for Ti-in-zircon thermometry. Chem Geol 280:33–46, doi:10.1016/j.chemgeo.2010.10.005

Tirone M, Ganguly J, Dohmen R, Langenhorst F, Hervig R, Becker H-W (2005) Rare earth diffusion kinetics in garnet: Experimental studies and applications. Geochim Cosmochim Acta 69:2385–2398

Tomkins HS, Powell R, Ellis DJ (2007) The pressure dependence of the zirconium-in-rutile thermometer. J Metamorph Geol 25:703–713

Turner S, Evans P, Hawkesworth C (2001) Ultrafast source-to-surface movement of melt at island arcs from Ra-226–Th-230 systematics. Science 292:1363–1366

Vielzeuf D, Saúl A (2011) Uphill diffusion, zero-flux planes and transient chemical solitary waves in garnet. Contrib Mineral Petrol 161:683–702, doi:10.1007/s00410-010-0557-4

Vielzeuf D, Veschambre M, Brunet F (2005) Oxygen isotope heterogeneities and diffusion profile in composite metamorphic-magmatic garnets from the Pyrenees. Am Mineral 90:463–472

Vielzeuf D, Baronnet A, Perchuk AL (2007) Calcium diffusivity in alumino-silicate garnets: an experimental and ATEM study. Contrib Mineral Petrol 154:153–170

Viete DR, Oliver GJH, Fraser GL, Forster MA, Lister GS (2013) Timing and heat sources for the Barrovian metamorphism, Scotland. Lithos 177:148–163, doi:10.1016/j.lithos.2013.06.009

Wang ZY, Hiraga T, Kohlstedt DL (2004) Effect of H+on Fe–Mg interdiffusion in olivine, $(FeMg)_2SiO_4$. Appl Phys Lett 85:209–211

Warren CJ, Beaumont C, Jamieson RA (2008) Formation and exhumation of ultra-high-pressure rocks during continental collision: Role of detachment in the subduction channel. Geochem Geophys Geosystem doi:10.1029/2007GC001839

Watson EB (1982) Basalt contamination by continental crust: some experiments and models. Contrib Mineral Petrol 80:73–87

Watson EB, Baxter EF (2007) Diffusion in solid-Earth systems. Earth Planet Sci Lett 253:307–327

Watson EB, Cherniak DJ (1997) Oxygen diffusion in zircon. Earth Planet Sci Lett 148:527–544

Watson EB, Wark DA, Thomas JB (2006) Crystallization thermometers for zircon and rutile. Contrib Mineral Petrol 151:413–433

Yund RA, Tullis J (1980) The effect of water, pressure, and strain on Al/Si order/disorder kinetics in feldspar. Contrib Mineral Petrol 72:297–302

Yund RA, Smith BM, Tullis J (1981) Dislocation-assisted diffusion of oxygen in albite. Phys Chem Miner 7:185–189

Yund RA, Quigley J, Tullis J (1989) The effect of dislocations on bulk diffusion in feldspars during metamorphism. J Metamorph Geol 7:337–341

Zack T, Moraes R, Kronz A (2004) Temperature dependence of Zr in rutile: empirical calibration of a rutile thermometer. Contrib Mineral Petrol 148:471–488

Zhang Y (2010) Diffusion in minerals and melts: theoretical background. Rev Mineral Geochem 72:5–59

Zhang Y, Cherniak DJ (2010) Diffusion in Minerals and Melts. Mineral Soc America, Washington, D. C.

Zhang Y, Ni H (2010) Diffusion of H, C, and O components in silicate melts. Rev Mineral Geochem 72:171–225

Zhang Y, Stolper EM, Wasserburg GJ (1991) Diffusion of a multi-species component and its role in oxygen and water transport in silicates. Earth Planet Sci Lett 103:228–240

Zhang Y, Xu Z, Zhu M, Wang H (2007) Silicate melt properties and volcanic eruptions. Rev Geophys 45, doi:10.1029/2006rg000216

Zhang Y, Ni H, Chen Y (2010) Diffusion data in silicate melts. Rev Mineral Geochem 72:311–408, doi:10.2138/rmg.2010.72.8

Reviews in Mineralogy & Geochemistry
Vol. 83 pp. 153–182, 2017
Copyright © Mineralogical Society of America

5

Electron Microprobe Petrochronology

Michael L. Williams, Michael J. Jercinovic

Department of Geosciences
University of Massachusetts
Amherst MA 01003
USA

mlw@geo.umass.edu

mjj@geo.umass.edu

Kevin H. Mahan

Department of Geological Sciences
University of Colorado
Boulder, CO 80302
USA

mahank@colorado.edu

Gregory Dumond

Department of Geosciences
University of Arkansas
Fayetteville, AR 72701
USA

gdumond@uark.edu

INTRODUCTION

The term petrochronology has increasingly appeared in publications and presentations over the past decade. The term has been defined in a somewhat narrow sense as "the interpretation of isotopic dates in the light of complementary elemental or isotopic information from the same mineral(s)" (Kylander-Clark et al. 2013). Although complementary isotopic and elementary information are certainly a central and critical part of most, if not all, petrochronology studies, the range of recent studies that might use the term covers a much broader scope. The term "petrochronology" might alternatively be defined as the detailed incorporation of chronometer phases into the petrologic (and tectonic) evolution of their host rocks, in order to place direct age constraints on petrologic and structural processes. As noted by Kylander-Clark et al. (2013), the linkage between geochronology and petrology can involve a variety of data including mineral textures and fabrics, the distribution of mineral modes or volume proportions, compositional zoning, mineral inclusion relationships, and certainly major element, trace element, and isotopic composition of the chronometer and all other phases.

Electron probe micro-analysis (EPMA) has a central and critical role to play in establishing the linkage between chronometer phases and their host assemblage. The basic instrument is an electron microscope which can be used in either scanning or fixed beam modes, with integrated wavelength dispersive spectrometers (WDS), energy dispersive spectrometers (EDS), electron detectors (to image secondary and backscattered signals) a light optical system, and optionally cathodoluminescence (CL) detection. The electron microprobe is used to investigate the distribution, composition, and compositional zonation of all mineral phases, the

1529-6466/17/0083-0005$05.00 (print)
1943-2666/17/0083-0005$05.00 (online)

data that underpin thermobarometric analysis and modeling of *P–T* histories. The microprobe, with μm-scale spatial resolution, can also characterize compositional zonation in very small accessory phases including monazite, xenotime, zircon, allanite, titanite, apatite, and others. This, as discussed below, can be a critical step in linking geochronology to petrology. Finally, in many circumstances, the microprobe can be used to determine or constrain the age of domains within certain chronometers, especially monazite and xenotime. Where the compositional domains are small (<5 μm), the microprobe may be the only feasible tool that can constrain the age. Narrow rim domains are commonly key constraints on the petro-tectonic history.

This chapter is focused on the role of EPMA in petrochronological studies. The early parts of the chapter highlight analytical considerations in using the electron microprobe: first, for compositional characterization and for establishing the linkage between chronometer phases and the petro-tectonic history, and second, for precise trace element analysis and dating. Although the process is relatively straightforward, special analytical methods must be adopted for high-precision trace element analysis by electron microprobe. The later parts of the chapter provide examples and illustrations of the different roles that the electron microprobe can play in petrochronological studies. Many of the examples, and much of the recent research, concern monazite and xenotime in deformed and metamorphosed rocks. Although zircon has been widely used to constrain age or provenance, and more rarely metamorphic history, monazite and xenotime can record, with high fidelity, multiple stages in the igneous and metamorphic history and can also provide some key constraints on the deformational history. Finally, it should be noted that neither the techniques nor the electron microprobe instrument itself have reached their ultimate potential for petrochronological analysis. We hope to shine some light on future directions and challenges that suggest the ability to more efficiently extract, and to more tightly constrain, the *P–T*–time–deformation history of rocks.

REACTION DATING

One of the major goals of igneous or metamorphic analysis is determining the sequence of minerals or mineral assemblages that were present during the evolution of the rocks of interest, and ultimately interpreting the sequence of chemical reactions that relate the minerals or assemblages. The characterization of mineral assemblages and reactions, based on petrographic, petrologic, and microstructural analysis allows the construction of *P–T–t* (±*D*=deformation) histories that are central to most petrologic investigations. Although in some previous studies, workers have used resetting of geochronometers during metamorphism as a means to constrain the "age of metamorphism", the very sluggish rate of diffusion in high-temperature chronometers such as zircon, monazite, or xenotime (Cherniak et al. 2004; Cherniak and Pyle 2008; Cherniak 2010), make resetting generally unlikely. Instead, new mineral grains (neocrystals), or new domains within chronometer minerals, probably grow during the reaction history. This is particularly true of monazite, which can serve as a source or sink of a wide variety of elements, including Rare Earths, actinides, and others, that are minor and trace elements in most silicates. As such, the essence of petrochronology involves integrating trace elements and trace-element-bearing chronometer phases into the silicate reaction history. Trace element partitioning may place additional constraints on the metamorphic equilibria (Hickmott and Spear 1992; Bea et al. 1997; Pyle and Spear 1999; Yang et al. 1999; Spear and Pyle 2002), and in the present context, dating specific generations of chronometer phases that have been associated with a particular reaction allows a date to be directly associated with a reaction. This process could be described as "reaction dating" and we note that it is increasingly becoming the focus of many researchers around the world (for example Larson et al. 2011; Dumond et al. 2015; Regis et al. 2016; and many others). Focusing on the term, "reaction dating" deemphasizes the goal of "dating metamorphism" and instead, emphasizes the goal of dating or constraining as many prograde and retrograde reactions as possible, in order to characterize the timing and hopefully duration of metamorphic or igneous events.

Most of the examples and illustrations in this chapter focus on petrochronology involving monazite and to a lesser extent xenotime and zircon. For many reasons, monazite may be the ideal chronometer for EPMA petrochronology. First, monazite has a wide stability field from diagenesis to high-grade metamorphism, and it occurs in a range of metamorphic and igneous rock compositions (Overstreet 1967; Williams et al. 2007; Catlos 2013). Further, its broad compositional range implies that monazite can, to some degree, participate in a variety of silicate reactions, serving as a source or sink of minor components that are also present in the silicate minerals. There are many examples of monazite or xenotime reacting to (or from) other phosphate or REE-bearing minerals (Finger et al. 1998; Janots et al. 2008; Budzyn et al. 2011), but as discussed below, monazite compositions can also be modified during many silicate reactions because of minor components either liberated or consumed during the reaction(s). Finally, monazite is particularly amenable to electron microprobe analysis. In addition to U, Th, and Pb, many components, including light REEs and actinides, are abundant enough to be analyzed by EPMA. The current analytical protocol at the University of Massachusetts includes 25 elements in a standard monazite analysis. Many other minerals can provide key petrochronological data including apatite, allanite, titanite, thorite, uraninite, and others (see also Vance et al. 2003). The methods described below are applicable to varying degrees to all of these phases. Examples of some representative studies are highlighted, but many other examples could certainly be included.

Monazite has been an important geologic chronometer since the classic study of Parrish (1990). However, starting in the late 1990s, a large number of studies documented multiple generations of metamorphic monazite (and xenotime) and to varying degrees attempted to integrate monazite or xenotime generations with silicate reactions. They illustrate the steadily evolving logic used to correlate monazite generations and silicate reactions and assemblages. Early studies tended to focus on Y-concentration because of the inverse relationship with garnet abundance (Pyle and Spear 1999, 2003; Wing et al. 2003; Foster et al. 2004; Gibson et al. 2004; Berger et al. 2005; Kohn et al. 2005; McFarlane et al. 2005; Yang and Pattison 2006). Along with heavy Rare Earth Elements, Y is strongly partitioned into garnet, and as such, Y in monazite (and the overall abundance of xenotime) typically decreases during garnet growth and increases on garnet break-down. This Y–garnet–monazite–xenotime connection is still one of the most powerful and widely used petrochronological tools, especially for microprobe-based studies.

More recently, workers have considered other trace components and proposed increasingly more specific reactions and mechanisms linking accessory phases with silicate assemblages (i.e., Rubatto et al. 2006; Buick et al. 2010; Dumond et al. 2015; Regis et al. 2016; Rocha et al. in press). In addition, new experimental data and studies of well-constrained natural assemblages have provided constraints on partitioning between accessory phases and silicate assemblages and melts (Hermann and Rubatto 2003; Krenn and Finger 2004; Rubatto and Hermann 2007; Stepanov et al. 2012), updating and expanding the classic work of Bea and coworkers (Bea et al. 1994; Bea and Montero 1999). One ultimate goal is to better constrain the thermodynamic properties of accessory phases such that the accessory chronometers can be incorporated into phase diagrams, particularly isochemical phase diagram models (see below). Several important steps have been made in modeling monazite, xenotime, and zircon abundance and composition in metamorphic rocks and in melt–silicate systems (Kelsey et al. 2008; Spear 2010; Spear and Pyle 2010; Kelsey and Powell 2011).

COMPOSITIONAL MAPPING, TRACE ELEMENT ANALYSIS, AND DATING BY ELECTRON MICROPROBE

Compositional mapping for petrochronology

Compositional maps have been used for many years to explore zoning in single minerals (Tracy 1982; c.f. Kohn 2013). Early maps made by contouring point analyses were time consuming and generally of low resolution. Modern maps made by rastering the electron microprobe (beam or stage) or by mosaicking a grid of smaller SEM images are simple and efficient to collect. Resolution can be as high as 1 μm per pixel (or better), although time constraints may limit the resolution for larger maps.

Full-thin-section or large-area compositional maps are much less common in the literature, but can be invaluable for petrochronological analysis (Fig. 1). Maps of major element abundances, commonly Mg, Ca, K etc. can show the distribution of major phases and also significant compositional zoning within the larger minerals. Maps of selected elements (Ce for monazite, Y for xenotime, Zr for zircon) can be used to locate accessory chronometer phases (see also Williams et al. 2006; Larson et al. 2011). Depending on the number of elements required and number of available spectrometers or compositional channels, multiple mapping acquisitions of the same area may be necessary in order to identify and characterize the major and accessory phases of interest. Background maps (maps collected in a background position for the desired element) or calculated backgrounds based on mean atomic number (Donovan et al. 2016) can be subtracted from peak-position compositional maps in order to perform a background correction. Also, peak pixel values can be calibrated in order to constrain phase compositions (see also Kohn and Spear 2000; Clarke et al. 2001; De Andrade et al. 2006), and image math can be used to calculate hybrid maps such as age maps (Goncalves et al. 2005). Commonly, a single set of maps can be used to successfully identify the major phases and to locate and evaluate the chronometer phases. Other instruments (and methods) besides electron microprobe WDS mapping are available for large-area compositional mapping, including scanning electron microscopy (SEM) with EDS (sometimes also WDS) and electron backscatter diffraction (EDS and EBSD) capabilities, and the related electron beam techniques offered by quantitative evaluation of minerals by scanning electron microscopy (QEMSCAN), and mineral liberation analysis (MLA). QEMSCAN in particular is capable of efficiently capturing compositional information from minerals in large-area maps, although these instruments, to date, have mainly been used in economic geology applications.

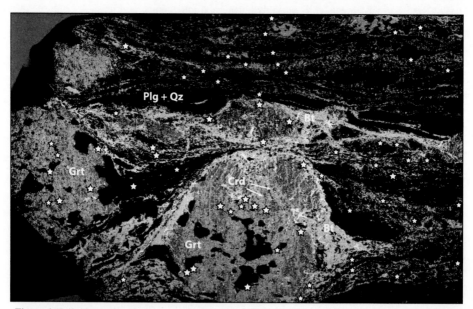

Figure 1. Full-thin-section Mg Kα compositional map of sample S32D. Lighter grey tone corresponds to great Mg content. The large central subhedral garnet has been partly replaced by biotite (outside) and cordierite on fractures within the crystal. Abbreviations from Whitney and Evans (2010). White stars show the location of Monazite grains; larger stars are larger grains. The stars represent (i.e., are placed on) high-Ce pixels (or clusters of pixels) on the full-section map. The map was run at 300 nA, 25 ms/pixel with a 35 μm step size and 25–30 μm beam. Modified from Mahan et al. (2006).

Large-area maps showing the location of chronometer phases are, in themselves, useful tools for illuminating the petrogenesis of the major and the accessory phases. For example xenotime grains specifically situated around the rim of garnet crystals commonly represent late-stage xenotime growth during garnet resorption (Fig. 2). Zircon in leucosome domains can represent zircon formed during melt crystallization (Flowers et al. 2006a). These maps are particularly useful when they are generated early in the petrographic/petrologic analysis cycle rather than later after the petrographic analysis and interpretation has been completed.

High-resolution (small-area) maps of chronometer phases, including WDS, cathodoluminescence, backscattered electron, etc., have been used extensively to characterize zoning in chronometer phases in order to plan a strategy for dating, and to aid in interpreting geochronologic results (Williams et al. 2006, 2007; Larson et al. 2011; Peterman et al. 2016; and many others). When placed into the context of full-section compositional maps, the high-resolution maps can be an even more powerful tool for petrochronology and reaction dating. One method for integrating the maps involves placing the high-resolution maps around or on the full-section image with links to the actual grain locations (Fig. 3a). This allows the zonation within a high-resolution map to be interpreted in the context of its setting within the thin section. Recent work has shown that even the most subtle compositional zoning in monazite can reflect the local setting in the thin section (i.e., porphyroblast inclusion relations, nearby phases, local structures, fractures). It is particularly important to process the high-resolution compositional maps with the same look-up table such that similar compositions have similar intensities on the images (Williams et al. 2006). One simple technique using Adobe software is summarized in Appendix-1 (deposited with MSA and available from the authors).

The combination of high-resolution and low-resolution (large area) compositional maps allows chronometer phases to be integrated into the petrographic analysis process. In the case of monazite, zircon, or xenotime, it is commonly possible to identify several key generations of the chronometer phase(s) and to relate the generations, at least in a qualitative way, to the silicate phases/assemblages and to fabric generations. This also changes the nature of the geochronologic analysis strategy. Rather than dating a number of grains and interpreting ages, the analytical strategy involves first, verifying that chronometer populations identified on compositional maps actually represent specific

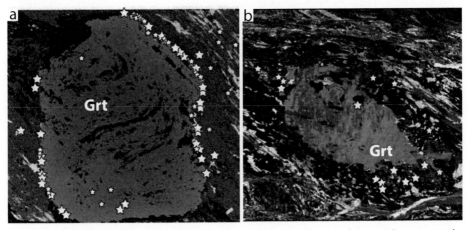

Figure 2. (a) Ca Kα compositional map of garnet from Vermont showing xenotime crystals concentrated on the margin of resorbed garnet. Modified from Gatewood et al. (2015). (b) Mg Ka map showing resorbed garnet from the Park Range, Colorado. Xenotime is concentrated in sinistral strain shadows of the garnet. The xenotime is interpreted to have been stabilized during garnet breakdown and release of Y.

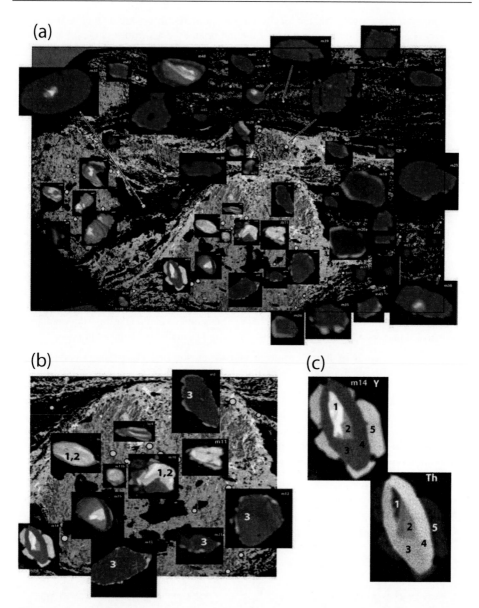

Figure 3. (a) Full-thin-section Mg Kα compositional map of sample S32D (Fig. 1) with superimposed high-resolution Y Lα monazite grain maps. (b) Close-up of central garnet. Note inclusions in inner core contain monazite generations Mz1 and Mz2. Outer inclusions are dominated by generation Mz3. (c) High-resolution Y Lα map of Mz14 showing all five generations of monazite. See text for discussion. Modified from Mahan et al. (2006) and Villa and Williams (2012).

compositional and geochronological populations and second, dating ("sampling") each population in order to constrain the age of the particular assemblages or fabrics.

Trace element maps of silicate phases can add additional insights. Trace-element maps can be acquired by EPMA, and are particularly straightforward in minerals such as garnet and olivine that are stable under high sample current (Spear and Kohn 1996; Pyle and Spear 1999;

Goodrich et al. 2013; Kohn 2013). Trace element mapping has also recently been done by LA-ICP-MS (Kylander-Clark 2017; Lanari and Engi 2017). With several notable exceptions, this is a largely untapped opportunity for relating chronometer phases to silicate assemblages. Pyle and Spear (1999) used high-current mapping of Y in garnet and documented the close coupling between Y in garnet and xenotime in metamorphic rocks over a range of metamorphic grades from garnet nucleation to partial melting. The xenotime–garnet relationships provide insights into parts of the *P–T* history that are not well constrained by the major elements and silicate phases alone; several thermometers have been calibrated for Y partitioning between garnet and xenotime, and the xenotime itself can be dated, commonly by electron probe, in order to place timing constraints on parts of the prograde and retrograde *P–T* path.

It is worth noting that many geochronological studies involve dating a large number of chronometer grains/domains, plotting dates on an "age histogram" and then interpreting populations. This "top-down" approach is useful for detrital mineral analysis, but is less appropriate for analysis of metamorphic or igneous rocks. The "bottom-up" approach (Williams et al. 2006) involves: (1) establishing populations, based on composition and texture; (2) developing hypotheses about chronometer-forming reactions and relative timing; and (3) constraining the age of the populations using the most appropriate analytical technique. For chronometers such as monazite and xenotime, narrow rim domains can be very important for constraining the petrologic history. For these domains, electron microprobe total Pb dating may be the only analytical option.

Trace-element analysis by electron microprobe—analytical considerations

The electron microprobe was initially developed for, and most applications involved, rapid non-destructive major and minor element analysis. However, even during the early development, some workers (e.g., Goldstein and Wood 1966; Goldstein 1967) recognized that the instrumentation had great utility in the trace element realm (below 1 wt.%). In the modern age, where LA-ICP-MS and Ion Microprobe instruments can analyze a broad suite of trace elements with low detection limits (e.g., Kylander-Clark 2017; Schmitt and Vazquez 2017), EPMA still has numerous applications in the measurement of minor and trace components, especially when non-destructive analysis and high spatial resolution are important. Further, during the past several decades, there have been a number of significant improvements in hardware, software, and analytical procedures that have increased EPMA precision and accuracy for trace element analysis.

U–Pb dating (i.e., total-Pb dating) by EPMA is a relatively new application (Parslow et al. 1985; Bowles 1990; Suzuki and Adachi 1991; Asami et al. 1996; Montel et al. 1996; Cocherie et al. 1998; Williams et al. 1999). It is based on the assumption that common Pb is insignificant compared to radiogenic Pb in U- and Th-rich (and Ca-poor) phases like monazite, xenotime, zircon, uraninite, or thorite. As such, accurate measurement of the total amount of Pb, U, and Th can be used to calculate a date from very small domains in the chronometer phases. As noted above, many compositional domains, especially rim domains and core domains, can be extremely small or narrow, but are critical for constraining *P–T* histories. Although Th can be present in monazite at the major or minor element level, U and Pb are almost always trace element measurements whose emission lines lie in a complicated part of the X-ray spectrum, particularly when REEs are present.

One lesson learned over the past decade is that it is not adequate to apply the major element analytical protocol to trace elements and simply increase the current and count time. Every aspect of the analytical routine, from sample (and standard) preparation to data processing, must be specifically designed for the trace elements involved and for the particular application (e.g., Scherrer et al. 2000; Jercinovic et al. 2008). Many aspects of the analytical problems and potential solutions have been described in other publications (Pyle et al. 2002; Williams et al. 2006; Jercinovic and Williams 2005; Jercinovic et al. 2008, 2012; Spear et al. 2009). The following paragraphs summarize some of the most recent improvements and some of the major

Table 1. Analytical Steps in EPMA Geochronology.

	Procedure	Instrument Setup
1	Find the accessory minerals in thin section (monazite, xenotime, zircon, thorite, uraninite, etc.) by full thin section mapping. Ce, Y, Zr, Mg, Ca.	15 kV, 300 nA, 25 ms/pixel, 35 mm beam diameter. 35 mm step, stage raster.
2	Process full section maps to overlay indicator elements on base map (reveals accessory mineral grain locations in relation to microstructure).	15 kV, 200 nA, 80 ms/pixel.
3	Micromap the accessory minerals to define compositional domains (e.g., Monazite: Y, U, Th, Ca, N).	focused beam. 0.5 mm step size.
4	Process maps: simultaneous and individual; superimpose maps onto full-section map for analysis.	
5	Select grains and domains for analysis based on compositional and microstructural significance.	
6	Recoat for quantitative analysis.	Al coat 25 nm, followed by C-coat 8 nm.
7	Quantitative analysis for full chemistry and age. Trace element methodology for Pb, U, and Th, as well as low concentration elements relevant to key reactions (see Table 3).	15 kV, 200 nA. Focused beam.

lessons learned especially with respect to trace element analysis for petrochronology. Key steps in the overall preparation and analytical procedure are summarized in Table 1.

Analytical strategy

An obvious goal in designing hardware, software, and analytical procedures for trace element analysis is to obtain the smallest possible detection limits within a practical setup/analytical timeframe. As formalized by Ziebold (1967), the propagated precision on the k-ratio is given by:

$$\sigma_k^2 = k^2 \left[\frac{\overline{N} + \overline{N}(B)}{n\left(\overline{N} - \overline{N}(B)\right)^2} + \frac{\overline{N}_s + \overline{N}_s(B)}{n'\left(\overline{N}_s - \overline{N}_s(B)\right)^2} \right] \qquad (1)$$

and similarly, the variance of the concentration is given by:

$$\sigma_C^2 = C^2 \left[\frac{\overline{N} + \overline{N}(B)}{n\left(\overline{N} - \overline{N}(B)\right)^2} + \frac{\overline{N}_s + \overline{N}_s(B)}{n'\left(\overline{N}_s - \overline{N}_s(B)\right)^2} \right] \times \left[1 - \frac{(a-1)C}{a} \right]^2 \qquad (2)$$

where σ_k = standard deviation, N = sample peak counts, $N(B)$ = sample bkg counts, N_s = std. peak counts, $N_s(B)$ are the std. bkg counts, $k = k$-ratio $= (N - N(B))/(N_s - Ns(B))$ for a pure element standard, n and n' = number of measured points on sample and standard, σ_c = concentration standard deviation, C = concentration, and a = correction factor that relates the k-ratio to concentration.

The relationships in Equations (1) and (2) describe the precision of the acquisition based on the Poisson statistics of X-ray emission. The equations also suggest where minor inaccuracies can strongly influence the result, particularly in the case of low-net-intensity unknowns such as trace elements. One clear implication is that the standard concentration for trace element analysis should be as high as possible to diminish the second term in equation 1. Then, for cases of diminishing peak/background, the precision of the result depends critically on: (1) maximizing peak counts (by increasing current, voltage, counting time, spectrometer collection efficiency, increasing the number of analysis points in an acquisition), and (2) the precision of the background, which ultimately determines the signal/noise level to be overcome in order for an element to be considered detectable. Each of these will be briefly discussed below. See also Ziebold (1967), Goldstein (1967), Ancey et al. (1978), Merlet and Bodinier (1990), and Lifshin et al. (1999) for formalisms of detection limit calculations.

The precision of an analysis in EPMA is clearly dependent on optimizing the total counts collected at the characteristic wavelength (or energy) for the element of interest. The simplest approach in the acquisition of more counts is to increase beam current and/or counting time, but as with many analytical variables, there are important trade-offs. High beam current with a small beam size results in high beam power density (Jercinovic et al. 2012) and subsequent beam damage and contamination. Long count time will also exacerbate time-dependent manifestations of beam damage and contamination, and also increase the potential for instrumental drift or internal charge effects. Another option is to acquire a population of peak acquisitions within a homogeneous compositional domain. However, the potential gain diminishes as the number of points increases above approximately 6. Jercinovic et al. (2012) showed that increasing the number of analysis points from 7 to 10 for a trace element analysis of Pb resulted in an increase the sensitivity of only 2 ppm. Other approaches are also very useful, including the use of multiple spectrometer simultaneous counting of the same line, with subsequent integration of counts. In this way, quite high sensitivity is possible with reasonable counting times, for example, a detection limit of 2–3 ppm is possible for Ti in quartz by integrating 5 spectrometers in 960 s acquisitions (Donovan et al. 2011). Additionally, the types of spectrometers employed can make a substantial difference, including the use of large and very-large monochromators (and carefully adjusted counter parameters, specific to the element being analyzed), and/or smaller radius focusing geometries (sacrificing spectral resolution). Higher voltage will also increase the ionization efficiency, but with the obvious trade-off of decreased spatial resolution and the possible introduction of increased spectral complexity resulting from higher energy ionizations (see Jercinovic et al. 2008).

One realization has been evident for decades—trace element sensitivity (lowering detection limit) results from the ability to characterize background at high precision as much as it does from maximizing peak counts (Reed 1993). Simple off-peak, two-point interpolation of background will not adequately account for curvature, particularly if positions are limited in complex matrices (Jercinovic et al. 2008, 2012). Background shape, a convolution of Bremsstrahlung emission vs. wavelength and spectrometer efficiency, must be directly assessed, and must also include the recognition of even the slightest component of interference in the background measurement regions of the spectrum. The combination of background curvature and interferences can produce disastrously inaccurate trace element results (Fig. 4) if not properly accounted for (Merlet and Bodinier 1990; Jercinovic et al. 2008). Further, because of the complexity of the spectrum in minerals such as monazite, xenotime, or zircon, is it extremely difficult to recognize small background interferences even on relatively high-resolution wavelength scans (see Williams et al. 2006; Jercinovic et al. 2008). Detailed WDS scanning is essential, but not completely satisfactory unless the precision of the WDS acquisition matches the precision of the quantitative analysis. As discussed below, errors due to background interference are systematic and reproducible for the domain being analyzed. The error could be severe even if secondary standards give acceptable results, and particularly if only a single secondary standard has been evaluated.

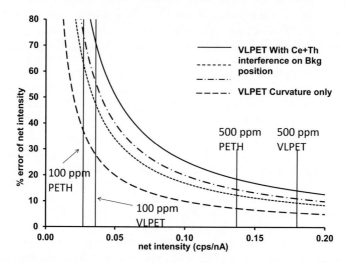

Figure 4. Growth of error (in %) on the net intensity (= Peak cps/nA – background cps/nA) as a function of net intensity for Pb Mα in monazite if a two-point linear interpolation is used. One bkg point was placed between Pb Mα and Pb Mβ, and the other at a suitable wavelength above the Pb Mα analytical line (above the interferences from first order Th Mζ_2, second order La Lα, and first order S Kα lines). There are two components to the error, one arises from the curvature of the background itself (dashed curves show this component only), and the other from the subtle interference that exists between the Pb Mα and Pb Mβ wavelengths. Two different monochromators are used for this estimation: PETH is the JEOL high intensity spectrometer, and VLPET is the Cameca very large PET monochromator + extended width detector. Note that at about 1500 ppm Pb, the error is a few percent, but below 500 ppm, the error on the estimate of the net intensity grows rapidly, approaching 100% at the 10 ppm level. Modified from Jercinovic et al. (2008) [Used by permission of Elsevier from Jercinovic et al. (2008) Chemical Geology, v, 254, Fig. 18, p. 213].

One approach to background characterization involves collecting a high-resolution wavelength scan of a broad region around the peak of interest and then selecting regions that are approximately background and regressing the background value at the peak position (see Williams et al. 2006). It is important to collect a very high-resolution scan and to avoid any fast-scan mode so that per-step count times are compatible with peak counts. This method has been successful when used for all reference materials and unknowns. However, there is a certain degree of subjectivity in choosing regions to be included in the regression analysis. An alternative method, called "multipoint background acquisition" (Allaz et al. 2011), involves acquisition of background intensities at multiple wavelength positions above and below the peak position. These points are exponentially regressed and evaluated for goodness of fit. When the regression meets certain criteria, the background value at the peak position is calculated. This method has the advantage of being objective and reproducible, and the regression statistics and background intensity can be evaluated and modified retroactively if necessary.

Analytical protocol

The protocol for quantitative trace element analysis is dynamic, depending on the questions being asked (see Pyle et al. 2005; Williams et al. 2006). Count times, background acquisition strategy, coating materials, etc. must be varied as the focus on key elements shifts to address particular reactions. Beam damage at high beam power density is problematic (Jercinovic et al. 2012), therefore the tradeoffs of beam diameter (plus scattering dimensions), beam current, and coating thickness/material all must be weighed. Table 2 lists the steps involved in the full quantitative EPMA characterization of phosphates such as monazite or xenotime. Additional comments and suggestions are included in Appendix 2 (deposited with MSA).

Table 2. Analytical Protocol for EPMA Trace Element Analysis.

Procedure	Setup
1 Coating. Samples and standards should be coated simultaneously for the best possible accuracy.	Plasma clean, then apply 20 nm Al, followed by 8 nm C.
2 Calibration. U, Th, Pb and, other major, minor, and interference calibrations as needed. After initial setup, routine recalibration is less routinely necessary as long a high quality secondary standards are available.	15 kV, 80 nA, 5 μm beam diameter.
3 Define the setup, including the use of multipoint background modeling, interference corrections, time-dependent intensity corrections, multiple spectrometer integration, and blank corrections as necessary. All elements should be acquired at each point, with count times adjusted to maximize precision for key elements.	
4 Analysis of secondary standards. This may include blanks to the protocol.	15 kV, 200 nA, focused beam.
5 Analyze unknown domain. A single homogeneous domain is analyzed at a time. Measure background (multipoint or WDS scanning). Accumulate peak measurements (typically 5–6 points) around the background location. Calculate weighted mean age, uncertainty, and MSWD.	15 kV, 200 nA, focused beam. Single background acquisition per domain.
6 Reanalyze consistency standard during and after session.	

The analytical philosophy for high-precision EPMA trace element analysis involves first, using compositional mapping to define compositionally homogeneous domains and second, independently constraining peak and background counts (count rates) at the highest possible precision. Because compositionally homogeneous domains are determined in advance by mapping key elements, it is possible to decouple peak and background measurement. For example, it has been noted that Th/LREE tends to control the relative background intensity in monazite domains, so a Th map is crucial in defining domains in monazite for background acquisition (Williams et al. 2007). Peak and background measurement strategy and protocol are developed independently for maximum precision considering sample stability. Optimally, background is measured using a multipoint regression scheme during a similar time window as peak measurements. Multiple peak measurements are made adjacent to the background position until uncertainty on the age calculations stabilizes at a minimum value (typically 5–7 peak measurements). The single background measurement is used with all peak measurements in order to calculate a single "date" for the domain. The MSWD (Mean Squared Weighted Deviation, Wendt and Carl 1991) value for the weighted mean of the multiple peak measurements provides a test of the compositional homogeneity; values significantly greater than unity generally indicate compositional heterogeneity within the analytical domain.

Error assessment

Propagated counting statistics in EPMA account for only a portion of the total uncertainty on a trace element analysis, or calculated physical parameter based on trace concentrations (i.e., temperature, pressure, or age, etc.). Systematic errors, arising particularly from instrumental factors or nonrandom analytical factors are particularly serious as they may not be detected by comparisons with secondary standards. Pyle et al. (2005) highlighted many of the factors contributing to the overall error, and in particular, suggested ways to minimize systematic errors. Williams et al. (2006) suggested that one might distinguish three components of error: (1) short-term random error—primarily counting statistics; (2) short-term systematic error—primarily from background methodology (regression models, etc.), but also coating variation and conductivity issues; (3) long-term systematic error—quality of standards and calibrations, interference correction algorithms, matrix corrections, current measurement, dead-time corrections, etc.

The first component includes primarily the propagated count statistical error. This error is readily calculated and is typically apparent from scatter in results. The third component produces errors that are fully reproducible when comparing results from the same chronometer measured on the same instrument. However, these errors may be reflected in systematic differences between the results of different dating techniques or instrument types. The second error component, as noted above, is much more problematical. It includes errors resulting from compositional effects such as peak or background interferences. They may seriously compromise results from one compositional domain but not from another, and unknowns may give erroneous results when secondary standards are accurately dated. As shown by Jercinovic et al. (2008), errors of this type could be on the order of tens of millions of years or several percent of the calculated age. Currently, relatively small, but significant, background interferences cannot be predicted from knowledge of the electromagnetic spectrum or sample composition. Thus, background estimation based on wavelength scanning or multipoint regression of all unknowns is essential for recognizing and accommodating this type of error. Careful background measurement by regression is commonly also necessary in measurements for overlap corrections on standards.

Results of EPMA geochronology have been published or presented in many venues. As with all types of geochronology, estimates of the precision and accuracy of the dates must be evaluated in light of the analytical protocol and specific components incorporated in the error estimation. In particular, systematic errors associated with background estimation (type-2 above) can be very large and are generally not included in published results. Background estimations based on any two-point interpolation have the potential for large systematic interference-related error that must be incorporated into estimates of accuracy or even comparison of age determinations. In addition, small uncertainties calculated simply by taking the mean of large numbers of measurements must be viewed with caution as this tends to yield a misleading average with inappropriate uncertainty, especially when the total range of dates is much larger than twice the calculated error estimate for a homogeneous dataset.

Because of the strong dependency on phase composition and analytical methodology, it is difficult to provide generalizations about the precision of EPMA geochronology. Uncertainties increase dramatically in U- or Th-poor chronometers and also increase in young samples because of low Pb abundance. Because trace element analysis by EPMA is essentially a measurement of minor signals above background, as peak heights diminish, minor inaccuracies in background can produce large errors in net intensity and calculated concentrations and ages. Spear et al. (2009) suggested that 2% uncertainties might be a best estimate for EPMA ages using standard hardware (and this estimate does not include background-related systematic error). Using optimized hardware, such as the SX-Ultrachron microprobe, and incorporating background regression procedures (i.e., Allaz et al. 2011), errors approaching 1% (and for older samples errors less than 1%) are possible and have been demonstrated using multiple standards analyzed by multiple techniques in multiple facilities.

The examples below show the preferred manner of presenting EPMA monazite data (see Fig. 7). Each probability density function (PDF) represents results from one compositional domain, that is, one background analysis and 5–7 peak analyses. The width of the PDF plot (the calculated 2-sigma error on the population of measurements) reflects the propagated error from peak and background analysis. The magnitude of uncertainty reflects the abundance of U, Th and Pb in the monazite domain, but large deviations from expected PDF uncertainty commonly indicate compositional heterogeneity within the domain. Weighted means can be calculated from individual dates, aggregating results from a particular generation of monazite.

APPLICATIONS AND EXAMPLES OF EPMA PETROCHRONOLOGY

Textural and compositional correlation between accessory and major phases

Compositional mapping by electron microprobe, at multiple scales, is an essential part of petrochronologic analysis regardless of the instrument or technique ultimately used for dating. As noted above, the mapping can provide the critical linkage between silicate assemblages and chronometers. For example, the linkage between Y in monazite and Y in garnet is exceedingly powerful and has been extensively used in petrochronology, especially EPMA-based studies (Pyle and Spear 2003; Foster et al. 2004; Mahan et al. 2006; Williams et al. 2007). This is partly because Y is abundant enough in monazite and garnet to be readily measured by electron probe. Also, in medium and high-grade metamorphic rocks (those without xenotime), monazite and garnet are the major hosts for Y, and thus, changes in the Y content of monazite commonly reflect modal changes in garnet, and because garnet is one of the key petrologic index minerals, constraining the timing of garnet growth and break-down events commonly provides some of the most critical timing constraints on the tectonic history.

Monazite–garnet–yttrium connection—Example 1: Legs Lake shear zone, Saskatchewan

One example of integrated compositional mapping and petrologic analysis, and the power of the Y–monazite–garnet connection, comes from Mahan et al. (2006) but has been expanded for use here. Mahan et al. (2006) investigated the timing and significance of the Legs lake shear zone, Athabasca Granulite terrane, Saskatchewan. The shear zone separates granulite facies (~1.1 GPa) rocks from amphibolite facies (~0.5 GPa) rocks. Kinematic analysis indicates oblique thrust-sense shearing of the high-grade rocks over lower-grade rocks. The granulite facies rocks contain evidence for two high-P–T tectonic events, one at ca. 2.6–2.55 Ga and one at ca 1.9 Ga. Thrusting/shearing occurred at ca. 1.85 Ga, and is interpreted to have been associated with exhumation of the granulite facies rocks. Mahan et al. (2006) calculated isochemical phase diagrams for rocks outside of and within the shear zone, and interpreted the P–T history and reaction history of the rocks during their evolution from dry deep-crustal conditions to hydrous mid-crustal conditions (Fig. 5).

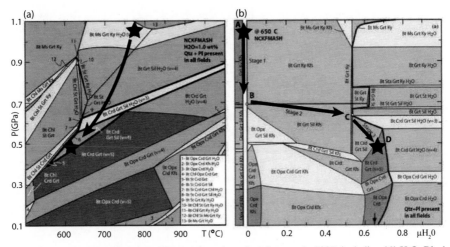

Figure 5. a) P–T isochemical phase diagram (pseudosection) for sample S32D including 1% H_2O. Black arrow shows inferred path during Legs Lake shear zone thrusting and exhumation based on peak and retrograde assemblages and reaction relationships (Mahan et al. 2006). b) P–M_{H_2O} pseudosection sample S32D. Black arrow shows path of decompression and hydration during Legs Lake shear zone thrusting. Thick black line marks H_2O saturation. Monazite population-5 is interpreted to reflect reaction B→C (Grt + Kfs + Pl1 + H_2O = Bt + Sil + Pl2 + Mz5 + Ap) [Used by permission of John Wiley and Sons from Mahan et al. (2006), Journal of Metamorphic Geology, v24, Figs. 6, 7, p.203,205].

Sample S32D was collected from within the Legs Lake shear zone. It preserves a particularly complete record of the metamorphic history including hydration associated with retrograde metamorphism. Figure 1 is a full-thin-section Mg Ka compositional map of sample S32D. The map shows several garnet porphyroblasts that have been fractured and partially replaced by cordierite (inside) and by biotite (outside). Stars mark the location of monazite grains, identified by full-thin section compositional mapping. Figure 3a shows the full-section map with high-resolution Y La maps of monazite grains superimposed. The maps were processed simultaneously so that intensities are comparable from grain to grain. Five monazite populations can be distinguished based on monazite composition (Fig. 6). Note that populations are here termed Mz1 through Mz5; Mahan et al. 2006 used the abbreviation pop-1 through pop-5. Monazite inclusions in garnet have been described by Mahan et al. (2006) and by Williams and Jercinovic (2012). Inclusions within innermost garnet cores are dominated by Mz1 and Mz2. Mz1 has high Y (up to several weight percent) and high Th content. Mz2 is characterized by distinctly lower Y content, reduced by at least 70%. Monazite inclusions in the outermost portions of garnet porphyroblasts are dominated by low-Y, low-Th Mz3 (Y contents of several hundred PPM or less). The very high Y content of Mz1 (comparable to monazite present in garnet absent rocks) suggests that this population probably grew before significant garnet growth. Because of the strong partitioning of Y into garnet, growth of even small amounts of garnet are associated with significant decreases of Y in monazite. The stepwise drop in Y from Mz1 to Mz2 to Mz3 (Fig. 6) is interpreted to represent two period(s) of garnet growth in the rock. This and the spatial separation of Mz2 and Mz3 within garnet suggests that these two populations may represent two distinct monazite producing reactions associated with two different garnet growth reactions. Mz3 contains the lowest Y content seen in the rock and is interpreted to represent the time of maximum garnet mode (i.e., volume percent). Th has been seen to increase from Mz2 to Mz3 in some monazite grains and to decrease in others. This may indicate the presence of limited amounts of partial melt where local access to melt yielded higher Th and isolation from melt yielded lower Th.

Most matrix monazite grains are relatively unzoned. Although rare cores of earlier monazite generations (Mz1 or Mz2) are locally present, most matrix grains are dominated by low-Y Mz4. Locally, monazite grains display high-Y rims (Mz5). These rims are generally restricted to monazite grains that are within several hundred μm of a garnet porphyroblast,

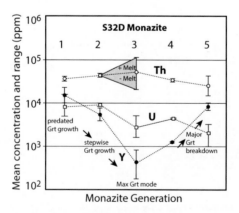

Figure 6. Plot showing variation in Y, Th and U among five monazite populations (Mz1–Mz5) in retrograde felsic granulite S32D from the Legs Lake shear zone, Athabasca area, Canada. [Used by permission of John Wiley and Sons from Mahan et al. (2006) *Journal of Metamorphic Geology*, v. 24, Fig. 8, p 208].

probably attesting to the limited mobility of Y in the matrix. In addition, monazite grains that are included in dynamically recrystallized plagioclase do not have the high-Y rims. This suggests that dynamic recrystallization of matrix plagioclase occurred after growth of Mz4 but before significant garnet break-down. That is, Mz4 monazite was trapped within recrystallizing plagioclase, and was shielded from later Mz5 overgrowths by the enclosing plagioclase.

Two monazite grains, of the 50 grains that were mapped, have all five populations (m14 and m48—Fig. 3). No other grain has more than three of the populations. We refer to these grains as "Rosetta-stone" grains (Dahl et al. 2005) because they are particularly important for petrochronology, i.e., linking chronology to tectonic events. Interestingly, the two grains are not among the largest monazite grains in the section, nor do they occupy the same microstructural setting. Monazite 14 occurs within the garnet rim and monazite 48 is in the matrix. Because of their small size, neither grain would probably have been recovered in a traditional mineral separation. Based on experience mapping a large number of samples from many localities, it is common to have a monazite population dominated by relatively simple one- or two-generation grains and then to have a small number of Rosetta-grains that capture important details of the history. This underscores the need to map a relatively large number of chronometer grains in a structurally and petrologically important sample.

Table 3 summarizes a set of observations that can be made from monazite compositional relationships in sample S32D. Some interpretations are rather speculative but they establish a set of hypotheses that can be tested and refined by quantitative monazite analysis and by analysis of other rocks. The next step is to document the composition and compositional variation within each monazite population and in the process, confirm that the map-defined populations are indeed homogeneous compositional populations. Then, the final step is to develop an analytical strategy to determine the age or age range for each population. This can be thought of as sampling each population, using the most appropriate geochronologic technique, in order to constrain the age.

Mahan et al. (2006) presented dates for monazite from the five populations using earlier EPMA techniques, and some new dates have been determined using techniques summarized above. Mz1, Mz2, and Mz3 are Archean (2.6–2.55 Ga) and confirm the Archean age of most garnet in the sample. Mz4 is Proterozoic (ca. 1.9 Ga), and corresponds to the second metamorphic event in the area. The presence of Mz4 inclusions in some garnet rims suggests that there may have been some Proterozoic garnet growth. Further, the very low Y content of both Mz3 and Mz4 suggests that there may not have been significant decompression (i.e., exhumation) between Mz3 and Mz4, i.e., between the Archean and the Proterozoic events. Garnet consumption associated with decompression is typically associated with growth of new Y-richer monazite. This is one piece of evidence supporting prolonged residence of the terrane in the deep crust, possibly from ca. 2.55 Ga to 1.9 Ga. Finally, high-Y Mz5 monazite constrains thrust-sense shearing and exhumation associated with the Legs Lake shear zone to have occurred between 1.9 and 1.85 Ga. The significantly increased Y in Mz5 is interpreted to reflect garnet breakdown during exhumation and decompression.

Sample S32D is an example of a "Rosetta-Stone" sample that preserves a particularly long and complete record of the tectonic history. Typical of many studies, once such a sample is discovered and characterized, other samples can fill in missing details of the *P–T–t–D* history. For example, the Legs Lake shear zone is locally cut by one other shear zone, the Grease River shear zone. Within this zone, the Legs Lake assemblages once again have been reequilibrated. Biotite plus cordierite pseudomorphs after garnet such as those in sample S32D (Fig. 1) have been flattened and deformed into the new shear-related foliation, and replacement of cordierite by chlorite and epidote documents greenschist facies conditions. One new population of monazite has developed in the late-stage structures, postdating Mz5 and constraining the age of this latest shearing event, and of greenschist facies metamorphism, to ca. 1.80 Ga (Mahan et al. 2006; Dumond et al. 2013).

Table 3. Interpretive Petrochronologic History of Sample S32D.

1. Garnet growth occurred in at least two phases, probably from two garnet-producing reactions. Monazite Mz1 predates significant garnet growth. Mz2 postdates the first garnet growth phase and Mz3 postdates the second growth phase.

2. Mz1 and Mz2 are abundant and closely spaced in all garnet cores, but few Mz1 or Mz2 remnants are present outside of the inner cores of garnet. Thus, early monazite was consumed or recrystallized before growth of Mz3. This may represent a stage of melting as suggested by the synchronous growth of new garnet and high-Th nature of monazite Mz3.

3. The efficient removal of early monazite may also indicate a deformation phase between Mz2 and Mz3.

4. Little new garnet growth occurred after Mz3. Mz3 records the lowest Y (and Gd) contents of all monazite populations.

5. The similar Y-content of Mz3 and Mz4 suggests that little garnet consumption occurred between these two monazite populations.

6. Matrix monazite is dominated by Mz4, with few cores or remnants of earlier populations. This suggests a second period of efficient monazite consumption after Mz3 and after garnet growth.

7. Dynamically recrystallized plagioclase wraps around garnet and contains only Mz4 monazite. Deformation and recrystallization occurred during or after Mz4 and before Mz5. Plagioclase is unzoned and thus the composition equilibrated with Mz4–Mz5 phases.

8. Mz5 was associated with garnet consumption as indicated by the major increase in Y. Mz5 rims are not aligned with dominant fabric; many occur on perpendicular sectors. This the dominant matrix deformation had terminated before Mz5. This is consistent with the persistence of abundant early monazite.

9. Cordierite in garnet fractures contains inclusions of xenotime. Thus, at least the later stages of cordierite replacement of garnet occurred below the monazite $\Rightarrow$ xenotime reaction.

Monazite–garnet–Y connection. Example-2: dating deformation

Figure 7 is one example of EMPA reaction dating following the protocol described above and modified from Williams and Jercinovic (2012). The basic question concerned the timing of folding and of garnet growth in a metamorphosed sediment from the Northwest Territories, Canada. Inclusion fabrics suggest that garnet grew during folding of an earlier fabric (S1) that was sub-parallel to compositional layering. Based on Y and Th content, three generations of monazite were distinguished, one pre-garnet, one syn-garnet, and one post garnet. Once the timing hypothesis was formulated, each generation was analyzed several times and dates calculated. Figure 7b show results for pre and post-garnet domains and Figure 7c shows syn-garnet results superimposed. Taken together the results show that garnet growth and folding occurred at ca. 1900 Ma. and retrogression began soon afterward. As noted above, Figure 7 also shows the preferred manner of presenting EPMA monazite data.

Other compositional/textural linkages with silicate assemblages

Recently, a broader suite of trace elements and element ratios have been used to link monazite with silicate assemblages in EPMA-based and also in LA-ICP-MS-based studies. Th content in monazite may be particularly useful as a monitor of melting reactions. Th is partitioned into the melt phase relative to the residual phases especially for haplogranite melts (Keppler and Wyllie 1990; Stepanov et al. 2012), but Th is partitioned into monazite relative to haplogranite melt, i.e., the monazite/melt partitioning coefficient is greater than unity (Rapp et al. 1987; Xing et al. 2013). In addition, the solubility of monazite in granitic melt has been interpreted to decrease with increasing pressure (Dumond et al. 2015). This suggests

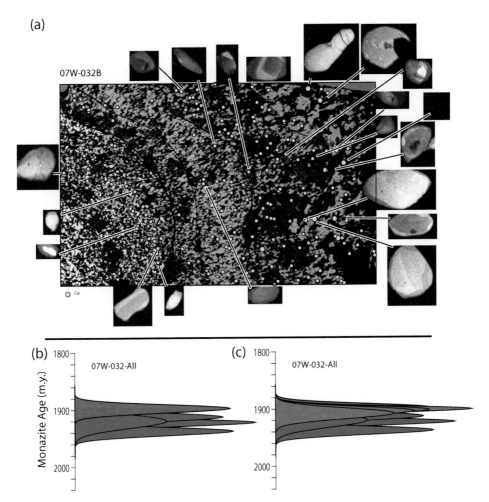

Figure 7. Monazite geochronology from west of Snowbird Lake, Northwest Territories, Canada. (a) Ca Kα WDS full-thin-section compositional map showing folded early compositional layering and axial plane S2 cleavage. High-resolution Y Lα compositional maps superimposed. Note, some grains are parallel to S1/S0 (also defined by inclusions in garnet), and some grains are parallel to S2 (upper left to lower right). Zoning in monazite is much more pronounced in Grt-rich layers. (b) Microprobe monazite geochronology. Each probability distribution plot represents one monazite date, including one background and 5–8 peak analyses (see text for discussion). Green curves are grains that are early with respect to S2 cleavage and garnet growth. Red curve is late with respect to S2 cleavage. (c) Same plot as (b) with addition of dates for syn-S2 and syn-garnet monazite. S2 cleavage and peak metamorphism (garnet growth) are constrained to be ca. 1900±8 Ma. Early stage of exhumation and garnet consumption are constrained to be 1893±12 Ma. Modified from Williams and Jercinovic (2012). [Used by permission of John Wiley and Sons, Williams and Jercinovic (2012) Journal of Metamorphic Geology, v30, Fig 5, p. 749].

that at higher pressures, monazite may provide key constraints on the timing and character of melt reactions. For example, Dumond et al. (2015) used phase equilibria to characterize melting reactions, adjust bulk compositions for melt loss, and interpret P–T–t–D histories for high-T granulite facies migmatites from the Athabasca Granulite Terrane (Fig. 8). They used multi-scale compositional mapping to integrate monazite into the reaction and melting history. The rocks contain multiple generations of monazite. The earliest generation is interpreted to represent detrital/inherited monazite. High-Th, low-Y monazite was interpreted to have been produced during biotite dehydration melting. Sharp drops in Th in later monazite generations were interpreted to reflect melt loss from the system, i.e., significant drops in bulk-rock Th content (Fig. 8). Thus, Th in monazite may not only serve as a monitor of melting but also of melt-extraction events in high-grade metamorphic rocks (Dumond et al. 2015).

More speculatively, Dumond et al. (2010, 2015) suggested that elevated Eu (and positive Eu anomalies) in monazite may be related to plagioclase break-down and to some degree, to sequestering of heavy REEs into garnet. In addition, light REE (especially Ce and La) enrichment has been suggested to be related to feldspar loss or recrystallization (Dumond et al. 2015; also see Bea 1996; Villaseca et al. 2003). These observations suggest a linkage between monazite and feldspar that may further allow monazite generations to be integrated into silicate reactions.

Because of the limited mobility of Rare Earth elements, especially in relatively dry rocks, it may be particularly useful to distinguish and compare different compositional layers or domains in rocks. For example, in the eastern Adirondack Mountains, monazite in garnet-rich gneiss has high-Y rims while in strongly recrystallized and lineated K-feldspar-rich layers monazite rims are subtly enriched in U. The increased U is interpreted to have been hosted by feldspar before recrystallization. The garnet rich and K-feldspar rich layers share a common fabric,

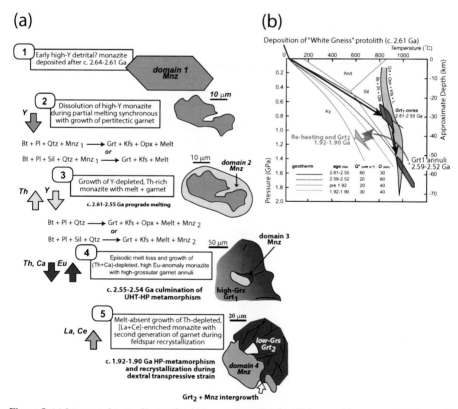

Figure 8. (a) Interpreted monazite reaction history and geochemical linkages with garnet generations and melting in high-*T* granulite facies migmatites from the Athabasca Granulite Terrane, Saskatchewan. (b) Interpreted *P–T* history based on petrologic modeling and monazite geochronology. See text for discussion. Modified from Dumond et al. (2015). [Used by permission of John Wiley and Sons from Dumond et al. (2015) Figures 15, 16, v.33, p 755–756].

mineral lineation, and kinematic shear sense, and both types of monazite rims have similar ages (Wong et al. 2012). Taken together, the two layer types suggest that garnet consumption (i.e., decompression), K-feldspar recrystallization, and extensional shearing were all synchronous. This allows the earliest phases of extensional shearing to be constrained and tied to early garnet resorption and exhumation in the eastern Adirondack Mountains (Wong et al. 2012).

In some samples, the limited mobility of trace elements is particularly apparent. An Al-rich gneissic sample from the Adirondack Mountains is distinctly layered with garnet-rich and garnet-poor layers. Monazite composition and dates are dramatically different from layer to layer. At least three distinct layer types can be recognized (Fig. 9). Layer-1 preserves only relatively young monazite that largely reflects the early stages of exhumation. This layer may have experienced more intense, late-stage strain, removing older monazite and promoting growth or recrystallization of new monazite. Layer-2 preserves older monazite (1180–1150 Ma). These are present as inclusions in garnet and also as matrix grains. It may be that these garnet-rich layers were stronger and thus less deformed during the late-stage deformation events. Layer-3 preserves a heterogeneous assemblage of monazite with dates that are more difficult to interpret. Some of these grains bear the unmistakable signature of dissolution-reprecipitation (see below) suggesting that this layer was affected by fluids of the right composition to interact with monazite (Harlov and Hetherington 2010; Williams et al. 2011). Taken together, monazite in this sample provides a very complete record of the petro-tectonic history from early high-grade metamorphism to late extensional collapse, but without careful multiscale mapping and in-situ analysis, accurate analyses would be impossible and the results would be very difficult to interpret.

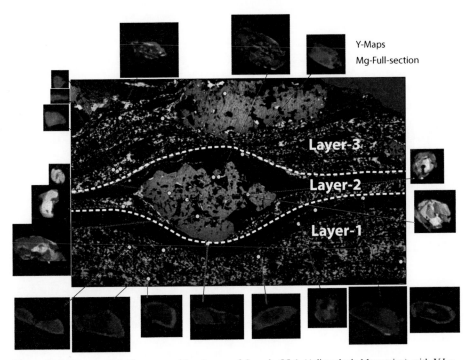

Figure 9. Full-section Mg Kα compositional map of Sample 85-1 (Adirondack Mountains) with Y Lα maps of monazite grains. Three different layer types have distinct monazite compositions and ages. Layer-1 grains contain exclusively 1050 Ma cores and 1030 Ma rims. Layer-2 grains have 1180 Ma core and 1150 Ma rims. Layer-3 grains have 1030 Ma domains and 980 Ma domains and complex textures characteristic of fluid-related recrystallization.

EPMA PETROCHRONOLOGY COMBINED WITH ISOTOPIC ANALYSIS

The electron microprobe also plays a major role in petrochronologic analysis in studies where the electron probe cannot be used for dating. Monazite and xenotime in relatively young metamorphic or igneous rocks (younger than approximately 100 m.y.) is generally not amenable to EPMA dating because of low Pb abundance. However, electron microprobe mapping and analysis of the chronometer phases and evaluation of the local reaction context are critical to drawing conclusions about the *P–T–t–D* history. Many studies in relatively young rocks have involved electron microprobe mapping and analysis followed by isotopic analysis by ion microprobe (i.e., see references in Vance et al. 2003; Kohn et al. 2005) or LA-ICP-MS (i.e., Larson et al. 2011; Regis et al. 2016). One of the disadvantages of this type of analysis is that many early core domains and late rim domains in chronometer phases are very small or narrow even for a focused laser or ion microprobe. Compositional mapping of a larger number of chronometer phases and identification of generations allow relatively larger domains to be dated and then correlated with smaller domains in distinctive structural or petrological setting.

Titanite is known to contain significant amounts of common Pb and thus cannot be readily dated using a total-Pb method such as electron microprobe dating. However, titanite compositions can vary widely and can be integrated into silicate reactions. Titanite is certainly amenable to the type of multi-scale compositional mapping and textural analysis as presented above for monazite. For example, Wintsch et al. (2005) used compositional mapping to identify several generations of titanite in granodioritic orthogneiss in Connecticut. The authors interpreted a reaction involving replacement of metastable magmatic K-feldspar, hastingsite, magnetite, and titanite by new metamorphic biotite, epidote, quartz and metamorphic titanite. Specifically, titanite was interpreted to have been produced by a reaction such as clinozoisite = plagioclase + Al-dominant titanite (Wintsch et al. 2005), and comparisons of REE composition of titanite and epidote were used to distinguish titanite generations. EPMA and SEM images were used to select Al-richer and Al-poorer titanite dating targets. Then, SHRIMP U–Pb dating was used to constrain ages. Although this is an example of petrochronology involving the electron microprobe, it is suspected that even more insight into the nature and relative timing of titanite producing reactions would result from integrated multiscale image analysis such as that described above.

One additional petrochronology application that involves an important component of EPMA analysis and also has great future potential is $^{40}Ar–^{39}Ar$ dating of mica, hornblende, and other K-bearing phases. The electron microprobe has been used to identify and distinguish compositional populations of micas or amphiboles and to generate phase diagrams incorporating the stability of different generations of these phases. Then, ages have been obtained by one of several possible methods: (1) in-situ laser techniques, (2) carefully micro-sampling distinct structural/petrologic domains, or (3) by interpreting distinct ages from step-heating results (e.g., Villa et al. 2000; Müller et al. 2002; White and Hodges 2003; Condon et al. 2006; Flowers et al. 2006b; Wells et al. 2008; Growdon et al. 2013; Schneider et al. 2013; Lanari et al. 2014; Chafe et al. 2014; Villa et al. 2014). One important strength of this approach is the possibility to exploit the $^{37}Ar–^{38}Ar–^{39}Ar$ correlation diagrams to fingerprint phases, and so to integrate deformation stages, as defined by mica fabrics and heterochemical generations, with *P–T* histories (Villa and Williams 2012).

Because K-feldspar, biotite, and especially white mica and amphibole can preserve microstructural relicts and the associated isotope inheritance even under medium and high-*T* conditions, all while new heterochemical mineral generations are growing, K–Ar chronometers are not substantially different from U–Pb chronometers such as monazite (Villa and Williams 2012), as both are "Class II" mineral chronometers (Villa 2016). $^{40}Ar–^{39}Ar$ chronology, in combination with detailed petrological characterization and dating of other geochronometer systems, has been particularly useful for constraining the retrogression history all the way from upper amphibolite facies to sub-greenschist facies. As new micas and feldspars tend to grow during retrograde events that take place relatively late, they have been used to constrain the exhumation stages of tectonic histories (i.e., White and Hodges 2003; Flowers et al. 2006b).

LOW-GRADE METAMORPHISM AND FLUID–ROCK INTERACTION

Many of the examples discussed above involve medium and high-grade metamorphism where reactions involving silicates and chronometer phases have been interpreted. Another rich avenue for petrochronologic study, and EPMA petrochronology, is in low-grade metamorphic rocks and hydrothermal processes. A number of reactions have been interpreted during diagenesis or low-grade metamorphism including: alteration of detrital monazite and growth of allanite, alteration of detrital monazite and growth of new monazite, growth of xenotime on detrital zircon, and growth of new generations of monazite, xenotime, or zircon during fluid infiltration events (Rasmussen 1996; Vallini et al. 2002; Rasmussen and Muhling 2007, 2009; Janots et al. 2007; Rasmussen et al. 2007; Allaz et al. 2013). In each case, detailed textural and compositional analysis of the chronometer phases have shed light on the nature and timing of digenetic and low-grade metamorphic reactions. Commonly, the authigenetic, digenetic, or low-grade metamorphic chronometer phases are very fine grained (less than 10 mm). The electron microprobe or SEM can be essential for simply finding and identifying the minerals and for quantifying compositions and ages. One of the most common applications of petrochronology in low-grade rocks is in dating and characterizing economic mineral and ore deposits (i.e., Rasmussen et al. 2006; Muhling et al. 2012; Zi et al. 2015).

Diffusional resetting of monazite, zircon, or xenotime ages is considered unlikely under most geological conditions because of extremely slow diffusion (Cherniak et al. 2004; Cherniak and Pyle 2008; Cherniak 2010). However, experimental work and empirical observations suggest that it is possible to reset zircon, monazite, and xenotime through dissolution-precipitation mechanisms (Seydoux-Guillaume et al. 2012; Villa and Williams 2012; Didier et al. 2013; Ruiz-Agudo et al. 2014), and that resetting may occur over a range of metamorphic grades (Harlov and Forster 2002; Harlov et al. 2007, 2011; Hetherington et al. 2008; Harlov and Hetherington 2010; Williams et al. 2011; Kelley et al. 2012). These observations have important implications for petrochronology because reset (recrystallized) chronometers may not be in equilibrium with the associated silicate assemblage and thus, may lead to incorrect timing constraints on the petrologic history (see Villa and Williams 2012). However, they also suggest a new avenue in petrochronology, that is, the accessory phases can place constraints on composition, character, and especially the timing of fluid–rock interactions (Grand'Homme et al. 2016).

The electron microprobe also has a role to play in identifying and characterizing the products of these dissolution-precipitation reactions. First, it is critical to recognize domains in zircon or monazite that have undergone dissolution-precipitation. The regions can have remarkably straight and crystallographically controlled boundaries and can be easily mistaken for overgrowth domains (see Harlov et al. 2011; Williams et al. 2011; Villa and Williams 2012). However, high-resolution compositional mapping and backscattered electron imaging can show narrow reset domains that follow inclusion trails, or cracks, and locally form delicate fingers intruding the core domains. Further, reset domains commonly have a distinctive composition, typically characterized by a more nearly end-member monazite, xenotime, or zircon composition than the host crystal (see discussion in Williams et al. 2011; also Didier et al. 2013). The electron microprobe can be used to investigate the composition of very small altered domains and for older chronometers to constrain the age of the alteration.

FUTURE TRENDS IN EPMA PETROCHRONOLOGY

Electron microprobe instrumental aspects

Electron Probe Micro-Analysis (EPMA) is a time-honored electron beam technique for the non-destructive quantitative analysis of micro-volumes in-situ. Since its inception in 1950, hardware, software, and physical theory have all advanced considerably, and continue to evolve to this day. Many advances are the result of technological improvements that have been inspired by the needs of the scientific community, and have primarily been aimed at improving sensitivity, accuracy, spatial resolution, and efficiency. For high spatial resolution trace element analysis and geochronology, a number of developments (some already in progress) can increase the sensitivity, spatial resolution, and especially, the efficiency of compositional mapping and quantitative analysis that are important for petrochronologic studies.

One recent emphasis centers around the improvement and refinement of low energy X-ray detection and quantification in order to exploit the low scattering volumes attainable at low accelerating voltage, resulting in higher analytical spatial resolution (e.g., McSwiggen et al. 2011; Armstrong et al. 2013; Hombourger and Outrequin 2013; Gopon et al. 2013; Susan et al. 2015). This development has been inspired in part by improvements in application of higher brightness sources (rare earth hexaboride and Schottky field emitters). However, the challenges are formidable particularly in trace element analysis, in which case high beam current and/or long counting time can compromise accuracy at high current density (Jercinovic et al. 2012), and more generally in evaluating soft X-ray spectra resulting from valence band transitions (Burgess et al. 2014; MacRae et al. 2016a,b). We expect significant progress over the next decade in the quantification of these difficult spectral emission regions, and regardless of difficulties in quantification, high spatial resolution mapping is extremely valuable. Quantification of WDS/EDS map data is also developing rapidly, even becoming useful in trace element applications (e.g., Donovan et al. 2016). The approach to petrochronologic studies with EPMA is mapping centric, and improvements in mapping efficiency, performance, signal integration (e.g., MacRae et al. 2016a) or processing will be advantageous, including rapid simultaneous processing (see Williams et al. 2006) of large sets of maps from multiple samples in characterizing geologic terrains.

The spectrometers currently deployed on WDS instruments cover just about any need, but there is room for improvement in terms of collection efficiency, specifically optimized monochromators, or detector gases and pressures for certain elements or element ranges and/or suppression of high order diffraction effects. These issues involve both high energy and low energy detection, for example, there are reasons to explore more efficient detection of the uranium L series using LIF [220], topaz [303], or α-quartz [20$\bar{2}$3], particularly in low Th monazite, xenotime, or zircon analysis. As discussed above, there is also renewed emphasis on low energy X-ray detection, including parallel beam optics (e.g., LEXS [EDAX, Inc.]), grating optics with CCD photon counters (Terauchi et al. 2012; MacRae et al. 2016b; Robertson and McSwiggen 2016), adaptation of silicon drift detector (SDD) technology to WDS (Moran and Wuhrer 2016), which may help both in improving collection efficiency as well as spectral resolution (e.g., Hombourger and Outrequin 2013; Robertson and McSwiggen, 2016). Obviously continued improvement in the energy resolution in energy dispersive detectors, and refinement of energy dispersive spectrometry for full integration into WDS based systems will be important particularly in improving efficiency for rapid analysis of many chronometer phases from multiple samples. So-called hyperspectral mapping datasets that combine signals for phase identification/classification offer a powerful approach to large scale mapping projects involving multiphase materials significant to petrochronology (e.g., MacRae et al. 2016c). This approach will undoubtedly gain significance in routine EPMA sample evaluation. In addition, EDS may be routinely employed in determination of actual accelerating potential in low kV systems, particularly where sample biasing is used to lower landing voltage and matrix corrections may be compromised at such low overvoltage if beam voltage is not accurately known.

Dual or multi-beam systems, and continued development of high performance sources will likewise be important in EPMA-based petrochronologic studies. Electron-laser dual beam integration may become tenable, allowing initial mapping and elemental analysis by EPMA, followed by high sensitivity LA mass spectrometry for further trace element evaluation and isotopic analysis. Continued development of electron sources for higher brightness at higher, more stable current, along with continued improvements in lowering energy spread for lowering chromatic aberration and improved performance at low kV. Multi-beam characterization machines are already finding use in life sciences, and may find utility in extremely high resolution geologic sample characterization as well, allowing detailed compositional analysis not accessible by X-ray tomography, and a scale not practicable by tomographic atom probe (Eberle et al. 2015).

As mentioned above, the very high current density required for high sensitivity—high spatial resolution trace element analysis is problematic. Therefore, continued exploration of surface conductive coatings and anti-contamination are an important frontier. In particular, as beam landing energy is lowered, more of the interaction volume involves interaction with the conductive coating along with enhanced contamination effects (e.g., Gopon et al. 2013). Advanced materials based on nanostructures such as graphene or stanine for high electrical and thermal conductivity may be developed that provide reliable surfaces for microanalysis (Park et al. 2016). Low contamination vacuum systems have improved remarkably over the past few decades, and improvements in vacuum technology will continue to provide even cleaner environments. Likewise, improvements in mounting media may replace epoxies that release carbon under vacuum and beam exposure. Anticontamination systems are already commonplace, and continue to be evaluated and improved. Advancements such as plasma cleaners are increasingly common in SEMs for low kV characterization, and are now being considered for EPMA instruments, as could other methods for sputtering or actively removing contaminants.

EPMA instrumentation includes electron detectors, specifically separate secondary and backscattered detectors. However, electron microscopy has benefited greatly from newer detection systems to better allow interrogation of the full electron spectrum at tunable energies to enhance phase contrast. These include in-lens designs that can offer energy discrimination as well as having the advantage of being less exposed to contaminants in the sample chamber. For petrochronology, any advancement in phase imaging is potentially useful.

Software development always plays an important role in advancing analytical technologies, and EPMA continues to see many exceptional improvements affecting all areas of data acquisition and processing. Improvements in mapping software that permit very rapid acquisition of a broad array of X-ray and other signals (electron, IR, visible, UV, EBSD, etc.), automatic identification and location of accessory phases, and even automated mapping of these phases at high resolution (e.g., MacRae et al. 2016c). Such mapping could potentially also take advantage of techniques devised to limit beam exposure to the grains of interest only (polygon or other shape-limited beam or stage motion), and potentially improve analytical efficiency significantly.

Thermochemical aspects

One major goal of future petrochronology research is to establish the thermodynamic properties and phase relationships of accessory chronometer phases. That is, to quantitatively incorporate accessory phases into phase diagrams. Thermodynamic modeling with modified thermodynamic databases such as that done by Spear and coworkers (Spear and Pyle 2010; Spear 2010) or Kelsey (i.e., Kelsey et al. 2008; Kelsey and Powell 2011) are pioneering studies in this context. In addition, experimental studies constraining phase relationships and trace-element partitioning (i.e., Rubatto and Hermann 2007; Stepanov et al. 2012) are needed to provide critical thermodynamic data and stability relationships.

New dissolution-precipitation experiments, such as those carried out by Harlov and coworkers (Harlov et al. 2007, 2011) are needed in order to better illuminate the controls on accessory phase recrystallization and replacement and importantly, the composition of the reactive fluids. Subtle differences in fluid composition can have a significant effect on the degree of recrystallization (Harlov and Hetherington 2010). This certainly can complicate petrochronological studies but also will provide new avenues for constraining fluid–rock interactions (Rasmussen et al. 2007; Rasmussen and Muhling 2009; Peterman et al. 2016).

A first step toward better integrating accessory phases into petrologic phase diagrams involves constraining, even qualitatively, the reactions that produce chronometer phases. The electron microprobe can quantify the more abundant trace elements in chronometers including monazite, titanite, xenotime, and zircon. However, the trend is clearly toward incorporating the broadest suite of trace elements possible. We see a future involving a combination of high-spatial resolution EPMA characterization with LA-ICP-MS analysis of larger domains for greater compositional resolution. It is important to stress that the electron microprobe is unsurpassed as a tool for characterizing spatial/compositional relationships by X-ray mapping in situ, the absolutely critical first step in petrochronology. Trace elements in associated silicate phases will generally need to be evaluated by LA-ICP-MS or ion microprobe. The classic studies of Bea are seminal (Bea et al. 1994; Bea 1996), but many more are needed, particularly those in which pre- and post-reaction compositions of silicate phases can be characterized. In all cases, these trace element studies need to be closely integrated with in-situ characterization and compositional mapping of accessory and silicate phases. As noted above, trace element mapping of silicate phases holds great potential, and in certain minerals, can be readily made by electron microprobe. For less abundant trace elements and less stable silicates, LA-ICP-MS (i.e., Koening 2010; Ubide et al. 2015; Kylander-Clark 2017; Lanari and Engi 2017), ion microprobe, or even synchrotron X-ray mapping (i.e., Dyl et al. 2014) may be necessary.

One major challenge for integrating accessory and silicate phases involves the different scales (and degrees) of equilibrium for trace elements and for major elements and even for different trace elements (Berger et al. 2005). Compositions of accessory phases may reflect a much smaller equilibrium domain and a more restricted assemblage compared to the silicate phases. Bulk-rock analyses from hand samples or thin-section chips have generally been extensively used for forward modeling of silicate assemblages. For trace element modeling, evaluation of local compositional subdomains may be more important (for example see Chernoff and Carlson 1999). Multiscale compositional mapping, including trace-element mapping, and analysis by electron microprobe is ever more essential for evaluating equilibrium domains and for selecting chronometer domains that reflect specific reactions (i.e., Fig. 9). New dating techniques and instrumentation will undoubtedly be developed allowing smaller and smaller domains to be dated more and more precisely. Because of its spatial resolution and increasing efficiency in major and trace element capability it is also clear that the EPMA will play a central role in petrologic and petrochronologic analysis for the foreseeable future.

ACKNOWLEDGMENTS

The authors sincerely than Harold Stowell, Philippe Goncalves, Igor Villa, and an anonymous reviewer for their careful and helpful review comments. In addition, we sincerely thank Pierre Lanari for excellent review comments and editorial handling. Jeffrey Webber and Sean Regan are thanked for their contributions to this research and for reading an earlier draft of the manuscript. The research presented in this manuscript was partly supported by NSF grants NSF/EAR-1419843 and NSF/EAR-1419876 to M.L. Williams for research in norther Saskatchewan and the Adirondack Mountains respectively.

REFERENCES

Allaz J, Williams ML, Jercinovic MJ, Donovan J (2011) A new technique for electron microprobe trace element analysis: the multipoint background method. EMAS Annual Meeting, Angers (France)

Allaz J, Selleck B, Williams ML, Jercinovic MJ (2013) Dating fluid events through the microprobe dating of detrital monazite from the Potsdam Sandstone, NY, USA. Am Mineral 98:1106–1119

Ancey M, Bastenaire F, Tixier R (1978) Application des methodes statistiques en microanalyse. *In*: Microanalyse, Microscopie Eléctronique à Balayage. Maurice F, Meny L, Tixier R, (eds). Les Exlitions du Physicien, Orsay, p 323–347

Armstrong JT, McSwiggen P, Nielsen3 C (2013) A thermal field-emission electron probe microanalyzer for improved analytical spatial resolution. Microsc Anal 27:18–22

Asami M, Suzuki K, Adachi M (1996) Monazite ages by the chemical Th-U-total Pb isochron method for pelitic gneisses from the eastern Sor Rondane Mountains, East Antarctica. Proc NIPR Symp Antarct Geosci 9:49–64

Bea F (1996) Residence of REE, Y, Th and U in granites and crustal protoliths; Implications for the chemistry of crustal melts. J Petrol 37:521–552, doi:10.1093/petrology/37.3.521

Bea F, Montero P (1999) Behavior of accessory phases and redistribution of Zr, REE, Y, Th, and U during metamorphism and partial melting of metapelites in the lower crust: an example from the Kinzigite Formation of Ivrea-Verbano, NW Italy. Geochim Cosmochim Acta 63:1133–1153, doi:10.1016/S0016-7037(98)00292–0

Bea F, Pereira MD, Stroh A (1994) Mineral/leucosome trace-element partitioning in a peraluminous migmatite (a laser ablation–ICP-MS study). Chem Geol 117:291–312, doi:10.1016/0009–2541(94)90133–3

Bea F, Montero P, Garuti G, Zacharini F (1997) Pressure-Dependence of Rare Earth Element Distribution in Amphibolite- and Granulite- Grade Garnets. A LA-ICP-MS Study. Geostand Newslett 21:253–270, doi:10.1111/j.1751-908X.1997.tb00674.x

Berger A, Scherrer NC, Bussy F (2005) Equilibration and disequilibration between monazite and garnet: indication from phase-composition and quantitative texture analysis. J Metamorph Geol 23:865–880, doi:10.1111/j.1525–1314.2005.00614.x

Bowles JFW (1990) Microanalytical methods in mineralogy and geochemistry age dating of individual grains of uraninite in rocks from electron microprobe analyses. Chem Geol 83:47–53, doi:10.1016/0009–2541(90)90139-X

Budzyń B, Harlov DE, Williams ML, Jercinovic MJ (2011) Experimental determination of stability relations between monazite, fluorapatite, allanite, and REE-epidote as a function of pressure, temperature, and fluid composition. Am Mineral 96:1547–1567, doi:10.2138/am.2011.3741

Buick IS, Clark C, Rubatto D, Hermann J, Pandit M, Hand M (2010) Constraints on the Proterozoic evolution of the Aravalli–Delhi Orogenic belt (NW India) from monazite geochronology and mineral trace element geochemistry. Lithos 120:511–528, doi:10.1016/j.lithos.2010.09.011

Burgess S, Li X, Holland J, Statham P, Bhadare S, Birtwistle D, Protheroe A (2014) Development of soft X-ray microanalysis using windowless SDD technology. Microsc Microanal 20 (suppl 3):1–2

Catlos EJ (2013) Versatile Monazite: resolving geological records and solving challenges in materials science. Generalizations about monazite: Implications for geochronologic studies. Am Mineral 98:819–832, doi:10.2138/am.2013.4336

Chafe AN, Villa IM, Hanchar JM, Wirth R (2014) A re-examination of petrogenesis and $^{40}Ar/^{39}Ar$ systematics in the Chain of Ponds K-feldspar: "diffusion domain" archetype versus polyphase hygrochronology. Contrib Mineral Petrol 167:1010, doi:10.1007/s00410-014-1010-x

Cherniak DJ (2010) Cation diffusion in feldspars. Rev Mineral Geochem 72:691–733

Cherniak DJ, Pyle JM (2008) Th diffusion in monazite. Chem Geol 256:52–61

Cherniak DJ, Watson EB, Grove M, Harrison TM (2004) Pb diffusion in monazite: a combined RBS/SIMS study. Geochim Cosmochim Acta 68:829–840

Chernoff CB, Carlson WD (1999) Trace element zoning as a record of chemical disequilibrium during garnet growth. Geology 27:555–558, doi:10.1130/0091–7613(1999)027<0555:tezaar>2.3.co;2

Clarke GL, Daczko NR, Nockolds C (2001) A method for applying matrix corrections to X-ray intensity maps using the Bence–Albee algorithm and Matlab. J Metamorph Geol 19:635–644, doi:10.1046/j.0263–4929.2001.00336.x

Cocherie A, Legendre O, Peucat JJ, Kouamelan A (1998) Geochronology of polygenetic monazites constrained by in situ electron microprobe Th–U–Total Pb determination: implications for Pb behavior in monazite. Geochim Cosmochim Acta 62:2475–2497

Condon DJ, Hodges KV, Alsop GI, White A (2006) Laser ablation $^{40}Ar/^{39}Ar$ dating of metamorphic fabrics in the Caledonides of North Ireland. J Geol Soc London 163:337–345, doi:10.1144/0016-764904-066

Dahl PS, Hamilton MA, Jercinovic MJ, Terry MP, Williams ML, Frei R (2005) Comparative isotopic and chemical geochronometry of monazite, with implications for U–Th–Pb dating by electron microprobe; an example from metamorphic rocks of the eastern Wyoming Craton (USA). Am Mineral 90:619–638

De Andrade V, Vidal O, Lewin E, O'Brien P, Agard P (2006) Quantification of electron microprobe compositional maps of rock thin sections: an optimized method and examples. J Metamorph Geol 24:655–668, doi:10.1111/j.1525-1314.2006.00660.x

Didier A, Bosse V, Boulvais P, Bouloton J, Paquette J-L, Montel J-M, Devidal J-L (2013) Disturbance versus preservation of U–Th–Pb ages in monazite during fluid–rock interaction: textural, chemical and isotopic in situ study in microgranites (Velay Dome, France). Contrib Mineral Petrol 165:1051–1072, doi:10.1007/s00410-012-0847-0

Donovan JJ, Lowers HA, Rusk BG (2011) Improved electron probe microanalysis of trace elements in quartz. Am Mineral 96:274–282, doi:10.2138/am.2011.3631

Donovan JJ, Singer JW, Armstrong JT (2016) EPMA method for fast trace element analysis in simple matrices. Am Mineral 101:1839–1853

Dumond, G. PG, Williams ML, Jercinovic MJ (2010) Subhorizontal fabric in exhumed continental lower crust and implications for lower crustal flow: Athabasca granulite terrane, western Canadian Shield. Tectonics 29 TC2006, doi:10.1029/2009TC002514

Dumond G, Mahan KH, Williams ML, Jercinovic MJ (2013) Transpressive uplift and exhumation of continental lower crust revealed by synkinematic monazite reactions. Lithosphere 5:507–512, doi:10.1130/l292.1

Dumond G, Goncalves P, Williams ML, Jercinovic MJ (2015) Monazite as a monitor of melting, garnet growth and feldspar recrystallization in continental lower crust. J Metamorph Geol 33:735–762, doi:10.1111/jmg.12150

Dyl KA, Cleverley JS, Bland PA, Ryan CG, Fisher LA, Hough RM (2014) Quantified, whole section trace element mapping of carbonaceous chondrites by synchrotron X-ray fluorescence microscopy: 1. CV meteorites. Geochim Cosmochim Acta 134:100–119, doi:10.1016/j.gca.2014.02.020

Eberle AL, Mikula S, Schalek R, Lichtman J, Tate MLK, Zeidler D (2015) High-resolution, high-throughput imaging with a multibeam scanning electron microscope. J Microsc 259:114–120, doi:10.1111/jmi.12224

Finger F, Broska I, Roberts MP, Schermaier A (1998) Replacement of primary monazite by apatite-allanite-epidote coronas in an amphibolite facies granite gneiss from the Eastern Alps. Am Mineral 83:248–258, doi:10.2138/am-1998-3-408

Flowers RM, Bowring SA, Williams ML (2006a) Timescales and significance of high-pressure, high-temperature metamorphism and mafic dike anatexis, Snowbird tectonic zone, Canada. Contrib Mineral Petrol 151:558–581

Flowers RM, Mahan KH, Bowring SA, Williams ML, Pringle MS, Hodges KV (2006b) Multistage exhumation and juxtaposition of lower continental crust in the western Canadian Shield; linking high-resolution U/Pb and $^{40}Ar/^{39}Ar$ thermochronometry with pressure–temperature–deformation paths. Tectonics 25, doi:10.1029/2005TC001912

Foster G, Parrish RR, Horstwood MSA, Chenery S, Pyle J, Gibson HD (2004) The generation of prograde *P–T–t* points and paths; a textural, compositional, and chronological study of metamorphic monazite. Earth Planet Sci Lett 228:125–142

Gatewood MP, Dragovic B, Stowell HH, Baxter EF, Hirsch DM, Bloom R (2015) Evaluating chemical equilibrium in metamorphic rocks using major element and Sm–Nd isotopic age zoning in garnet, Townshend Dam, Vermont, USA. Chem Geol 401:151–168, doi:10.1016/j.chemgeo.2015.02.017

Gibson HD, Carr SD, Brown RL, Hamilton MA (2004) Correlations between chemical and age domains in monazite, and metamorphic reactions involving major pelitic phases: an integration of ID-TIMS and SHRIMP geochronology with Y–Th–U X-ray mapping. Chem Geol 211:237–260

Goldstein JI, Wood F (1966) Experimental procedures for the determination of trace elements by electron probe microanalysis. First National Electron Probe Meeting, College Park, Md

Golostnn JI (1967) Distribution of germanium inthe metallic phases of some iron-meteorites. J Geophys Res 72:4689–4696

Goncalves P, Williams ML, Jercinovic MJ (2005) Electron-microprobe age mapping of monazite. Am Mineral 90:578–587

Goodrich CA, Treiman AH, Filiberto J, Gross J, Jercinovic M (2013) K_2O-rich trapped melt in olivine in the Nakhla meteorite: Implications for petrogenesis of nakhlites and evolution of the Martian mantle. Meteorit Planet Sci 48:2371–2405, doi:10.1111/maps.12226

Gopon P, Fournelle J, Sobol PE, Llovet X (2013) Low-voltage electron-probe microanalysis of Fe–Si compounds using soft X-rays. Microsc Microanal 19:1698–1708, doi:10.1017/S1431927613012695

Grand'Homme A, Janots E, Seydoux-Guillaume A-M, Guillaume D, Bosse V, Magnin V (2016) Partial resetting of the U–Th–Pb systems in experimentally altered monazite: Nanoscale evidence of incomplete replacement. Geology 44:431–434, doi:10.1130/g37770.1

Growdon ML, Kunk MJ, Wintsch RP, Walsh GJ (2013) Telescoping metamorphic isograds; evidence from $^{40}Ar/^{39}Ar$ dating in the Orange-Mulford Belt, southern Connecticut. Am J Sci 313:1017–1053

Harlov DE, Foerster H-J (2002) High-grade fluid metasomatism on both a local and a regional scale; the Seward Peninsula, Alaska, and the Val Strona di Omegna, Ivrea-Verbano Zone, northern Italy; Part II, Phosphate mineral chemistry. J Petrol 43:801–824

Harlov DE, Hetherington CJ (2010) Partial high-grade alteration of monazite using alkali-bearing fluids: Experiment and nature. Am Mineral 95:1105–1108

Harlov DE, Wirth R, Hetherington CJ (2007) The relative stability of monazite and huttonite at 300–900°C and 200–1000MPa: Metasomatism and the propagation of metastable mineral phases. Am Mineral 92:1652–1664

Harlov DE, Wirth R, Hetherington CJ (2011) Fluid-mediated partial alteration in monazite: the role of coupled dissolution–reprecipitation in element redistribution and mass transfer. Contrib Mineral Petrol 162:329–348, doi:10.1007/s00410-010-0599-7

Hermann J, Rubatto D (2003) Relating zircon and monazite domains to garnet growth zones: age and duration of granulite facies metamorphism in the Val Malenco lower crust. J Metamorph Geol 21:833–852, doi:10.1046/j.1525-1314.2003.00484.x

Hetherington CJ, Jercinovic MJ, Williams ML, Mahan KH (2008) Understanding geologic processes with xenotime: Composition, chronology, and a protocol for electron probe microanalysis. Chem Geol 254:133–147

Hickmott DD, Spear FS (1992) Major- and trace-element zoning in garnets from calcareous pelites in the NW Shelburne Falls Quadrangle, Massachusetts: Garnet growth histories in retrograded rocks. J Petrol 33:965–1005

Hombourger C, Outrequin M (2013) Quantitative analysis and high resolution X-ray mapping with a field emission electron microprobe. Microsc Today 21:10–15

Janots E, Brunet F, Goffé B, Poinssot C, Burchard M, Cemič L (2007) Thermochemistry of monazite-(La) and dissakisite-(La): implications for monazite and allanite stability in metapelites. Contrib Mineral Petrol 154:1–14, doi:10.1007/s00410-006-0176-2

Janots E, Engi M, Berger A, Allaz J, Schwarz JO, Spandler C (2008) Prograde metamorphic sequence of REE minerals in pelitic rocks of the Central Alps: implications for allanite–monazite–xenotime phase relations from 250 to 610 °C. J Metamorph Geol 26:509–526, doi:10.1111/j.1525-1314.2008.00774.x

Jercinovic MJ, Williams ML (2005) Analytical perils (and progress) in electron microprobe trace element analysis applied to geochronology: Background acquisition interferences, and beam irradiation effects. Am Mineral 90:526–546

Jercinovic MJ, Williams ML, Lane ED (2008) In-situ trace element analysis of monazite and other fine-grained accessory minerals by EPMA. Chem Geol 254:197–215

Jercinovic MJ, Williams ML, Allaz J, Donovan JJ (2012) Trace analysis in EPMA. IOP Conf Ser: Mater Sci Eng 32:012012

Kelly NM, Harley SL, Möller A (2012) Complexity in the behavior and recrystallization of monazite during high-*T* metamorphism and fluid infiltration. Chem Geol 322–323:192–208, doi:10.1016/j.chemgeo.2012.07.001

Kelsey DE, Powell R (2011) Progress in linking accessory mineral growth and breakdown to major mineral evolution in metamorphic rocks: a thermodynamic approach in the Na_2O–CaO–K_2O–FeO–MgO–Al_2O_3–SiO_2–H_2O–TiO_2–ZrO_2 system. J Metamorph Geol 29:151–166, doi:10.1111/j.1525-1314.2010.00910.x

Kelsey DE, Clark C, Hand M (2008) Thermobarometric modelling of zircon and monazite growth in melt-bearing systems; examples using model metapelitic and metapsammitic granulites. J Metamorph Geol 26:199–212, doi:10.1111/j.1525-1314.2007.00757.x

Keppler H, Wyllie PJ (1990) Role of fluids in transport and fractionation of uranium and thorium in magmatic processes. Nature 348:531–533

Koenig AE (2010) Methodology for detailed trace element mapping of garnet by laser ablation ICP-MS; a look at unraveling zoning and inclusions. Abstracts with Programs–Geological Society of America 42:627–628

Kohn M (2013) Geochemical zoning in metamorphic minerals. *In*: Treatise on Geochemistry, vol 3: The Crust. Rudnick R, (ed) Elsevier, p. 249–280

Kohn MJ, Spear F (2000) Retrograde net transfer reaction insurance for pressure–temperature estimates. Geology 28:1127–1130, doi:10.1130/0091-7613(2000)28<1127:rntrif>2.0.co;2

Kohn MJ, Wieland MS, Parkinson CD, Upreti BN (2005) Five generations of monazite in Langtang gneisses; implications for chronology of the Himalayan metamorphic core. J Metamorph Geol 23:399–406

Krenn E, Finger F (2004) Metamorphic formation of Sr-apatite and Sr-bearing monazite in a high-pressure rock from the Bohemian Massif. Am Mineral 89:1323–1329

Kylander–Clark ARC (2017) Petrochronology by laser–ablation inductively coupled plasma mass spectrometry. Rev Mineral Geochem 83:183–198

Kylander-Clark ARC, Hacker BR, Cottle JM (2013) Laser-ablation split-stream ICP petrochronology. Chem Geol 345:99–112, doi:10.1016/j.chemgeo.2013.02.019

Lanari P, Engi M (2017) Local bulk composition effects on metamorphic mineral assemblages. Rev Mineral Geochem 83:55–102

Lanari P, Rolland Y, Schwartz S, Vidal O, Guillot S, Tricart P, Dumont T (2014) *P–T–t* estimation of deformation in low-grade quartz–feldspar-bearing rocks using thermodynamic modelling and $^{40}Ar/^{39}Ar$ dating techniques: example of the Plan-de-Phasy shear zone unit (Briançonnais Zone, Western Alps). Terra Nova 26:130–138, doi:10.1111/ter.12079

Larson KP, Cottle JM, Godin L (2011) Petrochronologic record of metamorphism and melting in the upper Greater Himalayan sequence, Manaslu–Himal Chuli Himalaya, west-central Nepal. Lithosphere 3:379–392, doi:10.1130/l149.1

Lifshin E, Doganaksoy N, Sirois J, Gauvin R (1999) Statistical considerations in microanalysis by energy-dispersive spectrometry. Microsc Microanal 4:598–604

MacRae CM, Wilson NC, Torpy A, Bergmann J, Takahashi H (2016a) Holistic mapping—first results combining SXES, Windowless-SDD and CL spectrometry in an EPMA. MAS EPMA 2016 Topical Conference Program Guide with Abstracts 82–83

MacRae CM, Wilson NC, Torpy A, Bergmann J, Takahashi H (2016b) Collecting and analyzing 1.6eV–20keV emission spectra in an EPMA. Microsc Microanal 22 (Suppl 3):410–411

MacRae CM, Torpy A, Glenn AM, Pownceby MI, Grey IE, Wilson NC, Pundas P (2016c) Zircon zonation and trace chemistry characterized by mapping and analysis. MAS EPMA 2016 Topical Conference Program Guide with Abstracts 110–111

Mahan KH, Goncalves P, Williams ML, Jercinovic MJ (2006) Dating metamorphic reactions and fluid flow: application to exhumation of high-*P* granulites in a crustal-scale shear zone, western Canadian Shield. J Metamorph Geol 24:193–217

McFarlane CRM, Connelly JN, Carlson WD (2005) Monazite and xenotime petrogenesis in the contact aureole of the Makhavinekh Lake Pluton, northern Labrador. Contrib Mineral Petrol 148:524–541

McSwiggen P, Mori N, Takakura M, Nielsen C (2011) Improving analytical spatial resolution with the JEOL field emission electron microprobe. Microsc Microanal 17 (Suppl 2):624–625

Merlet C, Bodinier J-L (1990) Electron microprobe determination of minor and trace transition elements in silicate minerals: A method and its application to mineral zoning in the peridotite nodule PHN 1611. Chem Geol 83:55–69, doi:10.1016/0009–2541(90)90140–3

Montel J, Foret S, Veschambre M, Nicollet C, Provost A (1996) Electron microprobe dating of monazite. Chem Geol 131:37–53

Moran K, Wuhrer R (2016) Current state of combined EDS–WDS quantitative X-ray mapping. Microsc Microanal 22 (Suppl 3):92–93

Muhling JR, Fletcher IR, Rasmussen B (2012) Dating fluid flow and Mississippi Valley type base-metal mineralization in the Paleoproterozoic Earaheedy Basin, Western Australia. Precambrian Res 212–213:75–90, doi:10.1016/j.precamres.2012.04.016

Müller W, Kelley SP, Villa IM (2002) Dating fault-generated pseudotachylytes: comparison of [40]Ar/[39]Ar stepwise-heating, laser-ablation and Rb–Sr microsampling analyses. Contrib Mineral Petrol 144:57–77, doi:10.1007/s00410-002-0381-6

Overstreet WC (1967) The geologic occurrence of monazite. U S Geological Survey Professional Paper

Park JB, Kim YJ, Kim SM, Yoo JM, Kim Y, Gorbachev R, Barbolina II, Kim SJ, Kang S, Yoon MH, Cho SP (2016) Non-destructive electron microscopy imaging and analysis of biological samples with graphene coating. 2D Mater 3:045004

Parrish RR (1990) U–Pb dating of monazite and its application to geological problems. Can J Earth Sci 27:1431–1450, doi:10.1139/e90-152

Parslow GR, Brandstatter F, Kurat G, Thomas D (1985) Chemical ages and mobility of U and Th in anatectites of the Cree Lake Zone, Saskatchewan. Can Mineral 23:543–551

Peterman EM, Snoeyenbos DR, Jercinovic MJ, Kylander-Clark A (2016) Dissolution–reprecipitation metasomatism and growth of zircon within phosphatic garnet in metapelites from western Massachusetts. Am Mineral 101:1792–1806, doi:10.2138/am-2016-5524

Pyle JM, Spear FS (1999) Yttrium zoning in garnet: coupling of major and accessory phases during metamorphic reactions. Geol Mater Res 1:1–49

Pyle JM, Spear FS (2003) Four generations of accessory-phase growth in low-pressure migmatites from SW New Hampshire. Am Mineral 88:338–351

Pyle JM, Spear FS, Wark DA (2002) Electron microprobe analysis of REE in apatite, monazite and xenotime; protocols and pitfalls. Rev Mineral Geochem 48:337–362

Pyle JM, Spear FS, Wark DA, Daniel CG, Storm LC (2005) Contributions to precision and accuracy of monazite microprobe ages. Am Mineral 90:547–577, doi:10.2138/am.2005.1340

Rapp RP, Ryerson FJ, Miller CF (1987) Experimental evidence bearing on the stability of monazite during crustal anaatexis. Geophys Res Lett 14:307–310, doi:10.1029/GL014i003p00307

Rasmussen B (1996) Early-diagenetic REE-phosphate minerals (florencite, gorceixite, crandallite, and xenotime) in marine sandstones; a major sink for oceanic phosphorus. Am J Sci 296:601–632, doi:10.2475/ajs.296.6.601

Rasmussen B, Muhling JR (2007) Monazite begets monazite: evidence for dissolution of detrital monazite and reprecipitation of syntectonic monazite during low-grade regional metamorphism. Contrib Mineral Petrol 154:675–689, doi:10.1007/s00410-007-0216-6

Rasmussen B, Muhling JR (2009) Reactions destroying detrital monazite in greenschist-facies sandstones from the Witwatersrand basin, South Africa. {Stepanov, 2012 #1783}Chem Geol 264:311–327, doi:10.1016/j.chemgeo.2009.03.017

Rasmussen B, Sheppard S, Fletcher IR (2006) Testing ore deposit models using in situ U–Pb geochronology of hydrothermal monazite: Paleoproterozoic gold mineralization in northern Australia. Geology 34:77–80, doi:10.1130/g22058.1

Rasmussen B, Fletcher IR, Muhling JR (2007) In situ U–Pb dating and element mapping of three generations of monazite: Unravelling cryptic tectonothermal events in low-grade terranes. Geochim Cosmochim Acta 71:670–690, doi:10.1016/j.gca.2006.10.020

Reed SJB (1993) Electron Microprobe Analysis. Cambridge University Press, Cambridge

Regis D, Warren CJ, Mottram CM, Roberts NMW (2016) Using monazite and zircon petrochronology to constrain the *P–T–t* evolution of the middle crust in the Bhutan Himalaya. J Metamorph Geol 34:617–639, doi:10.1111/jmg.12196

Robertson VE, McSwiggen P (2016) Low Voltage, high spatial resolution, field emission SEM imaging coupled with new low energy X-ray spectrometers, a novel new technique—successes and challenges. MAS EPMA 2016 Topical Conference Program Guide with Abstracts:86–87

Rocha BC, Moraes R, Moller A, Cioffi CR, Jercinovic MJ (2017) Timing of anataxis and melt crystallization in the Socorro-Guaxupe Nappe, SE Brazil: Insights from trace element composition of zircon, monazite, and garnet coupled to U–Pb geochronology. Lithos 277:337–355, doi: 10.1016/j.lithos.2016.05.020

Rubatto D, Hermann J (2007) Experimental zircon/melt and zircon/garnet trace element partitioning and implications for the geochronology of crustal rocks. Chem Geol 241:38–61, doi:10.1016/j.chemgeo.2007.01.027

Rubatto D, Hermann J, Buick IS (2006) Temperature and bulk composition control on the growth of monazite and zircon during low-pressure anatexis (Mount Stafford, central Australia). J Petrol 47:1973–1996

Ruiz-Agudo E, Putnis CV, Putnis A (2014) Coupled dissolution and precipitation at mineral–fluid interfaces. Chem Geol 383:132–146, doi:10.1016/j.chemgeo.2014.06.007

Scherrer NC, Engi M, Gnos E, Jakob V, Liechti A (2000) Monazite analysis; from sample preparation to microprobe age dating and REE quantification. Schweiz Mineral Petrogr Mitt 80:93–105

Schoene B, Baxter EF (2017) Petrochronology and TIMS. Rev Mineral Geochem 83:231–260

Schneider S, Hammerschmidt K, Rosenberg CL (2013) Dating the longevity of ductile shear zones: Insight from $^{40}Ar/^{39}Ar$ in situ analyses. Earth Planet Sci Lett 369–370:43–58, doi:10.1016/j.epsl.2013.03.002

Seydoux-Guillaume A-M, Montel J-M, Bingen B, Bosse V, de Parseval P, Paquette J-L, Janots E, Wirth R (2012) Low-temperature alteration of monazite: Fluid mediated coupled dissolution–precipitation, irradiation damage, and disturbance of the U–Pb and Th–Pb chronometers. Chem Geol 330–331:140–158, doi:10.1016/j.chemgeo.2012.07.031

Spear FS (2010) Monazite-allanite phase relations in metapelites. Chem Geol 279:55–62, doi:10.1016/j.chemgeo.2010.10.004

Spear FS, Kohn MJ (1996) Trace element zoning in garnet as a monitor of crustal melting. Geology 24:1099–1102

Spear FS, Pyle JM (2002) Apatite, monazite, and xenotime in metamorphic rocks. Rev Mineral Geochem 48:293–335

Spear FS, Pyle JM (2010) Theoretical modeling of monazite growth in a low Ca metapelite. Chem Geol 273:111–119, doi:10.1016/j.chemgeo.2010.02.016

Spear FS, Pyle JM, Cherniak D (2009) Limitations of chemical dating of monazite. Chem Geol 266:218–230, doi:10.1016/j.chemgeo.2009.06.007

Stepanov AS, Hermann J, Rubatto D, Rapp RP (2012) Experimental study of monazite/melt partitioning with implications for the REE, Th and U geochemistry of crustal rocks. Chem Geol 300–301:200–220, doi:10.1016/j.chemgeo.2012.01.007

Susan D, Grant RP, Rodelas JM, Michael JR, Maguire MC (2015) Comparing field emission electron microprobe to traditional EPMA for analysis of metallurgical specimens. Microsc Microanal 21 (Suppl 3):2107–2108

Suzuki K, Adachi M (1991) Precambrian provenance and Silurian metamorphism of the Tsunosawa paragneiss in the South Kitakami terrane, Northeast Japan, revealed by the chemical Th–U–total Pb isochron ages of monazite, zircon and xenotime. J Geochem 25:357–376

Terauchi M, Takahashi H, Handa N, Murano T, Koike M, Kawachi T, Imazono T, Koeda M, Nagano T, Sasai H, Oue Y (2012) Ultrasoft X-ray emission spectroscopy using a newly designed wavelength-dispersive spectrometer attached to a transmission electron microscope. J Electron Microsc 61:1–8

Tracy RJ (1982) Compositional zoning and inclusions in metamorphic minerals. Rev Mineral 10:355–397

Ubide T, McKenna CA, Chew DM, Kamber BS (2015) High-resolution LA-ICP-MS trace element mapping of igneous minerals; in search of magma histories. Chem Geol 409:157–168, doi:10.1016/j.chemgeo.2015.05.020

Vallini D, Rasmussen B, Krapež B, Fletcher IR, McNaughton NJ (2002) Obtaining diagenetic ages from metamorphosed sedimentary rocks: U–Pb dating of unusually coarse xenotime cement in phosphatic sandstone. Geology 30:1083–1086, doi:10.1130/0091-7613(2002)030<1083:odafms>2.0.co;2

Vance D, Müller W, Villa IM (eds) (2003) Geochronology: linking the isotope record with petrology and textures

Villa IM (2016) Diffusion in mineral geochronometers: Present and absent. Chem Geol 420:1–10, doi:10.1016/j.chemgeo.2015.11.001

Villa IM, Williams ML (2012) Geochronology of metasomatic events. *In:*Metasomatism and the Chemical Transformation of Rock. Harlov DE, Austrheim H (eds). Springer-Verlag, Heidelberg, p. 171–202

Villa IM, Hermann J, Müntener O, Trommsdorff V (2000) $^{39}Ar–^{40}Ar$ dating of multiply zoned amphibole generations (Malenco, Italian Alps). Contrib Mineral Petrol 140:363–381, doi:10.1007/s004100000197

Villa IM, Bucher S, Bousquet R, Kleinhanns IC, Schmid SM (2014) Dating polygenetic metamorphic assemblages along a transect across the Western Alps. J Petrol 55:803–830, doi:10.1093/petrology/egu007

Villaseca C, Martín Romera C, De la Rosa J, Barbero L (2003) Residence and redistribution of REE, Y, Zr, Th and U during granulite-facies metamorphism: behaviour of accessory and major phases in peraluminous granulites of central Spain. Chem Geol 200:293–323, doi:10.1016/S0009-2541(03)00200–6

Wells ML, Spell TL, Hoisch TD, Arriola T, Zanetti KA (2008) Laser-probe ^{40}Ar/^{39}Ar dating of strain fringes: Mid-Cretaceous synconvergent orogen-parallel extension in the interior of the Sevier orogen. Tectonics 27: TC3012, doi:10.1029/2007TC002153. doi:10.1029/2007TC002153

Wendt I, Carl C (1991) The statistical distribution of the mean squared weighted deviation. Chem Geol; Isotope Geosci Sect 86:275–285

White AP, Hodges KV (2003) Pressure–temperature–time evolution of the Central East Greenland Caledonides: quantitative constraints on crustal thickening and synorogenic extension. J Metamorph Geol 21:875–897, doi:10.1046/j.1525-1314.2003.00489.x

Whitney DL, Evans BW (2010) Abbreviations for names of rock-forming minerals. Am Mineral 95:185–187, doi:10.2138/am.2010.3371

Williams ML, Jercinovic MJ (2012) Tectonic Interpretation of metamorphic tectonites: integrating compositional mapping, microstructural analysis, and in-situ monazite dating. J Metamorph Geol 30:739–732, doi:10.1111/j.1525-1314.2012.00995.x

Williams ML, Jercinovic MJ, Terry M (1999) High resolution "age" mapping, chemical analysis, and chemical dating of monazite using the electron microprobe: A new tool for tectonic analysis. Geology 27:1023–1026

Williams ML, Jercinovic MJ, Goncalves P, Mahan KH (2006) Format and philosophy for collecting, compiling, and reporting microprobe monazite ages. Chem Geol 225:1–15

Williams ML, Jercinovic MJ, Hetherington CJ (2007) Microprobe monazite geochronology: understanding geologic processes by integrating composition and chronology. Ann Rev Earth Planet Sci 35:137–175

Williams ML, Jercinovic MJ, Harlov DE, Budzyn B, Hetherington CJ (2011) Resetting monazite ages during fluid-related alteration. Chem Geol 283:218–225, doi:10.1016/j.chemgeo.2011.01.019

Wing BA, Ferry JM, Harrison TM (2003) Prograde destruction and formation of monazite and allanite during contact and regional metamorphism of pelites; petrology and geochronology. Contrib Mineral Petrol 145:228–250

Wintsch RP, Aleinikoff JN, Yi K (2005) Foliation development and reaction softening by dissolution and precipitation in the transformation of granodiorite to orthogneiss, Glastonbury Complex, Connecticut, U.S.A. Can Mineral 43:327–347, doi:10.2113/gscanmin.43.1.327

Wong MS, Williams ML, McLelland JM, Jercinovik MJ, Kowalkoski J (2012) Late Ottawan extension in the eastern Adirondack Highlands: Evidence from structural studies and zircon and monazite geochronology. Geol Soc Am Bull 124:857–869, doi:10.1130/b30481.1

Xing L, Trail D, Watson EB (2013) Th and U partitioning between monazite and felsic melt. Chem Geol 358:46–53, doi:10.1016/j.chemgeo.2013.07.009

Yang P, Pattison D (2006) Genesis of monazite and Y zoning in garnet from the Black Hills, South Dakota. Lithos 88:233–253, doi:10.1016/j.lithos.2005.08.012

Yang P, Rivers T, Jackson SJ (1999) Crystal-chemical and thermal controls on trace-element partitioning between coexisting garnet and biotite in metamorphic rocks from western Labrador. Can Mineral 37:443–468

Zi J-W, Rasmussen B, Muhling JR, Fletcher IR, Thorne AM, Johnson SP, Cutten HN, Dunkley DJ, Korhonen FJ (2015) In situ U–Pb geochronology of xenotime and monazite from the Abra polymetallic deposit in the Capricorn Orogen, Australia: Dating hydrothermal mineralization and fluid flow in a long-lived crustal structure. Precambrian Res 260:91–112, doi:10.1016/j.precamres.2015.01.010

Ziebold TO (1967) Precision and sensitivity in electron microprobe analysis. Anal Chem 36:858–861

Reviews in Mineralogy & Geochemistry
Vol. 83 pp. 183–198, 2017
Copyright © Mineralogical Society of America

6

Petrochronology by
Laser-Ablation Inductively Coupled Plasma
Mass Spectrometry

Andrew R. C. Kylander-Clark

Department of Earth Science
University of California, Santa Barbara
Santa Barbara, CA 93106-9630
U.S.A.

kylander@geol.ucsb.edu

INTRODUCTION

Petrochronology is a field of Earth science in which the isotopic and/or elemental composition of a mineral chronometer is interpreted in combination with its age, thus yielding a more synergistic combination of petrology and chronology that can be used to interpret geologic processes. It has recently attracted renewed interest as technologies for mineral analysis have improved. Examples are many, and continue to grow, from the early adoption of U/Th ratios in zircon as an indicator for magmatic vs. igneous crystallization (e.g., Ahrens 1965), to using the Nd isotopic composition in titanite to track source contribution over time (see *Applications*; B. R. Hacker, personal communication). Age and chemical information can be obtained by a variety of techniques: electron microprobe (age; major and minor elements; see Williams et al. 2017), secondary ion mass spectrometry (SIMS; age; trace elements; isotopic ratios; see Schmitt and Vazquez 2017), and laser-ablation inductively coupled plasma mass spectrometry (LA-ICPMS; age; trace elements; isotopic ratios).

Laser-ablation ICPMS instrumentation and techniques, the focus of this chapter, have been employed as a petrochronologic tool for decades, starting with separate analyses of ages and elemental and/or isotopic compositions, which were then combined and interpreted. For example, Zheng et al. (2009) employed LA-ICPMS to analyze the trace-element (TE) chemistry, Hf isotopic composition, and age of zircons from kimberlites by using three spots on each zircon grain, one for each type of analysis. This work was relatively time consuming and expensive, given the required number of analytical sessions, but yielded far better confidence in the conclusions, because of the link between physical conditions (petrology) and time (chronology).

Instrumentation and techniques which employ LA-ICPMS have continued to improve, particularly in the ease with which petrochronologic data can be obtained. A single LA-ICPMS instrument can now measure both the age and TE composition of one spot in a grain. Alternatively, the laser aerosol can be split for simultaneous measurement of U–Pb age or isotopic composition on one instrument and age and/or elemental concentration on another. One of the first examples of the latter was given by Chen et al. (2010), in which—using the first laser-ablation split-stream (LASS) setup, described by Yuan et al. (2008)—the authors distinguished 4 different types of metamorphic zircon (re)crystallization, based on linked U–Pb age, Hf isotope ratio and TE composition analyses from ultrahigh-pressure rocks in China.

The purpose of this paper is to describe the analytical setup and measurement of age and chemistry of mineral chronometers used for petrochronology. LA-ICPMS techniques that measure only age or chemistry have been previously described in detail; the focus

1529-6466/17/0083-0006$05.00 (print)
1943-2666/17/0083-0006$05.00 (online)

http://dx.doi.org/10.2138/rmg.2017.83.6

herein is on recent analytical developments specifically related to petrochronology that have occurred within the last decade. Extensive reviews of the basics of LA-ICPMS instrumentation and analytical techniques can be found in Jackson et al. (1992), Jackson et al. (2004), Longerich (2008), Schaltegger et al. (2015) and references therein.

VIRTUES OF LA-ICPMS

Laser ablation has advantages and disadvantages. First, analysis is relatively fast, inexpensive and straightforward. Preparation of samples requires only a moderately well-polished grain mount or thin section (grains or rock fragments can also be mounted); the laser is much more forgiving of surface topography than SIMS, though pits and μm-scale scratches can negatively affect precision and accuracy. Improvements in the technique have increased the accuracy for U–Pb spot dating—now limited to ~ 1–2%—and most LA-ICPMS systems can achieve this result with ≤60 second analyses (e.g., Košler et al. 2013); TE analyses are similar in duration, but the ppm-level precision necessary for most isotopic analyses requires analyses of up to 2 minutes on advanced systems (e.g., Kylander-Clark et al. 2013).

This short analysis time (an order of magnitude faster than SIMS), and the lower cost of instrumentation, yields data with unprecedented ease, but is not without drawbacks. First, a significant amount of material is removed from the sample—unlike electron-probe microanalysis (EPMA) sampling, and much more than SIMS. Typical spot sizes for U–Pb age analyses can be as small as 10 μm wide by a few μm deep (~1 ng of material), for samples with ppm levels of radiogenic Pb. The most recent advances in reducing ablated material show that even a single pulse of the laser (~100 nm depth) can yield U–Pb dates with a precision of 5% and yield ages that are within 2% of accepted values (Cottle et al. 2009a, 2012; Viete et al. 2015). Isotopic tracer analyses such as for Hf require larger spots with a ~100 ng minimum of ablated material (e.g., a 50×20 μm spot). Second, although the limit of detection for TE is much better than for EPMA (sub-ppm vs. 10s of ppm respectively, depending on element, spot size, etc.), ICPMS precision and accuracy is generally worse than EPMA. Thus, when performing analyses by LA-ICPMS, it is desirable to perform all EPMA major and minor element analyses required for interpretation of the data prior to ablation of the sample, and, if sample conservation is critical, the extra time and cost of SIMS may be appropriate. Nevertheless, use of LA-ICPMS for petrochronology has increased at a near-exponential rate for the last 20 years and is arguably becoming the primary means by which to acquire isotopic and elemental data on mineral chronometers (Fig. 1).

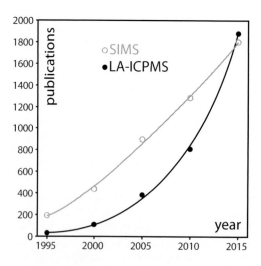

Figure 1. Publications involving petrochronology, with SIMS and LA-ICPMS, per year, summed over 5 year intervals. LA-ICPMS continues to grow at an exponential rate. Source: Google Scholar.

INSTRUMENTATION

Petrochronologic analyses by LA-ICPMS generally consist of two types: age+TE concentration, or age+isotopic composition. Three different mass spectrometers are generally employed for such analyses, and each instrument has its benefits and drawbacks. The following describes each system, and how it is typically used.

Multi-Collector (MC) ICPMS

A MC-ICPMS consists of an ICP source, a series of focusing lenses, an electrostatic analyzer (energy filter), a magnet (mass analyzer), and a detector array. The ion source of all ICP instruments is at atmospheric pressure, which therefore requires several step-down vacuum chambers, through which the ion beam travels on its way to the detector. Multi-collector instruments tend to be the largest, most precise instruments, and thus much effort is placed to reduce the pressure in the analyzer region to $\leq 10^{-8}$ mbar (running pressure). This has the effect of decreasing the possible number of collisions, thereby increasing sensitivity and improving abundance sensitivity. An MC-ICPMS uses multiple collectors to measure all isotopes of interest simultaneously. The advantage over a single-collector (SC) instrument (described below) is two-fold: first, sensitivity is increased linearly by measuring each isotope in a different collector. For example, for U–Pb measurements, 6 different isotopes—^{238}U, ^{232}Th, ^{208}Pb, ^{207}Pb, ^{206}Pb, and ^{204}Pb—are measured coincidentally on an MC-ICPMS, whereas each has to be measured sequentially on an SC-ICPMS. Given equal dwell time, 6 times as many counts are measured on the MC-ICPMS, increasing precision by ~2.5×. Second, because all isotopes are measured at the same time, any change in signal intensity affects all isotopic measurements at the same time, such that measured ratios remain constant. Multi-collector instruments are designed to be extremely stable, and as a result, it takes too long to switch the magnet (i.e., mass) during the course of a laser ablation analysis. Given their stability and sensitivity, MC-ICPMS is used exclusively for ratio measurements, both for U–Pb age determinations and single-element isotope ratio measurements such as for Sr (titanite, apatite, calcite), Nd (titanite, monazite), Hf (zircon), and Pb (apatite, titanite, calcite).

Single-collector (SC) ICPMS

SC-ICPMS instruments are inherently less sensitive than MC-ICPMS instruments. They are designed for rapid mass switching and therefore the instrument of choice in which a large number of masses are analyzed and precision is less important. Element concentration analysis fits into this category, but SC instruments can also be used for isotopic analyses, such as U–Pb, and in recent cases Rb–Sr (Zack and Hogmalm 2016). Two fundamental types of SC instruments exist, though both generally serve the same purpose: a quadrupole MS is less sensitive, and has less resolving power than a magnetic-sector MS, but is considerably less expensive and can be more flexible, especially if analysis across a large mass range is desirable. A magnetic-sector instrument requires relatively long settling times for magnet switching, so small mass jumps are obtained by steering the beam with a voltage potential. This minimizes the number of magnet positions during a single cycle.

Petrochronologic analyses can be performed solely by SC-ICPMS, whereby the TE composition and U–Pb age are determined entirely by one instrument, wherein each sweep of the mass range includes all necessary isotopes (e.g., Iizuka and Hirata 2004; Yuan et al. 2008; Kohn and Corrie 2011; Kylander-Clark et al. 2013). This technique can be sufficient in many cases, however, compared to LASS analyses in which a MC-ICPMS is used for U–Pb age determination, this yields relatively poor precision on the age, and requires larger pits, more analyses, or both, to yield the ultimate precision on an age of 1–2%.

Laser

The laser-ablation system may be the most critical component of any LA-ICPMS setup. Two main components of the LA system affect the quality of the data: 1) laser wavelength and pulse width, and 2) geometry of the sample cell. Absorbance of light by minerals increases as the wavelength of the light decreases. Lasers with very short (femtosecond) pulse widths have been shown to minimize melting of the sample (Russo et al. 2002), which produces volatility-controlled inter-element fractionation as the laser pit is progressively deepened—an effect referred to as "down-hole laser-induced elemental fractionation (LIEF)". Thus, instrumentation has moved towards employing lasers with shorter wavelength and shorter pulse widths. Both yield more consistent and cleaner ablation craters, producing particles with smaller sizes and more uniform size distributions (Guillong et al. 2003), which are in turn ionized more efficiently in the ICP. This increase in sensitivity allows smaller pit depths, reducing LIEF, which is a major factor limiting the application of analyses of standards to analyses of unknowns. Because LIEF varies least within a single mineral type, it is best to use a matrix-matched standard, i.e., a standard of the same mineral as the unknown. This applies to both geochronology and geochemistry; however, suitable homogenous reference materials (RMs) are not yet available for every type of analysis. Many matrix-matched reference materials exist for U–Pb geochronologic analyses for the most commonly employed geochronometers (zircon, monazite, titanite, apatite, and rutile), but fewer have been found/calibrated for TE analyses. Fortunately, internal standards—such as Ca, Si, and Ti—can be employed with non-matrix-matched reference materials, such that signal intensity can be partially corrected for variations in ablation rate between the reference material and the unknown. Because accuracy and precision of an age tend to be more important than the accuracy and precision on elemental abundance, the use of a short wavelength laser and matrix-matched standard is more critical in the acquisition of geochronologic data.

Some of the most important advances in improving accuracy, precision, and reproducibility of LA-ICPMS data have come from improving flow dynamics in the sample chamber. This is justly so, as the inter-element mass bias is strongly affected by the flow of the carrier gas across the sample surface (Bleiner and Günther 2001). Modern MC-ICPMS instruments can measure U–Pb ratios with a precision of much better than 1% for a single spot analysis, however, relatively large deviations (>2%) in inter-element fractionation can occur from spot to spot due to changes in flow dynamics in different parts of the cell (e.g., Bleiner and Bogaerts 2006; Muller et al. 2009), and thus cell design in recent years has been aimed at creating identical flow characteristics in every part of the cell. That way, samples and standards are subject to the same flow-induced fractionation—that is, the conditions under which the unknown is ablated are the same as the standard, and the true vs. measured ratio of each is the same.

The first sample chambers used for LA-ICPMS were simple single, small-volume cavities in blocks or discs. The newest cells involve a dual-volume design, employing a small-volume cup that resides inside a large-volume cell that contains several samples. The small-volume portion allows for rapid washout times (e.g., Bleiner and Günther 2001), and similar flow dynamics independent of position within the large-volume cell. The first conceptions of this dual-volume design—such as the HelEx cell (Eggins et al. 2005) found on the latest Photon Machines lasers—require an arm from inside the cell to the outside, and a moderately complicated system to keep the cell sealed as the arm position changes. A similar configuration exists for ESI lasers, however, other configurations, such as the RESOlution S155 cell, have a larger cell that includes the entire sample stage, allowing the small-volume cup to remain stationary in the large-volume cell minimizing the potential for changes in flow dynamics between sample and standard. If sealed and aligned properly, both designs work with the same efficiency and accuracy and are vast improvements over earlier cell designs.

TREATMENT OF UNCERTAINTIES

The treatment of uncertainties for all LA-ICPMS petrochronologic data invariably differs from lab to lab and is likely poorly understood by many users, such that uncertainty propagation of data in submitted manuscripts varies considerably. This variability is problematic, because it makes comparison of datasets from different sources quite difficult. Some geochronologic datasets from a single laboratory can be compared at a level better than 2% (2σ). However, it should be assumed that any two datasets from different LA-ICPMS laboratories cannot be distinguished, should their ages lie within 2% of one another (Košler et al. 2013). Furthermore, unless a study produces some confirmation of reproducibility, there can be little confidence in the data presented. Therefore, it is necessary to report the reproducibility of primary and secondary reference materials within each session.

Some attempts have been made to standardize the propagation of uncertainty in U–Pb data (Horstwood et al. 2016; McLean et al. 2016) and recommendations are published online (www.plasmage.org). Single element concentration and isotope ratio data should be treated in a similar manner with propagation of uncertainty using stated methods, including results for primary and secondary RMs. As noted below, uncertainties in TE data typically receive less scrutiny than isotopic data, possibly because composition is commonly compared in log space or as internally-derived ratios. Nevertheless, TE data can be collected and reduced with a wide variety of methods, and as such, those methods should be reported along with estimates on uncertainties for compositional data.

LASER ABLATION SPLIT STREAM (LASS)

Most recently, LA-ICPMS petrochronology has begun to take advantage of the virtues of each instrument: the MC-ICPMS, for its superior isotope-ratio measurements, and the SC-ICPMS, for its rapid mass switching. In such a configuration, the laser aerosol + He ± Ar carrier gas is split into two streams, each of which are sent to the two different mass spectrometers. Several configurations are possible, and will be discussed in sequence.

The first demonstrated LASS technique used a Nu Plasma MC-ICPMS coupled with an Elan 6100DRC quadrupole instrument (Yuan et al. 2008). This setup enabled the petrochronologic analysis of zircon, whereby Hf isotopic data were measured on the MC-ICPMS, and U–Pb age ± TE data were measured on the quadrupole. This type of analysis saves time and money in terms of laboratory operations, compared to separate sessions of analysis by each instrument. More importantly, it guarantees that the same volume of material is used for the ^{176}Hf/^{177}Hf ratio, the U–Pb age, and TE concentration. The downside is that Hf isotopic data requires a large signal intensity—for both acceptable signal:noise ratio and measurement precision—and thus, diverting part of the laser aerosol from the MC-ICPMS to the quadrupole instrument, requires the spot size to virtually double. As an example, Yuan et al. (2008) required a 44 µm diameter spot in the LASS configuration to get the same precision that they achieved using a 32 µm spot with only the MC-ICPMS; this is a nearly a 2-fold increase in spot area and volume. Furthermore, precision of the U–Pb age data is compromised using this technique: a SC-ICPMS has a much lower effective sensitivity than a MC-ICPMS, and is further reduced when TE data are also collected.

Another example is given in Figure 2. In this case, U–Pb zircon ages from a LASS analysis on the quadrupole are shown with U–Pb data from a MC-ICPMS; the quadrupole analyses included TE data (Si, Ti, Y, Zr, Nb, REE, Hf), but the signal intensity was increased by ~40× over the MC-ICPMS analyses by using a larger spot (65 µm vs 15 µm) and higher laser repetition rate (8 Hz vs. 4 Hz). Even with the increase in ablated volume for the SC-ICPMS, the precision of the U–Pb date using the SC-ICPMS is far inferior, and some samples with small, few, and complexly zoned petrochronometer minerals may not afford an abundance of large-spot analyses.

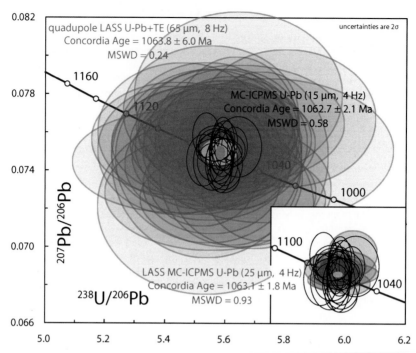

Figure 2. U–Pb zircon age data from two sessions, highlighting differences in analytical uncertainty between quadrupole and MC-ICPMS data. Data collected with the quadrupole was during a LASS session (approximately half the laser aerosol was diverted to a MC-ICPMS for Hf isotopic measurement) and also includes TE concentration data, employing a 65 μm spot and twice the laser repetition rate of U–Pb age analyses on the MC-ICPMS. Much better precision on U–Pb zircon ages can be obtained using the MC-ICPMS; a 25 μm spot using LASS (inset) yields approximately the same precision as a 15 μm spot using solely the MC-ICPMS.

One alternative to measuring all data of interest with just a single spot is to measure Hf isotopes in one session by MC-ICPMS, and use LASS in a second session to measure U–Pb age by MC-ICPMS and TE data by SC-ICPMS. If the second session is performed using LASS, a relatively small spot can be used for each session because one has the advantage of increased Hf isotopic precision in the first session and U–Pb precision in the second session; Figure 2 exemplifies this. This gives the user some freedom to place 2 smaller spots in roughly the same area as required by a single spot, which can be more advantageous in grains with complex zoning. This alternative method takes longer and is more costly; for simple, large grains, a single session of LASS, with isotopic composition measured on a MC-ICPMS and U–Pb age ± TE data measured on a SC-ICPMS is the most effective. At the time of this writing, LASS Hf-isotopic composition + U–Pb age ± TE has been employed only for metamorphic and igneous zircon (e.g., Chen et al. 2010; Wu et al. 2010).

If isotopic data are not required, LASS can be employed with MC-ICPMS for U–Pb age and SC-ICPMS for TE analysis (Kylander-Clark et al. 2013). As noted in the previous paragraph and shown in Figure 2, this technique was developed to take advantage of the superior stability and sensitivity of the MC-ICPMS to retrieve precise and accurate age data from single spots, while measuring elemental concentrations on a SC-ICPMS. As with the LASS Hf + U–Pb age ± TE technique described above, a larger spot for a single analysis takes the place of two single smaller spots. In this case, however, because measurements on a MC-ICPMS are so precise, sufficiently small spots can still be employed with excellent precision on the age. The precision and accuracy of element concentrations have not received the same scrutiny as U–Pb dates, and depend on many factors—including the matrix characteristics of the reference materials and unknowns—but LASS provides the same accuracy, albeit with a drop in precision commensurate with reduced signal intensity (Kylander-Clark et al. 2013). One important feature of LASS is the ability to match TE data with U–Pb data throughout the analysis to identify inclusions or changes in growth zoning within a mineral, down-hole during the analysis. This can be done without LASS (solely on a SC-ICPMS as described above), but the U–Pb analytical precision is generally prohibitively poor to recognize changes in U–Pb age with changes in chemistry. Two examples are given in Figures 3 and 4. The downside of LASS may be the larger spot size which increases the likelihood of intersecting a growth zone or inclusion; however, growth zoning or inclusions are difficult to identify without corresponding U–Pb age / TE information.

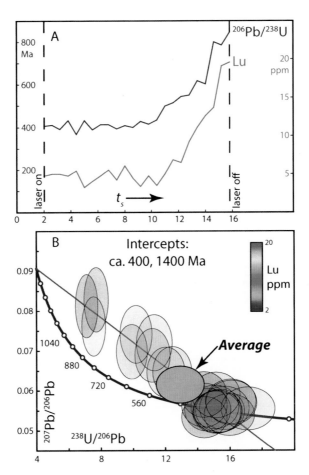

Figure 3. A) LASS trace of a single laser pit in a zircon; Lu data collected with other TE, using a SC-ICPMS and U–Pb age data was collected using a MC-ICPMS. The laser began the analysis sampling a garnet-stable (i.e., low Lu) ca. 400 Ma metamorphic rim, but eventually intersected the inherited, ca. 1400 Ma igneous, high-Lu core. B) Individual data points from A, plotted on a Tera-Wasserburg diagram. The time-resolved data (transparent ellipses shown with a uniform, 10% uncertainty) yield a discordia array between the metamorphic age and inherited age, with a corresponding increase in Lu concentration with apparent age. The average of the analysis (solid, 2SE, ellipse) is not nearly as informative, nevertheless, the entire example illustrates the benefits of collecting the age/TE data from one laser pit, rather than two adjacent pits.

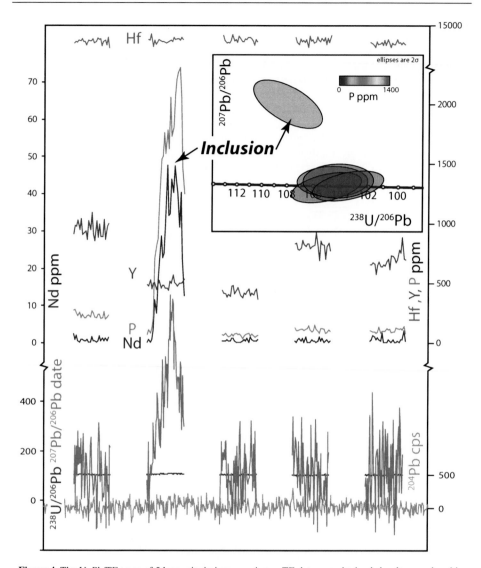

Figure 4. The U–Pb/TE trace of 5 laser pits in igneous zircon; TE data was obtained simultaneously with U–Pb data (quadrupole and MC-ICPMS, respectively). During the second analysis, the laser intersected an inclusion, which shows up in LREEs, such as Nd, and P (apatite?); elements that are elevated in zircon, such as Hf and Y, are unaffected. Though there is no measurable increase in [204]Pb, the age is significantly affected; most affected is the [207]Pb/[206]Pb age. The coincident TE data allows the user to confidently discard this data point when calculating an age for the sample.

LASS has advanced science by rapidly providing a wealth of information about both age and geochemistry (see Applications below for examples), however, LASS analyses should be used to complement, rather than replace the information gained by imaging and major elemental analysis prior to ablation. It is prudent to make images prior to analysis to guide spot placement—cathodoluminescence (CL) of zircon, or Y concentration maps of monazite, for example (e.g., Engi 2017; Lanari and Engi 2017; Williams et al. 2017). These images, and major- and minor-element analyses, used for thermobarometry (e.g., monazite–garnet; Pyle et al. 2001) for example, are easier to gather prior to ablation, which can remove substantial areas of grains of interest.

Once prior, non-destructive mapping is done, maps can also be made using LA(SS)-ICPMS, which can add significant insight to processes which formed/modified the geochronometers and/or coeval phases (Woodhead et al. 2007; Paul et al. 2014; Ubide et al. 2015). An example is shown in Figure 5; original igneous zircon is replaced and overgrown by rims and embayments of metamorphic zircon. Low HREE concentrations coincide with (neo/re)crystallized zircon, implying that metamorphism occurred in the garnet-stability field, whereas apparent age data is partly decoupled, implying that the U–Pb system may have been modified by a later, different process.

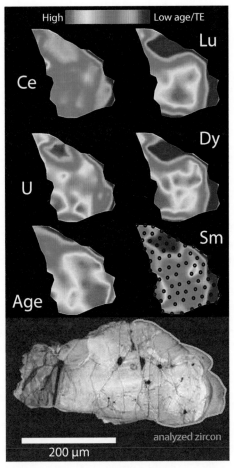

Figure 5. LASS maps (circles on Sm map show spot locations) of igneous zircon that grew at ca. 900 Ma, and underwent partial (re/neo)crystallization during a subsequent ca. 400 Ma garnet-stable metamorphic event. Rims and embayments contain considerably less HREE and moderately less MREE. Age map does not perfectly match that of REE, suggesting that late fissures (cutting both inherited and (neo/re)crystallized zircon) may have mobilized radiogenic Pb. The combination of age, TE and CL maps greatly increase the ability to interpret any single set of data.

Plumbing for LASS

There are a few different possible combinations to feed the He, Ar and ablated material into two different ICPMS instruments (Fig. 6). The simplest arrangement is a T- or Y-connector placed after the Ar+He mixing plumbing (be it a mixing bulb, 'T-connector,' etc.), sending

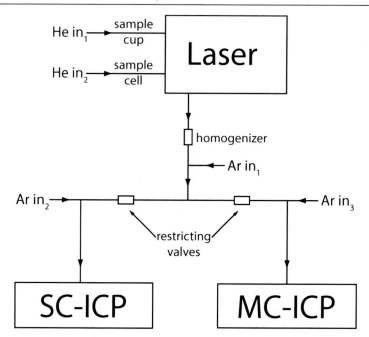

Figure 6. Configuration of instrumental setup for petrochronology as described in the text.

approximately half of the gas and aerosol to each ICPMS instrument. The advantage of this setup is its simplicity, but it leaves little room to tune each instrument separately by changing the Ar flow rate. Because each instrument will require a specific gas flow for maximum sensitivity, the backpressure of each line must be matched well enough so that tuning can take place by either changing the z-position of the torch, or by increasing or decreasing the auxiliary Ar flow to the torch. If the inner-torch inside diameter (ID) of each instrument is similar (and smaller than the tubing to the torch), the backpressure is likely to be the same in each line, and minor adjustments for maximum sensitivity of both instruments should be possible. If not, the backpressure in each line can be equalized by using similar length, identical ID tubing to both instruments, or by placing a restricting valve to the instrument with lower initial backpressure.

In theory, if an uneven distribution of signal intensity is desired, one can force more of the aerosol to one instrument by adding Ar to each ICPMS after the split (see the setup described in Ibanez-Mejia et al. 2015). An uneven split in Ar will force the aerosol to the line with the lower additional flow rate. This may be desirable for analyses in which one half of the split needs greater precision (i.e., count rates). An example of this would be a LASS setup in which Hf isotopes in zircon are measured on a MC-ICPMS and U–Pb age on a SC-ICPMS. As mentioned above, because the required precision of Hf isotopes is so much greater than that for U–Pb age, it would be advantageous to send a larger proportion to the MC-ICPMS.

APPLICATIONS

LA-ICPMS petrochronology, with and without LASS, has been applied to numerous minerals—both igneous and metamorphic—including, but not limited to zircon, monazite, titanite, rutile, and apatite. Sedimentary rocks have also received recent attention, as TE or Hf isotopic data can complement U–Pb detrital zircon ages to aid in provenance determination, or to better understand the processes leading to source terrane formation. To give the reader an idea of what is possible, a few case studies are described in detail, with an emphasis on the latest LASS techniques.

LASS of metamorphic zircon

Young and Kylander-Clark (2015) explored the transformation or lack thereof of felsic gneiss during subduction to (ultra)high-pressure (UHP) depths. In the Western Gneiss Region of western Norway, the bulk of the Caledonian (U)HP terrane consists of amphibolite-facies, plagioclase-bearing (i.e., low-pressure) assemblages. The extent of transformation is of great interest to those who model the density and strength of continental crust during subduction and exhumation. Transformation was assessed by comparing the observed mineral assemblage to that expected at high pressure and by using LASS petrochronology on zircon; a MC-ICPMS was employed for U–Pb geochronology, and a SC-ICPMS was used for TE compositions. It was deduced that most, but not all, of the continental crust did not transform at (U)HP conditions; the LASS data were essential in reaching this conclusion because the bulk of the data yielded inherited and mixed inherited-Caledonian dates with typical igneous zircon rare-earth-element (REE) patterns: a steep increase through the heavy REE (HREE), and a negative Eu anomaly. Mixed analyses could be recognized through matched U–Pb–TE traces. Caledonian rims yielded different patterns: lower REE concentrations and reduced HREE/MREE, indicate that the bulk rock underwent minor prograde, garnet-stable metamorphism early in the subduction/exhumation cycle. The disappearance of a negative Eu anomaly concomitant with the youngest dates indicate that some, if minor (re)crystallization, also occurred at near-peak metamorphic conditions.

Single-shot LASS (SS-LASS) analysis of thin metamorphic zircon rims

In an attempt to decipher the timing and conditions of minor metamorphic zircon growth in HP metamorphic rocks from the Cordillera de la Costa, Venezuela, Viete et al. (2015) used LASS—MC-ICPMS was used for U–Pb age and a quadrupole MS for TE analysis—to measure ages and TE on thin (<1 µm) rims of zircon. The authors mounted the zircons without polishing them, and ablated the rims in single-shot, 100nm pulses (Cottle et al. 2012) to retrieve pit-depth changes in age vs. TE concentration; Y was analyzed as a proxy for garnet stability, Ti as a proxy for temperature, and Eu as a proxy for plagioclase-stability. Because of the transient nature of the signal, only two TEs were measured in any given session; this resulted in ~20% long-term precision and accuracy on the analyses. Four discrete pulses of metamorphism were discovered, between 33–18 Ma, with a negative correlation between age and Y, Th, U, Eu concentration. In many grains, the first several pulses give the metamorphic crystallization age, and are followed by pulse-dates indicating an increasing mixture of inherited zircon.

Depth profiling of rutile

Thermally activated Pb diffusion in rutile occurs at sufficiently low temperatures such that rutile can be used to recover cooling paths (see Zack and Kooijman 2017). Smye and Stockli (2014) used a LASS configuration with two SC-ICPMS instruments to obtain a µm-scale depth profile of both U–Pb date (first instrument) and Zr concentration (second instrument). The authors found a monotonic decrease in age toward the rim; an inversion of that profile indicated rapid initial cooling in the Early Jurassic, followed by a period of slower cooling. The Zr profile showed a decrease in concentration from the core, but a sharp increase in the last few µm at the rim. These profiles indicated that the rutile, which originally crystallized during a Permian granulite-facies event, was reheated in the Early Jurassic. This conclusion was only possible because of the combined age + composition information.

Petrochronology of detrital zircon

Age combined with Th/U ratio may be the earliest petrochronologic tool used for evaluating detrital zircon data (e.g., Amelin 1998; Hartmann and Santos 2004), as Th/U ratio has long been known to indicate changes in petrogenetic conditions during zircon (re) crystallization (Ahrens 1965; Hoskin and Schaltegger 2003 and references therein). The

Th/U ratio in detrital zircon via standard LA-ICPMS is easily obtained and has therefore been used frequently (e.g., Bingen et al. 2005; Iizuka et al. 2005; Phillips et al. 2006). Recent advances have led to the ability to combine Hf isotopic data with age; for example a recent study by Ibanez-Mejia et al. (2015) used LASS data (MC-ICPMS for Hf and a SC-ICPMS for U–Pb age) from (meta)sediments in the northern Andes to better understand how the Amazon craton was involved in the assembly of Rodina. Using the combined age–isotopic data, the authors found an early interval of steeply increasing $^{176}Hf/^{177}Hf$ with decreasing age, followed by a later interval in which the trend shallows; these data were interpreted to indicate two contrasting phases of orogenesis. Other studies have combined TE composition with the age of detrital zircon data; Johnston and Kylander-Clark (2016) tracked the evolution of the Mesozoic California margin using detrital zircon age and chemistry (MC-ICPMS for U–Pb age and quadrupole for TE concentrations). The authors showed evidence for the rapid exhumation and denudation of a Jurassic ophiolite in the forearc from a distinct set of zircons found only in 102–97 Ma sediments, as well as present supporting evidence for forearc and intra-arc shortening during the Late Cretaceous via abrupt changes in HREE and Eu/Eu*.

Monazite petrochronology

Monazite has had an increasingly important role as a petrochronometer because of its affinity for both U and Th (and thus age determination), and a broad range of TE that can be linked to (re)crystallization conditions (Foster et al. 2002; Kohn and Malloy 2004; Kohn et al. 2005; Finger and Krenn 2007; Cottle et al. 2009b,c; Janots et al. 2009). Because of its large concentrations of Th and U (commonly 1000s of ppm) monazite grains can be analyzed with small laser spots, allowing for rapid acquisition of both U–Th/Pb and elemental composition data. For example, Stearns et al. (2013) used ages and HREE chemistry (MC-ICPMS + SC-ICPMS) of Himalayan and Pamir monazite to document synchronous prograde–retrograde garnet growth, in which 28–20 Ma monazite revealed decreasing HREE abundance, and 20–15 Ma monazite showed increasing HREE abundances. Štípská et al. (2015) documented monazite growth in several metamorphic grades along a *P–T–d* path; staurolite-grade monazite yielded older ages with high Y and HREE (garnet unstable), kyanite-grade monazite contained the same early zones in their cores, but with low Y, HREE mantles (garnet stable), and sillimanite-grade monazite contained both staurolite- and kyanite-grade monazite in their cores and mantles, respectively, but with high Y, HREE rims (retrograde garnet resorption). Holder et al. (2015) used LASS (MC-ICPMS + quadrupole) to examine the change of monazite chemistry during high-grade metamorphism, concluding that coincident increases in Sr, Pb, and Eu could be attributed to decrease in feldspar stability at high pressure. Hacker et al. (2015) applied monazite LASS (MC-ICPMS + quadrupole or SC-ICPMS) petrochronology to nearly 70 samples from the Scandinavian Caledonides to determine the response of monazite to widespread, high-grade metamorphism. In this case, the majority of samples revealed (re)crystallization of monazite during late-stage retrogression; both the age and chemistry were consistent. Both pre-Caledonian, and peak-Caledonian monazite was preserved in several cases, again evident in both age and composition.

Monazite also contains considerable concentrations of Nd, and favorable Sm/Nd ratios, affording the opportunity for in-situ isotopic Nd measurements. Poletti et al. (2017) used LA-ICPMS U–Th/Pb, elemental composition, and Nd isotopic composition of monazite from the Mountain Pass Intrusive suite, and were able to genetically link the monazite-bearing carbonatites to a nearby ultrapotassic suite of plutonic rocks.

Titanite Petrochronology

Titanite is a common trace phase in many quartzofeldspathic rocks that can persist through multiple metamorphic, igneous, and sedimentary transport events that span a wide range of pressures, temperatures, and deformation and fluid conditions. It often (re)crystallizes in response to changes in these parameters and can record these changes—in both age and chemistry—in distinct compositional domains. Though it can be more problematic to date, given its lower relative U and Th and higher relative common Pb contents than the other minerals mentioned in this section, it is widespread, has an affinity for a large range of TE, the latter of which can aid in the assignment of conditions during (re)crystallization (e.g., Franz and Spear 1985; Green and Pearson 1986; Frost et al. 2001; Prowatke and Klemme 2005; Hayden et al. 2008). Given the variable common-Pb content of titanite, the rapid analytical capability of LA-ICPMS is ideally suited for the mineral, because an isochronous discordia array between an upper intercept, common-Pb value and concordia (on a Tera-Wasserburg diagram) is easily achieved with a large number of spots (e.g., Storey et al. 2006). Perhaps the most common use of titanite in petrochronology by LA-ICPMS is that in which the age and temperature (Zr-in-titanite; Hayden et al. 2008) are linked: Kohn and Corrie (2011) used combined age–Zr profiles from titanite from the Himalaya to argue for long-lived thick and weak crustal channels; Gao et al. (2011) used LA-SC-ICPMS titanite data to distinguish between early titanite that crystallized at relatively cold temperatures, and late retrograde, high-temperature titanite; Schwartz et al. (2016) employed LASS (MC-ICPMS + quadrupole) for a persistent high-temperature lower crust from temperature and age data of titanites from Zealandia; Stearns et al. (2016) was the first to provide depth-profile Zr ppm + U–Pb date via single-shot LASS (MC-ICPMS + quadrupole), and argued for (re)crystallization rather than diffusional modification of Pamir titanites. Other LA-ICPMS studies have used other TE or isotopic analyses in titanite in concert with their age in order to solve geologic problems: Gao et al. (2011) also used REE composition of titanites to distinguish whether titanite grew in the presence of garnet and/or plagioclase; Stearns et al. (2015) presented TE + U–Pb age data (LASS; MC-ICPMS + quadrupole) from titanites in which age, Zr, Y, Sr covaried; B. R. Hacker (personal communication; Fig. 7) used *in-situ* Nd isotopic data to reveal changes in melt isotopic composition through time throughout the evolution of the Pamir plateau.

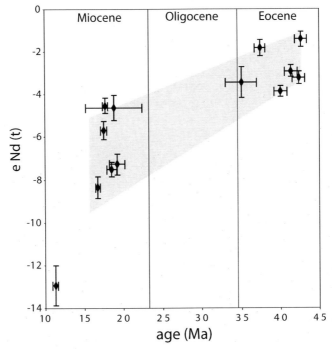

Figure 7. Petrochronologic (U–Pb age + Nd isotopes) data of Pamir titanites obtained by LA-ICPMS show that intrusions become more evolved over time.

CONCLUDING REMARKS

Petrochronologic data via LA-ICPMS, particularly by LASS, is more readily obtained and widely used than ever before, and new applications are continuing to be developed at an increasing rate. Improvements in instrument sensitivity, stability, laser coupling and gas-flow dynamics have broadened the applicability of LA-ICPMS petrochronology to a larger set of geologic problems, but limitations in necessary precision, sample size, composition still exist, and, in most cases LA(SS)-ICPMS petrochronology is most effective when combined with analytical data from other sources. A range of options exist, from single-instrument age–composition analysis to multiple-instrument, multiple session age–isotopic composition analysis, and choosing the right type of analysis requires careful considerations of all the requirements involved in producing interpretable results.

ACKNOWLEDGMENTS

This manuscript was significantly improved by inputs from reviewers D. Chew and P. Sylvester, editor P. Lanari, and B. Hacker.

REFERENCES

Ahrens LH (1965) Some observations on the uranium and thorium distributions in accessory zircon from granitic rocks. Geochim Cosmochim Acta 29:711–716, doi:10.1016/0016–7037(65)90064–5

Amelin YV (1998) Geochronology of the Jack Hills detrital zircons by precise U–Pb isotope dilution analysis of crystal fragments. Chem Geol 146:25–38, doi:10.1016/S0009-2541(97)00162–9

Bingen B, Griffin WL, Torsvik TH, Saeed A (2005) Timing of Late Neoproterozoic glaciation on Baltica constrained by detrital zircon geochronology in the Hedmark Group, south-east Norway. Terra Nova 17:250–258, doi:10.1111/j.1365–3121.2005.00609.x

Bleiner D, Günther D (2001) Theoretical description and experimental observation of aerosol transport processes in laser ablation inductively coupled plasma mass spectrometry. J Anal Atom Spectrom 16:449–456, doi:10.1039/b009729m

Bleiner D, Bogaerts A (2006) Computer simulations of laser ablation sample introduction for plasma-source elemental microanalysis. J Anal Atom Spectrom 21:1161–1174, doi:10.1039/b607627k

Chen R-X, Zheng Y-F, Xie L (2010) Metamorphic growth and recrystallization of zircon: Distinction by simultaneous in-situ analyses of trace elements, U–Th–Pb and Lu–Hf isotopes in zircons from eclogite-facies rocks in the Sulu orogen. Lithos 114:132–154, doi:10.1016/j.lithos.2009.08.006

Cottle JM, Horstwood MSA, Parrish RR (2009a) A new approach to single shot laser ablation analysis and its application to in situ Pb/U geochronology. J Anal Atom Spectom 24:1355–1363, doi:10.1039/b821899d

Cottle JM, Kylander-Clark ARC, Vrijmoed JC (2012) U–Th/Pb geochronology of detrital zircon and monazite by single shot laser ablation inductively coupled plasma mass spectrometry (SS-LA-ICPMS). Chem Geol 332–333:136–147, doi:10.1016/j.chemgeo.2012.09.035

Cottle JM, Searle MP, Horstwood MSA, Waters DJ (2009b) Timing of midcrustal metamorphism, melting, and deformation in the Mount Everest region of Southern Tibet revealed by U(–Th)–Pb geochronology. J Geol 117:643–664, doi:10.1086/605994

Cottle JM, Jessup MJ, Newell DL, Horstwood MSA, Noble SR, Parrish RR, Waters DJ, Searle MPCTC (2009c) Geochronology of granulitized eclogite from the Ama Drime Massif: Implications for the tectonic evolution of the South Tibetan Himalaya. Tectonics 28, doi:10.1029/2008tc002256

Eggins SM, Grün R, McCulloch MT, Pike AW, Chappell J, Kinsley L, Mortimer G, Shelley M, Murray-Wallace CV, Spötl C, Taylor L (2005) In situ U-series dating by laser-ablation multi-collector ICPMS: new prospects for Quaternary geochronology. Quat Sci Rev 24:2523–2538, doi:10.1016/j.quascirev.2005.07.006

Engi M (2017) Petrochronology based on REE–minerals: monazite, allanite, xenotime, apatite. Rev Mineral Geochem 83:365–418

Finger F, Krenn E (2007) Three metamorphic monazite generations in a high-pressure rock from the Bohemian Massif and the potentially important role of apatite in stimulating polyphase monazite growth along a *PT* loop. Lithos 95:103–115, doi:10.1016/j.lithos.2006.06.003

Foster G, Gibson HD, Parrish R, Horstwood M, Fraser J, Tindle A (2002) Textural, chemical and isotopic insights into the nature and behaviour of metamorphic monazite. Chem Geol 191:183–207

Franz G, Spear FS (1985) Aluminous titanite (sphene) from the Eclogite Zone, south-central Tauern Window, Austria. Chem Geol 50:33–46, doi:10.1016/0009–2541(85)90110-X

Frost BR, Chamberlain KR, Schumacher JC (2001) Sphene (titanite); phase relations and role as a geochronometer. Chem Geol 172:131–148

Gao XY, Zheng YF, Chen YX (2011) U–Pb ages and trace elements in metamorphic zircon and titanite from UHP eclogite in the Dabie orogen: constraints on *P–T–t* path. J Metamorph Geol 29:721–740, doi:10.1111/j.1525–1314.2011.00938.x

Green TH, Pearson NJ (1986) Rare-earth element partitioning between sphene and coexisting silicate liquid at high pressure and temperature. Chem Geol 55:105–119, doi:10.1016/0009–2541(86)90131–2

Guillong M, Horn I, Gunther D (2003) A comparison of 266 nm, 213 nm and 193 nm produced from a single solid state Nd:YAG laser for laser ablation ICP-MS. J Anal Atom Spectom 18:1224–1230, doi:10.1039/b305434a

Hacker BR, Kylander-Clark ARC, Holder R, Andersen TB, Peterman EM, Walsh EO, Munnikhuis JK (2015) Monazite response to ultrahigh-pressure subduction from U–Pb dating by laser ablation split stream. Chem Geol 409:28–41, doi:10.1016/j.chemgeo.2015.05.008

Hartmann LA, Santos JOS (2004) Predominance of high Th/U, magmatic zircon in Brazilian Shield sandstones. Geology 32:73–76, doi:10.1130/g20007.1

Hayden LA, Watson EB, Wark DA (2008) A thermobarometer for sphene (titanite). Contib Mineral Petr 155:529–540, doi:10.1007/s00410-007-0256-y

Holder RM, Hacker BR, Kylander-Clark ARC, Cottle JM (2015) Monazite trace-element and isotopic signatures of (ultra)high-pressure metamorphism: Examples from the Western Gneiss Region, Norway. Chem Geol 409:99–111, doi:10.1016/j.chemgeo.2015.04.021

Horstwood MS, Košler J, Gehrels G, Jackson SE, McLean NM, Paton C, Pearson NJ, Sircombe K, Sylvester P, Vermeesch P, Bowring JF (2016) Community-derived standards for LA-ICP-MS U–(Th–)Pb geochronology—Uncertainty propagation, age interpretation and data reporting. Geostandard Geoanal Res, doi:10.1111/j.1751-908X.2016.00379.x

Hoskin PWO, Schaltegger U (2003) The composition of zircon and igneous and metamorphic petrogenesis. Rev Mineral Geochem 53:27–62

Iizuka T, Hirata T (2004) Simultaneous determinations of U–Pb age and REE abundances for zircons using ArF excimer laser ablation-ICPMS. GEOCHEMICAL JOURNAL 38:229–241, doi:10.2343/geochemj.38.229

Iizuka T, Hirata T, Komiya T, Rino S, Katayama I, Motoki A, Maruyama S (2005) U–Pb and Lu–Hf isotope systematics of zircons from the Mississippi River sand: Implications for reworking and growth of continental crust. Geology 33:485–488, doi:10.1130/g21427.1

Jackson SE, Longerich HP, Dunning GR, Fryer BJ (1992) The application of laser ablation-microprobe-inductively coupled plasma-mass spectrometery (LAM-ICP-MS) to in situ trace-element analysis in minerals. Can Mineral 30:1049–1064

Jackson SE, Pearson NJ, Griffin WL, Belousova EA (2004) The application of laser ablation-inductively coupled plasma-mass spectrometry to in situ U–Pb zircon geochronology. Chem Geol 211:47–69, doi:10.1016/j.chemgeo.2004.06.017

Janots E, Engi M, Rubatto D, Berger A, Gregory C, Rahn M (2009) Metamorphic rates in collisional orogeny from in situ allanite and monazite dating. Geology 37:11–14, doi:10.1130/g25192a.1

Johnston SM, Kylander-Clark ARC (2016) Rapid ophiolite exhumation and arc thickening in the southern california late cretaceous convergent margin as defined by nacimiento block foreack detrital zircon geochronology and geochemistry. Geol Soc Am Abstracts with Programs 48:18–5

Kohn MJ, Malloy MA (2004) Formation of monazite via prograde metamorphic reactions among common silicates: Implications for age determinations. Geochim Cosmochim Ac 68:101–113

Kohn MJ, Corrie SL (2011) Preserved Zr-temperatures and U–Pb ages in high-grade metamorphic titanite: Evidence for a static hot channel in the Himalayan orogen. Earth Planet Sc Lett 311:136–143, doi:10.1016/j.epsl.2011.09.008

Kohn MJ, Wieland MS, Parkinson CD, Upreti BN (2005) Five generations of monazite in Langtang gneisses: implications for chronology of the Himalayan metamorphic core. J Metamorph Geol 23:399–406, doi:10.1111/j.1525-1314.2005.00584.x

Košler J, Sláma J, Belousova E, Corfu F, Gehrels GE, Gerdes A, Horstwood MS, Sircombe KN, Sylvester PJ, Tiepolo M, Whitehouse MJ (2013) U–Pb Detrital Zircon Analysis—Results of an inter-laboratory comparison. Geostandard Geoanal Res 37:243–259, doi:10.1111/j.1751-908X.2013.00245.x

Kylander-Clark ARC, Hacker BR, Cottle JM (2013) Laser-ablation split-stream ICP petrochronology. Chem Geol 345:99–112, doi:10.1016/j.chemgeo.2013.02.019

Lanari P, Engi M (2017) Local bulk composition effects on metamorphic mineral assemblages. Rev Mineral Geochem 83:55–102

Longerich H (2008) Laser Ablation-Inductively Coupled Plasma-Mass Spectrometry (LA-ICP-MS): An Introduction. Laser Ablation ICP-MS in the Earth Sciences: Current Practices and Outstanding Issues (P Sylvester, ed) Mineral Assoc Canada Short Course Ser 40:1–18

McLean N, Bowring J, Gehrels G (2016) Algorithms and software for U–Pb geochronology by LA-ICPMS. Geochem Geophy Geosy, doi:10.1002/2015gc006097

Muller W, Shelley M, Miller P, Broude S (2009) Initial performance metrics of a new custom-designed ArF excimer LA-ICPMS system coupled to a two-volume laser-ablation cell. J Anal Atom Spectom 24:209–214, doi:10.1039/b805995k

Paul B, Woodhead JD, Paton C, Hergt JM, Hellstrom J, Norris CA (2014) Towards a method for quantitative LA-ICP-MS imaging of multi-phase assemblages: mineral identification and analysis correction procedures. Geostandard Geoanal Res 38:253–263, doi:10.1111/j.1751-908X.2014.00270.x

Phillips G, Wilson CJL, Campbell IH, Allen CM (2006) U–Th–Pb detrital zircon geochronology from the southern Prince Charles Mountains, East Antarctica—Defining the Archaean to Neoproterozoic Ruker Province. Precambrian Res 148:292–306, doi:10.1016/j.precamres.2006.05.001

Poletti JE, Cottle JM, Hagen-Peter GA, Lackey JS (2017) Petrochronological constraints on the origin of the Mountain Pass ultrapotassic and carbonatite intrusive suite, California. J Petrol 57:1555–1598, doi:10.1093/petrology/egw050

Prowatke S, Klemme S (2005) Effect of melt composition on the partitioning of trace elements between titanite and silicate melt. Geochim Cosmochim Acta 69:695–709, doi:10.1016/j.gca.2004.06.037

Pyle JM, Spear FS, Rudnick RL, McDonough WF (2001) Monazite–xenotime and monazite–garnet equilibrium in a prograde pelite sequence. J Petrol 42:2083–2117

Russo RE, Mao X, Gonzalez JJ, Mao SS (2002) Femtosecond laser ablation ICP-MS. J Anal Atom Spectom 17:1072–1075, doi:10.1039/b202044k

Schaltegger U, Schmitt AK, Horstwood MSA (2015) U–Th–Pb zircon geochronology by ID-TIMS, SIMS, and laser ablation ICP-MS: Recipes, interpretations, and opportunities. Chem Geol 402:89–110, doi:10.1016/j.chemgeo.2015.02.028

Schoene B, Baxter EF (2017) Petrochronology and TIMS. Rev Mineral Geochem 83:231–260

Schwartz JJ, Stowell HH, Klepeis KA, Tulloch AJ, Kylander-Clark ARC, Hacker BR, Coble MA (2016) Thermochronology of extensional orogenic collapse in the deep crust of Zealandia. Geosphere 12:647–677, doi:10.1130/ges01232.1

Smye AJ, Stockli DF (2014) Rutile U–Pb age depth profiling: A continuous record of lithospheric thermal evolution. Earth Planet Sci Lett 408:171–182, doi:10.1016/j.epsl.2014.10.013

Stearns MA, Cottle JM, Hacker BR, Kylander-Clark ARC (2016) Extracting thermal histories from the near-rim zoning in titanite using coupled U–Pb and trace-element depth profiles by single-shot laser-ablation split stream (SS-LASS) ICP-MS. Chem Geol 422:13–24, doi:10.1016/j.chemgeo.2015.12.011

Stearns MA, Hacker BR, Ratschbacher L, Rutte D, Kylander-Clark ARC (2015) Titanite petrochronology of the Pamir gneiss domes: Implications for middle to deep crust exhumation and titanite closure to Pb and Zr diffusion. Tectonics 34:784–802, doi:10.1002/2014tc003774

Stearns MA, Hacker BR, Ratschbacher L, Lee J, Cottle JM, Kylander-Clark ARC (2013) Synchronous Oligocene–Miocene metamorphism of the Pamir and the north Himalaya driven by plate-scale dynamics. Geology 41:1071–1074, doi:10.1130/g34451.1

Štípská P, Hacker BR, Racek M, Holder R, Kylander-Clark ARC, Schulmann K, Hasalová P (2015) Monazite dating of prograde and retrograde *P–T–d* paths in the Barrovian terrane of the Thaya window, Bohemian Massif. J Petrol 56:1007–1035, doi:10.1093/petrology/egv026

Storey CD, Jeffries TE, Smith M (2006) Common lead-corrected laser ablation ICP-MS U–Pb systematics and geochronology of titanite. Chem Geol 227:37–52, doi:10.1016/j.chemgeo.2005.09.003

Ubide T, McKenna CA, Chew DM, Kamber BS (2015) High-resolution LA-ICP-MS trace element mapping of igneous minerals: In search of magma histories. Chem Geol 409:157–168, doi:10.1016/j.chemgeo.2015.05.020

Viete DR, Kylander-Clark ARC, Hacker BR (2015) Single-shot laser ablation split stream (SS-LASS) petrochronology deciphers multiple, short-duration metamorphic events. Chem Geol 415:70–86, doi:10.1016/j.chemgeo.2015.09.013

Williams ML, Jercinovic MJ, Mahan KH, Dumond G (2017) Electron microprobe petrochronology. Rev Mineral Geochem 83:153–182

Woodhead JD, Hellstrom J, Hergt JM, Greig A, Maas R (2007) Isotopic and elemental imaging of geological materials by laser ablation inductively coupled plasma-mass spectrometry. Geostandard Geoanal Res 31:331–343, doi:10.1111/j.1751-908X.2007.00104.x

Wu F-Y, Ji W-Q, Liu C-Z, Chung S-L (2010) Detrital zircon U–Pb and Hf isotopic data from the Xigaze fore-arc basin: Constraints on Transhimalayan magmatic evolution in southern Tibet. Chem Geol 271:13–25, doi:10.1016/j.chemgeo.2009.12.007

Young DJ, Kylander-Clark ARC (2015) Does continental crust transform during eclogite facies metamorphism? J Metamorph Geol 33:331–357, doi:10.1111/jmg.12123

Yuan H-L, Gao S, Dai M-N, Zong C-L, Günther D, Fontaine GH, Liu X-M, Diwu C (2008) Simultaneous determinations of U–Pb age, Hf isotopes and trace element compositions of zircon by excimer laser-ablation quadrupole and multiple-collector ICP-MS. Chem Geol 247:100–118, doi:10.1016/j.chemgeo.2007.10.003

Zack T, Hogmalm KJ (2016) Laser ablation Rb/Sr dating by online chemical separation of Rb and Sr in an oxygen-filled reaction cell. Chem Geol 437:120–133, doi:10.1016/j.chemgeo.2016.05.027

Zack T, Kooijman E (2017) Petrology and geochronology of rutile. Rev Mineral Geochem 83:443–467

Zheng JP, Griffin WL, O'Reilly SY, Zhao JH, Wu YB, Liu GL, Pearson N, Zhang M, Ma CQ, Zhang ZH, Yu CM (2009) Neoarchean (2.7 Ga–2.8 Ga) accretion beneath the North China Craton: U–Pb age, trace elements and Hf isotopes of zircons in diamondiferous kimberlites. Lithos 112:188–202, doi:10.1016/j.lithos.2009.02.003

Reviews in Mineralogy & Geochemistry
Vol. 83 pp. 199–230, 2017
Copyright © Mineralogical Society of America

7

Secondary Ionization Mass Spectrometry Analysis in Petrochronology

Axel K. Schmitt

Institute of Earth Sciences
Heidelberg University
69120 Heidelberg
Germany

axel.schmitt@geow.uni-heidelberg.de

Jorge A. Vazquez[1,2]

[1]*Stanford-USGS Ion Microprobe Laboratory*
U.S. Geological Survey
Menlo Park, CA 94025

[2]*Stanford University*
Stanford, CA 94305
USA

jvazquez@usgs.gov

INTRODUCTION AND SCOPE

The goal of petrochronology is to extract information about the rates and conditions at which rocks and magmas are transported through the Earth's crust. Garnering this information from the rock record greatly benefits from integrating textural and compositional data with radiometric dating of accessory minerals. Length scales of crystal growth and diffusive transport in accessory minerals under realistic geologic conditions are typically in the range of 1–10's of μm, and in some cases even substantially smaller, with zircon having among the lowest diffusion coefficients at a given temperature (e.g., Cherniak and Watson 2003). Intrinsic to the compartmentalization of geochemical and geochronologic information from intra-crystal domains is the requirement to determine accessory mineral compositions using techniques that sample at commensurate spatial scales so as to not convolute the geologic signals that are recorded within crystals, as may be the case with single grain or large grain fragment analysis by isotope dilution thermal ionization mass spectrometry (ID-TIMS; e.g., Schaltegger and Davies 2017, this volume; Schoene and Baxter 2017, this volume). Small crystals can also be difficult to extract by mineral separation techniques traditionally used in geochronology, which also lead to a loss of petrographic context. Secondary Ionization Mass Spectrometry, that is SIMS performed with an ion microprobe, is an analytical technique ideally suited to meet the high spatial resolution analysis requirements that are critical for petrochronology (Table 1).

In SIMS, bombardment of solid targets with an energetic ion beam removes atoms from the sample where primary ions are implanted into the target material to a depth of <5–10 nm. Lateral resolution is controlled by primary ion beam dimensions (sub-μm to few 10's of μm) with an upper limit set by the acceptance angle of the ion optical system which collects the secondary ions emitted from the source into the mass spectrometer. This upper limit is typically a few 100's of μm, provided sample charging can be mitigated, but only rarely are such large areas analyzed in geologic samples. Depth resolution is typically one order-of-magnitude smaller than lateral resolution, depending on conditions, and total crater depths for petrochronologic analysis are few μm at most. Analysis durations are typically a few to 10's of minutes.

1529-6466/17/0083-0007$05.00 (print)
1943-2666/17/0083-0007$05.00 (online)

http://dx.doi.org/10.2138/rmg.2017.83.7

Table 1. Advantages and limitations of *in-situ* SIMS analysis for petrochronology in comparison with other isotope selective methods

In-situ analysis capabilities	*Limitations*
Versatile for radiogenic and stable isotopes (−ID-TIMS, −LA-ICP-MS)	Quantitation where ultimate precision and accuracy is limited by sample volume and/or sputtering physics (−ID-TIMS, ±LA-ICP-MS)
Isotope ratios (e.g., complete U–Th–Pb system with common Pb correction) and trace elements (±ID-TIMS, ±LA-ICP-MS)	Dependence on reference materials (−ID-TIMS, +LA-ICP-MS)
Generates diverse data from spots (including depth profiling) or areas by ion imaging (−ID-TIMS, +LA-ICP-MS)	Trade-off between small sample volume and precision so that reducing sampling volumes by using smaller beams limits precision, also through higher background to signal ratios (+ID-TIMS, +LA-ICP-MS)
Spatial selectivity (−ID-TIMS, +LA-ICP-MS)	Complex mass spectrum (−ID-TIMS, −LA-ICP-MS)
Isotope analysis in petrographic context (−ID-TIMS, +LA-ICP-MS)	Complex and costly instrumentation required (−ID-TIMS, −LA-ICP-MS)
Reveals mixed populations (±ID-TIMS, +LA-ICP-MS)	
Can reveal diffusion in natural and experimental samples (−ID-TIMS, ±LA-ICP-MS)	
High sensitivity and low backgrounds (−ID-TIMS, −LA-ICP-MS)	
No sample chemistry (−ID-TIMS, +LA-ICP-MS)	
Individual particle analysis (−ID-TIMS, +LA-ICP-MS)	
Minimal sample consumption enabling correlated analysis (−ID-TIMS, −LA-ICP-MS)	

Note: + indicates property shared with other methods; - indicates property unique, and not shared with other methods; other methods: isotope dilution thermal ionization mass spectrometry ID-TIMS, laser ablation-inductively coupled plasma-mass spectrometry LA-ICP-MS.

In contrast to laser ablation-inductively coupled plasma-mass spectrometry (LA-ICP-MS; e.g., Kylander-Clark 2017, this volume) the principle of SIMS requires that sample removal ("sputtering") and ionization occur in the same location. Hence the sample surface is part of an ion optical system that controls the trajectories of the primary ion beam, and the secondary ion emission. While this results in high ion collection efficiencies, it also places strict requirements on the sample surface which must be flat and oriented parallel to the secondary ion extraction optical components. Moreover, the sample must be conductive, or coated with a conductive material, and sustain high vacuum. Sputtering generates complex mass spectra of atomic, molecular, and multiply charged ion species, which requires a careful spectral evaluation and the adoption of measures that can mitigate mass spectral interferences. These measures may result in a loss of transmission in the mass spectrometer.

This article summarizes the strengths of SIMS for petrochronology which include isotope specific analysis, superb spatial resolution, and high sensitivity. We also address some limitations of SIMS analysis which comprise potential pitfalls resulting from complex mass spectra and matrix-dependent variability in secondary ion formation that can compromise data accuracy if not properly accounted for (Table 1). There are many excellent summary publications of SIMS methodologies that have focused on key applications in the Earth Sciences: geochronology, trace element, and stable isotope analysis (e.g., Shimizu and Hart 1982; Ireland 1995; Williams and McKibben 1998; Ireland and Williams 2003; Stern 2009; Valley and Kita 2009; Ireland 2015). Although this literature is as old as 35 years, the principles of geochemical SIMS analysis and even some of the fundamental instrumental designs have changed only marginally, although reliability, routinely achieved precision (especially in multi-collection stable isotope analysis), and automation have substantially progressed since then. Our goal for this review is to focus on those aspects of SIMS analysis and data treatment that are of interest to igneous or metamorphic petrologists and that are unique to the requirements of optimizing the acquisition of spatially correlated geochronological and geochemical information from crystal domains and analyzing accessory minerals in a textural context (*in-situ* analysis sensu stricto). We only review basic aspects of SIMS geochronology, trace element, and stable isotopic analysis where it is necessary to provide a context for the integration of these methods; we refer to the excellent reviews listed above for more detail on individual methodologies.

INSTRUMENTATION AND SAMPLE PREPARATION

Large-magnet radius ion microprobes for petrochronology, and complementary instrumentation

The chronology aspect in petrochronology is almost exclusively the domain of large magnet radius SIMS instruments that can routinely resolve the complex mass spectra that result from sputtering chemically complex accessory minerals. We therefore focus below on currently used instruments of the CAMECA ims 1270/1280 and extended SHRIMP families (e.g., SHRIMP-II, and -RG). Smaller SIMS instruments can yield valuable information on the chemical composition and stable isotopes in certain minerals, but they are limited in resolving geochronologically relevant isotopes (e.g., ^{204}Pb, ^{206}Pb, ^{207}Pb, ^{208}Pb) for accessory minerals where isobaric interferences abound. The CAMECA NanoSIMS instruments that are specifically designed to utilize small primary ion beams ($\ll 1\,\mu m$) are widely applied in cosmochemistry and biogeochemistry (e.g., Hoppe et al. 2013). Because the ultra-high spatial resolution that results from a highly focused primary beam intrinsically comes at a loss of secondary ion intensity, the role of NanoSIMS in the analysis of trace species important for petrochronology is limited (Table 1). Where this instrument has been used successfully for dating purposes (e.g., Stern et al. 2005; Yang et al. 2012), primary beam dimensions are typically $>1\,\mu m$ and thus equivalent to those achieved in large magnet radius instruments (e.g., Harrison and Schmitt 2007). NanoSIMS has occasionally been used to investigate trace element compositions of accessory minerals (e.g., Hofmann et al. 2009, 2014), and for high-resolution geospeedometry to derive the durations, rather than absolute dates, over which diffusion occurred between the compositional zones in major minerals that record rapid petrologic evolution. When combined with petrologic information from the compositions of the mineral zoning, these diffusion studies have been used for "petrospeedometry" to resolve the durations of processes such as silicic magma generation (e.g., Till et al. 2015) and quartz growth and residence in rhyolitic magma that was eventually erupted (e.g., Seitz et al. 2016).

Large magnet radius instruments for geochronology and petrochronology

CAMECA-family. CAMECA has developed ion microprobes since the late 1960s, with the first commercially successful small radius magnet instrument, the IMS 3f, essentially earning the company monopoly status for magnetic sector SIMS in the semi-conductor industry (de Chambost 2011). Following the initial development of the SHRIMP instrument (see historical review by Ireland et al. 2008) in the 1970s at Australian National University (ANU), CAMECA decided to also branch out into geological applications (de Chambost 2011). The idea was to scale up the concurrently marketed IMS 4f to meet the transmission and mass resolution requirements for Pb isotope analysis of zircon, while at the same time maintaining the ion imaging capabilities of the IMS instruments that had proven extremely helpful for tuning and target location purposes. This design is also beneficial in enhancing lateral resolution, and in turn the resolution of depth profiling by means of inserting a field aperture into the secondary ion path (Fig. 1). The instrument's name IMS 1270 was in fact a misnomer which was based on a 10× scaling of the IMS 4f magnet radius assumed to be 127 mm (in fact it was only 117 mm), with the realized magnet radius being only 5 × 117 mm = 585 mm

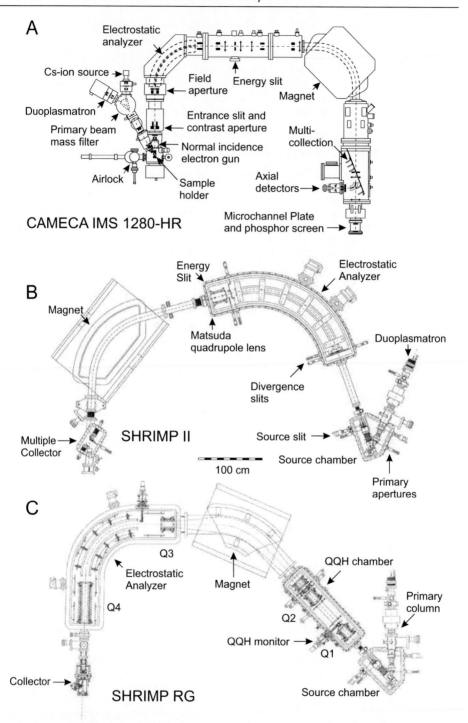

Figure 1. A) Schematic diagrams of large magnet radius SIMS instruments. A) CAMECA IMS 1280 with multi-collector. Double-arrows and black boxes indicate ion optical devices (electrostatic lenses and deflecting plates, respectively). B) SHRIMP II and C) SHRIMP-RG. Major differences include the magnet-electrostatic analyzer geometries, and the presence of four quadrupoles (Q1-Q4) in SHRIMP-RG.

(de Chambost 2011). The first instrument was installed at the University of California Los Angeles (UCLA) in December 1992. A multi-collector detection system became available in 1997. Following an upgrade of the instrument electronics, the IMS 1270 became the IMS 1280, the first of which was installed at University of Wisconsin in 2005, but the fundamental instrumental layout (Fig. 1) has experienced only incremental changes and remains largely constant to the present. There are currently about 30 instruments operating worldwide.

The IMS 1270/1280 instruments feature a split primary column where both duoplasmatron gas and Cs^+ surface ionization primary ion sources can be held simultaneously under vacuum (Fig. 1). The primary beam mass filter (PBMF; Fig. 1) generates an isotopically pure primary ion beam. This is critical for certain uses, such as for avoiding the implantation of OH^- together with O^- into the target which would generate difficult-to-resolve hydride interferences (see below). The PBMF also acts as a shunt magnet which allows for sequential use of both sources without breaking vacuum. Latest generation instruments can continuously monitor primary beam intensity through rapid beam switching.

Samples are introduced via an airlock into an antechamber which can hold multiple samples for rapid exchange. They are pumped to vacuum ($\sim 10^{-5}$ to 10^{-7} Pa) and then transferred into the sample chamber where the surface can be observed in reflected light through an off-axis microscope objective with illumination through an oblique vacuum-external light source. A unique design is the presence of a gas inlet through the extraction plate opposite the sample which can locally enhance the concentration of gas species in the sample chamber ("oxygen flooding"). Oxygen flooding enhances ionization of Pb^+ in some minerals (e.g., zircon, baddeleyite, rutile), but not others (e.g., monazite; Schuhmacher et al. 1994; Schmitt et al. 2010; Schmitt and Zack 2012). Mount holders can accommodate samples up to 25.4 mm in diameter (note: the useful area is smaller; see below) and a height of preferably <7 mm; smaller samples can be mounted in holders with smaller (e.g., 5 mm) cutouts in the top plate, but the limited useful surface area renders this type of mounting less preferable. To compensate for electron loss from the sputter area when the sample is at negative potential (analyzing negative ions, e.g., $^{18}O^-/^{16}O^-$), a normal incidence electron gun is used to prevent charging of insulators (Fig. 1). Positive or negative secondary ions are extracted in an electrostatic field between the biased sample surface and the grounded extraction plate at a potential gradient of up to 2 V/μm. The total secondary beam path is ~ 5 m resulting in a time-of-flight of 10's of μs.

The secondary ('Transfer') ion optical system permits adjustment of the secondary ion trajectory relative to the entrance slit and the co-located, and generally wide open contrast aperture, as well as the pre-electrostatic analyzer (ESA) field aperture. Tuning of the transfer system achieves the required transmission and limits spherical (angular) aberrations that could compromise the focusing capabilities of the mass spectrometer (Fig. 1). Moreover, the field aperture plane contains a magnified image of the ions emitted from the sample surface. The field aperture width can be continuously modified to achieve selection of a spatial subset of secondary ions. For analyses where surface contamination is of concern (as is the case for trace Pb), it can be adjusted for transmission of ions from the center of the analysis crater, thus blocking most ions emitted from the more slowly sputtered edges of the crater. While clipping of the beam in the field aperture reduces all secondary ion intensities, surface contaminants from the edges are preferentially excluded, thus enhancing depth resolution.

The mass spectrometer section of the instrument has a double-focusing geometry with an ESA followed by a magnetic prism (Fig. 1). Mass resolution can be enhanced by electrostatically modifying the secondary ion beam in a direction perpendicular to the mass focal plane in the so-called XY mode (Z being the direction of secondary ion travel in CAMECA coordinates); this is commonly done for U–Pb analysis. The energy slit between the ESA and the magnet (Fig. 1) provides a band-pass for secondary ion energies. The low-energy slit is set to a position that permits sampling of all ions below the nominal accelerating potential, whereas the high-energy slit is independently moved to typically admit ions within a few 10's of eV of the maximum. Analysis types where interferences are difficult to resolve benefit from sampling only the high-energy population of secondary ions. This "energy filtering" mode of operation takes advantage of the narrow energy distribution of most molecular ions (hydrides being a notable exception), whose intensities are significantly reduced at high (several 10's to 100's of eV) energy offsets (e.g., Shimizu 1978). Energy filtering reduces the relative contribution of molecular interferences on atomic ion species, for example that of light rare earth element (REE) oxides on heavy REE atomic ions (Zinner and Crozaz 1986), and mitigates matrix effects (e.g., Shimizu and Hart 1982).

Secondary ions are detected either by Faraday cup (FC) or electron multiplier detectors (EM). Single collection is achieved by electrostatically bending the secondary ion beam coming through the exit slit in a small ESA and steering it into the so-called axial EM, or one of two bracketing FCs (Fig. 1). Multi-collection instruments additionally have an array of five moveable detector carriers ("trolleys"), which are equipped with miniaturized EMs or FCs. On the current model of the IMS 1280-HR multi-collection is possible at 1 u dispersion from mass 7 to 240. Pre-amplifier circuits for FCs contain resistors between 10^9 and $10^{11} \, \Omega$; maximum currents (at $10^{10} \, \Omega$) are 6×10^9 counts per second (cps). Thermal drift at $\sim 4 \times 10^3$ cps/°C is mitigated by housing of the resistor circuit in a thermally stable environment, but count rates $<10^6$ cps are difficult to measure by these FCs because of Johnson-Nyquist noise. At low secondary ion intensities ($<10^6$ cps) electron multipliers in pulse-counting mode are currently used because they offer extremely low noise levels (<0.01 cps). One drawback of the EM is dead time which requires significant corrections at high count rates required for accurate analysis (i.e., 1‰ correction per nsec dead time at 10^6 cps). Moreover, aging of the EM detectors requires regular monitoring of the pulse voltage distribution to maintain sensitivity, and because the EM yield deteriorates with total accumulated counts, frequent cross-calibration between different EMs or EM and FC detectors is required if multiple detectors are used. Lastly, a terminal microchannel plate detector on the axis of the spectrometer allows for direct ion imaging of a mass-filtered secondary ion beam (Fig. 1). Direct ion imaging is not as sensitive as scanning ion imaging with EMs and at best semi-quantitative due to detector non-linearity and heterogeneous aging. Nevertheless, many practitioners value it as an indispensable aid for tuning, rapid beam location, and characterization of analytical targets. Direct ion imaging is a unique feature of all IMS (= ion microscope) instruments. Nevertheless, scanning ion imaging (analogous to scanning electron microscopy) where the primary beam is rastered over the analyte area and secondary ion signals are detected in fast-response EM detectors is the preferred method to create ion images and derivative spatially-correlated geochemical and isotope data on these instruments.

SHRIMP-family. The Sensitive High Resolution Ion Microprobes (SHRIMP) manufactured by Australian Scientific Instruments (ASI) are employed for geochronology, isotopic, and trace element analyses of geologic materials. These large-format, double-focusing mass spectrometers are descendants of the original SHRIMP (subsequently named SHRIMP I) instrument developed by William Compston and Steve Clement in the late 1970's at the Research School of Earth Sciences, ANU. SHRIMP I was developed to perform isotopic analyses with high sensitivity and at high mass resolution sufficient to resolve isobaric interferences when analyzing geologic samples. Many of the design features of SHRIMP I are employed in newer models of the SHRIMP family. SHRIMP I was designed based on the Matsuda (1974) ion optic configuration for a double-focusing mass spectrometer. To optimize transmission at high ($m/\Delta m > 5000$) mass resolution, SHRIMP I was designed with an astigmatic secondary ion extraction system, a 1270 mm radius electrostatic analyzer, and a 1000 mm turning radius magnet, all of which led to an instrument with an ~7 m beam path. These basic elements have been used in subsequent

SHRIMP models. Consequences of the astigmatic nature of the secondary ion optics are that SHRIMP instruments cannot be used as an ion microscopes like the CAMECA IMS family of ion probes, and that the depth profiling capability is limited because secondary ions from the bottom and the walls of the sputter pit are convoluted. After about three decades of service, SHRIMP I was decommissioned in late 2010.

In the early 1990's, SHRIMP II was introduced based on the successful design elements of SHRIMP I, and became commercially available through ASI. SHRIMP II featured a redesigned source (= sample) chamber, primary column and secondary ion optics system that reduced aberrations and doubled sensitivity. The redesigned primary column incorporated a lens and aperture system for Köhler illumination to produce flat-bottomed and steep-sided analysis spots, thereby reducing spherical aberrations and contamination from the non-sputtered surface of the target. The incorporation of a Schwarzschild microscope system, which allows coincident reflected light imaging of samples and alignment of the primary-secondary ion optics, has allowed user-friendly operation and precise sampling of target domains. The sample stage in the source chamber can hold two mounts for analysis; one position is typically reserved for a setup mount that contains a suite of concentration and secondary reference materials. A separate sample lock houses a motorized rack that holds up to four mounts under high vacuum in preparation for exchanges into the sample stage. Mount exchange is automated using a motorized hoist and pneumatic clamp system. Originally, SHRIMP II instruments were fitted with a single-collector system, but newer versions have featured a multi-collection system with ion counters and FCs (Holden et al. 2009). For most geochronology (e.g., U–Pb, U–Th) and trace element (e.g., REE, Ti) applications, a primary beam of negative oxygen ions that are generated in a cold-cathode duoplasmatron is used to bombard a sample and generate positive secondary ions. A Wien mass filter is used to select preferred primary beam species (e.g., O^- vs O_2^-) and exclude species, such as OH^- and NO_2^- that may generate interferences. For trace elements that more readily yield negative secondary ions during sputtering, such as halogens, the duoplasmatron is physically replaced by a cesium source to yield Cs^+ ions, with charge compensation on the sample surface provided by an electron gun. Helmholtz coils surround the source chamber and are used to moderate light isotope fractionation by canceling out the vertical component of the geomagnetic field.

After the introduction of SHRIMP II, a variant (SHRIMP-RG) was designed based on the reversely configured mass spectrometer of Matsuda (1990), in order to reach ultra-high mass resolution for analyses of intermediate-mass elements, such as the REE, that are plagued by interferences during SIMS analysis. Only two SHRIMP-RG instruments have been manufactured; one is located at ANU, and the other is at Stanford University where it is cooperatively operated with the U.S. Geological Survey. The reverse geometry of SHRIMP-RG, which employs the magnet before the ESA (Fig. 1), allows a mass resolution of up to four times higher than for other SHRIMPs (Compston and Clement 2006; Ireland et al. 2008). Consequently, the REE and other trace elements in accessory minerals can be resolved without the need to employ aggressive energy filtering that leads to a reduction of secondary ions transmitted to the collector, or where substantial reduction in molecular intensities cannot be achieved at higher energies, as is the case for hydrides of intermediate to high atomic number elements (Fig. 2). Energy filtering can be employed as needed using an energy selection slit located immediately before the ESA, and abundance sensitivity can be enhanced with a retardation lens located before the collector. An off-axis FC detector is available and can be employed in conjunction with an on-axis EM for measuring trace elements that yield count rates varying by several orders of magnitude, for example $^{139}La^+$ and $^{155}Gd^+$ in allanite or monazite. This approach has been employed by the Stanford-USGS SHRIMP-RG to perform simultaneous U–Th–Pb geochronology and analysis of REE in light REE-rich minerals. However, a key limitation of the SHRIMP-RG design is that multicollection of isotopes is not possible.

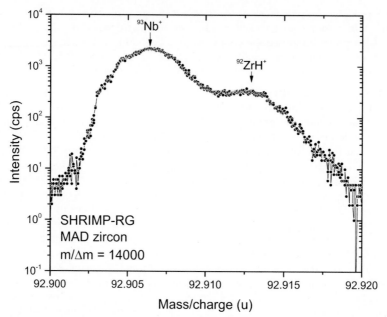

Figure 2. High mass resolution scan of MAD zircon on mass/charge 93 u ($^{93}Nb^+$) at a mass resolving power $m/\Delta m$ = 14000. Scan shows the $^{92}ZrH^+$ interference resolved on the the high-mass side of the Nb-peak. Data generated on the Stanford-USGS SHRIMP-RG, an instrument designed to achieve high mass resolution with minimal loss in transmission.

Recently, a variant (SHRIMP-SI, aka SHRIMP V) has been specifically designed for precise multi-collector measurement of stable isotopes including oxygen (Ireland et al. 2008; Ireland et al. 2014). SHRIMP-SI has been fitted with a new electrometer system for FCs with $10^{10-12}\,\Omega$ resistors and a capacitor charge mode capability optimized for count rates (10^5–10^6 cps) over the upper and lower effective performance limits of EM and FC detectors, respectively (Ireland et al. 2014).

Sample preparation for petrochronology

Because the sample surface is part of the ion extraction geometry in all SIMS instruments, stringent requirements exist for samples to be vacuum-proof, flat, and conductive. Sample preparation is thus a critical step for successful SIMS analysis. Furthermore, extensive sample characterization prior to SIMS analysis (e.g., by optical and/or electron beam imaging) and documentation is essential for efficient targeting. Key aspects of common preparation and imaging procedures used for petrochronological SIMS analysis are summarized here.

Grain mounts. SIMS analysis of geologic materials traditionally employs grain mounts with isolated crystals embedded into commercially available epoxy (Fig. 3). Grain mounts are typically composed of one mineral type to maximize the number of analyzed grains per sample mount, minimize the time spent changing mounts, and facilitate automated analysis, all of which are often important factors for research productivity. Mineral grains, either as whole crystals or fragments, are typically isolated via conventional mineral separation involving rock crushing and pulverizing, heavy liquid density separation, and magnetic separation. Individual mineral grains are then segregated and/or are hand-picked in preparation for mounting. As is performed before most ID-TIMS analyses, individual zircon crystals can be treated to "chemical abrasion" prior to mounting in order to remove domains affected by cryptic Pb-loss (Watts et al. 2016).

Grain mounts are typically prepared by placing individual or aggregations of grains onto double-sided tape. Individual grains or groups are commonly arranged into rows or other regular patterns in order to provide organization and later assist navigation using the ion microprobe's reflected light microscope (Fig. 3). It is important that mounts also include reference materials needed for the intended geochronology, trace element or stable isotope calibration. The CAMECA and SHRIMP ion microprobes have steel mount holders that accept 25.4 mm diameter round mounts. Because it has been demonstrated that instrumental mass fractionation can be exacerbated for analyses at the margins of the mounts near the steel-epoxy interface (e.g., Treble et al. 2007; Ickert et al. 2008), grains should be placed within the inner 15 mm diameter of the mount, away from the mount edge. After organization, the tape-mounted grains are surrounded with a 25.4 mm inner diameter plastic or teflon ring which is filled with a well-stirred mixture of resin and hardener. Choice of the epoxy is critical: the epoxy must meet quality criteria for categories such as minimal outgassing under vacuum, hardness and good bonding properties, resilience under electron beam bombardment, and preferably having low cathodoluminescence backgrounds. The University of Edinburgh Ion Microprobe facility has provided an excellent survey of the characteristics of numerous varieties of epoxy, available online (http://www.ed.ac.uk/geosciences/research/facilities/ionprobe/technical/epoxyresins). It is also important that the epoxy is free of bubbles, and that bubbles do not form on the sides of crystals, flaws that may result when the epoxy is overly viscous or poured too quickly. Sectioned bubbles can result in relief near or adjacent to grains, which can deform the local electrostatic field and affect secondary ion trajectories, possibly resulting in significant isotope fractionation and reduced precision (Kita et al. 2009). Curing the epoxy under pressure can mitigate bubble formation. After hardening, the epoxy mount may be milled to an appropriate overall thickness (~5 mm), and the side exposing the sample is ground with abrasives (e.g., diamond) to expose grain interiors at the required depth. Finally, the mount is polished to generate a flat surface across the entire mount with minimal relief.

Pre-analysis microscopic (transmitted and reflected light) and scanning electron microscope (SEM, using cathodoluminescence CL and/or back-scattered electrons BSE) imaging is essential because it provides a guide for navigation and targeting during analytical sessions and later provides a spatial context for organization, evaluation, and interpretation of combined geochronologic and trace element data. Electronic mount images may be uploaded and registered to stage coordinates for point-and-click recording of spot locations that are later referenced for automated analyses. For electron beam imaging, a conductive carbon (C) coating is usually applied. The C coating is then typically removed by gentle re-polishing and replaced prior to SIMS analysis by a gold (Au) coating. The Au coating has the advantages of being more conductive, monoisotopic, and rapidly removed due to its faster sputtering compared to C.

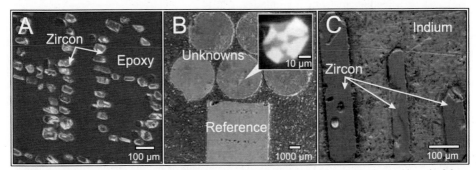

Figure 3. Examples for different mounting techniques. A) grain mount with zircon (imaged in cathodoluminescence CL) embedded in epoxy. B) optical image of a SIMS mount with diamond core petrographic thin-section disks mounted together with zircon reference grains prepared as a sectioned epoxy mount. Inset shows CL image of a small zircon from within the rock. C) zircon crystals pressed into indium showing unsectioned grains with their pristine crystal surfaces (back-scattered electron (BSE) image).

In-situ mounts. *In-situ* analysis using polished rock sections is a viable alternative to conventional procedures that aim at liberating accessory minerals from the rock and pre-concentrating them in a heavy mineral separate. The advantages of *in-situ* dating relative to mineral separation are partially offset by the requirement of preparing a larger number of mounts because of the typically low abundance of accessory grains exposed on the surface of a rock section. Consequently, the need for frequent sample changes and the time required to locate targets, especially if they are small, reduces analytical throughput when *in-situ* mounts are used.

Standard rectangle petrographic thin-sections where datable accessory minerals have been located and identified must be trimmed to dimensions that can fit ion microprobe holders (e.g., Rayner and Stern 2002). Even 25.4 mm round petrographic sections are best reformatted to permit co-mounting of reference materials. This is advantageous for throughput and data quality because swapping between mounts during the analysis is time-consuming and exact analytical conditions may not always be replicated after sample exchange. Trimming can be performed by cutting sections into smaller pieces containing the region of interest using a low-speed diamond saw, or by coring rounds using a diamond drill bit. The cut-outs are placed face-down on adhesive tape together with blocks of pre-sectioned and pre-polished reference material grains (Fig. 3). Subsequent preparation steps are identical to those described in the previous section, with or without the grinding and polishing steps.

Surface analysis and depth profiling mounts. Secondary ion emission in SIMS occurs from the top few nm of the target, making it ideally suited for the analysis of thin surface layers (e.g., Hunter 2009). SIMS is widely used to detect dopants in synthetic electronics materials via depth profiling where continuous sputtering produces a time-resolved signal that can be related to depth if the sputtering rate is known (Hunter 2009). Instrumental parameters that determine SIMS depth resolution include impact energy, beam density distribution, primary ion species, sample flatness, and secondary ion tuning. Especially the field aperture setting in CAMECA ion microprobes is critical as it serves to preferentially transmit secondary ions from the center of the crater and mitigate contribution of ions emitted from the more slowly sputtered crater edges (Hunter 2009). SIMS depth profiling of synthetic or experimentally treated natural samples has been successfully applied to quantify diffusion properties of geologic materials (e.g., Cherniak et al. 2004). In such samples, it is generally easier to control the geometric parameters (i.e., orientation of chemical/isotopic layers) that enhance depth resolution, whereas accessory minerals that are targets for petrochronology often lack these ideal properties. Nonetheless, SIMS surface analysis and depth profiling has potential to identify age or chemical variations at a scale that is inaccessible by other techniques. Crystals that have well developed flat crystal faces such as the (100) or (110) prisms in zircon lend themselves to depth profiling, provided that internal growth layers are oriented parallel to the surface (e.g., Breeding et al. 2004; Vorhies et al. 2013).

The shallow depth resolution characteristics of SIMS analyses thus can be leveraged to provide ultra-high spatial resolution sampling of these crystal faces, i.e., last crystallization layers of crystals that grew in a concentric fashion. The last interval of crystal growth by accessory minerals, often recording the final magmatic or metamorphic history, may archive important time-compositional information for petrochronology. In the case of accessory minerals like zircon, the outermost compositional zones may be only nano- to several micrometers in thickness, and thus beyond the spatial resolution of most other approaches for micro-geochronology and trace element quantification. When sampling crystal faces, lateral resolution is less critical because growth zones are often parallel to the plane of sputtering. Special techniques for depth profiling sample preparation where crystal faces are arranged parallel to the mount surface involve pressing crystals into soft and malleable metal such as indium (In; Fig. 3). Although the bonding of In is weaker than with epoxy, sectioning and polishing In-mounted crystals after depth profiling is possible and permits subsequent analysis of the crystal interiors. Alternatively, crystals can be extracted from the In mount, and recast into epoxy.

QUANTITATIVE SIMS ANALYSIS

Relative sensitivity and instrumental mass fractionation factors

The two major aspects of quantitative SIMS analysis are the interpretation of mass spectra that exist in the mass range of interest, and the quantification of secondary ion intensities of the species of concern with the goal to obtain accurate abundance information. Spectrum interpretation is thus a prerequisite to reliably extract the SIMS signal of interest, and isolate it from potentially overlapping signals such as interferences or backgrounds (e.g., abundance sensitivity). This is often not a trivial task because mass spectra in SIMS are intrinsically complicated, and comprise isobaric interferences composed of atoms from the target material, primary ion beam species, conductive coating, vacuum residua, and other potential contaminants (polishing materials, embedding medium, implanted atoms from previous SIMS analysis, etc.). Hydrides are particularly critical because they require high mass resolution (e.g., $m/\Delta m = 14000$ for $^{92}ZrH^+$ on $^{93}Nb^+$; Fig. 2), and are difficult to eliminate by energy-filtering. Achieving low H backgrounds, where critical, requires organics-free mounting media and the use of a liquid-N_2 cold trap or helium cryopump to pump hydrogen gas efficiently. Singly-charged ions dominate the mass spectrum, but dual positively charged ions (e.g., $^{90}Zr^{++}$ on $^{45}Sc^+$) can cause interferences with an analysis if unmitigated, e.g., through high mass resolution (in the case of Sc at $m/\Delta m = 13000$). EM backgrounds are typically negligibly low, but corrections for abundance sensitivity (the tailing of adjacent high-intensity peaks at a distance of 1 u; it is ~250 ppb for large magnet radius SIMS instruments) can be essential for ultra-trace analysis such as ^{230}Th or ^{231}Pa. Abundance sensitivity is often monitored adjacent to the peak of interest using an interference-free mass station, sometimes at the opposite mass side relative to the interfering peak, or a nearby mass with interference patterns similar to the peak of interest.

Secondary ion signals must then be accurately quantified as isotope ratios or concentrations. The fundamental equation that underlies quantitative SIMS analysis (Eqn. 1):

$$I_x = I_p \cdot Y_s \cdot \alpha_x^{\pm} \cdot f_x \cdot c_x \tag{1}$$

relates I_x = current of ion species x in terms of I_p = primary ion beam current, Y_s = the sputter yield (atoms removed per incoming primary ion), α_x = ionization probability of species x (as positive or secondary ions), f_x = fraction of x ions transmitted and detected, and c_x = concentration of x, with concentration being the quantity of interest. Sputter yield, ionization probability, transmission and detection efficiency can be combined into an element-specific useful yield (which is the ratio between ions of x detected divided by atoms x removed). For positive atomic secondary ions at low energies useful yields range between 10^{-3} and 10^{-1} for most elements (Hinton 1990; Hervig et al. 2006). For similar matrices (e.g., silicates) and analytical conditions (e.g., primary ion species, secondary ion energy), useful yields or absolute sensitivities follow basic chemical trends, and tabulated data can be used to make first-order predictions about the expected intensities of an ion species for a given concentration (Hinton 1990; Hervig et al. 2006).

Quantification of element abundances or relative abundances for analytical purposes require empirical corrections because the sputtering and instrument related unknowns in Equation (1) cannot be deduced theoretically, and they are variable with the material analyzed. These corrections are based on a comparison of unknowns with natural or synthetic measurement standards analyzed under reproducible conditions. Secondary ion intensities for the species of interest x are corrected for detector gain, dead time (in the case of EM detectors), and background (in particular for FC detectors which have much higher noise levels compared to EMs), and divided by the intensity of a normalizing species m corrected in the same manner. For a reference material, concentrations of x and m (or their ratio c_x/c_m) are known, and from the ratio "measured over true" a relative sensitivity factor (RSF) can be determined. This RSF combines all unknown parameters in Equation (1). The primary beam intensity cancels, provided it remains constant, or is corrected for drift if masses are analyzed sequentially (Ludwig 2009):

$$\frac{I_x}{I_m} = \frac{I_p \cdot Y_s \cdot \alpha_x^{\pm} \cdot f_x \cdot c_x}{I_p \cdot Y_s \cdot \alpha_m^{\pm} \cdot f_m \cdot c_m} = RSF \cdot \frac{c_x}{c_m} \tag{2}$$

The RSF represents the correction factor that needs to be applied to the measured quantity on the sample to obtain the "true" abundance (e.g., U ppm for trace element analysis) or ratio (e.g., $^{206}Pb/^{238}U$ for geochronology). Provided the RSF is identical for unknowns and references, the following relation holds (Eqn. 3 as an example for the quantification of U abundance in zircon):

$$RSF = \frac{I_U^{unknown}}{I_{Zr}^{unknown}} \cdot \frac{c_{Zr}^{unknown}}{c_U^{unknown}} = \frac{I_U^{reference}}{I_{Zr}^{reference}} \cdot \frac{c_{Zr}^{reference}}{c_U^{reference}} \tag{3}$$

where I_{Zr} is the intensity of a zirconium (Zr) species of interest (e.g., $^{94}Zr_2O^+$ at mass/charge = 203.8076). To estimate concentrations (e.g., U ppm), c_{Zr} needs to be quantified for the reference zircon and the unknown, unless it can be assumed that they are identical and cancel each other. This is—to a first order—the case for stoichiometric Zr in zircon. The U concentration (in the chosen units, e.g., in ppm) then simplifies to (Eqn. 4):

$$c_U^{unknown} = \frac{I_U^{unknown}}{I_{Zr}^{unknown}} \cdot \frac{1}{RSF} \tag{4}$$

with the RSF determined on a reference zircon (e.g., 91500 with ~80 ppm U; Wiedenbeck et al. 2004). If a glass (e.g., NIST SRM 610) is used to quantify U in zircon and a Si ion species is the normalizing species, then different SiO_2 abundances in unknown zircon and reference glass need to be carried through for quantification (e.g., $SiO_2 = 70$ wt% for NIST SRM 610 and 32 wt% for stoichiometric zircon). Multiple measurements of reference materials with different abundances of the trace element of interest can be used to establish a working curve, which helps to validate the linearity of the method and account for variations in the reliability of reference materials (e.g., for Ti-in-quartz; Fig. 4).

Before moving on to more complex calibrations, it should be mentioned that externally calibrated stable isotope ratios are quantified in an analogous way, whereby an instrumental mass fractionation factor (IMF) is determined on a reference, and then applied to the unknown, as expressed in Equation (5) and (6) for the example of oxygen isotopic ratios (where R_{true} is the "true" abundance ratio of $^{18}O/^{16}O$ and R_{meas} the measured ratio of the secondary ion intensities for $^{18}O^-$ and $^{16}O^-$):

$$IMF = R_{meas}^{reference} / R_{true}^{reference} = R_{meas}^{unknown} / R_{true}^{unknown} \tag{5}$$

$$R_{true}^{unknown} = R_{meas}^{unknown} \cdot \frac{1}{IMF} \tag{6}$$

Two- and three-dimensional relative sensitivity calibrations

It has long been recognized that Pb/U and Pb/Th RSF values in SIMS can vary strongly, even when analytical conditions are kept as constant as possible. The reasons for this behavior remain poorly understood, but covariations in Pb/U and Pb/Th RSF have been observed with U and Th oxide formation from which a RSF tailored to a given XO_m/XO_n of the unknown can be obtained (XO_m/XO_n, X = U or Th; $m = 1$–2; $n = 0$–1; $m \neq n$; Hinthorne et al. 1979, and reviews in

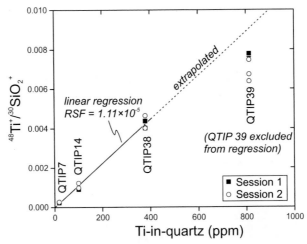

Figure 4. Example for a working curve for SIMS trace element analysis with the useful range shown as solid line and extrapolation (not used for quantification) as dashed line. Synthetic Ti-in-quartz reference materials ("QTIP") from Thomas et al. 2010). Secondary ion intensities for $^{48}Ti^+$ and $^{30}SiO_2^+$ were counted simultaneously using dual electron multiplier detection with a detector slit width of 150 μm (corresponding to $m/\Delta m = 8000$). The relative sensitivity factors (RSF) for two analytical sessions (May and August 2013) are indistinguishable. Note that the QTIP 39 reference with the highest Ti is heterogeneous, and therefore omitted from the calibration. Calibration data from Shulaker et al. (2015) and collected with multi-collection on a CAMECA IMS 1270.

Williams and McKibben 1998; Ireland and Williams 2003; Jeon and Whitehouse 2015). These empirically determined functions (Fig. 5) can have the form of a power law:

$$\frac{Pb}{XO_n} = a \left(\frac{XO_m}{XO_n} \right)^b \tag{7}$$

or, if spread is minor, reasonably be approximated by a linear fit:

$$\frac{Pb}{XO_n} = a \left(\frac{XO_m}{XO_n} \right) + b \tag{8}$$

Such two-dimensional (2-D, because they depend on two measured indicators) RSF calibrations have been developed for different accessory minerals (e.g., allanite, apatite, baddeleyite, monazite, rutile, titanite, xenotime, zircon), and also for calculating U abundances (e.g., Williams and McKibben 1998). The optimal choice of XO_m/XO_n will depend on the matrix and instrument parameters (e.g., O^- and O_2^- primary ion beam, or use of O_2-flooding), but differences in the resulting calibration uncertainties are generally minor. A long-term intercomparison of different calibration schemes for the same reference material (91500 zircon) shows that RSF values scattered within 0.79 and 1.1% relative uncertainty (Jeon and Whitehouse 2015), and monitoring this variability is a fundamental requirement for precise SIMS geochronology.

In some instances, three-dimensional calibrations (3-D) have been proposed, e.g., for Th–Pb geochronology of allanite where matrix effects could be monitored by also including FeO^+/SiO^+ in the calibration (Catlos et al. 2000). Other studies have subsequently established that conventional 2-D calibrations for allanite yielded sufficiently accurate results (Gregory et al. 2007).

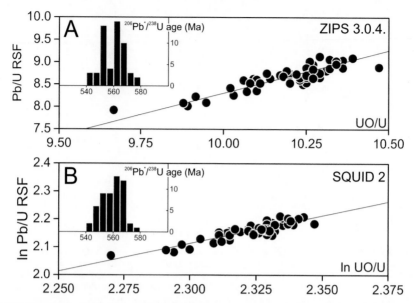

Figure 5. Example for U–Pb zircon calibration plots for zircon reference z6266 with an age of 559 Ma (Stern and Amelin 2003). Panels A and B show the same data, with A using a linear calibration and B a power law calibration. Data reduction was performed using ZIPS 3.0.4. (created by C. Coath), and SQUID 2 (K. Ludwig). Significant differences in the age and uncertainty calculations are absent, demonstrating the comparability of data reduced with different calibrations and software. Data collected using a CAMECA IMS 1270.

Another variation of a relative sensitivity calibration is used to determine accurate Th/U in accessory minerals. This is important as Th/U is often used as a first-order indicator for magmatic vs. metamorphic origins (where magmatic Th/U in zircon is ~0.3–0.7, whereas many metamorphic zircons have Th/U < 0.1; Rubatto 2002; Hoskin and Schaltegger 2003), and to implement accurate disequilibrium corrections (see below). Although Th/U RSF values are much closer to unity than those for Pb/U, quantification is hampered by Th and U heterogeneities in many reference zircons (with the notable exception of 91500 zircon). An elegant solution is a calibration, where "true" Th/U is proxied by $^{208}Pb^*/^{206}Pb^*$ which shows insignificant isotopic fractionation in SIMS. Assuming concordancy between Th–Pb and U–Pb decay systems, Th/U can be derived from measured $^{208}Pb^*/^{206}Pb^*$ (Hinthorne et al. 1979; Reid et al. 1997).

In all these cases, potential bias can arise if reference materials used for calibration behave differently during primary ion bombardment than the unknowns. Matrix effects are biases that can result from differences in chemical composition, which for zircon can be high (> 2500 ppm) U, although White and Ireland (2012) have proposed that bias in the analysis of high-U zircons may reflect incipient metamictization rather than a chemical matrix effect. Crystallographic orientation can also influence relative sensitivities (e.g., baddeleyite, Wingate and Compston 2000) and instrumental mass fractionation (e.g., oxygen isotopes in magnetite and rutile; Huberty et al. 2010; Shulaker et al. 2015). Adapting analytical conditions can mitigate both effects. Dispersion of U–Pb was reduced for randomly oriented baddeleyite and rutile crystals when O_2-flooding was applied (Li et al. 2010; Schmitt et al. 2010; Schmitt and Zack 2012), and oxygen isotope data showed less spread if low energy primary ions were used for magnetite and rutile (Huberty et al. 2010; Shulaker et al. 2015).

Specific analytical consideration and strategies for petrochronology

Geochronology. Coupling geochronology with textural and petrologic information requires adhering to an analytical strategy to optimize data quality and efficiency. Preparation for geochronology considers the availability of datable minerals in a rock, mineral compositions, and the range of expected dates. This also includes the availability of reference minerals for calibration. Zircon is probably the most-straightforward mineral for geochronology using SIMS because it is widespread in crustal rocks and has been employed as a geochronometer for decades (e.g., Rubatto 2017, this volume). Multiple reference materials exist for calibration of Pb/U ratios and trace elements in zircon. Most silicic plutonic rocks and many of their volcanic equivalents contain zircon, either as an early or late microphenocryst or as inclusions in major phases. Mafic intrusions often reach zircon saturation as they cool to their solidi. Metamorphic rocks typically contain zircon grains as crystals found in the matrix as well as in porphyroblasts. Other accessory minerals tend to be more restricted in their occurrences, for example baddeleyite is found mostly in mafic rocks, chevkinite in alkalic rocks, and monazite found predominantly in metamorphic rocks. Some accessory minerals may be preferable to others for petrochronology depending on compositions and project goals. For example, the complex composition of allanite facilitates the combination of geochronology with elemental concentrations that are reflective of the major-minor element evolution of the parent from which it grew, such as changes in Mn and Mg (e.g., Vazquez and Reid 2004), whereas zircon generally provides information about trace element evolution (e.g., REE, Hf) of its parent. In addition, accessory minerals differ in their closure behavior, and thus can be harnessed as thermochronometers.

The range of expected dates and mineral compositions, e.g., Tertiary monazite versus Proterozoic zircon, will dictate if dates are best derived by $^{207}Pb/^{206}Pb$, $^{206}Pb/^{238}U$, $^{207}Pb/^{235}U$ $^{208}Pb/^{232}Th$, etc., methods, or $^{238}U-^{230}Th$ disequilibrium in the case of young (<350 ka) accessory minerals. Because dates using any of these methods are derived from parent/daughter or daughter isotope ratios, a sufficient concentration of parent isotope is needed, especially when dating young minerals. High U concentrations have allowed U–Th dating of zircons as young as ca. 2.5 ka (Schmitt et al. 2013; Wright et al. 2015). However, there can be too much of a good thing: high U concentrations may be associated with metamictization, which compromises the assumption of matrix-matching between unknown and the reference material (see above), and which may be associated with Pb-loss and result in spurious crystallization ages. High U intensities can also compromise the EM detector and trigger safeties to protect it. High U concentrations and metamict domains are often recognizable in cathodoluminescence and backscattered electron images, and thus can be avoided via pre-analysis imaging. In general, instrumental conditions for U–Pb and U–Th dating are similar, with mass resolutions of $m/\Delta m > 5000$ needed to resolve $^{206}Pb^+$ from nearby $^{178}Hf^{28}Si^+$ and $^{174}Hf^{16}O_2^+$ in zircon. In CAMECA instruments, oxygen flooding is routinely used to enhance secondary ion yields of Pb^+ in zircon.

Analogous to the $^{40}Ar/^{39}Ar$ method, SIMS U–Pb dates are calculated relative to a reference that yields concordant U–Pb dates, typically independently constrained by conventional isotope dilution mass spectrometry (ID-TIMS). Several reference zircons are commonly used for U–Pb dating by SIMS, including Temora (418 Ma), R33 (419 Ma), SL13 (572 Ma), z6266 (559 Ma), 91500 (1065 Ma), AS3, and FC1 (both 1099 Ma). For some of these reference materials an approximately 1 million year range in the $^{206}Pb/^{238}U$ dates has been reported by different laboratories, e.g., for zircon R33 (Black et al. 2004; Mattinson 2010; Schaltegger et al. 2015). Through the use of commonly distributed spikes (e.g., EARTHTIME) for isotope dilution, the range of dates from different laboratories is likely to be reduced in the future. Minerals with high Th concentrations, such as monazite and allanite, are attractive for their potential to generate precise $^{208}Pb/^{232}Th$ dates. High Th minerals as reference materials for U–Pb and/or Th–Pb dating include: monazite 44069 (Aleinikoff et al. 2006), monazite 554 (Harrison et al. 1995), titanite BLR-1 (Aleinikoff et al. 2007), allanite Siss and allanite Bona (von Blanckenburg 1992), allanite Tara (Gregory et al. 2007), and xenotime MG-1 (Fletcher et al. 2004), although not all of these minerals have $^{208}Pb/^{232}Th$ dates that are confirmed by ID-TIMS (see below).

SIMS U–Pb dating also requires corrections for common Pb; the resulting radiogenic Pb* is used for the age calculation. For this, stable [204]Pb is analyzed and used for correction in combination with appropriate isotopic compositions of common Pb. For zircon, [204]Pb intensities are low, and often decreasing during the analysis, indicating that common Pb is surface derived. Consequently, anthropogenic common Pb compositions are adequate. Other minerals (e.g., allanite, apatite, monazite, titanite) have significant intrinsic common Pb, and common Pb compositions used for correction need to be carefully chosen. Accurate determination of [204]Pb is hampered by unresolvable interferences (e.g., $^{232}Th^{144}Nd^{16}O_2^{++}$ in monazite). Peak-stripping (using $^{232}Th^{143}Nd^{16}O_2^{++}$ at mass 203.5) or some amount of energy filtering and/or use of a pre-collector retardation lens is needed to reduce isobars that interfere with the Pb isotopes in monazite (Fletcher et al. 2010). Alternative corrections for common Pb are therefore often advantageous: for unradiogenic materials, or minerals with low Th (e.g., baddeleyite, rutile), a [208]Pb-based correction (Compston et al. 1984) is a viable alternative to the [204]Pb-based common Pb correction. If [204]Pb is difficult to measure and/or the dated material is sufficiently young that concordancy can be reasonably assumed, the common-Pb uncorrected data can be regressed to obtain a concordia intercept age (e.g., Baldwin and Ireland 1995). The effects of variable initial disequilibrium, however, can hamper the accuracy of this approach (e.g., Janots et al. 2012).

Accessory minerals in Quaternary (< 300 ka) rocks are amenable to U–Th dating. Due to the greater secondary ion yields for actinide oxides, U–Th dating by SIMS typically employs measurements of UO$^+$ and ThO$^+$ ions at $m/\Delta m = 5000$–6000 in order to resolve $^{230}Th^{16}O^+$ from any REE molecules and other species of the same unit mass (Schmitt 2011). Analyses for U–Th dating of zircon, allanite, or chevkinite typically require primary beam intensities that are 5–10 times greater than used for U–Pb dating. Count rates for [230]Th during analyses of zircons are typically 0.1–10 cps for U concentrations of 10–1000 ppm and a 50 nA O$^-$ primary beam (Schmitt 2011). Zircons with low U (< 200 ppm) can be challenging to date by U–Th or U–Pb methods, and may necessitate relatively large datasets to resolve statistically meaningful dates (e.g., Coombs and Vazquez 2014). The duration of analyses for ^{238}U–^{230}Th dating is typically 30–45 minutes, but this can be shortened by ~30% using multi-collection (e.g., Storm et al. 2011; Vazquez et al. 2014). Quantification of the relative ionization between U and Th is needed for U–Th dating, and can be accomplished using the observed versus true $^{208}Pb/^{206}Pb$–$^{232}Th/^{238}U$ relation for a reference mineral of known age or by precise measurement of a mineral in ^{238}U–^{230}Th secular equilibrium (Schmitt 2011).

Trace elements in accessory minerals. The combination of trace element analyses and geochronology using SIMS has been extensively used to understand the petrologic evolution of magmas and compositional evolution of metamorphic rocks. The use of zircon for petrochronology has been popular over the last decade, in particular due to the development and refinement of the Ti-in-zircon geothermometer (Watson and Harrison 2005; Ferry and Watson 2007) and advancements in the methodology for high-spatial resolution and high-precision dating of old and young zircons (e.g., Schaltegger et al. 2015). The combination of trace element geochemistry, Ti thermometry, stable isotope analysis (see below) and high-precision U–Th geochronology has been used to resolve the chronology of changing magma composition and temperature at young volcanoes, including the identification of crystals recycled by the thermal rejuvenation of stalled intrusions. This has contributed significantly to a better understanding of realistic crystallization conditions of zircon in magmas and their longevity (e.g., Claiborne et al. 2010; Storm et al. 2011).

In general, a mass resolution $m/\Delta m > 8000$ is required for interference-free analysis of REE in zircon, including the resolution of $^{48}Ti^+$ from $^{96}Zr^{2+}$. This level of resolution normally requires some energy filtering and/or closing of slits. Alternatively, high energy filtering (at ~100 eV offsets) and peak-stripping of independently determined atomic/oxide ion ratios can be applied for REE analysis at moderate mass resolution $m/\Delta m = 2000$. Significantly higher mass resolution

(>10,000) is needed to adequately resolve some other trace elements, such as $^{93}Nb^+$ from $^{92}ZrH^+$ (Fig. 2) that have been used for tracing the tectono-magmatic environments of zircon (e.g., Grimes et al. 2015). Allowing epoxy mounts to outgas at least overnight under high vacuum can minimize hydride interferences. Guide peaks are typically needed to autocenter the peaks of REE species that tend to be found at relatively low concentrations, such as Eu in zircon. In zircon, the ubiquitous presence of Zr molecules within the LREE spectrum provides a reliable set of guides (e.g., $^{91}Zr^{30}Si^{16}O_2^+$ for $^{153}Eu^+$) that can be employed for automated analysis.

Simultaneous measurement of selected trace elements and the isotopes used for geochronology is feasible (e.g., Abbott et al. 2012). For example, the Stanford-USGS SHRIMP–RG routinely performs U–Th–Pb dating at mass resolution of $m/\Delta m > 8000$, with simultaneous measurement of REE, Y, and Hf. Most studies, however, have used a two-step approach to combine dates and trace element concentrations, with trace element measurements occurring before or after those analyses utilized for dating, usually by placing an analysis spot within the same compositional domain sampled for geochronology with a smaller primary beam spot size and at lower intensity as for geochronology. Pre- and post-analysis imaging is of critical importance to confirm the link between the dating and compositional information. The correspondence between the domains sampled for trace elements and dating can be checked by evaluating the concordance of a single trace element, like uranium, that is measured during both analyses (e.g., Mattinson et al. 2006).

Trace element concentrations determined by SIMS are typically calculated by comparing the count rates measured in an unknown to those measured in a standard with known concentrations, typically normalized to a mutually stoichiometric element such as Si or Zr in the case of zircon. Matrix-dependent ionization of elemental and isotopic compositions is characteristic of SIMS analyses (e.g., Ireland 2015), and thus unknowns should ideally be matched to standards in terms of composition and crystal structure. Trace element concentrations for accessory minerals are generally calibrated to a single standard during an analytical session, but calibration curves (Fig. 3) that are derived from the relation between secondary ion yield and known concentrations for a suite of standards may be employed (e.g., Hiess et al. 2008). Various trace element reference zircons are employed by different SIMS laboratories, with 91500 (Wiedenbeck et al. 2004), z6266 (Stern and Amelin 2003), and MAD (Barth and Wooden 2010) as the most commonly used for calibration of REE and other elements in zircons. These are homogenous gem-quality materials and are derived from megacrysts. SL13 zircon is noteworthy because its Ti concentration is well characterized and homogenous (~6.3 ppm; Hiess et al. 2008), which is important for application of the Ti-in-zircon geothermometer. In contrast, the array of references that are commonly used for geochronology (e.g., Temora, R33, and AS3 for zircon, and 44069 for monazite) are mostly heterogeneous with respect to trace elements. References for calibrating SIMS measurements of trace elements in the other accessory minerals used for geochronology are less abundant than for zircon, but include xenotime BS-1 (Fletcher et al. 2004; Aleinikoff et al. 2012), monazite NAM (Aleinikoff et al. 2012), and titanite BLR (Mazdab 2009). For those accessory minerals with high REE concentrations, domains that have been dated via SIMS may be characterized using electron microprobe analyses, and their results combined to evaluate time-compositional evolution (e.g., Vazquez and Reid 2004; Vazquez et al. 2014).

The presence of micro-inclusions may hinder efforts to quantify the trace element concentrations of accessory minerals; applying fine beams (e.g., on instruments dedicated to high spatial resolution analysis such as NanoSIMS) may be required in such cases. Accessory minerals, in particular zircon, are often riddled with micro-inclusions trapped during crystal growth including other accessory minerals such as apatite, allanite, or oxides, as well as melt (glass) inclusions. Inadvertent sampling of these inclusions by the primary beam can dramatically skew trace element concentrations. For example, incorporation of apatite and glass inclusions, which are especially common in igneous zircons, may result in an REE pattern that is reversed from the high HREE/LREE pattern that typifies zircon. Cracks and other imperfections in zircons may contain elevated concentrations of trace elements such as Ti, which if sampled might skew calculated crystallization temperatures (Harrison and Schmitt 2007). To identify problematic analysis, it is recommended to always monitor elements that are indicative of inclusions (e.g., Mg, P, Fe) along with the trace elements of interest. Reproducibility for Ti in SL13 is <5% (2σ; Hiess et al. 2008), but contamination from beam overlap with crystal imperfections can be a major source of bias for the typically low abundances of Ti in zircon: at ~5 ppm Ti (corresponding to a crystallization temperature of ~680 °C) a 1 ppm excess would increase apparent temperatures by ~15 °C. RSF reproducibility is thus a minor source of uncertainty for the Ti-in-zircon thermometer compared to surface contamination and inclusion of non-zircon phases in the analysis volume. Because of the large sample volume of LA-ICP-MS compared to SIMS, the presence of subsurface inclusions can be problematic in LA-ICP-MS even when they are not visible at the surface.

Stable and long-lived radiogenic isotopes. Stable isotopes (specifically, O and Li) and long-lived radiogenic isotopes (Hf- and Nd-isotopes) have revealed important constraints on the origins of Hadean detrital zircons where rock context is lacking. These isotope systems are also relevant for the study of igneous and metamorphic zircon, and, in the case of oxygen isotopes, for accessory monazite and rutile. Because of the comparatively slow sputter rate of SIMS, secondary ion intensities are typically too low to be measured in FC detectors, and consequently data lack the precision that is required for a meaningful characterization of the Hf and Nd isotopic composition (e.g., Kinny et al. 1991), where LA-ICP-MS is utilized for zircon and monazite, respectively (e.g., McFarlane and McCulloch 2007; Fisher et al. 2014). In developing the optimal analytical strategy, it is, however, important to consider that SIMS is the most conservative analytical technique with regard to sample consumption, and that it is generally advantageous to start a sequence of petrochronological analyses with the least destructive methods, whereby SIMS can be considered as practically non-destructive when comparing the small amounts of materials sputtered to those removed during sample preparation using grinding and polishing abrasives. Because of the comparatively large amounts of material consumed in LA-ICP-MS Hf- and Nd-isotopic analysis (crater dimensions are several 10's of µm in width and depth), such data are best collected after all other required measurements have been completed on the target crystals.

Oxygen isotopes in zircon analysis protocols, including reference materials, have been established in different labs and using different instrumentation. Of these, oxygen isotopes in zircon have experienced the widest range of applications, and SIMS has practically become the method of choice for oxygen isotope ($\delta^{18}O_{VSMOW}$) analysis of zircon (Valley and Kita 2009). If oxygen isotopes cannot be analyzed first, then previously analyzed grains (e.g., for geochronology) require removal of existing craters or placing spots away from them because of ion implantation with the commonly used mass-filtered $^{16}O^-$ beam. Multicollection of $^{16}O^-$ and $^{18}O^-$ is performed whereby an intermediate detector is sometimes dedicated to $^{17}O^-$ (ca. 4-times lower in abundance than ^{18}O) when extraterrestrial materials are investigated for mass-independent fractionation effects which were inherited in the rocky planets from early solar system oxygen isotopic heterogeneities (e.g., Nemchin et al. 2006). For terrestrial zircon, simultaneous analysis of mass 17 along with 16 and 18 can also target $^{16}OH^-$ (requiring a mass resolution $m/\Delta m = 6000$ to be resolved from $^{17}O^-$) which can reveal the presence of water in nominally anhydrous zircon as an indication of metamictization (Pidgeon et al. 2013). For high precision analysis, shaped beam (Köhler) illumination or rastering of a focused Cs^+ beam over ~5 to 10 µm is applied to minimize downhole fractionation of isotope ratios. Charge build-up in the crater is prevented by re-supplying electrons via a normal incidence electron gun held at equipotential to the sample surface (CAMECA) or an oblique electron gun delivering a constant current (SHRIMP). Raw counts are corrected for FC backgrounds and detector yield. Bracketing and intermittent analyses of zircon references are performed to calibrate instrumental mass fractionation, and to correct isotopic ratios of the unknowns. External reproducibility of zircon $\delta^{18}O$ is ~0.2–0.3‰ (2σ), provided that the sample is flat (Kita et al. 2009), and targets are located at some distance from the edges of the crater (Treble et al. 2007). New mount designs (see below) have improved reproducibility of isotope analyses over an area up to ~15 mm in diameter. High spatial resolution analysis of $^{18}O^-$ and $^{16}O^-$ negative ions employs Cs^+ sources at crater dimensions (width and depth) of down to ~1 μm (Page et al. 2007). An ~0.5 μm Ga^+ beam emitted from a liquid metal ion source has also been successfully used in reconnaissance studies (Bindeman et al. 2014). For high spatial resolution analysis, secondary intensities are too low to measure $^{18}O^-$ in an FC detector. Instead, an EM is used for detection of $^{18}O^-$ which limits precision to ~1–2‰ (2σ; Page et al. 2007).

Accurate oxygen isotope analysis of monazite is more problematic than of zircon because of monazite's strong compositional variability, whereas zircon is essentially stoichiometric with only minor variability in Hf content causing generally negligible matrix effects. Several studies have demonstrated a matrix dependent variation in IMF for monazite where compositional variations of Th (via the huttonite substitution: $Th^{4+} + Si^{4+} = REE^{3+} + P^{5+}$) and Ca (via the brabantite substitution $Ca^{2+} + Th^{4+} = 2REE^{3+}$) cause IMF values to vary by several permil, depending on the instrumentation used, and the overall chemical variation delineated by the reference monazites analyzed (Breecker and Sharp 2007; Rubatto et al. 2014). Uncertainty in the bulk oxygen isotope analysis of reference monazite (Breecker and Sharp 2007; Rubatto et al. 2014) is another source of potential bias. Because of these uncertainties, reliable oxygen isotope analysis of monazite (e.g., for the purpose of monazite–quartz isotope exchange thermometry) requires careful consideration of all external and internal sources of error.

Lithium isotope ($^7Li/^6Li$ expressed as δ^7Li_{LSVEC}) analysis is emerging as a potentially useful indicator of magma sources and the thermal history of zircon (Ushikubo et al. 2008; Trail et al. 2016). Zircon incorporates Li as an interstitial cation through the coupled substitution $2(REE, Y)^{3+} = Li^+ + P^{5+}$ (Bouvier et al. 2012). Zircon in magma-derived continental crust has higher Li abundances compared to zircon from oceanic crustal plagiogranites (Grimes et al. 2011). Zircon from Archean felsic plutonic rocks mostly have near-mantle δ^7Li isotopic compositions of $+3.8 \pm 1.5$‰ (from the compilation in Bouvier et al. 2012), whereas Hadean Jack Hills (Australia) zircon has a significant population with negative δ^7Li characteristic of continentally-derived soils and sediments (Ushikubo et al. 2008; Bouvier et al. 2012). Li-in-zircon is a potentially useful diffusion thermochronometer, where the preservation of diffusion profiles can constrain the maximum temperature a crystal has experienced when timescales of high-temperature crystal storage are known or can be estimated (Rubin et al. 2014; Trail et al. 2016). Conversely, incomplete diffusive equilibration and uncertainties resulting from difficulties in identifying homogeneous (with regard to Li abundances and δ^7Li) reference zircon crystals hamper the interpretation of Li isotopic data for zircon.

Data reporting for U–Th–Pb geochronology

Shared attributes of SIMS with other U–Th–Pb dating techniques. SIMS is a highly versatile tool, and can produce diverse data such as trace element and isotope compositions from spot analyses, depth profiles, as well as 2-D and 3-D images. Here, we concentrate on general aspects of data reporting in SIMS geochronology. SIMS quantification includes assumptions and corrections that result from interelement fractionation in different matrices where true values rely on measurements of reference materials by other techniques (typically ID-TIMS), and as such, it shares many attributes with LA-ICP-MS (e.g., Horstwood et al. 2016).

As with LA-ICP-MS, uncertainties resulting from the determination of RSFs on reference materials are typically larger than internal analytical precision. Accurate dating also requires corrections for common Pb and initial disequilibrium. In combination, these corrections limit SIMS age uncertainties to 1–2% (relative error). Systematic errors from uranium isotopic abundances and decay constants (e.g., Boehnke and Harrison 2014) are common to all U–Th–Pb dating methods, but are frequently negligible for SIMS and LA-ICP-MS ages because of their comparatively large uncertainties, especially for unpooled data.

Proper calibration. The ratio of interest should always be calibrated on the same ratio of the reference material. This means that matching ID-TIMS values should be used: e.g., for $^{206}Pb/^{238}U$ RSF calibrations on a reference zircon, its reported $^{206}Pb/^{238}U$ is relevant, and not a ratio calculated from a $^{207}Pb/^{206}Pb$ age. Discordance, or corrections for disequilibrium in individual decay chains (e.g., ^{230}Th in the ^{238}U decay chain) applied to ID-TIMS dates need to be accounted for if published ages for reference materials used to calculate isotopic ratios. Monazite is particularly critical because of the scarcity of ID-TIMS $^{208}Pb/^{232}Th$ ages, and it is potentially problematic to derive $^{208}Pb/^{232}Th$ from U–Pb or Pb–Pb ages without independent verification.

Common Pb correction. Common Pb corrections are an additional source of uncertainty because ^{204}Pb counts are typically low and fraught with high uncertainties. Measurement uncertainties for the Pb species used for the common Pb correction should be propagated into the final age uncertainty, but uncertainties stemming from the choice of common Pb compositions should only be included in the data point uncertainty if they are likely to be variable, or are derived from a measurement pertinent to individual samples (Schoene 2014).

Disequilibrium corrections. Uranium series disequilibrium is a major source of age uncertainty for Quaternary zircons (Fig. 6), and together with pre-eruptive zircon residence can significantly bias cross-calibration of decay constants for ^{40}K- and U-based chronometers (Simon et al. 2008). This mainly affects the ^{238}U and ^{235}U decay chains whereas disequilibrium effects in the ^{232}Th-decay chain can be reasonably neglected due to the absence of long-lived intermediate daughter isotopes. Using ^{230}Th (the longest-lived non-uranium intermediate daughter isotope in the U-decay chains) as an example, the ingrowth of ^{206}Pb under conditions of initial disequilibrium is expressed as (Eqn. 9; modified from Wendt and Carl 1985):

$$\frac{^{206}Pb}{^{238}U} = \left(e^{\lambda_{238}t} - 1\right) + e^{\lambda_{238}t} \cdot \frac{\lambda_{238}}{\lambda_{230}} \cdot \left[\frac{\left(^{230}Th\right)_0}{\left(^{238}U\right)_0} - 1\right] \cdot \left(1 - e^{\lambda_{230}t}\right) \tag{9}$$

with $(^{230}Th)_0/(^{238}U)_0$ representing the initial (at the time of crystallization) activity ratio. The deviation from equilibrium can be predicted from the ratio of Th and U partitioning coefficients, assuming that the melt is in equilibrium (Eqn. 10; Schärer 1984):

$$f = \frac{\left(^{230}Th\right)_0}{\left(^{238}U\right)_0} = \frac{D_{Th}^{mineral-melt}}{D_U^{mineral-melt}} \cdot \frac{\left(^{230}Th\right)_0^{melt}}{\left(^{238}U\right)_0^{melt}} \tag{10}$$

D-values can be determined from ^{232}Th and ^{238}U concentrations measured in the mineral and glass, or if unavailable whole-rock. Alternatively, model D values (e.g., Blundy and Wood 2003) can be applied. Because $^{208}Pb/^{232}Th$ ages are not measurably affected by disequilibrium, they are preferred when D_U/D_{Th} is poorly constrained. Janots et al. (2012) have inverted the problem and used the difference between monazite $^{206}Pb/^{238}U$ and $^{208}Pb/^{232}Th$ ages to reconstruct the U/Th of hydrothermal fluids.

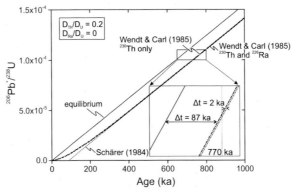

Figure 6. Effect of initial disequilibrium on $^{206}Pb^*/^{238}U$ zircon ages. Four different curves are shown between 0 and 1000 ka: secular equilibrium (solid line), disequilibrium for ^{230}Th according to Equation (9) ($D_{Th}/D_U=0.2$; dash-dot), a simplified estimate from Schärer (1984) using $f=0.2$ (short dash), and the unabridged calculation from Wendt and Carl (1985) which also accounts for ^{226}Ra disequilibrium ($D_{Ra}/D_U=0$). Inset illustrates the resulting age differences (Δt) for disequilibrium uncorrected and corrected ages for ca. 770 ka zircon (equivalent to the disequilibrium-corrected ID-TIMS zircon age of Bishop Tuff; Crowley et al. 2007). Ickert et al. (2015) emphasize that disequilibrium corrections introduce systematic uncertainty, and should not be propagated prior to calculation of weighted average ages (cf. Crowley et al. 2007).

By analogy to the recommended treatment of uncertainty of common Pb corrections, disequilibrium-correction uncertainties should be treated as systematic uncertainties if f (Eqn. 10) depends on model assumptions which is mostly the case because of the difficulties in reconstructing melt Th/U and the impossibility of defining $(^{230}Th)/(^{238}U)$ once equilibrium has been reached (Ickert et al. 2015; Boehnke et al. 2016). These considerations also hold for disequilibrium corrections in the ^{235}U-decay chain, where D_{Pa} can only be estimated from model (e.g., Blundy and Wood 2003) or empirical D_{Pa}/D_U partitioning coefficient ratios (Schmitt 2007, 2011; Rioux et al. 2015) because no long-lived or stable Pa isotope exists, unlike ^{230}Th where long-lived ^{232}Th can be used to estimate partitioning. Additional bias may result from initial disequilibrium in ^{234}U and ^{226}Ra (Fig. 6). Unity activity ratios for $(^{234}U)/(^{238}U)$ can be reasonably expected for melts, but geothermal fluids often have elevated values which are reflected by high initial $(^{234}U)/(^{238}U)$ in hydrogenic minerals such as opal (e.g., Paces 2015). Little is known on ^{226}Ra partitioning in zircon and other accessory minerals, but D values are expected to be very low (10^{-2}–10^{-3}) for zircon when using Ba as a proxy (Thomas et al. 2002).

Validation. Analyzing a secondary reference intermittently during an analytical session and processing the data as if it was an unknown is an essential check to detect instrumental malfunction and/or human error. This is important for both $^{206}Pb/^{238}U$ and $^{207}Pb/^{206}Pb$ ages. The secondary reference should be as similar to the unknown as possible (e.g., in age and/or U abundance) to facilitate the detection of potential analytical bias that could go undetected in other reference materials. Potential bias could occur from unmitigated surficial contamination (e.g., ^{204}Pb or ^{204}Hg), matrix effects (e.g., high-U zircon; White and Ireland 2012), or crystal orientation effects (e.g., baddeleyite, or rutile; Wingate and Compston 2000; Schmitt and Zack 2012; Taylor et al. 2012). Validation data should be included in the publications, either as a summary, or preferably together with data tabulations. To ensure comparability and possibly recalculation of data, it is essential that data sources for reference materials are adequately cited. When comparing data point analyses from the same analytical session (e.g., for calculating a population average, uncertainty, and values for the mean square of weighted deviates MSWD), systematic uncertainties should be excluded. Conversely, when comparing data from different sessions, or labs, systematic uncertainties for different reference materials should be included.

COMPARING SIMS TO OTHER TECHNIQUES

SIMS stands out amongst other techniques for micro-isotopic analysis for its small sampling volume and depth, and the preservation of samples for later analyses that may consume large domains or entire crystals. Other techniques, such as micro-milling or laser ablation, typically sample much larger volumes and often penetrate or consume entire crystals. The high sensitivity and relatively slow sputter rates (e.g., ~0.1 μm^3/s/nA for O$^-$ sputtering of zircon) characteristic of SIMS ensure that sampling depths are shallow and that experiments will be non-destructive relative to the typical size of accessory minerals. Analysts can have confidence that "what you see is what you get" when analyses are guided by cathodoluminescence or backscattered electron images. However, SIMS analysis times are significantly longer than those for comparable analyses by LA-ICP-MS, which is impractical for studies that require large datasets, such as the datasets that are standard for detrital mineral geochronology. Atom probe tomography (APT) has been used to obtain isotope-specific 3D reconstructions of individual atoms in samples shaped to tips < 10 nm wide and few 10's of nm long (e.g., Valley et al. 2015; Peterman et al. 2016). While this technique permits unique insights into the chemical structure of accessory minerals at the nanoscale, the expensive and time-consuming preparation of APT tips, and the small sample volume resulting in large uncertainties for trace components limit the petrochronologic applicability of APT.

Preservation of the textural context of datable minerals is one of the key advantages of microbeam methods, and SIMS excels for analysis of accessory mineral inclusions which are too small to be separated for ID-TIMS, or to yield sufficient signal in LA-ICP-MS analysis. *In-situ* dating of monazite, for example, has revealed age differences between crystals included in garnet vs. those hosted in the matrix, which was attributed to shielding of monazite by its garnet host (e.g., Foster et al. 2000; Catlos et al. 2002). By contrast, it is known that rock disintegration and conventional mineral separation using gravity separation techniques (e.g., through water-shaking, or heavy liquids) may fail in recovering the full spectrum of accessory minerals present (e.g., Sláma and Košler 2012), or even any at all if grains are small (especially < 10 μm) or included in mechanically strong host phases. Lab equipment used in mineral separation (e.g., crushers, shaking tables, magnetic separators, etc.) is also a potential source for contamination, and especially for samples with low accessory mineral abundances it is imperative to use new equipment and/or adhere to strict cleaning procedures to minimize the risk of crystal carry-over from previously processed rocks (e.g., Torsvik et al. 2013). Even when thorough measures are implemented in the mineral separation laboratory, there remains potential for unexpected, and difficult to mitigate, contamination as demonstrated by the recent discovery of abundant zircon in sepiolite drill mud used during ocean drilling (Andrews et al. 2016).

Lastly, SIMS is preferable when intact samples need to be preserved for later analyses. Several studies have demonstrated that a combination of geochronology, trace elements, and oxygen isotopes (measured by SIMS) and Hf isotopes (measured by LA-ICP-MS), all from the same zircons, can provide essential insights ranging from the secular evolution of Earth's crust and mantle (e.g., Harrison et al. 2008) to the evolution of young magma chambers beneath restive volcanoes (e.g., Stelten et al. 2015; Rubin et al. 2016). In these studies, SIMS analyses were used first to generate dates and measure trace elements, then the more penetrating LA-ICP-MS analyses were used to measure Hf isotope compositions. The relatively non-destructive nature of SIMS allows single zircon crystals to be "double-dated" (e.g., in volcanic rocks analyzed for their crystallization and eruption ages). For example, Danišík et al. (2012) used SIMS to generate ^{238}U–^{230}Th dates for zircons from a stratigraphically important tephra in New Zealand. The dated zircons were plucked from their mounts and individually heated to release and analyze their radiogenic helium. This "double-dating" approach was essential for generating a robust eruption age for the tephra, which was confirmed by ^{14}C dating of logs that were charred by the hot volcanic ash.

CASE STUDIES FOR SIMS PETROCHRONOLOGY

Other chapters in this volume cover case studies for SIMS in petrochronology in depth (e.g., Engi 2017, this volume; Lanari and Engi 2017, this volume; Zack and Kooijman 2017, this volume), and we therefore only briefly showcase some recent studies from the authors' respective laboratories where the strengths of SIMS (i.e., *in-situ* analysis capability at high spatial resolution, high-sensitivity, and versatility) proved essential for the success of the research project. One such study (Shulaker et al. 2015) was the first to utilize rutilated quartz as a thermogeochronometer. In rutilated quartz, low-temperature quartz (as evidenced from the presence of fluid inclusions) hosts rutile, often in the form of fine needles with diameters of few 10's of μm. Although the exact formation mechanisms remain unclear, the close spatial relationship between quartz and rutile offers the possibility to apply independent geothermometers: Ti-in-quartz (TITANIQ; Wark and Watson 2006; Thomas et al. 2010), Zr-in-rutile (Zack et al. 2004; Tomkins et al. 2007), and oxygen isotope exchange between quartz and rutile (Matthews 1994). Trace element and isotopic equilibrium between rutile and quartz is more likely for contact pairs, and analyzing rutile and quartz in petrographic context is therefore a prerequisite for testing consistency between these independent geothermometers. Shulaker et al. (2015) investigated a suite of six rutilated quartz samples from Alpine fissures from host rocks which experienced peak metamorphic temperatures between $\lesssim 350$ and $600\,°C$. Two of six samples yielded significant radiogenic ^{206}Pb and a concordia intercept age of ca. 15 Ma, consistent with radiometric ages for scarce monazite and titanite in Alpine clefts (e.g., Janots et al. 2012). Reproducibility of SIMS oxygen isotope data in reference quartz was ~0.3‰ (2σ) whereas randomly oriented rutile crystals (R10B; Luvizotto et al. 2009) were only reproducible to ~1‰ under the same analytical conditions. Shulaker et al. (2015) attributed this to a crystal orientation effect. Because of strong isotopic differences over the pertaining temperature interval ($\Delta^{18}O_{quartz-rutile} = \delta^{18}O_{quartz} - \delta^{18}O_{rutile} = +7$ to +15‰), uncertainties related to crystal orientation translate to only minor (20–60 °C) temperature uncertainties for the quartz–rutile oxygen isotope exchange thermometer. Interestingly, oxygen isotope exchange temperatures show no correlation with TITANIQ temperatures (using calibrations of Thomas et al. 2010; Huang and Audétat 2012) at low temperatures (Fig. 7), whereas such a correlation exists between oxygen isotope and Zr-in-rutile thermometers (Zack et al. 2004; Tomkins et al. 2007), albeit with a significant offset for the Tomkins et al. (2007) calibration. These results demonstrate how the strong oxygen isotopic fractionation between rutile and quartz can be harnessed as a geothermometer, but they also urge caution for the application of trace element geothermometers at low temperatures. The common occurrence of rutile and quartz in metamorphic and hydrothermal vein rocks offers the possibility to further exploit this mineral pair as a geothermometer with the bonus that rutile can be radiometrically dated.

Another recent study where the versatility and superb spatial resolution of SIMS came to bear is Matthews et al. (2015), who combined trace elements and U–Pb geochronology of Pleistocene zircons to resolve the eruption age and petrologic evolution of the magma chamber responsible for the Lava Creek Tuff super-eruption and the formation of Yellowstone caldera (Fig. 8). The Lava Creek eruption deposited a voluminous ignimbrite around Yellowstone, and generated an ash bed that blanketed much of the western and central United States. The Lava Creek ash has been a key marker bed for correlating glacial and pluvial deposits in North America (Sarna-Wojcicki et al. 1984), delimiting rates of uplift (e.g., Darling et al. 2012) and establishing faunal sequences (Bell et al. 2004). To constrain the eruption age of the Lava Creek magma and resolve the final stages of its thermal and compositional evolution, Matthews et al. (2015) sampled the unpolished faces of zircons from different members of the eruptive stratigraphy, and integrated the results with a comparable dataset for the interiors of sectioned crystals from the same populations of zircon. Matthews et al. (2015) derived a $^{206}Pb/^{238}U$ date of ca. 627 ± 6 ka (2σ error) from the zircon crystal faces, which included a ca. +95 ka

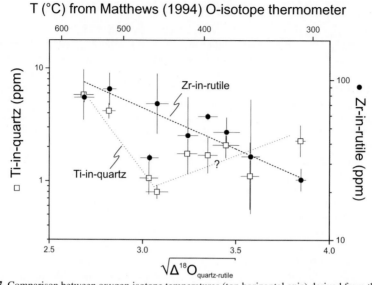

Figure 7. Comparison between oxygen isotope temperatures (top horizontal axis) derived from the rutile–quartz oxygen isotope exchange thermometer (Matthews 1994; bottom axis) and temperature-dependent Ti-in-quartz and Zr-in-rutile thermometers for rutilated quartz from Alpine fissures. Ti-in-quartz and Zr-in-rutile temperatures are not quantified, but shown as abundances. Zr-in-rutile and oxygen isotope temperatures correlate as expected, whereas no correlation exists between Ti-in-quarz and oxygen isotope temperatures. Data from Shulaker et al. (2015) collected using a CAMECA IMS 1270.

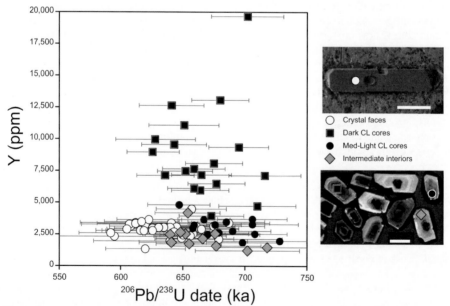

Figure 8. Petrochronology of zircons from Member B of the Lava Creek Tuff, erupted from Yellowstone caldera at ca. 630 ka. The ^{206}Pb/^{238}U dates for the cores and intermediate zones of sectioned crystals yield a mean crystallization age that is ca. 35 kyr older than the mean age derived from corresponding crystal faces (Matthews et al. 2015). Cores that are dark gray in cathodoluminescence images have elevated U, Y, and large negative europium anomalies, suggesting growth from low temperature and evolved, i.e., near solidus, rhyolitic magma prior to incorporation in less evolved rhyolite (Matthews et al. 2015). Data generated on the Stanford-USGS SHRIMP-RG. Scale bar is 100 μm.

correction for initial $^{230}Th/^{238}U$ disequilibrium. This zircon rim crystallization age effectively matched the eruption age apparent from the authors' $^{40}Ar/^{39}Ar$ dating of coexisting sanidines, as well as the stratigraphic position of the Lava Creek ash bed in marine sediments marking the Marine Isotope Stage 15–16 boundary (Dean et al. 2015). In addition, the petrochronologic approach used by Matthews et al. (2015) revealed that the Lava Creek magma evolved over a ca. 35 kyr interval, with reversely zoned rims (i.e., lower U, Th, REE, Hf, and higher Eu/Eu* compared to interior domains) indicating a change to a hotter and less evolved composition within about 10 kyr of eruption (Fig. 8). Independent U–Pb dating of Lava Creek zircons by ID-TIMS (Wotzlaw et al. 2015) confirmed the range of crystallization dates derived by Matthews et al. (2015) via SIMS, and has illustrated the ability of high-spatial resolution SIMS dating to provide reliable constraints on eruption ages with application to tephrochronology.

OUTLOOK

Improved detection

The fundamental design of large magnet radius SIMS instruments has experienced little modification since the prototypes were commissioned at ANU (SHRIMP I) and UCLA (IMS 1270) in 1981 and 1992, respectively. In the late 1990's, the reverse-geometry SIMS instrument (SHRIMP-RG) deviated significantly from earlier designs, but it was not commercially successful. Multi-detection capabilities were introduced for CAMECA and SHRIMP II instruments in the late 1990's and early 2000's, respectively. Since then, SIMS platforms have developed high-precision stable isotope analysis capabilities, which became only possible with the advent of multi-collection using FC detectors. Further improvements of the detection system remain highly desirable from a SIMS analyst's perspective. Although multi-collection of low intensity secondary ion signals is possible with discrete-dynode (CAMECA) and continuous (SHRIMP) EM detectors, dead-time uncertainties and ageing are intrinsic draw-backs of EM detectors. Multi-collection at low counts rates typical for the analysis of Pb and U-series intermediate daughter isotopes is hardly affected by this, as EM gain drift correlates with cumulative counts detected. For these applications, multi-collection can dramatically increase through-put, if gain changes can be monitored and corrected for (e.g., Holden et al. 2009). Much more problematic are count rates around the "Megahertz gap", the intensity range of 10^6 counts per second, where EMs age rapidly. Although software permits gain drift and dead time of EMs to be monitored in automated fashion, these parameters still result in significant (at the 1‰ level) uncertainty into isotope ratio measurements. In this regard, designing FC pre-amplifier circuits with high-impedance resistors is promising, as noise is expected to decrease by a factor of ~3 per order-of-magnitude increase in resistivity. For SIMS instruments, this has been successfully demonstrated for $10^{12}\,\Omega$ resistors (the iFlex Faraday counting system for SHRIMP), and even for $10^{13}\,\Omega$ resistors in TIMS instruments where FC detection yielded more precise measurements than EMs when currents were $>2\times10^4$ cps (Koornneef et al. 2014).

Ion source developments

Another field of current development is to improve the brightness, stability, and longevity of ion sources. For positive secondary ion analysis, duoplasmatron sources are traditionally applied. Maximum primary ion intensities of these sources are ~1 µA for O⁻, ~250 nA for O_2^-, and 10 µA for O_2^+. Poor stability and frequent replacement of metal parts exposed to the oxygen plasma (typically every 200–300 hours of use) are significant draw-backs of the duoplasmatron. Recently, plasma ion sources that can operate in dual polarity (producing O⁻ or O_2^+) have been commercialized (Hyperion II by Oregon Physics) and been installed on CAMECA instruments, including the large-magnet radius sources at UCLA. Initial tests showed a factor of 8 gain in beam density for O⁻ (Ming-Chang Liu, pers. comm). Higher brightness of the oxygen ion source will lead to improved spatial resolution, and/or higher sputtering rates at the same spot size compared to duoplasmatron sources, thus enhancing secondary ion signals. The higher secondary ion currents resulting from the use of the plasma ion source will often be in the range where they transcend the optimal intensity range for EM detection, which also calls for the implementation of low-noise FC detectors. In addition to increasing source brightness, there are developments in improving secondary ion yields through post-sputtering ionization. Resonance ionization mass spectrometry (RIMS) at much higher (~30–40%) useful yields than possible in normal SIMS is currently performed with experimental instrumentation (e.g., the Chicago Instrument for Laser Ionization CHILI; Stephan et al. 2016).

Sample holder design and automation

Precision of stable isotope analysis has benefited from the introduction of sample holders by CAMECA (Peres et al. 2013) with a larger annulus to hold the edge of the sample than previous holders. This has the advantage that a more uniform sample potential can be maintained across the central part of the holder than is possible with traditional sample holders, thus eliminating strong fractionation when approaching the edges of the sample (Treble et al. 2007). SHRIMP megamounts with 35 mm diameter were designed to achieve the same uniform IMF over the central 19 mm diameter bullseye (Ickert et al. 2008). Compared to commercial laser ablation cells, however, the useful surface area in SIMS is much smaller. Automated sample exchange improves throughput, and while this is available for SHRIMP instruments, it is currently under development for large magnet radius CAMECA instruments.

Final considerations

The past decade has seen a dramatic improvement of stable isotope analytical capabilities with SIMS, with 0.1–0.2‰ precision for replicate individual spot analyses for $\delta^{18}O$ in chemically simple materials such as zircon, quartz, or calcite. By contrast, analytical limitations for geochronologic SIMS have remained at the ~1–2% (2σ) level which was originally achieved about three decades ago. These limitations are imposed by the reproducibility of the 2-D calibration curve for U–Pb and Th–Pb, and are largely independent of the instrumental design. This does not imply complacency by SIMS geochronologists: significant instrument time was dedicated by all laboratories invested in petrochronology to gain a better understanding of the sources of uncertainty and bias in SIMS. However, no silver bullet has been identified to break through the 1% reproducibility barrier for U–Pb and Th–Pb despite many valuable insights gained into the understanding of how instrumental parameters (e.g., Magee et al. 2014) as well as matrix and crystal orientation effects (e.g., Taylor et al. 2012; White and Ireland 2012) can affect SIMS relative sensitivities for geochronologically relevant elements. In this regard, the high-spatial resolution methods, SIMS and LA-ICP-MS, remain limited in precision to ID-TIMS. For petrochronological analysis of complexly zoned accessory minerals the key advantage is accuracy albeit at lower precision compared to high-precision ID-TIMS ages where bulk analysis of crystal aliquots, and even single crystals and crystal fragments, would yield geologically meaningless averages (Table 1). This level of resolution is particularly valuable where crystals are analyzed *in situ*. Reconnaissance studies indicate a lack of bias if accessory minerals are analyzed by SIMS as grain separates conventionally embedded in epoxy vs. *in-situ* in a petrographic thin-section, even when the primary beam partially overlaps adjacent minerals in the thin section (Schmitt et al. 2010), but few such systematic comparisons have been conducted. Because *in-situ* analysis is such an important field of SIMS petrochronology, revisiting targeting, mounting, instrumental tuning, and calibration strategies for small accessory minerals hosted in a rocky matrix should become a major goal for improving throughput and bolstering confidence in the data. Another highly promising field is the combination of chemical (via SIMS) and structural information (via electron backscatter diffraction, or micro-Raman spectroscopy). These techniques are powerful in identifying phase changes, metamictization or overgrowths at the micrometer scales which can aid in the interpretation of petrochronologic data (e.g., for impact metamorphosed meteorites; Darling et al. 2016). At present, users of SIMS petrochronological data should stay aware of the non-routine nature of many of these analyses which can still provide some formidable challenges for the SIMS specialist.

ACKNOWLEDGMENTS

AKS expresses special thanks to all present and former members of the UCLA ion microprobe lab for their long-term support, in particular Chris Coath, Marty Grove, T. Mark Harrison, George Jarzebinski, Ming-Chang Liu, Oscar Lovera, Kevin McKeegan, and Lvcian Vltava. Insightful discussions with Thomas Ludwig (Heidelberg University) are acknowledged. JAV thanks Charlie Bacon, Matt Coble, Marty Grove, Brad Ito, Marsha Lidzbarski, and Joe Wooden for their help and support in the Stanford-USGS ion microprobe laboratory, and is grateful to the UCLA ion probe lab for his graduate training. We also thank John Aleinikoff, Brian Monteleone, and Richard Stern for their helpful reviews, and Pierre Lanari for careful editorial handling. The ion microprobe facility at UCLA is partly supported by a grant from the Instrumentation and Facilities Program, Division of Earth Sciences, National Science Foundation. Any use of trade, firm, or product names is for descriptive purposes only and does not imply endorsement by the U.S. Government.

REFERENCES

Abbott SS, Harrison TM, Schmitt AK, Mojzsis SJ (2012) A search for thermal excursions from ancient extraterrestrial impacts using Hadean zircon Ti–U–Th–Pb depth profiles. PNAS 109:13486–13492

Aleinikoff JN, Schenck WS, Plank MO, Srogi L, Fanning CM, Kamo SL, Bosbyshell H (2006) Deciphering igneous and metamorphic events in high-grade rocks of the Wilmington Complex, Delaware: Morphology, cathodoluminescence and backscattered electron zoning, and SHRIMP U–Pb geochronology of zircon and monazite. Geol Soc Am Bull 118:39–64

Aleinikoff JN, Wintsch RP, Tollo RP, Unruh DM, Fanning CM, Schmitz MD (2007) Ages and origins of rocks of the Killingworth dome, south-central Connecticut: Implications for the tectonic evolution of southern New England. Am J Sci 307:63–118

Aleinikoff JN, Grauch RI, Mazdab FK, Kwak L, Fanning CM, Kamo SL (2012) Origin of an unusual monazite–xenotime gneiss, Hudson Highlands, New York: SHRIMP U–Pb geochronology and trace element geochemistry. Am J Sci 312:723–765

Andrews GD, Schmitt AK, Busby CJ, Brown SR, Blum P, Harvey J (2016) Age and compositional data of zircon from sepiolite drilling mud to identify contamination of ocean drilling samples. Geochem Geophys Geosyst:3512–3526

Baldwin SL, Ireland TR (1995) A tale of two eras: Pliocene-Pleistocene unroofing of Cenozoic and late Archean zircons from active metamorphic core complexes, Solomon Sea, Papua New Guinea. Geology 23:1023–1026.

Barth AP, Wooden JL (2010) Coupled elemental and isotopic analyses of polygenetic zircons from granitic rocks by ion microprobe, with implications for melt evolution and the sources of granitic magmas. Chem Geol 277:149–159

Bell CJ, Lundelius Jr EL, Barnosky A, Graham R, Lindsay E, Ruez Jr D, Semken Jr H, Webb S, Zakrzewski R, Woodburne M (2004) The Blancan, Irvingtonian, and Rancholabrean mammal ages. Late Cretaceous and Cenozoic mammals of North America: Biostratigraphy and Geochronology:232–314

Bindeman I, Serebryakov N, Schmitt A, Vazquez J, Guan Y, Azimov PY, Astafiev BY, Palandri J, Dobrzhinetskaya L (2014) Field and microanalytical isotopic investigation of ultradepleted in ^{18}O Paleoproterozoic "Slushball Earth" rocks from Karelia, Russia. Geosphere 10:308–339

Black LP, Kamo SL, Allen CM, Davis DW, Aleinikoff JN, Valley JW, Mundil R, Campbell IH, Korsch RJ, Williams IS (2004) Improved $^{206}Pb/^{238}U$ microprobe geochronology by the monitoring of a trace-element-related matrix effect; SHRIMP, ID–TIMS, ELA–ICP–MS and oxygen isotope documentation for a series of zircon standards. Chem Geol 205:115–140

Blundy J, Wood B (2003) Mineral–melt partitioning of uranium, thorium and their daughters. Rev Mineral Geochem 52:59–123

Boehnke P, Harrison TM (2014) A meta-analysis of geochronologically relevant half-lives: what's the best decay constant? Int Geol Rev 56:905–914

Boehnke P, Barboni M, Bell E (2016) Zircon U/Th model ages in the presence of melt heterogeneity. Quat Geochronol 34:69–74

Bouvier A-S, Ushikubo T, Kita NT, Cavosie AJ, Kozdon R, Valley JW (2012) Li isotopes and trace elements as a petrogenetic tracer in zircon: insights from Archean TTGs and sanukitoids. Contrib Mineral Petrol 163:745–768

Breecker DO, Sharp ZD (2007) A monazite oxygen isotope thermometer. Am Mineral 92:1561–1572

Breeding CM, Ague JJ, Grove M, Rupke AL (2004) Isotopic and chemical alteration of zircon by metamorphic fluids: U–Pb age depth-profiling of zircon crystals from Barrow's garnet zone, northeast Scotland. Am Mineral 89:1067–1077

Catlos E, Sorensen SS, Harrison TM (2000) Th–Pb ion-microprobe dating of allanite. Am Mineral 85:633–648

Catlos E, Gilley L, Harrison TM (2002) Interpretation of monazite ages obtained via *in situ* analysis. Chem Geol 188:193–215

Cherniak DJ, Watson EB (2003) Diffusion in zircon. Rev Mineral Geochem 53:113–143

Cherniak D, Watson EB, Grove M, Harrison TM (2004) Pb diffusion in monazite: a combined RBS/SIMS study. Geochim Cosmochim Acta 68:829–840

Claiborne LL, Miller CF, Flanagan DM, Clynne MA, Wooden JL (2010) Zircon reveals protracted magma storage and recycling beneath Mount St. Helens. Geology 38:1011–1014

Compston W, Clement SWJ (2006) The geological microprobe: The first 25 years of dating zircons. Appl Surf Sci 252:7089–7095

Compston W, Williams I, Meyer C (1984) U-Pb geochronology of zircons from lunar breccia 73217 using a sensitive high mass-resolution ion microprobe. J Geophys Res Solid Earth 89(S02): B525-B534

Coombs ML, Vazquez JA (2014) Cogenetic late Pleistocene rhyolite and cumulate diorites from Augustine Volcano revealed by SIMS ^{238}U-^{230}Th dating of zircon, and implications for silicic magma generation by extraction from mush. Geochem Geophys Geosyst 15:4846–4865

Crowley J, Schoene B, Bowring S (2007) U–Pb dating of zircon in the Bishop Tuff at the millennial scale. Geology 35:1123–1126

Danišík M, Shane P, Schmitt AK, Hogg A, Santos GM, Storm S, Evans NJ, Fifield LK, Lindsay JM (2012) Re-anchoring the late Pleistocene tephrochronology of New Zealand based on concordant radiocarbon ages and combined ^{238}U/^{230}Th disequilibrium and (U–Th)/He zircon ages. Earth Planet Sci Lett 349:240–250

Darling AL, Karlstrom KE, Granger DE, Aslan A, Kriby E, Ouimet WB, Lazear GD, Coblentz DD, Cole RD (2012) New incision rates along the Colorado River system based on cosmogenic burial dating of terraces: Implications for regional controls on Quaternary incision. Geosphere:1020–1041

Darling JR, Moser DE, Barker IR, Tait KT, Chamberlain KR, Schmitt AK, Hyde BC (2016) Variable microstructural response of baddeleyite to shock metamorphism in young basaltic shergottite NWA 5298 and improved U–Pb dating of Solar System events. Earth Planet Sci Lett 444:1–12

de Chambost E (2011) A history of CAMECA (1954–2009). Adv Imaging Electron Phys 167:1–119

Dean WE, Kennett JP, Behl RJ, Nicholson C, Sorlien CC (2015) Abrupt termination of Marine Isotope Stage 16 (Termination VII) at 631.5 ka in Santa Barbara Basin, California. Paleoceanogr 30:1373–1390

Engi M (2017) Petrochronology based on REE–minerals: monazite, allanite, xenotime, apatite. Rev Mineral Geochem 83:365–418

Ferry JM, Watson EB (2007) New thermodynamic models and revised calibrations for the Ti-in-zircon and Zr-in-rutile thermometers. Contrib Mineral Petrol 154:429–37

Fisher CM, Vervoort JD, Hanchar JM (2014) Guidelines for reporting zircon Hf isotopic data by LA-MC-ICPMS and potential pitfalls in the interpretation of these data. Chem Geol 363:125–133

Fletcher IR, McNaughton NJ, Aleinikoff JA, Rasmussen B, Kamo SL (2004) Improved calibration procedures and new standards for U–Pb and Th–Pb dating of Phanerozoic xenotime by ion microprobe. Chem Geol 209:295–314

Fletcher IR, McNaughton NJ, Davis WJ, Rasmussen B. (2010) Matrix effects and calibration limitations in ion probe U–Pb and Th–Pb dating of monazite. Chem Geol 270:31–44.

Foster G, Kinny P, Vance D, Prince C, Harris N (2000) The significance of monazite U–Th–Pb age data in metamorphic assemblages; a combined study of monazite and garnet chronometry. Earth Planet Sci Lett 181:327–340

Gregory CJ, Rubatto D, Allen CM, Williams IS, Hermann J, Ireland T (2007) Allanite micro-geochronology: a LA-ICP-MS and SHRIMP U–Th–Pb study. Chem Geol 245:162–182

Grimes CB, Ushikubo T, John BE, Valley JW (2011) Uniformly mantle-like δ^{18}O in zircons from oceanic plagiogranites and gabbros. Contrib Mineral Petrol 161:13–33

Grimes C, Wooden J, Cheadle M, John B (2015) "Fingerprinting" tectono-magmatic provenance using trace elements in igneous zircon. Contrib Mineral Petrol 170:1–26

Harrison TM, Schmitt AK (2007) High sensitivity mapping of Ti distributions in Hadean zircons. Earth Planet Sci Lett 261:9–19

Harrison TM, McKeegan K, LeFort P (1995) Detection of inherited monazite in the Manaslu leucogranite by [208]Pb–[232]Th ion microprobe dating: crystallization age and tectonic implications. Earth Planet Sci Lett 133:271–282

Harrison TM, Schmitt AK, McCulloch MT, Lovera OM (2008) Early (≥4.5 Ga) formation of terrestrial crust: Lu–Hf, δ^{18}O, and Ti thermometry results for Hadean zircons. Earth Planet Sci Lett 268:476–486

Hervig RL, Mazdab FK, Williams P, Guan Y, Huss GR, Leshin LA (2006) Useful ion yields for Cameca IMS 3f and 6f SIMS: Limits on quantitative analysis. Chem Geol 227:83–99

Hiess J, Nutman AP, Bennett VC, Holden P (2008) Ti-in-zircon thermometry applied to contrasting Archean metamorphic and igneous systems. Chem Geol 247:323–338

Hinthorne J, Andersen C, Conrad R, Lovering J (1979) Single-grain [207]Pb/[206]Pb and U/Pb age determinations with a 10-µm spatial resolution using the ion microprobe mass analyzer (IMMA). Chem Geol 25:271–303

Hinton R (1990) Ion microprobe trace-element analysis of silicates: Measurement of multi-element glasses. Chem Geol 83:11–25

Hofmann AE, Valley JW, Watson EB, Cavosie AJ, Eiler JM (2009) Sub-micron scale distributions of trace elements in zircon. Contrib Mineral Petrol 158:317–335

Hofmann AE, Baker MB, Eiler JM (2014) Sub-micron-scale trace-element distributions in natural zircons of known provenance: implications for Ti-in-zircon thermometry. Contrib Mineral Petrol 168:1–21

Holden P, Lanc P, Ireland TR, Harrison TM, Foster JJ, Bruce Z (2009) Mass-spectrometric mining of Hadean zircons by automated SHRIMP multi-collector and single-collector U/Pb zircon age dating: the first 100,000 grains. Int J Mass Spectrom 286:53–63

Hoppe P, Cohen S, Meibom A (2013) NanoSIMS: technical aspects and applications in cosmochemistry and biological geochemistry. Geostand Geoanal Res 37:111–154

Horstwood MS, Košler J, Gehrels G, Jackson SE, McLean NM, Paton C, Pearson NJ, Sircombe K, Sylvester P, Vermeesch P (2016) Community-derived standards for LA-ICP-MS U-(Th-) Pb geochronology–uncertainty propagation, age interpretation and data reporting. Geostand Geoanal Res 40:311–332

Hoskin PW, Schaltegger U (2003) The composition of zircon and igneous and metamorphic petrogenesis. Rev Mineral Geochem 53:27–62

Huang R, Audétat A (2012) The titanium-in-quartz (TitaniQ) thermobarometer: a critical examination and re-calibration. Geochim Cosmochim Acta 84:75–89

Huberty JM, Kita NT, Kozdon R, Heck PR, Fournelle JH, Spicuzza MJ, Xu H, Valley JW (2010) Crystal orientation effects in δ^{18}O for magnetite and hematite by SIMS. Chem Geol 276:269–283

Hunter J (2009) Improving depth profile measurements of natural materials: Lessons learned from electronic materials depth-profiling. Secondary Ion Mass Spectrometry in the Earth Sciences: Gleaning the Big Picture from a Small Spot. Mineral Assoc Can Short Course 41:133–148

Ickert R, Hiess J, Williams I, Holden P, Ireland T, Lanc P, Schram N, Foster J, Clement S (2008) Determining high precision, *in situ*, oxygen isotope ratios with a SHRIMP II: analyses of MPI-DING silicate-glass reference materials and zircon from contrasting granites. Chem Geol 257:114–128

Ickert RB, Mundil R, Magee CW, Mulcahy SR (2015) The U–Th–Pb systematics of zircon from the Bishop Tuff: A case study in challenges to high-precision Pb/U geochronology at the millennial scale. Geochim Cosmochim Acta 168:88–110

Ireland TR (1995) Ion microprobe mass spectrometry: techniques and applications in cosmochemistry, geochemistry, and geochronology. Adv Anal Geochem 2:1–118

Ireland TR (2015) Secondary ion mass spectrometry (SIMS). *In:* Encyclopedia of Scientific Dating Methods. Rink WJ, Thompson JW (eds) Springer, Berlin, p 739–740

Ireland TR, Williams IS (2003) Considerations in zircon geochronology by SIMS. Rev Mineral Geochem 53:215–241

Ireland T, Clement S, Compston W, Foster J, Holden P, Jenkins B, Lanc P, Schram N, Williams I (2008) Development of SHRIMP. Australas J Earth Sci 55:937–954

Ireland T, Schram N, Holden P, Lanc P, Avila J, Armstrong R, Amelin Y, Latimore A, Corrigan D, Clement S (2014) Charge-mode electrometer measurements of S-isotopic compositions on SHRIMP-SI. Int J Mass Spectrom 359:26–37

Janots E, Berger A, Gnos E, Whitehouse M, Lewin E, Pettke T (2012) Constraints on fluid evolution during metamorphism from U–Th–Pb systematics in Alpine hydrothermal monazite. Chem Geol 326:61–71

Jeon H, Whitehouse MJ (2015) A critical evaluation of U–Pb calibration schemes used in SIMS zircon geochronology. Geostand Geoanal Res 39:443–452

Kinny PD, Compston W, Williams IS (1991) A reconnaissance ion-probe study of hafnium isotopes in zircons. Geochim Cosmochim Acta 55:849–859

Kita NT, Ushikubo T, Fu B, Valley JW (2009) High precision SIMS oxygen isotope analysis and the effect of sample topography. Chem Geol 264:43–57

Koornneef J, Bouman C, Schwieters J, Davies G (2014) Measurement of small ion beams by thermal ionisation mass spectrometry using new 10^{13} Ohm resistors. Analytica Chim Acta 819:49–55

Kylander–Clark ARC (2017) Petrochronology by laser–ablation inductively coupled plasma mass spectrometry. Rev Mineral Geochem 83:183–198

Lanari P, Engi M (2017) Local bulk composition effects on metamorphic mineral assemblages. Rev Mineral Geochem 83:55–102

Li Q-L, Li X-H, Liu Y, Tang G-Q, Yang J-H, Zhu W-G (2010) Precise U–Pb and Pb–Pb dating of Phanerozoic baddeleyite by SIMS with oxygen flooding technique. J Anal At Spectrom 25:1107–1113

Ludwig K (2009) Errors of isotope ratios acquired by double interpolation. Chem Geol 268:24–26

Luvizotto G, Zack T, Meyer H, Ludwig T, Triebold S, Kronz A, Münker C, Stockli D, Prowatke S, Klemme S (2009) Rutile crystals as potential trace element and isotope mineral standards for microanalysis. Chem Geol 261:346–369

Matsuda H (1974) Double focusing mass spectrometers of second order. Int J Mass Spectrom Ion Phys 14:219–233

Matsuda H (1990) High performance mass spectrometers of magnetic sector type. Int J Mass Spectrom Ion Processes 100:31–39

Matthews A (1994) Oxygen isotope geothermometers for metamorphic rocks. J Metamorph Geol 12:211–219

Matthews NE, Vazquez JA, Calvert AT (2015) Age of the Lava Creek supereruption and magma chamber assembly at Yellowstone based on $^{40}Ar/^{39}Ar$ and U-Pb dating of sanidine and zircon crystals. Geochem Geophys Geosyst 16:2508–2528

Mattinson JM (2010) Analysis of the relative decay constants of ^{235}U and ^{238}U by multi-step CA-TIMS measurements of closed-system natural zircon samples. Chem Geol 275:186–198

Mattinson CG, Wooden JL, Liou JG, Bird D, Wu C (2006) Age and duration of eclogite-facies metamorphism, North Qaidam HP/UHP terrane, Western China. Am J Sci 306:683–711

Magee C Jr, Ferris J, Magee C Sr (2014) Effect of impact energy on SIMS U–Pb zircon geochronology. Surf Interface Anal 46(S1):322–5.

Mazdab FK (2009) Characterization of flux-grown trace-element-doped titanite using the high-mass-resolution ion microprobe (SHRIMP–RG). Can Mineral 47:813–831

McFarlane C, McCulloch M (2007) Coupling of *in-situ* Sm–Nd systematics and U–Pb dating of monazite and allanite with applications to crustal evolution studies. Chem Geol 245:45–60

Nemchin A, Whitehouse M, Pidgeon R, Meyer C (2006) Oxygen isotopic signature of 4.4–3.9 Ga zircons as a monitor of differentiation processes on the Moon. Geochim Cosmochim Acta 70:1864–1872

Paces JB (2015) Uranium Series, Opal. Encyclopedia of Scientific Dating Methods:837–843

Page F, Ushikubo T, Kita NT, Riciputi L, Valley JW (2007) High-precision oxygen isotope analysis of picogram samples reveals 2 µm gradients and slow diffusion in zircon. Am Mineral 92:1772–1775

Peres P, Kita N, Valley J, Fernandes F, Schuhmacher M (2013) New sample holder geometry for high precision isotope analyses. Surf Interface Anal 45:553–556

Peterman EM, Reddy SM, Saxey DW, Snoeyenbos DR, Rickard WD, Fougerouse D, Kylander-Clark AR (2016) Nanogeochronology of discordant zircon measured by atom probe microscopy of Pb-enriched dislocation loops. Sci Adv 2:e1601318

Pidgeon R, Nemchin A, Cliff J (2013) Interaction of weathering solutions with oxygen and U–Pb isotopic systems of radiation-damaged zircon from an Archean granite, Darling Range Batholith, Western Australia. Contrib Mineral Petrol 166:511–523

Rayner NM, Stern RA (2002) Improved sample preparation method for SHRIMP analysis of delicate mineral grains exposed in thin sections. Natural Resources Canada, Geological Survey of Canada, http://www. publications.gc.ca/collections/collection_2011/rncan-nrcan/M44-2002-F10-eng.pdf

Reid MR, Coath CD, Harrison TM, McKeegan KD (1997) Prolonged residence times for the youngest rhyolites associated with Long Valley Caldera: $^{230}Th–^{238}U$ ion microprobe dating of young zircons. Earth Planet Sci Lett 150:27–39

Rioux M, Bowring S, Cheadle M, John B (2015) Evidence for initial excess ^{231}Pa in mid-ocean ridge zircons. Chem Geol 397:143–156

Rubatto D (2002) Zircon trace element geochemistry: partitioning with garnet and the link between U–Pb ages and metamorphism. Chem Geol 184:123–138

Rubatto D (2017) Zircon: The metamorphic mineral. Rev Mineral Geochem 83:261–295

Rubatto D, Putlitz B, Gauthiez-Putallaz L, Crépisson C, Buick IS, Zheng Y-F (2014) Measurement of *in-situ* oxygen isotope ratios in monazite by SHRIMP ion microprobe: Standards, protocols and implications. Chem Geol 380:84–96

Rubin A, Cooper K, Kent A, Costa Rodriguez F, Till C (2014) Using Li diffusion to track thermal histories within single zircon crystals. AGU Abstract Fall 2014.

Rubin A, Cooper KM, Leever M, Wimpenny J, Deering C, Rooney T, Gravley D, Yin Q-Z (2016) Changes in magma storage conditions following caldera collapse at Okataina Volcanic Center, New Zealand. Contrib Mineral Petrol 171:1–18

Sarna-Wojcicki AM, Bowman H, Meyer CE, Russell P, Woodward M, McCoy G, Rowe Jr J, Baedecker P, Asaro F, Michael H (1984) Chemical analyses, correlations, and ages of upper Pliocene and Pleistocene ash layers of east-central and southern California. *In:* Chemical Analyses, Correlations, and Ages of Ipper Pliocene and Pleistocene Ash Layers of East-Central and Southern California. USGS Prof P 1293:1–40.

Schaltegger U, Davies JHFL (2017) Petrochronology of zircon and baddeleyite in igneous rocks: Reconstructing magmatic processes at high temporal resolution. Rev Mineral Geochem 83:297–328

Schaltegger U, Schmitt A, Horstwood M (2015) U–Th–Pb zircon geochronology by ID-TIMS, SIMS, and laser ablation ICP-MS: recipes, interpretations, and opportunities. Chem Geol 402:89–110

Schärer U (1984) The effect of initial ^{230}Th disequilibrium on young U–Pb ages: the Makalu case, Himalaya. Earth Planet Sci Lett 67:191–204

Schmitt AK (2007) Letter: Ion microprobe analysis of $(^{231}$Pa)/$(^{235}$U) and an appraisal of protactinium partitioning in igneous zircon. Am Mineral 92:691–694

Schmitt AK (2011) Uranium series accessory crystal dating of magmatic processes. Annu Rev Earth Planet Sci 39:321–349

Schmitt AK, Zack T (2012) High-sensitivity U–Pb rutile dating by secondary ion mass spectrometry (SIMS) with an O$_2^+$ primary beam. Chem Geol 332:65–73

Schmitt AK, Chamberlain KR, Swapp SM, Harrison TM (2010) *In situ* U–Pb dating of micro-baddeleyite by secondary ion mass spectrometry. Chem Geol 269:386–395

Schmitt AK, Martín A, Stockli DF, Farley KA, Lovera OM (2013) (U–Th)/He zircon and archaeological ages for a late prehistoric eruption in the Salton Trough (California, USA). Geology 41:7–10

Schoene B (2014) 4.10-U–Th–Pb Geochronology. Treatise on Geochemistry, Second Edition, Elsevier, Oxford:341–378

Schoene B, Baxter EF (2017) Petrochronology and TIMS. Rev Mineral Geochem 83:231–260

Schuhmacher M, De Chambost E, McKeegan K, Harrison TM, Migeon H (1994) *In situ* dating of zircon with the CAMECA ims 1270. Secondary Ion Mass Spectrometry SIMS IX:919–922

Seitz S, Putlitz B, Baumgartner LP, Escrig S, Meibom A, Bouvier AS (2016) Short magmatic residence times of quartz phenocrysts in Patagonian rhyolites associated with Gondwana breakup. Geology 44:67–70

Shimizu N (1978) Analysis of the zoned Plagioclase of different magmatic environments: A Preliminary ion microprobe study, Earth Planet Sci Lett 39:398–406.

Shimizu N, Hart S (1982) Isotope fractionation in secondary ion mass spectrometry. J Appl Phys 53:1303–1311

Shulaker DZ, Schmitt AK, Zack T, Bindeman I (2015) *In-situ* oxygen isotope and trace element geothermometry of rutilated quartz from Alpine fissures. Am Mineral 100:915–925

Simon JI, Renne PR, Mundil R (2008) Implications of pre-eruptive magmatic histories of zircons for U–Pb geochronology of silicic extrusions. Earth Planet Sci Lett 266:182–194

Sláma J, Košler J (2012) Effects of sampling and mineral separation on accuracy of detrital zircon studies. Geochem Geophys Geosyst 13:Q05007, doi:10.1029/2012GC004106

Stelten ME, Cooper KM, Vazquez JA, Calvert AT, Glessner JJ (2015) Mechanisms and timescales of generating eruptible rhyolitic magmas at Yellowstone caldera from zircon and sanidine geochronology and geochemistry. J Petrol 56:1607–1642

Stephan T, Trappitsch R, Davis AM, Pellin MJ, Rost D, Savina MR, Yokochi R, Liu N (2016) CHILI–the Chicago Instrument for Laser Ionization–a new tool for isotope measurements in cosmochemistry. Int J Mass Spectrom 407:1–15

Stern R (2009) An introduction to secondary ion mass spectrometry (SIMS) In: Geology. Secondary Ion Mass Spectrometry in the Earth Sciences: Gleaning the big picture from a small spot. Mineral Assoc Can Short Course 41:1–18

Stern RA, Amelin Y (2003) Assessment of errors in SIMS zircon U–Pb geochronology using a natural zircon standard and NIST SRM 610 glass. Chem Geol 197:111–142

Stern RA, Fletcher IR, Rasmussen B, McNaughton NJ, Griffin BJ (2005) Ion microprobe (NanoSIMS 50) Pb-isotope geochronology at <5μm scale. Int J Mass Spectrom 244:125–134

Storm S, Shane P, Schmitt AK, Lindsay JM (2011) Contrasting punctuated zircon growth in two syn-erupted rhyolite magmas from Tarawera volcano: Insights to crystal diversity in magmatic systems. Earth Planet Sci Lett 301:511–520

Taylor R, Clark C, Reddy SM (2012) The effect of grain orientation on secondary ion mass spectrometry (SIMS) analysis of rutile. Chem Geol 300:81–87

Thomas J, Bodnar R, Shimizu N, Sinha A (2002) Determination of zircon/melt trace element partition coefficients from SIMS analysis of melt inclusions in zircon. Geochim Cosmochim Acta 66:2887–2901

Thomas JB, Watson EB, Spear FS, Shemella PT, Nayak SK, Lanzirotti A (2010) TitaniQ under pressure: the effect of pressure and temperature on the solubility of Ti in quartz. Contrib Mineral Petrol 160:743–759

Till CB, Vazquez, JA, Boyce JW (2015) Months between rejuvenation and volcanic eruption at Yellowstone caldera, Wyoming. Geology 43:695–698

Tomkins H, Powell R, Ellis D (2007) The pressure dependence of the zirconium-in-rutile thermometer. J Metamorph Geol 25:703–713

Torsvik TH, Amundsen H, Hartz EH, Corfu F, Kusznir N, Gaina C, Doubrovine PV, Steinberger B, Ashwal LD, Jamtveit B (2013) A Precambrian microcontinent in the Indian Ocean. Nat Geosci 6:223–227

Trail D, Cherniak DJ, Watson EB, Harrison TM, Weiss BP, Szumila I (2016) Li zoning in zircon as a potential geospeedometer and peak temperature indicator. Contrib Mineral Petrol 171:1–15

Treble P, Schmitt AK, Edwards R, McKeegan KD, Harrison T, Grove M, Cheng H, Wang Y (2007) High resolution Secondary Ionisation Mass Spectrometry (SIMS) $\delta^{18}O$ analyses of Hulu Cave speleothem at the time of Heinrich Event 1. Chem Geol 238:197–212

Ushikubo T, Kita NT, Cavosie AJ, Wilde SA, Rudnick RL, Valley JW (2008) Lithium in Jack Hills zircons: Evidence for extensive weathering of Earth's earliest crust. Earth Planet Sci Lett 272:666–676

Valley JW, Kita NT (2009) *In situ* oxygen isotope geochemistry by ion microprobe. Secondary Ion Mass Spectrometry in the Earth Sciences: Gleaning the big picture from a small spot. Mineral Assoc Can Short Course 41:19–63

Valley JW., Reinhard DA, Cavosie AJ, Ushikubo T, Lawrence DF, Larson DJ, Kelly TF, Snoeyenbos DR, Strickland A (2015) Presidential Address. Nano-and micro-geochronology in Hadean and Archean zircons by atom-probe tomography and SIMS: New tools for old minerals. Am Mineral 100:1355–1377

Vazquez JA, Reid MR (2004) Probing the accumulation history of the voluminous Toba magma. Sci 305:991–994

Vazquez JA, Velasco NO, Schmitt AK, Bleick HA, Stelten ME (2014) ^{238}U–^{230}Th dating of chevkinite in high-silica rhyolites from La Primavera and Yellowstone calderas. Chem Geol 390:109–118

von Blanckenburg F (1992) Combined high-precision chronometry and geochemical tracing using accessory minerals: applied to the Central-Alpine Bergell intrusion (central Europe). Chem Geol 100:19–40

Vorhies SH, Ague JJ, Schmitt AK (2013) Zircon growth and recrystallization during progressive metamorphism, Barrovian zones, Scotland. Am Mineral 98:219–230

Wark DA, Watson EB (2006) TitaniQ: a titanium-in-quartz geothermometer. Contrib Mineral Petrol 152:743–754

Watson E, Harrison T (2005) Zircon thermometer reveals minimum melting conditions on earliest Earth. Sci 308:841–844

Watts KE, Coble MA, Vazquez JA, Henry CD, Colgan JP, John DA (2016) Chemical abrasion-SIMS (CA-SIMS) U–Pb dating of zircon from the late Eocene Caetano caldera, Nevada. Chem Geol 439:139–151

Wendt I, Carl C (1985) U/Pb dating of discordant 0.1 Ma old secondary U minerals. Earth Planet Sci Lett 73:278–284

White L, Ireland T (2012) High-uranium matrix effect in zircon and its implications for SHRIMP U–Pb age determinations. Chem Geol 306:78–91

Wiedenbeck M, Hanchar JM, Peck WH, Sylvester P, Valley J, Whitehouse M, Kronz A, Morishita Y, Nasdala L, Fiebig J (2004) Further characterisation of the 91500 zircon crystal. Geostand Geoanal Res 28:9–39

Williams I, McKibben M (1998) Applications of microanalytical techniques to understanding mineralizing processes. Rev Econ Geol 7:1–35

Wingate M, Compston W (2000) Crystal orientation effects during ion microprobe U–Pb analysis of baddeleyite. Chem Geol 168:75–97

Wotzlaw J-F, Bindeman IN, Stern RA, D'Abzac F-X, Schaltegger U (2015) Rapid heterogeneous assembly of multiple magma reservoirs prior to Yellowstone supereruptions. Sci Rep 5, doi:10.1038/srep14026

Wright HM, Vazquez JA, Champion DE, Calvert AT, Mangan MT, Stelten M, Cooper KM, Herzig C, Schriener A (2015) Episodic Holocene eruption of the Salton Buttes rhyolites, California, from paleomagnetic, U-Th, and Ar/Ar dating. Geochem Geophys Geosyst 16:1198–1210

Yang W, Lin Y-T, Zhang J-C, Hao J-L, Shen W-J, Hu S (2012) Precise micrometre-sized Pb–Pb and U–Pb dating with NanoSIMS. J Anal At Spectrom 27:479–487

Zack T, Kooijman E (2017) Petrology and geochronology of rutile. Rev Mineral Geochem 83:443–467

Zack T, Moraes R, Kronz A (2004) Temperature dependence of Zr in rutile: empirical calibration of a rutile thermometer. Contrib Mineral Petrol 148:471–488

Zinner E, Crozaz G (1986) A method for the quantitative measurement of rare earth elements in the ion microprobe. Int J Mass Spectrom Ion Processes 69:17–38

Reviews in Mineralogy & Geochemistry
Vol. 83 pp. 231–260, 2017
Copyright © Mineralogical Society of America

8

Petrochronology and TIMS

Blair Schoene

Department of Geosciences
Princeton University
Princeton, NJ 08544
USA

bschoene@princeton.edu

Ethan F. Baxter

Department of Earth and Environmental Sciences
Boston College
Chestnut Hill, MA 02467
USA

baxteret@bc.edu

INTRODUCTION

Thermal ionization mass spectrometers, or TIMS, were developed by the pioneers of mass spectrometry in the mid-20th century, and have since been workhorses for generating isotopic data for a wide range of elements. Later-developed mass spectrometric techniques have many advantages over TIMS, including higher spatial resolution with *in situ* techniques, such as secondary ion mass spectrometry (SIMS) and laser ablation inductively coupled plasma mass spectrometry (LA-ICPMS), and greater versatility in terms the elements that can be easily- and well-measured. The reason TIMS persists as an important method for geochronology is that for some key parent-daughter systems (e.g., U–Pb, Sm–Nd), it can produce isotopic data and resultant dates with 10–100 times higher precision and more quantifiable accuracy than *in situ* techniques, even when sample sizes are very small (such as those that might result from single crystals, or even small portions of zoned crystals). For many questions in the geosciences, the highest achievable precision and accuracy are required to resolve the timescales of processes and/or correlate events globally. As an example, modern TIMS U–Pb geochronology is capable of producing dates with precision and accuracy better than 0.1% of the age for single crystals with only a few picograms (pg) of Pb. Therefore, it is possible to constrain the durations of single zircon crystal growth in magmatic systems over tens to hundreds of kyr in Mesozoic and younger rocks. If these dates and rates can be connected with other igneous processes such as magma transfer, emplacement and crystallization, then it becomes possible to calibrate thermal and mass budgets in magmatic systems and evaluate competing models for pluton assembly and subvolcanic magma storage. As another example, Sm–Nd geochronology of garnet permits dates with precision better than ±1 million years for garnets of any age, including multiple concentric growth zones in single crystals. Such zoned mineral chronology of this common porphyroblastic mineral in metamorphic rocks can be linked with thermodynamic modeling, permitting constraints on rates, durations, and pulses of heating, burial, and devolatilization during tectonometamorphic evolution.

As with other techniques used in geochronology, it is important by TIMS to combine the parent and daughter isotope ratios with complementary geochemical, isotopic, and textural information in order to interpret those dates within the context of the geologic processes being investigated. In this volume, focus is directed to the science of understanding the rates of mineral- and rock-

http://dx.doi.org/10.2138/rmg.2017.83.8

forming processes, or petrochronology. Generally the goal of petrochronology is to capitalize on a thermodynamic, geochemical, or mechanical framework to understand metamorphic or igneous processes. This goal is no different in TIMS compared to other techniques, but TIMS poses a different set of challenges that make harmonizing textural and geochemical information with dates more difficult than with *in situ* techniques. This chapter outlines some of the aspects of TIMS that are important in doing good petrochronology, and then highlights some examples from the recent literature that exemplify both conceptual advances and creative workflows. We focus primarily on the U–Pb and Sm–Nd systems, but acknowledge that other systems, for example Rb–Sr and Re–Os, have also been exploited to obtain petrochronologic data with the same goal of understanding the rates of rock-forming and geodynamic processes.

A BRIEF REVIEW OF TIMS GEOCHRONOLOGY

While the details of thermal ionization mass spectrometry will not be covered in this chapter (see Carlson 2014, for a recent review), we mention the mechanism of sample ionization because it is the major difference between TIMS and other mass spectrometers used in geochronology: in short, it is the primary control on both the positive and negative aspects of TIMS in petrochronology. Sample ionization in TIMS, as the name suggests, is carried out by heating a sample on a thin metal filament until the target element in question is either volatilized or ionized (preferably the latter, but ionization efficiencies are typically < 5%; although recent advances approach or exceed 25% ion yield for NdO^+ analysis, e.g., Harvey and Baxter 2009), and these ions are accelerated through a focusing lens and into a mass separator such as a magnet. In contrast, ICPMS provides 100% ionization for all elements. However, ICPMS provides very low ion transmission from the plasma source to the analyzer, typically < 1%, whereas TIMS ion transmission is much higher, probably > 50%. Either method can result in ion beams that are stable for several hours for typical sample sizes, and through simple counting statistics this can result in high precision isotopic ratios. Once the ion beam is established, run duration is dictated by the desired analytical precision and is identical via TIMS or MC-ICPMS. Generally, TIMS outperforms MC-ICPMS for isotopic analysis when ionization efficiency for a given element exceeds ~2%, especially for smaller sample sizes. As will be highlighted below, for small sample sizes (defined below) targeted in petrochronology, minimizing laboratory contamination, or *blank*, is also important and often easier by TIMS given complex sample introduction systems associated with solution ICPMS.

There are, however, several drawbacks to creating ions by thermal ionization. One is that the sample must be transformed into a substance that can be adhered onto, and easily ionized on, a filament that is a few millimeters wide. The other is that TIMS are generally low mass resolution instruments, meaning they cannot resolve isobaric or polyatomic interferences of the same nominal mass (e.g., cannot differentiate between masses 205.9 and 206.0). The first issue is addressed by dissolving the sample in acid and concentrating it into a small volume of liquid substrate, selected based on its ability to enhance ionization, then drying that volume on the filament before loading the sample into the TIMS source. Sample size is thus limited in this way to that which the geochronologist can physically manipulate with tweezers or a pipette to transfer the sample into beakers for dissolution.

The issue of low mass resolution is remediated by putting the dissolved sample through various types of ion separation chemistry that use organic resins capable of separating different elements based on their chemical properties and the type/concentration of reagent used (Schönbächler and Fehr 2014). The exact type and volume of resin, in addition to the reagent recipe, for the procedure depends on the elements requiring separation. For example, high-precision Sm–Nd geochronology requires separation and isolation of the rare earth elements (REEs) prior to mass spectrometry due to isobaric interferences that would obfuscate the measured isotopic ratios (Wasserburg et al. 1981). TIMS also carries the benefit of differentiating certain interfering elements from the desired element due to the different temperatures at which each element ionizes. This can help to recognize and mitigate the effects of interfering elements during an analysis though in general the goal of column chemistry is to remove interfering elements as much as possible.

Because the whole purpose of ion chromatography is to separate elements in the sample (including the parent from the daughter), the measurement of parent/daughter ratio required for geochronology is not possible during a single analysis. Instead, in order to still calculate accurate ages, a technique called isotope dilution (ID) is used. Isotope dilution is the process by which a tracer (or "spike") solution is added to the sample, usually prior to dissolution, that contains concentrated isotopes of the parent and daughter elements (Stracke et al. 2014). Because the ratio of those isotopes in the tracer solution are very well known through a series of calibration experiments, the concentration of parent and daughter isotopes can be backed out by measuring the sample isotopes relative to spike isotopes (Wasserburg et al. 1981; Condon et al. 2015; McLean et al. 2015). For some applications, synthetic isotopes such as ^{205}Pb and ^{233}U and ^{236}U are used, whereas for others natural isotopes are used. The mathematics for calculating parent/daughter ratios using the differing approaches are well-defined (Compston and Oversby 1969; Galer and Abouchami 1998; Thirlwall 2000; Schmitz and Schoene 2007; McLean et al. 2011; Stracke et al. 2014a). An added benefit of isotope dilution as a means of calculating parent/daughter ratios is that the tracer solution is homogeneous and analyzed at the exact same time as the sample, thereby eliminating concerns about variable parent/daughter fractionation between sample and standard. This point is an important source of uncertainty using *in situ* techniques that correct both isotopic and elemental fractionation through sample/standard bracketing (Williams 1998; Horstwood et al. 2016). In other words, the accuracy of a sample's date can be tracked directly to the accuracy to which one knows the composition of the tracer solution.

The process of sample preparation is briefly described above, but in practice is quite time-intensive. When compounded with long analysis times required for high precision data, TIMS geochronology is an extremely time consuming process. Furthermore, because the steps listed above require extensive sample handling and use of reagents, blank is an issue and for some methods can be a limiting factor in the sample size analyzed. Therefore, sample dissolution and ion separation chemistry are all done in clean labs, further adding to the cost and complexity of TIMS geochronology.

There are obvious and not-so-obvious difficulties in doing TIMS petrochronology. By way of comparison, with *in situ* techniques sample preparation can be as simple as making polished thin sections or grain mounts as a way of obtaining petrographic and textural information of dated minerals. Following thorough optical and geochemical sample characterization, the target minerals can be directly dated with spatial resolution of < 30 μm (see Kylander-Clark 2017; Schmitt and Vazquez 2017, both this volume). Developments of both conventional and split-stream LA-ICPMS techniques permit analysis of both U–Pb date and geochemistry simultaneously, further simplifying the work-flow and removing uncertainties about comparing dates and geochemistry from different volumes of material (Yuan et al. 2008; Kohn and Corrie 2011; Kylander-Clark et al. 2013). In TIMS geochronology, the sample preparation follows some variation of that used for *in situ* analysis, with the caveat that the sample to be dated must be physically removed from grain mount or thin section or bulk rock and dissolved, which requires steady hands and/or precise tools (such as a MicroMill). The specifics of the workflow needed to achieve petrochronologic characterization vary depending on the isotopic system and minerals being targeted. As such, the rest of this chapter is divided into the two isotopic dating methods most commonly used for petrochronology by TIMS, and within each section the narrative focuses mostly on the workflow considerations while presenting some results from the literature. The reader is referred to those publications for more detail on the interpretations and geologic significance of the data.

U–Pb ID-TIMS PETROCHRONOLOGY

Workflows in petrochronology

As broadly outlined above, there are many important steps prior to analysis by isotope dilution (ID) TIMS geochronology (see Davis et al. 2003; Corfu 2013; Schoene 2014, for additional details and references). Because most high-U minerals targeted for U–Pb geochronology are accessory phases (i.e., <1% volume of rock), typical sample processing begins with crushing followed by density+magnetic separations to concentrate the mineral(s) of interest, e.g., zircon, apatite, titanite, monazite, etc. This was an essential step several decades ago when a pure aliquot of several milligrams of high-U mineral was required for a single analysis. Comparatively, modern ID-TIMS U–Pb geochronology focuses on single accessory crystals (mostly <300 μm in diameter) or fragments of single crystals. Bulk crushing of rocks to obtain large mineral separates is therefore not necessarily required unless the concentration of accessory minerals is very low, as may be the case with crystal-poor volcanic rocks. This emphasis on smaller sample size is driven largely by increased awareness of the complexity of mineral growth in igneous and metamorphic systems. In theory, the ideal TIMS workflow would not differ much from that commonly pursued by in situ techniques, namely: 1) identify high-U minerals in thin section, 2) characterize their petrographic context and geochemistry in relation to rock-forming minerals that contain complementary information about the system's petrogenetic history, and 3) remove those same minerals from thin section for ID-TIMS analysis. However, such an approach is rare in reality due to the challenge of physically manipulating micrometer-scale mineral domains. Despite these difficulties, the clear advantage of TIMS temporal precision relative to in situ dating methods underscores its utility in petrologic studies (e.g., Fig. 1), and researchers have increasingly developed workflows to complement TIMS geochronology with additional textural and geochemical information (Fig. 2). Nonetheless, bulk crushing remains an important method of mineral separation for three reasons: 1) it is not uncommon to identify high-U minerals through bulk crushing that are either absent or rare in thin section, 2) to ensure more accurate representation of the total crystal population of a sample, from which to then characterize and select crystals for analysis, and 3) it is difficult to extract small minerals from thin section. Examples from the recent literature of creative workflows in ID-TIMS petrochronology are given below in the section *Linking geochemistry with U–Pb ID-TIMS geochronology*, but first a discussion of limitations imposed by sample size is presented, as this factor dictates what workflows are possible.

Limits on sample size and precision

ID-TIMS U–Pb geochronology of high-U minerals with low amounts of initial Pb (e.g., zircon) is the most precise available geochronological method. Reported age precision for ID-TIMS has increased dramatically over the past few decades, from the percent level on large milligram size fractions of minerals to better than 0.1% (2σ) on fragments of single crystals (Schoene 2014). Advances in mass spectrometry have played a pivotal role in this development, particularly in relation to ion generation, transmission sensitivity, and ion beam measurement. However, other sources of uncertainty can be far greater than associated with analytical precision, depending on the age of the sample, the ratio of initial U to Pb, and laboratory blank. Several recent contributions have discussed these sources in detail (Schmitz and Schoene 2007; McLean et al. 2011; Schoene 2014); examples are depicted in Fig. 3 for zircons with different ages and U contents. Because one goal of ID-TIMS petrochronology is to explore potentially complex and protracted mineral growth histories through dating increasingly smaller samples, those uncertainties that pose the greatest limitations to sample size are discussed below. Other sources of uncertainty, such as mass-dependent isotope fractionation during mass spectrometric measurements (Amelin and Davis 2006; McLean et al. 2015), corrections for initial daughter product disequilibrium (Parrish 1990; Schärer 1984; Crowley et al. 2007; Schmitz and Schoene 2007), and oxygen isotopic composition of UO_2 (Condon et al. 2015) are independent of sample size but can be the dominant source of uncertainty for cases where sample size is not a limit on precision (Fig. 3).

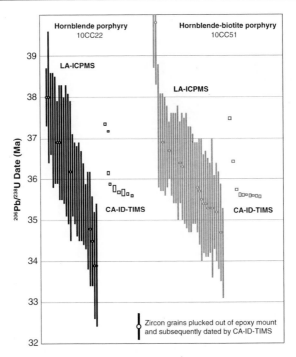

Figure 1. Rank order plot illustrating the precision of Chemical Abrasion (CA-)ID-TIMS geochronology relative to LA-ICPMS U–Pb geochronology. Shown are two examples of $^{206}Pb/^{238}U$ dates on zircon from the same rocks, and in some cases for the exact same zircon grains. These data illustrate both the ability of chemical abrasion to remove zones of Pb-loss and also that higher precision reveals complexity in zircon growth that is unresolvable with lower precision data. The height of symbols represents 2σ uncertainties. Data from Chelle-Michou et al. (2014).

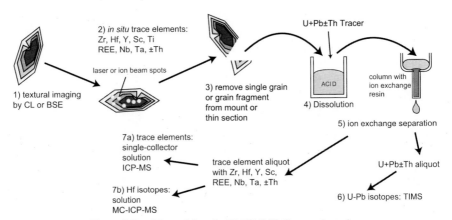

Figure 2. Possible workflow for ID-TIMS U–Pb petrochronology.

Pb-blank. The greatest limitation on sample size is the amount of radiogenic Pb in a sample (Pb^*) compared to the non-radiogenic, or common, Pb (Pb_c). Pb_c integrates Pb taken up by the grain during crystallization and the laboratory blank. For zircon, there is no quantifiable Pb included in the crystal structure during crystallization, so laboratory blank is the only source of Pb_c (although some Pb_c may be incorporated into zircon during metamictization and

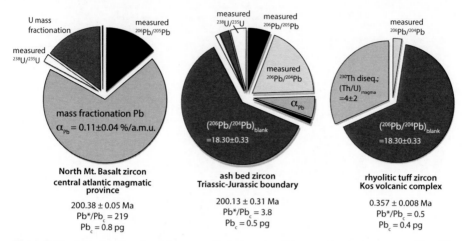

Figure 3. Pie charts detailing the variance contributions to the total uncertainty budget of some $^{206}Pb/^{238}U$ zircon dates. Modified from Schoene (2014). See text for discussion.

recrystallization (Geisler et al. 2003)). For other minerals, such as monazite, small amounts of Pb$_c$ (typically < 1 ppm) may be incorporated into the crystal lattice. For titanite, apatite and allanite, among others, 10s to 100s ppm Pb$_c$ may be incorporated during crystallization.

Figure 3 shows that for young and/or low U zircons, the correction for Pb$_c$ is by far the largest source of uncertainty for a given U–Pb date (manifested in the correction for the $^{206}Pb/^{204}Pb$ ratio of the Pb blank). Figure 4 further quantifies this by showing a compilation of zircon data from a single study where the Pb*/Pb$_c$ varies from ~1 to > 150 as a function of the sample size (all analyses are < 1 zircon grain), U and Pb* content, and amount of blank Pb. This figure clearly demonstrates that the precision of an analysis scales with Pb*/Pb$_c$, with dates becoming exponentially less precise when Pb*/Pb$_c$ <~15–20. Laboratories that perform high precision geochronology of small and/or young zircons typically have Pb blanks < 1 pg, but strive to remain < 0.5 pg.

Comparatively, in the case of Archean zircons with tens, if not hundreds, of picograms Pb*, sample size is essentially limited by one's skill with a set of tweezers. For most applications, however, there is a trade-off between the volume of sample selected, age, U content, and routinely-achievable blanks for a given laboratory. The average uncertainty in the Oligocene zircon dataset in Figure 3 of 0.07% on a $^{206}Pb/^{238}U$ date translates into about ±20 ka for a single zircon fragment, which is far smaller than the average range of zircon crystallization ages for a granitoid sample from that study (i.e., ~500 kyr spread in dates for a given hand sample). While this implies that even smaller sample sizes could elucidate further details of zircon growth histories, it also requires lower Pb blanks. The difference between a 0.6 and 0.3 pg Pb blank in such studies can determine whether one can successfully resolve temporal variability *between* grains or detect intra-crystal growth histories by subsampling and dating multiple fragments from *single* grains.

A large part of the uncertainty associated with Pb$_c$ in dated minerals comes from the difficulty in constraining the Pb$_c$ isotopic composition. Because ^{204}Pb is non-radiogenic and stable, in ID-TIMS U–Pb geochronology, the moles of ^{204}Pb can be determined through isotope dilution and the amount of ^{206}Pb, ^{207}Pb and ^{208}Pb from blank can be subtracted from the radiogenic Pb isotopes by calculating their abundance relative to ^{204}Pb. In the case of minerals that incorporate relatively high Pb$_c$ into their mineral structure (e.g., titanite), the composition of this initial Pb has been estimated in several ways, including 1) assuming a composition

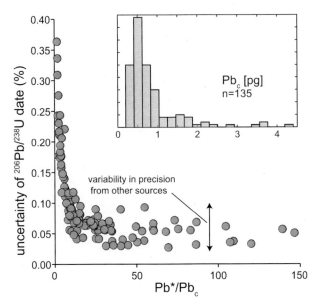

Figure 4. Precision of ID-TIMS U–Pb date as a function of the ratio of radiogenic Pb (Pb*) to unradiogenic Pb (laboratory blank in this case; Pbc). Data are for single zircon fragments from (Samperton et al., 2015), and given 2σ uncertainties, as a percent of the age. Inset shows histogram of Pb_c values from the same dataset.

based on whole-earth Pb evolution models (Stacey and Kramers 1975), 2) measuring the Pb isotopic composition of a coexisting low-U mineral, e.g., feldspar (Chamberlain and Bowring 2001), and 3) application of standard isochron methods. The benefits and drawbacks of each of these approaches are reviewed in detail by Schoene (2014) and the references therein.

For minerals such as zircon where blank Pb is the dominant source of Pb_c, knowledge of that composition can be a limiting factor in the precision of a date, even if blank amounts are very low. The blank isotopic composition depends on mixtures between the different sources of blank Pb in the lab. The obvious sources are the reagents added to the sample during grain dissolution, ion separation chemistry, and filament loading, in addition to Pb_c ionized from the filament, diffused from Teflon during dissolution, or not removed from the surface of the sample during pre-cleaning steps. Because the amounts of each source will all vary slightly between analyses, a typical approach to estimating the composition of these mixtures is to analyze "total procedural blanks", where a dissolution capsule is left empty, spiked with tracer solution, and treated identically to normal samples. The standard deviation of these measurements should characterize the range of possible blank compositions for samples, and this uncertainty can be propagated into dates (Schmitz and Schoene 2007; McLean et al. 2011). Figure 5 shows the measured isotopic compositions of ~600 Pb blanks in the Princeton ID-TIMS laboratory over a 3 year period. Most of these analyses were measured with the intention of calculating an accurate mass of Pb, not the Pb isotopic composition, and thus emphasized speed over quality. For example, it is observed in many TIMS labs that isobaric interferences are present under all Pb isotopes at low temperature, and that as the filament is ramped to higher temperatures these interferences "burn off". This effect takes time, and while ensuring that interferences are burned off is not crucial for determining the amount of Pb blank in an analysis, it is crucial for determining the Pb blank isotopic composition. Therefore the accuracy of the isotopic composition is questionable in the majority of measurements shown in Figure 5. Two datasets

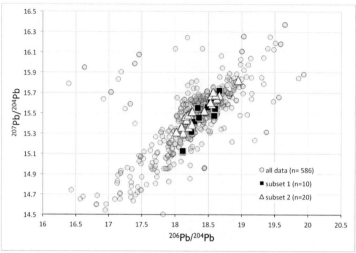

Figure 5. Compilation of measurements of the isotopic composition of Pb from 2013–2016 in the Princeton ID-TIMS laboratory. Gray circles represent all data, including measurements not meant to target the blank isotopic composition. These analyses were done quickly with less care taken to allow low temperature isobaric interference "burn off", noting these interferences are observed to decrease through an analysis until high quality data is obtained. Black squares (Samperton et al. 2015) and white triangles (Barboni et al. 2017) are two examples of total procedural blanks measured by a single user over the course of one or two studies. Mean values and 1-standard deviation are: $^{206}Pb/^{204}Pb = 18.42 \pm 0.18$, $^{207}Pb/^{204}Pb = 15.48 \pm 0.17$ for (Samperton et al. 2015) and $^{206}Pb/^{204}Pb = 18.44 \pm 0.18$, $^{207}Pb/^{204}Pb = 15.53 \pm 0.18$ for (Barboni et al. 2017). Uncertainties (1RSE) for individual data points are approximately the symbol size.

are also shown in Figure 5 wherein analyses were carefully performed over individual studies and run identically to typical samples; these analyses were used to estimate the Pb blank composition. Typically the uncertainty of a given blank analysis is far less than the variance observed in blank isotopic composition of multiple measurements; thus, while measuring more blanks results in a more accurate description of that variance, it does not reduce the uncertainty in the correction. In other words, measuring more blanks over the course of a study increases the accuracy, but not necessarily the precision of the resulting U–Pb dates. Reducing the variance in Pb blank composition, and therefore reducing the uncertainty in the blank subtraction, is an important way to increase precision on dates for samples with low Pb^*/Pb_c.

Analytical precision. Another limitation on sample size relates to the ionization of Pb and U in the mass spectrometer. While analytical precision would appear to be of secondary importance based on Figure 3, this is primarily because those examples were chosen to highlight the importance of other sources of uncertainty. For the zircon to the left in Figure 3, using a double-Pb, double-U isotopic tracer (e.g., the EARTHTIME $^{202}Pb–^{205}Pb–^{233}U–^{235}U$ tracer; Condon et al. 2015; McLean et al. 2015) would eliminate the uncertainty of Pb mass fractionation during mass spectrometry, and in that case the measured $^{206}Pb/^{205}Pb$ would be the largest contributor to the uncertainty in the date. Figure 6 compiles data from the Princeton ID-TIMS lab between 2012–2016 for zircons with a range of intensities of ^{205}Pb and variable $^{206}Pb/^{205}Pb$ ratios. The data were measured on an IsotopX PhoeniX62 TIMS on a Daly photomultiplier ion counter, all with ~10 pg ^{205}Pb. While these data encompass a range of N-values (i.e., number of ratios measured, or total integration time) and total ion yield (average intensity is plotted), they form a roughly linear cloud in log–log space, where precision increases as a function of increasing intensity and increasing sample/spike Pb (represented as the $^{206}Pb/^{205}Pb$, where ^{205}Pb is the tracer isotope). The plot also shows the precision of uranium analyses for the same grains. Uranium was measured as UO_2^+ on three static Faraday cups

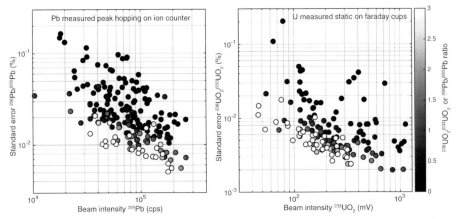

Figure 6. Demonstration of analytical precision of Pb and U isotopic ratios as a function of sample size. Data are from routine measurements with the EARTHTIME ET535 or ET2535 tracer in the Princeton ID-TIMS laboratory between 2012–2016, and were chosen to include a wide range in sample/spike ratio and beam intensities. All measurements performed on an IsotopX PhoeniX TIMS by peak-hopping on a Daly photomultiplier ion counter for Pb and as static measurements with 10^{12} ohm amplifier boards on Faraday cups for U. Each sample has ~1–1.5 ng of ^{233}U and 10–15 pg of ^{205}Pb. Datasets include ~50–100 ratios (250–500 second integration on each isotope) for Pb and 100–300 ratios (500–1500 second integration) for U. Grayscale on datapoints represent the sample/spike ratio, with key at right. Beam intensities are reported as the average intensity over the entire analysis. Uncertainties (vertical axes) are reported as 1RSE. See text for discussion.

on 10^{12} ohm resistors. Although each measurement contains ~1 ng of ^{233}U, larger scatter in the measured precision as a function of intensity and $^{238}UO_2/^{233}UO_2$ compared to Pb reflects less predictable ionization for U (i.e., more variable N or total integration time). For very low $^{238}UO_2/^{233}UO_2$ or particularly poorly-ionizing samples, UO_2 can also be measured on an ion counter with similar results to Pb. A couple take-home points can be made from Figure 6. One, not surprisingly, is that precision is improved as more ions are extracted from the filament and measured. Most labs mix their sample with a slurry of colloidal silicic acid and phosphoric acid, which has been shown to increase ionization (Gerstenberger and Haase 1997) compared to loading the sample on the filament with other materials. Variability in ionization using this technique is large from sample-to-sample and user-to-user, resulting in much of the dispersion in Figure 6. Regardless, ionization for both Pb and U (typically calculated by summing the total number of ions measured relative to the calculated sample size) is typically ≤5% of the total sample on the filament. There is, therefore, an opportunity for drastic improvements in ionization, which would decrease the amount of sample needed for ID-TIMS U–Pb geochronology and permit examination of increasingly smaller mineral domains.

Chemical abrasion. As with any geochronological technique, assessing open-system behavior is of primary importance. In the U–Pb system, for minerals older than a couple hundred millions of years (Ma), discordance between $^{206}Pb/^{238}U$ and $^{207}Pb/^{235}U$ dates can be used as a means of identifying whether initial Pb, or loss/gain of Pb or U, have affected a mineral's isotopic systematics (Wetherill 1956; Corfu 2013). This information can be used to assess the accuracy of a given date. Of all the possible sources of discordance in the U–Pb system, Pb-loss has been the prime suspect, and great measures have been undertaken to both understand and remediate this problem (Corfu 2013; Schoene 2014). In U–Pb thermochronometers such as titanite, apatite, and rutile, Pb-loss can be used advantageously to constrain the thermal history of rocks (Blackburn et al. 2011; Smye and Stockli 2014; note however that for titanite and rutile there is ongoing discussion about diffusion kinetics; e.g., Kohn 2016; Kohn and Penniston-Dorland 2017, this volume). In minerals with negligible rates of Pb diffusion under most crustal conditions, however, little has been learned in attempts to interpret Pb-loss; instead, the overwhelming goal has been to exclude from analysis crystal domains affected by Pb-loss. Following decades of attempts, Mattinson (2005) presented a technique called chemical abrasion TIMS (CA-TIMS), in which zircon grains are annealed at >900 °C for 48–60 hours and then leached with HF acid in order to remove domains compromised by Pb-loss. In short, this technique has revolutionized ID-TIMS zircon U–Pb geochronology such that almost every lab has adopted some form of this method and adapted it for single- or sub-grain zircon analysis (see Mundil et al. 2004, Schoene et al. 2006) for early examples, but note that all recent U–Pb ID-TIMS papers employ CA). CA-TIMS has taken huge steps in eliminating discordance in Archean zircon and improved accuracy in dating Phanerozoic zircons as well. Notably, attempts to extend CA-TIMS to minerals other than zircon have not yet been successful (Rioux et al. 2010; Peterman et al. 2012). In CA-TIMS, an analyst will typically leach single pre-selected grains, perform several cleaning steps, spike with an isotopic tracer, and then dissolve the whole grain. An important point about this workflow for petrochronology is that, due to the leaching step, no information is available about what parts of the grains are dissolved. Polishing and imaging of zircon that has experienced chemical abrasion shows that the leaching procedure is complicated and does not simply attack the outermost zircon layers (Mundil et al. 2004; Mattinson 2011). Hydrofluoric acid apparently penetrates zircon along fractures and attacks discrete domains, often burrowing into grains in unpredictable, asymmetric ways. It is therefore not straightforward to connect zircon textural or geochemical information obtained prior to chemical abrasion with the age that is eventually measured. This "information decoupling" is similar to the practice of many *in situ* workflows that, e.g., involve measuring age from one laser spot and geochemistry ± Hf isotopes from a nearby spot in the grain. While it is not unreasonable to associate such data, there is potentially nontrivial uncertainty in such a correlation. For *in situ* methods, the LASS method helps remediate this problem (Kylander-Clark 2017, this volume); for TIMS, the TIMS-TEA method, described below, aims to achieve similar goals.

Linking textures with dates in ID-TIMS U–Pb geochronology

In situ attempts. In situ geochronological methods have revealed that linking dates to internal mineral textures and petrographic context can be highly useful in interpreting such dates in terms of *P–T* paths in metamorphic rocks and mineral growth histories in both metamorphic and igneous rocks (see Rubatto 2017; Schaltegger and Davies 2017, both this volume). ID-TIMS has benefited from these insights but is faced with the obvious challenge of physically isolating grains targeted for geochronology prior to dissolution. There have been some notable examples where grains have been removed from thin sections and dated by ID-TIMS following characterization in thin section.

One such example is that of (Lanzirotti and Hanson 1996), who characterized two populations of monazite in metapelites from the Appalachian-aged Wepawaug Schist in Connecticut, USA. They were able to characterize multiple populations of monazite on the basis of morphology and geochemistry, and also link their growth to peak versus retrograde metamorphic conditions. By simply plucking the monazites out of thin section and dating them by TIMS, they were able to show that the populations differed in age by ~30 Ma, thereby placing robust time constraints on the retrograde *P–T* path.

More recently, Corrie and Kohn (2007) used a microdrill to remove cores of monazite grains from thin section and then date them by TIMS. The monazite forms part of a Barrovian-type assemblage in the Great Smoky Mountains, USA, and was demonstrated to have grown near the staurolite-in isograd during the Taconian orogeny. Because the monazites were relatively large (>200 µm diameter) and high-U, Corrie and Kohn (2007) were able to drill out and analyze <50 µm diameter pieces of monazite that had been characterized petrographically. One of the challenges of this is the possibility of contamination during drilling, since monazite has a high acid solubility and is difficult to clean. Regardless, this approach to linking TIMS dates to metamorphic reactions remains underexploited.

Although not carried out in thin section, Barboni and Schoene (2014) were able to link zircon dates to their setting in the rock by extracting zircon inclusions from K-feldspar megacrysts in the 7 Ma Mt. Capanne pluton, Elba, Italy. By cutting megacrysts into core and rim pieces and then crushing them, they were able to build a stratigraphy through the K-feldspar with uncertainties on U–Pb dates of only a few thousand years (Fig. 7). Using logic similar to that of a detrital zircon study in sedimentary rocks, they showed that youngest zircons from the megacryst cores are tens of thousands of years older than the rims, and this duration is therefore the maximum growth interval of the K-feldspar. By combining U–Pb data with observations of K-feldspar inclusion suites and thermodynamic phase modeling, they argued that megacryst growth occurred in an upper crustal magmatic mush, thus placing time constraints on upper crustal cooling and pluton crystallization.

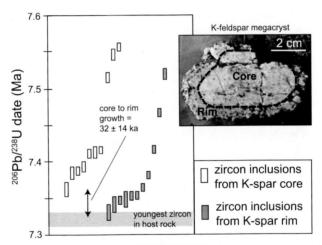

Figure 7. Determining crystal growth rates with ID-TIMS U–Pb zircon geochronology. In this example, K-feldspar megacrysts were removed from the host granotoid, the Sant Andrea phase of the Mt. Capanne pluton, Elba, Italy. Zircon inclusions from within K-feldspar megacrysts were removed by bulk crushing after slicing up K-feldspar into core and rim domains. ^{206}Pb/^{238}U dates from individual zircons are plotted in a rank order plot. The youngest zircon date from the host granitoid is shown to be identical to the youngest K-feldpsar rim zircon date; core to rim growth duration is thus a maximum. Data are from Barboni and Schoene (2014).

Ex situ attempts. It is now common to characterize internal textures of minerals following bulk mineral separation. This can give insight into whether minerals have igneous or metamorphic origins, or whether igneous or metamorphic cores of grains are overgrown by later episodes of mineral growth. Interpreting these textures has been important for attaching significance to dates but remains a qualitative and imperfect tool. Nonetheless, knowledge regarding whether a date isolates or mixes visible growth domains within a mineral is an important first-order petrochronologic observation. Textures within these domains can give information about the igneous or metamorphic origin of mineral growth; the nature of the contacts between crystal domains also provides information regarding periods of dissolution, reprecipitation, or recrystallization.

In U–Pb ID-TIMS geochronology, linking internal mineral textures with dates is most commonly done by imaging minerals in grain mount and then removing and dating them as whole grains or fragments (Crowley et al. 2006; Matzel et al. 2006; Gordon et al. 2010; Rivera et al. 2013). Although admittedly crude, it is possible to break grains near or along boundaries between different growth domains and date them in order to isolate the timing of growth of different sub-domains. There are, of course, large uncertainties in how observed domains project into the third dimension of the grain and therefore into the sampled volume. Microsampling previously imaged minerals is not yet routine in most labs, but is becoming increasingly common and in every case having textural information that can be linked to dates is better than the alternative.

Linking geochemistry with U–Pb ID-TIMS geochronology

In situ geochemical analysis prior to TIMS. An extension of the petrographic and textural characterization techniques described above involves carrying out geochemical analyses in thin section or grain mount and then microsampling grains for geochronology. Depending on the concentration of the elements or isotopes measured, the analyses can be performed by electron probe microanalysis (EPMA), LA-ICPMS or SIMS. In the case of LA-ICPMS care must be taken not to ablate so much sample such that the remainder is too small to analyze by TIMS. The goal is to combine geochemical information with age information in a way that can be related back to the composition of the liquid from which the minerals crystallized or identifying equilibrium mineral assemblages through application of partition coefficients.

There are not many examples of this approach, but the increasing accessibility of *in situ* techniques and the number of TIMS labs capable of measuring small amounts of Pb will surely lead to more common implementation. Crowley et al. (Crowley et al. 2007), working on zircon from the Bishop Tuff, California, measured U and Th concentrations and produced CL images by EPMA prior to removing grains from grain mount for U–Pb analysis. U and Th increased by up to a factor of 5 across grain transects. In addition to providing information about magma evolution during zircon growth, this approach provided a means to compare bulk U and Th concentrations obtained from the TIMS analysis with finer scale zonation that is obscured by sampling whole grains or fragments.

Rivera et al. (2014) used LA-ICPMS followed by TIMS U–Pb zircon geochronology on the same grains to investigate the magmatic history of the precursor magma that fed the eruption of the Huckleberry Ridge Tuff at ca. 2.1 Ma. By measuring the europium anomaly in zircons (a proxy for fractional crystallization of plagioclase and sanidine within the magmatic system), and combining this with Ti-in-zircon measurements to investigate temperature, they were able to model the crystallinity in the magma as a function of temperature (Fig. 8). These same zircons were then removed from grain mount and dated by TIMS. The results were used to texturally and geochemically identify cores of zircons that predated the main phase of emplacement and residence of the magma in the upper crust by up to 120 ka. The combination of Ti-in-zircon, geochemistry, and geochronology showed that magma residence and differentiation in the upper crust predated eruption by only tens of ka, arguing for high magma flux and short upper crustal residence leading to the Huckleberry Ridge supereruption.

Similarly, Samperton et al. (2015) conducted trace element transects across zircon grains to look at the relative geochemical evolution during zircon growth in a suite of samples from the Bergell Intrusion, Italy/Switzerland. These data were used to argue that relatively gradual changes in zircon trace elements reflected evolving magma composition over several hundred kyr due to fractionation of both major and accessory phases in the magma. Microsampling of the grains provided up to 5 analyses from single zircons with individual uncertainties as low as ±20 ka, permitting evaluation of growth timescales of individual zircons within a geochemical framework. A remaining uncertainty in quantitatively relating trace elements in zircon to

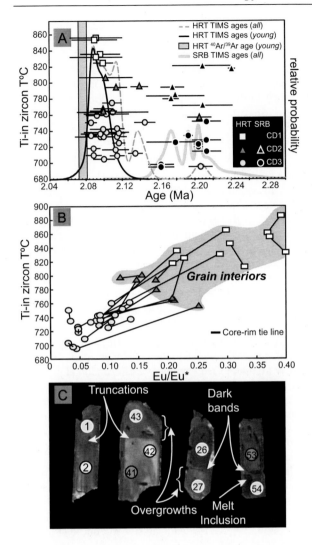

Figure 8. LA-ICPMS + ID-TIMS U–Pb petrochronology in the Huckleberry Ridge Tuff (HRT) and Snake River Butte rhyolite (SRB) (Rivera et al. 2014). (A) Temperature versus $^{206}Pb/^{238}U$ age from zircon fragments that were first measured for trace elements by LA-ICPMS in situ, then removed from grain mount and dated by ID-TIMS. CD1, etc., are manual groupings based on zircon geochemistry. (B) Geochemistry of zircon core–rim pairs that were subsequently dated. Eu/Eu* is the europium anomaly. (C) Examples of zircon CL images, textural interpretations, and LA spots, color coded by geochemical domain to those in A.

that of the liquid, while not unique to TIMS petrochronology, derives from limited partition coefficient data under a range of pressures, temperatures, and magma compositions. More experiments are needed to help address the paucity of partitioning data with respect to such intensive variables for zircon and other high-U minerals used for petrochronology.

Another important uncertainty remains in relating trace element geochemistry determined *in situ* on the surface of a polished grain mount to dates derived by ID-TIMS on grains removed from that mount: in other words, one is left comparing two-dimensional geochemical data to three-dimensional chronological data. In every type of analysis, be it *in situ* or whole-grain, the trace element and age information obtained are integrations of the analyzed volume. If those volumes are not identical, then it is difficult to quantitatively compare the results.

Indirect in situ analysis. One of the parameters that falls out of U–Pb geochronology is the Th/U of the grain. While in U–Th–Pb analysis this can be directly determined by $^{232}Th/^{238}U$ measurement, a model Th/U can be calculated without measuring Th through measurement of the $^{208}Pb/^{206}Pb$ and assuming concordance between $^{206}Pb/^{238}U$ and $^{208}Pb/^{232}Th$ dates. Therefore, it is common to plot the Th/U of a suite of minerals versus the measured dates, and this has been used to distinguish between metamorphic versus igneous grains and/or evaluate the petrologic implications of evolving Th/U during crystallization. Given this information, it is also possible to measure Th/U of datable minerals in thin section and correlate those data indirectly with Th/U derived from the isotopic measurements. As an example of this approach, Oberli et al. (2004) measured major and trace element geochemistry of magmatic allanite by EPMA. They were able to relate the Th/U measured by EPMA with that determined through U–Th–Pb isotopic analysis, and by inference could then relate the major element geochemistry to the calculated U–Th–Pb dates. They conclude that allanite grew in the presence of an evolving silicate liquid composition over a period of 5 Ma following intrusion of the studied magma pulse.

Geochemistry and dates from the same dissolved minerals. As described in the section *A brief review of TIMS geochronology*, an essential part of TIMS geochronology is that target elements in the sample must be separated from other elements that may interfere with the masses measured. In the cases discussed in this section, U, Th and Pb can be separated quantitatively from the other elements in the dissolved minerals. Though often overlooked, it is possible to retain the portions of the sample that are not analyzed for geochronology—the so-called "wash" solution—and these can then be measured for geochemistry or isotopes. The benefit of this approach over *in situ* analysis prior to geochronology is that it compares the dates and geochemistry and/or isotopes from the exact same volume of dated material. Thus, the average geochemical composition should in most cases correspond to the moment in time measured for average growth age over that volume of mineral (Samperton et al. 2015).

A relatively common example is that of Hf in zircon. Given that Hf is present in zircon at the weight percent level, it is possible to measure the Hf isotopic composition of the wash solution by multicollector ICPMS. The first published example was in Amelin et al. (1999), carried out on a set of Hadean Jack Hills zircons, in order to combine high-precision dates with their corresponding Hf isotopic values to gain insight into the presence or absence of depleted early-Earth reservoirs. Subsequent investigations have used measured Hf isotopic values to look at the importance of magma mixing and assimilation during the crystallization history of igneous zircons (Schaltegger et al. 2002, 2009; Schoene et al. 2012; Broderick et al. 2015), as well as the sources of silicate fluids in metamorphic zircon growth (Crowley et al. 2006). Because recovery of Hf in the wash solution is very close to 100%, isotopic fractionation during ion separation chemistry is not a concern. In some cases, the wash solution has been put through a second stage separation to remove REEs from the Hf fraction, given isobaric interferences such as ^{176}Yb on ^{176}Hf. Alternatively, this interference can be corrected for by measuring ^{175}Yb, as is routinely done during LA-ICPMS Hf analysis (Woodhead et al. 2004; Fisher et al. 2014; D'Abzac et al. 2016).

Similarly, given the sensitivity of modern ICPMS instruments, it is possible to measure the trace element geochemistry of the wash solutions as well for single minerals or fragments of minerals (Schoene et al. 2010). This technique, dubbed TIMS-TEA (for Trace Element Analysis), is simply a downsizing of previous work on larger aliquots of dated minerals (Heaman et al. 1990; Root et al. 2004), and thus usable for modern ID-TIMS U–Pb geochronology where an emphasis is placed on shrinking sample size to isolate heterogeneities. Schoene et al. (2010, 2012) applied this method to zircon and titanite from igneous rocks ranging in composition from gabbroic to granitic and also to a granulite-derived zircon. Those studies showed that elemental fractionation during ion separation chemistry was less than a few percent and that Hf isotopes could be measured from the same aliquot as the other trace elements. Trace element data from Schoene et al. (2012) served several purposes. The first was to identify zircons that are similar in age but inherited from other batches of magma (sometimes called antecrysts; Miller et al. 2007)). The second was to show that zircon and titanite geochemistry reflects the geochemistry of the liquid from which these minerals crystallized. The third was to show that changes in trace element signatures of the accessory minerals as a function of time record the timescales of magma differentiation processes such as assimilation, magma mixing and fractional crystallization. Those data, from the Adamello Batholith, Italy, show that the timescales associated with these processes are from 10^4–10^5 years, and therefore require the precision of TIMS U–Pb geochronology to resolve them.

Samperton et al. (2015) used TIMS-TEA of imaged and microsampled zircons to investigate the timescales of magma differentiation in the Bergell Intrusion, Italy/Switzerland. They combined this approach with in situ determination of trace elements by LA-ICPMS to compare analytical approaches and to learn more about both high spatial resolution relative changes in trace element geochemistry and lower spatial resolution (i.e., zircon crystal- and fragment-scale) trace element evolution in absolute time (Fig. 9). One of the purposes of obtaining in situ trace elements is to answer the question of whether the TEA data record mixing of core versus rim signatures of zircon, given the crude nature of microsampling zircons prior to TIMS analysis, or whether trends in time exhibited by TEA data record protracted evolution of magma geochemistry. In this particular case, it was shown that tonalites from the Bergell recorded zircon crystallization in a differentiating liquid during post-emplacement cooling; granodioritic zircon preserved inheritance from slightly older tonalitic magma pulses, which was subsequently overgrown by zircon that tracked liquid compositions. The main point to make here is that it seems necessary to combine CL imaging with both in situ and TEA geochemistry to fully understand of zircon growth and magma evolution in these systems. However, the agreement observed in studies with both TEA data and in situ geochemical analyses in both igneous and metamorphic rocks (DesOrmeau et al. 2015; Samperton et al. 2015) show that either approach may be appropriate depending on the questions that are asked (Schoene et al. 2010; DesOrmeau et al. 2014; Broderick et al. 2015; Deering et al. 2016).

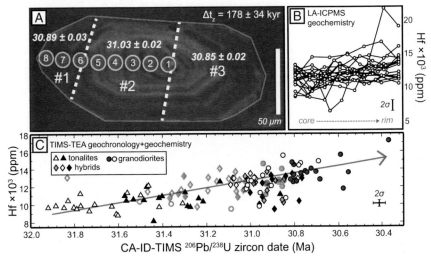

Figure 9. U–Pb TIMS-TEA, illustrating combined LA-ICPMS/TIMS workflow. (A) CL image of zircon. Circles show where geochemical data was obtained by LA-ICPMS. Grain fragments #1–#3 were then extracted by breaking grain along dotted lines. Fragments were then dated by CA-ID-TIMS, and dates are indicated with 2σ uncertainties. Δt_z is the minimum growth duration of the crystal. (B) LA-ICPMS geochemical transects across multiple zircons, including that shown in A. Plotted as relative core to rim transects. (C) Geochemical and age data determined by TIMS-TEA for seven hand samples ranging in composition from tonalite to granodiorite. Datapoints represent dates from single zircon fragments as in A, and geochemistry was measured on the same volume of dissolved zircon by solution ICPMS. Data come from the Bergell Intrusion, Italy/Switzerland, and are from Samperton et al. (2015). See text for discussion.

SAMARIUM–NEODYMIUM ID-TIMS PETROCHRONOLOGY

While the first Nd isotopic measurements were conducted in the mid-1970's (e.g., DePaolo and Wasserburg 1976; Lugmair et al. 1976; Richard et al. 1976), the earliest mineral-specific geochronology using Sm–Nd can probably be traced to 1980 focusing on the mineral garnet (van Bremen and Hawkesworth 1980; Griffin and Brueckner 1980). Like U–Pb in zircon, garnet has an unusually high Sm–Nd (parent to daughter) ratio making it amenable to Sm–Nd geochronology. Since then, the Sm–Nd system has also been used to constrain the mineralization of other relatively high Sm/Nd minerals such as scheelite (e.g., Bell et al. 1989), fluorite (e.g., Chesley et al. 1991), apatite (e.g., Rakovan et al. 1997), carbonate (e.g., Peng et al. 2003; Henjes-Kunst et al. 2014), and rare earth rich xenotime (e.g., Thoni et al. 2008) and eudialyte (e.g., Sjöqvist et al. 2016). Unlike U–Pb in zircon (where U/Pb ratio often exceeds 100), the Sm–Nd ratio in garnet (or any high Sm/Nd mineral) is too low (clean garnet has $^{147}Sm/^{144}Nd$ generally between 1.0 and 5.0; other minerals rarely exceed 1.0) for us to ignore "common neodymium" or initial $^{143}Nd/^{144}Nd$. Rather, all Sm–Nd petrochronology is isochron geochronology, requiring precise analysis not just of the mineral of interest but also of a second point (or points) with low Sm/Nd to pin the isochron and calculate an age. The reader is referred to other papers describing the choice of second point(s) on the isochron and associated age derivation (e.g., Baxter and Scherer 2013; Baxter et al. 2017, this volume), though most often it is appropriate to use the whole rock or matrix surrounding the garnet (or other mineral) as the anchor point for an isochron. Here, we review the use of TIMS to extract precise and accurate Sm–Nd isotopic data needed for precise geochronology, with potential application not just to garnet, but to any relatively high Sm/Nd mineral. Then, we turn our focus to garnet-specific workflows and analytical advances, including the advent of chemically contoured micromill-TIMS of zoned crystals. Most garnet petrochronology combines an Sm–Nd age (and/or Lu–Hf age, which would be gained through MC-ICPMS analysis) with any number of complementary petrologic, geochemical, or textural constraints (see Baxter et al. 2017, this volume). On the other hand, whether for garnet or for other dateable minerals such as zircon or monazite, Nd-isotopic measurements (i.e., $\varepsilon Nd_{(init)}$) can themselves provide valuable petrologic or tectonic context for "petrochronology" (e.g., zircon, monazite, allanite; Amelin 2004; McFarlane and McCulloch 2007; Iizuka et al. 2011). Such work has utilized LA-ICPMS when Nd concentrations are high enough, though TIMS can provide data on materials yielding much smaller amounts of Nd (especially zircon; e.g., Amelin 2004). This review of TIMS Sm–Nd methodologies and capabilities is thus motivated by both the petrologic and chronologic aspects of petrochronology.

Why ID-TIMS Sm–Nd for Petrochronology?

Sm–Nd can be analyzed with high precision by TIMS or by MC-ICPMS. However, the main advantage of TIMS is the ability to maintain such high precision (and accuracy) even as sample size decreases. Part of the reason involves the option to ionize Nd as the oxide, NdO^+, first recognized in the earliest days of Nd-isotope geochemistry (e.g., Lugmair et al. 1976; DePaolo and Wasserburg 1976). Numerous methods have been developed to maximize ionization including the use of an oxygen bleed valve (e.g., Sharma et al. 1995), and various loading activator solutions (e.g., Thirlwall 1991; Griselin et al. 2001; Amelin 2004; Chu et al. 2009; Harvey and Baxter 2009). For example, Harvey and Baxter (2009) employed a Ta_2O_5 powder slurry in dilute phosphoric acid as an activator solution. Without the need for an oxygen bleed valve, source pressures remain low during analysis (about two orders of magnitude lower than when using the bleed valve). Calculated total ionization efficiency using the Ta_2O_5 method are as high as ~30%, representing a major boost over more common TIMS (or MC-ICPMS) analysis of Nd metal where ion yields are < 10%. Given the unavoidable sample attrition during focusing, this total ionization efficiency is very close to the theoretical maximum possible value. $^{143}Nd/^{144}Nd$ precision via TIMS NdO^+ analysis using the Harvey and Baxter (2009) loading solution yields 10 ppm 2RSD for 4 ng loads, and 28–37 ppm 2RSD for 400 pg loads of a standard solution (Fig. 10). In practice, most natural samples that are dissolved and run through columns yield precision (internal 2RSE) 2 to 5 times worse than pure standards (external 2RSD; Fig. 10) for loads of at least 100 pg. For loads smaller than 100 pg, performance of natural samples worsens further in most cases. This familiar and frustrating difference between standard solution and natural sample performance in any lab likely results from loading technique, and additional non-Nd material coming from the column chemistry that may inhibit or destabilize ionization of Nd. Reducing the performance gap between standard solutions and dissolved samples remains a frontier for TIMS analysis.

NdO$^+$ analysis also requires correction for interferences from the various oxide species formed when a particular Nd isotope combines with ^{16}O, ^{17}O, or ^{18}O. In practice, this interference correction is quite straightforward. Oxygen isotope corrections should be made first, beginning with lowest mass Nd isotope (^{142}Nd) up to the highest mass (^{150}Nd). This therefore requires measurement of ^{140}Ce and ^{141}Pr during analysis (even though good column chemistry should have eliminated these interfering elements, see below). Harvey and Baxter (2009) ignored the ^{140}Ce correction as it has negligible effect on the crucial measurement of ^{143}Nd/^{144}Nd needed for traditional ^{147}Sm–^{143}Nd geochronology. Note, however, that failure to measure and correct for ^{140}Ce oxide interference can lead to spurious ^{142}Nd abundances; as petrochronologic applications have now arisen that may require ultrahigh precision on ^{142}Nd/^{144}Nd as a tracer of fractionated reservoirs in the early Earth (e.g., O'Neil et al. 2008), the analytical method would need to be modified to include measurement of ^{140}Ce. The oxide correction also requires an accurate and precise knowledge of the oxygen isotopic composition during analysis. Published "natural" oxygen isotope compositions are not sufficient; rather this must be carefully calibrated on the TIMS using monoisotopic REE solutions such as pure ^{141}Pr or ^{150}Nd spike (e.g., Baxter and Harvey 2009). Baxter and Harvey (2009) also showed that the oxygen isotope reservoir was large enough not to be fractionated during typical runs.

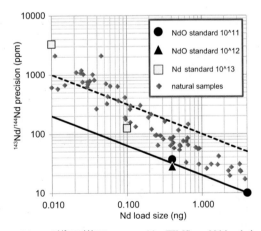

Figure 10. Analytical precision of ^{143}Nd/^{144}Nd measured by TIMS vs. Nd load size. Large symbols represent external precision (2RSD) of repeat loads of a pure Nd standard solution. Filled symbols are for samples run as NdO$^+$ following Baxter and Harvey (2009). Filled circles and triangle were run in dynamic analysis on an Isotopx Phoenix TIMS with 10^{11} and 10^{12} ohm amplifiers, respectively. Open symbols are for samples run as Nd-metal in static analysis with 10^{13} ohm amplifiers on a Triton TIMS (from Koornneef et al. 2014). Small diamond symbols represent internal analytical precision (2RSE) of natural samples (mostly from tiny garnets) from which Nd was extracted via column chemistry and run as NdO$^+$ with 10^{11} ohm amplifiers on a Triton TIMS following Baxter and Harvey (2009) by Kathryn Maneiro (Maneiro 2016). Solid line is a simple statistical projection of expected precision based on the 10 ppm precision of the 4 ng Nd standard solution. Dashed line is five times higher than the statistical projection of the standard.

Final Nd isotope ratios must be normalized to account for TIMS mass-dependent fractionation. The accepted normalizing ratio is $^{146}Nd/^{144}Nd = 0.7219$. Normalization should be done after oxygen isotope and interference corrections have been performed. As discussed above, ID-TIMS involves the addition of an enriched isotopic tracer or "spike" for precise measurements of Sm and Nd. Typically, ^{150}Nd is the Nd spike, and ^{147}Sm or ^{149}Sm is used for Sm spike. By using a well-calibrated mixed spike (e.g., ^{147}Sm and ^{150}Nd) added to and fully equilibrated with a dissolved sample before column chemistry, any inaccuracies in absolute concentration of Sm and Nd will perfectly correlate thus resulting in accurate Sm/Nd ratios. With spiked samples, normalization and spike subtraction (to achieve both accurate concentration and accurate $^{143}Nd/^{144}Nd$ ratios) must be done iteratively. Sm ID-TIMS analysis is done separately on its own filament; either a double/triple Re filament, or on a single Re or Ta filament run as the metal.

Recent years have witnessed attempts by TIMS manufacturers to utilize high resistance amplifiers which should theoretically boost signal to noise ratio for smaller beams, further improving precision for the smallest sample sizes (e.g., Koornneef et al. 2013, 2014; Tappa et al. 2016). However, these higher resistance amplifiers (10^{12} and 10^{13} ohm) may require a longer decay time during analysis leading ultimately to loss of much of the analyzable signal as compared to typical 10^{11} ohm resistors. While still few published data from higher ohm amplifiers exists, it remains unclear how much improvement will ultimately manifest from such amplifiers (Fig. 10). This is particularly true when considering the other limitations and sources of error (such as the blank, see below) that may dwarf TIMS analytical errors at small sample sizes when the potential benefits of 10^{12} or 10^{13} ohm amplifiers might be maximized.

Sample Preparation for Sm–Nd TIMS Petrochronology

Recall that all Sm–Nd geochronology is isochron geochronology. Thus, a whole rock or matrix sample (prepared using traditional methods of powdering and dissolution) is required to pair with the mineral of interest. In the next section we will address the challenges of extracting and preparing the specific mineral of interest for dissolution. Once any sample (a whole rock, matrix, or mineral powder) has been acquired, the next step is full dissolution. This is followed by addition of the mixed calibrated Sm–Nd spike solution. The sample is then ready for column chromatography. There are several methods for clean separation of Nd and Sm for analysis, each generally involving a pre-concentration step (or steps) to extract the REEs. For traditional Nd-metal analysis, most labs use LN-Spec resin for the final separation of Sm from Nd (e.g., Pin and Santos Zalduegui 1997). For NdO analysis, column chemistry must do a good job of also eliminating Pr from the Nd; LN-Spec resin generally does a poor job at that. So, methyl-lactic acid (also called alpha-hydroxy-isobutyric acid) is frequently used for final separation of all the REEs (e.g., DePaolo and Wasserburg 1976; Harvey and Baxter 2009). Care must be taken to calibrate these columns as they are pH and also temperature sensitive. Even with good column chemistry, interferences from Pr and Sm (on Nd) and from Gd (on Sm) should be monitored during TIMS analysis.

Garnet Petrochronology. Sm–Nd petrochronology is most widely applied to garnet because it is a common porphyroblastic mineral in a wide array of metamorphic rocks, rather than an accessory phase. The use of garnet as a monitor of tectonometamorphic evolution—via its well preserved concentric chemical, textural, and isotopic zonation—is well established after many decades of fundamental thermodynamic, geochemical, and analytical investments (see Baxter et al. 2017, this volume, for a review). Relatively speaking, the chronology of garnet is still being refined. Garnet petrochronology can be done either via bulk mineral analysis of whole grains, or via analysis of multiple growth zones within a single crystal via chemically contoured microsampling (see Pollington and Baxter 2011, Baxter and Scherer 2013, or Baxter et al. 2017, this volume for further review). We note that one of the very first pioneering studies of zoned garnet geochronology used the Rb–Sr system (Christensen et al. 1989) though subsequent work on garnets has moved much more towards Sm–Nd (via TIMS) and Lu–Hf (via MC-ICPMS). Lu–Hf geochronology is not discussed in this TIMS-focused chapter; for a full discussion of Lu–Hf garnet geochronology see Baxter et al. 2017, this volume and references therein. Sample preparation for bulk garnet analysis requires standard techniques which may include heavy liquids, magnetic separation, and handpicking to extract an optically pure garnet separate. Lapen et al. (2003) notes the possibility of unintentionally fractionating different growth zones in garnet during magnetic separation if magnetite or ilmenite inclusions occur heterogeneously. But, garnet petrochronology via TIMS is at its most powerful when pairing high precision, chemically contoured, concentric age zonation in single garnet crystals to the commonly zoned major element, trace element, isotopic, and textural information that can also be extracted from the same growth zones prior to their consumption during Sm–Nd analysis. Zoned garnet geochronology first requires a thin (1–2 mm) disc-shaped slice through the geometric center of a garnet from which a map of the zoned garnet can be produced to guide the sampling. The adjacent garnet surface may also be preserved for a thin section and to add petrologic and structural context. The map could be based on simple textures, colors, or concentric geometries visible with naked eye, or it may be a chemical map of the disc's surface acquired via electron microprobe or LA-ICPMS. In any case, once the map has been created, the petrochronologist's next task is to physically extract different portions of the garnet. Different methods have been developed over the years (stick it on tape and whack it with a hammer–Cohen et al. 1988; rectangular saw cuts–Christensen et al. 1989; cylindrical drill cores–Stowell et al. 2001) culminating in the use of a high spatial resolution MicroMill device either to extract microdrilled garnet powders (Ducea et al. 2003) or intact solid garnet annuli (Pollington and Baxter 2011). Once garnet separates have been prepared, they must then be cleaned of their mineral inclusions. Avoiding visible inclusions during micromilling and hand picking can go a long way but most of the time an additional "partial dissolution" step is required to eliminate microscopic inclusions (see Baxter et al. 2017, this volume for further discussion). Once the garnet has been properly cleaned, it is ready for full dissolution and column chemistry described above. Figure 11 shows the typical workflow for chemically contoured Sm–Nd garnet geochronology via TIMS.

Numerous studies have now paired zoned garnet geochronology with related petrologic, geochemical, or textural observations in the same garnets; the possibilities for future garnet petrochronology are vast. Figure 12 illustrates four examples of chemically contoured garnet microsampling conducted following the workflow of Figure 11. 5–6 cm diameter garnet crystals (rare as they may be) have produced concentric age information from up to 13 sampled annuli revealing not just when the garnet grew, and not just how long it grew, but how the rate of growth changed during that timespan. Figure 13 highlights examples of the

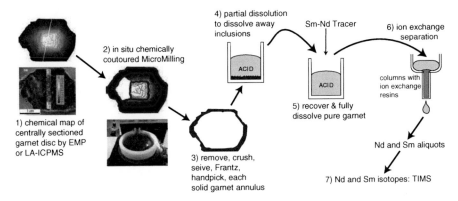

Figure 11. Workflow involved in chemically contoured Sm–Nd ID-TIMS garnet geochronology. Example garnet shown in #1,2,3 is from Gatewood et al. (2015). Micromill image is from Pollington and Baxter (2010).

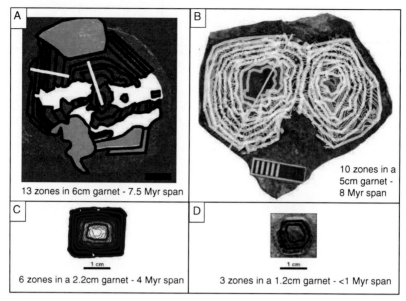

Figure 12. Four examples of chemically contoured microsampling for garnet geochronology via TIMS. Panels are shown at the same scale; 1 cm scale bars indicated. A. garnet from a chlorite schist in a local shear zone, Tauern Window, Austria (from Pollington and Baxter 2010); B. garnet from a felsic blueschist-facies rock from Sifnos, Greece (from Dragovic et al. 2015); C. garnet from a regional metamorphic schist at Townshend Dam, Vermont, USA (from Gatewood et al. 2015); D. garnet from a mafic blueschist from Sifnos, Greece (from Dragovic et al. 2012). Dark or colored concentric lines on each figure represent the drill trenches from which garnet powder was discarded. Instead, solid annuli between each trench was kept for analysis.

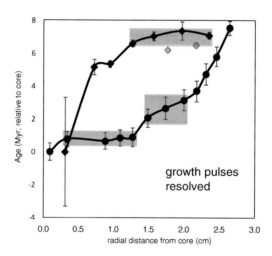

Figure 13. Resolving power of zoned garnet geochronology. Circles are data from Pollington and Baxter (2010) as pictured in Figure 12a. Diamonds are data from Dragovic et al. (2015) as pictured in Figure 12b. Note individual age precision is generally <±1 Myr. Two gray diamonds are age data not included in interpretations due to a likely artifact from sample preparation (see Dragovic et al. 2015 for discussion). Shaded bars highlight growth pulses resolved by the data. Note the 2-order of magnitude acceleration in volumetric growth rate of the garnet from Dragovic et al. (2015). The Pollington and Baxter (2010) garnet is consistent with a constant volumetric growth rate with the two rapid pulses superimposed.

geochronologic resolution that can be gained from this method for the two largest garnets. Pollington and Baxter (2010) showed geochronologic evidence for two rapid pulses of garnet growth that correlate with garnet chemical zonation; these pulses reflect evolving kinetic and thermodynamic drivers during the waning stages of the Alpine Orogeny 20–28 million years ago. Dragovic et al. (2015) showed that while this 5 cm garnet crystal from Sifnos, Greece grew over 8 million years, the majority of its growth happened in a final rapid pulse lasting just a few hundred thousand years. Using thermodynamic analysis of the garnet forming reaction, this pulse was linked to rapid dehydration from this lithology during subduction 45 million years ago. Smaller garnets 1–2 cm in diameter are also suitable for this approach. Gatewood et al. (2015) conducted zoned garnet chronology on ten different crystals from the same rock volume (one of these is picture in Fig. 12c) of a regional metamorphic schist from Townshend Dam Vermont. Comparison of Mn chemical zonation showed a correlation between Mn and age indicating first-order rock-wide chemical equilibrium for major elements in the rock volume. Dragovic et al. (2012) conducted zoned chronology on two crystals in a mafic blueschist from Sifnos Greece providing evidence again of a very rapid pulse of garnet growth and related dehydration. The reader is directed to these papers for a full accounting of the petrochronologic results, but we present them here as recent examples of what is already possible with chemically contoured garnet Sm–Nd petrochronology. Given available methodologies, chemically contoured garnet petrochronology on a single crystal requires a grain diameter of at least ~5 mm (Pollington and Baxter 2011), though smaller crystals may be drilled and lumped together in some cases. On the one hand, the scale of this "microsampling" seems enormous compared to the tiny zircons of U–Pb geochronology; however the sample size of Nd extracted from the garnet zones itself is still quite small due to the low concentration of Nd (<< 1 ppm) in most garnets. Fortunately, unlike accessory minerals, garnet crystals often grow large enough to permit zoned chronology and high resolution petrochronology.

Internal Sm–Nd Isochron Petrochronology. It is also theoretically possible to construct an isochron solely from the mineral to be dated, called an "internal mineral" isochron. This only works if the mineral exhibits some natural zonation in Sm/Nd, which allows a reasonable spread along the isochron and worthwhile age precision (e.g., Rakovan et al. 1997), and a physical means of separating and analyzing the different zones (e.g., Sjoqvist et al. 2016). Sector zoned minerals are good candidates for this approach as they can have easily recognizable zones with strongly variable chemistry, but also grew together at the same time (thus conforming to the fundamental isochron assumption). The elegance of an internal isochron is that it no longer requires the use of points on the isochron besides the mineral of interest, nor all of the (sometimes fraught) assumptions that must be made about the suitability of any second (or third, fourth, etc) point added to an isochron besides the mineral of interest. Rakovan et al. (1997) used sector zoned apatite and cut small chunks from different crystal faces to construct a 2 point isochron of 43.8 ± 4.7 Ma based on a spread in $^{147}Sm/^{144}Nd$ from 0.16 to 0.52. Somewhat analogously, Sjövqist et al. (2016) dated the REE ore mineral eudialyte from the Norra Kärr deposit, Sweden, via an internal isochron (Fig. 14). In this case sector zonation revealed much more subtle variations in $^{147}Sm/^{144}Nd$ ratio. Given the high Nd concentrations in the mineral (several thousand ppm) tiny 150 um diameter microdrilled pits provided enough material for precise analysis, and permitted careful spatial control to sample pits from different sector zones as guided by in situ LA-ICPMS mapping. Figure 14 shows the resulting five point internal isochron of 1040 ± 44 Ma based. The relatively poor precision is reflective of the small spread in $^{147}Sm/^{144}Nd$ from 0.141 to 0.186. The petrologic value of this age is that it directly constrains the primary ore mineral's formation even as evidence for complex multistage open system processes abounds. The age provides important context for the specific processes that led to ore formation within this complex system. While few minerals will be amenable to such internal isochron petrochronology, it is an elegant approach that can yield valuable petrochronologic constraints when possible.

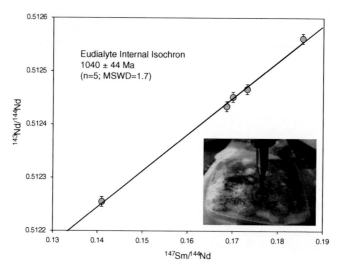

Figure 14. Eudialyte internal isochron. Each datapoint represents a different ~75 μm radius microdrilled pit within a single sector zoned eudialyte crystal. Inset shows in situ micromilling of the sample in progress. Data from Sjövqist et al. (2016).

Sm–Nd Age Precision

The precision of an Sm–Nd isochron age depends on the following factors: (1) Maximum spread in $^{147}Sm/^{144}Nd$ between points on the isochron; (2) Analytical precision of the $^{143}Nd/^{144}Nd$ data; (3) Analytical precision of the $^{147}Sm/^{144}Nd$ data; (4) Number of points on the isochron and the MSWD.

If we consider a simple two-point isochron between whole rock and garnet, in general, the garnet (or other high Sm/Nd phase) controls the precision and the age. The higher the garnet's $^{147}Sm/^{144}Nd$ the more precise the isochron age can potentially be. After that, the analytical precision of the $^{143}Nd/^{144}Nd$ and $^{147}Sm/^{144}Nd$ determine two-point isochron age precision, though their relative importance varies with the age of the garnet and the absolute $^{147}Sm/^{144}Nd$ ratio (Fig. 14). The $^{147}Sm/^{144}Nd$ precision matters most to age precision when the garnet is very old (i.e., Proterozoic to Archean) and has a higher $^{147}Sm/^{144}Nd$ ratio. For young garnet and with lower absolute $^{147}Sm/^{144}Nd$ ratio, it is the $^{143}Nd/^{144}Nd$ precision that matter most to age precision. Baxter and Scherer (2013) and Baxter et al. (2017, this volume) show how age precision varies as a function of $^{147}Sm/^{144}Nd$ ratio in garnet, and for different $^{143}Nd/^{144}Nd$ analytical precisions. Here, Figure 15 shows how age precision varies as a function of $^{147}Sm/^{144}Nd$ ratio in garnet, both for different $^{147}Sm/^{144}Nd$ analytical precision and for differently aged garnet. An important nuance to appreciate here, which differs in the way we typically think about U–Pb ages, is that the absolute precision of a garnet age generally does not increase proportionally with increasing age of the garnet, except when Sm/Nd ratio is highest and/or $^{147}Sm/^{144}Nd$ analytical precision is poorest. That is, whereas U–Pb geochronologists typically speak about % age uncertainty, a Sm–Nd geochronologist would speak about absolute uncertainty. For example, as shown in Figure 15B, all things being equal, if $^{147}Sm/^{144}Nd$ analytical precision is very good (e.g., 0.02%) one could date a 10 million year old garnet at ±1 Ma absolute precision (±10%) and also a 1000 million year old garnet at the same ±1 Ma absolute precision (±0.1%). The reason is the slow decay of ^{147}Sm, and the relatively low parent/daughter ratio, as compared to most minerals dated via U–Pb.

The major limitation in analytical precision, as already discussed, is sample size. The concentration of Nd in truly clean garnet is very low, usually between 0.01 and 1.0 ppm (Baxter et al. 2017, this volume). Thus, as the petrochronologist seeks to microsample smaller and smaller zones, smaller amounts of Nd (and Sm) will need to be analyzed. Or, we may be working with a different mineral rich in Nd (e.g., monazite, xenotime, eudialyte) but we may be limited in sample size simply by natural grain size or by the complexity and resolution of internal zonation we wish to microsample, for example via tiny MicroMill pits just 50–150 μm wide (e.g., Sjövqist et al. 2016; Fig. 14). Figure 10 provides a sense of the analytical precision one can expect from different load sizes of Nd. $^{147}Sm/^{144}Nd$ can be measured with external reproducibility no worse than 0.1% to 0.5% with advances towards 0.01% precision underway (e.g., Dragovic et al. 2015 reports 0.023% external precision on $^{147}Sm/^{144}Nd$ on repeat analyses of a mixed gravimetric Sm–Nd standard, spiked and run through columns). Only at sub-nanogram sample sizes does internal run precision of $^{147}Sm/^{144}Nd$ exceed the external precision and thus come into play. In practice, one should always use the greater (poorer) of the internal run precision or the long term external precision (based on repeat runs of standards of similar sample size) in age error propagation.

The final aspect of isochron age precision comes from additional points on the isochron and the degree to which they fit on the same isochron. The choice of additional points on an isochron is a delicate matter (see Baxter et al. 2017, this volume) so one must resist the urge to populate an "isochron" with as many points as possible just to meet some statistical criterion. Indeed the business of high-resolution petrochronology seeks to resolve real geological differences and subtleties in age, not smear them out and obfuscate them with the brush of statistics. Still, if multiple preparations of the same generation of garnet (or mineral) growth (as determined by microdrilling of similar age zones, or by chemical correlation, or textural arguments) can be collected, a multipoint isochron is always preferred. The MSWD is a statistical measure of the goodness of fit accounting for the scatter that would be expected given the reported analytical uncertainties (Wendt and Carl 1991). When the fit is good, an MSWD is close to 1.0 and age precision will improve as compared to a simple two-point isochron age precision. When fit is poor, an MSWD is $\gg 1$ and age precision can worsen as compared to a simple two-point isochron. As discussed in Baxter et al. (2017, this volume) and Kohn (2009) a poor MSWD should not be viewed as a failure; rather a high MSWD can be an indication that the samples on the isochron did not in fact grow at a single time, but rather that a more complex geochronologic story is there, waiting for a clever petrochronologist to dissect.

Another crucial variable that can affect the isochron age precision—and accuracy—is contamination from blank. As discussed for U–Pb geochronology, characterization and correction for the blank can often be the primary limitation to age precision. While blank corrections have always been a major concern for U–Pb dating, Sm–Nd geochronology has only recently pushed into small enough sample sizes where blank now becomes a major limitation. Blanks affect both the $^{143}Nd/^{144}Nd$ and the $^{147}Sm/^{144}Nd$ measurements. Thus, a blank correction should be made (and may become significant) whenever sample/blank ratio falls below ~1000:1. To accomplish this, a good constraint on the $^{147}Sm/^{144}Nd$ and $^{143}Nd/^{144}Nd$ of the blank is required. This can pose a major challenge given that good lab blanks are < 10 pg of Nd. The good news is that the $^{147}Sm/^{144}Nd$ and $^{143}Nd/^{144}Nd$ blank corrections are correlated, and generally pull each datapoint down along the isochron. This is because most blanks should approximate most common rocks, thus plotting near the whole rock point on the isochron for most samples. This problem is exacerbated for the oldest samples that deviate the most from modern crustal average values. Just as is common practice in U–Pb geochronology, rigorous statistical propagation of correlated errors in blank corrections are needed especially when sample/blank is $\ll 1000:1$. For Sm–Nd, especially given the somewhat more reagent intensive NdO column chemistry, the best

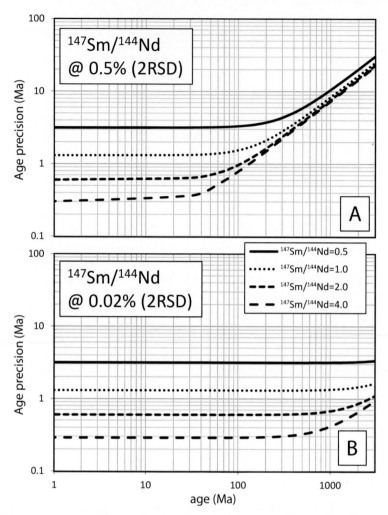

Figure 15. Sm–Nd age precision vs. age of sample. All calculations are for a two-point garnet–matrix Sm–Nd isochron where matrix has a $^{147}Sm/^{144}Nd = 0.15$, and the analytical precision for $^{143}Nd/^{144}Nd$ is 10 ppm (2RSD) for both datapoints. Each curve is for a different absolute $^{147}Sm/^{144}Nd$ of the garnet ranging from 0.5 (solid) to 4.0 (long dashed). Panel A is for a $^{147}Sm/^{144}Nd$ analytical precision of 0.5% (2RSD) whereas Panel B is for a $^{147}Sm/^{144}Nd$ analytical precision of 0.02%. Note that the importance of $^{147}Sm/^{144}Nd$ analytical precision is greatest at older ages.

Nd blanks are in the 1–10 pg range. Thus, as we dive into the sub-nanogram range of analysis, blank begins to represent the largest source of uncertainty, dwarfing the improvements that could be made by higher ohm resistors or new-and-improved loading solutions to further enhance ion yield (e.g., Baxter et al. 2014). The authors are not aware of any reports of labs having achieved total chemistry blanks below 1 pg of Nd; that will be an important frontier as we seek to push the barriers of petrochronologic age resolution further. In this regard, Sm–Nd geochronologists have much to learn from U–Pb geochronologists as the two fields have advanced at different paces.

THE FUTURE

As geochronologists continue to calibrate the rates of geologic processes at a finer and finer scale, there is an inevitable result that geologists and geochronologists are driven to ask questions that then require higher precision dates. This cyclic process is why TIMS geochronology continues to thrive despite its time-consuming and relatively cumbersome nature. While three decades ago it was sufficient to ask what the age of a pluton was, igneous petrologists now want to know over what timescales a pluton is constructed. While it was once common to classify a metamorphic "event", it is now more interesting, and also possible, to look at the rates of metamorphic reactions within an event that occurred over millions of years. As we learn more about these processes and about how they are recorded by minerals we can actually date, it becomes even more important to place geochronologic data within the context of petrogenesis. The challenges and opportunities that face ID-TIMS petrochronology outlined above must be met by sustained innovation, starting with sample collection in the field and ending with mass spectrometry and data reduction and analysis.

While this review focuses on U–Pb and Sm–Nd petrochronology, other systems including Rb–Sr and Re–Os are also amenable linking dates with rock forming processes, and numerous examples from these systems exist. Because the Rb–Sr system is useful in dating the growth or evolution of numerous rock forming minerals, this tool has been used to link such minerals to growth or deformation textures some of which may relate to larger scale tectonic or geochemical forcing (e.g., Christensen et al. 1989; Muller et al. 2000; Cliff and Meffan-Main 2003; Charlier et al. 2006; Glodny et al. 2008; de Meyer et al. 2014; Walker et al. 2016). Generally, in situ measurements of $^{87}Sr/^{86}Sr$ are more robust with MicroMill-TIMS (followed by column chemistry to remove Rb interference) than with LA-ICPMS due to the challenges of Rb and Kr interferences that can exist in LA-ICPMS analysis (though reaction-cell methods now allow direct in-situ LA-ICPMS Rb–Sr geochronology; e.g. Zack and Hogmalm 2016). Petrochronologic applications employing high resolution MicroMill sampling of individual growth zones in feldspars (e.g., Charlier et al. 2006) or polymineralic porphyroblast strain fringes (e.g., Muller et al. 2000) represent examples of what is possible with in situ MicroMill-TIMS methods and Rb–Sr. Re–Os geochronology has proven useful in calibrating the timescales of ore-forming processes through molybdenite geochronology (Selby and Creaser 2001; Stein et al. 2001; Bingen and Stein 2003), and continued interest in understanding these deposits and their links to igneous and hydrothermal systems will continue to motivate work in Re–Os petrochronology (Zimmerman et al. 2014).

It is unlikely that ID-TIMS geochronology will meet the spatial resolution of *in situ* techniques (though in situ MicroMill pits may be as small as ~50–100 μm diameter using fine tipped tungsten-carbide bits; Charlier et al. 2006), but it has yet to be seen whether *in situ* geochronology will match the precision and accuracy currently afforded by ID-TIMS. The goal in ID-TIMS geochronology, therefore, is to bridge the gap between spatial resolution and petrographic and geochemical context that is more easily attained by *in situ* techniques. Many of the tools to do this are in place but underutilized. Sample characterization via in situ techniques such as petrography, SIMS, LA-ICPMS and EPMA should be a rule rather than an exception for those wishing to understand high-precision geochronologic data and attach petrologic significance to it. Although ID-TIMS is rightly touted as the most precise and accurate technique for some geochronologic systems, increasing accuracy of reported ages remains dependent on correct data interpretation. Developing workflows that benefit most from multiple analytical and theoretical tools for interpreting geochronologic data in terms of petrogenesis will therefore become even more important in the future.

ACKNOWLEDGMENTS

EFB gratefully acknowledges support from NSF grants EAR-1250497/1561882 and PIRE-1545903. EFB thanks Kathryn Maneiro for sharing otherwise unpublished data from her PhD thesis at Boston University, and for her pioneering efforts in sub-ng Nd analysis. BS would like to thank Kyle Samperton and Michael Eddy for feedback on the manuscript, and Kyle Samperton for help compiling data from the Princeton lab. The authors are grateful for the careful reviews from Urs Schaltegger and Randy Parrish, in addition to constructive feedback, patience, and editorial handling by Matt Kohn.

REFERENCES

Amelin Y (2004) Sm–Nd systematics of zircon. Chem Geol 211:375–387

Amelin Y, Davis WJ (2006) Isotopic analysis of lead in sub-nanogram quantities by TIMS using a $^{202}Pb–^{205}Pb$ spike. J Anal At Spectrom 21:1053–1061

Amelin Y, Lee D-C, Halliday AN, Pidgeon RT (1999) Nature of the Earth's earliest crust from hafnium isotopes in single detrital zircons. Nature 399:252–255

Barboni M, Schoene B (2014) Short eruption window revealed by absolute crystal growth rates in a granitic magma. Nat Geosci 7:524–528

Barboni M, Boehnke P, Keller CB, Kohl I, Schoene B, Young ED, McKeegan KD (2017) Early formation of the moon 4.51 billion years ago. SciAdv 3.1:e1602365

Baxter EF, Scherer EE (2013) Garnet geochronology: timekeeper of tectonometamorphic processes. Elements 9:433–438

Baxter EF, Honn DK, Sullivan NS, Eccles KA (2014) Sub-nanogram Nd isotope analysis via TIMS: Magic potions, fancy resistors, but don't forget the blank. Goldschmidt Meeting, Sacramento CA

Baxter EF, Caddick MJ, Dragovic B (2017) Garnet: A rock–forming mineral petrochronometer. Rev Mineral Geochem 83:469–533

Bell K, Anglin CD, Franklin JM (1989) Sm–Nd and Rb–Sr isotope systematics of scheelites: Possible implications for the age and genesis of vein-hosted gold deposits. Geology 17:500–504

Bingen B, Stein H (2003) Molybdenite Re–Os dating of biotite dehydration melting in the Rogaland high-temperature granulites, S. Norway. Earth Planet Sci Lett 208:181–195

Blackburn T, Bowring S, Schoene B, Mahan K, Dudas F (2011) U–Pb thermochronology: creating a temporal record of lithosphere thermal evolution. Contrib Mineral Petrol 162:479–500

Broderick C, Wotzlaw JF, Frick DA, Gerdes A, Ulianov A, Günther D, Schaltegger U (2015) Linking the thermal evolution and emplacement history of an upper-crustal pluton to its lower-crustal roots using zircon geochronology and geochemistry (southern Adamello batholith N. Italy). Contrib Mineral Petrol 170:1–17

Carlson RW (2014) 15.18 - Thermal Ionization Mass Spectrometry A2. *In:*Treatise on Geochemistry (Second Edition) Holland, Heinrich D, Turekian KK (ed.) Oxford, Elsevier, p. 337–354

Chamberlain KR, Bowring SA (2001) Apatite–feldspar U–Pb thermochronometer: a reliable mid-range (~450 °C), diffusion controlled system. Chem Geol 172:173–200

Charlier BLA, Ginibre C,Morgan D,Nowell GM, Pearson,DG, Davidson JP, Ottley CJ (2006) Methods for the microsampling and high-precision analysis of strontium and rubidium isotopes at single crystal scale for petrological and geochronological applications. Chem Geol 232:114–133

Chelle-Michou C, Chiaradia M, Ovtcharova M, Ulianov A, Wotzlaw J-F (2014) Zircon petrochronology reveals the temporal link between porphyry systems and the magmatic evolution of their hidden plutonic roots (the Eocene Coroccohuayco deposit, Peru). Lithos 198:129–140

Chesley JT, Halliday AN, Scrivener RC (1991) Samarium–neodymium direct dating of fluorite mineralization. Science 252:949–951

Christensen JN, Rosenfeld JL, DePaolo DJ (1989) Rates of tectonometamorphic processes from rubidium and strontium isotopes in garnet. Science 244:1465–1469

Chu ZY, Chen FK, Yang YH, Guo JH (2009) Precise determination of Sm, Nd concentrations and Nd isotopic compositions at the nanogram level in geological samples by thermal ionization mass spectrometry. J Anal At Spectrom 24:1534–1544

Cliff RA, Meffan-Main S (2003) Evidence from Rb–Sr microsampling geochronology for the timing of Alpine deformation in the Sonnblick Dome, SETauern Window, Austria. Geol Soc Spec Publ 220:159–172

Cohen AS, O'Nions RK, Siegenthaler R, Griffin WL (1988) Chronology of the pressure–temperature history recorded by a granulite terrain. Contrib Mineral Petrol 98:303–311

Compston W, Oversby VM (1969) Lead isotopic analysis using a double spike. J Geophys Res 74:4338–4348

Condon D, Schoene B, McLean N, Bowring S, Parrish R (2015) Metrology and traceability of U–Pb isotope dilution geochronology (EARTHTIMETracer Calibration Part I). Geochim Cosmochim Acta 164:464–480

Corfu F (2013) A century of U–Pb geochronology: The long quest towards concordance. Geol Soc Am Bull 125:33–47

Corrie SL, Kohn MJ (2007) Resolving the timing of orogenesis in the Western Blue Ridge, southern Appalachians, via in situ ID-TIMS monazite geochronology. Geology 35:627–630, DOI: 610.1130/G23601A23601

Crowley JL, Schmitz MD, Bowring SA, Williams ML, Karlstrom KE (2006) U–Pb and Hf isotopic analysis of zircon in lower crustal xenoliths from the Navajo volcanic field: 1.4 Ga mafic magmatism and metamorphism beneath the Colorado Plateau. Contrib Mineral Petrol 151:313–330, doi: 310.1007/s00410-00006-00061-z

Crowley JL, Schoene B, Bowring SA (2007) U–Pb dating of zircon in the Bishop Tuff at the millennial scale. Geology 35:1123–1126; doi: 1110.1130/G24017A

D'Abzac F-X, Davies JH, Wotzlaw J-F, Schaltegger U (2016) Hf isotope analysis of small zircon and baddeleyite grains by conventional multi collector-inductively coupled plasma-mass spectrometry. Chem Geol 433:12–23

Davis DW, Williams IS, Krogh TE (2003) Historical development of zircon geochronology. Rev Mineral 53:145–181

Deering CD, Keller B, Schoene B, Bachmann O, Beane R, Ovtcharova M (2016) Zircon record of the plutonic-volcanic connection and protracted rhyolite melt evolution. Geology 44:267–270

de Meyer CMC, Baumgartner LP, Beard BL, Johnson CM (2014) Rb–Sr ages from phengite inclusions in garnets from high pressure rocks of the Swiss Western Alps. Earth Planet Sci Lett 395:205–216

DePaolo DJ, Wasserburg GJ (1976) Nd isotopic variations and petrogenetic models. Geophys Res Lett 3:249–252

DesOrmeau JW, Gordon SM, Little TA, Bowring SA (2014) Tracking the exhumation of a Pliocene (U) HP terrane: U–Pb and trace-element constraints from zircon, D'Entrecasteaux Islands, Papua New Guinea. Geochem Geophys Geosystems 15:3945–3964

DesOrmeau JW, Gordon SM, Kylander-Clark AR C, Hacker BR, Bowring SA, Schoene B, Samperton KM (2015) Insights into (U)HP metamorphism of the Western Gneiss Region, Norway: A high-spatial resolution and high-precision zircon study. Chem Geol 414:138–155

Dragovic B, Samanta LM, Baxter EF, Selverstone J (2012) Using garnet to constrain the duration and rate of water-releasing metamorphic reactions during subduction: An example from Sifnos, Greece. Chem Geol 314–317:9–22

Dragovic B, Baxter EF and Caddick MJ (2015) Pulsed dehydration and garnet growth during subduction revealed by zoned garnet geochronology and thermodynamic modeling, Sifnos, Greece. Earth Planet Sci Lett 413:111–122

Ducea MN, Ganguly J, Rosenberg EJ, Patchett PJ, Cheng WJ and Isachsen C (2003) Sm–Nd dating of spatially controlled domains of garnet single crystals: a new method of high-temperature thermochronology. Earth Planet Sci Lett 213:31–42

Fisher CM, Vervoort JD, DuFrane SA (2014) Accurate Hf isotope determinations of complex zircons using the "laser ablation split stream" method. Geochem Geophys Geosystems 15:121–139

Galer S, Abouchami W (1998) Practical application of lead triple spiking for correction of instrumental mass discrimination. Mineral Mag A 62:491–492

Gatewood MP, Dragovic B, Stowell HH, Baxter EF, Hirsch DM, Bloom R (2015) Evaluating chemical equilibrium in metamorphic rocks using major element and Sm–Nd isotopic age zoning in garnet, Townshend Dam, Vermont, USA. Chem Geol 401:151–168

Geisler T, Pidgeon RT, Kurtz R, van Bronswijk W, Schleicher H (2003) Experimental hydrothermal alteration of partially metamict zircon. Am Mineral 88:1496–1513

Gerstenberger H, Haase G (1997) A highly effective emitter substance for mass spectrometric Pb isotope ratio determinations. Chem Geol 136:309–312

Glodny J, Kühn A, Austrheim H (2008) Geochronology of fluid-induced eclogite and amphibolite facies metamorphic reactions in a subduction—collision system, Bergen Arcs, Norway. Contrib Mineral Petrol, 156:27–48

Gordon SM, Bowring SA, Whitney DL, Miller RB, McLean N (2010) Time scales of metamorphism, deformation, crustal melting in a continental arc, North Cascades, USA. Geol Soc Am Bull: B30060-1

Griffin WL and Brueckner HK (1980) Caledonian Sm–Nd ages and a crustal origin for Norwegian eclogites. Nature 285:319–321

Griselin M, van Belle JC, Pomies C, Vroon PZ, van Soest MC, Davies GR (2001) An improved chromatographic separation of Nd with application to NdO+ isotope analysis. Chem Geol 172:347–359

Harvey J, Baxter EF (2009) An improved method for TIMS high precision neodymium isotope analysis of very small aliquots (1–10ng). Chem Geol 258:251–257

Heaman LM, Bowins R, Crocket J (1990) The chemical composition of igneous zircon suites: implications for geochemical tracer studies. Geochim Cosmochim Acta 54:1597–1607

Henjes-Kunst F, Prochaska,W, Niedermayr A, Sullivan N, Baxter E (2014) Sm–Nd dating of hydrothermal carbonate formation: the case of the Breitenau magnesite deposit (Styria, Austria). Chem Geol 387:184–201

Horstwood MSA, Košler J, Gehrels G, Jackson SE, McLean NM, Paton C, Pearson NJ, Sircombe K, Sylvester P, Vermeesch P, Bowring JF, Condon DJ, Schoene B (2016) Community-derived standards for LA-ICP-MSU-Th–Pb geochronology—uncertainty propagation, age interpretation and data reporting. Geostand Geoanal Res, doi: 10.1111/j.1751-908X2016.00379.x

Iizuka T, Nebel O, McCulloch MT (2011) Tracing the provenance and recrystallization processes of the Earth's oldest detritus at Mt. Narryer and Jack Hills, Western Australia: An in situ Sm–Nd isotopic study of monazite. Earth Planet Sci Lett 308:350–358

Kohn MJ (2009) Models of garnet differential geochronology. Geochim Cosmochim Acta 73:170–182

Kohn MJ (2016) Metamorphic chronology—a tool for all ages: Past achievements and future prospects. Am Mineral 101:25–42

Kohn MJ, Corrie SL (2011) Preserved Zr-temperatures and U–Pb ages in high-grade metamorphic titanite: Evidence for a static hot channel in the Himalayan orogen. Earth Planet Sci Lett 311:136–143

Kohn MJ, Penniston–Dorland SC (2017) Diffusion: Obstacles and opportunities in petrochronology. Rev Mineral Geochem 83:103–152

Koornneef JM, Bouman C, Schwieters JB, Davies GR (2013) Use of 10^{12} ohm current amplifiers in Sr and Nd isotope analyses by TIMS for application to sub-nanogram samples. J Anal At Spectrom 28:749–754

Koornneef JM, Bouman C, Schwieters JB, Davies GR (2014) Measurement of small ion beams by thermal ionisation mass spectrometry using new 10^{13} ohm resistors. Anal Chim Acta 819:49–55

Kylander–Clark ARC (2017) Petrochronology by laser–ablation inductively coupled plasma mass spectrometry. Rev Mineral Geochem 83:183–198

Kylander-Clark ARC, Hacker BR, Cottle JM (2013) Laser-ablation split-stream ICP petrochronology. Chem Geol 345:99–112

Lanzirotti A, Hanson GN (1996) Geochronology and geochemistry of multiple generations of monazite from the Wepawaug Schist, Connecticut, USA: implications for monazite stability in metamorphic rocks. Contrib Mineral Petrol 125:332–340

Lapen TJ, Johnson CM, Baumgartner LP, Mahlen NJ, Beard BL, Amato JM (2003) Burial rates during prograde metamorphism of an ultra-high-pressure terrane: an example from Lago di Cignana, western Alps, Italy. Earth Planet Sci Lett 215:57–72

Lugmair GW, Marti K, Kurtz JP, Scheinin NB (1976) History and genesis of lunar troctolite 76535 or: how old is old? Proc 7th Lunar Sci Conf:2009–2033

Maneiro, KA (2016) Development of a Detrital Garnet Geochronometer and the Search for Earth's Oldest Garnet. PhD Thesis Boston University

Mattinson JM (2005) Zircon U–Pb chemical-abrasion ("CA-TIMS") method: combined annealing and multi-step dissolution analysis for improved precision and accuracy of zircon ages. Chem Geol 220:47–56

Mattinson JM (2011) Extending the Krogh legacy: development of the CATIMS method for zircon U–Pb geochronology. Can J Earth Sci 48:95–105

Matzel JP, Bowring SA, Miller RB (2006) Timescales of pluton construction at differing crustal levels: examples from the Mount Stuart and Tenpeak intrusions, North Cascades, WA: Geol Soc Am Bull 118:1412–1430 doi: 1410.1130/B25923.25921

McFarlane CRM, McCulloch MT (2007) Coupling of in-situ Sm–Nd systematics and U–Pb dating of monazite and allanite with applications to crustal evolution. Chem Geol 245:45–60

McLean NM, Bowring JF, Bowring SA (2011) An algorithm for U–Pb isotope dilution data reduction and uncertainty propagation. Geochem Geophys Geosystem 12:Q0AA18

McLean NM, Condon DJ, Schoene B, Bowring SA (2015) Evaluating uncertainties in the calibration of isotopic reference materials and multi-element isotopic tracers (EARTHTIME Tracer Calibration Part II). Geochim Cosmochim Acta 164:481–501

Miller JS, Matzel JP, Miller CF, Burgess SD, Miller RB (2007) Zircon growth and recycling during the assembly of large, composite arc plutons. J Volcanol Geotherm Res 167:282–299

Muller W, Aerden D, Halliday AN (2000) Isotopic dating of strain fringe increments: duration and rates of deformation in shear zones. Science 288:2195–2198

Mundil R, Ludwig KR, Metcalfe I, Renne PR (2004) Age and timing of the Permian mass extinctions: U/Pb dating of closed-system zircons. Science 305:1760–1763

Oberli F, Meier M, Berger A, Rosenberg CL, Giere R (2004) U–Th–Pb and ^{230}Th/^{238}U disequilibrium isotope systematics: Precise accessory mineral chronology and melt evolution tracing in the Alpine Bergell intrusion. Geochim Cosmochim Acta 68:2543–2560

O'Neil J, Carlson RW, Francis D, Stevenson RK (2008) Neodymium-142 evidence for Hadean mafic crust. Science 321:1828–1831

Parrish RR (1990) U–Pb dating of monazite and its application to geological problems. Can J Earth Sci 27:1431–1450

Peng J-T, Hu R-Z, Burnard PG (2003) Samarium–neodymium isotope systematics of hydrothermal calcites from the Xikuangshan antimony deposit (Hunan, China): the potential of calcite as a geochronometer. Chem Geol 200:129–136

Peterman EM, Mattinson JM, Hacker BR (2012) Multi-step TIMS and CA-TIMS monazite U–Pb geochronology. Chem Geol 312–313:58–73

Pin C, Santos Zalduegui JF (1997) Sequential separation of light-rare-earth elements, thorium and uranium by miniaturization extraction chromatography: application to isotopic analyses of silicate rocks. Anal Chim Acta 339:79–89

Pollington AD, Baxter EF (2010) High resolution Sm–Nd garnet geochronology reveals the uneven pace of tectonometamorphic processes. Earth Planet Sci Lett 293:63–71

Pollington AD, Baxter EF (2011) High precision microsampling and preparation of zoned garnet porphyroblasts for Sm–Nd geochronology. Chem Geol 281:270–282

Rakovan,J., McDaiel, D.K., Reeder R.J., 1997, Use of surface-controlled REE sectoral zoning in apatite from Llallagua, Bolivia, to determine a single-crystal Sm–Nd age. Earth Planet Sci Lett 146:329–336

Richard P, Shimizu N, Allegre CJ (1976) ^{143}Nd/^{146}Nd, a natural tracer: an application to oceanic basalts. Earth Planet Sci Lett 31:269–278

Rioux M, Bowring S, Dudás F, Hanson R (2010) Characterizing the U–Pb systematics of baddeleyite through chemical abrasion: application of multi-step digestion methods to baddeleyite geochronology. Contrib Mineral Petrol 160:777–801

Rivera TA, Storey M, Schmitz MD, Crowley JL (2013) Age intercalibration of ^{40}Ar/^{39}Ar sanidine and chemically distinct U/Pb zircon populations from the Alder Creek Rhyolite quaternary geochronology standard. Chem Geol 345:87–98

Rivera TA, Schmitz MD, Crowley JL, Storey M (2014) Rapid magma evolution constrained by zircon petrochronology and 40Ar/39Ar sanidine ages for the Huckleberry Ridge Tuff, Yellowstone, USA. Geology 42:643–646

Root DB, Hacker BR, Mattinson JM, Wooden JL (2004) Zircon geochronology and ca. 400 Ma exhumation of Norwegian ultrahigh-pressure rocks: an ion microprobe and chemical abrasion study. Earth Planet Sci Lett 228:325–341

Schaltegger U, Davies JHFL (2017) Petrochronology of zircon and baddeleyite in igneous rocks: Reconstructing magmatic processes at high temporal resolution. Rev Mineral Geochem 83:297–328

Samperton KM, Schoene B, Cottle JM, Brenhin Keller C, Crowley JL, Schmitz MD (2015) Magma emplacement, differentiation and cooling in the middle crust: Integrated zircon geochronological–geochemical constraints from the Bergell Intrusion, Central Alps. Chem Geol 417:322–340

Schaltegger U, Zeilinger G, Frank M, Burg JP (2002) Multiple mantle sources during island arc magmatism: U–Pb and Hf isotopic evidence from the Kohistan arc complex, Pakistan. Terra Nova 14:461–468

Schaltegger U, Brack PB, Ovtcharova M, Peytcheva I, Schoene B, Stracke A, Bargossi GM (2009) Zircon U, Pb, Th, Hf isotopes record up to 700 kyrs of magma fractionation and crystallization in a composite pluton (Adamello batholith, N. Italy). Earth Planet Sci Lett 286:208–218

Schärer U (1984) The effect of initial ^{230}Th disequilibrium on young U–Pb ages: the Makalu case, Himalaya. Earth Planet Sci Lett 67:191–204

Schmitt AK, Vazquez JA (2017) Secondary ionization mass spectrometry analysis in petrochronology. Rev Mineral Geochem 83:199–230

Schmitz MD, Schoene B (2007) Derivation of isotope ratios, errors, error correlations for U–Pb geochronology using ^{205}Pb-^{235}U-(^{233}U)-spiked isotope dilution thermal ionization mass spectrometric data. Geochem Geophys Geosystem 8:Q08006

Schoene B (2014) U–Th–Pb geochronology. *In:*Treatise on Geochemistry, Volume 4.10, Rudnick R (ed.) Oxford UK, Elsevier, p. 341–378

Schoene B, Crowley JL, Condon DC, Schmitz MD, Bowring SA (2006) Reassessing the uranium decay constants for geochronology using ID-TIMS U–Pb data. Geochim Cosmochim Acta 70:426–445

Schoene B, Latkoczy C, Schaltegger U, Gunther D (2010) A new method integrating high-precision U–Pb geochronology with zircon trace element analysis (U–Pb TIMS-TEA). Geochim Cosmochim Acta 74:7144–7159

Schoene B, Schaltegger U, Brack P, Latkoczy C, Stracke A, Günther D (2012) Rates of magma differentiation and emplacement in a ballooning pluton recorded by U–Pb TIMS-TEA, Adamello batholith, Italy. Earth Planet Sci Lett 355–356:162–173

Schönbächler M, Fehr MA (2014) 15.7—Basics of ion exchange chromatography for selected geological applications A2. *In:*Treatise on Geochemistry (Second Edition), Holland, Heinrich D, Turekian KK (eds.) Oxford, Elsevier, p. 123–146

Selby D, Creaser RA (2001) Re–Os geochronology and systematics in molybdenite from the Endako porphyry molybdenum deposit, British Columbia, Canada. Econ Geol 96:197–204

Sharma M, Wasserburg GJ, Papanastassiou DA, Quick JE, Sharkov EV, Laz'ko EE (1995) High ^{143}Nd/ ^{144}Nd in extremely depleted mantle rocks. Earth Planet Sci Lett 135:101–114

Sjöqvist ASL, Zack T, Baxter EF, Honn DK (2016) Post-magmatic implications for rare-earth element mineralisation from a microgeochemical in situ ID-TIMS Sm–Nd isochron from a single magmatic eudialyte crystal from the Norra Kärr alkaline complex. IGC Conference, Cape Town, South Africa

Smye AJ, Stockli DF (2014) Rutile U–Pb age depth profiling: A continuous record of lithospheric thermal evolution. Earth Planet Sci Lett 408:171–182

Stacey JC, Kramers JD (1975) Approximation of terrestrial lead isotope evolution by a two-stage model. Earth Planet Sci Lett 26:207–221

Stein H, Markey R, Morgan J, Hannah J, Scherstén A (2001) The remarkable Re–Os chronometer in molybdenite: how and why it works. Terra Nova 13:479–486

Stowell HH, Taylor DL, Tinkham DL, Goldberg SA and Ouderkirk KA (2001) Contact metamorphic *P–T–t* paths from Sm–Nd garnet ages, phase equilibria modelling and thermobarometry: Garnet Ledge, south-eastern Alaska, USA. J Metamorph Geol 19:645–660

Stracke A, Scherer EE, Reynolds BC (2014) 15.4 –Application of Isotope Dilution in Geochemistry A2 - Holland, *In:* Heinrich D, Turekian KK (eds) Treatise on Geochemistry (Second Edition): Oxford, Elsevier:71–86

Tappa MJ, Baxter EF, Maneiro KA, Guest RE (2016) Sub-nanogram neodymium isotope measurements on a New Isotopx Phoenix TIMS using 10^{11} and 10^{12} ohm resistors: GSA Fall Conference, Denver CO, USA

Thirlwall MF (1991) High-precision multicollector isotopic analysis of low levels of Nd as oxide. Chem Geol 94:13–22

Thirlwall MF (2000) Inter-laboratory and other errors in Pb isotope analyses investigated using a $^{207}Pb–^{204}Pb$ double spike. Chem Geol 163:299–322

Thoni M, Miller C, Blichert-Toft J, Whitehouse MJ, Konzett J, Zanetta A (2008) Timing of high-pressure metamorphism and exhumation of the eclogite type-locality (Kupplerbrunn–Prickler Halt, Saualpe, south-eastern Austria): constraints from correlations of the Sm–Nd, Lu–Hf, U–Pb and Rb–Sr isotopic systems. J Metamorph Geol 26:561–81, doi:10.1111/j.1525–1314.2008.00778.x

van Breemen O and Hawkesworth CJ (1980) Sm–Nd isotopic study of garnets and their metamorphic host rocks. Trans R Soc Edinburgh: Earth Sci 71:97–102

Walker S, Thirlwall MF, Strachan RA, Bird AF (2016) Evidence from Rb–Sr mineral ages for multiple orogenic events in the Caledonides of Shetland, Scotland. J Geol Soc London 173:489–503

Wasserburg GJ, Jacousen SB, DePaolo DJ, McCulloch MT, Wen T (1981) Precise determinations of Sm/Nd ratios, Sm and Nd isotopic abundances in standard solutions. Geochim Cosmochim Acta 45:2311–2323

Wendt I, Carl C (1991) The statistical distribution of the mean squared weighted deviation. Chem Geol 86:275–285

Wetherill GW (1956) Discordant uranium-lead ages. Trans Am Geophys Union 37:320–326

Williams IS (1998) U–Th–Pb geochronology by ion microprobe. *In:* Applications of Microanalytical Techniques to Understanding Mineralizing Processes. McKibben MA, Shanks WC III, Ridley WI (eds.) Volume 7, p. 1–35

Woodhead J, Hergt J, Shelley M, Eggins S, Kemp R (2004) Zircon Hf-isotope analysis with an excimer laser, depth profiling, ablation of complex geometries, concomitant age estimation. Chem Geol 209:121–135

Yuan H-L, Gao S, Dai M-N, Zong C-L, Günther D, Fontaine GH, Liu X-M, Diwu C (2008) Simultaneous determinations of U–Pb age, Hf isotopes and trace element compositions of zircon by excimer laser-ablation quadrupole and multiple-collector ICP-MS. Chem Geol 247:100–118

Zack T, Hogmalm KJ. Laser ablation Rb/Sr dating by online chemical separation of Rb and Sr in an oxygen-filled reaction cell. Chem Geol 437:120–133

Zimmerman A, Stein HJ, Morgan JW, Markey RJ, Watanabe Y (2014) Re–Os geochronology of the El Salvador porphyry Cu–Mo deposit, Chile: tracking analytical improvements in accuracy and precision over the past decade. Geochim Cosmochim Acta 131:13–32

Reviews in Mineralogy & Geochemistry
Vol. 83 pp. 261–295, 2017
Copyright © Mineralogical Society of America

9

Zircon: The Metamorphic Mineral

Daniela Rubatto

Institute of Geological Sciences
University of Bern
Baltzerstrasse 1–3
CH-3012, Switzerland
and
Research School of Earth Sciences
Australian National University
Acton 2601, ACT
Australia

daniela.rubatto@geo.unibe.ch

INTRODUCTION

A mineral that forms under conditions as variable as diagenesis to deep subduction, melt crystallization to low temperature alteration, and that retains information on time, temperature, trace element and isotopic signatures is bound to be a useful petrogenetic tool. The variety of conditions under which zircon forms and reacts during metamorphism is a great asset, but also a challenge as interpretation of any geochemical data obtained from zircon must be placed in pressure–temperature–deformation–fluid context. Under which condition and by which process zircon forms in metamorphic rocks remains a crucial question to answer for the correct interpretation of its precious geochemical information.

In the last 20 years there has been a dramatic evolution in the use of zircon in metamorphic petrology. With the advent of in situ dating techniques zircon became relevant as a mineral for age determinations in high-grade metamorphic rocks. Since then, there has been incredible progress in our understanding of metamorphic zircon with the documentation of growth and alteration textures, its capacity to protect mineral inclusions, zircon thermometry, trace element patterns and their relation to main mineral assemblages, solubility of zircon in melt and fluids, and isotopic systematics in single domains that go beyond U–Pb age determinations. Metamorphic zircon is no longer an impediment to precise geochronology of protolith rocks, but has become a truly indispensable mineral in reconstructing pressure–temperature–time–fluid-paths over a wide range of settings. An obvious consequence of its wide use, is the rapid increase of literature on metamorphic zircon and any attempt to summarize it can only be partial: in this chapter, reference to published works are intended as examples and not as a compilation.

This chapter approaches zircon as a metamorphic mineral reporting on its petrography and texture, deformation structure and mineral chemistry, including trace element and isotopic systematics. Linking this information together highlights the potential of zircon as a key mineral in petrochronology.

PREAMBLE: THE MANY FACES AND NAMES OF METAMORPHIC ZIRCON

Various terms have been used, more or less loosely, to describe features and processes that form zircon in metamorphic conditions. Different authors may have used the same term differently, causing additional confusion. Any classification has to be based on texture,

1529-6466/17/0083-0009$05.00 (print)
1943-2666/17/0083-0009$05.00 (online)
http://dx.doi.org/10.2138/rmg.2017.83.09

crystal structure and zircon composition and requires understanding of the formation process, which is not always the case in published studies. Within this contribution a generic terminology is adopted that uses three principal terms for metamorphic zircon: alteration, replacement/recrystallization and new growth. Examples of texture of these three categories are shown in Figure 1 and an attempt to summarize zircon features distinctive of particular metamorphic environments and processes is presented in Table 1.

Alteration is a process that partly overprints and disturbs a relict mineral and where textural and chemical vestiges of the relict are preserved. Alteration is often, but not necessarily, aided by prior metamictization as the damaged zircon domains are particularly prone to alteration by

Table 1. Summary of metamorphic zircon characteristics.

	Characteristic	Metamorphic conditions	Process and/or cause
Zoning	Regular polygonal zoning, oscillatory or sector, generally weak, and mostly euhedral external shape	Anatexis, granulite facies and hydrothermal conditions	Crystallization from a melt or precipitation from a fluid
	Patchy, mosaic zoning	Subsolidus	Metamictization, fluid alteratioin, initial stages of replacement
	Unzoned or weak convolute zoning	Subsolidus above greenschist facies	Replacement including dissolution-precipitation
	Sawtooth overgrowths	Diagenesis to low greenschist facies < 400°C	Dissolution-precipitation
	Intragrain crystallographic missorientation and formation of subgrains	Amphibolite to UHT	High strain rates and milonitization
Microstructure	Microzircon around major minerals (rutile, ilmenite, garnet)	Cooling from high temperatures	Expulsion of Zr during mineral breakdown or recrystallization
	Porosity and inclusions of Th and U phases	Subsolidus	Dissolution–precipitation
	Presence of non-formula elements (Ca, Al...)	From diagenesis to extreme conditions	Metamictization and fluid alteratioin
Anomalous composition	Pb nuggets	Ultra high temperature >900°C	Pb mobilization
	Low Th/U	Subsolidus to migmatites, less common in UHT and mafic compositions	Coexistence with Th-rich phase such as monazite or allanite
	Flat HREE pattern	Amphibolite, eclogite and granulite facies to extreme $P–T$	Coexistence with garnet
Chemistry	Strong LREE depletion and steep REE pattern	Amphibolite to granulite facies	Coexistence with abundant LREE-rich phases such as titanite, allanite, monazite
	Absence of negative Eu-anomaly	Eclogite facies (or assemblages lacking feldspar)	Lack of significant amount of feldspars in the assemblage
	Decoupling of U–Pb and Hf systematics	Subsolidus to granulite	Alteration and incomplete replacement

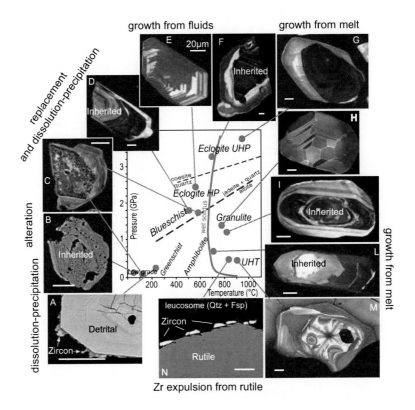

Figure 1. Typical internal zoning and textures of zircons from different metamorphic grades. Images in A, B and N are BSE, others are CL images. Horizontal scale bar in all images is 20 microns. Labels on the outside of images indicate the main process responsible for zircon growth or disturbance. The central diagram summarizes the main metamorphic facies. A) Zircon overgrowth on detrital core in greenschist facies shale [used by permission of Springer, license 3930810283470, from Rasmussen et al. (2005), Contributions to Mineralogy and Petrology, Vol. 150, Fig. 3k, p. 149]. B) Altered inherited zircon in digenetic sandstone [used by permission of John Wiley and Sons, license 3930080281612, from Hay and Dempster (2009), Sedimentology, Vol. 56, Fig. 4A, p. 2181]. C) Zircon with core altered during sea-floor alteration and rim formed during high-pressure metamorphism, same sample as described in Spandler et al. (2004). D) Zircon with inherited core and two metamorphic rims from eclogitic micaschist [used by permission of Nature Publishing Group, from Rubatto et al. (2011), Nature Geoscience, Vol. 4, Fig. 2a, p. 339]. E) Zircon formed in a fluid vein within eclogite [used by permission of Elsevier, from Rubatto and Hermann (2003), Geochimica et Cosmochimica Acta, Vol. 67, Fig. 4b, p. 2179]. F) Zircon with inherited core and two metamorphic rims from UHP whiteschist [used by permission of Springer, license 3930090407326, from Gauthiez-Putallaz et al. (2016), Contributions to Mineralogy and Petrology, Vol. 171, Fig. 3a, p. 15]. G) Zircon grown under UHP to granulite facies metamorphic conditions in a Kokchetav gneiss, courtesy of A. Stepanov. H) Fir-tree sector zoning in metamorphic zircon from eclogite [used by permission of Elsevier, license 3930810027220, Root et al. (2004), Earth and Planetary Science Letters, Vol. 228, Fig. 3a, p. 330]. In this case the process for zircon formation is unclear. I) Zircon from granulite with two metamorphic overgrowths around inherited core [used by permission of John Wiley and Sons, license 3930090711788, from Hermann and Rubatto (2003), Journal of Metamorphic Geology, Vol. 171, Fig. 3a, p. 15]. L) Zircon from low temperature migmatite with two metamorphic overgrowths around inherited core [used by permission of Springer, license 3930090864124, from Rubatto et al. (2009), Contributions to Mineralogy and Petrology, Vol. 158, Fig. 3l, p. 708]. M) Zircon from a leucocratic vein that records the age of UHT metamorphism [used by permission of Oxford University Press, license 930800811618, from Harley and Nandakumar (2014), Journal of Petrology, Vol. 55, Fig. 8b, p. 1978]. Note the feathered texture and a multiphase inclusion in the core. N) Microzircons around rutile grain formed by expulsion of Zr during recrystallization of rutile upon cooling from UHT metamorphism [used by permission of Springer, license 3930091093780, from Ewing et al. (2013), Contributions to Mineralogy and Petrology, Vol. 165, Fig. 5d, p. 766].

interaction with fluids. Alteration is the term used, for example, for old relict cores that have only partly lost their U–Pb and other isotopic signatures, and may have ghost or fuzzy zoning.

Replacement is an in situ process that changes the chemical composition of an existing domain, occurs at sub-solidus conditions and is commonly aided by fluids. Recrystallization is another term that is often used to describe this process. Replaced/recrystallized zircon domains show evidence of complete resetting of the chemical/isotopic system, a sharp boundary with inherited domains, and lack regular growth textures (shape or internal zoning), suggesting the domain formed in replacement of pre-existing zircon. One of the best investigated and widely recognized replacement processes that forms metamorphic zircon in sub-solidus conditions is in situ dissolution–precipitation (Geisler et al. 2003a, 2007; Tomaschek et al. 2003; Rubatto et al. 2008; Putnis 2009).

Overgrowth or new growth indicates a new crystal domain that shows sharp boundaries with any existing relict core and regular growth textures (shape or internal zoning). These domains have distinct chemical and isotopic compositions that are generally homogenous within a single domain. This type of zircon is commonly found in melt or fluid-rich systems and is caused by crystallization from a melt or precipitation from a fluid. Overgrowths on detrital zircon grains at very low grade would also be included in this category.

This terminology has to be taken as zircon specific, and may not apply to other minerals. While many zircons in metamorphic rocks can be described with these three categories, it is evident that individual cases exist that do not clearly fit a single category as multiple processes may affect the same zircon population, and the distinction between one and another may be blurred (Spandler et al. 2004; Tichomirowa et al. 2005; Zheng et al. 2005; Chen et al. 2011; Gao et al. 2015).

PETROGRAPHY OF ZIRCON

Textural relationships and inclusions

Metamorphic petrology is grounded on careful textural observations to establish mutual relationships between minerals. Defining which mineral is coexistent with others (paragenesis) based on textural equilibrium, inclusion relationships and composition is the backbone of metamorphic petrology. Textural criteria are often questionable when small accessory and refractive minerals like zircon are involved (Fig. 2). Small zircons are commonly included in larger grains, but this does not mean that the host and the inclusion are in equilibrium. A common observation under the optical microscope is the presence of inherited zircons included in key metamorphic minerals, that despite the apparent equilibrium texture with straight grain boundaries, have no petrological relationship with the host mineral (Fig. 2B). Zircon in gneisses, eclogitic metagabbros, and amphibolites that are included in metamorphic garnet, pyroxene or amphibole may be inherited and unrelated to metamorphism. Textural relationships between major minerals and zircon (and refractory accessory minerals in general) alone are not a robust criterion for age interpretation, particularly at low to medium metamorphic grade. A more robust link between the stability of major minerals and zircon can be established based on mineral inclusions in zircon (see below).

Petrography has proven a powerful and necessary tool to identify low-grade metamorphic zircon that overgrows detrital grains (Rasmussen 2005; Hay and Dempster 2009b). The small size and sawtooth-shape of these overgrowths is so characteristic that the texture alone is a strong evidence of metamorphic growth (Fig. 1A). Similarly, petrography is crucial in identifying micro-zircon that may form by exsolution or metamorphic reactions during cooling and breakdown of Zr-rich minerals. Examples are micro zircons in cordierite coronas around garnet (Fig. 2C, Degeling et al. 2001) or micro zircon around ilmenite and rutile (Bingen et

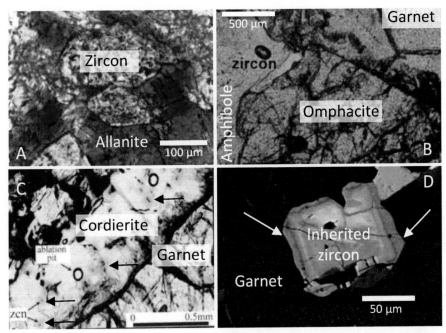

Figure 2. Textural relationships between zircon and other metamorphic minerals. A) Zircon in apparent textural equilibrium with allanite (brown) in an eclogite facies rock; zircon and monazite are both metamorphic but differ in age by 10 Ma [used by permission of Mineralogical Society of America, from Rubatto et al. (2007), American Mineralogist, Vol. 94, Fig. 1, p. 1521]. B) Inherited magmatic zircon included in amphibole in an eclogite. C) Micro-zircons (indicated by arrows) in the cordierite corona around garnet, which formed during decompression from granulite facies [used by permission of Mineralogical Society of Great Britain and Ireland, license 3930121477614, from Degeling et al. (2001), Mineralogical Magazine, Vol. 65, Fig. 2b, p. 752]. D) Zircon included in garnet in amphibolite; most of the zircon is inherited and only the thin rim (indicated by arrows) is metamorphic, same sample as described in Buick et al. (2006).

al. 2001; Ewing et al. 2013). Unfortunately the size of these zircons is commonly below the common spatial resolution of microbeam dating techniques (10–30 μm, Fig. 1N and 2C), but their presence still provides important information on the petrogenesis of the rocks.

Particularly important is the use of petrography to recognize and characterize mineral inclusions in zircon. These inclusions are not only valuable for relating zircon ages to metamorphic assemblages, but may also provide unique petrological information on the *P–T* evolution of the host sample. The most striking example is from high and ultra-high pressure rocks (UHP), were prograde to peak mineralogy is easily replaced during decompression. Coesite is a key indicator mineral for ultra-high pressure metamorphism, but a robust container, such as garnet or zircon, is often needed to preserve coesite in natural rocks. Additionally, during decompression of felsic UHP rocks at *T* > 700 °C, phengite melting occurs leading to a pervasive recrystallization of the main rock-forming minerals (Hermann et al. 2006c). For this reason, relicts of UHP metamorphism in felsic rocks are most commonly found in refractory minerals such as garnet or zircon. In the subducted continental rocks of the Kokchetav Metamorphic Complex, diamonds in gneisses and marbles are mainly found as inclusions in zircon and garnet (Shatsky and Sobolev 2003). The spectacular record of peak to retrograde inclusions in zircon has proven a key tool for age interpretation (Hermann et al. 2001; Katayama and Maruyama 2009). In the vast Dabie-Sulu orogen, the extent of the crust subducted to UHP conditions could only be demonstrated through the widespread occurrence of coesite inclusions in zircon from gneisses (Ye et al. 2000) that otherwise preserve little or no relict of UHP assemblages.

Petrography of mineral inclusions in zircon must however deal with the possibility of secondary inclusions (Fig. 3). Altered and metamict zones in inherited zircon cores have been proven to contain secondary inclusions, as for example high pressure minerals in "magmatic" zircon (Gebauer et al. 1997; Zhang et al. 2009a; Gauthiez-Putallaz et al. 2016). These secondary inclusions may be identified because of disturbed cathodoluminescence zoning, fractures, porosity and evidence of metamictization in the zircon around the inclusions. More controversial is the finding of white mica inclusions (Hopkins et al. 2008; Rasmussen et al. 2011) and carbon-phase inclusions (Menneken et al. 2007) in Early Archean zircons from the Jack Hills quartzite, where the debate is ongoing to what extent these inclusions are primary or secondary (see discussion in Harrison et al. 2017).

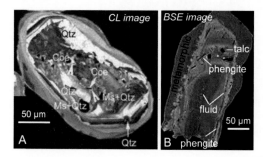

Figure 3. Secondary metamorphic inclusions in inherited zircon cores that underwent high pressure metamorphism. The cores also show disturbance of the original zoning likely due to alteration by metamorphic fluids. A) Zircon from micaschist recovered from the Chinese Continental Scientific Drilling Main Hole in the Sulu orogeny, China [used by permission of John Wiley and Sons, license 3930800210763, from Zhang et al. (2009a), Journal of Metamorphic Geology, Vol. 27, Fig. 3c, p. 321]. B) Zircon from a whiteschist of the Dora Maira unit, Western Alps, Italy [used by permission of Springer, license 3930130984661, from Gauthiez-Putallaz et al. (2016), Contributions to Mineralogy and Petrology, Vol. 171, Fig. 3b, p. 15].

Internal zoning

It is impossible to overstate the importance of characterizing internal zoning in metamorphic zircon: recognizing the presence of detrital cores, relict protolith magmatic zircon, and multiple domains formed during metamorphism is a fundamental step in the correct interpretation of any zircon geochemical information and eventually age. This is most commonly achieved with panchromatic cathodoluminescence (CL) or back-scattered-electron (BSE) imaging, which have become ubiquitous tools in zircon studies. Since the early applications to metamorphic zircon it became clear that CL and BSE images reveal internal structure otherwise invisible in light microscopy or etching (e.g., Hanchar and Miller 1993; Hanchar and Rudnick 1995; Vavra et al. 1996). CL and BSE emission are both proxies for chemical signals and are broadly anti-correlated because intrinsic CL attributed primarily to Dy is suppressed by the heavy element U, which increases the BSE signal (Rubatto and Gebauer 2000; Poller et al. 2001). Composition is however not the only player in CL emission, which is also controlled by structural parameters such as crystallinity or the presence of defect centers (Nasdala et al. 2002). The application of CL to metamorphic zircon has revolutionized its use in metamorphic petrology, making it easy to identify distinct growth zones and, to some degree, deformation features.

There have been a few attempts to categorize zoning (particularly CL zoning) of metamorphic zircon (Rubatto and Gebauer 2000; Corfu et al. 2003; Rubatto and Hermann 2007b), but more commonly every case has been presented in separate studies. The variety of textures in zircon is extremely wide, but some general systematics exist (Fig. 1, Table 1). It is commonly recognized that metamorphic zircon domains (including zircons forming in anatectic melts, which are considered metamorphic in this paper) have weak zoning when compared with the marked oscillatory and sector zoning of magmatic zircon (see a review in Rubatto and Gebauer 2000; Corfu et al. 2003). Zircon formed in sub-solidus conditions most commonly shows no regular zoning, having either a homogeneous CL/BSE emission or a cloudy and irregular zonation. However exceptions exist particularly for zircons attributed to fluid-related processes like in metamorphic veins or jadeitites. Weak internal zoning also characterizes metamorphic zircon that crystallized in high-grade rocks, likely from anatectic melts. In this case, weak oscillatory, sector, or fir-tree zoning is more common (Vavra et al. 1996; Schaltegger et al. 1999; Corfu et al. 2003; Claesson et al. 2016). High grade metamorphic zircon often displays a relatively low CL emission, possibly related to a high U content (see also section on Th–U composition), but the opposite has also been observed (Vavra et al. 1996; Corfu et al. 2003; Fu et al. 2008). Notably low-U zircon domains that are irregular in shape and form embayments into magmatic zircon, and thus look quite similar to metamorphic domains, are also formed during late-magmatic processes (Corfu et al. 2003). While CL, and to a lesser extent BSE imaging remains a powerful tool to recognize different growth domains, additional data are often required to determine the growth environment of metamorphic zircon.

Zircon imaging by more advanced techniques such as Electron Back Scattered Diffraction (EBSD), Transmission Electron Microscopy (TEM), and element mapping is more time consuming than CL and BSE imaging. These advanced techniques are applied to cases where deformation or particular chemical information is targeted (Reddy et al. 2008, 2010; Austrheim and Corfu 2009; Timms et al. 2011; Piazolo et al. 2012; Vonlanthen et al. 2012). It has been proposed that magmatic and metamorphic zircon can be distinguished based on Raman spectra (Xian et al. 2004), but this application remains limited and widely untested. The analytical approach for Raman identification has indeed been contested (Nasdala and Hanchar 2005) because it is based on a laser-induced photoluminescence peak. Given the possible difference in trace element composition between magmatic and metamorphic (i.e., lower in rare earth elements—REE—and Th) zircon, it is plausible that in some cases a distinction based on spectroscopy may work. The greatest use of Raman spectroscopy is in the documentation of completely to partly metamict zircon, where the amorphisation process leads to changes in the wavenumber and half-width of the Raman bands (Nasdala et al. 2003).

The superior spatial resolution of TEM analyses (McLaren et al. 1994; Hay and Dempster 2009b; Hay et al. 2010; Vonlanthen et al. 2012) and, more recently, atom probe (Valley et al. 2014) are promising investigative tools, but they require advanced sample preparation and are partly destructive. Their future application to recrystallization fronts and domain boundaries within metamorphic zircon may be particularly interesting for the understanding of processes of metamorphic zircon formation.

Deformation

There is no regional metamorphism without deformation and thus the effect of deformation on metamorphic zircon systematics must be taken into account. Evidence of crystal-plastic deformation of zircon crystals affecting composition and most importantly U–Pb systematics mainly come from high temperature shear zones (Reddy et al. 2006; Timms et al. 2006; Austrheim and Corfu 2009; Piazolo et al. 2012), although reports from unfoliated rocks also exist (Timms et al. 2011). Large zircon crystals (mm in size) in rocks that deformed under amphibolite- to granulite-facies conditions ($>700\,^{\circ}C$) display intragrain crystallographic misorientation of 2–20° at the crystal tips (Reddy et al. 2006, 2007; Timms et al. 2006; Piazolo et al. 2012). This miss-orientation correlates with panchromatic CL emission (reduced CL at the loci of low angle boundaries), REE composition (increase in total REE and of middle-REE—MREE—with respect to heavy-REE—HREE) or increase in Th/U (Timms et al. 2006; Piazolo et al. 2012). In some cases, deformation results in microfractures that define small subgrains that are misoriented by up to 10° and contain less Ti than the original crystal (Timms et al. 2011; Piazolo et al. 2012). Planar deformation features have been observed in zircon from pseudotachylytes (Austrheim and Corfu 2009). While these features are common in impact-related minerals, in this case they have been attributed to extreme strain rate during seismic deformation.

Studies agree that the dislocations and deformation features act as fast diffusion pathways for trace elements, U, Th and Pb. The creation of subgrains may also enhance chemical exchanges with any alteration fluid due to high surface area (Piazolo et al. 2012). How much this deformation disturbs the age is not always clear because its detection depends on the relative timing of crystallization versus deformation. Partial to complete resetting of U–Pb ages in the deformed domain is observed in some cases (Timms et al. 2011; Piazolo et al. 2012).

While full characterization of deformation features (best done by EBSD) and degree of chemical and isotopic resetting may not always be possible, panchromatic CL images can give a first hint on the presence of deformation. In all cases reported, a general correspondence between low-angle boundaries and low CL emission exists and such features should be a warning for any isotopic analysis. This correspondence is in agreement with observed recovery of CL emission by annealing of crystal defects (Nasdala et al. 2002).

MINERAL CHEMISTRY

A few chemical indicators have been commonly used in identifying metamorphic zircon and to create links between measured ages and metamorphic conditions (see also Table 1). Chemical criteria that can relate zircon composition to metamorphic assemblages are particularly useful for age interpretation. This however requires that the chemical (trace elements including Th and U) and isotopic (Pb) systems are equally robust. Experimental studies indicate that diffusion of the large divalent Pb^{2+} ion is comparable to that of trivalent REE and orders of magnitude faster than tetravalent ions like Th and U, which are essentially immobile under most geological conditions (Cherniak and Watson 2003). Additionally, radiogenic Pb is internally produced and may not be bonded in the crystal lattice in a structural site, and this might enhance its capacity to escape the crystal. Decoupling of U–Pb ages and element abundances has been reported for samples that have experienced high temperature (Kusiak et al. 2013a), metamictization or intense deformation (Reddy et al. 2006; Timms et al. 2006, 2011). Kaczmarek et al. (2008) reported zircon from deformed metagabbros that preserved magmatic REE patterns, but whose apparent ages varied between that of the protolith and of later metamorphism. Zircons from the Dabie-Sulu high pressure rocks commonly contain relicts of magmatic zircon, that may still have their original high Th/U or steep REE patterns, but whose U–Pb system has been reset to the age of metamorphism (Zheng et al. 2005; Xia et al. 2009). Studies of natural samples that suggest diffusional re-equilibration of REE, but not of Th and U, as predicted by diffusion experiments, are lacking.

Th/U systematics

The Th–U composition of zircon is routinely measured during dating and thus has become an easy-to-acquire and widely used criteria for zircon classification. The Th/U of metamorphic zircons is generally <0.1, but exceptions do exist (see below). This criteria was proposed based on the study of low temperature, high pressure zircons (Rubatto and Gebauer 2000), and it has been proven valid in countless cases. Exemplary are numerous studies of eclogite-facies zircon, and zircon in migmatites and granulites (Fig. 4). In general terms, the robustness of this simple chemical criteria appears to be independent of the process that led to zircon formation, from solid-sate replacement to crystallization from anatectic melts (Zhao et al. 2015). It is also important to note that the opposite is also true: most magmatic zircons have Th/U > 0.1 unless altered (e.g., Belousova et al. 2002; Grimes et al. 2015).

It has been demonstrated that metamorphic zircon does not always have low Th/U. The most occurrences of metamorphic zircon with Th/U > 0.1 are from high and ultra high temperature (>900 °C) samples [(Vavra et al. 1996; Schaltegger et al. 1999; Möller et al. 2003; Kelly and Harley 2005) see also a discussion in (Harley et al. 2007)]. The incorporation

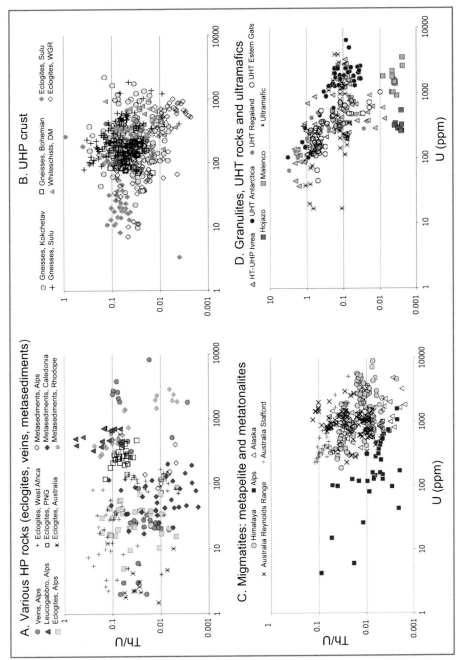

Figure 4. Th/U versus U content (in ppm) of metamorphic zircon from different tectonic settings. The compilation is based on circa 1400 published analyses (Vavra et al. 1996; Gebauer et al. 1997; Rubatto and Gebauer 2000; Hermann et al. 2001, 2006b; López Sánchez-Vizcaíno et al. 2001; Rubatto et al. 2001, 2006, 2008, 2009, 2013; Möller et al. 2002; Cesare et al. 2003; Hermann and Rubatto 2003; Rubatto and Hermann 2003; Root et al. 2004; Kelly and Harley 2005; Spandler et al. 2005; Bauer et al. 2007; Zhang et al. 2009b; Gasser et al. 2012; Gordon et al. 2012; Ewing et al. 2013; Korhonen et al. 2013; Kylander-Clark et al. 2013; Ganade de Araujo et al. 2014; Phillips et al. 2015; Rubatto and Angiboust 2015; Gauthiez-Putallaz et al. 2016; Stepanov et al. 2016b). See text for discussion.

of Th in zircon is primarily controlled by the availability of Th and U in the system and partitioning with other phases. The common presence in crustal metamorphic rocks of Th-rich phases such as monazite and allanite is an obvious reason for low Th/U in coexisting metamorphic zircon in eclogite, amphibolite and granulite facies rocks. The absence of these phases in some crustal rocks (either by melting under ultra-high temperature—UHT, or because of composition) should produce metamorphic zircon with high Th/U.

A compilation of Th/U versus U content of metamorphic zircon in different tectonic settings shows some interesting systematics (Fig. 4, ca. 1400 analyses). Data are grouped in 4 categories representing different metamorphic conditions.

(A) Relatively low temperature, high-pressure rocks of various compositions from mafic eclogites and leucogabbros, to micaschists and metamorphic veins (Rubatto and Gebauer 2000; Rubatto and Hermann 2003; Spandler et al. 2005; Bauer et al. 2007; Rubatto et al. 2008; Gordon et al. 2012; Ganade de Araujo et al. 2014; Phillips et al. 2015; Rubatto and Angiboust 2015). In all cases metamorphic temperatures are below the solidus and there is no evidence of melting in the rocks. According to the studies, most metamorphic zircon in these samples formed under HP conditions. In such "cold" eclogites Th/U of metamorphic zircon is mostly, but not restricted to <0.1, with values between 0.001 and 0.6. U contents are also variable from a few to 1000s of ppm, but mostly below 1000 ppm. The variability in zircon Th–U composition of HP rocks overall is large compared to any other category. The large range reflects the variety of rock types but also the lack of a dominant buffering phase. Monazite is not a common mineral in these rocks, but allanite is present in many samples. Another secondary effect that influences Th/U in these samples may be the temperature dependence of Th incorporation in zircon: the relatively larger Th^{+4} ion may fit proportionally less in a lower-T crystal structure than the smaller U^{+4} ion (Rubatto and Gebauer 2000).

(B) In mafic and felsic crustal rocks that underwent UHP conditions and thus higher temperatures of re-equilibration, metamorphic zircon Th–U composition is more restricted. Uranium ranges between 10–2500 ppm (at least in the selected samples) and Th/U is mainly below 0.2, with less than 10% of data (total analyses 524) higher than this value. As documented by Stepanov et al. (2016b) for some UHP–T gneisses of the Kokchetav metamorphic complex, the relatively high solubility of monazite in ultrahigh temperature melts (Stepanov et al. 2012; Stepanov et al. 2014) will allow zircon crystallization with high Th/U in some rocks. Indeed, in one Kokchetav sample it has been documented that prograde metamorphic zircon cores with low Ti-contents have Th/U<0.1, consistent with coexisting monazite, whereas peak metamorphic zircon domains with high Ti contents have Th/U of 0.4–0.6 and formed at 1000 °C, 5 GPa when monazite was completely dissolved in the partial melt. Zircon rims with low Ti that formed during retrograde crystallization of melts, when monazite is again present in the assemblage, show a low Th/U<0.1 (Stepanov et al. 2016a). Zircons in gneisses from the Bohemian Massif UHP unit have a restricted composition (Th/U 0.02–0.1, Kylander-Clark et al. 2013). UHP mafic eclogites from the Western Gneiss Region (Root et al. 2004; Kylander-Clark et al. 2013) and from the Dora Maira whiteschists (Gebauer et al. 1997; Gauthiez-Putallaz et al. 2016) also have zircon Th/U consistently below 0.1. Zircon rims from the Sulu UHP mafic and felsic rocks show higher values up to 0.4 (Zhang et al. 2009b).

(C) Zircon in migmatites, where a significant amount of leucosome is preserved, consistently has Th/U at 0.1 or below. This is independent of metamorphic temperature or pressure, from the 800 °C and 9 kbar of the Himalayan Higher Crystalline (Rubatto et al. 2013), to the water assisted melting at 650–700 °C, 5–13 kbar in the Central Alps (Rubatto et al. 2009) and Alaska Chugach complex (Gasser et al. 2012), including low pressure migmatites of central Australia. Most of these samples are metapelitic migmatites, where monazite is always an abundant accessory mineral in both paleosome and leucosome. In metatonalites from the central Alps, allanite is a nearly ubiquitous accessory (Rubatto et al. 2009). Notably, in the metapelites the U content of metamorphic zircon is also quite restricted, never below ~100 ppm, whereas it can be as low as 10 ppm in the migmatitic tonalites. In migmatites of intermediate composition where neither allanite nor monazite are stable then higher Th/U are expected (see an example in the Lewisian granulites of Norhern Scotland Whitehouse and Kemp 2010). The remarkably consistent Th–U composition of zircons in many migmatites may be also related to the presence of partial melts, which are a Th- and U-bearing phase. Experimental studies show that the relative partitioning of Th and U between zircon and granitic melt does not significantly change with T in the range 800–1050 °C (Rubatto and Hermann 2007a). In very oxidized environments, some of the U might occur as 5+ or even 6+ cation, which are significantly more incompatible than U^{4+} (Burnham and Berry 2012). As Th remains as 4+ cation this might potentially contribute to high Th/U in highly oxidizing environments.

(D) UHT rocks ($T > 900$ °C) are a distinct case in Th–U metamorphic zircon composition, as the majority of data plot above Th/U of 0.1 with values as high as 3. Samples include the pigeonite-bearing granulites from the Rogaland anorthosite complex (Möller et al. 2002), the saffirine-bearing orthogneiss and charnokite of the Napier Complex in Antarctica (Kelly and Harley 2005), enderbite and migmatitic gneisses of the Eastern Ghats belt in India (Korhonen et al. 2013), and metapelitic rocks of the lower crustal section of the Ivrea Zone (Vavra et al. 1996; Ewing et al. 2013). Note that some of the UHP samples from the Kokchetav massif plotted in category B also recorded T of 900–1000 °C and Th/U can be > 0.1. Lower crustal metapelites that did not reach temperatures > 850 °C and where monazite is stable are plotted in Figure 4D for comparison (Hojazo and Malenco, Cesare et al. 2003; Hermann and Rubatto 2003); these relatively lower T granulites have very low Th/U (0.001–0.01) and U content is above 100 ppm. Zircon from ultramafic rocks where there is no stable Th-phase are also included in this plot and indeed they show Th/U of 0.1–1. Examples are metamorphic zircons from a HP metapyroxenite (López Sánchez-Vizcaíno et al. 2001) and zircon from the Duria garnet peridotite (Hermann et al. 2006b).

Rare earth elements

Rare earth elements (REE) and particularly mid to heavy REE (M-HREE, Sm–Lu) can also be used for recognizing metamorphic zircons. The principle is based on partitioning with co-existing minerals that sequester M-HREE in the metamorphic assemblage (Fig. 5). Garnet is commonly a main host of HREE in medium to high-grade mafic to pelitic metamorphic rocks. Zircon that grows in a garnet-rich assemblage, where HREE are sequestered in garnet, will show a relatively flat HREE pattern compared to the HREE enrichment in magmatic zircon. This low HREE signature has been widely reported for metamorphic zircon in garnet-bearing eclogitic and granulitic rocks (Schaltegger et al. 1999; Rubatto 2002; Hermann and Rubatto 2003; Rubatto and Hermann 2003; Whitehouse and Platt 2003; Bingen et al. 2004; Gilotti et al. 2004; Hokada and Harley 2004; Root et al. 2004; Kelly and Harley 2005; Hermann et al. 2006a; Rubatto et al. 2006, 2013; Wu et al. 2008a; Wu et al. 2008b; Fornelli et al. 2014; Whitehouse et al. 2014; Gauthiez-Putallaz et al. 2016).

Similarly, Eu deficiency relative to other REE in zircon (i.e., negative Eu anomaly) has been attributed to the presence of feldspars that sequester Eu (Schaltegger et al. 1999; Rubatto 2002). The fact that zircon mainly incorporates Eu^{3+}, whereas feldspars take up mainly Eu^{2+} is a further complication (see also a discussion in Kohn 2016). Two main metamorphic conditions have been related to changing Eu anomaly in metamorphic zircon. (i) In eclogite facies assemblages, where albite breaks down to jadeite and quartz, zircon commonly has a weak or no Eu anomaly, at least in rocks that have no strong bulk Eu anomaly. As eclogitic assemblages also commonly yield garnet, the lack of a negative Eu-anomaly is coupled to a relatively flat HREE pattern (Fig. 5A) (Rubatto 2002; Rubatto and Hermann 2003; Baldwin et al. 2004; Gilotti et al. 2004; Wu et al. 2008a,b; Gauthiez-Putallaz et al. 2016) (ii) In migmatites, melting reactions involving micas produce

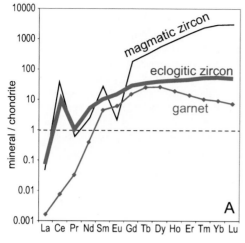

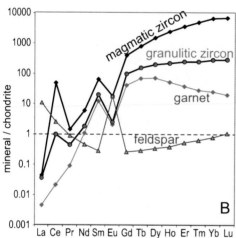

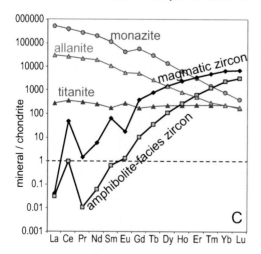

peritectic K-feldspar that incorporates all the available Eu^{2+}. Metamorphic zircon (and monazite) growing from anatectic melts acquires a stronger negative Eu anomaly relative to their protolith or sub-solidus counterpart (Fig. 5B, Schaltegger et al. 1999; Rubatto et al. 2006, 2013). Similarly, metamorphic zircon growing in an assemblage rich in L-MREE phases such as titanite, allanite or monazite can develop a particularly light-REE (La–Nd) depleted pattern (Fig. 5C). This has been observed for example in zircon from the amphibolite-facies migmatites of the Central Alps, which are rich in accessory allanite and titanite (Rubatto et al. 2009).

It is important to bear in mind that, as already stated for Th/U, these REE signatures are not absolute and depend on a number of factors: (i) bulk rock composition, e.g., in rocks strongly enriched in HREE, both garnet and metamorphic zircon will have relatively high HREE; zircon growing in a feldspar-free, HP assemblage may still have a negative Eu anomaly if the bulk rock is Eu-depleted; e.g., HP zircon in some Kokchetav gneisses (Hermann et al. 2001) and Dora Maira whiteschists (Gauthiez-Putallaz et al. 2016). (ii) The volume percent of the HREE or Eu controlling phase: rocks in which garnet is only a minor phase may still have metamorphic zircon with high HREE contents. (iii) The presence of other phases controlling the HREE or Eu budget, as for example abundant orthopyroxene that can accommodate significant HREE (Fornelli et al. 2014) and will increase the HREE depletion in granulite facies zircon.

Figure 5. Representative REE patterns of zircon types and other relevant minerals, normalized to Chondrite values. Magmatic zircon is plotted for reference in all diagrams. Relevant REE-bearing minerals coexisting with metamorphic zircon in eclogite (A), granulite (B) and amphibolite facies (C) assemblages are represented in the respective diagrams to illustrate the competition for REE and the consequent REE signature of metamorphic zircon in each assemblage. See text for discussion.

The use of HREE in linking metamorphic zircon to garnet in the co-existing assemblage can be exploited further if the equilibrium partitioning between zircon and garnet is known for different temperatures and garnet compositions. In samples with zoned garnet and multiple metamorphic zircon growth this could lead to identifying which specific garnet and zircon growth are in equilibrium (Hermann and Rubatto 2003). Experimental and empirical trace element zircon/garnet equilibrium partitioning vary over an order of magnitude, particularly for the HREE (Table 2) (Rubatto 2002; Hermann and Rubatto 2003; Rubatto and Hermann 2003, 2007a; Hokada and Harley 2004; Kelly and Harley 2005; Buick et al. 2006; Rubatto et al. 2006; Taylor et al. 2014). Element partitioning between two phases is firstly controlled by temperature, and secondarily by composition, whereas pressure is likely to have a negligible effect (Rubatto and Hermann 2007a). The published values cover metamorphic temperatures from 550–1000 °C for natural samples and 800–1000 °C for experiments, and variable garnet compositions from 0 to 8 wt% CaO (subset of studies for which the garnet composition is provided). In this wide range of conditions, variations are expected. Taking as an example Yb, the most abundant HREE in both minerals and thus relatively easy to measure, correlations between these parameters can be seen (Fig. 6). The zircon/garnet partition coefficient for Yb shows a negative correlation with T when the experimental studies are considered. This T-dependence is confirmed by some natural samples at 1000 °C that have a low zircon/garnet partition coefficient for Yb of 0.4–1.2, whereas granulites at ~800 °C have a Yb partition coefficient of 5–17. Both experimental studies and natural samples at these T are equilibrated with melts. Notably, the three samples that recorded lower metamorphic temperatures, where melt was not present, fall off the trend defined by melt-present samples/experiments. For the subset of studies that report garnet composition, most data also show a correlation between zircon/garnet Yb partitioning and grossular component in garnet. This highlights an additional complexity in the application of equilibrium partitioning as garnet major element composition varies widely, whereas zircon major composition remains constant. More systematic studies are needed to fully map out the effects of temperature and garnet composition on HREE partitioning.

In samples where garnet is zoned and zircon has multiple growth zones partition coefficients could be used to recognize equilibrium versus disequilibrium growth. In general, it is easier to detect when garnet and zircon are clearly not in equilibrium than when they potentially are. The partitioning can be strongly affected by lack of preservation of growth zones and original REE composition. Resorption of garnet or zircon during decompression or melting would impede correct partitioning determination. In rocks that reach relatively high temperatures, garnet commonly grows during the prograde evolution, whereas zircon is more likely to form during melt crystallization upon cooling, and thus the two phases are

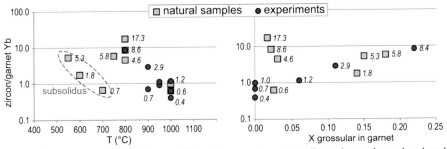

Figure 6. Zircon/garnet Yb partition coefficient from experimental studies and natural samples plotted against T (A) and garnet composition represented by molar proportion of grossular (B). The dotted ellipse in A groups natural samples that did not reach melting conditions. The number next to each symbol is the partition coefficient. Data are from Table 2 that also contains the references. See text for discussion.

Table 2. Zircon/garnet partition coefficient for middle to heavy REE obtained from natural rocks and experiments. The major element garnet composition for each sample is reported when available.

Rock type	Reference	Gd	Tb	Dy	Ho	Er	Tm	Yb	Lu	Y	T (°C)	X_{Grs}	X_{Alm}	X_{Prp}	X_{Spess}
HP vein	Rubatto and Hermann 2003	1.6	1.6	2.1	2.9	3.6	4.5	5.3	6.0	3.4	550	0.15	0.55	0.30	0.001
HP schist	Rubatto 2002	0.3	0.4	0.7	1.0	1.2	1.5	1.8	2.1	1.3	600	0.14	0.68	0.16	0.018
Gneiss	Buick et al. 2006	2.3	2.5	3.8	5.6	8.2	10.9		15.5		800	0.016	0.63	0.35	0.006
Granulite	Hermann and Rubatto 2003	0.8	1.0	1.3	1.7	2.6	3.9	5.8	8.2	2.1	750	0.18	0.63	0.18	0.008
Granulite	Rubatto 2002	1.6	1.8	2.6	4.3	7.1	11.1	17.3	23.9	4.8	800	0.017	0.81	0.16	0.015
Granulite	Rubatto 2002	0.9	0.9	1.3	2.0	3.2	5.2	8.6	12.1	2.3	800	0.023	0.82	0.13	0.026
Granulite	Rubatto et al. 2006	0.7	0.8	1.1	1.6	2.4	3.4	4.6	6.3	1.9	800	0.031	0.84	0.1	0.027
Granulite	Whitehouse and Platt 2003	0.5	0.7	0.8	0.8	1.0	0.7	0.7	0.8		700				
UHT gneiss	Hokada and Harley 2004	1.3	0.9	0.7	0.7	0.7	0.7	0.6	0.7		1000	0.024	0.52	0.46	0.003
UHT gneiss	Kelly and Harley 2005	0.7	0.7	0.7	0.8	0.9	0.9	1.0	1.2		1000				
Experiment	Rubatto and Hermann 2007	1.0		2.8		5.5		8.4	11.6	3.9	800	0.22	0.5	0.3	0.0
Experiment	Rubatto and Hermann 2007	1.0		1.3		2.1		2.9	3.5	1.8	900	0.11	0.52	0.37	0.0
Experiment	Rubatto and Hermann 2007	0.8		0.6		0.8		1.2	1.4	0.7	1000	0.06	0.47	0.47	0.0
Experiment	Taylor et al. 2014	0.6	0.5	0.5	0.5	0.5	0.6	0.7	0.8		900	0.0	0.53	0.47	0.0
Experiment	Taylor et al. 2014	0.7	0.6	0.6	0.6	0.7	0.8	1.0	1.2		950	0.0	0.56	0.44	0.0
Experiment	Taylor et al. 2014	0.7	0.6	0.6	0.6	0.7	0.8	0.9	1.1		950				
Experiment	Taylor et al. 2014	0.9	0.8	0.7	0.8	0.9	1.0	1.1	1.5		950				
Experiment	Taylor et al. 2014	0.7	0.5	0.5	0.4	0.4	0.4	0.4	0.5		1000	0.0	0.55	0.45	0.0
Experiment	Taylor et al. 2014	0.8	0.6	0.5	0.5	0.5	0.6	0.6	0.7		1000				

not necessarily in trace element equilibrium. An additional complication is diffusional re-equilibration, which can affect particularly the garnet composition: it has been shown that above 700–900 °C, diffusional equilibration of trace elements in garnet occurs over geological timescales, whereas zircon still preserves prograde growth stages (Stepanov et al. 2016b). As it is not trivial to detect resorption or diffusion in garnet and zircon, and textural coexistence is not a valid argument to prove chemical equilibrium, the use of HREE zircon/garnet partitioning for a direct link between metamorphic zircon and garnet growth zones must be carefully evaluated case by case. However, the general rule that metamorphic zircon has flat HREE in garnet-rich assemblages, and variable Eu depending on the presence of feldspar has been observed in numerous studies and remains a useful tool for relating metamorphic zircon to assemblages and ultimately metamorphic conditions.

Ti-in-zircon thermometry

The Ti-in-zircon thermometer is based on the principle that, in a buffered assemblage, the incorporation of Ti in zircon depends on T (Watson and Harrison 2005; Ferry and Watson 2007). One of the main attractions of this single mineral thermometer is that it can yield temperatures from isolated zircon crystals because in most crustal rocks the activity of SiO_2 is high and a Ti-phase is present. First applied to Early Archean detrital zircons of magmatic origin (Watson and Harrison 2005), the Ti-in-zircon thermometer has since proved a seemingly simple tool for petrology and its application has quickly spread from detrital to magmatic and metamorphic zircon. For temperature-dominated metamorphism the capacity to relate an age to a temperature in the P–T path is certainly appealing. Countless studies have applied this thermometer obtaining reasonable metamorphic temperatures, especially in upper amphibolite facies to lower granulite facies conditions, where zircon starts to be reactive and full buffering by quartz and a Ti-phase is achieved. Under lower and higher metamorphic grade, caution is needed in the application of the Ti-in-zircon thermometer as discussed below.

Metamorphic zircons that have formed under low temperature in sub-solidus conditions (<600 °C) have low Ti as predicted by the thermometer (<2 ppm, unpublished data, Tailby et al. 2011). Tailby et al. (2011) have shown that different sectors in a fir-tree zoned zircon from a vein containing rutile and quartz have variation in Ti contents that would correspond to variations of ~40 °C. This cannot be reconciled with T oscillations during growth and indicates that crystallographic sectors have some influence on Ti content rather than solely crystallization temperature. This effect may be particularly relevant at very low Ti contents, i.e., low temperature, but zircon is hard to react under such conditions and thus very few cases exist to test this issue. For example Ti-thermometry of diagenetic or low-grade zircon has never been achieved. At low Ti concentrations, any contamination during analysis from Ti-bearing micro-inclusions, neighboring phases or fractures could be serious (Harrison and Schmitt 2007). This demands particular cautions during analysis of Ti in potentially low temperature zircons. Additionally, it is much more difficult to prove that a buffer assemblage was reactive at low metamorphic temperatures.

On the other end, under extreme metamorphic temperatures, Ti content in zircon is higher and easier to measure (18–90 ppm at 800–1000 °C). The Ti-in-zircon thermometer can indeed record T of 900–1000 °C, as shown in some Kaapvaal xenoliths containing two stages of zircon growth (Baldwin et al. 2007). Extreme T is recorded by zircons from the Kokchetav diamondiferous rocks where Ti-in-zircon thermometry returns T of 910–1080 °C (Stepanov et al. 2016b), corroborating peak conditions of UHP metamorphism. In high grade rocks, which commonly undergo partial melting, a major issue can instead be that zircon crystallizes upon crystallization of melt during decompression and cooling and not at the T peak. For example, in the Anápolis-Itauçu Complex in central Brazil, samples with UHT assemblages that recorded metamorphic temperature of ~1000 °C, contain metamorphic zircon that record Ti-in-zircon temperatures mainly in the range 800–950 °C (Baldwin et al. 2007). Obtained T and textural relationships between zircon and other minerals were used to conclude that the zircon formed mainly during prograde and retrograde reactions and not at the UHT peak. The lower crustal section of the Ivrea Zone, Italy, contains metapelitic septa within gabbros that recorded metamorphic T of 900–1050 °C according to Zr-in-rutile thermometry (Ewing et al. 2013). In these rocks Ti-in-zircon thermometry records lower temperatures around 750–800 °C, likely because zircon crystallized only upon cooling. Ti-in-zircon temperatures below the peak T (700–850 °C versus ~900 °C) are also reported for metamorphic zircon in leucosomes of the Bohemian Massif (Kotkova and Harley 2010). Therefore, in the case of zircon crystallization from a partial melt, the T of Zr saturation and thus zircon crystallization can be significantly lower than the peak T experienced by the rocks, but also well below the T of melting (see details in thermodynamic modeling by Kelsey et al. 2008; Yakymchuk and Brown 2014; Kohn et al. 2015).

As any successful petrological tool, Ti-in-zircon thermometry has its limitations. Hofmann et al. (2009) pointed out possible problems of this thermometer due to non-equilibrium effects in Ti incorporation in zircon, the effect of other substituting elements, and the contamination of Ti analyses from edges and fractures (see also Harrison and Schmitt 2007; Hiess et al. 2008). Others have highlighted the issue of underestimation of crystallization temperatures in magmatic zircon (Fu et al. 2008). In applying Ti-in-zircon thermometry to metamorphic rocks there are other potential limitations to be considered. (1) Equilibrium with a Ti-phase is a prerequisite for the correct application of the thermometer (Ferry and Watson 2007). In metamorphic rocks where more than one metamorphic assemblage is preserved establishing the presence of a Ti-phase at the stage when zircon formed may not be trivial. Most commonly this leads to an underestimation of the real Ti-activity and thus the temperature of zircon crystallization. (2) The effect of pressure on Ti incorporation in zircon remains unconstrained. The Ti-in-zircon thermometer (Ferry and Watson 2007) was calibrated for pressures close to 10 kbar (Ferry and Watson 2007). Tailby et al. (2011) showed that Ti substitutes for Si at these pressures and proposed that at high pressure the solubility of Ti should decrease. As a result, the Ti-in-zircon thermometer may underestimate temperature. On the other hand, it is expected that under low pressure conditions (<5 kbar), the thermometer likely overestimates temperatures. Ferriss et al. (2008) suggested that with increasing pressure, Ti might additionally substitute into the Zr site, in turn increasing Ti solubility in zircon. Testing on natural samples is limited, but Ti-in-zircon temperature estimates for at least the UHP Kokchetav rocks are close to the peak T estimated by other thermometers (Stepanov et al. 2016b), indicating that, with pressure, decreasing amounts of Ti on the Si site might be compensated by increasing Ti concentration on the Zr site. Until a better understanding of the effect of pressure on this thermometer is gained, it is safer to use Ti-in-zircon temperatures in (U)HP rocks as a relative T indicator, as suggested by Stepanov et al. (2016b). (3) Deformation under high temperature can modify zircon chemical composition, generally inducing loss of Pb, REE and Ti or redistribution of elements (Reddy et al. 2009); application of the thermometer to highly deformed rocks has to be done with caution or supported by EBSD analysis and chemical mapping of the zircon crystals (see section on Deformation).

ISOTOPE SYSTEMATICS

Since the very early days of geochronology, zircon has been a prime target for U–Pb isotopic investigations in order to obtain crystallization ages. In the last decades Lu–Hf and oxygen isotope investigations in zircon domain have been developed and primarily applied to magmatic or detrital crystals. In metamorphic settings, the use of isotopic tracers to understand zircon petrogenesis and assist in U–Pb age interpretation is less widespread, but increasing. Correlations between different isotopic system (U–Pb, Lu–Hf and oxygen) and chemical signatures acquired from the same growth domain remain underexplored, but have already demonstrated some valid concepts that are summarized below. For any isotopic systematic it is important to consider diffusivity (see also Kohn and Penniston-Dorland 2017), robustness to metamorphic resetting/alteration and thus possible decoupling of different systems.

U–Pb isotopes

Zircon is the most widely used mineral for U–Pb age determination also in metamorphic rocks. U–Pb geochronology (Th is not sufficiently abundant in zircon to be a useful chronometer) in metamorphic zircon has increased dramatically since the development of micro-beam techniques that can measure U–Pb isotopic ratios at the 10–50 μm scale, namely high resolution ion microprobes and Laser Ablation ICP-MS (Kylander-Clark 2017; Schmitt and Vazquez 2017). Whereas these methods may not reach the sub-1% age precision of thermal ionization mass spectrometry (TIMS, Schoene and Baxter 2017) they can spatially resolve the internal growth zones typical of metamorphic zircon.

Zircon alteration and replacement under sub-solidus conditions are difficult to date accurately. In altered zircon domains the U–Pb system is disturbed and measured dates are commonly a mix between the age of the inherited grain and that of metamorphic disturbance. Several examples of these systematics exist in eclogite of the Dabie-Sulu orogenic belt, where altered zircon cores yield a range of spurious dates (Zheng et al. 2005; Xia et al. 2009; Chen et al. 2011; Sheng et al. 2012). Textures also suggests that alteration affects zircon at a micron scale below the spatial resolution of micro-beam geochronology (10–50 μm) and thus mixing of domains is a common problem in analyzing altered zircon. Alteration that produces spurious dates is not limited to sub-solidus conditions, but can also occur in higher grade samples (Tichomirowa et al. 2005). Metamictization of zircon domains commonly results in dark, mottled CL emission (Table 1) and zircons with these features are preserved in all metamorphic environment from diagenesis (Hay and Dempster 2009a) to high grade (Tichomirowa et al. 2005). Reversely discordant dates are commonly measured in metamict, altered domains where there has been Pb mobility (Tichomirowa et al. 2005; Kusiak et al. 2013a). Additional indications of zircon alteration are (i) high initial Pb contents and (ii) homogeneous Lu–Hf and oxygen isotopic composition in domains whose ages scatter (see sections on Lu–Hf and oxygen isotopes). Complete metamorphic replacement of zircon domains does achieve resetting of the U–Pb system: such domains (commonly zircon rims) have poor zoning, homogeneous chemistry and yield accurate and reproducible ages. In sub-solidus rocks, metamorphic zircon commonly has low U content and relatively high initial Pb content, which can limit age precision. Eclogite-facies rocks that were subducted along a cold-geotherm yield this type of zircon (Rubatto et al. 1999, 2011; Baldwin et al. 2004; Spandler et al. 2005). Replacement/recrystallization may proceed along a well defined front (Hoskin and Black 2000), but it can also be more localized and randomly distributed to generate sub-domains where CL-zoning is chaotic (Rubatto et al. 2008). Where pristine, altered and replaced domains are combined at a micron scale, age determination results in mixed ages. Most other chemical signals are analyses at the same scale as ages and thus will also be mixed.

Metamorphic zircon forms more readily in high-grade rocks where it commonly constitutes overgrowths on inherited magmatic or detrital cores. Chronology of migmatites and granulites that crystallized zircon from melts is relatively straightforward: metamorphic zircon domains are texturally and chemically distinct from the inherited cores and record faithfully the age of crystallization, at least below UHT conditions. Common Pb is generally not an issue in such zircon as Pb strongly partitions in the melt during crystallization. Reproducible and concordant dates are commonly achieved for zircon overgrowths and their chemistry can assist in age interpretation. However, even under conditions where Pb diffusion/mobility, strong deformation or later fluid alteration do not disturb the zircon formation age, it is possible to obtain a spread in zircon ages. This is not necessarily poor geochronology, as this spread can have a geological significance indicating a long-lasting process. Indeed, in several terranes it has been demonstrated that melting and high-grade metamorphism can persist for 10s of million years and thus zircon can form over a period of time. However, during protracted metamorphism the environment of zircon growth changes because of metamorphic reactions or new melt injections and thus zircon grows metamorphic domains that are distinct in chemistry and internal texture (Vavra et al. 1996; Hacker et al. 1998; Schaltegger et al. 1999; Rubatto et al. 2001, 2009, 2013; Möller et al. 2002; Hermann and Rubatto 2003; Montero et al. 2004; Root et al. 2004; Tichomirowa et al. 2005; Hermann et al. 2006a; Gerdes and Zeh 2009; Kotkova and Harley 2010; Gordon et al. 2012, 2013; Korhonen et al. 2013; Kylander-Clark et al. 2013; Harley and Nandakumar 2014; Young and Kylander-Clark 2015). In such samples, the presence of distinct metamorphic domains that are internally homogeneous, suggests that zircon grows in discrete, relatively brief episodes.

Extreme conditions with $T > 1000\,°C$ and P up to 50 kbar are documented in metamorphic rocks and such conditions might persist for long geological times (10s Ma). In order to understand ages obtained from zircon in UHT rocks it is thus important to discuss the robustness of the U–Pb system in zircon under these conditions. Experimental studies indicate that Pb diffusion in zircon is not significant at $T < 900\,°C$ even for geological times of several million years (see a review in Cherniak 2010). Metamorphism, however, can reach extreme temperatures up to $1050\,°C$, where even a perfectly crystalline zircon could diffuse Pb over geological time scales (Cherniak and Watson 2003). Inherited zircon (detrital or magmatic cores/grains) that still retain pre-metamorphic ages are preserved in various terranes that underwent extreme T for long periods of time. Examples are the Rogaland sapphirine-granulites (peak and decompression at $950–1000\,°C$ between 7.5 and 5.5 kbar, Möller et al. 2003; Drüppel et al. 2013), and the lower crustal rocks of the Ivrea Zone ($900–1000\,°C$, 6–10 kbar, Vavra et al. 1996; Ewing et al. 2013). Rare inherited cores with pre-metamorphic ages are also preserved in crustal rocks from the Kokchetav metamorphic belt in Kazakhstan, where peak T was around $1000\,°C$ at elevated P of 40–50 kbar. The existence of inherited zircons in some UHP–T samples imply that zircon was a stable phase even under extreme conditions and that U–Pb ages can at least partially survive such high T, even when associated with deformation and melting. In the Kokchetav belt, the duration of UHP–T metamorphism was only a few million years (Claoué-Long et al. 1991; Hermann et al. 2001; Katayama et al. 2001; Stepanov et al. 2016b). Zircons in some Kokchetav samples have dates scattering over 20 Ma that do not correlate with the zircon internal zoning texture or composition (Stepanov et al. 2016b). This observation was interpreted as partially reset during peak temperature of $1000\,°C$ in zircon that do not show any evident sign of metamictization. What it is hard to establish is if any age disturbance under these extreme metamorphic conditions is due to diffusional Pb loss or dissolution–precipitation of zircon in several stages during metamorphism.

A particular case of Pb mobility is documented in the Archean zircons of the Napier Complex, Antarctica, which underwent UHT metamorphism at $>900\,°C$. The discovery of reversely discordant ages ($^{206}Pb/^{238}U$ age older than the $^{207}Pb/^{206}Pb$ age) led to postulate the presence of unsupported radiogenic Pb in some zircon domains (Williams et al. 1984). More recent studies imaged in detail the distribution of Pb isotopes in these zircons and proved the patchy distribution of radiogenic Pb at the micron scale. Pb forms nuggets of Pb metal and its distribution is mostly unrelated to U content (Kusiak et al. 2013a; Kusiak et al. 2013b; Kusiak et al. 2015) and has no correlation with Th/U, REE or oxygen isotopes either. The mobility of radiogenic Pb caused apparent $^{207}Pb/^{206}Pb$ dates within a modified domain to vary over 500 Ma, with the oldest dates being spurious and older than the formation of the crystal. Kusiak et al. (2013a) proposed that extreme metamorphic temperatures in a dry environment caused Pb mobility within the crystal, without net Pb loss. This process of within-crystal Pb mobility has been also identified in the UHT rocks of southern India (Whitehouse et al. 2014). Such reports will likely increase with the development of atom probe analysis and ion microprobe mapping. Deformation under relatively high temperature has also been found to cause Pb mobility (see section on deformation).

Pb diffusion profiles in pristine zircon (not metamict, nor deformed) have so far never been measured, but increased spatial resolution in modern analytical techniques (atom probe, NanoSIMS analysis and SIMS depth profiling) may eventually resolve diffusion at a submicron scale. The bulk of available evidence indicates that loss of Pb by volume diffusion in non-metamict zircon is not an important factor even under extreme metamorphic conditions.

Lu–Hf isotopes

Hafnium isotopes in zircon have become a widely used petrogenetic tool in magmatic and sedimentary rocks for crustal evolution studies. For this isotopic system, the role of metamorphism is primarily a negative one, as metamorphism can cause Pb loss and thus compromise the veracity of calculated epsilon Hf values and model ages (see a review in Vervoort and Kemp 2016). The application of Hf isotopes to metamorphic zircon has been directed to understand zircon petrogenesis and resetting of the U–Pb system. Decoupling of the U–Pb and Lu–Hf systems in zircon during metamorphism and alteration has been documented in natural samples (Zheng et al. 2005; Gerdes and Zeh 2009; Xia et al. 2009; Zhao et al. 2015) and by experimental work (Lenting et al. 2010). While variable U–Pb dates can be the product of alteration/resetting and be unrelated to geological events (see above), distinct $^{176}Hf/^{177}Hf$ signature are expected if zircon domains formed during different episodes of the rock evolution (Gerdes and Zeh 2009). The plot in Figure 7A illustrates the different evolution in $^{176}Hf/^{177}Hf$ over time of components (bulk rock, magmatic zircon, metamorphic zircon, garnet) of a metagabbro over a 160 Ma period. Because of the high Hf contents, there is insignificant ingrowth of radiogenic Hf into the magmatic zircon. For other minerals and the bulk rock, radiogenic ingrowth depends on the Lu/Hf ratio. As most of the Hf is locked away in zircon, the "reactive" bulk rock Hf isotopes will evolve much faster than the bulk Lu/Hf. Let's consider a prograde metamorphic event producing garnet 100 Ma after the protolith crystallization, when magmatic zircon was not reactive. Given the high Lu/Hf of garnet, even within a few million years of a metamorphic cycle, garnet can evolve highly radiogenic $^{176}Hf/^{177}Hf$. Metamorphic zircon that forms at peak metamorphic conditions a few million year after prograde garnet can potentially acquire any $^{176}Hf/^{177}Hf$ in between the value of the protolith zircon and the highly radiogenic value of the garnet depending on which reservoir (protolith zircon, bulk rock, bulk without zircon, or garnet) the new zircon equilibrated with. Zircon that forms after dissolution of some protolith zircon and some garnet will acquire an intermediate $^{176}Hf/^{177}Hf$, higher than the protolith zircon. The same effect can be produced by dissolution of other high Lu/Hf phases such as apatite (Valley et al. 2010).

In metamorphic gneisses and mafic eclogites of the Dabie-Sulu HP terrane, inherited magmatic cores, whose ages were partly reset during metamorphism, still preserve high $^{176}Lu/^{177}Hf$ and Th/U relative to metamorphic zircon (Zheng et al. 2005; Xia et al. 2009; Gao et al. 2015). Even when metamorphic disturbance of the U–Pb age and Th/U system is nearly complete (spongy textures, modified CL, apparent ages close to lower intercept and low Th/U) the Lu/Hf system of the protolith zircon remains undisturbed (Xia et al. 2009). On the contrary, zircon grown during metamorphism has not only new U–Pb age and Th–U compositions but also consistently lower $^{176}Lu/^{177}Hf$ (2–10 times lower) than the protolith zircon value (Fig. 7B). Some of the zircon domains that crystallized during metamorphism also have higher $^{176}Lu/^{177}Hf$, which points to release of radiogenic Hf from other sources, most likely garnet. The low $^{176}Lu/^{177}Hf$ and high $^{176}Hf/^{177}Hf$ in some metamorphic zircon implies that, at the time of zircon growth, garnet was present as a Lu sink (see section on REE), but recrystallized releasing radiogenic $^{176}Hf/^{177}Hf$

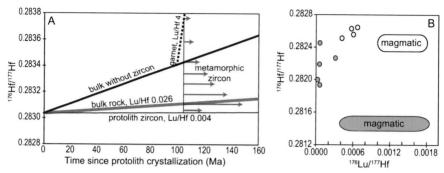

Figure 7. Lu–Hf systematics in metamorphic zircon. A) $^{176}Hf/^{177}Hf$ isotopic evolution over time for different components of a gabbroic rock that crystallized magmatic zircon at time 0 and metamorphic zircon 100 Ma later, in a garnet bearing assemblage. Because zircon is the main host of Hf in the rock, the evolution of the bulk rock where protolith zircon remains isotopically isolated (bulk without zircon) is significantly more radiogenic than that of the protolith zircon. Garnet that forms during prograde metamorphism has a much higher Lu/Hf than the bulk (4 versus 0.026) or the magmatic zircon (Lu/Hf 0.004) and rapidly increases its $^{176}Hf/^{177}Hf$ with time. The $^{176}Hf/^{177}Hf$ ratio of metamorphic zircon that forms a few Ma after garnet (peak to retrograde path) will depend on which component(s) of the system the zircon equilibrates with: only the magmatic zircon, the bulk rock fully equilibrated, the bulk rock without participation of the magmatic zircon, or even only the garnet. The grey arrows indicate the trajectory of new metamorphic zircon depending on its acquired $^{176}Hf/^{177}Hf$, and their variable length represents the likelihood of that composition occurring. Model based on data from Monviso eclogite described in Spandler et al. (2011). B) Difference in $^{176}Hf/^{177}Hf$ and $^{176}Lu/^{177}Hf$ systematics between magmatic protolith zircons (large fields) and new metamorphic zircon (small circles) in two HP samples from the Dabie orogeny. Data are from Zheng et al. (2005), white symbols, and Gao et al. (2015), grey symbols. Note the tendency of the metamorphic zircon to higher $^{176}Hf/^{177}Hf$ and lower $^{176}Lu/^{177}Hf$ due to the likely effect of garnet in the metamorphic assemblage.

(Zheng et al. 2005; Xia et al. 2009). Even more complicated $^{176}Lu/^{177}Hf$ versus $^{176}Hf/^{177}Hf$ trends can be created in metamorphic zircon if multiple Hf sources are identified, such as dissolution in anatectic melts of much older detrital zircon of various age and multiple metamorphic events overprinting each other (Gerdes and Zeh 2009; Zhao et al. 2015). In the process of U–Pb and Lu–Hf decoupling, there is a fundamental role for aqueous fluids and melts in redistributing and transporting these elements. The decoupling of the U–Pb and Lu–Hf systems can also be used to identify analyses mixing inherited cores and metamorphic rims, which produces a correlation between the two systems (Xia et al. 2009; Zhao et al. 2015).

Oxygen isotopes

Oxygen isotopes can be readily measured in single zircon growth zones by ion microprobe to a precision of 0.2‰ (Valley 2003). A common application of this isotopic system is tracing the source of zircon grains, particularly in magmas and sediments, to reconstruct crustal reworking. This strategy applies equally to inherited and detrital zircons in metamorphic rocks (Rumble et al. 2002; Zhao et al. 2008; Rubatto and Angiboust 2015), and is particularly powerful in recrystallized rocks where relic zircon cores may be the only remnant of the protolith (Zheng et al. 2008; Fu et al. 2010; Chen et al. 2011; Sheng et al. 2012; Gauthiez-Putallaz et al. 2016).

Similarly to the Lu–Hf system, oxygen isotopes can give insight into metamorphic replacement and modification of zircons, as well as partial resetting of the U–Pb system and mixed ages. Natural samples indicate that, at least in some environments, the U–Pb system is easier to reset than the oxygen isotopes and thus partly reset zircon domains may still preserve protolith oxygen isotopic composition (Wu et al. 2006; Martin et al. 2008). This may seem at odds with experimental data that report diffusion of oxygen isotopes faster than that of Pb (Cherniak and Watson 2003), but age resetting during metamorphism is generally not dominated by simple solid state diffusion. In other cases, metamorphic resetting of zircon age and Th/U weakly correlates with changes in $\delta^{18}O$ suggesting that metamorphism eventually affects oxygen isotopes (Petersson et al. 2015; Rubatto and Angiboust 2015). This difference may reflect different processes and degree of resetting. Therefore the use of oxygen isotopes to detect metamorphic disturbance of the age may have to be considered case by case.

While replacement/recrystallization can be an effective way to reset the oxygen isotopic composition of zircon, natural and experimental studies indicate that diffusion of oxygen does not play a significant role (Watson and Cherniak 1997; Page et al. 2007). Oxygen isotope diffusion in zircon is still debated (see also Kohn and Penniston-Dorland 2017) because of disagreement whether "dry" or "wet" diffusion applies in natural rocks and the general lack of data (i.e., measurable profiles in natural samples). The point has been made that even in apparently "dry" metamorphic rocks where free fluids may be lacking, the activity of H_2O is still buffered by mineral phases and significant, and thus "wet" diffusion generally applies (Kohn 1999). The potential retentivity of $^{18}O/^{16}O$ isotopic ratio in zircon has been modeled by Cherniak and Watson (2003) using the only experimentally determined wet diffusion data. Compared to Pb, U or other trace elements, oxygen diffusion in zircon is much faster so that oxygen isotope signature of a ~100 μm domain should survive less than 0.1 Ma at temperature of 700 °C, and only 5700 years at 900 °C. Natural cases, however, suggest higher retentivity of oxygen isotopes in zircon. Inherited, relict zircon cores with very low $\delta^{18}O$ that survived UHP metamorphism at ~750 °C are common in the Dabie-Sulu gneiss (Chen et al. 2011; Sheng et al. 2012). Bowman et al. (2011) have shown that even zircon that underwent relatively high-grade metamorphism for several 10s of million years preserve sharp $\delta^{18}O$ changes (within the 5–10 μm resolution of the analyses) between inherited cores (~6 ‰) and metamorphic rims (8–10‰). Claesson et al. (2016) suggest that the oxygen isotopic composition of non-metamict zircons that still preserve magmatic-type zoning were modified under high temperature (~700 °C) in hydrothermal conditions. They base their conclusion on the finding of unusually high $\delta^{18}O$ in Archean zircons and on previous experimental determination of relatively fast oxygen diffusion under wet conditions (Cherniak and Watson 2003). However, also in this case, no oxygen isotope diffusion profile could be measured in zircon. These examples show that oxygen isotope diffusion in zircon is still far from being resolved and requires future work.

Oxygen isotopic composition of zircon can assist in the challenging task to link zircon ages to metamorphic assemblages. Oxygen fractionation factors between zircon and some major minerals are reasonably well known (Valley 2003). For example, given known metamorphic temperatures, oxygen isotopic equilibrium between zircon and garnet can support a case for metamorphic zircon formation (Martin et al. 2006). Conversely, isotopic disequilibrium between zircon and major phases would suggest that inherited zircons are preserved within metamorphic assemblages, where major minerals are isotopically equilibrated (Zheng et al. 2003). Fu et al. (2010) demonstrated that the mantle-like $\delta^{18}O$ (4.8–5.2 ‰) of zircon and Ti-in-zircon thermometry (600–800 °C) in numerous jadeitites are not supportive of hydrothermal zircon formation in a vein, but rather of their oceanic crust protolith. The systematics of oxygen isotopes is more complicated during sub-solidus alteration and partial to complete replacement of zircon. Does a zircon formed by in situ dissolution–precipitation equilibrate with the bulk rock oxygen isotopic signature or inherit the $\delta^{18}O$ of the protolith zircon it replaces? Two examples of metamorphic zircon that did not re-equilibrate with the bulk oxygen signature are presented in Figure 8. In a metabasite from Naxos that experienced metamorphism at 500 °C, a number of metamorphic zircon rims yield variable $\delta^{18}O$ that in each grain is identical to the values of the magmatic core (Martin et al. 2006), suggesting that the rims inherited the core isotopic signature. In an eclogitic metasediment from the Sesia Zone in the Italian Alps (Rubatto et al. 1999), metamorphic zircon rims yield different $\delta^{18}O$ than the detrital cores, but still differing one from another. In this case it is postulated that the isotopic composition of the metamorphic zircon equilibrated only locally and not rock-wide. On the other hand, zircon overgrowths that formed in a metamorphic vein, hosted by the eclogitic metasediment, displays a homogenous oxygen isotope composition, as expected for newly formed zircon that equilibrated with the bulk rock.

Particularly insightful is the use of oxygen isotopes in zircon and co-existing minerals to trace metasomatism and open versus closed system metamorphism. In orthogneisses, significant shifts in $\delta^{18}O$ between relict magmatic cores and metamorphic rims have been interpreted as open system behavior and crystallization of zircons from metamorphic fluids (Chen et al. 2011; Sheng et al. 2012). Zircon within metamorphic veins in the Monviso eclogite, have significantly lower $\delta^{18}O$ than inherited magmatic zircon in the country rock. The zircon oxygen composition and age, together with the variable $\delta^{18}O$ in garnet, reveal deep sea floor alteration as well as metasomatism during high pressure metamorphism (Rubatto and Angiboust 2015). As this example shows, a more convincing argument for metasomatism can be made when the isotopic variation in metamorphic zircon is correlated to the composition of major minerals, such as garnet (Martin et al. 2006; Page et al. 2014; Rubatto and Angiboust 2015).

Changes in the oxygen isotopic composition of minerals can change according to temperature, assemblages and external fluids. Therefore the interpretation of oxygen isotope signatures in metamorphic systems in terms of external fluids requires a control on the change in assemblages and temperatures between the protolith and metamorphic stages. Large variations in mineral crystallization and volume of quartz and plagioclase can, for example, shift the $\delta^{18}O$ of metamorphic minerals of a few ‰ within a constant bulk $\delta^{18}O$. This was modeled in metamorphic rocks from Alpine Corsica (France) where Permian granulites (T 650–800 °C) composed mainly of garnet, quartz and feldspars were transformed by Eocene high-pressure metamorphism (T 400–500 °C) into garnet, amphibole, lawsonite, quartz and phengite (Martin et al. 2014). The shift in $\delta^{18}O$ between the granulitic garnet core (9.9‰) and the high-pressure garnet domains (7.2‰) is significant, but it can be largely reconciled with changes in assemblage and T alone. In that sample, a further stage of metamorphism under high pressure

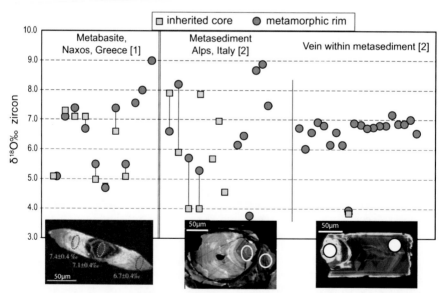

Figure 8. Oxygen isotopic composition of zircon core and rim from two samples that underwent meta-morphism at 500–600 °C and where zircon rims record the age of metamorphism. Analyses of rim and core from the same crystal are linked by thin vertical lines. In both samples, the zircon rim oxygen isotope composition did not equilibrate with the bulk rock. In contrast, zircon in the metamorphic vein within the micaschist (right) have a uniform oxygen isotope composition. Example of the zircon texture is shown in the CL image below the data. Errors on $\delta^{18}O$ values are not plotted. Data and images are from [1] Martin et al. (2006) and [2] unpublished data from sample MST2 and MST1 described in Rubatto et al. (1999) [Images are used by permission of Elsevier, license 3930790667589, in Martin et al. (2006), Vol. 87, Fig. 7, p.183; license 3930131166344, from Rubatto et al. (1999), Vol. 167, Fig. 2, p. 146]. See text for discussion.

was related to metasomatism by external fluids. While the Corsica rocks may represent a rather extreme case (large variation in *T* and assemblages between metamorphic cycles), the case highlights that the interpretation of $\delta^{18}O$ in zircon or any metamorphic mineral must be supported by accurate *P–T* constraints, and that changes in $\delta^{18}O$ between growth zones of up to a couple of ‰ do not necessarily imply external fluids.

Particularly low $\delta^{18}O$ have been reported for metamorphic zircons (Fig. 9) that grew in high-pressure rocks such as mafic eclogites and veins (Fu et al. 2010, 2012; Chen et al. 2011; Sheng et al. 2012; Rubatto and Angiboust 2015). The $\delta^{18}O$ value of these zircons is well below the value of the relict magmatic zircons present in the same or nearby rocks and below the mantle value of 5.3±0.3 ‰, 1SD (Valley 2003). In some cases the low zircon values are also found in coexisting garnet (Rubatto and Angiboust 2015). Two explanations have been proposed for these low $\delta^{18}O$ zircons. In the case of values between 1–5 ‰ in mafic rocks and included veins the low $\delta^{18}O$ has been attributed to ocean floor alteration of the protolith and/or presence of low $\delta^{18}O$ metasomatic fluids generated from nearby altered crust (ultramafic or mafic) during zircon growth. Indeed, bulk rock analyses have proven that altered oceanic crust, even when subducted to high pressure conditions, generally preserves the $\delta^{18}O$ typical of ocean floor alteration (Putlitz et al. 2000; Miller et al. 2001). However profiles through oceanic crust in Oman report minimum $\delta^{18}O$ values of 3‰ for serpentinization under relatively high temperature (Gregory and Taylor 1981); these values are still higher of what is measured in some metamorphic zircon in eclogites and veins (Fig. 9). Internal fractionation of oxygen isotopes during metamorphism by crystallization of high $\delta^{18}O$ phases could additionally shift the zircon oxygen composition to lighter values. Alternatively, the protolith of subducted oceanic crust could have experienced more extreme high *T* alteration than found in the Oman sequence.

A particularly intriguing case is that of negative $\delta^{18}O$ values (as low as -10 ‰) in metamorphic zircons of the Dabie–Sulu eclogites, veins and gneisses, where inherited zircon are -4 to 3.5‰ in the gneiss and 2–10‰ in the eclogite (Chen et al. 2011; Sheng et al. 2012). These negative values have been interpreted as metamorphic zircon growth from externally derived negative $\delta^{18}O$ fluids produced by dehydration of gneissic protoliths that were glacial-hydrothermally altered, in a particularly cold clima (the so called "Snowball Earth"). Such extremely light oxygen signature has been extensively documented in the inherited magmatic zircon over a wide section of the South China Craton (Zheng et al. 2008).

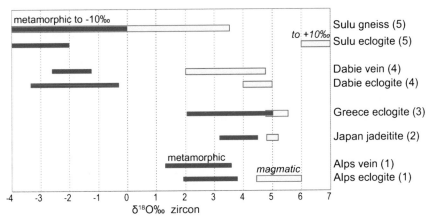

Figure 9. Compilation of low $\delta^{18}O$ values in metamorphic zircon (black bars) compared to protolith zircon of the same sample (empty bars). Data are from (1) Rubatto and Angiboust (2015), (2) Fu et al. (2010), (3) Fu et al. (2012), (4) Sheng et al. (2012) and (5) Chen et al. (2011). See text for discussion.

PETROGENESIS

Compared to the wealth of petrological knowledge on major metamorphic minerals, understanding the behavior of accessory phases is still in its infancy. Relatively few studies have dealt with the stability of accessory minerals. Zircon is a robust accessory mineral and case studies show that its stability is wide and its reactivity can vary case by case. Only a few reactions that form new metamorphic zircon from major or accessory minerals have been proposed. Processes that produce metamorphic zircon from precursor zircon by replacement/recrystallization at the solid state, dissolution (re)precipitation in situ aided by fluids, or melting and crystallization in leucosomes have been recognized as significant.

In this section the complex topic of the petrogenesis of zircon in metamorphic rocks is discussed for four main processes/environments roughly corresponding to increasing metamorphic grade: (i) diagenesis and low-T metamorphism, (ii) zircon replacement or dissolution–precipitation in sub-solidus conditions and in the presence of aqueous fluids, (iii) zircon in melt-dominated metamorphic systems and (iv) zircon-forming reactions involving major/minor minerals under amphibolite to granulite facies conditions. An overview of typical zircon types under variable metamorphic conditions is shown in Figure 1.

Diagenesis and low-T metamorphism

What may happen to zircon in the early stages of metamorphism is best shown by the study of lightly metamorphosed sandstones (Hay and Dempster 2009a). During burial at $<100\,°C$, detrital zircons in these sediments show signs of alteration and new metamorphic zircon is formed. Delicate textures of zircon showing alteration have been documented under such conditions (Fig. 1A): they are unlikely to have survived the depositional process, and thus represent diagenetic/metamorphic modifications. The modification is driven by fluids and affects particularly radiation-damaged zones (Fig. 1B), as predicted by experimental studies (Geisler et al. 2003a). The altered zones are enriched in elements such as Ca, Al, Fe, Mn and LREE that do not enter the zircon structure (Geisler et al. 2003b; Hay and Dempster 2009a), and show porosity and fractures. These zones are similar to what observed in zircons in hydrothermal altered granites (Geisler et al. 2003b). The altered zones are progressively replaced by new crystalline zircon that also forms jagged, sawtooth-shaped zircon overgrowths of a few microns in thickness (Hay and Dempster 2009a), which become relatively more abundant in lower greenschist facies samples (200–400 °C) (Rasmussen 2005; Hay and Dempster 2009b). The formation of low-grade metamorphic xenotime, even finely intergrown with zircon, is commonly associated to these first metamorphic zircon overgrowths (Rasmussen 2005; Hay and Dempster 2009a; Hay et al. 2010). The mechanisms that have been proposed to be dominant at this low grade are dissolution–precipitation and solid-state reaction (Hay and Dempster 2009a).

The sawtooth-shaped zircon overgrowths are generally absent (see an exception in Franz et al. 2015) from higher grade rocks (upper greenschist to lower amphibolite facies, Rubatto et al. 2001; Hay and Dempster 2009b) and thus it is likely that later dissolution or Ostwald ripening is erasing this early record. Zr released by dissolution of small zircon overgrowths during prograde metamorphism could be accommodated in other growing metamorphic minerals (Kohn et al. 2015) explaining why metamorphic zircon is generally absent in greenschist to lower amphibolite facies rocks. Partly altered, porous and metamict zircon domains, which are the likely source of Zr for low-grade zircon growth, are partly dissolved during early metamorphism, but can also be preserved at higher grade (see below).

Metamorphic zircon and fluids at sub-solidus conditions

Zircon replacement and modification by fluids has been identified in a variety of metamorphic settings from diagenetic environments (see above) to granulite facies metamorphism, and particularly in rocks that record high pressure metamorphism (see also a review in Rubatto and

Hermann 2007b). A variety of textures have been attributed to fluid-driven alteration/replacement of zircon (Fig. 1, Table 1, see also Corfu et al. 2003). One of the lowest T reported for zircon alteration in metamorphic rocks is from Spandler et al. (2004), who proposed that relict zircon cores preserved in schist from New Caledonia underwent alteration by fluids during sea floor alteration of the mafic protolith at $T < 100\,°C$, based on their inclusion assemblage and texture (Fig. 1C). Such process expelled trace elements from the original zircon, increased porosity, and produced mottled and patchy zoning. These features have many similarities to the dissolution–precipitation process described in zircon from rocks at $400–600\,°C$ (Tomaschek et al. 2003; Rubatto et al. 2008). Similar alteration and replacement processes have been described for other accessory minerals such as apatite and monazite and reproduced in controlled experiments (Hetherington et al. 2010; Harlov et al. 2011; Seydoux-Guillaume et al. 2012). For a more general description of mineral replacement see also Putnis (2009).

More extensive alteration leads to full replacement/recrystallization of domains that are texturally discordant with the original zoning and often marked by a recrystallization front (Hoskin and Black 2000; Vonlanthen et al. 2012). These replaced domains are chemically and microstructurally different from the cores: they commonly have a lower trace element content, they lack any elements typical for alteration zones such as LREE, Ca, Al and Fe, and they are inclusion poor. TEM investigation of these domains shows that they are relatively free of defects (Vonlanthen et al. 2012). Most studies of natural samples report that during this process the chemical composition and isotopic systematics of the replaced zircon is totally reset. However, remnants of the chemical and isotopic composition of protolith zircon have been reported in such domains when analyses were performed with a spatial resolution of $\sim 20\,\mu m$ (Hoskin and Black 2000). This indicates that the replacement process is not always complete and that "islands" of altered zircon are preserved at the micro to nanoscale. Such micro-relicts have been identified in zircon where the replacement/recrystallization process led to micro-zircon formation (Rubatto et al. 2008).

It has been proposed that dissolution–precipitation can proceed efficiently even with minimum free fluid, and with low Zr solubility in that fluid (Tomaschek et al. 2003; Geisler et al. 2007). The process does not require transport of Zr and other zircon-forming elements outside the zircon itself, and can proceed in a virtually closed system (Hoskin and Black 2000). This implies that this process, that cannibalizes inherited zircon and resets its isotopic and chemical composition to form metamorphic zircon, is not necessarily communicating with the reactive bulk rock and can occur independently of Zr solubility or metamorphic reactions. It requires aggressive fluids and it is enhanced by temperature and the presence of metamict zircon (Geisler et al. 2003a).

Zircon with textures suggestive of a metamict state is preserved up to high metamorphic conditions. For example, porous zircon domains that contain very low-grade ($< 100\,°C$) mineral inclusions have been recovered in blueschist- to eclogite-facies rocks (Spandler et al. 2004). A whiteschists that recorded ultra high pressure metamorphism ($35\,kbar$ and $750\,°C$) preserves altered zircon cores from the precursor granite that contain prograde to peak metamorphic mineral inclusions in healed alteration zones (Gauthiez-Putallaz et al. 2016). These reports suggest that dissolution of metamict zircon domains by metamorphic fluids that provides Zr and other essential elements for metamorphic zircon formation can occur under diverse conditions.

Preservation of metamict zircon up to high grade conditions is supported by experimental work and calculations. Zircon is predicted to stop accumulating radiation damage at $\sim 250\,°C$, the T above which amorphisation will no longer occur, because annealing becomes faster than damage accumulation (Meldrum et al. 1999; Ewing et al. 2003). The process of recovery and structural re-organisation of zircon, as described in Ewing et al. (2003, and references therein) occurs in stages over a T range and is particularly low in zircon compared to any related phase. In experiments, the first stage of recovery occurs below ~700 °C (recovery of point defects and short length damage). The second stage of re-organisation occurs above ~700 °C and, full structural re-organisation has been documented at 900 °C or above (Geisler et al. 2001b). During this process, islands of recovered material are surrounded by amorphous domains until damaged material is no longer detectable (by Raman or TEM investigation). Diffusion dominates the recovery process which is thus *T*- and time-dependent (Ewing et al. 2003). Calculations indicate that it would take up to 370 Ma to fully recover the structure of metamict zircon at 700 °C (Geisler et al. 2001a), but the process is significantly more efficient under hydrothermal or wet geological conditions (Geisler et al. 2001b).

Experimental investigations have shown very low solubility of Zr in aqueous fluids, that increases with increasing Si contents and alkalinity of the fluids (Ayers et al. 2012). Both, Si contents and alkalinity in aqueous fluids increase significantly with increasing pressure (Hermann and Rubatto 2014). This might explain why abundant metamorphic zircon occurs in high-pressure crustal rocks that record metamorphic temperatures below the solidus <700 °C, as for example in Alpine eclogites. Examples are large metamorphic zircon in the omphacite–garnet–rutile veins within the Monviso eclogites (Rubatto and Hermann 2003) and the intense zircon dissolution–precipitation process in the jadeite–leucogabbro of the Lanzo Massif (Rubatto et al. 2008). Metamorphic zircon in jedeitites could also form by precipitation from alkaline-rich fluids (e.g., Mori et al. 2011). The abundance of metamorphic zircon in the HP rocks of the Dabie-Sulu region has also been largely attributed to the activity of aggressive fluids (Zheng et al. 2005; Zhang et al. 2009b; Zhao et al. 2015). In contrast, metamorphic zircon is very rare in greenschist to amphibolite facies crustal rocks that record Barrovian metamorphism at similar *T*.

If fluids are the driver of zircon dissolution–precipitation, zircon reactivity would be much less dependent on *T*, but rather enhanced by the release of fluids in the rock system. Thus, metamorphic zircon formation can occur even while *T* is increasing, unlike what indicated by thermodynamic models that require equilibrium among all phases (Kohn et al. 2015). Prograde metamorphic zircon have been for example documented in the Dora Maira schists (Gauthiez-Putallaz et al. 2016), where zircon domains have been related to episode of fluid release by dehydration reactions (see texture in Fig. 1F). Such cases of prograde zircon must be considered when interpreting zircon ages.

Metamorphic zircon and melts

The most common form of metamorphic zircon is overgrowth on detrital or inherited magmatic grains during partial melting. Field studies that looked at prograde metamorphic sequences, mainly of pelitic/arcosic compositions, show that new zircon formation under sub-solidus conditions is virtually absent, but becomes abundant as soon as partial melting is observed (Rubatto et al. 2001; Williams 2001). Such metamorphic overgrowths on zircon cores are more abundant in leucosomes than melanosomes (Rubatto et al. 2001).

Zr solubility in anatectic melts ranges from 10s to 100s of ppm according to temperature and melt composition (Boehnke et al. 2013). A few studies have attempted to model the behavior of zircon and monazite in migmatites using thermodynamic databases (Kelsey et al. 2008; Kelsey and Powell 2011; Yakymchuk and Brown 2014; Kohn et al. 2015). These models agree that in migmatites, most of the bulk Zr will be either stored in the melt or locked in undissolved inherited zircon. They also predict that zircon crystallization will occur significantly during cooling when Zr saturation is reached in leucosomes or in interstitial melt. A common conclusion is that most of the new zircon will crystallize in leucosomes compared to melanosomes or restitic portions. The solubility models also predict that, in felsic compositions, zircon dissolution and thus crystallization from a melt is less effective than that of monazite, and thus relict zircon will survive to higher metamorphic grade than monazite. The models consistently predict that anatectic zircon should grow during cooling when Zr solubility decreases in the melt or the solidus is reached (Kelsey et al. 2008; Kelsey and Powell 2011; Kohn et al. 2015).

The thermodynamic models all make significant assumptions that may differ from natural cases. A common assumption to all models is that all zircon and monazite crystals are in contact with the melt over the entire *P–T* evolution (Kelsey et al. 2008; Kelsey and Powell 2011; Yakymchuk and Brown 2014; Kohn et al. 2015). In natural rocks, shielding of inherited grains from contact with the melt (or fluids) is often the case. The observation that metamorphic zircon rims even in migmatites are variable in size from grain to grain, with the common presence of grains that are lacking overgrowths, support this scenario. Therefore any attempt to model the Zr budget in metamorphic rocks has the limitation that a variable but significant proportion of Zr may remain shielded from the melt. A second assumption in some models is that no melt loss from the system occurs. The pioneering modeling of Kelsey et al. (2008) mainly considered closed system behavior, with one episode of melt loss. Melt loss is undoubtedly a complication in natural rocks and successive models demonstrated that such behavior will reduce the amount of melt generated, and thus the solubility of Zr and the production of new zircon (Yakymchuk and Brown 2014). Ignoring the effect of Zr-release and uptake from other Zr-bearing phases in thermodynamic models (Kelsey et al. 2008; Yakymchuk and Brown 2014) does not seems to affect the general conclusions compared to studies that budgeted for Zr in minerals such as garnet, rutile and amphibole (Kelsey and Powell 2011; Kohn et al. 2015).

Crystallization from a Zr saturated melt upon cooling, as predicted by the models, is not the only process to form zircon overgrowth at high-grade metamorphic conditions. Some studies concluded that zircon in migmatites or granulites can form during prograde or peak metamorphism (Hermann and Rubatto 2003; Baldwin et al. 2007; Gordon et al. 2013; Rubatto et al. 2013). Arguments are based on zircon inclusions, trace element composition, Ti-in-zircon-temperatures and different zircon ages from continuous sequences. In the most compelling cases, a change in REE composition has been observed between relatively older and younger metamorphic zircon domains (Hermann and Rubatto 2003; Gordon et al. 2013; Rubatto et al. 2013). When the change in zircon REE composition has been related to the abundance of coexisting phases (particularly garnet and feldspars, see discussion above) and in turn to the *P–T* path, prograde to peak zircon growth was proposed. Such examples span different tectonic setting and *P–T* evolutions from collisional Barrovian metamorphism (Rubatto et al. 2013), UHP metamorphism (Gordon et al. 2013) and lower crustal melting in extensional settings (Hermann and Rubatto 2003). In the case of fluid-induced melting in the central Alps, Rubatto et al. (2009) observed that zircon growth from melt occurred at different times (million of years apart) in segregated leucosomes sampled only meters from each other. Some migmatites contain multiple growth zones (Fig. 1) that have distinct age and composition, within single samples or even zircon grains. The intermittent availability of water for fluid-induced melting has been proposed by Rubatto et al. (2009) as a mechanism to explain multiple zircon growth zones in the same crystal and with age differences of several million years. These observation provide strong evidence that dissolution–precipitation of zircon in the presence of a melt can occur at any stage—prograde or retrograde—as long as melt is present.

Zircon forming reactions

Textural observations in natural samples that support metamorphic zircon growth from the breakdown of another phase have been reported for ilmenite (Bingen et al. 2004), rutile (Ewing et al. 2013; Pape et al. 2016) and garnet (Fraser et al. 1997; Degeling et al. 2001); see textures in Fig. 1N and Fig. 2C. In some cases the textural observations are supported by trace element mass balance considerations (Degeling et al. 2001; Ewing et al. 2013). The main idea is that, in a metamorphic reaction, there is a decrease of Zr solubility in source minerals (rutile, garnet etc...) and the expelled Zr results in the precipitation of metamorphic zircon. It has been suggested that garnet plays a minor role in the Zr budget of crustal rocks (Kelsey and Powell 2011), and thus garnet breakdown reactions might result in the formation of new zircon. A recent compilation of Zr content in major minerals (Kohn et al. 2015) identified rutile, ilmenite, titanite, garnet and hornblende as carriers for Zr: such minerals contain a few to 100s of ppm of Zr, and even 1000s of ppm in the case of rutile. A review of own data on Zr content in garnet (320 analyses, 20 samples) in rocks metamorphosed from 500 to 1000 °C where metamorphic zircon is found, show Zr contents of only 5–20 ppm Zr in garnet. Zr concentrations in minerals increase with increasing temperature. Thus, the reaction of magmatic minerals such as pyroxene, amphibole and ilmenite, but also volcanic glass to lower temperature metamorphic minerals provides a mechanism to form metamorphic zircon.

The storage capacity of Zr in rock forming minerals can be considered in a simple example: A garnet-amphibolite with 20% garnet, 20% hornblende with contents of 3, 30 or 100 ppm Zr, that also contains 2 % of rutile with 40, 1500 or 3500 ppm Zr (T of 500, 700 and 900 °C, respectively) would provide a maximum of 2, 42 or 110 ppm Zr for the bulk, respectively. Thus, Zr release at low temperature of 500 °C from these minerals is irrelevant, whereas is more significant when very high temperature minerals are affected. As modeled by Kohn et al. (2015) for a basaltic and metapelitic composition, Zr released from major and accessory minerals during prograde metamorphism is expected to be entirely taken up by other minerals, at least as long as the T is increasing. Particularly in rutile-bearing rocks, any Zr release in the reactive bulk can be taken up by growing rutile (Kohn et al. 2015). Therefore zircon-forming reactions will be mainly related to decompression when rutile transforms to ilmenite or titanite, or related to retrograde replacement of garnet by minerals that have a lower capacity to store Zr (cordierite, biotite or chlorite).

CONCLUSIONS AND OUTLOOK

The investigation of metamorphic zircon has dramatically increased since the development of in situ analytical methods that allow measuring diverse chemical and isotopic signals in distinct zircon domains. Combined with essential imaging of internal textures, geochemical information has provided the necessary base for metamorphic petrology of zircon.

Metamorphic zircon forms by a series of processes from the lowest grade to extreme metamorphism. At low grade, fluid alteration and solid-state replacement are dominant mechanisms affecting zircon. During partial melting zircon is particularly reactive with high solubility in the melts and crystallization of overgrowths. Under extreme conditions metamorphic zircon remains stable, but different elements and isotope systems are likely to be affected in different ways. The most prominent process is the loss of incompatible Pb that occurs in altered, deformed or metamict zircon, whereas the REE and HFSE trace element chemistry and Hf systematics are generally preserved. The behavior of oxygen isotopes in zircon under extreme conditions remains uncertain.

Linking U–Pb ages to metamorphic conditions for correct age interpretation requires the combination of multiple information, including internal zoning, deformation features, inclusions, Ti-thermometry, trace element patterns, Lu–Hf and/or oxygen isotopes. These different systems may have different retentivity and thus are not always coupled, and their comparison provides additional information. Zircon is not only a robust geochronometer, but also a mineral relevant for the petrogenesis of metamorphic rocks that can provide details on protolith, temperature evolution, deformation, fluids and melts and assists the reconstruction of crustal processes.

Current and future analytical developments will increase our capability to collect geochemical information with a greater spatial resolution, that in turn will allow resolving element and isotopic diffusion, fine scale alteration and replacement/recrystallization. The systematics of numerous trace elements hosted in zircons remains underexplored (e.g., H, Li, Nb, Ta, Sc) and requires systemic studies. Trace element partitioning with other phases and melts of different compositions have to be further investigated over a range of $P-T$ to fully exploit the capacity of zircon to monitor geochemical differentiation. Ti-in-zircon thermometry is lacking studies on the effect of pressure and Ti activity. Diffusion of crucial elements, for example oxygen, requires further investigation both experimentally and of natural samples. Petrogenesis of metamorphic zircon will gain from additional knowledge of Zr and Hf distribution in metamorphic minerals and fluids, and of zircon forming reactions. The complex processes of zircon alteration, replacement, dissolution, precipitation and modification in general should be approached through simulations, experiments and systematic studies in natural samples.

ACKNOWLEDGMENTS

A wide thanks goes to all the students and collaborators that I have worked with while investigating the many faces and lives of metamorphic zircon. Their constant supply of interesting samples has much benefited my research.

I thank Matt Kohn, Martin Engi and Pierre Lanari for inviting me to contribute to this volume and keeping up the information flow and discussion during manuscript preparation. This contribution has benefited from the constructive reviews of Martin Whitehouse, Bradley Hacker and Matt Kohn. A special thanks goes to Jörg Hermann for discussion and comments on this chapter.

REFERENCES

Austrheim H, Corfu F (2009) Formation of planar deformation features (PDFs) in zircon during coseismic faulting and an evaluation of potential effects on U–Pb systematics. Chem Geol 261:24–30

Ayers JC, Zhang L, Luo Y, Peters TJ (2012) Zircon solubility in alkaline aqueous fluids at upper crustal conditions. Geochim Cosmochim Acta 96:18–28

Baldwin SL, Monteleone B, Webb LE, Fitzgerald PG, Grove M, Hill EJ (2004) Pliocene eclogite exhumation at plate tectonic rates in eastern Papua New Guinea. Nature 431:263–267

Baldwin JA, Brown M, Schmitz MD (2007) First application of titanium-in-zircon thermometry to ultrahigh-temperature metamorphism. Geology 35:295–298

Bauer C, Rubatto D, Krenn K, Proyer A, Hoinkes G (2007) A zircon study from the Rhodope Metamorphic Complex, N-Greece: Time record of a multistage evolution. Lithos 99:207–228

Belousova E, Griffin W, O'Reilly SY, Fisher N (2002) Igneous zircon: trace element composition as an indicator of source rock type. Contrib Mineral Petrol 143:602–622

Bingen B, Austrheim H, Whitehouse M (2001) Ilmenite as a source of zirconium during high-grade metamorphism? Textural evidence from the Caledonides of Western Norway and implications for zircon geochronology. J Petrol 42:355–375

Bingen B, Austrheim H, Whitehouse MJ, Davis WJ (2004) Trace element signature and U–Pb geochronology of eclogite-facies zircon, Bergen Arcs, Caledonides of W Norway. Contrib Mineral Petrol 147:671–683

Boehnke P, Watson EB, Trail D, Harrison TM, Schmitt AK (2013) Zircon saturation re-visited. Chem Geol 351:324–334

Bowman JR, Moser DE, Valley JW, Wooden JL, Kita NT, Mazdab FK (2011) Zircon U–Pb isotope, $\delta^{18}O$ and trace element response to 80 m.y. of high temperature metamorphism in the lower crust: Sluggish diffusion and new records of Archean craton formation. Am J Sci 311:719–772

Buick IS, Hermann J, Williams IS, Gibson R, Rubatto D (2006) A SHRIMP U–Pb and LA-ICP-MS trace element study of the petrogenesis of garnet–cordierite–orthoamphibole gneisses from the Central Zone of the Limpopo Belt, South Africa. Lithos 88:150–172

Burnham AD, Berry AJ (2012) An experimental study of trace element partitioning between zircon and melt as a function of oxygen fugacity. Geochim Cosmochim Acta 95:196–212

Cesare B, Gómez-Pugnaire MT, Rubatto D (2003) Residence time of S-type anatectic magmas beneath the Neogene Volcanic Province of SE Spain: a zircon and monazite SHRIMP study. Contrib Mineral Petrol 146:28–43

Chen Y-X, Zheng Y-F, Chen R-X, Zhang S-B, Li Q, Dai M, Chen L (2011) Metamorphic growth and recrystallization of zircons in extremely 18O-depleted rocks during eclogite-facies metamorphism: Evidence from U–Pb ages, trace elements, and O–Hf isotopes. Geochim Cosmochim Acta 75:4877–4898

Cherniak DJ (2010) Diffusion in accessory minerals: Zircon, titanite, apatite, monazite and xenotime. Rev Mineral Geochem 72:827–869

Cherniak DJ, Watson BE (2003) Diffusion in zircon. Rev Mineral Geochem 53:113–143

Claesson S, Bibikova EV, Shumlyanskyy L, Whitehouse MJ, Billström K (2016) Can oxygen isotopes in magmatic zircon be modified by metamorphism? A case study from the Eoarchean Dniester-Bug Series, Ukrainian Shield. Precambr Res 273:1–11

Claoué-Long JC, Sobolev NV, Shatsky VS, Sobolev AV (1991) Zircon response to diamond-pressure metamorphism in the Kokchetav massif, USSR. Geology 19:710–713

Corfu F, Hanchar JM, Hoskin PWO, Kinny P (2003) Atlas of zircon textures. Rev Mineral Geochem 53:469–500

Degeling H, Eggins S, Ellis DJ (2001) Zr budget for metamorphic reactions, and the formation of zircon from garnet breakdown. Mineral Mag 65:749–758

Drüppel K, Elsäßer L, Brandt S, Gerdes A (2013) Sveconorwegian mid-crustal ultrahigh-temperature metamorphism in Rogaland, Norway: U–Pb LA-ICP-MS geochronology and pseudosections of sapphirine granulites and associated paragneisses. J Petrol 54:305–350

Ewing RC, Meldrum A, Wang L, Weber WJ, Corrales LR (2003) Radiation effects in zircon. Rev Mineral Geochem 53:387–425

Ewing T, Hermann J, Rubatto D (2013) The robustness of the Zr-in-rutile and Ti-in-zircon thermometers during high-temperature metamorphism (Ivrea-Verbano Zone, northern Italy). Contrib Mineral Petrol 165:757–779

Ferriss EDA, Essene EJ, Becker U (2008) Computational study of the effect of pressure on the Ti-in-zircon geothermometer. Eur J Mineral 20:745–755

Ferry JM, Watson EB (2007) New thermodynamic models and revised calibrations for the Ti-in-zircon and Zr-in-rutile thermometers Contrib Mineral Petrol 154:429–437

Fornelli A, Langone A, Micheletti F, Pascazio A, Piccarreta G (2014) The role of trace element partitioning between garnet, zircon and orthopyroxene on the interpretation of zircon U–Pb ages: An example from high-grade basement in Calabria (Southern Italy). Int J Earth Sci 103:487–507

Franz G, Morteani G, Rhede D (2015) Xenotime-(Y) formation from zircon dissolution–precipitation and HREE fractionation: an example from a metamorphosed phosphatic sandstone, Espinhaço fold belt (Brazil). Contrib Mineral Petrol 170

Fraser G, Ellis D, Eggins S (1997) Zirconium abundance in granulite-facies minerals, with implications for zircon geochronology in high-grade rocks. Geology 25:607–610

Fu B, Page FZ, Cavosie AJ, Fournelle J, Kita NT, Star Lackey J, Wilde SA, Valley JW (2008) Ti-in-zircon thermometry: applications and limitations. Contrib Mineral Petrol 156:197–215

Fu B, Valley JW, Kita NT, Spicuzza MJ, Paton C, Tsujimori T, Bröcker M, Harlow GE (2010) Multiple origins of zircons in jadeitite. Contrib Mineral Petrol 159:769–780

Fu B, Paul B, Cliff J, Bröcker M, Bulle F (2012) O–Hf isotope constraints on the origin of zircon in high-pressure mélange blocks and associated matrix rocks from Tinos and Syros, Greece. Eur J Mineral 24:277–287

Ganade de Araujo CE, Rubatto D, Hermann J, Cordani UG, Caby R, Basei MAS (2014) Ediacaran 2,500-km-long synchronous deep continental subduction in the West Gondwana Orogen. Nature Commun 5:5198

Gao X-Y, Zheng Y-F, Chen Y-X, Tang H-L, Li W-C (2015) Zircon geochemistry records the action of metamorphic fluid on the formation of ultrahigh-pressure jadeite quartzite in the Dabie orogen. Chem Geol 419:158–175

Gasser D, Rubatto D, Bruand E, Stüwe K (2012) Large-scale, short-lived metamorphism, deformation, and magmatism in the Chugach metamorphic complex, southern Alaska: A SHRIMP U–Pb study of zircons. Geol Soc Am Bull 124:886–905

Gauthiez-Putallaz L, Rubatto D, Hermann J (2016) Dating prograde fluid pulses during subduction by in situ U–Pb and oxygen isotope analysis. Contrib Mineral Petrol 171:15

Gebauer D, Schertl H-P, Brix M, Schreyer W (1997) 35 Ma old ultrahigh-pressure metamorphism and evidence for very rapid exhumation in the Dora Maira Massif, Western Alps. Lithos 41:5–24

Geisler T, Pidgeon RT, van Bronswijk W, Pleysier R (2001a) Kinetics of thermal recovery and recrystallization of partially metamict zircon: a Raman spectroscopic study. Eur J Mineral 13:1163–1176

Geisler T, Ulonska M, Schleicher H, Pidgeon RT, van Bronswijk W (2001b) Leaching and differential recrystallization of metamict zircon under experimental hydrothermal conditions. Contrib Mineral Petrol 141:53–65

Geisler T, Pidgeon RT, Kurtz R, van Bronswijk W, Schleicher H (2003a) Experimental hydrothermal alteration of partially metamict zircon. Am Mineral 88:1496–1543

Geisler T, Rashwan AA, Rahn MKW, Poller U, Zwingmann H, Pidgeon RT, Schleicher H, Tomaschek F (2003b) Low-temperature hydrothermal alteration of natural metamict zircons from the Eastern Desert, Egypt. Mineral Mag 67:485–508

Geisler T, Schaltegger U, Tomaschek F (2007) Re-equilibration of zircon in acqueous fluids and melts. Elements 3:43–50

Gerdes A, Zeh A (2009) Zircon formation versus zircon alteration—New insights from combined U–Pb and Lu–Hf in-situ LA-ICP-MS analyses, and consequences for the interpretation of Archean zircon from the Central Zone of the Limpopo Belt. Chem Geol 261:230–243

Gilotti JA, Nutman AP, Brueckner HK (2004) Devonian to Carboniferous collision in the Greenland Caledonides: U–Pb zircon and Sm–Nd ages of high-pressure and ultrahigh-pressure metamorphism. Contrib Mineral Petrol 148:216–235

Gordon SM, Little TA, Hacker BR, Bowring SA, Korchinski M, Baldwin SL, Kylander-Clark ARC (2012) Multi-stage exhumation of young UHP–HP rocks: Timescales of melt crystallization in the D'Entrecasteaux Islands, southeastern Papua New Guinea. Earth Planet Sci Lett 351–352:237–246

Gordon SM, Whitney DL, Teyssier C, Fossen H (2013) U–Pb dates and trace-element geochemistry of zircon from migmatite, Western Gneiss Region, Norway: Significance for history of partial melting in continental subduction. Lithos 170–171:35–53

Gregory RT, Taylor HP (1981) An oxygen isotope profile in a section of Cretaceous oceanic crust, Samail Ophiolite, Oman: Evidence for $\delta^{18}O$ buffering of the oceans by deep (>5 km) seawater-hydrothermal circulation at mid-ocean ridges. J Geophys Res 86:2737–2755

Grimes CB, Wooden JL, Cheadle MJ, John BE (2015) "Fingerprinting" tectono-magmatic provenance using trace elements in igneous zircon. Contrib Mineral Petrol 170:1–26

Hacker RB, Ratschbacher L, Webb L, Ireland T, Walker D, Dong S (1998) U/Pb zircon ages constrain the architecture of the ultrahigh-pressure Qinling–Dabie Orogen. Earth Planet Sci Lett 161:215–230

Hanchar JM, Miller CF (1993) Zircon zonation patterns as revealed by cathodoluminescence and backscattered electron images: Implications for interpretation of complex crustal histories. Chem Geol 110:1–13

Hanchar JM, Rudnick RL (1995) Revealing hidden structures: The application of cathodoluminescence and back-scattered electron imaging to dating zircons from lower crust xenoliths. Lithos 36:289–303

Harley SL, Nandakumar V (2014) Accessory mineral behaviour in granulite migmatites: a case study from the Kerala Khondalite Belt, India. J Petrol 55:1965–2002

Harley SL, Kelly NM, Möller A (2007) Zircon behaviour and the thermal histories of mountain chains. Elements 3:25–30

Harlov DE, Wirth R, Hetherington CJ (2011) Fluid-mediated partial alteration in monazite: the role of coupled dissolution–reprecipitation in element redistribution and mass transfer. Contrib Mineral Petrol 162:329–348

Harrison TM, Bell EA, Boehnke P (2017) Hadean zircon petrochronology. Rev Mineral Geochem 83:329–363

Harrison TM, Schmitt AK (2007) High sensitivity mapping of Ti distributions in Hadean zircons. Earth Planet Sci Lett 261:9–19

Hay DC, Dempster TJ (2009a) Zircon alteration, formation and preservation in sandstones. Sedimentology 56:2175–2191

Hay DC, Dempster TJ (2009b) Zircon behaviour during low-temperature metamorphism J Petrol 50:571–589

Hay DC, Dempster TJ, Lee MR, Brown DJ (2010) Anatomy of a low temperature zircon outgrowth. Contrib Mineral Petrol 159:81–92

Hermann J, Rubatto D (2003) Relating zircon and monazite domains to garnet growth zones: age and duration of granulite facies metamorphism in the Val Malenco lower crust. J Metamorph Geol 21:833–852

Hermann J, Rubatto D (2014) Subduction of continental crust to mantle depth: Geochemistry of ultrahigh-pressure rocks. *In:* Rudnick R (ed) The Crust vol 4. Elsevier Amsterdam pp 309–340

Hermann J, Rubatto D, Korsakov A, Shatsky VS (2001) Multiple zircon growth during fast exhumation of diamondiferous, deeply subducted continental crust (Kokchetav massif, Kazakhstan). Contrib Mineral Petrol 141:66–82

Hermann J, Rubatto D, Korsakov A, Shatsky VS (2006a) The age of metamorphism of diamondiferous rocks determined with SHRIMP dating of zircon. Russ Geol Geophys 47:513–520

Hermann J, Rubatto D, Trommsdorff V (2006b) Sub-solidus Oligocene zircon formation in garnet peridotite during fast decompression and fluid infiltration (Duria, Central Alps). Mineral Petrol 88:181–206

Hermann J, Spandler C, Hack A, Korsakov AV (2006c) Aqueous fluids and hydrous melts in high-pressure and ultra-high pressure rocks: Implications for element transfer in subduction zones. Lithos 92:399–417

Hetherington CJ, Harlov DE, Budzyn B (2010) Experimental metasomatism of monazite and xenotime: Mineral stability, REE mobility and fluid composition. Mineral Petrol 99:165–184

Hiess J, Nutman AP, Bennett VC, Holden P (2008) Ti-in-zircon thermometry applied to contrasting Archean metamorphic and igneous systems. Chem Geol 247:323–338

Hofmann AE, Valley JW, Watson EB, Cavosie AJ, Eiler JM (2009) Sub-micron scale distributions of trace elements in zircon. Contrib Mineral Petrol 158:317–335

Hokada T, Harley SL (2004) Zircon growth in UHT leucosome: constraints from zircon–garnet rare earth elements (REE) relations in Napier Complex, East Antarctica. J Mineral Petrol Sci 99:180–190

Hopkins M, Harrison TM, Manning CE (2008) Low heat flow inferred from > 4 Gyr zircons suggests Hadean plate boundary interactions. Nature 456:493–496

Hoskin PWO, Black LP (2000) Metamorphic zircon formation by solid-state recrystallization of protolith igneous zircon. J Metamorph Geol 18:423–439

Kaczmarek M-A, Müntener O, Rubatto D (2008) Trace element chemistry and U–Pb dating of zircons from oceanic gabbros and their relationship with whole rock composition (Lanzo, Italian Alps). Contrib Mineral Petrol 155:295–312

Katayama I, Maruyama S (2009) Inclusion study in zircon from ultrahigh-pressure metamorphic rocks in the Kokchetav massif: an excellent tracer of metamorphic history. J Geol Soc 166:783–796

Katayama I, Maruyama S, Parkinson CD, Terada K, Sano Y (2001) Ion micro-probe U–Pb zircon geochronology of peak and retrograde stages of ultrahigh-pressure metamorphic rocks from the Kokchetav massif, northern Kazakhstan. Earth Planet Sci Lett 188:185–198

Kelly N, Harley S (2005) An integrated microtextural and chemical approach to zircon geochronology: refining the Archean history of the Napier Complex, east Antarctica. Contrib Mineral Petrol 149:57–84

Kelsey DE, Powell R (2011) Progress in linking accessory mineral growth and breakdown to major mineral evolution in metamorphic rocks: a thermodynamic approach in the Na_2O–CaO–K_2O–FeO–MgO–Al_2O_3–SiO_2–H_2O–TiO_2–ZrO_2 system. J Metamorph Geol 29:151–166

Kelsey DE, Clark C, Hand M (2008) Thermobarometric modelling of zircon and monazite growth in melt-bearing systems: examples using model metapelitic and metapsammitic granulites. J Metamorph Geol 26:199–212

Kohn MJ (1999) Why most "dry" rocks should cool "wet". Am Mineral 84:570–580

Kohn MJ (2016) Metamorphic chronology—a tool for all ages: Past achievements and future prospects. Am Mineral 101:25–42

Kohn MJ, Penniston–Dorland SC (2017) Diffusion: Obstacles and opportunities in petrochronology. Rev Mineral Geochem 83:103–152

Kohn MJ, Corrie SL, Markley C (2015) The fall and rise of metamorphic zircon. Am Mineral 100:897–908

Korhonen FJ, Clark C, Brown M, Bhattacharya S, Taylor R (2013) How long-lived is ultrahigh temperature (UHT) metamorphism? Constraints from zircon and monazite geochronology in the Eastern Ghats orogenic belt, India. Precambr Res 234:322–350

Kotkova J, Harley SL (2010) Anatexis during high-pressure crustal metamorphism: evidence from garnet–whole-rock REE relationships and zircon–rutile Ti–Zr thermometry in leucogranulites from the Bohemian Massif. J Petrol 51:1967–2001

Kusiak MA, Whitehouse MJ, Wilde SA, Dunkley DJ, Menneken M, Nemchin AA, Clark C (2013a) Changes in zircon chemistry during archean UHT metamorphism in the Napier Complex, Antarctica. Am J Sci 313:933–967

Kusiak MA, Whitehouse MJ, Wilde SA, Nemchin AA, Clark C (2013b) Mobilization of radiogenic Pb in zircon revealed by ion imaging: Implications for early Earth geochronology. Geology 41:291–294

Kusiak MA, Dunkley DJ, Wirth R, Whitehouse MJ, Wilde SA, Marquardt K (2015) Metallic lead nanospheres discovered in ancient zircons. PNAS 112:4958–4963

Kylander–Clark ARC (2017) Petrochronology by laser–ablation inductively coupled plasma mass spectrometry. Rev Mineral Geochem 83:183–198

Kylander-Clark ARC, Hacker BR, Cottle JM (2013) Laser-ablation split-stream ICP petrochronology Chem Geol 345:99–112

Lenting C, Geisler T, Gerdes A, Kooijman E, Scherer EE, Zeh A (2010) The behavior of the Hf isotope system in radiation-damaged zircon during experimental hydrothermal alteration. Am Mineral 95:1343–1348

López Sánchez-Vizcaíno V, Rubatto D, Gómez-Pugnaire MT, Trommsdorff V, Müntener O (2001) Middle Miocene HP metamorphism and fast exhumation of the Nevado Filabride Complex, SE Spain. Terra Nova 13:327–332

Martin L, Duchêne S, Deloule E, Vanderhaeghe O (2006) The isotopic composition of zircon and garnet: a record of the metamorphic history of Naxos (Greece). Lithos 87:174–192

Martin LAJ, Duchêne S, Deloule E, Vanderhaeghe O (2008) Mobility of trace elements and oxygen in zircon during metamorphism: Consequences for geochemical tracing. Earth Planet Sci Lett 267:161–174

Martin L, Rubatto D, Crepisson C, Hermann J, Putlitz B, Vitale-Brovarone A (2014) Garnet oxygen analysis by SHRIMP-SI: matrix corrections and application to high pressure metasomatic rocks from Alpine Corsica. Chem Geol 374–375:25–36

McLaren AC, Fitz Gerald JD, Williams IS (1994) The microstructure of zircon and its influence on the age determiantion from Pb/U isotopic ratios measured by ion microprobe. Geochim Cosmochim Acta 58:993–1005

Meldrum A, Boatner LA, Zinkle SJ, Wang S-X, Wang L-M, Ewing RC (1999) Effects of dose rate and temperature on the crystalline-to-metamict transformation in the ABO_4 orthosilicates. Can Mineral 37:207–221

Menneken M, Nemchin AA, Geisler T, Pidgeon RT, Wilde SA (2007) Hadean diamonds in zircon from Jack Hills, Western Australia. Nature 448:917–U915

Miller JA, Cartwright I, Buick I, Barnicoat A (2001) An O-isotope profile through the HP-LT Corsican ophiolite, France and its implications for fluid flow during subduction. Chem Geol 178:43–69

Möller A, O'Brien PJ, Kennedy A, Kröner A (2002) Polyphase zircon in ultrahigh-temperature granulites (Rogaland, SW Norway): Constraints for Pb diffusion in zircon. J Metamorph Geol 20:727–740

Möller A, O'Brien PJ, Kennedy A, Kröner A (2003) Linking growth episodes of zircon and metamorphic textures to zircon chemistry: an example from the ultrahigh-temperature granulites of Rogaland (SW Norway). Geol Soc, London, Spec Publ 220:65–81

Montero P, Bea F, Zinger TF, Scarrow JH, Molina JF, Whitehouse M (2004) 55 million years of continuous anatexis in Central Iberia: Single-zircon dating of the Peña Negra Complex. J Geol Soc 161:255–263

Mori Y, Orihashi Y, Miyamoto T, Shimada K, Shigeno M, Nishiyama T (2011) Origin of zircon in jadeitite from the Nishisonogi metamorphic rocks, Kyushu, Japan. J Metamorph Geol 29:673–684

Nasdala L, Hanchar JM (2005) Comment on: Application of Raman spectroscopy to distinguish metamorphic and igneous zircon (Xian et al., Anal Lett 2004, v. 37, p. 119). Anal Lett 38:727–734

Nasdala L, Lengauer CL, Hanchar JM, Kronz A, Wirth R, Blanc P, Kennedy AK, Seydoux-Guillaume AM (2002) Annealing radiation damage and the recovery of cathodoluminescence. Chem Geol 191:121–140

Nasdala L, Zhang M, Kempe U, Panczer G, Gaft M, Andrut M, Plötze M (2003) Spectroscopic methods applied to zircon. Rev Mineral Geochem 53:427–467

Page FZ, Essene EJ, Mukasa SB, Valley JW (2014) A garnet-zircon oxygen isotope record of subduction and exhumation fluids from the Franciscan complex, California. J Petrol 55:103–131

Page FZ, Ushikubo T, Kita NT, Riciputi LR, Valley JW (2007) High-precision oxygen isotope analysis of picogram samples reveals 2 µm gradients and slow diffusion in zircon. Am Mineral 92:1772–1775

Pape J, Mezger K, Robyr M (2016) A systematic evaluation of the Zr-in-rutile thermometer in ultra-high temperature (UHT) rocks. Contrib Mineral Petrol 171:1–20

Petersson A, Scherstén A, Andersson J, Whitehouse MJ, Baranoski MT (2015) Zircon U–Pb, Hf and O isotope constraints on growth versus reworking of continental crust in the subsurface Grenville orogen, Ohio, USA. Precambrian Res 265:313–327

Phillips G, Rubatto D, Phillips D, Offler R (2015) High-pressure metamorphism in the southern New England Orogen: implications for long-lived accretionary orogenesis in eastern Australia. Tectonics 34:1979–2010

Piazolo S, Austrheim H, Whitehouse M (2012) Brittle–ductile microfabrics in naturally deformed zircon: Deformation mechanisms and consequences for U–Pb dating. Am Mineral 97:1544–1563

Poller U, Huth J, Hoppe P, Williams IS (2001) REE, U, TH, and HF distribution in zircon from Western Carpathian Variscan granitoids: A combined cathodoluminescence and ion microprobe study. Am J Sci 301:858–876

Putlitz B, Matthews A, Valley JW (2000) Oxygen and hydrogen isotope study of high-pressure metagabbros and metabasalts (Cyclades, Greece): implications for the subduction of oceanic crust. Contrib Mineral Petrol 138:114–126

Putnis A (2009) Mineral replacement reactions. Rev Mineral Geochem 70:87–124

Rasmussen B (2005) Zircon growth in very low grade metasedimentary rocks: evidence for zirconium mobility at ~250 °C. Contrib Mineral Petrol 150:146–155

Rasmussen B, Fletcher IR, Muhling JR, Gregory CJ, Wilde SA (2011) Metamorphic replacement of mineral inclusions in detrital zircon from Jack Hills, Australia: Implications for the Hadean Earth. Geology 39:1143–1146

Reddy SM, Timms NE, Trimby P, Kinny PD, Buchan C, Blake K (2006) Crystal-plastic deformation of zircon: A defect in the assumption of chemical robustness. Geology 34:257–260

Reddy SM, Timms NE, Pantleon W, Trimby P (2007) Quantitative characterization of plastic deformation of zircon and geological implications. Contrib Mineral Petrol 153:625–645

Reddy SM, Timms NE, Eglington BM (2008) Electron backscatter diffraction analysis of zircon: A systematic assessment of match unit characteristics and pattern indexing optimization. Am Mineral 93:187–197

Reddy SM, Timms NE, Hamilton PJ, Smyth HR (2009) Deformation-related microstructures in magmatic zircon and implications for diffusion. Contrib Mineral Petrol 157:231–244

Reddy SM, Clark C, Timms NE, Eglington BM (2010) Electron backscatter diffraction analysis and orientation mapping of monazite. Mineral Mag 74:493–506

Root DB, Hacker BR, Mattinson JM, Wooden JL (2004) Zircon geochronology and ca. 400 Ma exhumation of Norwegian ultrahigh-pressure rocks: an ion microprobe and chemical abrasion study. Earth Planet Sci Lett 228:325–341

Rubatto D (2002) Zircon trace element geochemistry: distribution coefficients and the link between U–Pb ages and metamorphism. Chem Geol 184:123–138

Rubatto D, Angiboust S (2015) Oxygen isotope record of oceanic and high-pressure metasomatism: a *P–T*–time–fluid path for the Monviso eclogites (Italy). Contrib Mineral Petrol 170:44

Rubatto D, Chakraborty S, Dasgupta S (2013) Timescales of crustal melting in the Higher Himalayan Crystallines (Sikkim, Eastern Himalaya) inferred from trace element-constrained monazite and zircon chronology. Contrib Mineral Petrol 165:349–372

Rubatto D, Gebauer D (2000) Use of cathodoluminescence for U–Pb zircon dating by ion microprobe: some examples from the Western Alps. *In:* Pagel M, Barbin V, Blanc P, Ohnenstetter D (eds) Cathodoluminescence in geosciences, vol. Springer, Berlin Heidelberg New York, pp 373–400

Rubatto D, Gebauer D, Compagnoni R (1999) Dating of eclogite-facies zircons: the age of Alpine metamorphism in the Sesia-Lanzo Zone (Western Alps). Earth Planet Sci Lett 167:141–158

Rubatto D, Hermann J (2003) Zircon formation during fluid circulation in eclogites (Monviso, Western Alps): implications for Zr and Hf budget in subduction zones. Geochim Cosmochim Acta 67:2173–2187

Rubatto D, Hermann J (2007a) Experimental zircon/melt and zircon/garnet trace element partitioning and implications for the geochronology of crustal rocks. Chem Geol 241:62–87

Rubatto D, Hermann J (2007b) Zircon behaviour in deeply subducted rocks. Elements 3:31–35

Rubatto D, Williams IS, Buick IS (2001) Zircon and monazite response to prograde metamorphism in the Reynolds Range, central Australia. Contrib Mineral Petrol 140:458–468

Rubatto D, Hermann J, Buick IS (2006) Temperature and bulk composition control on the growth of monazite and zircon during low-pressure anatexis (Mount Stafford, central Australia). J Petrol 47:1973–1996

Rubatto D, Müntener O, Barnhorn A, Gregory C (2008) Dissolution–reprecipitation of zircon at low-temperature, high-pressure conditions (Lanzo Massif, Italy). Am Mineral 93:1519–1529

Rubatto D, Hermann J, Berger A, Engi M (2009) Protracted fluid-induced melting during Barrovian metamorphism in the Central Alps. Contrib Mineral Petrol 158:703–722

Rubatto D, Regis D, Hermann J, Boston K, Engi M, Beltrando M, McAlpine SRB (2011) Yo-Yo subduction recorded by accessory minerals in the Sesia Zone, Western Alps. Nature Geosci 4:338–342

Rumble D, Giorgis D, Ireland T, Zhang Z, Xu H, Yui TF, Yang J, Xu Z, Liou JG (2002) Low $\delta^{18}O$ zircons, U–Pb dating, and the age of the Qinglongshan oxygen and hydrogen isotope anomaly near Donghai in Jiangsu Province, China. Geochim Cosmochim Acta 66:2299–2306

Schaltegger U, Fanning M, Günther D, Maurin JC, Schulmann K, Gebauer D (1999) Growth, annealing and recrystallization of zircon and preservation of monazite in high-grade metamorphism: conventional and in-situ U–Pb isotope, cathodoluminescence and microchemical evidence. Contrib Mineral Petrol 134:186–201

Schmitt AK, Vazquez JA (2017) Secondary ionization mass spectrometry analysis in petrochronology. Rev Mineral Geochem 83:199–230

Schoene B, Baxter EF (2017) Petrochronology and TIMS. Rev Mineral Geochem 83:231–260

Seydoux-Guillaume A-M, Montel J-M, Bingen B, Bosse V, de Parseval P, Paquette J-L, Janots E, Wirth R (2012) Low-temperature alteration of monazite: Fluid mediated coupled dissolution–precipitation, irradiation damage, and disturbance of the U–Pb and Th–Pb chronometers. Chem Geol 330–331:140–158

Shatsky VS, Sobolev AV (2003) The Kokchetav massif, Kazakhstan. *In:* EMU Notes in Mineralogy Vol. 5: Ultrahigh Pressure Metamorphism, p. 75–103

Sheng Y-M, Zheng Y-F, Chen R-X, Li Q, Dai M (2012) Fluid action on zircon growth and recrystallization during quartz veining under UHP eclogite: Insights from U–Pb ages, O–Hf isotopes and trace elements. Lithos 136–139:126–144

Spandler C, Hermann J, Rubatto D (2004) Exsolution of thortveitite, yttrialite and xenotime during low temperature recrystallization of zircon from New Caledonia, and their significance for trace element incorporation in zircon. Am Mineral 89:1795–1806

Spandler C, Rubatto D, Hermann J (2005) Late Cretaceous–Tertiary tectonics of the southern Pacific; insight from U–Pb SHRIMP dating of eclogite-facies rocks from New Caledonia. Tectonics 24:TC3003, doi:3010.1029/2004TC001709

Spandler C, Pettke T, Rubatto D (2011) Internal and external fluid sources for eclogite-facies veins in the Monviso meta-ophiolite, Western Alps: Implications for fluid flow in subduction zones. J Petrol 52:1207–1236

Stepanov AS, Hermann J, Rubatto D, Rapp RP (2012) Experimental study of monazite/melt partitioning with implications for the REE, Th and U geochemistry of crustal rocks. Chem Geol 300–301:200–220

Stepanov A, Hermann J, Korsakov AV, Rubatto D (2014) Geochemistry of ultrahigh-pressure anatexis: fractionation of elements in the Kokchetav gneisses during melting at diamond-facies conditions. Contrib Mineral Petrol 167:1002

Stepanov A, Hermann J, Rubatto D, Korsakov AV, Danyushevsky (2016a) Melting history of an ultrahigh-pressure paragneiss revealed by multiphase solid inclusions in garnet, Kokchetav massif, Kazakhstan. J Petrol 57:1531–1554

Stepanov A, Rubatto D, Hermann J, Korsakov AV (2016b) Contrasting *P–T* paths within the Barchi-Kol UHP terrain (Kokchetav Complex): Implications for subduction and exhumation of continental crust. Am Mineral 101:788

Tailby ND, Walker AM, Berry AJ, Hermann J, Evans KA, Mavrogenes JA, O'Neill HSC, Rodina IS, Soldatov AV, Rubatto D, Newville M, Sutton SR (2011) Ti site occupancy in zircon. Geochim Cosmochim Acta 75:905–921

Taylor RJM, Harley SL, Hinton RW, Elphick S, Clark C, Kelly NM (2014) Experimental determination of REE partition coefficients between zircon, garnet and melt: A key to understanding high-*T* crustal processes. J Metamorph Geol 33:231–248

Tichomirowa M, Whitehouse MJ, Nasdala L (2005) Resorption, growth, solid state recrystallisation, and annealing of granulite facies zircon—a case study from the Central Erzgebirge, Bohemian Massif. Lithos 82:25–50

Timms NE, Kinny PD, Reddy SM (2006) Deformation-related modification of U and Th in zircon. Geochim Cosmochim Acta 70:A651

Timms NE, Kinny PD, Reddy SM, Evans K, Clark C, Healy D (2011) Relationship among titanium, rare earth elements, U–Pb ages and deformation microstructures in zircon: Implications for Ti-in-zircon thermometry. Chem Geol 280:33–46

Tomaschek F, Kennedy AK, Villa IM, Lagos M, Ballhaus C (2003) Zircons from Syros, Cyclades, Greece—recrystallization and mobilization of zircon during high-pressure metamorphism. J Petrol 44:1977–2002

Valley JW (2003) Oxygen isotopes in zircon. Rev Mineral Geochem 53:343–385

Valley PM, Fisher CM, Hanchar JM, Lam R, Tubrett M (2010) Hafnium isotopes in zircon: A tracer of fluid–rock interaction during magnetite–apatite ("Kiruna-type") mineralization. Chem Geol 275:208–220

Valley JW, Cavosie AJ, Ushikubo T, Reinhard DA, Lawrence DF, Larson DJ, Clifton PH, Kelly TF, Wilde SA, Moser DE, Spicuzza MJ (2014) Hadean age for a post-magma-ocean zircon confirmed by atom-probe tomography. Nature Geosci 7:219–223

Vavra G, Gebauer D, Schmidt R, Compston W (1996) Multiple zircon growth and recrystallization during polyphase Late Carboniferous to Triassic metamorphism in granulites of the Ivrea Zone (Southern Alps): an ion microprobe (SHRIMP) study. Contrib Mineral Petrol 122:337–358

Vervoort JD, Kemp AIS (2016) Clarifying the zircon Hf isotope record of crust–mantle evolution. Chem Geol 425:65–75

Vonlanthen P, Fitz Gerald JD, Rubatto D, Hermann J (2012) Recrystallization rims in zircon (Valle d'Arbedo. Switzerland): An integrated cathodoluminescence, LA-ICP-MS, SHRIMP, and TEM study. Am Mineral 97:369–377

Watson BE, Cherniak DJ (1997) Oxygen diffusion in zircon. Earth Planet Sci Lett 148:527–544

Watson EB, Harrison TM (2005) Zircon thermometer reveals minimum melting conditions on earliest Earth. Science 308:841–844

Whitehouse M, Kemp AIS (2010) On the difficulty of assigning crustal residence, magmatic protolith and metamorphic ages to Lewisian granulites: constraints from combined in situ U–Pb and Lu–Hf isotopes. Geological Society, London, Special Pubblications 335:81–101

Whitehouse MJ, Platt JP (2003) Dating high-grade metamorphism: constraints from rare-earth elements in zircon and garnet. Contrib Mineral Petrol 145:61–74

Whitehouse MJ, Ravindra Kumar GR, Rimša A (2014) Behaviour of radiogenic Pb in zircon during ultrahigh-temperature metamorphism: An ion imaging and ion tomography case study from the Kerala Khondalite Belt, southern India. Contrib Mineral Petrol 168:1–18

Williams I, Compston W, Black L, Ireland T, Foster J (1984) Unsupported radiogenic Pb in zircon: a cause of anomalously high Pb–Pb, U–Pb and Th–Pb ages. Contrib Mineral Petrol 88:322–327

Williams IS (2001) Response of detrital zircon and monazite, and their U–Pb isotopic systems, to regional metamorphism and host-rock partial melting, Cooma Complex, southeastern Australia. Aust J Earth Sci 48:557–580

Wu Y-B, Zheng Y-F, Zhao Z-F, Gong B, Liu XM, Wu F-Y (2006) U–Pb, Hf and O isotope evidence for two episodes of fluid-assisted zircon growth in marble-hosted eclogites from the Dabie orogen. Geochim Cosmochim Acta 70:3743–3761

Wu YB, Gao S, Zhang HF, Yang SH, Jiao WF, Liu YS, Yuan HL (2008a) Timing of UHP metamorphism in the Hong'an area, western Dabie Mountains, China: Evidence from zircon U–Pb age, trace element and Hf isotope composition. Contrib Mineral Petrol 155:123–133

Wu YB, Zheng YF, Gao S, Jiao WF, Liu YS (2008b) Zircon U–Pb age and trace element evidence for Paleoproterozoic granulite-facies metamorphism and Archean crustal rocks in the Dabie Orogen. Lithos 101:308–322

Xia QX, Zheng YF, Yuan H, Wu FY (2009) Contrasting Lu–Hf and U–Th–Pb isotope systematics between metamorphic growth and recrystallization of zircon from eclogite-facies metagranites in the Dabie orogen, China. Lithos 112:477–496

Xian WS, Sun M, Malpas J, Zhao GC, Zhou MF, Ye K, Liu JB, Phillips DL (2004) Application of Raman spectroscopy to distinguish metamorphic and igneous zircons. Anal Lett 37:119–130

Yakymchuk C, Brown M (2014) Behaviour of zircon and monazite during crustal melting. J Geol Soc 171:465–479

Ye K, Yao Y, Katayama I, Cong B, Wang Q, Maruyama S (2000) Large areal extent of ultrahigh-pressure metamorphism in the Sulu ultrahigh-pressure terrane of East China: new implications from coesite and omphacite inclusions in zircon of granitic gneiss. Lithos 52:157–164

Young DJ, Kylander-Clark ARC (2015) Does continental crust transform during eclogite facies metamorphism? J Metamorph Geol 33:331–357

Zhang ZM, Schertl HP, Wang JL, Shen K, Liou JG (2009a) Source of coesite inclusions within inherited magmatic zircon from Sulu UHP rocks, eastern China, and their bearing for fluid–rock interaction and SHRIMP dating. J Metamorph Geol 27:317–333

Zhang ZM, Shen K, Wang JL, Dong HL (2009b) Petrological and geochronological constraints on the formation, subduction and exhumation of the continental crust in the southern Sulu orogen, eastern-central China. Tectonophysics 475:291–307

Zhao ZF, Zheng YF, Wei CS, Chen FK, Liu X, Wu FY (2008) Zircon U–Pb ages, Hf and O isotopes constrain the crustal architecture of the ultrahigh-pressure Dabie orogen in China. Chem Geol 253:222–242

Zhao L, Li T, Peng P, Guo J, Wang W, Wang H, Santosh M, Zhai M (2015) Anatomy of zircon growth in high pressure granulites: SIMS U–Pb geochronology and Lu–Hf isotopes from the Jiaobei Terrane, eastern North China Craton. Gondwana Res 28:1373–1390

Zheng YF, Fu B, Gong B, Li L (2003) Stable isotope geochemistry of ultra-high pressure metamorphic rocks from the Dabie-Sulu orogen in China; implications for geodynamics and fluid regime. Earth-Sci Rev 62:105–161

Zheng YF, Wu Y-B, Zhao Z-F, Zhang S-B, Xu P, Wu F-Y (2005) Metamorphic effect on zircon Lu–Hf and U–Pb isotope systems in ultrahigh-pressure eclogite-facies metagranite and metabasalt. Earth Planet Sci Lett 240:378–400

Zheng YF, Gong B, Zhao ZF, Wu YB, Chen FK (2008) Zircon U–Pb age and O isotope evidence for Neoproterozoic low-^{18}O magmatism during supercontinental rifting in South China: Implications for the snowball earth event. Am J Sci 308:484–516

Reviews in Mineralogy & Geochemistry
Vol. 83 pp. 297–328, 2017
Copyright © Mineralogical Society of America

10

Petrochronology of Zircon and Baddeleyite in Igneous Rocks: Reconstructing Magmatic Processes at High Temporal Resolution

Urs Schaltegger

Department of Earth Sciences
University of Geneva
1205 Geneva,
Switzerland

urs.schaltegger@unige.ch

Joshua H.F.L. Davies

Department of Earth Sciences
University of Geneva
1205 Geneva
Switzerland

joshua.davies@unige.ch

INTRODUCTION

Zircon ($ZrSiO_4$) and baddeleyite (ZrO_2) are common accessory minerals in igneous rocks of felsic to   composition. Both minerals host trace elements substituting for Zr, among them Hf, Th, U, Y, REEs and many more. The excellent chemical and physical resistivity of zircon makes this mineral a perfect archive of chemical and temporal information to trace geological processes in the past, utilizing the outstanding power and temporal resolution of the U–Pb decay schemes. Baddeleyite is a chemically and physically much more fragile mineral. It preserves similar information only where it is shielded from dissolution and physical fragmentation as an inclusion in other minerals or in a   or non-reactive rock matrix. It   the potential for dating the   of   rocks with high-precision through its crystallization in small pockets of Zr-enriched melt, after extensive olivine and pyroxene fractionation. Zircon and baddelelyite U–Pb dates are, for an overwhelming majority of cases

The development of the U–Pb dating tool CA-ID-TIMS (chemical abrasion-isotope dilution-thermal ionization mass spectrometry) since 2005 has led to unprecedented precision of better than 0.1% in $^{206}Pb/^{238}U$ dates (Bowring et al. 2005). Increased sensitivity of mass spectrometers and low laboratory blanks due to reduction of acid volumes allow routine U–Pb age determinations of micrograms of material at   high radiogenic / common lead ratios (see Schoene and Baxter 2017, this volume).

In situ U–Pb age analysis using laser ablation or primary ion beam sputtering allows analysis of sub-microgram quantities of zircon material from polished internal sections or zircon surfaces with spot diameters ranging from ~30 μm for laser-ablation, inductively coupled plasma mass spectrometry (LA-ICP-MS) to 10 μm for secondary ion mass spectrometry (SIMS), lateral resolutions of 2–5 μm for NanoSIMS (Yang et al. 2012), or from high-voltage pulse induced evaporation of needle-tips of 200 nm diameter through

1529-6466/17/0083-0010$05.00 (print)
1943-2666/17/0083-0010$05.00 (online)

http://dx.doi.org/10.2138/rmg.2017.83.10

atom probe tomography (APT; Valley et al. 2015). Increased spatial resolution is logically linked to a drastic decrease of analyzed volume and to increased analytical uncertainty. In contrast, ID-TIMS analysis integrates the chemical and isotopic information over the analyzed volume of an entire grain or parts of a fragmented grain, which yields integrated compositional and temporal information at high precision. The dilemma of low-precision at high spatial resolution and high-precision at low spatial resolution is in many cases asking for an intelligent combination of the two approaches, both for elemental and isotopic analysis (Schaltegger et al. 2015). However, to resolve the timescale of processes that are active in magmatic systems, we need temporal information in the 10^3–10^5 years range, precluding the use of *in situ* dating techniques for rocks that are older than ~10 Ma; ID-TIMS analysis can provide the necessary temporal resolution up to ages of ~500 Ma. In rare cases, precise $^{207}Pb/^{206}Pb$ dates from Proterozoic and Archean zircon may provide   resolution to trace magmatic processes (e.g., Schoene and Bowring 2010; Zeh et al. 2015).

In this chapter we concentrate mainly on the approach of combining high-precision CA-ID-TIMS dating of zircon and baddeleyite with compositional imaging and analysis to obtain the maximum amount of information on the petrogenesis of the host rock as well as the physio-chemical conditions during crystallization at maximum temporal resolution. We also give credit to low-precision U–Pb petrochronology utilizing LA-ICP-MS, as well as SIMS, for the latter with a focus on disequilibrium dating of young volcanic zircon.

> *understanding the petrological evolution of a given rock and a given mineral assemblage in an absolute time frame that has been established by radio-isotopic dating.*

WORKFLOW FOR ZIRCON PETROCHRONOLOGY

The aim of petrochronology is to relate an age determination as closely as possible with elemental and isotopic analysis of the dated minerals. Looking beyond the horizon of zircon geochronology, we may want to analyze trace element concentrations and/or Hf, Nd, Pb or Sr isotopes in a variety of U-bearing minerals amenable to U–Pb dating (titanite, monazite, xenotime, rutile and potentially others). Concentrating on the mineral zircon, the analytical   are   epending on the choice of the dating method – high temporal resolution U–Pb dating using CA-ID-TIMS, or high spatial resolution LA-ICP-MS or SIMS techniques.

The concept of petrochronology was initially developed for studying the growth of minerals in metamorphic systems, where a combination of composition and age helped to determine the timing of growth as well as the controlling metamorphic reactions (e.g., Fraser et al. 1997). For igneous systems, petrochronology developed later, partially in combination with laser-ablation, split-stream analysis (LASS; Yuan et al. 2008; Kylander-Clark et al. 2013). LASS refers to the simultaneous analysis of chemical or isotopic composition and age on the exact same volume of a previously imaged grain (backscatterd electrons, BSE, or cathodo-luminescence, CL) mounted in epoxy resin and polished, or directly from a polished thin section. Initial chemical analysis could be done by electron microprobe (EMP), a required step for minerals other than zircon to quantify one element as an internal standard in LA-analysis. The LASS approach involves splitting the ablated particle stream, suspended in a Ar–He–N_2 gas mixture, and deviating, e.g., one half into a sector   ICP-MS (SF-ICP-MS) for U–Pb dating, the other half into a multicollector ICP-MS (MC-ICP-MS) for Hf isotope analysis. Such setups have proven suitable for precise and accurate age determination and Hf isotopic analysis of zircon (e.g., Fisher et al. 2014; Reimink et al. 2016). Another setup may be designed using a quadrupole ICP-MS (Q-ICP-MS) to determine elemental concentrations, and a SF or MC-ICP-MS for dating, when relating, e.g., heavy rare earth element (HREE) fractionation during magmatic or metamorphic crystallization to zircon ages. If LASS capabilities are not present in a given laboratory, the analytical steps can be done sequentially; one may   analyse the U–Pb age in a 30 μm diameter UV-laser pit, which is later "overdrilled" by a 70 μm UV-laser pit for Hf isotope analysis. However, age and Hf isotope information cannot always be considered identical since the volumes analysed   . This can lead to   Hf isotope variation in Archean and Hadean zircon due to inaccurate correction for the proportion of radiogenic Hf produced *in situ* from the radioactive decay of the parent isotope ^{176}Lu (e.g., Kemp et al. 2010).

LASS was successfully used to address the question of zircon growth in the stability of garnet through HREE analysis in high-pressure metamorphic terranes (e.g., Yuan et al. 2008; Kylander-Clark et al. 2013, and references therein). An example for sequential use of SIMS for U–Pb age determination, REE and Y elemental concentration analysis and O isotope analysis, followed by LA-MC-ICP-MS analysis of Hf isotope composition in diamond bearing zircon may be found in Kotková et al. (2016). Metamorphic reactions or short periods of zircon saturation in a magma may lead to thin overgrowths of zircon on pre-existing grains; such overgrowths may be analysed for age and elemental/isotopic composition using depth   on a SIMS (e.g., Reid and Coath 2000; Schmitt 2011), or by single shot analysis with a LA-ICP-MS equipment (e.g., Cottle et al. 2012).

SIMS analysis is the method of choice when applying ^{238}U–^{232}Th–^{230}Th dating techniqes to zircons from young volcanic rocks, utilizing initial ^{230}Th disequilibrium during the crystallization of zircon. However, the ages produced are model ages that rely on an estimated initial ^{232}Th/^{230}Th ratio, usually obtained from the whole rock, and therefore may not be acurate (Reid et al. 1997). The major   hattcan be obtained from disequilibrium zircon dating by SIMS is depth   of individual zircon grains for age, trace elements and isotopic composition (e.g., Storm et al. 2014; Rubin et al. 2016). Using these tools, the geochemical evolution of the zircon can be linked to ages in the 10's of ka to reveal magmatic processes, protracted zircon growth or recycling of old grains. In a general sense, U-series petrochronology has uncovered identical magmatic processes at young time scales to conventional U–Pb petrochronology. However, this technique is limited to ages younger than ~500 ka, and the ages produced have to be considered as model ages. In the following we focus our discussion mostly to examples of conventional U–Pb petrochronology.

All of the approaches descrived above are inadequate for resolving rapid geological processes in rocks that are older than ~100 Ma. For these we need to utilize high-precision U–Pb dating techniques involving CA-ID-TIMS. A complete   for zircon petrochronology involving CA-ID-TIMS high-precision U–Pb age determination is complex and consists of a combination of *in situ* and bulk-dissolution techniques (Fig. 1).

A) Imaging of the textural relationships from a thin section, using BSE images from a scanning electron microscope (SEM) or BSE and X-ray imaging on a QEMScan (Quantitative Evaluation of Minerals by Scanning electron microscopy). This analysis provides detailed information on the textural position of zircon and baddeleyite in the rock, thus allowing selective sampling of an accessory phase according to its paragenetic position. In addition, the QEMScan data set yield grain size distribution statistics for zircon if the imaging resolution is appropriately selected. This textural characterization prior to sampling of zircon directly from the imaged section will become increasingly important, since it allows us to directly link a paragenetic sequence and information with chemical composition and date.

B) Selected zircon grains may be extracted from thin section using a microdrill and individually annealed in quartz crucibles for later chemical abrasion treatment (e.g., Broderick 2013). It is also possible to mount grains after annealing in epoxy resin, thus preserving the initially polished surface from the thin section.

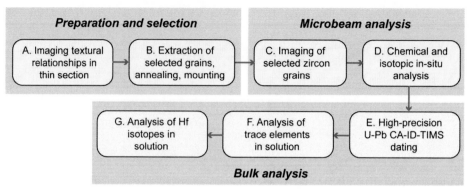

Figure 1.

C) After a very light repolishing, imaging of the growth structures of zircon using optical, or panchromatic or wavelength-resolved cathodo-luminescence (CL) on an SEM. This information is essential for relating zircon textures with U–Pb dates in order to establish a relative crystallization sequence and produce high quality images to guide *in situ* work.

D) Carrying out chemical or isotopic *in situ* analyses, such as trace element, oxygen or hafnium isotopic analysis by SIMS, or trace element and hafnium isotopic analysis through LA-ICP-MS. These *in situ* procedures allow separate and selective analysis of single growth zones of zircon for their chemical and isotopic composition. We note that ID-TIMS U–Pb age determinations performed at the University of Geneva laboratory of zircon grains that were previously sampled for laser ablation analysis did not reveal any   from prior laser analysis on the U–Pb system of zircon (e.g., Chelle-Michou et al. 2014).   , melt rims around laser craters are removed entirely during the partial dissolution step of the chemical abrasion procedure.

E) After careful extraction of selected, imaged and chemically and isotopically characterized zircon crystals, partial dissolution is undertaken to remove   of radiation-related lattice damage (Mattinson 2005). The grains are then subjected to dissolution after adding a tracer solution, such as, the (^{202}Pb–) ^{205}Pb–^{233}U–^{235}U EARTHTIME tracer (Condon et al. 2015), see details of analytical techniques in Schoene and Baxter (2017, this volume) or in Schoene (2014). Lead and uranium are isolated using anion-exchange chromatographic separation techniques and their isotopic compositions are measured on a thermal ionization mass spectrometer (TIMS). The resulting ^{206}Pb/^{238}U dates are very precise (at $\pm 0.05\%$ or better) and since many labs use the same EARTHTIME tracer solutions, the individual ages can be compared between these labs using only the internal (analytical, random) uncertainties, rather than including tracer calibration and decay constant uncertainties. These later uncertainties should be included when comparing ages between non EAR

F) Chromatography also isolates the trace element and REE fraction that can be analyzed for elemental concentrations and ratios using a SF-ICP-MS in dry plasma mode with a desolvating nebulizer, a technique described as TIMS-TEA (Schoene et al. 2010). This allows determination of chemical information from the same volume of the grain that is dated.

G) From the same trace element solution Hf isotope analyses can be carried out (if the Hf concentration is   high) using a MC-ICP-MS in dry plasma mode. At low Hf concentrations, standard bracketing with matrix-matched natural reference materials is crucial to obtain accurate results (D'Abzac et al. 2016).

WORKFLOW FOR BADDELEYITE PETROCHRONOLOGY

Baddeleyite petrochronology has been applied very rarely. The traditional chemical abrasion method applied to zircon has not proven to be   for baddeleyite (Rioux et al. 2010). Hence the problem of secondary lead loss in baddeleyite, which   biases the accuracy of U–Pb dates, remains an issue. Furthermore, the small size of baddeleyite, typically <70 μm, and its monoclinic platy habit typically prohibit a complex   involving extraction from thin section, remounting for trace elements/isotopic analysis and imaging, and then extraction from the mount again for U–Pb dating. However, a typical   for baddeleyite petrochronology proceeds similarly to zircon:

A) Electron imaging of thin sections to identify baddeleyite and determine its host or co-crystallizing minerals. This is the best way to determine the baddeleyite content of the rock and its potential for U–Pb geochronology. From the authors' personal experience, samples without baddeleyite in thin section usually do not have this mineral in their mineral separates. Since extraction from whole rock powder is trickier and more time consuming for baddeleyite than zircon, checking thin sections for the presence of baddeleyite and determining the average size of the grains before separation is advised.

B) Coring of the thin section to date the   baddeleyite crystals '*in situ*' by SIMS can be conducted at this point (e.g., Schmitt et al. 2010). Alternatively, traditional zircon separation techniques, or   hereof (Soderlund and Johansson 2002) are employed on whole rock powder crushed to <250 μm to extract the baddeleyite crystals. To our knowledge, no studies have employed extraction of baddeleyite directly from thin section for TIMS dating.

C) Large grains (> 100 μm) can be mounted for further electron imaging by SEM or *in situ* analysis by electron microprobe or LA-ICP-MS for Hf isotopes (Ibanez-Mejia et al. 2014).

D) Separated grains are either removed from the grain mount and dissolved for ID-TIMS U–Pb analysis or dissolved without mounting and imaging if the grains are small. Dissolution procedures are the same as for zircon, but without chemical abrasion pre-treatment. Uranium and Pb   column chemistry washes can also be collected and analyzed for Hf isotopic composition on an MC-ICP-MS if the grains are large enough to provide a high Hf signal. Special analytical procedures may be required to measure the Hf isotopic composition of small grains (e.g., D'Abzac et al. 2016).

PETROCHRONOLOGY OF ZIRCON IN INTERMEDIATE TO FELSIC SYSTEMS

In the following section we discuss the information that can be obtained from careful chemical and isotopic analysis of zircon. We show that zircon can record information on the temporal evolution of the chemical composition of a magmatic system and how it changes as a function of the physical processes such as crystallisation, and magma extraction and recharge. We   discuss how zircon in intermediate to felsic rocks   the evolution of a calc-alkaline system, then, in the following chapter, we discuss how baddeleyite and zircon

Crystallization of zircon in intermediate–felsic, calc-alkaline melts

First consider crystallization in a   *magmatic system* along a monotonic cooling path without any thermal disturbance from magma recharge. Temperature decreases over a period of time, before arriving at the solidus of a given chemical system.

Saturation of a given magmatic liquid with respect to zircon can be expressed as the zircon saturation temperature (T_{sat}; Watson and Harrison 1983). The zircon saturation temperature

T_{sat} is a function of the magma composition, depending on the concentration of Zr and the concentration ratio of network-forming elements Si+Al over K+Na+Ca in the melt, expressed as $M=[(Na+K+2Ca)/(Si+Al)]$, and for compositions of $M=1.0$ to 2.1. This range covers granitic to granodioritic composition, but does not include more melts. The dependence of the zircon saturation temperature on parameters other than the ones listed above (such as the variation in H_2O or changing melt composition through crystallization of concurrent silicate phases) leads to an uncertainty on T_{sat} of tens of degrees at least, see discussion in Boehnke et al. (2013). We also here the "saturation interval", which extends between the saturation temperature (T_{sat}) and the solidus temperature, i.e., the temperature range during which we expect zircon to grow.

Alternatively, the crystallization temperature can be approximated by measuring the concentration of Ti in zircon, since the incorporation of Ti into the lattice of cyrstallizing zircon has been shown to be dependent on temperature (Watson and Harrison 1983; Ferry and Watson 2007). However, further discussion of this approach by Fu et al. (2008) and Hofmann et al. (2009) suggested that Ti partitioning into zircon is strongly by factors other than just a_{TiO_2} and a_{SiO_2}. Those include pressure, trace element composition of the parent melt, relative rates of Ti, Si and Zr in the melt surrounding the growing zircon, and $Zr^{4+}_{+1} Hf^{4+}_{-1}$ and $(Si^{4+}+Zr^{4+})_1 (P^{5+}+Y^{3+})_{-1}$ exchange vectors (hafnon and xenotime solid solution vectors).

At the saturation temperature (T_{sat}), zircon nucleates either on pre-existing older zircon crystals through so-called heterogeneous nucleation, or spontaneously through homogeneous nucleation from a liquid that is (over)saturated with respect to zircon, which requires a lower activation energy. Zircon then continues growing through an Ostwald-ripening process. Heterogeneous nucleation takes place at the surface of older, resorbed zircon grains as well as on grains that have crystallized earlier in the magmatic system, in crystal mushes that were present at lower levels of the system ("antecrystic" zircon; Miller et al. 2007). Any magma may contain recycled zircon crystals from earlier phases of crystallization within the same magmatic system. In some cases (e.g., in Charlier et al. 2004), what looked like an "inherited core" in a zircon grain was shown to be only a few 10's to 100 ka older than the mostly oscillatory zone magmatic rim.

Obviously, zircon crystallizes over a certain period of time, during which the magma is cooling and the percentage of the total volume of crystals (zircon and others) is increasing. In intermediate magma, crystallization of large volumes of amphibole and plagioclase occur together with or prior to zircon precipitation. Zircon may thus occur as inclusions in the major minerals as well as in matrix minerals that crystallized subsequently from the interstitial liquid (or it may be fortouitously included in secondary minerals). In particular, the crystallization of plagioclase will fractionate Zr as an incompatible element into the residual liquid, where zircon will crystallize. The change of melt composition through fractional crystallization of major and accessory phases will also alter T_{sat}. Pressure is not known to zircon saturation in the liquid, but will the crystallization sequence of plagioclase and amphibole, and these in turn the trace element partitioning into zircon. Trace element partitioning into zircon is therefore a petrogenetic tool that allows reconstruction of the fractional crystallization path of a magmatic system.

In case of monotonic cooling and crystallization of a very small model pluton (*1000 °C melt temperature, 2 km thick cylindrical pluton with a volume of 12 km³, 10 km intrusion depth; 30 °C/km geothermal gradient; solidus temperature 680 °C*) zircon will crystallize over 7000 years, between 5 and 16 ka after intrusion, in a portion of the pluton that resides at an average temperature, along the solid thin line in Fig. 2A. Note that near the top and sides of the pluton zircon will crystallize over a shorter interval as cooling is faster than in interior and deep portions. Therefore, at the margin of the intrusion zircon crystallization is instant (the stippled curve "minimum temperature" remains below the solidus temperature at all times), whereas

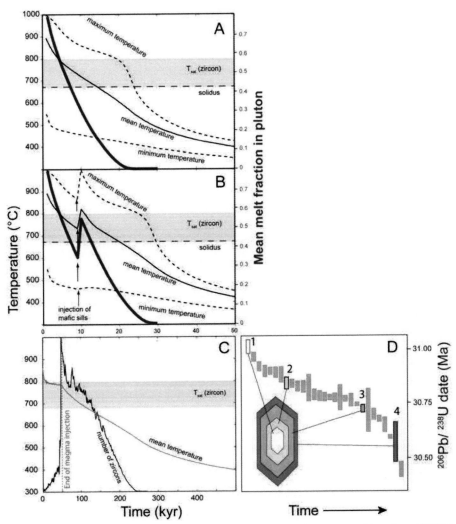

Figure 2. Modelled thermal decay curves and melt fractions for a cooling and crystallizing pluton (1000 °C melt temperature, 2 km thick cylindrical pluton = 12 km³, 10 km intrusion depth; 30 °C/km geothermal gradient; solidus temperature 680 °C); the grey horizontal band indicates the range of zircon saturation temperatures (T_{sat}) in intermediate to felsic magmatic systems; A: monotonous cooling; B: with one recharge after 13 ka (melt temperature 1100 °C). Stippled lines account for minimum temperature at the outermost border of the pluton, and maximum temperature in the core, the solid blue line is the melt fraction measured on the right axis. The model pluton is entirely solid at 20 and 27 ka, respectively. C) Theoretical dimensionless distribution of newly formed zircon crystals in a crystallizing magma, assuming intrusion over 50 ka at a   of 10⁻² km³/a and a   volume of 500 km³ (calculated and redrawn after Caricchi et al. 2014); D) theoretical, sigmoidal age distribution of zircon growth bands crystallizing from a monotonously cooling magma. Stages 1–4 discussed in text.

in the hottest part the liquid remains within the zircon saturation interval until ~25 ka after emplacement. This time span corresponds to the "magmatic residence time" and to the period of "autocrystic" zircon growth (Miller et al. 2007). The expected distribution of $^{206}Pb/^{238}U$ dates recorded by zircon crystals in a sample of this pluton will describe a unimodal, skewed distribution pattern, assuming linear nucleation and crystallization rates, as anticipated by the crystallization model of Caricchi et al. (2014; see Fig. 2C;). A zircon crystallized in such a   system would exhibit undisturbed oscillatory zoning visible in cathodo-luminescence (CL; Fig. 3A). If we could analyze each growth segment of an oscillatory zoned zircon separately, zones 1 to 4 (as an example) would   a systematic decrease of the date of crystallization (Fig. 2D). If a linear crystallization rate is assumed, the temporal distribution of individual growth zones would produce a curved, sigmoidal pattern with a majority of dates concentrated between dates equivalent to zone 2 and zone 3 (Fig. 2D); the curvature of the distribution of individual dates is a function of the changing proportion of magma within the zircon saturation interval, relative to magma at temperatures below or above during progressive crystallization.

For whole-grain ID-TIMS dates it is necessary to evaluate how a single zircon grain   these processes because it potentially integrates the entire, or at least part of the growth history including growth segments 1 to 4. Zircon single-grain dates from plutons or volcanic units show dispersions that can vary between near-zero (Wotzlaw et al. 2014) and up to several 10 to 100 ka (Leuthold et al. 2012; Schoene et al. 2012; Broderick et al. 2015). Despite the fact

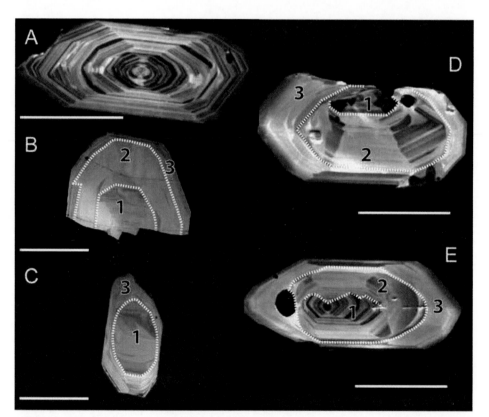

Figure 3. Cathodo-luminescence images of magmatic zircon crystals. A: oscillatory zoning from undisturbed growth in a magmatic liquid (leucotonalite, S. Adamello, N. Italy); B: magmatic zircon showing three growth zones—1) sector zoning, with truncated boundary versus 2) sector zoning, 3) oscillatory zoning (granite porphyry, Mongolia); C: zircon crystal from same rock sample as B, but devoid of growth zone 2, evidencing that not all zircon crystals of the same rock sample share the same growth history; D: complexly zoned zircon crystal with three growth periods—1) combined oscillatory and sector zoning, 2) sector zoning with somewhat subordinate oscillatory zoning/banding, 3) outer zone with faint oscillatory and sector zoning –growth zones are separated by intermittent resorption events (tonalite, S. Adamello, N. Italy); E: zircon from the same sample as D, showing very similar characteristics; note the non-planar zoning in zone

that zircon in a real magmatic system grows at   rates, and it nucleates homogeneously as well as heterogeneously, the age distribution of single-grain ID-TIMS dates is always expected to be sigmoidal in shape when a   number ($N > 15$) of grains is analyzed (e.g., Samperton et al. 2015). Other processes, such as magma recharge and inheritance from re-melting of partially or entirely crystallized precursor systems, will blur the initial distribution pattern. It has been demonstrated that a large part of the history of a magmatic plumbing system can be reconstructed from the zircon population of a single hand-sample (e.g., Broderick et al. 2015; Samperton et al. 2015). This fact invokes homogenization processes that transports zircon between   portions of a crystallizing magma reservoir and/or pass melts of composition through the porous framework of a crystal mush. If thermally driven convection is strong enough, initial small-scale, local and ephemeral equilibrium states get averaged out; by contrast, in low-volume and rapidly cooling systems such states will be preserved (Broderick et al. 2015). As a result, a U–Pb date of a zircon is to be considered as a chemical signal and cannot be readily related to a physical process other than its crystallization.

The more realistic case of a *complex magmatic system*, with periods of recharge from   and hot magma batches will lead to   emperature oscillations within the crystallizing magma, and to repeated intermittent periods of zircon resorption and dissolution during the growth history.

Consider a crystal-free, water-saturated basaltic melt ($50\% \, SiO_2$) that experiences monotonic crystallization from 1300 °C (Caricchi and Blundy 2015) and   olivine, clinopyroxene, plagioclase and amphibole fractionation at a deep level of the magmatic plumbing system (the "hot zone", following Annen 2011). A small volume of zircon forms during this period of rapid crystallization (rapid decrease of the melt fraction M_f; interval I on Fig. 4) in small, marginal volumes of the magmatic reservoir. In most magmatic systems, zircon would preferentially crystallize during a long period of slow crystallization (interval II), when the biggest part of the magma reservoir is within the T_{sat} window. Zircon growth during this time interval has been termed "equilibrium crystallization" (Watson and Liang 1995; Wang et al. 2011) and is commonly associated with sector-zoned growth (see, e.g., Corfu et al. 2003; Kotková et al. 2010; Tapster et al. 2016). Such early growth zones with sector zoning are displayed in Figs. 3B, C (zone 1) and Figs. 3 D, E (zone 2). Crystallization is terminated by rapid precipitation of oscillatory zoned zircon when approaching the solidus (interval III in Fig. 4). However, several additional processes may occur within the system (stippled lines in Fig. 4): i) andesitic magma may be extracted after interval I from the deep part of the plumbing system and rises into the middle crust; ii) subsequently, more batches of increasingly evolved melts are extracted from the crystallizing magma (A, B, C); iii) crystallizing magma batches ("crystal mushes") in the middle crust are injected and thermally rejuvenated by incoming hot (1100 °C) andesitic magma, or, iv) volatile loss shifts the crystallization curve at a given melt fraction to higher temperature.

Considering a single injection of 1100 °C hot andesite magma 10 ka after the initial emplacement (Fig. 2B), we observe several   compared to the simple case shown in Fig. 2A: (i) the time the magma takes to arrive at the solidus is prolonged to 20 ka in the part of the intrusion at average temperature (Fig. 2B); (ii) if   eltnis mixing with re-juvenated mush, the change in chemical composition will lead to a change in T_{sat}; (iii) withdrawal of Zr through zircon crystallization results in a decrease of T_{sat}; (iv) the zircon population will record a bimodal distribution of $^{206}Pb/^{238}U$ dates centered around ~8 and ~15 ka, due to crystallization after the recharge events; (v) the peak temperature during this thermal rejuvenation may exceed T_{sat} for the given magma composition and thus lead to a short period of zircon resorption (Fig. 2B).

The models show that recharge and rejuvenation of a magmatic system through hotter, incoming magma leads to thermal rejuvenation (decrease of the crystal proportion) and to repeated periods of enhanced zircon crystallization. And indeed, multi-peak zircon age distributions over 200 to 300 ka from high-precision U–Pb dates are a common feature of most magmatic systems,

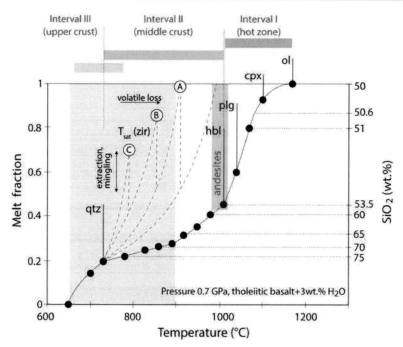

Figure 4. Fractional crystallization of major minerals and zircon from a tholeiitic basalt as a function of crystal content and temperature (after Caricchi et al. 2014, 2015). A, B and C denote extraction events of interstitial melt from the crystallizing mush. Dashed lines trace the melt fraction during extraction and subsequent crystallization of melt batches. The grey band delineates the temperature interval between arrival at T_{sat} (zircon) at 900 °C and the solidus, i.e., the interval during which zircon forms. Decreasing water content of melts due to volatile loss would shift the melt fraction curve towards higher temperatures (Caricchi and Blundy 2015).

suggesting that temperature oscillations are a common result of magma recharge events. Figures 5 A to C show three examples of complex magmatic systems (taken from Broderick et al. 2015; Samperton et al. 2015, and Wotzlaw et al. 2013, respectively). Other examples with high-precision geochronology can be found in Barboni et al. (2015; see distribution curves in Caricchi et al. 2016), Schoene et al. (2012), or Wotzlaw et al. (2012). Temperature cycling often involves periods of resorption, visible in CL or BSE images, during which temperature exceeded T_{sat} (zir). Sector and/or oscillating growth zones are truncated along rounded internal surfaces that are epitaxially overgrown by subsequent euhedral growth zones (Figs. 3B, C truncated zone 1; Figs. 3 D, E, truncated zones 1 and 2). The sequence of zircon growth zones represents a temporal "stratigraphy" of a thermally decaying magmatic system, where the more internal zones of a zircon crystal grew during earlier stages of the magmatic system. In case of mushes that were > 50% crystalline, these were subsequently rejuvenated due to temperature oscillations during magma recharge. Internal zones that formed during an early stage, at deeper levels of a magmatic plumbing systems, often show sector zoning or a combination of sector and oscillatory zoning (Fig. 3) and may be termed "antecrystic" in the sense of Miller et al. (2007). The onset of magmatic zircon crystallization (t=0) in the example of the Val Fredda Complex (Fig. 5A) was set so as to ignore two ca. 150 ka older zircon analyses from the same hand sample. Such zircons (several 100 ka older) may be derived from cannibalized "proto-plutons" (e.g., Chelle-Michou et al. 2014; Reubi and Blundy 2008; Wotzlaw et al. 2013) that were reactivated and rejuvenated during injection of new magma. A distinction between these types of "antecrysts" may be made through chemical and isotopic (O, Hf) analysis, see below. Not all zircon crystals of one hand sample need to show the same "growth stratigraphy". For example, the crystals shown in

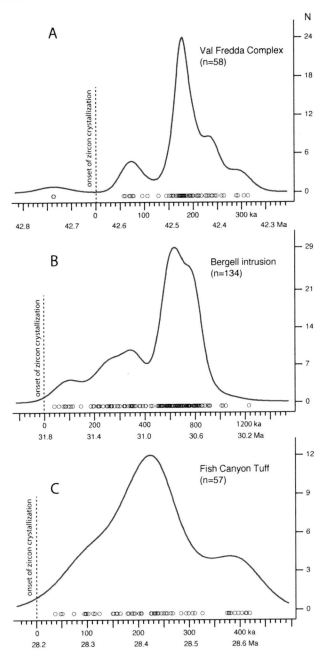

Figure 5. Examples of multi-peak zircon date distributions from magmatic systems of increasing volume (kernel density plots; Vermeesch et al. 2012); A: Val Fredda Complex, S. Adamello intrusion, N. Italy (Broderick et al. 2015); B: Bergell intrusives, N. Italy (Samperton et al. 2015); C: Fish Canyon T   Wotzlaw et al. 2013).

Figs. 3 B and C are separated from the same granite porphyry sample, but C did not experience the growth event 2, intermediate in age between 1 and 3, or, alternatively did not preserve it during the resorption event between 2 and 3.

Age distribution curves as shown in Figure 5 also carry information about the physical state of a magma reservoir. A small reservoir of magma (tens of km^3) that has been periodically replenished will   ephemeral, short episodes during which melt crystallized within 10^5 year timescales (Fig. 5A). Increasing the volume of a magmatic system to hundreds of km^3, longer periods at higher (>50% vol.) melt proportion can be produced, creating reservoir-wide melt convection and exchange of crystals (Fig. 5B). These features are shown in the zircon age distributions as numerous single peaks that get smoothed out, then the principal and secondary maxima become wider. Zircon that is crystallizing in equilibrium in a larger volume of melt over longer periods of time has to be considered "autocrystic", as proposed in Samperton et al. (2015). In a giant magmatic system (thousands of km^3), these homogenizing   re a even more important and lead to a large age dispersion with a single mode and a large standard deviation (Fig. 5C). Using the model approach of Caricchi et al. (2014, 2016), mode, median and standard deviation of populations of zircon U–Pb dates can be tentatively employed to estimate the integrated magma   and the volume of a magmatic system, even where it is not known a priori whether the resulting data derive from a lower or upper crustal magma reservoir.

Distributions of zircon dates over several 100 ka often are not in agreement with the short   times required for small magma volumes. To explain this apparent contradiction, crystallization of zircon at deeper crustal levels, followed by transport in suspension, in small magma batches, into the upper crust is typically invoked. Sequential accumulation of small magma batches into larger plutons at low magma   rates is a characteristic of many mid to upper crustal plutons (e.g., Tappa et al. 2011; Schoene et al. 2012; Rosera et al. 2013; Chelle-Michou et al. 2014; Barboni et al. 2015; Broderick et al. 2015). Incremental accumulation of small magma batches became popular as the major paradigm for the buildup of large plutons (after Glazner et al. 2004, and Coleman et al. 2004), as an alternative concept to "big tank" models that involve magma emplacement through diapirism.

Finally, a word of caution is due when referring to a number of dispersed (antecrystic) U–Pb single zircon dates grains as "magmatic residence time", or "period of protracted zircon growth": The age dispersion   by the antecrysts (Miller et al. 2007) is the result of processes, involving interaction with previously partly crystallized magmas during melt ascent and recycling of zircon from rejuvenated mushes. The oldest antecrysts may be many 100 ka older than the latest crystallization. It is therefore implied that "magmatic residence" includes in many cases long periods of storage of previously crystallized zircon in cold mushes close to the solidus and not as suspended crystals in a melt (Claiborne et al. 2010; Klemetti et al. 2014).

What does zircon chemistry tell us about magmatic processes?

Co-precipitating mineral assemblages. Zircon crystallizes together with a series of major and accessory minerals, which all compete for some critical trace elements. Analyses of these elements in dated zircon allow the calculation of time-resolved crystallization paths and the evolution of equilibrium melt compositions, via known partition   In practical terms, only a small series of elements in zircon is accessible to   through electron microprobe and laser ablation ICP-MS analyses, such as Zr, Hf, Ti, REE (mostly heavy REE, HREE), Y, Th, and U. The Th/U ratio may also be obtained from the radiogenic $^{208}Pb/^{206}Pb$ ratio from ID-TIMS age determination. In the following, we inspect the chemical variation of selected elements in zircon crystallizing in calc-alkaline magmas in continental arc settings,   excluding those melts in which monazite and xenotime are also stable, because no data are available yet for the chemical and isotopic composition of precisely (ID-TIMS) dated zircon from these.

Partitioning of rare earth elements. Detailed chemical investigation of the major and accessory minerals in a quartz–plagioclase diorite (PQD) and a tonalite (VFT) of the Val Fredda Complex (S. Adamello, N. Italy; Broderick 2013) reveals the relative importance of amphibole, plagioclase, apatite, titanite, allanite and zircon for the partitioning of REE's between the melt and the mineral phases (Fig. 6). The   f zircon is restricted to the HREE Dy to Lu; the main competitors for this element group are amphibole, apatite and titanite, in decreasing importance and modal abundance. The Dy concentration as well as the heavy/middle rare earth element ratio (HREE/MREE) Yb/Dy can be used as a chemical proxy for melt crystallinity, and also allow a qualitative assessment of the importance of titanite, allanite and apatite crystallization on the melt via the $D_{(Yb/Dy)}$ of these minerals. Allanite can be ubiquitous in dioritic to tonalitic melts although usually in low abundance. It preferentially scavenges light rare earth elements (LREE) and thus is of limited importance for the HREE/MREE ratio. For the examples shown in Fig. 6, Gd to Lu abundances in plagioclase were below the limit of detection, so this mineral is of subordinate importance for HREE/MREE fractionation. The REE patterns of zircon in Fig. 6 reveal the outstanding capacity of zircon to scavenge Ce^{4+} from the melt, leading to very high Ce^{4+}/Ce^{3+} ratios (Hoskin and Schaltegger 2003) that have been shown experimentally to be related to the oxidation state of the melt (Trail et al. 2012). The oxidation state may   the composition of source rocks and/or the oxidation state of assimilated material at higher levels of the crust. The Ce anomaly (Ce/Ce*) is commonly   y linear interpolation between La and Pr. But the concentration of these two elements in zircon are often below or close to the limit of detection, so the Ce/Nd ratio is used instead (Chelle-Michou et al. 2014).

The trace element composition of zircon (especially its MREE and HREE concentrations) may thus be used to model the chemical composition of the melt in equilibrium with zircon, as follows: (1) select partition   (K_D values) for REE in zircon from Rubatto and Hermann (2007), (2) use the whole-rock composition as the chemical starting composition in a closed system, (3) use published trace element partition   for the major and accessory minerals and the modal composition from thin section observations. Subsequently, the trace element composition of the fractionating bulk assemblage and the evolving melt may be reconstructed. As a result, the chemical composition of the zircon can be used, in combination with the other minerals, to reconstruct the magmatic evolution, for example, yielding a percentage of crystals as a function of time (e.g., Wotzlaw et al. 2013).

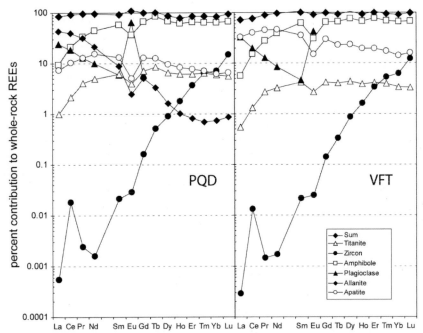

Figure 6. Rare earth element mineral budgets for a plagioclase–quartz–diorite (PQD) and a tonalite (VFT; Val Fredda Complex, Southern Adamello batholith, N. Italy), after Broderick (2013).

However, this model relies on single, published sets of REE partition for   accessory minerals and does not allow any   of accuracy and uncertainty. Theoretically, Monte Carlo simulations of the bulk partition   from the fractionating assemblage (from the published range of experimentally determined partition   of major and accessory minerals) could help to narrow down the range of possible K_D values by inverting them from the melt composition, back to measured mineral compositions. In addition, to validate the result, the calculated modal compositions can be compared to the modal abundance of mineral assemblages determined experimentally from Piwinskii and Wyllie (1968) and Nandedkar et al. (2014).

The large scatter of Dy concentration plotted versus the Yb/Dy ratio in zircon from   continental arc plutonic and volcanic rocks (Fig. 7A; all data from dated zircon grains or populations) suggests that zircon records the degree of fractionation of the LREE and MREE scavenging minerals (apatite, titanite, zircon) during crystallization. The trajectory y the Fish Canyon T   ircon can be recreated using a model with ~0.6 vol% titanite crystallization from a granodioritic initial magma (Wotzlaw et al. 2013). Zircon derived from Adamello and Bergell intrusions (N. Italy and S. Switzerland, respectively) align grossly along the same trend. Two well-dated intrusive units of the Southern Adamello, the Lago della Vacca Complex (Schoene et al. 2012) and the Val Fredda Complex (Broderick et al. 2015) show hardly any overlap, pointing to   tarting compositions and/or crystallization pathways. REE concentrations in zircon from gabbros and dolerite dykes of the Central Atlantic Magmatic province (Davies et al., in press; Schoene et al. 2010) show   scatter in Dy (and probably HREE) concentration up to 1800 ppm Dy (Fig. 7A). These zircon grains are interpreted as growing in small pockets of trace element-enriched, residual melt that arrived at zircon saturation just above the solidus, after abundant fractionation of olivine, pyroxene and plagioclase, but none of the REE-scavenging accessory minerals, thus representing local equilibria at small (possibly cm-to-dm) scale.

Partitioning of Th, U, Zr and Hf. Very similar systematics are displayed by Th/U vs. Zr/Hf ratios of the same zircon populations (Fig. 7B; for D values see overview in Table 1). Zircon incorporates U^{4+} in preference to Th^{4+} because the smaller ionic radius of U^{4+} better into the zircon lattice (Shannon 1976), leading to estimates of D_{Th}/D_U (zir) of between 0.15 and 0.39 for a temperature range from 700 to 900 °C (Rubatto and Hermann 2007). Th/U (zircon/rock) ratios show an even larger spread of 0.19–0.65 (Kirkland et al. 2015), but these values are in   with values for zircon D_{Th} and D_U derived from the lattice strain model (Blundy and Wood 2003). Knowing the Th/U of the melt is of paramount importance for obtaining accurate $^{206}Pb/^{238}U$ zircon dates because of the correction for initial ^{230}Th disequilibrium that leads to a   of radiogenic ^{206}Pb in the zircon lattice relative to the abundance of the parent ^{238}U (Schärer 1984). The D_{Zr}/D_{Hf} ratio of zircon has been estimated to be around 2, leading to a higher compatibility of Zr in the zircon lattice than Hf in agreement with the lattice strain model (Blundy and Wood 2003; no data provided). This value is in disagreement with the experimental data of Rubatto and Hermann (2007), who propose D_{Zr}/D_{Hf} values of 0.58–1.37 (Table 1), neither of which are fully supported by the lattice strain model (or the observed trends in real data, see below). These systematics imply that for Th, U, Zr and Hf, fractionation is solely due to the respective zircon/melt partition   Crystallization of zircon itself will leave a residual liquid with higher Th/U and lower Zr/Hf ratios, thus subsequently forming zircon will show increasing Th/U and decreasing Zr/Hf (Table 1).

To explain the co-variation of Th/U and Zr/Hf ratios in zircon (Fig. 7B) we need to infer co-precipitation of a Th-scavenging phase such as apatite, titanite or allanite. Contemporaneous crystallization of these mineral phases can indeed explain the correlated

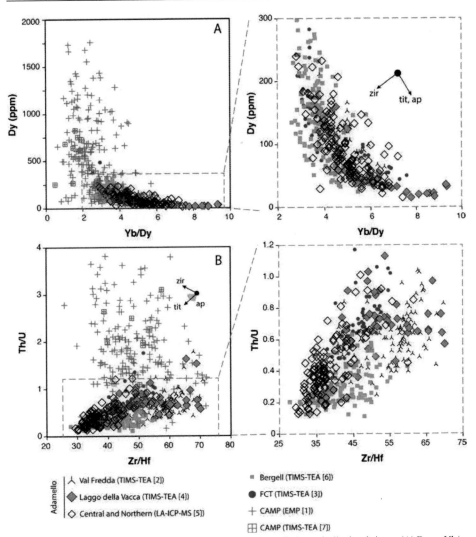

Figure 7. Trace element concentrations and ratios measured in radio-isotopically dated zircon. (A) Dy vs. Yb/Dy; (B) Th/U vs. Zr/Hf. CAMP = Central Atlantic Magmatic Province, dotted lines show blown up version of plots without CAMP data for clarity. Sources of data: [1] Davies et al., in press.; [2] Broderick et al. 2015; [3] Wotzlaw et al. 2013; [4] Schoene et al. 2012; [5] Skopelitis 2014; [6] Samperton et al. 2015; [7] Schoene et al. 2010. Schematic fractionation vectors of zircon, titanite and apatite are indicated, based on values in Table 1.

Table 1. Selected Δ-values for zircon, titanite and apatite.

Mineral	Yb/Dy	Th/U	Zr/Hf
Zircon	1.81 [1]	0.15 [3]	~2 [3]
	3.45–4.33 [2]	0.29–0.25 [2]	0.58–1.37 [2]
Titanite	0.12* [4]	2 [4]	5 [4]
Apatite	0.19–0.14* [5]	0.94–12.8 [5]	3.2–4.6 [5]

* Lu/Gd ratio; [1] Thomas et al. (2002); [2] Rubatto and Hermann (2007), temperature dependent; [3] Blundy and Wood (2003); [4] Prowatke and Klemme (2005), dacite; [5] Prowatke and Klemme (2006).

decrease of Th/U and Zr/Hf in Val Fredda and Lago della Vacca units of Southern Adamello, as well as for the Fish Canyon T   over time (as seen in Fig. 7B). This implies that the crystallization of accessory minerals is controlling the Th/U in the melt. Titanite and apatite are ubiquitous minerals in calc-alkaline rocks, and allanite has been reported as an importantly fractionating phase from the Bergell intrusion (Oberli et al. 2004; Samperton et al. 2015), for example. Titanite from the Fish Canyon T   as a high Th/U ratio of 6.7 (Bachmann et al. 2005), titanite of the Val Fredda Complex between 2.1 and 0.5 (Broderick 2014), whereas the Th/U ratio in titanite from the Lago della Vacca unit varies over several orders of magnitude between 0.02 and 11.8 as a function of host rock composition and in part due to subsolidus reactions (Schoene et al. 2012). Experimental data of Prowatke and Klemme (2005) point to a D_{Th}/D_U of ~2 (Table 1) for titanite. The fractionation of apatite with a D_{Th}/D_U of 0.94–12.78 (Prowatke and Klemme 2006) can have an even bigger impact on the Th/U budget of a crystallizing melt due to its higher modal abundance.

It would, however, be strange if titanite, apatite and allanite rather than zircon were controlling the Zr/Hf of a melt. Adopting a D_{Zr}/D_{Hf} ratio in zircon below one (e.g., a value of 0.58 from the 800 °C run of Rubatto and Hermann 2007, instead of the value of ~2 from Blundy and Wood 2003) would lead to increasing Zr/Hf over time through zircon crystallization alone, contrary to the trend in Figure 7B. This may be taken as an indication that the D_{Th}/D_U value in Rubatto and Hermann (2007) is not representative of these melt compositions.

Zircon crystallizing in small pockets of highly evolved residual melts in gabbros and dolerite dykes of the Central Atlantic Magmatic province show highly elevated Th/U up to 4, which may be explained by the lack of Th-scavenging minerals in the crystallizing assemblage (Fig. 7B).

Simple versus complex magmatic evolution. Temporal trends in the chemical composition of zircon are monotonic and continuous for the case of closed-system crystallization and cooling. In a magmatic system with a more complex history, involving periodic magma recharge/rejuvenation events that result in temperature oscillation and lead to magma mingling/mixing are unlikely to yield smooth and monotonous chemical trends. The   scatter in Dy and Yb data from the central and northern part of the Adamello intrusive suite (Skopelitis 2014), from the Val Fredda complex of S. Adamello (Broderick et al. 2013) or from the Bergell pluton (Samperton et al. 2015) are likely to indicate some of these processes (Fig. 7A).

The temporally resolved changes in zircon chemistry in the complex evolution of a large and homogeneous magma reservoir, such as represented by the Fish Canyon T   Wotzlaw et al. 2013), provide good insight into the power of zircon petrochronology. In a monotonously cooling granodioritic magma, chemical fractionation related to crystallization of titanite as the sole major carrier of HREE beside zircon would lead to an apparent monotonous increase in Yb/Dy ratios of consecutively crystallizing zircon grains 1 to 4 (Fig. 8). The case of the Fish Canyon T   hows, however, that the Yb/Dy ratios from sequentially crystallizing zircon grains 1' to 4' follow a   more complex fractionation trend over 450 ka. The geochemical model can be inverted to calculate hypothetical melt/crystal proportions,   a sequence of increasing crystal proportion, up to ~80%, followed by thermal rejuvenation, re-melting, a decrease of the crystal content, to ~45%, prior to violent eruption at 28.19 Ma (Fig. 8; see Wotzlaw et al. 2013 for details of the numerical model).

This example impressively shows how high-precision zircon petrochronology can yield a temporally resolved chemical record of magma evolution, from which parameters such as equilibrium melt composition, percentage of crystals, modal composition of the crystallizing mineral assemblage or even magma rheology may be computed, such as in Wotzlaw et al. (2013), or in Barboni and Schoene (2014).

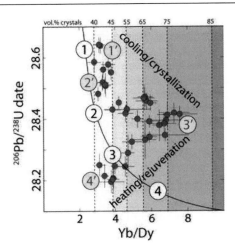

Figure 8. Temporal evolution of Yb/Dy as a function of crystallizing titanite in the granodioritic magma of the Fish Canyon T reservoir; titanite is the only HREE scavenging mineral beside zircon. Sequentially crystallized zircons 1 to 4 would delineate the evolution of an undisturbed, closed-system magmatic system; zircons 1' to 4' crystallize during complex system evolution involving thermal rejuvenation between 28.4 and 28.2 Ma of a nearly   mush followed by eruption at 28.19 Ma (redrawn after Wotzlaw et al. 2013).

Mingling magma batches of   ***rigin.*** When dating an igneous rock through zircon U–Pb geochronology, one underlying assumption is that zircon crystallized in the last melt batch at or at least close to the level of emplacement. This a priori assumption is wrong in some cases, as shown by the disagreement between apparent zircon residence times of several 100 ka and thermal models (as e.g., in Barboni and Schoene 2014, or Chelle-Michou et al. 2014). As an alternative it has been proposed that zircon crystallizes in compositionally   magma batches at   temperatures and at   depths of the magmatic plumbing system that have been sequentially and rapidly assembled in the upper crust after most of the zircon formed. Very short timescales of 10^2–10^3 years for merging smaller compositionally   agma reservoirs into large ones, and to subsequent supereruption have been suggested for Yellowstone-type magmatic systems (e.g., Wotzlaw et al. 2014, 2015). Cores of zircon   compositional   f the smaller-scale reservoirs shown by O and Hf isotope heterogeneity, whereas rim compositions may be homogeneous and in equilibrium with the last melt. However, for the Kilgore T   agma it has been shown that these processes were too rapid for the outermost rims of zircon to reach isotopic equilibrium with the interstitial glass (e.g., Wotzlaw et al. 2014). Mingling of distinct magma batches - from   ources or variably contaminated by   crustal materials - will be by heterogeneous oxygen and hafnium isotope composition of zircon, beside excess scatter in trace element abundances in zircon and other accessories.

The oxygen isotopic composition of zircon has been shown to be an excellent indicator for the assimilation of hydrothermally altered material, because remelting of upper crustal material that has been in contact with meteoric water ($\delta^{18}O = -14‰$; Taylor and Forester 1979) will produce zircon with $\delta^{18}O$ values   below the canonical value of 5.3 ± 0.3 ‰ for mantle zircon (Valley 2003). Low-$\delta^{18}O$ zircon has been discovered in volcanic rocks related to the Snake River Plain – Yellowstone plume (e.g., Bindeman and Valley 2001; Bindeman 2008; Wotzlaw et al. 2015), indicating assimilation of hydrothermally altered juvenile caldera material shortly before eruption. Similarly, late-stage   of the Skaergaard complex apparently digested previously emplaced, meteorically-altered and ^{18}O-depleted material (Wotzlaw et al. 2012).

Incorporation of old crustal components into a melt may also lead to changes in the $\delta^{18}O$ values, but a far more dramatic   can be seen in the Hf isotopic composition of zircon from such hybrid melts. For example, Wotzlaw et al. (2015) demonstrated that 45 vol% assimilation of Archean crust ($\varepsilon_{Hf} = -40$ to -60; 4–8 ppm Hf; $\delta^{18}O = +6$ to +9) into juvenile mantle melt ($\varepsilon_{Hf} = +5$ to +15; 2–6 ppm Hf; $\delta^{18}O = +5$ to +15) would lead to an increase of 1‰ for the

$\delta^{18}O$ value, but a marked decrease of 23 ε_{Hf} units in analyzed zircon. Oxygen and Hf isotope compositions of precisely-dated magmatic zircon are therefore perfectly suitable to trace the timing of the assembly of magma batches from   parts of the magmatic system, with crystallization/assimilation histories and/or   ources. Young magmatic systems provide   temporal resolution to demonstrate the ephemeral character of melt batches and the high speed at which the mingling between them must be acting, which cannot be resolved even by high-precision dating techniques. Therefore, zircon in a given portion of melt can crystallize over short time periods in equilibrium with a small-scale chemical system and may be   to zircons crystallized in other parts of the same magmatic system (e.g, Farina et al. 2014, Wotzlaw et al. 2015). During the assembly of melt portions with   history and/or source, suspended zircon can become physically mingled, which is recorded as non-systematic trace element and isotopic scatter.

Broderick et al. (2015) documented a case of small and ephemeral magma batches in the Val Fredda Complex in the Southern Adamello that were emplaced in the upper crust over 10^4 years' timescales and represent variably contaminated hybrid melts. Individual zircon crystals within a sample document variations of 4–6 ε_{Hf} units, unrelated to the actual high-precision U–Pb date (Fig. 9B) and pointing to mechanical mixing of zircon from   chemical environments into the   melt. The co-variation of ε_{Hf} with Th/U suggests, however, that contamination by SiO_2-rich crustal melts and fractional crystallization of possibly apatite and titanite were contemporaneous (Fig. 9A). The lacking correlation with time in Fig. 9B suggests that these chemical characteristics were, however, neither acquired in a coherent magma batch, nor in the last melt at the moment of emplacement, but in   mal melt

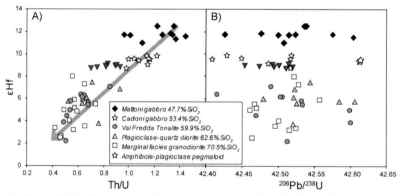

Figure 9. Zircon Hf, Th, U chemical and isotopic characteristics from the Val Fredda complex, southern Adamello. The lacking temporal correlation between crustal contamination and crystal fractionation processes points to zircon growth in distinct and small magma portions.

Incremental assembly of magma batches in the upper crust—wrapping up what we have learned

One of the essential messages emerging from the   of petrochronology is that magma is a suspension of crystals in a liquid and zircon is a part of this suspended crystal cargo. The suspended zircon crystals have formed over several 10 to 100 ka, as directly demonstrated by SIMS dating of volcanic zircon (e.g., Cooper and Kent 2014; Klemetti and Clynne 2014) or through zircon inclusions in K-feldspar phenocrysts (e.g., Barboni and Schoene 2014). As shown by the Fish Canyon T   example of Wotzlaw et al. (2013), zircon may even record crystallization in a precursor mush or "proto-pluton" (Annen 2011) up to 500 ka prior to   eruption, despite rejuvenation and subsequent eruption of the mush. This is considered a typical case for "cold" arc type magmatic settings, where a short magma recharge event leads to the eruption of crystal mushes of granodioritic composition, the so-called "monotonous intermediates" (e.g., Bachmann et al. 2007), but the temperature does not exceed T_{sat} over a long period of time, i.e., pre-existing zircons are not dissolved. Quantitative thermal models demonstrate that zircon dissolution is rather in calc-alkaline magmatic systems with magma temperatures below 800 °C (Watson 1996; Frazer et al. 2014). Zircon in such a case is thus recording a part of the entire lifespan of a magmatic system, from the initiation of crystallization in the lower part of the crust, through phases of intermediate storage and

At the other extreme, almost crystal mushes may be almost entirely rejuvenated by incoming, very hot basaltic magma in plume-type settings such as shown by the Kilgore T Wotzlaw et al. 2014), which leads to substantial re-melting before eruption. The pre-eruptive zircon memory may be entirely erased because the majority of the magma volume became thermally rejuvenated at temperatures above T_{sat} (zir), and additionally by magma that is undersaturated in Zr. Zircon geochronology will thus solely reveal the very short timescales of crystallization during rapid cooling from this rejuvenation event until eruption, i.e., the lifetime of the very last magmatic liquid. Nearly complete extraction of this high-temperature rhyolitic liquid containing very limited crystal cargo may be seen as the process leading to high-silica, phenocryst-free rhyolite, which apparently formed over very short lifetimes in hot appa et al. 2011; Wotzlaw et al. 2014).

Tilton et al. (1955) and Silver and Deutsch (1963) were among the geochronologists to apply U–Pb zircon dating to date the intrusion of granitoid plutons – what has remained from this initial motivation? With the advances of high-precision CA-ID-TIMS dating, the uncertainties of individual analyses of single grains or parts of grains have reached 0.05% in their $^{206}Pb/^{238}U$ date. These high precision ages inform us on the timescales of some of the above discussed processes, but with our improved understanding we that the ages and compositions do not necessarily relate to the physical emplacement of the magma in the crust. Zircon U–Pb dates provide ample information about the evolution of a magmatic plumbing system from its roots in the deep crust to its eventual emplacement in the upper crust or its eruption. The information we can obtain from high-precision U–Pb dating includes:

Zircon U–Pb dates very commonly display multi-peak distributions, describing a history of thermal oscillation over periods of 10^4 to 10^5 years. Zircon grains from one hand-sample do not necessarily share the same history but were assembled into the same melt batch. They represent snapshots from arts of a thermally, physically and chemically evolving magma reservoir. Zircon crystals are moving around in interstitial liquids percolating through crystal mushes, and/or record the passage of liquids over time via their growth history. Epitaxial

The trace element characteristics of zircon from small (tens of km^3) magma volumes may not show any coherent chemical trends, since every grain and every growth zone may represent a local and ephemeral equilibrium. Any diagram plotting two trace element ratios against each other, or plotting them against crystallization date, will yield non-systematic or chaotic trends (e.g., Broderick et al. 2015). If the magma volume is large enough and the heat content high, such a magma reservoir may start to convect and homogenize, which is evidenced by chemical trends from zircon that exhibit systematic variations (e.g., Schoene et al. 2012). In case of large-volume, agma systems, the trace element chemistry of zircon may represent pluton-wide chemical trends that can be modelled in terms of fractional crystallization, (e.g., shown for the Bergell intrusion by Samperton et al. 2015, for Zr and Hf, or for the Fish Canyon T Wotzlaw et al. 2013, for HREE and Th /U).

The mode in the distribution of zircon dates (such as in Figs. 5 A to C)   the fact that most of the zircon crystals form when the major part of the magma volume was within the zircon saturation interval. However, a small number of crystals will be preserved that formed earlier in cooler portions, already at lower levels of the crust, another population of zircon will form in small residual melt pockets when the magma approaches the solidus, after emplacement at shallow levels in the crust. Temperature oscillations due to multiple recharge events of hot   agma would not alter these systematics, but prolong the duration of the main peak (or peaks) of zircon crystallization. High-precision U–Pb zircon dates thus   the integrated crystallization history from zircon growth in precursor mushes to   In most cases it is therefore impossible to make a direct link to the physical movement of magma, i.e., to emplacement. Development of time-resolved thermal-physical models may help to quantify the moment when a magma has reached ca. 50% crystallinity, which would correspond to an approximate minimum age of emplacement. After emplacement, zircon may continue to crystallize in a stagnant interstitial liquid.

These considerations imply that zircon may not at all record the age relationships we can deduce in the   because it has crystallized mostly at deeper crustal levels than the present-day outcrop level. This may be suspected based on excessive dispersion of zircon dates up to 200-300 ka, in contradiction with thermal models. The U–Pb ages and the incoherent trace element systematics of zircon suggest injection of small melt batches into the upper crust, emplacing and crystallizing over no more than 10 ka (e.g., Chelle-Michou et al. 2014; Barboni et al. 2015), eventually building large plutons by sequential accretion over millions of years.

PETROCHRONOLOGY OF BADDELEYITE AND ZIRCON IN MAFIC SYSTEMS

Crystallization of zir   □

In this section, we concentrate our   on characterizing zircon and baddeleyite in tholeiitic rocks since recent geochemical and geochronological work on zircon from mid ocean ridge (MOR) gabbro and large igneous provinces (LIPs) suggests subtle between zircon in these environments and those found in calc-alkaline magmas. Zircon and baddeleyite are not uncommon in tholeiitic rocks, especially in the more evolved portions of coarse grained intrusions or in coarse mesostasis. Zircon is commonly documented from tonalites and trondhjemites associated with oxide-rich MOR gabbros (Coogan and Hinton 2006; Grimes et al. 2007, 2009; Lissenberg et al. 2009; Rioux et al. 2015a,b, 2016) and less commonly from thick sills, dykes and   in LIPs (Svensen et al. 2009; Schoene et al. 2010; Blackburn et al. 2013; Sell et al. 2014; Burgess and Bowring 2015; Davies et al. in press).

The common occurrence of zircon and baddeleyite in tholeiitic rocks may lead to the speculation that zircon can crystallize directly from tholeiitic melts at high temperature. However, application of the zircon saturation thermometry equations, with an extrapolation to higher M values for   elts (e.g., $M > 2.5$, DeLong and Chatelain 1990) suggests that zircon will not crystalize from basaltic magmas unless they have Zr concentrations of $> \sim 7400$ ppm at $> 1000\,°C$ (Boehnke et al. 2013), whereas average mid ocean ridge gabbro only has ~ 20 ppm Zr (Niu and O'Hara 2003). This simplistic application of the zircon saturation equations is consistent with Ti-in-zircon thermometry in MOR gabbro, although the absence of quartz or rutile from many of these magmas limits the application of titanium thermometry in   ircon. However, the observation that most silicic rocks have a_{TiO_2} between 0.6–0.9 and $a_{SiO_2} > 0.5$ (Ferry and Watson 2007) allows us to make estimates for the Ti and Si activities that result in temperature estimates with uncertainties of $\sim 30\,°C$. Compilations of   Ti-in-zircon temperatures indicate zircon crystallization at temperatures of ~ 950–$700\,°C$ (Fu et al. 2008; Grimes et al. 2009; Jöns et al. 2009; Rioux et al. 2015a, 2016). These temperatures can be compared to models of the liquid lines of descent for tholeiitic magmas which show that concentrations of Ti and Fe increase in the liquid as the basaltic liquid cools to ~1080–1120 °C and crystallizes until ~85% fractional crystallization, where the saturation point for Fe-Ti oxides is reached (depending on f_{O_2} and H_2O). From this point on in the crystallization sequence, SiO_2 increases in the melt along with other incompatible elements, causing zircon and other accessory minerals (apatite, quartz) to saturate (Niu et al. 2002). The point at which zircon crystallizes in these late-stage, evolved melts determines to what extent zircon chemistry can be used to trace petrogenetic processes. Zircon crystallization can also be modeled by partial re-melting of hydrothermally altered gabbro rather than by a fractional crystallization process directly from a tholeiitic melt (Koepke et al. 2007), but the geochemistry of zircon produced either way should be similar.

Chemical characteristics of zir

In a general sense, the trace element geochemistry of zircon crystallized from tholeiitic magmas is quite distinctive compared with zircon in calc-alkaline rocks. Grimes et al. (2007) showed that, because U and Yb have similar partition   between zircon and melt (226 ± 64 for Yb and 157 ± 51 for U, experiments at 850 °C, Rubatto and Hermann 2007), their ratio in zircon should   he ratio in the melt they crystallize from. Arc magmas have high U/Yb ratios > 1 due to high concentrations of U, whereas MORB has low U and is typically more enriched in Yb producing U/Yb < 1. When these ratios are plotted vs. HREE or Hf concentrations, tholeiitic zircon can be   distinguished from zircon grown in other magmas. However,   of zircon (especially tholeiitic zircon) on the basis of trace elements is hampered by the fact that variation within a single sample can match the variation shown by all samples in a group. Trace element variations within MOR gabbroic zircon can reveal fractional crystallization processes, similar to zircon from calc-alkaline settings, with negative correlations between Ti and Hf concentrations, which likely   incompatible behavior of Hf in the melt, and that zircon is the main Hf host (Fig. 10A; Grimes et al. 2009; Rioux et al. 2015a,b). At lower temperatures (<~750 °C), and more elevated Hf concentrations (~1.7 wt% Hf), the relationship between Hf and Ti appears to break, with increases in Hf occurring at constant Ti, which can be interpreted as   eutectic crystallization (Grimes et al. 2009). Within the same sample set, most REE display more scattered patterns compared with Hf when plotted against Ti (or temperature), suggesting that saturation of other accessory minerals (apatite, titanite, etc.) plays a role in controlling the budget of these elements in fractionated tholeiitic melts. Other inferences may be made based on Yb/Dy ratios, which in zircon from acidic rocks has been shown to record titanite + zircon crystallization (see above, Fig. 7A). In tholeiitic zircon, Yb/Dy appears to be extremely consistent over a range of temperatures, suggesting that the partition   of these elements in tholeiitic melts may be close to unity, which is quite   from what is expected ($D_{(Yb/Dy)} \approx 1.81$ from Thomas et al. 2002; or 3.45–4.3 from Rubatto and Hermann, 2007; see Table 1). After Ti concentrations drop below ~10 ppm (~725 °C), the consistent behavior of Yb/Dy stops, possibly indicating that a LREE-MREE scavenging phase starts to sequester Dy, increasing the Yb/Dy ratios of zircon (Fig. 10B). This transition to non-zircon controlled HREE partitioning occurs simultaneously between Yb/Dy and the possibly eutectic growth observed with Hf.

One interesting   in tholeiitic zircon geochemistry between MOR and LIP magmas is shown in Fig. 10C where the Th/U of LIP zircon is consistently elevated relative to MOR zircon. This   is easily modeled in LIP samples since zircon and baddeleyite are the only minerals that strongly partition U and Th into their crystal structure, whereas evolved melts that crystallize zircon from MOR magmas may contain apatite, titanite, and other U and Th scavaging trace phases. Both zircon and baddeleyite have $D_{Th/U}$ mineral–melt below 1, ~0.2 for zircon (Rubatto and Hermann, 2007), and < 0.18 for baddeleyite (Klemme and Meyer 2003), but since almost all analyses of baddeleyite have Th/U ~0.01, $D_{Th/U}$ is likely to be very low. Th/U in a melt should increase with fractional crystallization of zircon or baddeleyite, and consequently, zircon or baddeleyite grown at later stages should have elevated Th/U, assuming constant

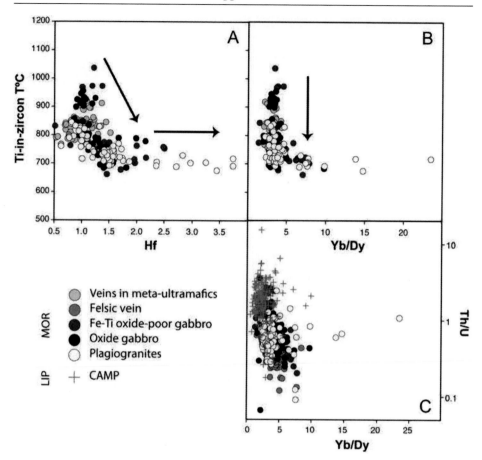

Figure 10. Zircon trace element ratios vs Ti-in-zircon temperatures for zircon from tholeiitic melts. A) Titanium temperature vs Hf (wt %), B) Titanium Temperature vs Yb/Dy, C) Th/U vs Yb/Dy. MOR = Mid Ocean Ridge, data from Grimes et al. 2009; Rioux et al. 2015a,b, 2016, CAMP = Central Atlantic Magmatic Province, data from Davies et al. in press and Schoene et al. (2010). The arrows denote general trends in the data.

partitioning (Barboni and Schoene 2014; Wotzlaw et al. 2014). However, the high Th/U ratios in zircon shown in the LIP dataset from the CAMP, suggest Th/U in the magma of up to 30. Such high Th/U values for whole rocks have been found for some granites (see Kirkland et al. 2015) but not for   ocks, which are typically <~6. To create these extreme Th/U magmatic values, zircon and/or baddeleyite need to make up a large proportion of the fractionally crystallizing assemblage, up to ~20%, however simple mass balance calculations indicate that this can not be the case. For example, the Zr concentrations of CAMP whole rocks are < 160 ppm (Marzoli et al. 2014), and crystallizing zircon and baddeleyite in large proportions (20% of the crystallizing phases) would reduce the Zr concentration of the melt to ~60 ppb after only 4.5% fractional crystallization while only increasing the Th/U of the melt by ~2, nowhere near the required Th/U of 30 even if the starting Th/U of the melt is 6 (assuming partition   of Rubatto and Herman 2007; and Klemme and Meyer 2003). Such extreme and apparently contradictory compositions required for these melts indicate an alternative explanation for the high Th/U in the LIP zircon such as the partition   for Th/U may be closer to 1 than 0.2. Highly fractionated   elts have not   been investigated in zircon–melt, or baddeleyite–melt trace element partitioning experiments, so this possibility remains currently speculative.

Tholeiitic zircon petrochronology has been applied much less than petrochronology in calc-alkaline rocks, and since zircon in tholeiitic rocks are typically crystallizing in late-stage evolved, to extremely fractionated melts, the recordable temporal history is likely to be short. Few geochronological studies from tholeiitic zircon have had the resolution to record protracted crystallization histories (Grimes et al. 2008; Rioux et al. 2012, 2015a,b), and until very recently (Rioux et al. 2016) high-resolution geochronological studies have not also collected geochemical information on the dated grains to directly compare chemistry to age. This comparison would be especially interesting in LIP magmas due to the apparently more frequent occurrence of baddeleyite than in MOR gabbros. Since baddeleyite petrochronology so far is underdeveloped, it is not clear if the   in baddeleyite abundance for MOR and LIP   merely a sampling   as more studies have targeted baddeleyite in LIP samples than in MOR, or if it is petrologically controlled.

Baddeleyite geochronology

Part of the problem with high precision U–Pb geochronology in   ocks is the limited abundance of zircon. But when it is extracted, chemical abrasion techniques can be applied to ensure that the   of Pb loss on the grain are removed (Mattinson 2005), resulting in reliable and accurate ages. For baddeleyite, there are no currently accepted techniques to remove the   of Pb loss (see Rioux et al. 2010), and therefore geochronology with this mineral results in scattered ages and   interpretations. Another datable mineral in iO_2O_7 presents similar analytical drawbacks (Wu et al. 2010). There have been numerous studies highlighting the power of baddeleyite for dating rock types. Recent examples and topics include meteorite impact events (Moser et al. 2013, Darling et al. 2016), alkaline magmas (Heaman and LeCheminant 2000; Heaman et al. 2009; Ibáñez-Mejía et al. 2014), diabase dykes and gabbros (Olsson et al. 2011; Davies and Heaman 2014; Ernst et al. 2016), (anorthosites, Wall and Scoates 2016; Wall et al. 2016), and vesicle-rich segregations in silica undersaturated lavas (Wu et al. 2015). Here we concentrate on some of the unresolved issues with U–Pb geochronology in baddeleyite.

U–Pb discordance in baddeleyite. Theoretically, in a cooling   elt with high Zr concentrations, baddeleyite saturation may be reached before silica saturation and zircon crystallization. Such relationships have been   qualitatively in thin section (Fig. 11A) and quantitatively though geochronology, where baddeleyite is shown to crystallize 0.16 ± 0.07 Ma before zircon in a gabbroic intrusion in the Canary Islands, followed by a period of baddeleyite and zircon co-precipitation (Allibon et al. 2011). In the Allibon study, zircon was chemically abraded to remove Pb loss, whereas if baddeleyite and zircon are dated from the same sample without any chemical abrasion applied to the zircon, baddeleyite ages are usually more concordant than zircon (Davies and Heaman 2014), suggesting that baddeleyite and zircon lose Pb in   ay. In detailed, high-resolution studies involving zircon chemical abrasion ages, baddeleyite frequently records ages younger than zircon from the same rock. In a dyke from the Moroccan CAMP, baddeleyite ages are up to 1.5 ± 0.4 Ma younger than the zircon, despite petrographic evidence that they crystallized before (Fig. 11B; Davies et al. in press). Discordance in baddeleyite is often attributed to mixtures between igneous (and presumably concordant) baddeleyite, with later hydrothermal or metamorphic zircon, producing a linear array between the igneous baddeleyite age, and the later zircon age (Heaman and LeCheminant 1993). In the CAMP example (Fig. 11), none of the dated baddeleyite grains contained zircon overgrowths or alteration suggesting that the discordance is attributable to baddeleyite itself. Attempts to air-abrade baddeleyite before TIMS U–Pb analysis to remove zircon rims have resulted in reduced discordance, but it has not been completely (e.g., Heaman and LeCheminant 2000; Wall et al. 2016). Part of the reason for only partial removal of discordance through air abrasion is that alteration of baddeleyite to produce zircon occurs both at the rim of the crystals and internally, where it will remain   by air abrasion (Rioux et al. 2010; Wall et al.

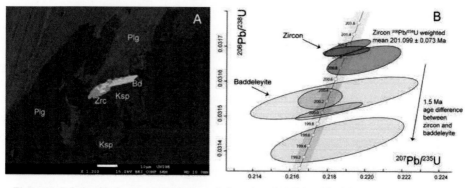

Figure 11. Relationship between baddeleyite and zircon in a CAMP basaltic dyke. A) BSE image from a thin section showing baddeleyite with zircon overgrowth, indicating that baddeleyite crystallized before zircon, B) Concordia diagram for the same sample, chemically abraded zircon analyses are shaded red, baddeleyite data without chemical abrasion are shaded grey. Note that the zircon analyses overlap within uncertainty and the baddeleyite ages are all younger, but still concordant. Data from Davies et al. in press.

2016). Also the small size of typical crystals < 100 µm, and their fragility means that air abrasion is not suitable in cases where only a few grains are extracted from a sample. Multi-step digestion has been demonstrated to isolate baddeleyite from secondary zircon where each component, i.e., baddeleyite and zircon, can be dated using high precision TIMS techniques independently (Rioux et al. 2010). This technique results in more concordant results than air abrasion alone,

On top of the Pb loss problems, baddeleyite is known to record reverse discordance, i.e., the U–Pb uncertainty ellipse plots to the left of the Concordia curve (Schoene and Bowring 2006; Nilsson et al. 2010; Söderlund et al. 2010). The cause of reverse discordance is enigmatic, although reverse discordance data commonly display linear arrays with normally discordant analyses, suggesting a common origin both physically and in time, for example U loss, Pb gain or isotopic fractionation of Pb during Pb loss. Baddeleyite can be forced to produce reverse discordance in a lab setting through partial HCl dissolution between 125 and 210 °C, similar to zircon (Mattinson 2005; Rioux et al. 2010), but again it is not clear what the exact mechanism is that leads to reverse discordance. With increased dissolution, the reverse Pb loss disappears, suggesting that it is associated with more disturbed areas of the crystal; however, 'normal' Pb loss then becomes dominant (Rioux et al. 2010), i.e., analyses fall to the right of the Concordia line.

Part of the problem in understanding discordance in baddeleyite is that there are no available experimental   data for Pb in baddeleyite. This area is receiving some attention (Bloch et al. 2014) but the observation that it preserves concordant, or slightly discordant ages though granulite grade metamorphism suggests that the closure temperature for Pb   is high (Soderlund et al. 2008; Beckman et al. 2014), hence thermally activated volume   may not be a viable mechanism for Pb loss. Lead disturbance was   during laser ablation analysis of baddeleyite from Duluth gabbro, where increased Pb counts were   next to high U zones (Ibáñez-Mejía et al. 2014). These anomalous Pb enrichments were interpreted to   intracrystalline Pb* migration from the surrounding U rich zones, possibly due to   of radiation damage. Baddeleyite responds to radiation damage   than zircon and maintains its crystallinity through high levels of radiation (see review by Trachenko 2004). However, ion bombardment experiments on monoclinic zirconia (a.k.a. baddeleyite) have shown that baddeleyite can undergo phase transformations due to the radiation damage: baddeleyite becomes tetragonal at high ion radiation   and this phase survives at atmospheric pressures and temperatures, unlike tetragonal zirconia produced thermally (Phillippi and Mazdiyasni 1971; Sickafus et al. 1999; Simeone et al. 2006). The phase change is a recrystallization of polymerized domains created during the collision cascade of the incoming ion (in terms of geochronological processes this would correspond to damage associated with alpha recoil), and in a natural baddeleyite crystal, recrystallization would likely cause Pb migration and fast pathway   These phase changes have not been   in natural baddeleyite ☐ crystals, but Raman spectra from the Phalaborwa baddeleyite do suggest the presence of a tetragonal phase (Fig. 12). Phalaborwa baddeleyite (2060 Ma) is known to record small amounts of Pb loss (Heaman 2009; Rioux et al. 2010), and the   of a tetragonal band in the Raman spectra leads to the speculation that Pb loss may   Pb mobility and   induced by a phase transformation (Davies 2014). Further study is required to show whether the   Raman spectra are due to the presence of a tetragonal phase, since Raman spectral bands can be created in a number of ways (Lenz et al. 2015). Once the mechanism for Pb loss in baddeleyite is   targeted methods can be developed to

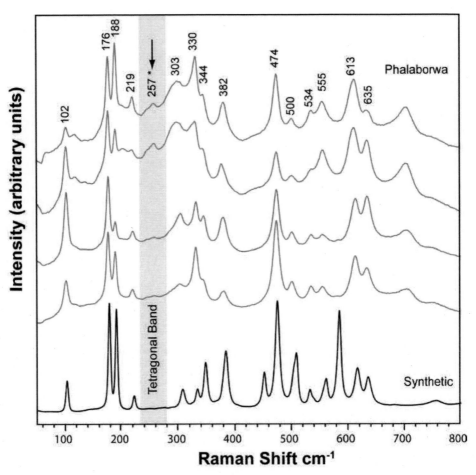

Figure 12. Baddeleyite Raman spectra. Spectrum for synthetic monoclinic baddeleyite (black line) compared with spectra for Phalaborwa natural baddeleyite (grey lines) which has an age of 2060 Ma (Heaman 2009). The location of the dominant tetragonal band at 257 cm^{-1} is highlighted. Note the appearance of a signal in this location in the Phalaborwa baddeleyite, but not the synthetic baddeleyite.

OUTLOOK

Zircon (and to a modest extent baddeleyite) petrochronology has revealed a great deal about magmatic processes through combining techniques such as geochronology, petrology and geochemistry. The   is, however, still rapidly moving ahead, through   of currently used techniques, more extensive use of chemical and structural mapping of mineral grains prior to isotopic analysis, or through the development of novel tools. For example, Li   much faster in zircon than most other detectable cations (Cherniak and Watson 2010). Recent experimental results suggest that relaxation of Li   from sector or oscillatory zones may be used as a geospeedometer and potentially a peak temperature indicator (Trail et al. 2016). The combination of Li   with Ti temperatures may provide interesting new information either on the thermal history of plutons after zircon saturation, or on zircon recycling events, depending on the temperatures involved. Another avenue for future work, so far under-utilized, is petrographically controlled sampling—context matters: whereas zircon crystallizing from the mesostasis between minerals in late-crystallizing melts should have a chemistry and age that   this, zircon inclusions in pyroxene, or in cores of plagioclase may have   age/chemical information. Even in a single pluton hand sample, it is clear that many zircons do not record the age of pluton emplacement but some protracted pre-emplacement history, but it is not so clear to what extent the other minerals in plutonic rocks record corresponding information. Targeting zircon inclusions with minerals could be used to add constraints on the temporal history of other phases.

A problem in urgent need of development relates to element partition   between melt of   ompositions and zircon and baddeleyite. The usefulness of chemical information in zircon and baddeleyite partially depends on such data, and as highlighted throughout this chapter, inconsistencies between   studies reduce the ability to extract robust petrochronological histories.

Reducing sample size has been a target of the TIMS U–Pb geochronology community for a long time and, with further reductions in laboratory blank below the 0.2 pg level, analyte sizes will continue to shrink allowing access to   and   spatial resolution while maintaining temporal precision. Obvious targets are zircons with multiple overgrowths that are too close in age for ion probe analysis to resolve (e.g., Reimink et al. 2016). However, recent high-resolution ion imaging studies have shown that Pb^* distribution is not correlated with U concentrations, instead it is concentrated in areas of the crystal (Kusiak et al. 2013, 2015). This may cause problems as sample sizes decrease, although, on the other hand it may open new avenues for future developments.

ACKNOWLEDGMENTS

Members of the Earth Science department at University of Geneva have contributed ideas and data to this review, especially C.A. Broderick, F. Martenot, G. Simpson, and L. Caricchi. The support from the Swiss National Science Foundation for the research of both authors and the isotope laboratory at University of Geneva is highly appreciated. An early version of the manuscript   from comments of M. Ovtcharova and F. Farina (Geneva), which are both acknowledged. The reviews of Drew Coleman, Matt Rioux and Blair Schoene helped to improve the text further, as well as the comments and careful editorial handling by Martin Engi. No research is possible without skilled and competent technical help at all stages of work—a special "Thank you!" therefore goes to the technical personnel, who work in the background and make the work possible.

REFERENCES

Allibon J, Ovtcharova M, Bussy F, Cosca M, Schaltegger U, Bussien D, Lewin E (2011) Lifetime of an ocean island volcano feeder zone: constraints from U–Pb dating on coexisting zircon and baddeleyite, and $^{40}Ar/^{39}Ar$ age determinations, Fuerteventura, Canary Islands. Can J Earth Sci 48:567–592, doi.org/10.1139/E10-032

Annen C (2011) Implications of incremental emplacement of magma bodies for magma differentiation, thermal aureole dimensions and plutonism–volcanism relationships. Tectonophysics 500:3–10, doi.org/10.1016/j.tecto.2009.04.010

Bachmann O, Dungan MA, Bussy F (2005) Insights into shallow magmatic processes in large silicic magma bodies: the trace element record in the Fish Canyon magma body, Colorado. Contrib Mineral Petrol 149:338–349, doi.org/10.1007/s00410-005-0653-z

Bachmann O, Miller C, de Silva S (2007) The volcanic–plutonic connection as a stage for understanding crustal magmatism. J Volcanol Geotherm Res 167:1–23

Barboni M, Schoene B (2014) Short eruption window revealed by absolute crystal growth rates in a granitic magma. Nat Geosci 7:524–528, doi.org/10.1038/ngeo2185

Barboni M, Annen C, Schoene B (2015) Evaluating the construction and evolution of upper crustal magma reservoirs with coupled U/Pb zircon geochronology and thermal modeling: A case study from the Mt. Capanne pluton (Elba, Italy). Earth Planet Sci Lett 432:436–448, doi.org/10.1016/j.epsl.2015.09.043

Beckman V, Möller C, Söderlund U, Corfu F, Pallon J, Chamberlain K (2014) Metamorphic zircon formation at the transition from gabbro to eclogite in Trollheimen–Surnadalen, Norwegian Caledonides. *In:* New Perspectives on the Caledonides of Scandinavia and Related Areas. Corfu F, Gasser D, Chew D (eds) Geol Soc Lon Spec Publ 390

Bindeman I (2008) Oxygen isotopes in mantle and crustal magmas as revealed by single crystal analysis. Rev Mineral Geochem 69:445–478, doi.org/10.1016/j.epsl.2015.09.043

Bindeman IN, Valley JW (2001) Low-delta O-18 rhyolites from Yellowstone: Magmatic evolution based on analyses of zircons and individual phenocrysts. J Petrol 42:1491–1517

Blackburn TJ, Olsen PE, Bowring SA, Mclean NM, Kent DV, Puffer J, McHone G, Rasbury TE, Et-Touhami M (2013) Zircon U–Pb geochronology links the End-Triassic extinction with the Central Atlantic magmatic province. Science 340:941–945, doi.org/10.1126/science.1234204

Bloch M, Watkins J, Van Orman J (2014) Diffusion kinetics of geochronologically relevant species in baddeleyite. EOS Trans, Am Geophys Union #V43B-4827

Blundy J, Wood B (2003) Mineral–melt partitioning of uranium, thorium and their daughters. Rev Mineral Geochem 52:59–123

Boehnke P, Watson EB, Trail D, Harrison TM, Schmitt AK (2013) Zircon saturation re-revisited. Chem Geol 351:324–334, doi.org/10.1016/j.chemgeo.2013.05.028

Broderick C (2013) Timescales and petrologic processes during incremental pluton assembly: a case study from the Val Fredda Complex, Adamello Batholith N Italy. PhD thesis nr. 4612, University of Geneva, Terre & Environnement 125, 169 pp

Bowring SA, Erwin D, Parrish RR, Renne P (2005) EARTHTIME: A community-based effort towards high-precision calibration of earth history. Geochim Cosmochim Acta 69:A316

Broderick C, Wotzlaw JF, Frick DA, Gerdes A, Ulianov A, Günther D, Schaltegger U (2015) Linking the thermal evolution and emplacement history of an upper-crustal pluton to its lower-crustal roots using zircon geochronology and geochemistry (southern Adamello batholith, N. Italy). Contrib Mineral Petrol 170:28, doi.org/10.1007/s00410-015-1184-x

Burgess SD, Bowring SA (2015) High-precision geochronology confirms voluminous magmatism before, during, and after Earth's most severe extinction. Sci Adv 1:e1500470–e1500470, doi.org/10.1126/sciadv.1500470

Caricchi L, Blundy J (2015) The temporal evolution of chemical and physical properties of magmatic systems. Geol Soc London, Spec Publ 422:1–15, doi.org/10.1144/SP422.11

Caricchi L, Simpson G, Schaltegger U (2014) Zircons reveal magma fluxes in the Earth's crust. Nature 511:457–461, doi.org/10.1038/nature13532

Caricchi L, Simpson G, Schaltegger U (2016) Estimates of volume and magma input in crustal magmatic systems from zircon geochronology: the effect of modeling assumptions and system variables. Frontiers Earth Sci 4:409, doi.org/10.1016/j.epsl.2015.02.035

Charlier BLA, Wilson C, Lowenstern J, Blake S, van Calsteren P, Davidson, J (2004) Magma generation at a large, hyperactive silicic volcano (Taupo, New Zealand) revealed by U–Th and U–Pb systematics in zircons. J Petrol 46:3–32, http://doi.org/10.1093/petrology/egh060

Chelle-Michou C, Chiaradia M, Ovtcharova M, Ulianov A, Wotzlaw JF (2014) Zircon petrochronology reveals the temporal link between porphyry systems and the magmatic evolution of their hidden plutonic roots (the Eocene Coroccohuayco deposit, Peru). Lithos 198–199:129–140, doi.org/10.1016/j.lithos.2014.03.017

Cherniak DJ, Watson EB (2010) Li diffusion in zircon. Contrib Mineral Petrol 160:383–390

Claiborne LL, Miller CF, Flanagan DM, Clynne MA, Wooden JL (2010) Zircon reveals protracted magma storage and recycling beneath Mount St. Helens. Geology 38:1011–1014, doi.org/10.1130/G31285.1

Coleman D, Gray W, Glazner A (2004) Rethinking the emplacement and evolution of zoned plutons: Geochronologic evidence for incremental assembly of the Tuolumne Intrusive Suite, California. Geology 32:433–436

Condon D, Schoene B, Mclean NM, Bowring SA, Parrish RR (2015) Metrology and traceability of U–Pb isotope dilution geochronology (EARTHTIME Tracer Calibration Part I). Geochim Cosmochim Acta 164:464–480, doi.org/10.1016/j.gca.2015.05.026

Coogan LA, Hinton RW (2006) Do the trace element compositions of detrital zircons require Hadean continental crust? Geology 34:633–636

Cooper K, Kent A (2014) Rapid remobilization of magmatic crystals kept in cold storage. Nature 506:480–483. http://doi.org/10.1038/nature12991

Corfu F, Hanchar JM, Hoskin PWO, Kinny P (2003) Atlas of zircon textures. Rev Mineral Geochem 53:468–500

Cottle JM, Kylander-Clark AR, Vrijmoed JC (2012) U–Th/Pb geochronology of detrital zircon and monazite by single shot laser ablation inductively coupled plasma mass spectrometry (SS-LA-ICPMS). Chem Geol 332–333: 136–147, doi.org/10.1016/j.chemgeo.2012.09.035

D'Abzac F-X, Davies JHFL, Wotzlaw JF, Schaltegger U (2016) Hf isotope analysis of small zircon and baddeleyite grains by conventional Multi Collector-Inductively Coupled Plasma-Mass Spectrometry. Chem Geol 433:12–23, doi.org/10.1016/j.chemgeo.2016.03.025

Darling JR, Moser DE, Barker IR, Tait KT, Chamberlain KR, Schmitt AK, Hyde BC (2016) Variable microstructural response of baddeleyite to shock metamorphism in young basaltic shergottite NWA 5298 and improved U–Pb dating of Solar Sytem events. Earth Planet Sci Lett 444:1–12

Davies JHFL (2014) Insights into the origin of the Scourie Dykes from geochemistry and geochronology. PhD Dissertation. University of Alberta, Edmonton, Canada

Davies JHFL, Heaman LM (2014) New U–Pb baddeleyite and zircon ages for the Scourie dyke swarm: A long-lived large igneous province with implications for the Paleoproterozoic evolution of NW Scotland. Precamb Res 249:180–198

Davies JHFL, Marzoli A, Bertrand H, Youbi N, Schaltegger U (2017) End-Triassic mass extinction started by intrusive CAMP activity. Nat Comm (in press)

DeLong SE, Chatelain C (1990) Trace element constraints on accessory-phase saturation in evolved MORB magma. Earth Planet Sci Lett 101:206–215

Ernst RA, Hamilton MA, Söderlund U, Hanes JA, Gladkochub DP, Okrugin AV, Kolotilina T, Mekhonoskin AS, Bleeker W, LeCheminant AN, Buchan KL, Chamberlain KR, Didenko AN (2016) Long-lived connection between southern Siberia and northern Laurentia in the Proterozoica. Nat Geosci 9:464–469, doi.org/10.1038/NGEO2700

Farina F, Stevens G, Gerdes A, Frei D (2014) Small-scale Hf isotopic variability in the Peninsula pluton (South Africa): the processes that control inheritance of source $^{176}Hf/^{177}Hf$ diversity in S-type granites. Contrib Mineral Petrol 168:1065, doi:10.1007/s00410-014-1065-8

Ferry J, Watson E (2007) New thermodynamic models and revised calibrations for the Ti-in-zircon and Zr-in-rutile thermometers. Contrib Mineral Petrol 154:429–437

Fisher CM, Vervoort JD, DuFrane SA (2014) Accurate Hf isotope determinations of complex zircons using the "laser ablation split stream" method. Geochem Geophys Geosystem 15:121–139, doi.org/10.1002/2013GC004962

Fraser G, Ellis D, Eggins S (1997) Zirconium abundance in granulite-facies minerals, with implications for zircon geochronology in high-grade rocks. Geology 25:607–610

Frazer RE, Coleman DS, Mills RD (2014) Zircon U–Pb geochronology of the Mount Givens Granodiorite: Implications for the genesis of large volumes of eruptible magma. J Geophys Res-Solid Earth 119:2907–2924, doi.org/10.1002/2013JB010716

Fu B, Page FZ, Cavosie AJ, Fournelle J, Kita NT, Lackey JS, Wilde SA, Valley JW (2008) Ti-in-zircon thermometry: applications and limitations. Contrib Mineral Petrol 156:197–215

Geisler T, Schaltegger U, Tomaschek F (2007) Re-equilibration of zircon in aqueous fluids and melts. Elements 3:43–50

Glazner A, Bartley J, Coleman D, Gray W, Taylor R (2004) Are plutons assembled over millions of years by amalgamation from small magma chambers? GSA Today 14:4–11, doi: 10.1130/1052–5173(2004)014<0004:APAOMO>2.0.CO;2

Grimes CB, John BE, Kelemen PB, Mazdab FK, Wooden JL, Cheadle MJ, Hanghøj K, Schwartz JJ (2007) Trace element chemistry of zircons from oceanic crust: A method for distinguishing detrital zircon provenance. Geology 35:643–646

Grimes CB, John BE, Cheadle MJ, Wooden JL (2008) Protracted construction of gabbroic crust at a slow spreading ridge: Constraints from $^{206}Pb/^{238}U$ zircon ages from Atlantis Massif and IODP Hole U1309D (30°N, MAR). Geochem Geophys Geosyst 9, doi:10.1029/2008GC002063

Grimes CB, John BE, Cheadle MJ, Mazdab FK, Wooden JL, Swapp S, Schwartz JJ (2009) On the occurrence, trace element geochemistry, and crystallization history of zircon from in situ ocean lithosphere. Contrib Mineral Petrol 158:757–783, doi.org/10.1007/s00410-009-0409-2

Heaman LM, LeCheminant AN (1993) Paragenesis and U–Pb systematics of baddeleyite (ZrO_2). Chem Geol 110:95–126

Heaman LM, LeCheminant AN (2000) Anomalous U–Pb systematics in mantle-derived baddeleyite xenocrysts from Ile Bizard: evidence for high temperature radon diffusion? Chem Geol 172:77–93

Heaman LM (2009) The application of U–Pb geochronology to mafic, ultramafic and alkaline rocks: An evaluation of three mineral standards. Chem Geol 261:43–52

Hofmann AE, Valley JW, Watson EB, Cavosie AJ, Eiler JM (2009) Sub-micron scale distributions of trace elements in zircon. Contrib Mineral Petrol 158:317–335

Hoskin PWO, Schaltegger U (2003) The composition of zircon and igneous and metamorphic petrogenesis. Rev Mineral Geochem 53:27–62

Ibanez-Meija M, Gehrels, GE, Ruiz J, Vervoort JD, Eddy MP, Li C (2014) Small-volume baddeleyite (ZrO_2) U–Pb geochronology and Lu–Hf isotope geochemistry by LA-ICP-MS Techniques and applications. Chem Geol 348:149–167

Jöns N, Bach W, Schroeder T (2009) Formation and alteration of plagiogranites in an ultramafic-hosted detachment fault at the Mid-Atlantic Ridge (ODP Leg 209). Contrib Mineral Petrol. 157:625–639 doi: 10.1007/s00410-008-0357-2

Kemp AIS, Wilde SA, Hawkesworth CJ, Coath CD, Nemchin A, Pidgeon RT, Vervoort JD, DuFrane AS (2010) Hadean crustal evolution revisited: New constraints from Pb–Hf isotope systematics of the Jack Hills zircons. Earth Planet Sci Lett 296:45–56, doi.org/10.1016/j.epsl.2010.04.043

Kirkland C L, Smithies RH, Taylor RJM, Evans N, McDonald B (2015) Zircon Th/U ratios in magmatic environs. Lithos 212–215:397–414, doi.org/10.1016/j.lithos.2014.11.021

Klemetti EW, Clynne MA (2014) Localized rejuvenation of a crystal mush recorded in zircon temporal and compositional variation at the Lassen volcanic center, Northern California. PLoS ONE 9:e113157, doi.org/10.1371/journal.pone.0113157.s002

Klemme S, Meyer H-P (2003) Trace element partitioning between baddeleyite and carbonatite melt at high pressures and high temperatures. Chem Geol 199:233–242

Koepke J, Berndt J, Feig ST, Holtz F (2007) The formation of SiO_2 rich melts within the deep oceanic crust by hydrous partial melting of gabbros. Contrib Mineral Petrol 153:67–84

Kotková J, Schaltegger U, Leichmann J (2010) Two types of ultrapotassic plutonic rocks in the Bohemian Massif—Coeval intrusions at different crustal levels. Lithos 115:163–176, doi.org/10.1016/j.lithos.2009.11.016

Kotková J, Whitehouse M, Schaltegger U, D'Abzac FX (2016) The fate of zircon during UHT-UHP metamorphism: isotopic (U/Pb, $\delta^{18}O$, Hf) and trace element constraints. J Metamorph Geol 34:719–739, doi.org/10.1111/jmg.12206

Kusiak MA, Whitehouse MJ, Wilde SA, Nemchin AA, Clark C (2013) Mobilization of radiogenic Pb in zircon revealed by ion imaging: Implications for early Earth geochronology. Geology 41:291–294

Kusiak MA, Dunkley DJ, Wirth R, Whitehouse MJ, Wilde SA, Marquardt K (2015) Metallic lead nanospheres discovered in ancient zircons. PNAS 112:4958–4963

Kylander-Clark AR, Hacker BR, Cottle JM (2013) Laser-ablation split-stream ICP petrochronology. Chem Geol 345:99–112, doi.org/10.1016/j.chemgeo.2013.02.019

Lenz C, Nasdala L, Talla D, Hauzenberger C, Seitz R, Kolitsch U (2015) Laser-induced REE^{3+} photoluminescence of selected accessory minerals—An "advantageous artefact" in Raman spectroscopy. Chem Geol 415:1–16

Leuthold J, Müntener O, Baumgartner L P, Putlitz B, Ovtcharova M, Schaltegger U (2012) Time resolved construction of a bimodal laccolith (Torres del Paine, Patagonia). Earth Plan Sci Lett 325–326:1–8, doi.org/10.1016/j.epsl.2012.01.032

Lissenberg CJ, Rioux M, Shimizu N, Bowrin, SA, Mével C (2009) Zircon dating of oceanic crustal accretion. Science 323:1048–1050

Marzoli A, Jourdan F, Bussy F, Chiaradia M, Costa F (2014) Petrogenesis of tholeiitic basalts from the Central Atlantic magmatic province as revealed by mineral major and trace elements and Sr isotopes. Lithos 188:44–59

Mattinson J (2005) Zircon U–Pb chemical abrasion ("CA-TIMS") method: combined annealing and multi-step partial dissolution analysis for improved precision and accuracy of zircon ages. Chem Geol 220:47–66

Miller J, Matzel J, Miller C, Burgess S, Miller R (2007) Zircon growth and recycling during the assembly of large, composite arc plutons. J Volcanol Geotherm Res 167:282–299

Moser DE, Chamberlain KR, Tait KT, Schmitt AK, Darling JR, Barker IR, Hyde BC (2013) Solving the Martian meteorite age conundrum using micro-baddeleyite and launch-generated zircon. Nature 499:454–457

Nandedkar RH, Ulmer P, Müntener O (2014) Fractional crystallization of primitive, hydrous arc magmas: an experimental study at 0.7 GPa. Contrib Mineral Petrol 167:1015, doi.org/10.1007/s00410-014-1015-5

Nilsson MKM, Söderlund U, Ernst RE, Hamilton MA, Scherstén A, Armitage PEB (2010) Precise U–Pb baddeleyite ages of mafic dykes and intrusions in southern West Greenland and implications for a possible reconstruction with the Superior craton. Precamb Res 183:399–415

Niu Y, O'Hara MJ (2003) Origin of ocean island basalts: a new perspective from petrology, geochemistry, and mineral physics considerations. J Geophys Res 108:1–19

Niu Y, Gilmore T, Mackie S, Greig A, Bach W (2002) Mineral chemistry, whole-rock compositions, and petrogenesis of Leg 176 gabbros: data and discussion. Proc Ocean Drill Prog Sci Results 176:1–60

Oberli F, Meier M, Berger A, Rosenberg CL, Gieré R (2004) U–Th–Pb and $^{230}Th/^{238}U$ disequilibrium isotope systematics: Precise accessory mineral chronology and melt evolution tracing in the Alpine Bergell intrusion. Geochim Cosmochim Acta 68:2543–2560, doi.org/10.1016/j.gca.2003.10.017

Olsson JR, Söderlund U, Hamilton MA, Klausen MB, Helffrich GR (2011) A late Archaean radiating dyke swarm as possible clue to the origin of the Bushveld Complex. Nat Geosci 4:865–869

Phillippi CM, Mazdiyasni KS (1971) Infrared and Raman spectra of zirconia polymorphs. J Am Ceram Soc 54:254–258

Piwinskii AJ, Wyllie PJ (1968) Experimental studies of igneous rock series: A zoned pluton in the Wallowa batholith, Oregon. J Geol 76: 205–234

Prowatke S, Klemme S (2005) Effect of melt composition on the partitioning of trace elements between titanite and silicate melt. Geochim Cosmochim Acta 69:695–709, doi.org/10.1016/j.gca.2004.06.037

Prowatke S, Klemme S (2006) Trace element partitioning between apatite and silicate melts. Geochim Cosmochim Acta 70:4513–4527, doi.org/10.1016/j.gca.2006.06.162

Reid MR, Coath CD (2000) In situ U–Pb ages of zircons from the Bishop Tuff: No evidence for long crystal residence times. Geology 28:443–446

Reid MR, Coath, Ca.D, Harrison TM, McKeegan KD (1997) Prolonged residence times for the youngest rhyolites associated with Long Valley Caldera: $^{230}Th–^{238}U$ ion microprobe dating of young zircons. Earth Planet. Sci. Lett., 150:27–39

Reimink JR, Davies JHFL, Chacko T, Stern RA, Heaman LM, Sarkar C, Schaltegger U, Creaser RA, Pearson DG (2016) No evidence for Hadean continental crust within Earth's oldest evolved rock unit. Nat Geosci, doi. org/10.1038/ngeo2786

Reubi O, Blundy J (2008) Assimilation of Plutonic Roots, Formation of High-K "Exotic" Melt Inclusions and Genesis of Andesitic Magmas at Volcan De Colima, Mexico. J Petrol 49:2221–2243, doi.org/10.1093/petrology/egn066

Rioux M, Bowring S, Dudás F, Hanson R (2010) Characterizing the U–Pb systematics of baddeleyite through chemical abrasion: application of multi-step digestion methods to baddeleyite geochronology. Contrib Mineral Petrol 160:777–801, doi.org/10.1007/s00410-010-0507-1

Rioux M, Lissenberg CJ, McLean NM, Bowring SA, MacLeod CJ, Hellebrand E, Shimizu N (2012) Protracted timescales of lower crustal growth at the fast-spreading East Pacific Rise. Nat Geosci 5:275–278, doi.org/10.1038/ngeo1378

Rioux M, Bowring S, Cheadle M, John B (2015a) Evidence for initial excess 321 Pa in mid-ocean ridge zircons. Chem Geol 397:134–156

Rioux M, Jöns N, Bowring S, Lissenberg CJ, Bach W, Kylander-Clark A, Hacker B, Dudás F (2015b) U–Pb dating of interspersed gabbroic magmatism and hydrothermal metamorphism during lower crustal accretion, Vema lithospheric section, Mid-Atlantic Ridge. J Geophys Res: Solid Earth 120:2093–2118

Rioux M, Cheadle M, John B, Bowring S (2016) The temporal and spatial distribution of magmatism during lower crustal accretion at an ultraslow-spreading ridge: High-precision U–Pb zircon dating of ODP Holes 735B and 1105A, Atlantis Bank, Southwest Indian Ridge. Earth Planet Sci Let doi: 10.1016/j.epsl.2016.05.047

Rosera JM, Coleman DS, Stein HJ (2013) Re-evaluating genetic models for porphyry Mo mineralization at Questa, New Mexico: Implications for ore deposition following silicic ignimbrite eruption. Geochem Geophys Geosystem 14:787–805, doi.org/10.1002/ggge.20048

Rubatto D, Hermann J (2007) Experimental zircon/melt and zircon/garnet trace element partitioning and implications for the geochronology of crustal rocks. Chem Geol 241:38–61, doi.org/10.1016/j.chemgeo.2007.01.027

Rubin A, Cooper KM, Leever M, Wimpenny J, Deering C, Rooney T, Gravley D, Yin QZ (2016) Changes in magma storage conditions following caldera collapse at Okataina Volcanic Center, New Zealand. Contrib Mineral Petrol 171:1–18

Samperton KM, Schoene B, Cottle JM, Keller CB, Crowley JL, Schmitz MD (2015) Magma emplacement, differentiation and cooling in the middle crust: Integrated zircon geochronological–geochemical constraints from the Bergell Intrusion, Central Alps. Chem Geol 417:322–340, doi.org/10.1016/j.chemgeo.2015.10.024

Simakin A, Bindeman I (2008) Evolution of crystal sizes in the series of dissolution and precipitation events in open magma systems. J Volcanol Geothermal Res 177:997–1010, doi.org/10.1016/j.jvolgeores.2008.07.012

Schaltegger U, Schmitt AK, Horstwood MSA (2015) U–Th–Pb zircon geochronology by ID-TIMS, SIMS, and laser ablation ICP-MS: Recipes, interpretations, and opportunities. Chem Geol 402:89–110, doi. org/10.1016/j.chemgeo.2015.02.028

Schärer U (1984) The effect of initial ^{230}Th disequilibrium on young U–Pb ages; the Makalu case, Himalaya. Earth Planet Sci Lett 67:191–204

Schmitt AK (2011) Uranium series accessory crystal dating of magmatic processes. Ann Rev Earth Planet Sci 39:321–349, doi.org/10.1146/annurev-earth-040610-133330

Schmitt AK, Vazquez JA (2017) Secondary ionization mass spectrometry analysis in petrochronology. Rev Mineral Geochem 83:199–230

Schmitt AK, Chamberlain KR, Swapp SM, Harrison TM (2010) In situ U–Pb dating of micro-baddeleyite by secondary ion mass spectrometry. Chem Geol 269:386–395

Schoene B (2014) U–Th–Pb Geochronology. Treatise of Geochemistry, The Crust (2nd ed., Vol. 4, pp. 341–378). Elsevier Ltd., doi.org/10.1016/B978-0-08-095975-7.00310–7

Schoene B, Baxter EF (2017) Petrochronology and TIMS. Rev Mineral Geochem 83:231–260

Schoene, B, Bowring S (2006) U–Pb systematics of the McClure Mountain syenite: thermochronological constraints on the age of the $^{40}Ar/^{39}Ar$ standard MMhb. Contrib Mineral Petrol 151:615–630

Schoene B, Bowring SA (2010) Rates and mechanisms of Mesoarchean magmatic arc construction, eastern kaapval craton, Swaziland. Geol Soc Amer Bull 122:408–429, doi.org/10.1016/0012-821X(96)00049–0

Schoene B, Latkoczy C, Schaltegger U, Günther D (2010) A new method integrating high-precision U–Pb geochronology with zircon trace element analysis (U–Pb TIMS-TEA). Geochim Cosmochim Acta 74:7144–7159, doi.org/10.1016/j.gca.2010.09.016

Schoene B, Schaltegger U, Brack P, Latkoczy C, Stracke A, Günther D (2012) Rates of magma differentiation and emplacement in a ballooning pluton recorded by U–Pb TIMS-TEA, Adamello batholith, Italy. Earth Planet Sci Lett 355–356:162–173, doi.org/10.1016/j.epsl.2012.08.019

Sell B, Ovtcharova M, Guex J, Bartolini A, Jourdan F, Spangenberg JE, Vicente JC, Schaltegger U (2014) Evaluating the temporal link between the karoo LIP and climatic–biologic events of the Toarcian Stage with high-precision U–Pb geochronology. Earth Plan Sci Lett 408:48–56 doi.org/10.1016/j.epsl.2014.10.008

Shannon RD (1976) Revised effective ionic radii and systematic studies of interatomic distances in halides and chalcogenides. Acta Crystallogr A32:751–767

Sickafus KE, Matzke H, Hartmann T, Yasuda K, Valdez JA, Chodak P, Nastasi M, Verrall RA (1999) Radiation damage effects in zirconia. J Nucl Mat 274:66–77

Silver LT, Deutsch S (1963) Uranium-lead isotopic variations in zircons: a case study. J Geol 71:721–758

Simeone D, Baldinozzi G, Gosset D, Caër SLe (2006) Phase transition of pure zirconia under irradiation: A textbook example. Nucl Inst Methods Phys Res B 250:95–100

Skopelitis A (2014) Formation of a tonalitic batholith through sequential accretion of magma batches: a study of chemical composition, age and emplacement mechanisms of the Adamello Batholith N Italy. Unpubl PhD thesis nr. 4660, University of Geneva

Söderlund U, Johanson L (2002) A simple way to extract baddeleyite (ZrO_2). Geochem Geophys Geosystem 3, doi.org/10.1029/2001GC000212

Söderlund U, Hellström FA, Kamo SL (2008) Geochronology of high-pressure mafic granulite dykes in SW Sweden: tracking the $P–T–t$ path of metamorphism using Hf isotopes in zircon and baddeleyite. J Metamorph Geol 26:539–560

Söderlund U, Hofmann A, Klausen MB, Olsson JR, Ernst RE, Persson P-O (2010) Towards a complete magmatic barcode for the Zimbabwe craton: Baddeleyite U–Pb dating of regional dolerite dyke swarms and sill complexes. Precam Res 183:388–398

Storm S, Schmitt AK, Shane P, Lindsay JM (2014) Zircon trace element chemistry at sub-micrometer resolution for Tarawera volcano, New Zealand, and implications for rhyolite magma evolution. Contrib Mineral Petrol 167:1–19

Svensen H, Planke S, Polozov AG, Schmidbauer N, Corfu F, Podladchikov YY, Jamtviet B (2009) Siberian gas venting and the end-Permian environmental crisis. Earth Planet Sci Lett 277:490–500

Tappa MJ, Coleman DS, Mills RD, Samperton KM (2011) The plutonic record of a silicic ignimbrite from the Latir volcanic field, New Mexico. Geochem Geophys Geosystem 12, doi.org/10.1029/2011GC003700

Tapster S, Condon DJ, Naden J, Noble SR, Petterson MG, Roberts NMW, Saunders AD, Smith JD (2016) Rapid thermal rejuvenation of high-crystallinity magma linked to porphyry copper deposit formation; evidence from the Koloula Porphyry Prospect, Solomon Islands. Earth Planet Sci Lett 442:206–217, doi.org/10.1016/j.epsl.2016.02.046

Taylor HP, Forester RW (1979) An oxygen and hydrogen isotope study of the Skaergaard intrusion and its country rocks: A description of a 55 My-old fossil hydrothermal system. J Petrol 20:355–419

Tilton GR, Patterson C, Brown H, Inghram M, Hayden R, Hess D, Larsen E (1955) Isotopic composition and distribution of lead, uranium and thorium in a Precambrian granite. Geol Soc Am Bull 66:1131–1148

Trachenko K (2004) Understanding resistance to amorphiziation by radiation damage. J Phys Condens Matter 16:R1491–R1515

Trail D, Watson EB, Tailby ND (2012) Ce and Eu anomalies in zircon as proxies for the oxidation state of magmas. Geochim Cosmochim Acta 97:70–87, doi.org/10.1016/j.gca.2012.08.032

Trail D, Cherniak DJ, Watson EB, Harrison TM, Weiss BP, Szumila I (2016) Li zoning in zircon as a potential geospeedometer and peak temperature indicator. Contrib Mineral Petrol 171, doi 10.1007/s00410-016-1238-8

Valley JW (2003) Oxygen isotopes in zircon. Rev Mineral Geochem 53:343–385

Valley JW, Reinhard DA, Cavosie AJ, Ushikubo T, Lawrence DF, Larson DJ, Kelly TF, Snoeyenbos DR, Strickland A (2015) Nano- and micro-geochronology in Hadean and Archean zircons by atom-probe tomography and SIMS: New tools for old minerals. Am Mineral 100:1355–1377, doi.org/10.2138/am-2015-5134

Vermeesch P (2012) On the visualisation of detrital age distributions. Chem Geol 312–313:190–194, doi. org/10.1016/j.chemgeo.2012.04.021

Wall CJ, Scoates JS (2016) High precision U–Pb zircon-baddeleyite dating of the J-M Reef platinum group element deposit in the Still Water complex, Montana (USA). Econ Geol 111:771–782

Wall CJ, Scoates JS, Weis D (2016) Zircon from the Anorthosite zone II of the Stillwater Complex as a U–Pb geochronological reference material for Archean rocks. Chem Geol 436:54–71

Wang X, Griffin WL, Chen J, Huang P, Li X (2011) U and Th contents and Th/U ratios of zircon in felsic and mafic magmatic rocks: improved zircon–melt distribution coefficients. Acta Geol Sinica 85:11164–17411

Watson EB (1996) Dissolution, growth and survival of zircons during crustal fusion: kinetic principles, geological models and implications for isotopic inheritance. Trans R Soc Edinburgh 87:43–56, doi.org/10.1017/ S0263593300006465

Watson EB, Harrison TM (1983) Zircon saturation revisited: temperature and composition effects in a variety of crustal magma types. Earth Planet Sci Lett 64:295–304

Watson EB, Liang Y (1995) A simple model for sector zoning in slowly grown crystals: Implications for growth rate and lattice diffusion, with emphasis on accessory minerals in crustal rocks. Am Mineral 80:1179–1187

Wotzlaw J-F, Bindeman IN, Schaltegger U, Brooks, Ca.K, Naslund HR (2012) High-resolution insights into episodes of crystallization, hydrothermal alteration and remelting in the Skaergaard intrusive complex. Earth Planet Sci Lett, 355–356:199–212

Wotzlaw JF, Schaltegger U, Frick DA, Dungan MA, Gerdes A, Günther D (2013) Tracking the evolution of large-volume silicic magma reservoirs from assembly to supereruption. Geology 41:867–870, doi.org/10.1130/ G34366.1

Wotzlaw JF, Bindeman IN, Watts KE, Schmitt AK, Caricchi L, Schaltegger U (2014) Linking rapid magma reservoir assembly and eruption trigger mechanisms at evolved Yellowstone-type supervolcanoes. Geology 42:807–810, doi.org/10.1130/G35979.1

Wotzlaw JF, Bindeman IN, Stern RA, D'Abzac FX, Schaltegger U (2015) Rapid heterogeneous assembly of multiple magma reservoirs prior to Yellowstone supereruptions. Sci Rep 1–10, doi.org/10.1038/srep14026

Wu F-Y, Yang Y-H, Mitchell RH, Bellatreccia F, Li Q-L, Zhao Z-F (2010) In situ U–Pb and Nd–Hf–(Sr) isotopic investigations of zirconolite and calzirtite. Chem Geol 277:178–195

Wu WN, Schmitt AK, Pappalardo L (2015) U–Th baddeleyite geochronology and its significance to date the emplacement of silica undersaturated magmas. Am Mineral 100:2082–2090

Yang W, Lin YT, Zhang JC, Hao JL, Shen WJ, Hu S (2012) Precise micrometre-sized Pb–Pb and U–Pb dating with NanoSIMS J Anal Atom Spectrom 27:479, doi.org/10.1039/c2ja10303f

Yuan HL, Gao S, Dai MN, Zong Ca.L, Gunther D, Fontaine GH, Liu XM, Diwu Ca.R,. (2008) Simultaneous determinations of U–Pb age, Hf isotopes and trace element compositions of zircon by excimer laser-ablation quadrupole and multiple-collector ICP-MS Chem Geol 247:100–118

Zeh A, Ovtcharova M, Wilson AH, Schaltegger U (2015) The Bushveld Complex was emplaced and cooled in less than one million years—results of zirconology, and geotectonic implications. Earth Planet Sci Lett 418:103–114, doi.org/10.1016/j.epsl.2015.02.035

Reviews in Mineralogy & Geochemistry
Vol. 83 pp. 329–363, 2017
Copyright © Mineralogical Society of America

11

Hadean Zircon Petrochronology

T. Mark Harrison, Elizabeth A. Bell, Patrick Boehnke

Department of Earth, Planetary and Space Sciences
University of California, Los Angeles
Los Angeles, CA 90095
USA

tmark.harrison@gmail.com

ebell21@ucla.edu

pboehnke@gmail.com

INTRODUCTION

The inspiration for this volume arose in part from a shift in perception among U–Pb geochronologists that began to develop in the late 1980s. Prior to then, analytical geochronology emphasized progressively lower blank analysis of separated accessory mineral aggregates (e.g., Krogh 1982; Parrish 1987), with results generally interpreted to reflect a singular moment in time. For example, a widespread measure of confidence in intra-analytical reliability was conformity to an MSWD (a form of χ^2 test; Wendt and Carl 1991) of unity. This approach implicitly assumed that geological processes act on timescales that are short with respect to analytical errors (e.g., Schoene et al. 2015). As *in situ* methodologies (e.g., Compston and Pidgeon 1986; Harrison et al. 1997; Griffin et al. 2000) and increasingly well-calibrated double spikes (e.g., Amelin and Davis 2006; McLean et al. 2015) emerged, geochronologists began to move away from interpreting geological processes as a series of instantaneous episodes (e.g., Rubatto 2002). At about the same time, petrologists developed techniques that permitted in situ chemical analyses to be interpreted in terms of continuously changing pressure–temperature– time histories (e.g., Spear 1988). The recognition followed that specific mineral reactions yielded products that could be directly dated or interpreted in terms of protracted petrogenetic processes. Part of this shift was due to an appreciation that trace elements in accessory phases could identify the changing nature of modal mineralogy during crystal growth (e.g., Pyle et al. 2001; Kohn and Malloy 2004) and thus potentially relate petrogenesis to absolute time. The transition to petrochronology was complete upon recognition that high MSWDs were in fact the expected case for most metamorphic minerals (Kohn 2009).

One of the great frontiers for fundamental discovery in the geosciences is earliest Earth (DePaolo et al. 2008). However, investigations of the first five hundred million years of Earth history—known as the Hadean eon (Cloud 1972, 1976)—are limited by the lack of a rock record older than 4.02 Ga (Bowring and Williams 1999; Mojzsis et al. 2014; Reimink et al. 2016; cf. O'Neil et al. 2008). This potentially leaves only a single strategy—the examination of Hadean detrital or inherited minerals—to directly assess the geophysical conditions, and therefore habitability, of early Earth. Not having access to the rock context in which a mineral geochronometer formed does reduce opportunities to understand its growth medium and the forces acting on it during crystallization. However, all is not lost. Virtually every accessory phase contains both trace element signatures and inclusions of coexisting minerals that were incorporated during its formation. Thus we can stand what seems to be a limitation on its head by viewing, for example, a detrital zircon as both a micro-rock encapsulation system and an elemental partition mirror of the magma from which it grew.

1529-6466/17/0083-0011$05.00 (print)
1943-2666/17/0083-0011$05.00 (online)

http://dx.doi.org/10.2138/rmg.2017.83.11

In this chapter, we review the petrochemical and -chronological systems in zircon and their application to detrital grains from the Jack Hills regions, enumerate the thirteen presently known localities from which Hadean zircons have been documented, review their age and geochemical properties, and discuss the results in context of possible and unlikely sources. For our purposes, we arbitrarily define the Hadean as the period of Earth history prior to formation of the oldest documented rock (i.e., older than the age of the Acasta metatonalite at 4.02 Ga). We conclude that, in contrast to the longstanding paradigm of a hellish early Earth devoid of oceans, continents and life, the Hadean zircon record, and the micro-rocks that they encapsulate, largely grew under a range of conditions far more similar to the present than once imagined.

WHY STUDY HADEAN ZIRCONS?

Due to zircon's inherent resistance to alteration by weathering, dissolution, shock, and diffusive exchange, and its enrichment in U and Th relative to daughter product Pb (Hanchar and Hoskin 2003), the U–Pb zircon system has long been regarded as the premier crustal geochronometer. While highly valued in that role, the trace element and isotopic compositions of zircon have become recognized as valuable probes of environmental conditions experienced during crystallization. Even in cases where zircon has been removed from its original rock context, such as detrital grains in clastic rocks, inclusions, trace element patterns and isotopic signatures can yield important information regarding source conditions if the record is undisturbed. Having emphasized the remarkably refractory nature and resistance to diffusive exchange of zircon (Cherniak and Watson 2003), it is important to note its Achilles heel. Zircon is sensitive to radiation damage and can degrade into heterogeneous microcrystalline zones encompassed by amorphous material (Ewing et al. 2003). Nature has, to some degree, already weeded out those grains most susceptible to metamictization from detrital zircon populations as high U and Th grains are unlikely to survive sediment transport (e.g., Hadean Jack Hills zircons with original U concentrations > 600 ppm are exceedingly rare). Thus care must be taken to ensure that effects of post-crystallization alteration are not mistaken as primary features.

The importance of Hadean zircons is then in the coupling of their great antiquity with their amenability to U–Pb dating and capacity to retain geochemical information. Although first documented at nearby Mt. Narryer (Froude et al. 1983), the vast majority of investigations of > 4 Ga zircons sampled heavy-mineral-rich quartz–pebble conglomerates from a locality in the Erawondoo region of the Jack Hills (Fig. 1) (Compston and Pidgeon 1986; Maas et al. 1992; Spaggiari et al. 2007). Zircons are typically extracted from these rocks using standard separatory methods based on their high density and low magnetic susceptibility, handpicked and secured in an epoxy mount which is then polished and analyzed using the $^{207}Pb/^{206}Pb$ ion microprobe dating approach (see Holden et al. 2009).

Over 200,000 of these grains have been $^{207}Pb/^{206}Pb$ dated in this fashion with over 6,000 yielding ages older than 4 Ga. Most of the ~3% of the analyzed grains that are > 4 Ga were then U–Pb dated using an ion microprobe, with several found to be as old as 4.38 Ga (Compston and Pidgeon 1986; Holden et al. 2009; Valley et al. 2014).

Although zircon is dominantly a mineral of the continental crust, its formation is not restricted to that environment nor, for that matter, to Earth. However, zircons of continental affinity can be readily distinguished from those derived from the mantle or oceanic crust by trace element characteristics (e.g., U/Y vs. Y) and significantly lower crystallization temperatures (Grimes et al. 2007; Hellebrand et al. 2007). Lunar and meteoritic zircons can be distinguished from terrestrial counterparts by their REE signature (e.g., lack of a Ce anomaly; Hoskin and Schaltegger 2003). Furthermore, apparent crystallization temperatures for lunar zircons range from 900 to 1100 °C (Taylor et al. 2009) in contrast to terrestrial Hadean zircons which are restricted to 600–780 °C (Harrison et al. 2007; Fu et al. 2008). Thus it is amply clear that the vast majority of Hadean

Figure 1. Quartz–pebble conglomerates on Erawondoo Hill, Jack Hills region of Western Australia. Samples collected within ~100 m of this site have produced > 95% of all Hadean zircons yet documented (Photo credit: Bruce Watson).

zircons are derived from terrestrial continental lithologies. Furthermore, textural characteristics of Hadean zircons from Jack Hills (e.g., growth zoning, inclusion mineralogy) indicate that most are derived from igneous sources (e.g., Cavosie et al. 2004; Hopkins et al. 2008).

Geochemical studies using upwards of half of the total Hadean grains thus far documented have inspired a variety of interpretations. However, there is a broad consensus that evidence derived from these ancient zircons implies abundant water at or near Earth's surface during that era (e.g., Wilde et al. 2001; Mojzsis et al. 2001; Rollinson 2008; Shirey et al. 2008; Harrison 2009). This represents a dramatic reversal from the conception of an uninhabitable, hellish world from which this time period gets its name (Solomon 1980; Smith 1981; Maher and Stevenson 1988; Abe 1993; Ward and Brownlee 2000).

We note at the outset that the intrinsic limitations (including preservation bias) of the Hadean zircon record could prevent 'smoking gun' conclusions about earliest Earth from ever being drawn. For example, the 70% of the Earth's surface that is today covered by MORB contributes essentially nothing to the archive of detrital or xenocrystic zircons. This concern is sure to diminish as Hadean zircons are documented from a growing number of globally diverse locations.

Modes of investigation

Most investigations of Hadean zircons to date have emphasized ion microprobe analysis to minimize the volume of mineral excavated during age surveys, thus maximizing the signals in our subsequent analyses (i.e., $\delta^{18}O$, $^{176}Hf/^{177}Hf$, Ti, etc.). This made sense for early studies of Jack Hills zircons (e.g., Compston and Pidgeon 1986; Mojzsis et al. 2001) as there were then no serious alternatives to the ion microprobe. Typical practice was to handpick individual zircons and mount them on double-sided adhesive tape in systematic grids together with zircon standards. This enabled the most ancient grains identified by the $^{207}Pb/^{206}Pb$ age survey to be easily located for subsequent analysis. When we began the program to date over 100,000 Jack Hills zircons in 2001, this laborious mounting process was not the rate limiting step in creating a large archive of Hadean zircons as then no ion microprobe yet had automated analysis capability. Indeed, that project led to the development of the automated stage on the SHRIMP instruments, followed shortly thereafter by CAMECA's 'chain analysis' tool. Arguably the most remarkable analytical development over the subsequent 15 years has been the development and refinement of laser ablation, inductively-coupled mass spectrometry (LA-ICP-MS). Effective yields have increased by over an order of magnitude dropping both analytical time and the mass of material needed to attain a specified precision. This in turn has decreased costs dramatically. While the ion microprobe remains the ultimate tool for *in situ* U–Pb dating, it is perhaps no longer cost effective in undertaking large (i.e., $\geq 5,000$ U–Pb zircon) age surveys. Where once zircons were largely evaporated to attain an LA-ICP-MS U–Pb age, modern multicollector instruments can obtain a U–Pb with $\pm 2\%$ precision from a $\sim 1000\,\mu m^3$ volume (Ibanez-Mejia et al. 2014) which begins to compare favorably with the $\sim 150\,\mu m^3$ volume long attainable using the ion microprobe. Even lower cost, quadrupole LA-ICP systems can attain similar precision from $\sim 6,000\,\mu m^3$ ablation craters, which still represents only $\sim 4\%$ of a typical Hadean zircon mass (i.e., $\sim 1\,\mu g$). However, a common Pb correction based on ^{204}Pb is problematic for LA-ICP instruments owing to near ubiquitous background at ^{204}Hg leading to seemingly precise but potentially inaccurate U–Pb ages.

Age distributions. Numerous age studies of Jack Hills detrital zircons all show a characteristic bimodal distribution with peaks close to 3.4 and 4.1 Ga with some grains as old as nearly 4.4 Ga (Compston and Pidgeon 1986; Maas et al. 1992; Amelin 1998; Amelin et al. 1999; Mojzsis et al. 2001; Cavoise et al. 2004; Trail et al. 2007; Holden et al. 2009; Bell et al. 2011, 2014; Bell and Harrison 2013).

How abundant are Hadean zircons on Earth? Most of the dozen or so localities for which at least one $>4\,Ga$ zircon has been documented were not targeted for that purpose but rather discovered serendipitously. Ancient metasediments and orthogneisses for which 10s to 100s of zircons have been U–Pb dated without identifying at least one Hadean zircon must fall into one of two categories: 1) those in which $>4\,Ga$ zircons are present at a level of less than $\sim 1\%$ but have not yet been detected, and 2) those in which they are simply absent. How many zircons should be dated to ascertain to which category a sample belongs? The probability of detecting at least a single $>4\,Ga$ zircon as a function of abundance is shown in Fig. 2. A reasonable assumption is that the ca. 3% $>4\,Ga$ zircon abundance in the Jack Hills (Holden et al. 2009) is anomalously high for most Archean quartzites and thus we examine the detection probabilities where abundances are between one and two orders of magnitude lower (i.e., 0.2–0.02%). From Fig. 2, we can see that at the 95% confidence level (bright red band), diminishing returns are achieved following analysis of ~ 5000 zircons, corresponding to an effective abundance limit of $\sim 0.05\%$. To our knowledge, only the Erawondoo locality in the Jack Hills has had more than 5000 zircons dated and thus intercomparisons are as yet of limited value.

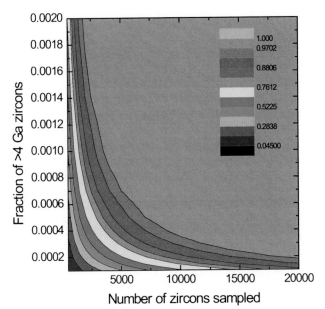

Figure 2. Plot showing the probability of identifying at least one Hadean zircon from populations with a range of assumed occurrence rates of >4 Ga grains as a function of number of dated grains. Note that achieving 89–93% confidence in detecting a Hadean zircon from a population in which 1-in-2000 are >4 Ga requires U–Pb dating ~5000 grains.

Indeed, the significance of Hadean zircons is sometimes dismissed by their seeming rarity—with less than a handful of these grains (the total mass acquired is less than 6 g), how can one begin to articulate the nature of early Earth? Our view is that this type of question is akin to asking how the Big Bang could possibly be characterized by capture of a vanishingly small fraction ($<10^{-70}$) of the photons in the observable universe. That is, observations on rare materials can lead to profound insights. There are two principal factors that inform this view. The first is, we do not expect to preserve a significant proportion of early formed crust on a dynamic planet. Assuming that Earth has recycled crust since formation at least as efficiently as we recognize it has throughout the Phanerozoic, only a few percent at most would likely remain (Armstrong 1981). The second is that we simply have not tried hard enough to find these remnants; we have sampled considerably less than 10^{-17} of the continental crust for geochronology, and support for reconnaissance dating surveys (i.e., "fishing trips") is notoriously difficult to obtain.

Isotope geochemistry. Several elements abundant in zircon comprise isotopic systems relevant to petrogenesis and have a significant role in defining conditions not only during the Hadean but throughout Earth history. The $^{18}O/^{16}O$ of magmas contains information regarding their sources with primary variations often reflecting incorporation of aqueously altered materials. Mantle-derived magmas display a narrow range of $^{18}O/^{16}O$, corresponding to zircons with an average $\delta^{18}O_{SMOW}$ of 5.3 ± 0.3 (1σ) (Valley et al. 1998). Aqueous alteration at low temperatures results in clay-rich sediments with higher $\delta^{18}O$, whereas hydrothermal alteration generally imparts lower $\delta^{18}O$ values. Incorporation of these altered materials into later magmas results in significant deviation from the mantle average value (e.g., O'Neil and Chappell 1977) which is reflected in the compositions of all silicate and oxide phases present, including zircon. That some Hadean Jack Hills zircons are significantly above the mantle value suggests abundant liquid water in the surface or near-surface environment as early as ca. 4.3 Ga (e.g., Mojzsis et al. 2001; Peck et al. 2001).

The Lu–Hf system is based on the decay of ^{176}Lu to ^{176}Hf ($t_{\frac{1}{2}} = 37$ Ga; Söderlund et al. 2004). Lu and Hf are fractionated during partial melting, such that higher Lu/Hf ratios form in depleted mantle and lower Lu/Hf ratios form in continental crust. Over Earth's history, this leads to differences in ^{176}Hf relative to the stable, primordial isotope ^{177}Hf (represented by ε_{Hf}) in these two reservoirs. The spread in ε_{Hf} between the continental crust and depleted mantle allows for the calculation of a model age of mantle extraction for igneous rocks. This isotopic system is useful on the level of individual zircons due to the incorporation of abundant (up to several weight percent) Hf in zircon and the lesser, ca. 100 ppm-level incorporation of Lu (e.g., Hoskin and Schaltegger 2003), which allows zircon to preserve the original magma ε_{Hf} with very little age-correction for ^{176} Lu ingrowth. Hadean Jack Hills zircons show dominantly negative (i.e., old crustal) ε_{Hf} with some grains requiring separation of very low-Lu/Hf (i.e., felsic) reservoirs by ca. 4.5 Ga (Harrison et al. 2008; Bell et al. 2014).

Mineral inclusions. As zircon crystallizes in the solid state or from magmas, it almost invariably traps exotic phases such as crystals and melt or other fluids (e.g., Maas et al. 1992; Chopin and Sobolev 1995; Tabata et al. 1998; Liu et al. 2001; Fig. 3). Coupled with the capacity for accurate U–Pb dating of the zircon host, these inclusions are a potentially rich source of information about petrogenetic conditions of formation and/or provenance.

Although mineral inclusions in magmatic zircon can record diagnostic information about the environment in which they formed, the extent to which the mineralogy and chemistry of zircons and their inclusion can be used to reconstruct petrogenesis and provenance is only now becoming clear. Darling et al. (2009) concluded that mineral inclusions in zircons grown in intermediate to felsic melts within the Sudbury impact melt sheet imply somewhat more felsic melt conditions than the associated whole rock in terms of the modal proportions of quartz, alkali feldspar, and plagioclase. However, Jennings et al. (2011) showed that the chemistry of igneous apatite and mafic phases is typically similar between crystals included in zircon and those in the whole rock.

Establishing the primary nature of inclusions and contamination introduced during sample preparation are potentially serious concerns that need to be explicitly addressed. For example, reports of abundant diamonds and graphite in Hadean Jack Hills zircons (Menneken et al. 2007; Nemchin et al. 2008) were later determined to be contaminants introduced during sample preparation (Dobrzhinetskaya et al. 2014).

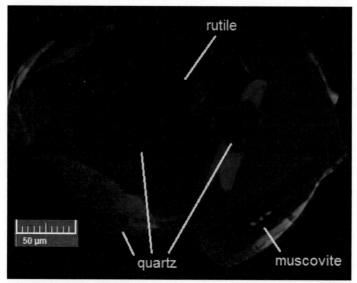

Figure 3. Cathodoluminescence image of Hadean Jack Hills zircon RSES77-5.7. This 4.06±0.1 Ga concordant grain contains likely primary inclusions of quartz, rutile and muscovite permitting reliable Ti thermometry and phengite barometry which can then be used to infer near surface thermal structure.

Mineral inclusions coupled with the chemistry of their host zircon are an underexploited resource for establishing internally consistent evidence for host rock character. The advent of the Ti-in-zircon thermometer, for instance, underscored the potential for thermodynamic relationships between included phases and elements partitioned into the zircon structure. Similarly, the incorporation of aluminous and carbonaceous inclusions into zircon (Hopkins et al. 2008; Rasmussen et al. 2011; Bell et al. 2015b; Harrison and Wielicki 2015) raises the possibility of calibrating trace elements in zircon as an indicator of host melt chemistry or volatile content.

Zircon geochemistry. Zircon incorporates many elements at the trace or minor level during crystallization, some of which are useful petrologic indicators. For example, the content of Ti in zircon serves as a crystallization thermometer given knowledge of the melt a_{SiO_2} and a_{TiO_2} (Watson and Harrison 2005; Ferry and Watson 2007). Ce/Ce*, or the excess in Ce over the other light rare earth elements (LREE) La and Pr, is a proxy for magma f_{O_2} (Trail et al. 2011a). Th/U can generally be used to distinguish magmatic from metamorphic zircons, with metamorphic zircons typically <0.07 and igneous zircon at higher values (Rubatto 2002, 2017). Other trace element concentrations have less quantitative ties to petrogenesis but may have the potential to yield important information.

Rare earth elements (REE) occur in terrestrial zircon with a characteristic chondrite-normalized abundance pattern characterized by relatively low LREE and increasingly abundant REE with increasing Z (e.g., Hoskin and Schaltegger 2003). Two exceptions to this rule include the aforementioned excess in Ce (Ce/Ce*) which is seen among virtually all unaltered terrestrial zircons and a deficit in Eu (Eu/Eu*). The steady increase in compatibility with increasing atomic mass for most REE in the zircon lattice results from the steady decrease in ionic radius coupled with the trivalent oxidation state in which most REE are found in the crustal and surficial environment. Significant amounts of tetravalent Ce (which is more compatible in zircon) and divalent Eu (largely taken up by plagioclase) lead to their respective anomalous contents. However, interpreting REE patterns in terms of zircon petrogenesis requires distinguishing pristine from altered zircon chemistry, and hydrothermal alteration of zircon is usually accompanied by an increase in LREE relative to the other REE and a flattening of the LREE pattern, obscuring the Ce/Ce* and potentially the Eu/Eu*.

JACK HILLS ZIRCONS

Isotopic results

U–Pb age. Various age surveys of detrital zircons from the Erawondoo Hill discovery site conglomerate (e.g., Crowley et al. 2005; Holden et al. 2009) generally show the zircons to have a bimodal age distribution with major peaks at ca. 3.4 and 4.1 Ga. Concordant zircons older than ca. 3.8 Ga make up approximately 5% of the population, and zircons become much less abundant with age older than ca. 4.2 Ga (Holden et al. 2009; Fig. 4). The remaining 95% of the population is mostly concentrated between 3.3 and 3.6 Ga, with a deficit of zircon ages between 3.6 and 3.8 Ga (Bell and Harrison 2013).

The confidence with which one can interpret the meaning of a U–Pb date of a >4 Ga zircon is challenged by the potential for later fluid alteration and thermal disturbances. While the concordance of U–Pb analyses (or lack thereof) can be used to assess the robustness of an interpreted age, this is generally insensitive to early Pb loss. Therefore, assessing the general reliability of ion microprobe U–Pb ages is an open challenge, especially given that most Hadean Jack Hills zircons contain multiple age domains. Valley et al. (2014) examined the possibility of Pb redistribution in a 4.38 Ga zircon core, imaged using atom probe tomography, encompassed by a ca. 3.4 Ga, 10-20 μm rim. In this analytical technique, a zircon sliver extracted using a focused ion beam was field evaporated and the emergent ions mass analyzed with a spatial

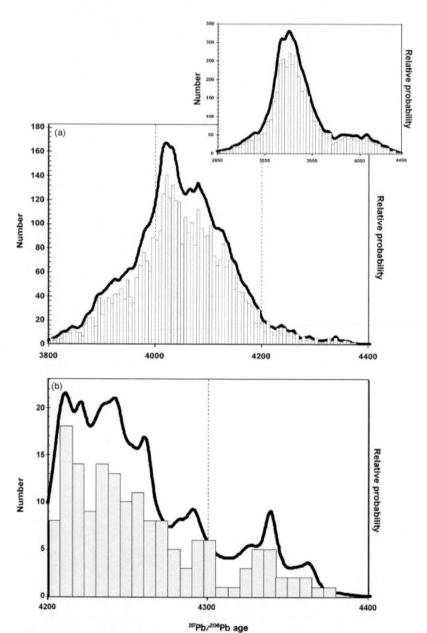

Figure 4. Histograms and probability-density curves for concordant Jack Hills zircons. (a) Histogram of rapid initial survey of individual $^{207}Pb/^{206}Pb$ ages undertaken to identify the >3.9 Ga population. Inset shows the whole population of 4500 rapidly scanned $^{207}Pb/^{206}Pb$ ages. (b) Histogram and probability density for the concordant >4.2 Ga zircons. The small peak at 4.35 Ga may be the oldest surviving crustal remnant. Reprinted fromPeter Holden et al. Mass spectrometric mining of Hadean zircons by automated SHRIMPmulticollector and singlecollector U/Pb zircon age dating: The first 100,000 grains, International Journal of Mass Spectrometry 286(2–3):53–63 (2009), with permission from Elsevier.

resolution of <1 nm. Results from their analysis show Pb redistribution into "nanoclusters" with ~10 nm diameter and spacing of ~10-50 nm. The $^{207}Pb/^{206}Pb$ age of the zircon outside of the "nanoclusters" is ~3.4 Ga and ~4.4 Ga for the entire analyzed volume. These data are consistent with an event at 3.4 Ga that mobilized radiogenic Pb that had accumulated since the zircon's crystallization at 4.38 Ga. The mobilized Pb migrated into "nano clusters" on a length scale of <50 nm, which is below the lateral spatial resolution of an ion microprobe. This finding supports the view that due to the generally slow diffusion of Pb in zircon and zircon's resistance to alteration, concordant U–Pb analyses likely record actual zircon crystallization ages.

Oxygen isotopes. Elevated values of $\delta^{18}O_{SMOW}$ observed in Hadean Jack Hills zircons led two independent groups to simultaneously propose (Mojzsis et al. 2001, Wilde et al. 2001) that the protolith of these grains contained ^{18}O-enriched clay minerals, in turn implying that liquid water was present at or near the Earth's surface by ~4.3 Ga. Numerous follow-up measurements (e.g., Cavosie et al. 2005, Trail et al. 2007, Harrison et al. 2008; Bell et al. 2016) confirmed that a significant fraction of Hadean Jack Hills zircons contain ^{18}O-enrichments 2 to 3‰ above the mantle zircon value of 5.3‰ (Valley et al. 1998). As the oxygen isotope fractionation between zircon and granitoid melt is approximately −2‰ (Valley et al. 1994, Trail et al. 2009), $\delta^{18}O$ values of the melt from which the zircons crystallized are inferred to have been up to +9 ‰.

Phanerozoic granitoids derived largely from orthogneiss protoliths (I types) tend to have $\delta^{18}O$ between 8-9‰, whereas those derived by melting of clay-rich (i.e., ^{18}O enriched) metasedimentary rocks (S types) have higher $\delta^{18}O$ (O'Neil and Chappell 1977). Granitoids with $\delta^{18}O$ values significantly less than 5‰ likely reflect hydrothermal interaction with meteoric water (Taylor and Sheppard 1986) rather than weathering. In general, S-type granitoids form by anatexis of metasediments enriched in ^{18}O, compared with I-type granitoids that form directly or indirectly from arc processes (Chappell and White 1974). Jack Hills zircons enriched in ^{18}O thus provide evidence indicating the presence in the protolith of recycled crustal material that had interacted with liquid water under surface, or near surface, conditions (i.e., at low temperature).

A limitation to this interpretation is the possibility of oxygen isotope exchange under hydrous conditions, even at post-depositional temperatures experienced by Jack Hills zircons (i.e., ~450 °C). For example, the characteristic diffusion distance for oxygen in zircon at 500 °C for 1 Ma is ~1 μm, assuming a high water activity (Watson and Cherniak 1997). Thus it is conceivable that oxygen isotope exchange during protracted thermal events could have introduced the heavy oxygen signature. This concern is somewhat mitigated by the relative improbability that hydrothermal fluids were highly $\delta^{18}O$ enriched. However, it does not preclude isotopic equilibration from having occurred prior to deposition at ca. 3 Ga.

Lutetium–Hafnium. Studies of initial $^{176}Hf/^{177}Hf$ in >4 Ga Jack Hills zircons show large deviations in $\varepsilon_{Hf(T)}$ from bulk silicate Earth (Kinny et al. 1991; Amelin et al. 1999; Harrison et al. 2005, 2008; Blichert-Toft and Albarède 2008; Bell et al. 2011, 2014; Kemp et al. 2010) that have been generally interpreted to reflect an early major differentiation of the silicate Earth (Fig. 5). Modeling these data by associating $\varepsilon_{Hf(T)}$ with the range of $^{176}Lu/^{177}Hf$ observed in large datasets of analyzed crustal rocks are consistent with the formation of crust occurring essentially continuously since 4.5 Ga. Several data (Harrison et al. 2008; Bell et al. 2014) yield $\varepsilon_{Hf(T)}$ within uncertainty of the solar system initial ratio (Iizuka et al. 2015) requiring that the zircon protoliths had been removed from a chondritic uniform reservoir (CHUR) by 4.5 Ga. Harrison et al. (2005) initially reported several Hadean Jack Hills zircons with positive $\varepsilon_{Hf(T)}$, but subsequent *in situ* studies have not confirmed significantly positive values. This likely reflects complications arising from the lack of simultaneous age and Hf isotope analysis, as described by Harrison et al. (2005).

The most robust aspect of this now large dataset is the cluster of results along a line corresponding to a Lu/Hf ≈ 0.01, a value characteristic of continental crust. Such a low-Lu/Hf reservoir at ~4 Ga is consistent with either early extraction of this very felsic crust or its

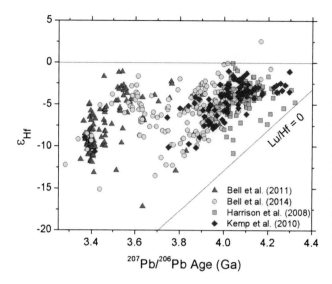

Figure 5. $\varepsilon_{Hf}(T)$ vs. [207]Pb/[206]Pb age of Jack Hills zircons. Reference [176]Lu/[177]Hf ratios for continental crust (0.01) is shown along with value for Bulk Earth (0.034) and primordial [176]Hf/[177]Hf (i.e., Lu/Hf=0). These data are consistent with the formation of continental crust occurring essentially continuously since 4.5 Ga (modified from Bell et al. 2014).

generation by remelting of a primordial more basaltic reservoir, but in either case extrapolation of this trend yields a present-day $\varepsilon_{Hf(T)}$ of approximately -100. This is substantially lower than the most negative value yet measured ($\varepsilon_{Hf(T)} = -35$; Guitreau et al. 2012). The lack of such a signal suggests substantial recycling of crust into the mantle during the early Archean (Bell et al. 2011, 2014). More specifically, zircons with $\varepsilon_{Hf(T)}$ consistent with continuing evolution of this reservoir appear absent from the Jack Hills record after 3.7 Ga (Bell et al. 2014). Combined with Hf isotopic evidence for juvenile mantle melts at ca. 3.9–3.7 Ga at both Jack Hills (Bell et al. 2014) and the nearby Mt. Narryer site with similarly aged zircon (Nebel-Jacobsen et al. 2010), these observations likely point to a recycling event ca. 3.9–3.7 Ga which resembles the Hf isotopic evolution of modern subduction-related orogens (e.g., Collins et al. 2011) and so may have additional tectonic significance.

Plutonium–Xenon. The meteorite record reveals that [244]Pu was present in the early solar system with an initial Pu/U abundance of ~0.007 (Ozima and Podosek 2002). However, its use as a geochemical tracer is restricted by its relatively short half-life ($t\frac{1}{2}$ = 82 Ma). As the only known relics of the Earth's earliest crust, analysis of Xe in Hadean zircons offers a way to determine terrestrial Pu/U ratios and potentially investigate Pu geochemistry during early crust forming events. Because these ancient zircons are detrital and of unknown provenance, it is essential that individual grains be analyzed. Turner et al. (2004) discovered the first evidence of extinct terrestrial [244]Pu in individual 4.15–4.22 Ga Jack Hills zircons. These measurements yielded initial Pu/U ratios ranging from chondritic (~0.007) to essentially zero. The latter results were first interpreted to be due to Xe loss during later metamorphism. This assumption was tested by irradiating 3.98–4.16 Ga zircons with thermal neutrons to generate Xe from [235]U neutron fission to determine Pu/U simultaneously with U–Xe apparent ages. Comparison of U–Pb and U–Xe ages showed varying degrees of Xe loss, but about a third of the zircons yield [207]Pb/[206]Pb and U-Xe ages that are concordant within uncertainty (Turner et al. 2007).

Given that U becomes oxidized to the soluble uranyl ion (UO_2^{2+}) under even mildly oxidized aqueous conditions while the solubilities of essentially all Pu species are generally much lower, variations in Pu/U has been suggested as a potential indicator of aqueous alteration in the Jack Hills zircon protoliths (Harrison 2009). To test this hypothesis, Bell (2013) collected a multivariate dataset on eleven zircons, including analysis of Xe isotopic ratios, U–Pb age, trace element contents, and $\delta^{18}O$, to look for correlations (e.g., $\delta^{18}O$ vs. Pu/U) expected from aqueous processes. With the exception of Nd/U, none were found. High-Nd/U zircons display only low Pu/U, while low Nd/U zircons show more heterogeneous Pu/U. The high-Nd/U group appears less magmatically evolved than other Hadean zircons, has REE patterns suggestive of some degree of alteration, either by hydrothermal fluid interaction or phosphate replacement, and consists of solely low-Pu/U zircons with a range of Hadean to Proterozoic U–Xe ages. The higher diversity of Pu/U among the rest of the population may reflect more heterogeneous processes, including possible primary Pu/U variations from a variety of processes that were not well-constrained. Thus the early promise that Pu/U variations might record aqueous fractionation events in the Hadean may not be realized.

Lithium. $\delta^7 Li$ analyses of Hadean Jack Hills zircons range from -19 to $+13‰$ (Ushikubo et al. 2008). These authors interpreted highly negative values to reflect zircon crystallization from a source that experienced intense weathering, thus placing the protolith at one time at Earth's surface. A limitation of this interpretation is that Li diffuses readily in zircons at relatively low temperatures (Cherniak and Watson 2010) and thus could have exchanged with hydrogen species during metamorphism (Trail et al. 2011b). Ushikubo et al. (2008) speculated that Li migration might be limited by coupling with the very slow REE diffusion in zircons thus limiting its geological transport rate. Recently Trail et al. (2016) examined just this relationship and found no detectable link between Li and REE diffusion.

Inclusions in zircon

The plentiful mineral inclusions preserved in detrital zircons from the Jack Hills, western Australia, have been the subject of several studies beginning with Maas et al. (1992) who recognized their dominantly granitic character.

Muscovite. Hopkins et al. (2008, 2010) followed up the Maas et al. (1992) study by examining > 1700 inclusion bearing zircons from Jack Hills. Their examination revealed that quartz and muscovite are the principal inclusion phases, potentially pointing to aluminous granitic sources; see example in Fig. 3). Hopkins et al. (2010) used a thermodynamic solution model for celadonite substitution in muscovite (White et al. 2001) to estimate pressures for muscovite inclusions in magmatic zircons. In all cases, pressures greater than 5 kbar (unsurprising given the presence of magmatic muscovite) were obtained which, coupled with the relatively low host zircons crystallization temperature (ca. 700 °C), implies remarkably low near surface heat flows (≤ 80 mW/m^2). This stands in stark contrast to previous model estimates of 160–400 mW/m^2 (Smith 1981; Sleep 2000). By analogy to modern Earth, this led them to suggest formation in an underthrust, or subduction-like, environment (Hopkins et al. 2008, 2010). However, the primary nature of these inclusions was brought into question by Rasmussen et al. (2011), who surveyed 1000 Jack Hills zircons from 4.2 to 3.0 Ga and showed that some inclusions fall on cracks in their host zircons and that phosphate inclusions generally record post-depositional U–Pb ages. They suggested that much of the mineral inclusion record was due to secondary mineralization.

However, a closer look at the Jack Hills mineral inclusion record reveals complexities not well explained by a largely secondary origin and argues for the preservation of many primary inclusions. Inclusions that intersect cracks in their host zircons display a different modal mineralogy than those isolated from cracks (Bell et al. 2015a). Muscovite inclusions record a wide range of silica substitution with Si-per-formula-unit (12 oxygen basis) ranging from 2.9 to 3.4, unlikely to all form from the same metamorphic fluid. The assemblage that intersects cracks is roughly intermediate between the isolated assemblage and the assemblage of secondary phases seen filling void space along cracks, probably showing partial replacement (Bell et al. 2015a). The isolated and likely primary assemblage is muscovite-dominated with abundant quartz, still suggestive of aluminous granitic protoliths, and minor phases such as biotite, apatite, and feldspars vary in abundance with zircon age (Bell et al.

2015a). Certain phases present in the isolated assemblage and absent in the crack-intersecting assemblage probably point to selective destruction of the minerals apatite and feldspar. Because of the relatively low numbers of identified rare phases (e.g., aluminosilicates), it is difficult at present to determine their significance for zircon provenance or for identifying the nature of the altering fluids that invaded the zircons along cracks over geologic time.

In many instances where a sound case for a preserved primary inclusion assemblage can be made, analyses cannot currently be effected due to size limitations. Indeed, only 6 of the 31 muscovites documented in the Hopkins et al. (2008) study could be reliably analyzed using EMPA due to their small (<2 μm on shortest dimension) size and the effects of secondary fluorescence. Typically, the oldest zircons (>4.2) contain the smallest white mica inclusions. For example, we have identified a zircon as old as 4.34 Ga containing white mica that, except for its size, is a candidate for thermobarometric analysis.

Fe oxides. The development of textural criteria for identifying primary inclusions (Bell et al. 2015a) opens up possibilities for recognizing zircons' changing provenance with time and investigating their post-depositional alteration history. One intriguing aspect of zircon provenance that could be further understood through the inclusion record is that of protolith magma f_{O_2} and its evolution. As described in more detail in the next section, Trail et al. (2011a) demonstrated that the Ce anomaly of a zircon (Ce/Ce*) is a quantitative estimate for host magma f_{O_2}, and furthermore that Hadean Jack Hills zircons show a range in f_{O_2} with an average near the fayalite–magnetite–quartz (FMQ) buffer, i.e., similar to the modern upper mantle. Granitoids form at a range of f_{O_2}, controlled both by source region and assimilation of wall rock material during ascent. Characteristic series of granites with contrasting f_{O_2} in accretionary environments are identified by their accessory Fe-Ti oxides, with more oxidized granites dominated by magnetite and more reduced granites dominated by ilmenite (Ishihara 1977).

Fe-Ti oxides occur commonly as inclusions in zircon (Rasmussen et al. 2011) and appear to be a robust if minor component of the Hadean primary assemblage (Bell et al. 2015a). Primary Fe-oxide inclusions may preserve geomagnetic information and multiple groups are currently investigating whether such signals are the oldest known records of a Hadean dynamo. Knowing when the geodynamo arose potentially constrains the Earth's early thermal structure and potential for atmospheric loss, as well as when compositionally-driven core convection began. At present, the oldest reliable determination of the terrestrial magnetic field is 3.45 Ga (Biggin et al. 2011).

Tarduno et al. (2015) interpreted Jack Hills zircons as containing magnetite inclusions that retained primary remanent magnetization as old as 4.2 Ga (cf. Weiss et al. 2015) but failed to demonstrate whether they had been remagnetized by thermal processes subsequent to formation. Tarduno et al. (2015) argued that their zircons had not experienced high-temperature metamorphism, as Pb would be redistributed in an inhomogeneous fashion at the nm-scale (Valley et al. 2014). This process would result in non-systematic Pb/U variations during SIMS depth profiling, which they did not observe. That view misrepresents their ion microprobe capability in three ways: 1) the sputtering process mixes near surface atoms at the ~10 nm-scale, 2) the SHRIMP instrument they used cannot truly depth profile as sputtered atoms from both crater bottom and surface are simultaneously accelerated into the mass spectrometer, and 3) the 10–20 μm diameter spot they used is three orders of magnitude larger than would be needed to reveal such heterogeneities, even if they existed. As described below, Trail et al. (2016) suggested that zircons exhibiting Li concentration heterogeneities, including oscillatory zoning, could be calibrated in this role as a peak temperature geothermometer for paleomagnetic studies.

Biotite. Biotite inclusions in magmatic zircon are relatively common (Rasmussen et al. 2011) and appear to reflect the composition of biotite in the host (Jennings et al. 2011), which varies considerably among granitoids (e.g., Buda et al. 2004; Abdel-Rahman 1994). Biotite shows characteristic variations in FeO, MgO, and Al_2O_3 contents that can discriminate among calc-alkaline, peraluminous, and alkaline anorogenic granitoids (Abdel-Rahman 1994). Thus, identifying and analyzing primary Hadean biotite inclusions could better constrain the nature of Hadean melt compositions that may have tectonic implications. In addition, rare sulfide (Mojzsis 2007) and carbonaceous (see next section) phases have also been identified in Hadean zircons. A systematic survey for these and other rare phases will further illuminate the volatile contents of Hadean magmas and their source materials.

Graphite. A key challenge in pondering the existence of life elsewhere is that we know of only one occurrence. While Earth is the only planet on which life is known to have emerged, we remain largely ignorant of the conditions, timing and mechanisms by which this occurred. A broad array of morphological and isotopic evidence supports the view that by 3.8 to 3.5 billion years (Ga) ago our planet hosted microbiota, including some with relatively sophisticated metabolisms (e.g., Mojzsis et al. 1996; Rosing 1999; McKeegan et al. 2007; Schopf 2014; Brasier et al. 2015).

As noted earlier, geochemical studies of Hadean Jack Hills zircons have led several authors to suggest relatively clement conditions on earliest Earth (e.g., Mojzsis et al. 2001; Wilde et al. 2001; Harrison 2009). This leaves open the possibility that our planet became habitable, and life emerged, during the first 500 million years of Earth history. Knowing when and under what conditions life emerged could tell us a great deal about the likelihood of life elsewhere. Were conditions clement or hellacious? Did life emerge virtually immediately or only after a half billion years of planetary preparation?

Thus reports of abundant diamond and graphite inclusions in the Jack Hills zircons (4% of each in the zircons investigated) and the spectrum of light carbon isotopic compositions they contained (Menneken et al. 2007; Nemchin et al. 2008) was met with both excitement and skepticism; the latter reflecting the seeming inconsistency of the presence of diamonds with the many inferences drawn from other zircon inclusions (e.g., their derivation from crustal melts; Mojzsis et al. 2001; Peck et al. 2001; Watson and Harrison 2005; Hopkins et al. 2010; Bell et al. 2015). Recently, the diamonds were shown definitively to be contamination from the polishing compound that was used during sample preparation (Dobrzhinetskaya et al. 2014). The origin of the graphite was less certain but deemed also likely due to contamination. This left the true occurrence rate and nature of carbonaceous materials in the Jack Hills zircons uncertain.

Bell et al. (2015b) optically examined a large number of >3.8 Ga Jack Hills zircons and found ~25% contain opaque inclusions. Imaging these selected grains by Raman spectroscopy revealed two isolated carbonaceous inclusions in a concordant, 4.10 Ga zircon (RSES 61-18.8; Fig. 6). To ensure that these inclusions were never in contact with the laboratory environment prior to structural and isotopic analyses, Bell et al. (2015b) extracted a ~160 ng sliver of the zircon containing the two carbonaceous phases via focused ion beam milling and examined it using X-ray nanotomography (Fig. 6). The 40 nm spatial resolution of this imaging method revealed no cracks associated with the graphite inclusions. Their isolation within the zircon crystal and from cracks indicated a primary origin. Carbon isotopic measurements using SIMS yielded an average $\delta^{13}C_{PDB}$ of -24 ± 5‰. As carbon isotopic fractionation between gaseous and condensed species in magmas is expected to be relatively small (e.g., ≤ 4‰; Javoy et al. 1978), such a low $\delta^{13}C$ value is consistent with a biogenic origin. While there are possible inorganic mechanisms that could also produce such a signal, they require what we see as an unlikely chain of geologic events (Bell et al. 2015b). Alternatively, House (2015) offered the "wild" suggestion that a high carbon content in Earth's core could have resulted in an initially highly ^{13}C-depleted mantle. If the Bell et al. (2015b) result does indeed represent an isotopic signal of biologic activity, it would extend our knowledge of the timing of terrestrial life back to at least 4.1 Ga, or ≥ 300 Ma earlier than the previously suggested and coincident with estimates derived from molecular divergence among prokaryotes (Battistuzzi et al. 2004).

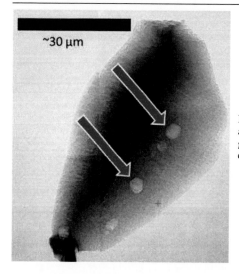

Figure 6. High resolution transmission X-ray image of RSES 61-18.8 with arrows pointing to the two graphite inclusions analyzed for carbon isotopes (Bell et al. 2015b).

Reports of graphite in S-type granites are relatively rare but cases have been documented in which it was inherited from the source (Seifert at al. 2010; Zeng et al. 2001), incorporated via wallrock assimilation (Duke and Rumble 1986), or precipitated during subsolidus interactions with CO_2 (Frezzotti et al. 1994; also see Carroll and Wyllie 1989). Graphite inclusions have been reported in metamorphic zircon (Song et al. 2005) but, to our knowledge, ours is the first documented case of primary graphite in magmatic zircon. Given the relative paucity of investigations of zircon inclusion populations, it is difficult to know whether this reflects their low abundance or simply the lack of a concerted search.

The oxygen fugacity over which graphite can be stable in a granitic magma depends on H_2O and H_2 activities (Ohmoto and Kerrick 1977), but relatively reducing redox conditions (i.e., below FMQ) are implied.

Zircon geochemistry

Titanium. Because the abundance of a trace element partitioned between mineral and melt is temperature dependent, crystallization temperatures can in principle be estimated from knowledge of the concentration of that element in the solid phase if the magma is appropriately buffered. The advent of the Ti-in-zircon thermometer permitted zircon crystallization temperatures to be assessed provided the activities of quartz and rutile can be estimated (Watson and Harrison 2005; Watson et al. 2006; Ferry and Watson 2007). The diffusion of Ti in zircon is vanishingly slow under crustal conditions (Cherniak and Watson 2007) and thus the potential for re-equilibration of the thermometer is very low. In the case in which zircon co-exists with both quartz and rutile (i.e., $a_{SiO_2} \approx a_{TiO_2} \approx 1$), an accurate and precise temperature (i.e., ±15 °C) can routinely be determined.

The first application of the Ti-in-zircon thermometer was to Hadean zircons from Jack Hills. Watson and Harrison (2005) measured Ti in zircons ranging from 3.91 to 4.35 Ga, and the vast majority of these plotting in a normal distribution. Excluding high temperature outliers yielded an average temperature of 680±25 °C (data shown in Fig. 7). However, a limitation in applying this thermometer to detrital zircons is the unknown a_{TiO_2} of the parent magma. In the case of the zircon shown in Fig. 3, which contains both primary quartz and rutile, an accurate crystallization temperature is expected. However, unless co-crystallization with rutile is known, the calculated temperature it is a minimum estimate. In the absence of rutile inclusions, Watson and Harrison (2005) argued that a_{TiO_2} is largely restricted to between

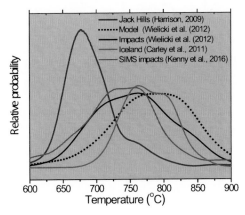

Figure 7. Probability plot of apparent zircon crystallization temperature comparing Hadean data (blue) with data for Icelandic (magenta) and impact formed (black) zircons. The dashed curve shows the distribution predicted by a model incorporating impact thermal effects, continental rock chemistry, and zircon saturation behavior (Modified from Wielicki et al. 2012).

~0.5 and 1 in continental igneous rocks as the general nature of evolving magmas leads to high a_{TiO_2} prior to zircon saturation. Thus for Hadean zircons of magmatic origin, it would be a rare case in which zircon formed in the absence of a Ti-rich phase (e.g., rutile, ilmenite, titanite), thus generally restricting a_{TiO_2} to ≥ 0.5. In case of $a_{TiO_2} \approx 0.5$, calculated temperatures in the range 650–700 °C would be underestimated by 40–50 °C, although this is entirely compensated for if $a_{SiO_2} = a_{TiO_2}$ (Ferry and Watson 2007). Hofmann et al. (2009) inferred that enhanced Ti contents could be incorporated during non-equilibrium crystallization resulting in higher than actual calculated temperatures. If this effect were significant in the generation of granitic Jack Hills zircons this would further support their low-temperature origin.

While it is widely acknowledged that water saturation in intracrustal magmas is rare and that the vast majority of intermediate to siliceous magmas form by dehydration melting under vapor absent conditions (Clemens 1984), Watson and Harrison (2005) concluded that the tight cluster of Hadean zircon crystallization temperatures at 680 ± 25 °C (Fig. 7) reflects prograde melting under conditions at or near water saturation. They arrived at this interpretation because prograde, vapor-absent melting of metapelites and orthogneisses containing typical crustal Zr concentrations (i.e., 150–200 ppm; Harrison et al. 2007) at 5 to 10 kbar would be expected to record zircon crystallization temperature peaks corresponding to the relevant dehydration melting equilibria (e.g., muscovite at ca. 740 °C, biotite at ca. 770–800 °C, amphibole at ≥ 800 °C; Spear 1993). Thus, for example, Hamilton's (2007) assertion that Hadean zircons were derived solely through melting resulting from hornblende breakdown is fundamentally inconsistent with all thermometric results to date.

Rock porosities in the middle and deep crust are typically <0.1% (Ingebritsen and Manning 2002) and thus <0.03 wt.% free H_2O is available to flux melting. During metamorphism, water is progressively lost from rocks via discontinuous, subsolidus dehydration reactions through the greenschist and amphibolite facies (Spear 1993). Structural water is stored in hydrous minerals (e.g., ~4% in muscovite, ~3% in biotite, ~2% in hornblende, ~2–4% in altered basalt at greenschist facies; Clemens and Vielzeuf 1987; Franzson et al. 2010). The correspondingly low water contents of pelitic (~1.2%) and quartzofeldspathic rocks (~0.6%) are expected to produce only small amounts of melt at temperatures close to 700 °C (Clemens and Vielzeuf 1987). Figure 8 (White et al. 2001) underscores the limited melting potential of a metapelite (represented by the Na+CaO–K_2O–FeO–MgO–Al_2O_3–SiO_2–H_2O system) for the temperature range (655–705 °C)

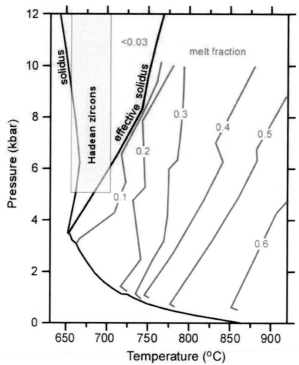

Figure 8. *P–T* pseudosection for a model pelite in the NCKFMASH system containing added 20 mol% added H₂O (modified from White et al. 2001). Note that even in the presence of this free water, the *P–T* region populated by Hadean zircons would result in essentially melt-free conditions indicating that very high water contents (>9 wt%; Burnham 1975) would be required to create significant, mobile magmas.

and pressures (>6 kbars) inferred for Hadean Jack Hills zircons which fall below the "effective solidus" melt fraction of 0.03 (even in the presence of 20 mol% added H₂O). That is, under vapor absent conditions at the pressure–temperature range documented for Hadean zircons (Hopkins et al. 2010), both pelitic and quartzofeldspathic source rocks would be effectively melt free.

Concluding that Hadean Jack Hills zircons largely formed under water-saturated conditions sharply limits the possible tectonic settings in which they formed. The key issue is that silicate magmas at pressures above 6 kbar dissolve much more H₂O than is available in rocks (up to 70 mol% at 10 kbar; Burnham 1975; Clemens 1984) and thus requires an external (e.g., dehydrating underthrust sediments) source of water for saturation to be achieved.

Rare earths. As previously noted, the abundance ratio of Ce in zircon relative to that interpolated from the light rare earth pattern (Ce/Ce*) has been developed as a quantitative estimate for host magma f_{O_2} (Trail et al. 2011a). Most Hadean Jack Hills zircons are within error of FMQ (similar present-day upper mantle) but range as low as IW suggesting a diversity of source materials (Trail et al. 2011a). Another, qualitative estimate for magma f_{O_2} involves the mineralogy of Fe-Ti oxide phases. Ishihara (1977) observed that both oxidized and reduced series of granitoids occur in accretionary settings, with the oxide mineralogy of high f_{O_2} granites dominated by magnetite and that of the reduced granites dominated by ilmenite. Since oxide inclusions are often a minor constituent of igneous zircon inclusion suites (e.g., Rasmussen et al. 2011), the coupled investigation of oxide inclusion mineralogy with Ce/Ce* in the host zircon provides the ability to check for internal consistency between these two estimates of Hadean magma redox conditions.

It will be helpful to establish both the characteristic ranges of zircon Ce/Ce* for magnetite vs. ilmenite series granitoids and whether the oxide mineralogy of the whole rock is accurately reflected by the mineralogy included in zircons. Reconnaissance EDS analysis (Hopkins et al. 2010; reported by Bell et al. 2015a) suggests that the Hadean opaque inclusions are dominated by Fe oxides, potentially magnetite. Further investigation of the mineralogy of Hadean oxide inclusions, coupled with their host zircon Ce/Ce*, may help to better classify the granitoids they derive from or potentially to diagnose alteration affecting the inclusions or host zircon. This coupled approach will better illuminate the redox conditions in the Hadean crust and any potential complexities that igneous zircon may record.

However, petrologically important trace element characteristics such as Ce/Ce* and Ti content can be obscured by alteration or contamination (e.g., cracks, inclusions, etc.). Hydrothermal alteration of zircon is often diagnosed by a high, flat light rare earth element (LREE) pattern. Among Jack Hills zircons, such alteration is dominantly characterized by anomalously high Ti, Fe, P, U, and LREE contents (Bell et al. 2016). To remediate this issue, Bell et al. (2016) developed a trace element indicator (i.e., the LREE-Index; Fig. 9) which permits altered and hydrothermal zircons to be clearly identified.

Lithium. As noted earlier, Trail et al. (2016) proposed the use of Li zoning in zircon as a peak temperature indicator, particularly for use in ascertaining the retention of primary remanent magnetic signals. Figure 9 shows a direct ion image of $^7Li^+$ of the surface of a sectioned 4.02 Ga Jack Hills zircon containing a ~5-μm-wide Li concentration band (Fig. 10). The general preservation of this band requires that peak heating temperature(s) for this detrital zircon did not exceed ~500 °C for million-year timescales. Thus this grain did not exceed the Curie temperature for magnetite of 585 °C and would be a viable candidate for study of primary magnetism. Of course each detrital zircon in a population may have a different pre-depositional thermal history, but this result indicates that, post deposition, the metaconglomerates at Erawondoo Hill have not experienced temperatures greater than 500 °C, consistent with other thermometric determinations (Rasmussen et al. 2010).

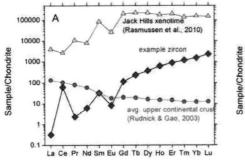 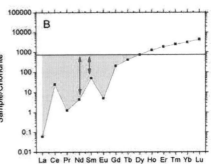

Figure 9. A) Typical zircon trace element pattern vs. that for metamorphic xenotime in Jack Hills discovery site quartzite and average upper continental crust. Most potential contaminating materials would have higher LREE and/or a higher LREE/HREE slope than primary magmatic zircon. B) Definition of the Light Rare Earth Element Index (LREE-I), based on the ratio of the LREE (represented by Nd and Sm) to the MREE (represented by Dy). Reprinted from Bell et al. (2016) Recovering the primary geochemistry of Jack Hills zircons through quantitative estimates of chemical alteration Geochimica et Cosmochimica Acta 191:287–292., with permission from Elsevier.

Aluminum. Trail et al. (2017) found that zircons from peraluminous granitoids contain average Al concentrations of ~10 ppm (with a range from 0 to 23 ppm), in contrast to I- and A-type zircons, which average ~1.3 ppm. Although alumina activity appears not to be

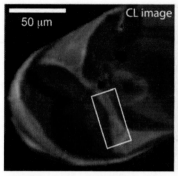 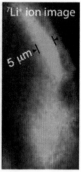

Figure 10. Li-in-zircon geospeedometry applied to a 4.03 Ga Jack Hills zircon. (Left) CL image showing growth zoning. (Right) Li+ image in boxed region at left. The 5 μm scale of Li banding requires peak heating temperatures to have never exceed 500 °C for Ma-timescales. This constrains the Jack Hills Hadean zircons to have never been heated above the magnetite Curie point since 4.03 Ga. Reprinted from Trail (2016) Li zoning in zircon as a potential geospeedometer and peak temperature indicator. Contributions to Mineralogy and Petrology 171:1–15, with permission from Springer.

a simple function of the degree of the peraluminosity, zircon Al concentration could be calibrated as a proxy for melt $Al_2O_3/(CaO + Na_2O + K_2O)$ where molar values > 1 reflect an origin from recycled pelitic material (Trail et al. 2017). They applied this approach to Hadean Jack Hills zircons and found both metalunimous and peraluminous sources, albeit with the former apparently dominating the population. Although the scarcity of high Al contents from Hadean zircons suggests that metaluminous crustal rocks may have been more common than peraluminous rocks in the Hadean, the ~20% overlap of low Al (i.e., < 5 ppm) in S-type zircons somewhat obscures this inference.

Carbon. As carbon is long known to dissolve in silicates at trace levels (e.g., Freund et al. 1980; Oberheuser et al. 1983; Mathez et al. 1984; Tingle et al. 1988; Keppler et al. 2003; Rosenthal et al. 2015), the coexistence of zircon and graphite raises the possibility that C could be present at measurable levels in zircon. As SIMS has the potential for detection levels of C as low as ~1 ppb, it is ideally suited for such a search.

The lack of dependency of carbon solubility in silicates on oxygen fugacity suggested to Shcheka et al. (2006) that C^{4+} substitutes for Si^{4+}, with increased levels as the volume of the SiO_4 tetrahedron decreases. In this regard, they emphasized that the relatively small volume of the SiO_4 tetrahedron in zircon should enhance carbon solubility. Alternatively, Sen et al. (2013) found evidence that C sbustitutes for nonbridging oxygen in synthesized silicate nanodomains. As such they hypothesized that trace carbon could be incorporated into silicates across a broader range of f_{O_2} than previously thought and speculated that this incorporation mechanism might have been preferentially important during the Hadean eon.

In the same way that a zircon co-crystallizing with rutile contains a predictable temperature-dependent Ti concentration, zircons growing in the presence of a carbonaceous species appear to partition C in a fashion that could be calibrated as a magma volatile probe. Having an approach with which to detect Hadean crustal C could potentially reconcile the disparate views regarding the magnitude of carbon in the crust during that eon. Some authors argued for a net increase in crustal carbon from essentially zero at 4 Ga (e.g., Hayes and Waldbauer 2006; Kelemen and Manning 2015) to its present day inventory in a broadly linear fashion. Marty et al. (2013) envisioned an essentially continuous transfer of carbon from undegassed mantle reservoirs implying a net increase to the crust over time. In contrast, Dasgupta (2013) advocated for higher than present day concentration on early Earth. The development of a proxy to detect the presence of carbon in Hadean (and younger) melts may eventually permit selection among these models.

OTHER WESTERN AUSTRALIAN HADEAN ZIRCON OCCURRENCES

Mt. Narryer

Ion microprobe dating of detrital zircons from several quartzites at Mt. Narryer in Western Australia (Fig. 11) have revealed a minor Hadean component ranging from 2% (Froude et al. 1983) to 12% grains >4.0 Ga (Pidgeon and Nemchin 2006), with younger zircons ranging to ca. 3 Ga. A LA-ICP-MS study of Mt. Narryer zircons of all ages suggested that they generally display higher U contents and lower Ce/Ce* than Jack Hills zircons (Crowley et al. 2005). Our preliminary ion microprobe data for 80 zircons between ca. 3 and 3.75 Ga in age from two Mt. Narryer quartzites suggests that zircons with unaltered magmatic chemistry (i.e., via the LREE-Index) do indeed show slightly higher U contents and lower Ce/Ce* than Jack Hills zircons of similar age, although there is significant overlap between the populations. However, crystallization temperatures for these zircons is higher than at Jack Hills, averaging ~750±50°C. All but one of our studied zircons has Th/U > 0.2, indicative of magmatic origins. Calculated f_{O_2} for these zircons suggests values on average several log units below the FMQ buffer, which overlaps the range of many less-oxidized Hadean Jack Hills zircons. Although we have not yet identified >4 Ga zircons from Mt. Narryer in our preliminary survey, these differences in chemistry likely point to a diversity of Eoarchean-Hadean source rocks represented in Western Australia, as also suggested by Crowley et al. (2005) and by Hf isotopic compositions (Nebel-Jacobsen et al. 2010).

Churla Wells

Ion microprobe dating of a zircon from an orthogneiss from near Churla Well,~25 km west of the Mt. Narryer site (Fig. 11), yielded grains with $^{207}Pb/^{206}Pb$ ages of 4.14 to 4.18 Ga (Nelson et al. 2000). Electron microprobe traverses show that the core containing the oldest ages has much lower Hf, REE, U and Th than the outer regions. Nonetheless, several observations—U contents in the core ranging up to 666 ppm, Th/U as high as 0.6, and trace element concentrations and ratios—strongly suggest its origin in a granitic magma.

Maynard Hills

In the Southern Cross Granite–Greenstone Terrane, Western Australia (Fig. 11), ion microprobe dating of a single zircon from a quartzite within the Maynard Hills greenstone belt (Wyche 2007) yielded a mean $^{207}Pb/^{206}Pb$ age of 4.35 ±0.01 Ga.

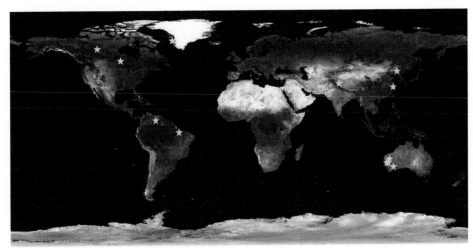

Figure 11. Location map of the 13 sites from which >4 Ga zircons have been documented.

Mt Alfred

At the Mt. Alfred locality of the Illaara Greenstone Belt further along strike, Nelson (2005) documented a concordant zircon with an age of 4.17 ± 0.01 Ga. Thern and Nelson (2012) reported three additional Hadean zircon ages from this sample ranging from 4.23 to 4.34 Ga. To our knowledge, no geochemistry for Hadean zircons from this sample have been published.

NORTH AMERICAN HADEAN ZIRCON OCCURRENCES

Northwest Territory, Canada

The Acasta tonalite orthogneiss from the Western Slave craton (Fig. 11) yields a range of U–Pb zircon ages interpreted to date protolith crystallization at 3.96 Ga (Bowring and Williams 1999; Stern and Bleeker 1998; Mojzsis et al. 2014). However, Iizuka et al. (2006) documented a 4.20 ± 0.06 Ga zircon grain using LA-ICP-MS. This apparent xenocryst has a LREE pattern (Bell et al. 2016) within the field associated with unaltered zircon. Its Th/U suggests a magmatic origin and, along with other trace element concentrations and ratios, derivation from a felsic melt by a process other than differentiation of a mafic magma. Pronounced Ce and Eu anomalies correspond, respectively, to an f_{O_2} close to the FMQ buffer (assuming a crystallization temperature of 750 °C) and a crustal, as opposed to mantle, origin.

Greenland

Detailed ion microprobe dating of a tonalitic orthogneiss from Akilia Island, West Greenland (Fig. 11), that crosscuts the oldest known marine sediment (Manning et al. 2006) established a crystallization age of 3.83 ± 0.01 Ga (Mojzsis and Harrison 2002). A U–Pb survey of zircons identified in thin section documented a single zircon, concordant within uncertainty, with an age of 4.08 ± 0.02 Ga.

ASIAN HADEAN ZIRCON OCCURRENCES

Tibet

Duo et al. (2007) report a 4.1 Ga ion microprobe age for a zircon from a quartzite in Buring County, western Tibet (Fig. 11). Th/U ratios greater than 0.7 suggest a magmatic origin for this detrital grain.

North Qinling

Wang et al. (2007) reported a LA-ICP-MS age of 4.08 ± 0.01 Ga for a xenocrystic zircon from Ordovician volcanics of the Caotangou Group, North Qinling Orogenic Belt (Fig. 11). Subsequent ion microprobe and LA-ICP-MS analyses identified additional Hadean grains with ages ranging from 4.03 to 4.08 Ga (Diwu et al. 2010, 2013). Hafnium isotope analyses of these grains are consistent with origin in crust extracted between 4.0 to 4.4 Ga (Diwu et al. 2013).

North China Craton

Cui et al. (2013) reported a LA-ICP-MS U–Pb date of 4.17 ± 0.05 Ga, concordant within uncertainly, for a xenocrystic zircon from the Anshan–Benxi Archaean supracrustal greenstone belt (Fig. 11). Correction of common Pb was made using [208]Pb and assumed concordancy between the U–Pb and Th–Pb systems. The zircon was separated from fine-grained amphibolites intruded into banded iron formation and bedded coarse-grained amphibolites. Its Th/U of 0.46 suggests a magmatic origin.

Southern China

Using ion microprobe U–Pb dating, two Hadean detrital zircons were documented from a quartzite within Neoproterozoic metasediments from the Cathaysia Block in southwestern

Zhejiang (Fig. 11; Xing et al. 2014). One zircon core yielded a $^{207}Pb/^{206}Pb$ age of 4.13 ± 0.01 Ga with a $\delta^{18}O = 5.9 \pm 0.1‰$. The other zircon grain has a 4.12 ± 0.01 Ga magmatic core, a $\delta^{18}O$ of $7.2 \pm 0.2‰$, a positive Ce anomaly indicative of highly oxidizing conditions, and a high apparent Ti-in-zircon crystallization temperature of 910 °C. While the authors interpreted these results to suggest the zircons originated via dry melting of oxidized and hydrothermally altered supracrustal rocks, closer examination of the high apparent Ti content may be warranted due to Ti contamination effects (Harrison and Schmitt 2007). Trace element discrimination diagrams and REE patterns place the zircon core within the continentally-derived field.

SOUTH AMERICAN HADEAN ZIRCON OCCURRENCES

Southern Guyana

A xenocrystic zircon from a felsic volcanic unit of the Iwokrama Formation, Guyana Shield (Fig. 11), yielded a concordant LA-ICP-MS U–Pb age of 4.22 ± 0.02 (Nadeau et al. 2013). No other geochemical analyses of this zircon have been reported.

Eastern Brazil

The Archean core of the São Francisco Craton, northeastern Brazil (Fig. 11), contains meta-volcanosedimentary supracrustal rocks including the Ibitira–Ubirac greenstone belt. Paquette et al. (2015) analyzed a zircon from an amphibolite facies pelite from this belt by LA-ICP-MS yielding a distribution of U–Pb ages that intersected concordia at 4.22 ± 0.02 Ga (four $^{207}Pb/^{206}Pb$ ages > 4.01 Ga). The core Th/U ratios of 0.8 and high U contents (up to 1400 ppm) suggest a felsic magmatic origin of this probably detrital grain.

The above discussion is complicated by limitations comparing data generated by different analytical methodologies. For example, the LA-ICP-MS approach lacks the capacity to measure common ^{204}Pb and thus the possibility exists that single zircon occurrences with apparent ages of > 4 Ga could be due to inclusion of non-radiogenic Pb.

OTHER PROPOSED MECHANISMS FOR FORMING HADEAN JACK HILLS ZIRCONS

The collective data obtained over the last 15 years broadly support an origin of many Hadean zircons as crystallizing from relatively cool, relatively wet felsic melts sourced at least in part from sedimentary protoliths at a plate boundary. Nonetheless, numerous other models have been proposed, and we evaluate the consistency of the data with these hypotheses.

Icelandic rhyolites

Iceland's unusual geochemical character and thick basaltic crust couple to produce an unusually high proportion (~10%) of silicic magmatism. These rocks in turn host abundant zircon making Iceland a seemingly attractive model to explain the production of Hadean zircons (Taylor and McLennan 1985; Galer and Goldstein 1991; Valley et al. 2002). However, comprehensive investigations of the trace element and oxygen isotope composition of Icelandic zircons show that the two populations to be distinctively different (Carley et al. 2011, 2014; cf. Reimink et al. 2014). In contrast to the elevated ^{18}O signature in some Hadean zircons, Icelandic zircons are characterized by ^{18}O-depleted values, likely the result of recycling of altered basaltic crust rather than direct melting of sedimentary rocks (Bindeman et al. 2012). Zircon crystallization temperatures are similarly different, with Icelandic zircons yielding an average of 780 °C compared to the 680 °C average of Jack Hills Hadean zircons.

As noted in the previous section, higher temperature data from the Mt. Narryer and southwestern Zhejiang sites may indicate their formation in different environments relative to the Jack Hills population, possibly similar to an Iceland-like source.

Intermediate igneous rocks

Several authors have argued that the low temperature Hadean peak could reflect zircon saturation at low temperatures in rocks of the tonalite-trondhjemite-granodiorite (TTG) suite (Glikson 2006; Nutman 2006). Glikson (2006) proposed that Hadean zircons could have originated in TTG's that formed at high-temperatures but did not crystallize zircon until near eutectic temperatures were reached. Similarly, Nutman (2006) argued on the basis of calculated saturation temperatures (Watson and Harrison 1983) for TTG's that high-temperature melts do not crystallize zircon until they cool to temperatures near that of minimum melting (i.e., zircons from both wet tonalite and minimum melts yield similarly low crystallization temperatures). However, Harrison et al. (2007) showed bulk rock saturation thermometry to be inapplicable to zircons crystallizing from TTGs. Rather, a cooling magma system first crystallizes modally abundant phases, increasing the Zr concentration in the residual melt while moving the melt towards compositions with much lower capacities to dissolve zircon. Thus temperatures calculated from bulk rock chemistry significantly underestimate the onset temperature of zircon crystallization in TTGs. The expected and observed result is that zircons crystallized from TTG melts (as documented by electron imaging studies) yield significantly higher temperatures than seen in the Hadean Jack Hills population (e.g., Harrison et al. 2007).

Mafic igneous rocks

A variety of authors have suggested that the > 4 Ga zircon temperature distribution could be derived from zircons originating in mafic magmas (Coogan and Hinton 2006; Valley et al. 2006; Rollinson 2008). However, zircon formation temperatures in these environments are significantly higher (>750 °C) than the Hadean peak (e.g., Harrison et al. 2007; Hellebrand et al. 2007). As a case in point, Rollinson (2008) argued that the $\delta^{18}O$ and trace element signatures in Hadean Jack Hills zircons were consistent with an origin in ophiolitic trondhjemites rather than continental crust. The author pointed to water-saturated, low pressure melting experiments on oceanic gabbros at > 900 °C that yielded trondhjemitic melts. While the origin of the excess water is potentially explicable in this scenario, the origin of muscovite, a mineral uncharacteristic of trondhjemite but the most common inclusion in Jack Hills Hadean zircons (Hopkins et al. 2008), was not addressed. Although we noted earlier the clear separation between Hadean and MORB zircons on a plot of U/Yb vs. Y plot (Grimes et al. 2007), Rollinson (2008) argued that data on such discrimination diagrams showed a ~20% overlap and were thus permissive of such an origin. While this is true when plotting present U concentrations, the separation becomes essentially complete once an appropriate correction for U decay has been made (e.g., a 4.3 Ga zircon presently containing 100 ppm U originally crystallized with 244 ppm U).

As noted earlier, zircons derived from a wide range of mafic rocks yield much higher average temperatures (~770 °C; Valley et al. 2006; Fu et al. 2008) than the Hadean population (Harrison et al. 2007). In the absence of a natural selection mechanism that preferentially excludes zircons formed at high temperature (the opposite of what is expected from preservation effects on high radioactivity zircons), intermediate to mafic sources are unlikely to have contributed significantly to the Hadean Jack Hills population. However, sparse, higher temperature zircons from the Mt. Narryer and southwestern Zhejiang locations may well be consistent with such an origin and could be tested by studies of their inclusion populations.

Sagduction

A feature of modern plate tectonics is that oceanic lithosphere older than about 20 Ma is negatively buoyant and thus can be underthrust beneath adjacent plates. At 80–120 km depths, the basaltic crust undergoes a transformation to much denser eclogite and the resulting pull on the downgoing slab provides a first-order contribution to the global plate tectonic energy budget. In the hotter mantle of early Earth, it is assumed that oceanic lithosphere would have been thicker and thus may not have been able to achieve the neutral buoyancy required for subduction making plate tectonic-type behavior uncertain (Davies 1992; cf., Korenaga 2013). Under such conditions, a low apparent geotherm could be achieved locally where thermally and/ or compositionally dense crust sinks into the mantle as downward moving drips (sagduction; Macgregor 1951). While this can insulate the descending mass from reaching melting temperature until high pressures are attained (e.g., Davies 1992), more nuanced scenarios are also possible (e.g., François et al. 2014). Such a mechanism was invoked by several authors (Williams 2007; Nemchin et al. 2008) to explain the anomalously low (<10 °C/km) geotherms required by the apparent occurrence of diamond in Hadean Jack Hills zircons, although recognition that the diamonds were contamination (Dobrzhinetskaya et al. 2014) obviated the need for such models.

The sagduction model shares similar limitations to those discussed above, the source of the needed water and, in the case of blocks delaminated into the mantle, the lack of a mechanism to return zircons formed by this mechanism to the surface. Consider the case of a sagducting block of mafic eclogite. As noted earlier, below the brittle-ductile transition rock porosities are typically <0.1% (Ingebritsen and Manning 2002). Structural water stored in hydrous minerals is limited to ≤2% of virtually all rock types and is lost progressively via discontinuous, subsolidus dehydration reactions through the greenschist and amphibolite facies (Spear 1993).

Any water liberated by dehydration is likely to ascend from the sagducting drip into colder, overlying rocks. Thus fusion is likely to be forestalled until temperatures greatly exceeding that of minimum melting are reached. In the case of complete devolatilization, temperatures of >900 °C would be required for melting of dry rock. As noted earlier, the absence of peaks in the Hadean zircon crystallization spectrum corresponding to dehydration melting does not support such a mechanism and such melts are unlikely to be characterized by quartz and muscovite inclusions. Even the most appealing scenario involving eclogitized pelite containing a 50:50 mixture of muscovite and quartz contains only ~2 wt.% water and vapour absent melting of such a protolith produces highly water-undersaturated melts (e.g., Patiño Douce and Harris 1998).

Sagduction models lack a mechanism to introduce water-rich fluids into fertile source rocks capable of yielding both peraluminous and metaluminous magmas at temperatures close to minimum melting (as required by Ti thermometry) and then sustain the supply of water until the rock's melt fertility is essentially exhausted (thus resulting in the single Hadean zircon peak at ca. 680 °C). The twofold appeal of a plate boundary environment is the continuous source of water available in the hangingwall of a submarine underthrust and the potential for long-term (i.e., >4 Ga) preservation of any zircons created by water-fluxed melting. In contrast, how zircons formed during dehydration melting in a block sagducting into the mantle will reappear to be preserved at the Earth's surface is unclear.

Impact melts

Given the likelihood of high bolide fluxes to early Earth, the potential for impact melts to be a source of Hadean zircons requires investigation. Studies of neo-formed zircon in preserved terrestrial basins large enough to have created melt sheets (e.g., Sudbury, Morokweng, Manicouagan, Vredefort) show that their crystallization temperatures average more than 100 °C greater than that of >4 Ga Jack Hills zircons and thus impacts do not represent a dominant source for that Hadean population (Darling et al. 2009; Wielicki et al. 2012). This observation is supported by models that relate expected impact thermal anomalies with early crustal rock chemistry (Wielicki et al. 2012; Fig. 7), which confirm that observations from the handful of known impact melt sheets is indeed globally representative. Recently, Kenny et al. (2016) argued that zircon crystallization temperatures for the granophyre layer at the Sudbury impact crater had been underemphasized and proposed this rock type as a source of at least a portion of Hadean Jack Hills zircons (Fig. 7). They raised the prospect of an unspecified selection process that had preferentially destroyed high temperature Hadean zircons and thus biased the detrital record to low temperatures. Nature does tend to bias the detrital zircon record but that mechanism operates in exactly the opposite sense. Late crystallizing, thus low temperature, granitoid zircons are known to contain elevated U and Th concentrations which lead to metamictization (Claiborne et al. 2010) and thus they are more likely lost from the detrital record, resulting in preferential preservation of higher temperature zircons (Harrison and Schmitt 2007). Wielicki et al. (2016) tested the Kenny et al. (2016) hypothesis statistically and showed that the probability of extracting the Hadean Jack Hills Ti-in-zircon temperature distribution from their data, or any permutation of the published dataset of impact-produced zircons, is vanishingly small.

Marchi et al. (2014) proposed that an intense bombardment event at ~4.1 Ga covered the planet with ca. 20 km of flood basalt in which the Hadean Jack Hills zircons were formed. In brief, this model is fundamentally incompatible with virtually every geochemical record obtained from that population (i.e., hydrous melting conditions, low geotherm, peraluminous compositions, etc.).

The above statements are specifically relevant to the Hadean Jack Hills zircons and conclusions should be tempered by the reconnaissance-scale data obtained for samples from Mt. Narryer and southwestern Zhejiang, which yield apparent zircon crystallization temperatures of 750 and 910 °C, respectively. While both results are subject to potential Ti contamination effects, developing a strong geochemical database from the dozen or so locations for which >4 Ga zircons have been documented should be a research priority.

Heat pipe tectonics

Moore and Webb (2013) investigated the thermal effects of "heat-pipe" magmatism in which volcanism dominates the near surface thermal structure early in planetary evolution. Their simulations showed that low geotherms could develop in response to frequent volcanic eruptions that advect surface material downwards. They argued that Hadean zircons arose in ascending TTG plutons within the diamond stability field produced at the intersection of the wet basalt solidus and their exceedingly low calculated geotherms. Unfortunately, this constraint was predicated on a report that diamonds had been included in these grains during formation (Menneken et al. 2007). This was subsequently shown to be due to contamination during sample preparation (Dobrzhinetskaya et al. 2014). Even putting this issue aside, left unaddressed is the source of sufficient water to saturate an intermediate melt (>25 wt.%; Mysen and Wheeler 2000) at depths of >100 km. As noted earlier, the geochemistry of Hadean Jack Hills zircons is inconsistent with low water activity melting and the inclusion assemblage is unlikely to arise from a basaltic source.

Terrestrial KREEP

KREEP is an acronym reflecting K-, REE- and P-enriched materials on Earth's moon, which are thought to reflect progressive crystallization of a magma ocean (Warren and Wasson 1979). As noted earlier, initial $^{176}Hf/^{177}Hf$ of Jack Hills zircons show large deviations in $\varepsilon_{Hf(T)}$ from bulk silicate Earth (see summary in Bell et al. 2014). The initial report of Harrison et al. (2005) of positive ε_{Hf} results utilizing ion microprobe U–Pb age spots with LA-ICP-MS Lu–Hf results on differing portions of the analyzed zircon were not reproduced in a follow-up study (Harrison et al. 2008) in which age and Hf isotopes were measured on the same volume. This was ascribed to non-linear mixing effects between zircon rims and cores (see Harrison et al. 2005) that almost certainly also affected the bulk results of Blichert-Toft and Albarède (2008).

Kemp et al. (2010) chose a small subset of the Jack Hills ε_{Hf} data that aligned along a subchondritic array extrapolating back to 4.4–4.5 Ga. They interpreted the coherence of this limited dataset to reflect formation of an incompatible element enriched reservoir during solidification of a magma ocean—in effect, a terrestrial KREEP layer. In their model, ~400 Ma of subsequent intra-crustal melting of basalt hydrated by interaction with an early atmosphere/hydrosphere produced the Hadean Jack Hills zircons including those with high ^{18}O. They interpreted the results of an experimental study of the simple system $CaO+MgO+Al_2O_3 + SiO_2 + H_2O$ (Ellis and Thompson 1986), which produced peraluminous melts at $\geq 800\,°C$ under water-saturation, as explaining the presence of muscovite inclusions in Jack Hills zircons and thus obviating the requirement for a metasedimentary source. While true that corundum-normative melts are produced under these conditions, muscovite was of course not present in the K-free experimental system and would have been an unlikely modal phase to form from a basaltic protolith. As with most of the above hypotheses, the principal problem with the Kemp et al. (2010) model is the lack of a source for the copious amounts of water required to saturate melts at high pressure and the implied high zircon formation temperatures. As described in the Sagduction and Heat pipe sections, carrying water from surface through a continuous series of dehydration reactions during burial would result in production of inextractable amounts of melt below 750 °C.

Multi-stage scenarios

Shirey et al. (2008) interpreted the origin of Hadean zircons through a multi-part model that included: 1) global separation of an early (>4.4 Ga) enriched reservoir, 2) deep mantle fractionation of Ca-silicate and Mg-silicate perovskite from a terrestrial magma ocean following lunar formation, 3) formation of a mafic to ultramafic crust, and 4) repeated cycles of remelting of that crust under "wet" conditions to produce progressively more silica-oversaturated TTGs. Their arguments against Hadean Jack Hills zircons forming in a dominantly granitic crust are the occurrence of zircons in MORB and Icelandic settings (see *Icelandic rhyolites* and *Mafic igneous rocks* sections as well as the requirement of water-saturated melting for counter arguments). Shirey et al. (2008) note that an Iceland-like environment would permit hydrothermally altered basalt to be buried to the depths of wet melting to produce zircons of similar character to the Jack Hills zircons. However, no such population has been documented in extensive studies of Icelandic zircons (Carley et al. 2011, 2014).

Summary

Although considered separately above, most of the alternative explanations for the geochemical characteristics of the Jack Hills zircons share the assumption that melting occurred intracrustally in the absence of an external source of water (i.e., sagduction, Davies 1992; burial beneath impact melts, Marchi et al. 2014; heat pipe burial, Moore and Webb 2013; terrestrial KREEP, Kemp et al. 2010). As noted previously (see Fig. 8), the *P–T* conditions indicated by these zircons puts them outside conditions that would produce extractable melts fractions thus requiring addition of significant water from an exotic source, such as dehydration of a downgoing slab. Despite the attractions of a plate boundary-type model, as noted at the outset, it is not possible to ascribe unique conditions to Hadean Earth from geochemical records preserved in >4 Ga zircons. Rather, our goal should be to identify a parsimonious, internally-consistent model of Earth evolution that explains robust aspects of the zircon record. We acknowledge that our perspective is strongly biased towards Jack Hills zircons because the vast majority of information about Hadean conditions has been derived from them. Increasingly comprehensive studies of other zircon populations, e.g., from Narryer Hills, might reveal other aspects of Hadean geology. Nonetheless, given data available to date, plate boundary interactions not only provide a setting that explains the nature of Hadean zircon geochemistry and the inclusions they host, but also invoke the simplest dynamical mode that is clearly plausible for this planet. Granted, internal consistency is not smoking gun proof and most accumulated evidence is indirect and open to alternate interpretations. Overly elaborate scenarios, especially those that stand in contradiction to important aspects of the geochemical evidence, do not help advance our understanding of the first five hundred million years of Earth history. Rather, they muddle a discussion that is generally poorly understood by those outside this somewhat narrow field. Furthermore, although it should go without saying, mantle-derived rocks do not possess a record that rocks of continental affinity do not exist elsewhere on the planet (e.g., Kamber et al. 2005; Kemp et al. 2015; Reimink et al. 2016). This is the equivalent of concluding that no cratons exist on the planet today from analysis of a rock from Samoa.

Even if we ultimately conclude that plate boundary interactions were the dominant source of the Hadean zircon geochemical signals, numerous questions remain unanswered. Was the process continuous throughout the Hadean or did it repeatedly start and stop? Is the inferred convergent boundary an island arc, a continent-continent collision, a mixture of the two, or an entirely different kind of setting unique to early Earth?

A LINK TO THE LATE HEAVY BOMBARDMENT?

As noted previously, Jack Hills detrital zircons show a characteristic bimodal age distribution with peaks at about 3.4 and 4.1 Ga, and ages as old as 4.38 Ga (Holden et al. 2009). A curious feature of this distribution is the relative rarity of zircons between 3.9 and 3.6 Ga (Bell and Harrison 2013), a period which includes a hypothesized spike in impacts to the Earth-Moon system (termed the Late Heavy Bombardment; LHB). Evidence of such an event was first seen in ca. 3.9 Ga isotopic disturbances of lunar samples (Tera et al. 1974), although others (e.g., Hartmann 1975) interpreted this as the tail of a decreasing bolide flux. The lack of an identifiable signature in the fragmentary terrestrial rock record from the LHB era (ca. 3.9 Ga) has limited the study of this period of solar system history almost entirely to extraterrestrial samples. Given its scaling to the Moon in terms of gravitational cross section and surface area, the Earth likely experienced ~20 times the impact flux to the Moon causing a widespread crustal thermal disturbance. Thus it is somewhat surprising that the Jack Hills zircon population does not contain a significant proportion grown in impact melt sheets (Wielicki et al. 2012, 2016). Because of their crustal origin, all Hadean Jack Hills zircons share one feature in common—they all must have resided within 10s of km of the Earth surface during the LHB era. Thermal perturbations in the crust during this time, perhaps due to impacts, could mobilize Zr to form epitaxial growths on Hadean-age zircons. Trail et al. (2007) U–Pb depth profiled four Hadean zircons and found that they preserved 3.94–3.97 Ga rims. While they could not rule out endogenic processes as the precipitating event, they speculated that this common trait might be the terrestrial evidence of the LHB. Abbott et al. (2012) followed up this study by simultaneously depth profiling U–Pb age and crystallization temperature of overgrowths on Hadean zircons. Of the eight grains examined, four had 3.85–3.95 Ga rims that yield significantly higher formation temperatures (>840 °C) than either younger rims or older cores. This was again seen as suggestive of an LHB link.

Bell and Harrison (2013) undertook an intensive age survey to archive a large number (>100) of Jack Hills zircons formed in the age range 3.9 to 3.6 Ga. Geochemical analyses on this population showed surprising differences. Specifically, zircons between ca. 3.91 and 3.84 Ga were found to be unique in the >3.6 Ga Jack Hills zircon record in having two distinct trace element groupings. The existence of a distinct high-U (and Hf), low-Ti (and Ce, P, Th/U) zircon provenance (they termed "Group II") is specific to this ca. 70 million year period. The remaining 3.91–3.84 Ga zircons (termed "Group I") resemble the majority of Hadean zircons both in apparent crystallization temperature and numerous other trace elements. These patterns in trace element depletion and enrichment, the seemingly paradoxical coincidence of the highest U contents with high degrees of concordance, and the homogeneous nature or very faint zoning found in many Group II grains, were interpreted to result from thermally-driven, transgressive recrystallization (Hoskin and Schaltegger 2003) at 3.91–3.84 Ga. The persistent coincidence of an apparent thermal event within the period postulated for the LHB suggests that the terrestrial archive of Hadean material may potentially be a superior to lunar samples for establishing the timing, and even existence (Boehnke and Harrison 2016), of a Late Heavy Bombardment.

BROADER IMPACTS OF HADEAN ZIRCONS

The role of Hadean zircons in geochemical innovation

Geochemical studies of Hadean zircons have collectively led to a paradigm shift in our concept of early Earth—there is general agreement that evidence derived from these zircons implies abundant water at or near Earth's surface during the Hadean. Almost as interesting as the paradigm shift itself is the manner by which it occurred and what that says about the role of early Earth research in driving geochemical innovation. As previously noted, it was the need to date large ($> 10^5$) numbers of zircons to create an archive of $> 10^3$ Hadean grains that drove development of the first automated ion microprobe stage.

Zircon has long been appreciated as the leading crustal geochronometer for its robust U–Pb system and resistance to physical and chemical alteration in most geologic environments. Thus knowing the temperature at which magmatic zircon crystallizes sharpens both interpretations of U–Pb dates and permits a range of new petrochronologic investigations. Although this appeal has existed since at least the 1970s, it was through the challenge of the unknown provenance of Hadean zircons that the Ti-in-zircon thermometer was realized (Watson and Harrison 2005; Watson et al. 2006; Ferry and Watson 2007). This application subsequently eruptedacross geochemistry and petrology (as attested by the over 1,700 citations these three papers have attracted since 2005; Google Scholar). Subsequent developments, including terrestrial Pu/U tracing (Turner et al. 2007), zircon f_{O_2} barometry (Trail et al. 2011a), and a magma aluminosity proxy (Trail et al. 2016), underscore the degree to which a vanishingly small lithic record has inspired innovation.

The role of Hadean zircons in scientific thought

Perhaps the most remarkable feature of inferences drawn from > 4 Ga zircons is that none were gleaned from theory. Rather, generations of models innocent of observational constraints fed a paradigm of a hellish, desiccated, uninhabitable Earth (e.g., Cloud 1976; Smith 1981; Sleep et al. 1989; Collerson and Kamber 1999; Ward and Brownlee 2000; O'Neill and Debaille 2014) for which there is no empirical evidence. What compelled the scientific community to create an origin myth in the absence of direct evidence? While science is distinguished from mythology by its emphasis on verification, its practitioners may be as subject to the same existential need for explanations as any primitive society. In context with high expected Hadean heat production and impact flux, it proved irresistible to explain the lack of ancient continental crust by its non-existence rather than the equally or more plausible notion that it has been largely consumed by the same processes we see operating on the planet today.

Whether or not this episode represents a scientific anomaly is, to us, a matter of debate. It is at least arguable that such behavior has been a feature of geophysical modelers approach to Earth evolution since Kelvin's "certain truth" that the planet was less than 2% of its actual age (Thompson 1897) or Jeffreys' denial of the fit between South America and Africa (Jeffreys 1924). For reasons that remain obscure, calculations carried out in the absence of observational constraints can take on an edifice-like character in the geo- and planetary sciences that slows progress and distracts the community from fresh, and possibly better, ideas. Perhaps David Stevenson (1983) said it best when referring to speculations of Hadean dynamics: "Basic physical principles need to be understood but detailed scenarios or predictions based upon them are best regarded as 'convenient fictions' worthy of discussion but not enshrinement".

SUMMARY

Advances in geochemical microanalysis and innovative applications of zircon geochemistry have made this mineral perhaps the only currently known probe of the time predating Earth's known rock record. Zircon's resistance to mechanical breakup and chemical weathering allow it to preserve chemical, isotopic, and mineral inclusion information with an associated U–Pb timestamp through later metamorphism and sedimentary cycling. Although the Jack Hills locality in Western Australia is the best known and most highly studied source of Hadean zircons, worldwide twelve additional localities are known to have yielded at least one zircon older than 4 Ga. Geochemical investigations carried out on Jack Hills zircons have yielded a view of Hadean Earth fundamentally at odds with traditional notions of a dry, impact-disrupted, certainly uninhabitable environment, and these lines of investigation can serve as a useful guide as attention turns to these additional Hadean zircon-bearing localities. Isotopic investigations reveal the likelihood of a surface or near-surface hydrosphere as early as 4.3 Ga ($\delta^{18}O$; e.g., Peck et al. 2001; Mojzsis et al. 2001) and of felsic crust as early as 4.5 Ga (Lu–Hf system; e.g., Harrison et al. 2008). Magmatic Th/U ratios (e.g., Cavosie et al. 2006; Bell and Harrison 2013; Bell et al. 2014), dominantly granitic mineral inclusions (Hopkins et al. 2008, 2010; Bell et al. 2015a), and innovations such as the Ti-in-zircon thermometer (Watson and Harrison 2005) and Ce/Ce^* as f_{O_2} barometer (Trail et al. 2011a) further suggest minimum melt conditions and redox conditions near the present-day upper mantle for the Hadean magmas that produced the Jack Hills zircons. One zircon containing primary graphite with a light isotopic value reminiscent of biologic carbon fixation may point to a terrestrial biosphere as early as 4.1 Ga (Bell et al. 2015b). Preliminary geochemical investigation of zircons from the other localities suggest higher-temperature origins, and further study of these materials will doubtless shed light on the diversity of preserved Hadean magmatic environments. Epitaxial rims on Hadean zircon cores tend toward ca. 3.9 Ga ages and their anomalously high formation temperatures have led some workers to speculate that this could be a terrestrial signal of the Late Heavy Bombardment.

Although a complete picture of Hadean Earth will likely not be possible from detrital zircon alone, they have thus far provided the only known empirical data on pre-4 Ga Earth. Models for this period must consider all of the data from the various isotopic, trace chemical, and mineral inclusion lines of evidence in the zircons. Increasing the geographic (and possibly petrogenetic) diversity of Hadean samples available for study will undoubtedly continue to drive geochemical innovation and new insights into this obscure yet important period of our planet's history.

ACKNOWLEDGMENTS

We thank Dustin Trail, an anonymous reviewer, and editor Matt Kohn for helpful reviews, and we thank Matthew Wielicki, Ellen Alexander, Bruce Watson, Stephen Mojzsis, and Ben Weiss for discussions that helped refine many of the arguments made in this paper. This work was supported by NSF grant EAR-1551437. The ion microprobe facility at the University of California, Los Angeles, is partly supported by a grant from the Instrumentation and Facilities Program, Division of Earth Sciences, National Science Foundation.

REFERENCES

Abdel-Rahman AFM (1994) Nature of biotites from alkaline calc-alkaline and peraluminous magmas. J Petrol 35:525–541
Abe Y (1993) Physical state of the very early Earth. Lithos 30:223–235
Abbott SS, Harrison TM, Schmitt AK, Mojzsis SJ (2012) A search for terrestrial evidence of the Late Heavy Bombardment in Ti–U–Th–Pb depth profiles of ancient zircons. PNAS 109:13,486–13,492
Amelin YV (1998) Geochronology of the Jack Hills detrital zircons by precise U–Pb isotope dilution analysis of crystal fragments. Chem Geol 146:25–38

Amelin Y, Davis WJ (2006) Isotopic analysis of lead in sub-nanogram quantities by TIMS using a ^{202}Pb–^{205}Pb spike. J Anal Atomic Spectrom 21:1053–1061

Amelin YV, Lee DC, Halliday AN, Pidgeon RT (1999) Nature of the Earth's earliest crust from hafnium isotopes in single detrital zircons. Nature 399:252–55

Armstrong RL (1981) Radiogenic isotopes: the case for crustal recycling on a near-steady-state no-continental growth Earth. Phil Trans R Soc London Ser A 301:443–71

Battistuzzi FU, Feijao A, Hedges SB (2004) A genomic timescale of prokaryote evolution: insights into the origin of methanogenesis phototrophy and the colonization of land. BMC Evol Biol 4:44–51

Bell EA (2013) Hadean–Archean transitions: Constraints from the Jack Hills detrital zircon record. PhD thesis, University of California, Los Angeles, 285 pp

Bell EA, Harrison TM (2013) Post-Hadean transitions in Jack Hills zircon provenance: A signal of the Late Heavy Bombardment? Earth Planet Sci Lett 364:1–11

Bell EA, Harrison TM, McCulloch MT, Young ED (2011) Early Archean crustal evolution of the Jack Hills Zircon source terrane inferred from Lu–Hf, ^{207}Pb/^{206}Pb, and δ^{18}O systematics of Jack Hills zircons. Geochim Cosmochim Acta 75:4816–4829

Bell EA, Harrison TM, Kohl IE, Young ED (2014) Eoarchean evolution of the Jack Hills zircon source and loss of Hadean crust. Geochim Cosmochim Acta 146:27–42

Bell EA, Boehnke P, Hopkins-Wielicki MD, Harrison TM (2015a) Distinguishing primary and secondary inclusion assemblages in Jack Hills zircons. Lithos 234:15–26

Bell EA, Boehnke P, Harrison TM, Mao W (2015b) Potentially biogenic carbon preserved in a 4.1 Ga zircon. PNAS 112:14518–14521

Bell EA, Boehnke P, Harrison TM (2016) Recovering the primary geochemistry of Jack Hills zircons through quantitative estimates of chemical alteration. Geochim Cosmochim Acta 191:187–202

Biggin AJ, de Wit MJ, Langereis CG, Zegers TE, Voûte S, Dekkers MJ, Drost K (2011) Palaeomagnetism of Archaean rocks of the Onverwacht Group Barberton Greenstone Belt (southern Africa): Evidence for a stable and potentially reversing geomagnetic field at ca. 3.5 Ga. Earth Planet Sci Lett 302:314–328

Bindeman I, Gurenko A, Carley T, Miller C, Martin E, Sigmarsson O (2012) Silicic magma petrogenesis in Iceland by remelting of hydrothermally altered crust based on oxygen isotope diversity and disequilibria between zircon and magma with implications for MORB. Terra Nova 24:227–232

Blichert-Toft J, Albarède F (2008) Hafnium isotopes in Jack Hills zircons and the formation of the Hadean crust. Earth Planet Sci Lett 265:686702

Boehnke P, Harrison TM (2016) Illusory Late Heavy Bombardments. PNAS 201611535

Bowring SA, Williams IS (1999) Priscoan (4.00–4.03 Ga) orthogneisses from northwestern Canada. Contrib Mineral Petrol 134:3–16

Brasier MD, Antcliffe J, Saunders M, Wacey, D (2015) Changing the picture of Earth's earliest fossils (3.5–1.9 Ga) with new approaches and new discoveries. PNAS 112:4859–4864

Buda G, Koller F, Kovács J, Ulrych J (2004) Compositional variation of biotite from Variscan granitoids in Central Europe: a statistical evaluation. Acta Mineral Petrograph Szeged 45:21–37

Burnham, CW (1975) Water and magmas; a mixing model. Geochim Cosmochim Acta 39:1077–1084

Carley TL, Miller CF, Wooden JL, Padilla AJ, Schmitt AK, Economos RC, Bindeman IN, Jordan BT (2014) Iceland is not a magmatic analog for the Hadean: Evidence from the zircon record. Earth Planet Sci Lett 405:85–97

Carley TL, Miller CF, Wooden JL, Bindeman IN, Barth AP (2011) Zircon from historic eruptions in Iceland: reconstructing storage and evolution of silicic magmas. Mineral Petrol 102:135–161

Cavosie AJ, Wilde SA, Liu D, Weiblen PW, Valley JW (2004) Internal zoning and U–Th–Pb chemistry of Jack Hills detrital zircons: a mineral record of early Archean to Mesoproterozoic (4348–1576 Ma) magmatism. Precambrian Res 135:251–279

Cavosie AJ, Valley JW, Wilde SA, EIMF (2005) Magmatic δ^{18}O in 4400–3900 Ma detrital zircons: A record of the alteration and recycling of crust in the Early Archean. Earth Planet Sci Lett 235:663–681

Cavosie AJ, Valley JW, Wilde SA, EIMF (2006) Correlated microanalysis of zircon: Trace element, δ^{18}O, and U–Th–Pb isotopic constraints on the igneous origin of complex >3900 Ma detrital grains. Geochim Cosmochim Acta 70:5601–5616

Chappell BW, White AJR (1974) Two contrasting granite types. Pac Geol 8:173–174

Cherniak DJ, Watson EB (2003) Diffusion in zircon. Rev Mineral Geochem 53:113–143

Cherniak DJ, Watson EB (2007) Ti diffusion in zircon. Chem Geol 242:470–483

Cherniak DJ, Watson EB (2010) Li diffusion in zircon. Contrib Mineral Petrol 160:383–390

Chopin C, Sobolev NV (1995) Principal mineralogic indicators of UHP in crustal rocks. *In:* Ultrahigh-Pressure Metamorphism. Coleman RG, Wang XM (Eds.) Cambridge University Press. England, p. 96–133

Claiborne LL, Miller CF, Wooden JL (2010) Trace element composition of igneous zircon: a thermal and compositional record of the accumulation and evolution of a large silicic batholith, Spirit Mountain, Nevada. Contrib Mineral Petrol 160:511–531

Clemens JD (1984) Water contents of silicic to intermediate magmas. Lithos 17:273–287

Clemens JD, Vielzeuf D (1987) Constraints on melting and magma production in the crust. Earth Planet Sci Lett 86:287–306

Cloud P (1972) A working model of the primitive Earth. Am J Sci 272:537–548

Cloud P (1976) Major features of crustal evolution. De Toit Lecture, Geol Soc S Afr:1–33

Collerson KD, Kamber BS (1999) Evolution of the continents and the atmosphere inferred from Th–U–Nb systematics of the depleted mantle. Science 283:1519–1522

Collins WJ, Belousova EA, Kemp AIS, Murphy JB (2011) Two contrasting Phanerozoic orogenic systems revealed by Hf isotopic data. Nat Geosci. 4:333–337

Compston W, Pidgeon RT (1986) Jack Hills Evidence of more very old detrital zircons in Western Australia. Nature 321:766–769

Coogan LA, Hinton RW (2006) Do the trace element compositions of detrital zircons require Hadean continental crust? Geology 34:633–636

Crowley JL, Myers JS, Sylvester PJ, Cox RA (2005) Detrital zircon from the Jack Hills and Mount Narryer, Western Australia: Evidence for diverse >4.0 Ga source rocks. J Geol 113:239–263

Cui PL, Sun JG, Sha DM, Wang XJ, Zhang P, Gu AL, Wang ZY (2013) Oldest zircon xenocryst (4.17 Ga) from the North China Craton. Int Geol Rev 55:1902–1908

Darling J, Storey C, Hawkesworth C (2009) Impact melt sheet zircons and their implications for the Hadean crust. Geology 37:927–930

Dasgupta R (2013) Ingassing storage and outgassing of terrestrial carbon through geologic time. Rev Mineral Geochem 75:183–229

Davies GF (1992) On the emergence of plate tectonics. Geology 20:963–966

DePaolo DJ, Cerling TE, Hemming SR, Knoll AH, Richter FM, Royden LH, Rudnick RL, Stixrude L, Trefil JS (2008) Origin and evolution of Earth: Research questions for a changing planet. Nat Acad Press 152 pp

Diwu CR, Sun Y, Dong ZC, Wang HL, Chen DL, Chen L, Zhang H (2010) In situ U–Pb geochronology of Hadean zircon xenocryst (4.1–3.9 Ga) from the western of the Northern Qinling Orogenic Belt. Acta Petrol Sin 26:1171–1174

Diwu C, Sun Y, Wilde SA, Wang H, Dong Z, Zhang H, Wang Q (2013) New evidence for~ 4.45 Ga terrestrial crust from zircon xenocrysts in Ordovician ignimbrite in the North Qinling Orogenic Belt China. Gondwana Res 23:1484–1490

Dobrzhinetskaya L, Wirth R, Green H (2014) Diamonds in Earth's oldest zircons from Jack Hills conglomerate Australia are contamination. Earth Planet Sci Lett 387:212–218

Duke EF, Rumble D (1986) Textural and isotopic variations in graphite from plutonic rocks South-Central New Hampshire. Contrib Mineral Petrol 93:409–419

Duo J, Wen CQ, Guo JC, Fan XP, Li XW (2007) 4.1 Ga old detrital zircon in western Tibet of China. China Sci Bull 52:23–26

Ellis DJ, Thompson AB (1986) Subsolidus and partial melting reactions in the quartz-excess CaO +MgO+Al$_2$O$_3$+SiO$_2$+H$_2$O system under water-excess and water-deficient conditions to 10 kb: some implications for the origin of peraluminous melts from mafic rocks. J Petrol 27:91–121

Ewing RC, Meldrum A, Wang L, Weber WJ, Corrales LR (2003) Radiation effects in zircon. Rev Mineral Geochem 53:387–425

Ferry JM, Watson EB (2007) New thermodynamic models and revised calibrations for the Ti-in-zircon and Zr-in-rutile thermometers. Contrib Mineral Petrol 154:429–437

François C, Philippot P, Rey P, Rubatto D (2014) Burial and exhumation during Archean sagduction in the east Pilbara granite–greenstone terrane. Earth Planet Sci Lett 396:235–251

Franzson H, Guðfinnsson GH, Helgadóttir HM, Frolova J (2010) Porosity, density and chemical composition relationships in altered Icelandic hyaloclastites. *In:* Water–Rock Interaction. Birkle P, Torres-Alvarado IS (eds.) London: Taylor & Francis Group, p. 199–202

Freund F, Kathrein H, Wengeler H, Knobel R, Heinen HJ (1980) Carbon in solid solution in forsterite—A key to the intractable nature of reduced carbon in terrestrial and cosmogenic rocks. Geochim Cosmochim Acta 44:1319–1333

Frezzotti ML, Di Vincenzo G, Ghezzo C, Burke EA (1994) Evidence of magmatic CO$_2$-rich fluids in peraluminous graphite-bearing leucogranites from Deep Freeze Range (northern Victoria Land Antarctica). Contrib Mineral Petrol 117:111–123

Froude DO, Ireland TR, Kinny PD, Williams IS, Compston W (1983) Ion microprobe identification of 4,100–4,200 Myr-old terrestrial zircons. Nature 304:616–618

Fu B, Page FZ, Cavosie AJ, Fournelle J, Kita NT, Lackey JS, Wilde SA, Valley JW (2008) Ti-in-Zircon thermometry: applications and limitations. Contrib Mineral Petrol 156:197–215

Galer SJG, Goldstein SL (1991) Early mantle differentiation and its thermal consequences. Geochim Cosmochim Acta 55:227–239

Glikson A (2006) Comment on "Zircon thermometer reveals minimum melting conditions on earliest Earth". Science 311:779a

Griffin WL, Pearson NJ, Belousova E, Jackson SE, Van Achterbergh E, O'Reilly SY, Shee SR (2000) The Hf isotope composition of cratonic mantle: LAM-MC-ICPMS analysis of zircon megacrysts in kimberlites. Geochim Cosmochim Acta 64:133–147

Grimes CB, John BE, Kelemen PB, Mazdab FK, Wooden JL, Cheadle MJ, Hanghøj K, Schwartz JJ (2007) Trace element chemistry of zircons from oceanic crust: A method for distinguishing detrital zircon provenance. Geology 35:643–646

Guitreau M, Blichert-Toft J, Martin H, Mojzsis SJ, Albarède F (2012) Hafnium isotope evidence from Archean granitic rocks for deep-mantle origin of continental crust. Earth Planet Sci Lett 337:211–223

Hamilton WB (2007) Earth's first two billion years—The era of internally mobile crust. Geol Soc Am Memr 200:233–296

Hanchar JM, Hoskin PWO (eds.) (2003) Zircon. Rev Mineral Geochem 53, Mineral Soc Am.

Harrison TM (2009) The Hadean crust: Evidence from >4 Ga zircons. Ann Rev Earth Planet Sci 37:479–505

Harrison TM, Schmitt AK (2007) High sensitivity mapping of Ti distributions in Hadean zircons. Earth Planet Sci Lett 261:9–19

Harrison TM, Wielicki MM (2015) From the Himalaya to the Hadean. Am Mineral 101:1348–1359

Harrison TM, Ryerson FJ, Le Fort P, Yin A, Lovera OM, Catlos EJ (1997) A Late Miocene–Pliocene origin for the Central Himalayan inverted metamorphism. Earth Planet Sci Lett 146:E1–E8

Harrison TM, Blichert-Toft J, Müller W, Albarede F, Holden P, Mojzsis SJ (2005) Heterogeneous Hadean hafnium: Evidence of continental crust by 4.4–4.5 Ga. Science 310:1947–1950

Harrison TM, Watson EB, Aikman AK (2007) Temperature spectra of zircon crystallization in plutonic rocks. Geology 35:635–638

Harrison TM, Schmitt AK, McCulloch MT, Lovera OM (2008) Early (≥4.5 Ga) formation of terrestrial crust: Lu–Hf, δ18O, and Ti thermometry results for Hadean zircons. Earth Planet Sci Lett 268:476–486

Hartmann WK (1975) Lunar "cataclysm": a misconception? Icarus 24:181–187

Hayes JM, Waldbauer JR (2006) The carbon cycle and associated redox processes through time. Phil Trans R Soc B: Biol Sci 361:931–950

Hellebrand E, Möller A, Whitehouse M, Cannat M (2007) Formation of oceanic zircons. Geochim Cosmochim Acta 71:A391–A391

Hofmann AE, Valley JW, Watson EB, Cavosie AJ, Eiler JM (2009) Sub-micron scale distributions of trace elements in zircon. Contrib Mineral Petrology 158:317–335

Holden P, Lanc P, Ireland TR, Harrison TM, Foster JJ, Bruce ZP (2009) Mass-spectrometric mining of Hadean zircons by automated SHRIMP multi-collector and single-collector U/Pb zircon age dating: The first 100 000 grains. Int J Mass Spectrom 286:53–63

Hopkins M, Harrison TM, Manning CE (2008) Low heat flow inferred from >4 Gyr zircons suggests Hadean plate boundary interactions. Nature 456:493–496

Hopkins M, Harrison TM, Manning CE (2010) Constraints on Hadean geodynamics from mineral inclusions in >4 Ga zircons. Earth Planet Sci Lett 298:367–376

Hoskin PWO, Schaltegger U (2003) The composition of zircon and igneous and metamorphic petrogenesis. Rev Mineral Geochem 53:27–62

House CH (2015) Penciling in details of the Hadean. PNAS 112:14410–14411

Ibanez-Mejia M, Gehrels GE, Ruiz J, Vervoort JD, Eddy MP, Li C (2014) Small-volume baddeleyite (ZrO2) U–Pb geochronology and Lu–Hf isotope geochemistry by LA-ICP-MS. Techniques and applications. Chem Geol 384:149–167

Ishihara S (1977) The magnetite-series and ilmenite-series granitic rocks. Min Geol 27:293–305

Iizuka T, Horie K, Komiya T, Maruyama S, Hirata T, Hidaka H, Windley BF (2006) 4.2 Ga zircon xenocryst in an Acasta gneiss from northwestern Canada: Evidence for early continental crust. Geology 34:245–248

Iizuka T, Yamaguchi T, Hibiya Y, Amelin Y (2015) Meteorite zircon constraints on the bulk Lu–Hf isotope composition and early differentiation of the Earth. PNAS 112:5331–5336

Ingebritsen SE, Manning CE (2002) Diffuse fluid flux through orogenic belts: Implications for the world ocean. PNAS 99:9113–9116

Javoy M, Pineau F, Iiyama I (1978) Experimental determination of the isotopic fractionation between gaseous CO2 and carbon dissolved in tholeiitic magma. Contrib Mineral Petrol 67:35–39

Jeffreys H (1924) The earth. Its origin, history and physical constitution. Cambridge University Press

Jennings ES, Marschall HR, Hawkesworth CJ, Storey CD (2011) Characterization of magma from inclusions in zircon: Apatite and biotite work well feldspar less so. Geology 39:863–866

Kamber BS, Whitehouse MJ, Bolhar R, Moorbath S (2005) Volcanic resurfacing and the early terrestrial crust: zircon U–Pb and REE constraints from the Isua Greenstone Belt, southern West Greenland. Earth Planet Sci Lett 240:276–290

Kelemen PB, Manning CE (2015) Reevaluating carbon fluxes in subduction zones what goes down mostly comes up. PNAS 112:E3997–E4006

Kemp AIS, Wilde SA, Hawkesworth CJ, Coath CD, Nemchin A, Pidgeon RT, Vervoort JD, DuFrane SA (2010) Hadean crustal evolution revisited: new constraints from Pb–Hf isotope systematics of the Jack Hills zircons. Earth Planet Sci Lett 296:45–56

Kemp AIS, Hickman AH, Kirkland CL, Vervoort JD (2015) Hf isotopes in detrital and inherited zircons of the Pilbara Craton provide no evidence for Hadean continents. Precambrian Res 261:112–126

Kenny GG, Whitehouse MJ, Kamber BS (2016) Differentiated impact melt sheets may be a potential source of Hadean detrital zircon. Geology 44:435–438

Keppler H, Wiedenbeck M, Shcheka SS (2003) Carbon solubility in olivine and the mode of carbon storage in the Earth's mantle. Nature 424:414–416

Kinny PD, Compston W, Williams IS (1991) A reconnaissance ion-probe study of hafnium isotopes in zircons. Geochim Cosmochim Acta 55:849–859

Kohn MJ (2009) Models of garnet differential geochronology. Geochim Cosmochim Acta 73:170–182

Kohn MJ, Malloy MA (2004) Formation of monazite via prograde metamorphic reactions among common silicates: implications for age determinations. Geochim Cosmochim Acta 68:101–113

Korenaga J (2013) Initiation and evolution of plate tectonics on Earth: theories and observations. Ann Rev Earth Planet Sci 41:117–151

Krogh TE (1982) Improved accuracy of U–Pb zircon ages by the creation of more concordant systems using an air abrasion technique. Geochim Cosmochim Acta 46:637–649

Liu J, Ye K, Maruyama S, Cong B, Fan H (2001) Mineral inclusions in zircon from gneisses in the ultrahigh-pressure zone of the Dabie Mountains China. J Geol 109:523–535

Maas R, Kinny PD, Williams IS, Froude DO, Compston W (1992) The Earth's oldest known crust: A geochronological and geochemical study of 3900–4200 Ma old detrital zircons from Mt. Narryer and Jack Hills Western Australia. Geochim Cosmochim Acta 56:1281–1300

Macgregor AM (1951) Some milestones in the Precambrian of Southern Rhodesia. Proc Geol Soc S Afr 54:27–71

McLean NM, Condon DJ, Schoene B, Bowring SA (2015) Evaluating uncertainties in the calibration of isotopic reference materials and multi-element isotopic tracers (EARTHTIME Tracer Calibration Part II). Geochim Cosmochim Acta 164:481–501

Maher KA, Stevenson DJ (1988) Impact frustration of the origin of life. Nature 331:612–614

Manning CE, Mojzsis SJ, Harrison TM (2006) Geology, age and origin of supracrustal rocks at Akilia, West Greenland. Am J Sci 306:303–366

Marchi S, Bottke WF, Elkins-Tanton LT, Bierhaus M, Wuennemann K, Morbidelli A, Kring DA (2014) Widespread mixing and burial of Earth's Hadean crust by asteroid . Nature 511:578–82

McKeegan KD, Kudryavtsev AB, Schopf JW (2007) Raman and ion microscopic imagery of graphitic inclusions in apatite from older than 3830 Ma Akiliasupracrustal rocks west Greenland. Geology 35:591–594

Marty B, Alexander CMD, Raymond SN (2013) Primordial origins of Earth's carbon. Rev Mineral Geochem 75:149–181

Mathez EA, Blacic JD, Beery J, Maggiore C, Hollander M (1984) Carbon abundances in mantle minerals determined by nuclear reaction analysis. Geophys Res Lett 11:947–950

Menneken M, Nemchin AA, Geisler T, Pidgeon RT, Wilde SA (2007) Hadean diamonds in zircon from Jack Hills Western Australia. Nature 448:917–920

Moore WB, Webb AAG (2013) Heat-pipe earth. Nature 501:501–505

Mojzsis SJ, Harrison TM (2002) Establishment of a 3.83-Ga magmatic age for the Akiliatonalite (southern West Greenland). Earth Planet Sci Lett 202:563–576

Mojzsis SJ, Arrhenius G, McKeegan KD, Harrison TM, Nutman AP, Friend CRL (1996) Evidence for life on Earth by 3800 Myr. Nature 384:55–59

Mojzsis SJ, Harrison TM, Pidgeon RT (2001) Oxygen-isotope evidence from ancient zircons for liquid water at the Earth's surface 4300 Myr ago. Nature 409:178–181

Mojzsis SJ, Cates NL, Caro G, Trail D, Abramov O, Guitreau M, Blichert-Toft J, Hopkins MD, Bleeker W (2014) Component geochronology in the polyphase ca. 3920 Ma Acasta Gneiss. Geochim Cosmochim Acta 133:68–96

Moore WB, Webb AAG (2013) Heat-pipe earth. Nature 501:501–505

Mysen BO, Wheeler K (2000) Solubility behavior of water in haploandesitic melts at high pressure and high temperature. Am Mineral 85:1128–1142

Nadeau S, Chen W, Reece J, Lachhman D, Ault R, Faraco MTL, Fraga LM, Reis NJ, Betiollo LM (2013) Guyana: the Lost Hadean crust of South America? Braz J Geol 43:601–606

Nebel-Jacobsen Y, Munker C, Nebel O, Gerdes A, Mezger K, Nelson DR (2010) Reworking of Earth's first crust: constraints from Hf isotopes in Archean zircons from Mt. Narryer Australia. Precambrian Res 182:175–186

Nelson DR (2002) Compilation of Geochronology Data 2001.No. 2002/2 Western Australia Geological Survey

Nelson DR (2005) Compilation of Geochronology Data 2003. No. 2005/2 Western Australia Geological Survey

Nelson DR, Robinson BW, Myers JS (2000) Complex geological histories extending for ≥4.0 Ga deciphered from xenocryst zircon microstructures Earth Planet Sci Lett 181:89–102

Nemchin AA, Whitehouse MJ, Menneken M, Geisler T, Pidgeon RT, Wilde SA (2008) A light carbon reservoir recorded in zircon-hosted diamond from the Jack Hills. Nature 454:92–95

Nutman AP (2006) Comments on "Zircon thermometer reveals minimum melting conditions on earliest Earth": Science 311:779b

Oberheuser G, Kathrein H, Demortier G, Gonska H, Freund F (1983) Carbon in olivine single crystals analyzed by the $^{12}C(d,p)^{13}C$ method and by photoelectron spectroscopy. Geochim Cosmochim Acta 47:1117–1129

Ohmoto H, Kerrick DM (1977) Devolatilization equilibria in graphitic systems. Am J Sci 277:1013–1044

O'Neil JR, Chappell BW (1977) Oxygen and hydrogen isotope relations in the Berridale Batholith, Southeastern Australia. J Geol Soc London 133:559–571

O'Neil C, Debaille V (2014) The evolution of Hadean–Eoarchaean geodynamics. Earth Planet Sci Lett 406:49–58

O'Neil J, Carlson RW, Francis D, Stevenson RK (2008) Neodymium-142 evidence for Hadean mafic crust. Science 321:1828–1831

Ozima M, Podosek FA (2002) Noble Gas Geochemistry. Cambridge Univ. Press. 291 pp

Patiño Douce A, Harris N (1998) Experimental constraints on Himalayan anatexis. J Petrol 39:689–710

Paquette JL, Barbosa JSF, Rohais S, Cruz SC, Goncalves P, Peucat JJ, Leal ABM, Santos-Pinto M, Martin H (2015) The geological roots of South America: 4.1 Ga and 3.7 Ga zircon crystals discovered in NE Brazil and NW Argentina. Precambrian Res 271:49–55

Parrish RR (1987) An improved micro-capsule for zircon dissolution in U–Pb geochronology; Chem Geol (Isotope Geosci Sec) 66:99–102

Peck WH, Valley JW, Wilde SA, Graham CM (2001) Oxygen isotope ratios and rare earth elements in 3.3 to 4.4 Ga zircons: Ion microprobe evidence for high $\delta^{18}O$ continental crust and oceans in the Early Archean. Geochim Cosmochim Acta 65:4215–4229

Pidgeon RT, Nemchin AA (2006) High abundance of early Archaean grains and the age distribution of detrital zircons in a sillimanite-bearing quartzite from Mt Narryer, Western Australia. Precambrian Res 150:201–220

Pyle JM, Spear FS, Rudnick RL, McDonough WF (2001) Monazite–xenotime–garnet equilibrium in metapelites and a new monazite–garnet thermometer. J Petrol 42:2083–2107

Rasmussen B, Fletcher IR, Muhling JR, Wilde SA (2010) In situ U–Th–Pb geochronology of monazite and xenotime from the Jack Hills belt: Implications for the age of deposition and metamorphism of Hadean zircons. Precambrian Res 180:26–46

Rasmussen B, Fletcher IR, Muhling JR, Gregory CJ, Wilde SA (2011) Metamorphic replacement of mineral inclusions in detrital zircon from Jack Hills Australia: Implications for the Hadean Earth. Geology 39:1143–1146

Reimink JR, Chacko T, Stern RA, Heaman LM (2014) Earth's earliest evolved crust generated in an Iceland-like setting. Nat Geosci 7:529–533

Reimink JR, Davies JHFL, Chacko T, Stern RA, Heaman LM, Sarkar C, Schaltegger U, Creaser RA, Pearson DG (2016) No evidence for Hadean continental crust within Earth's oldest evolved rock unit. Nat Geosci: doi:10.1038/ngeo2786

Rollinson H (2008) Ophiolitic trondhjemites: a possible analogue for Hadean felsic 'crust'. Terra Nova 20:364–369

Rosenthal A, Hauri EH, Hirschmann MM (2015) Experimental determination of C, F and H partitioning between mantle minerals and carbonated basalt CO_2/Ba and CO_2/Nb systematics of partial melting and the CO_2 contents of basaltic source regions. Earth Planet Sci Lett 412:77–87

Rosing MT (1999) ^{13}C-depleted carbon microparticles in >3700-Ma sea-floor sedimentary rocks from West Greenland. Science 283:674–676

Rubatto D (2002) Zircon trace element geochemistry: partitioning with garnet and the link between U–Pb ages and metamorphism. Chem Geol 184:123–138

Rubatto D (2017) Zircon: The metamorphic mineral. Rev Mineral Geochem 83:261–295

Schoene B, Samperton KM, Eddy MP, Keller G, Adatte T, Bowring SA, Khadri SF, Gertsch B (2015) U–Pb geochronology of the Deccan Traps and relation to the end-Cretaceous mass extinction. Science 347:182–184

Schopf JW (2014) Geological evidence of oxygenic photosynthesis and the biotic response to the 2400–2200 Ma "Great Oxidation Event". Biochem (Moscow)79:165–177

Sen S, Widgeon SJ, Navrotsky A, Mera G, Tavakoli A, Ionescu E, Riedel R (2013) Carbon substitution for oxygen in silicates in planetary interiors. PNAS 110:15904–15907

Seifert W, Thomas R, Rhede D, Förster HJ (2010) Origin of coexisting wüstite Mg-Fe and REE phosphate minerals in graphite-bearing fluorapatite from the Rumburk granite. Eur J Mineral 22:495–507

Shcheka SS, Wiedenbeck M, Frost DJ, Keppler H (2006) Carbon solubility in mantle minerals. Earth Planet Sci Lett 245:730–742

Shirey SB, Kamber BS, Whitehouse MJ, Mueller PA, Basu AR (2008) A review of the isotopic and trace element evidence for mantle and crustal processes in the Hadean and Archean: Implications for the onset of plate tectonic subduction. Geol Soc Am Spec Pap 440:1–29

Sleep NH (2000) Evolution of the mode of convection within terrestrial planets. J Geophys Res 105:17563–17578

Sleep NH, Zahnle KJ, Kasting JF, Morowitz HJ (1989) Annihilation of ecosystems by large asteroid impacts on the early Earth. Nature 342:139–142

Smith JV (1981) The 1st 800 million years of earth's history. Philos Trans R Soc London Ser A 301:401–422

Söderlund U, Patchett PJ, Vervoort JD, Isachsen CE (2004) The ^{176}Lu decay constant determined by Lu–Hf and U–Pb isotope systematics of Precambrian mafic intrusions. Earth Planet Sci Lett 219:311–324

Solomon SC (1980) Differentiation of crusts and cores of the terrestrial planets: Lessons for the early Earth? Precambr Res 10:177–194

Song S, Zhang L, Niu Y, Su L, Jian P, Liu D (2005) Geochronology of diamond-bearing zircons from garnet peridotite in the North Qaidam UHPM belt Northern Tibetan Plateau: a record of complex histories from oceanic lithosphere subduction to continental collision. Earth Planet Sci Lett 234:99–118

Spaggiari CV, Pidgeon RT, Wilde SA (2007) The Jack Hills greenstone belt, Western Australia: part 2: lithological relationships and implications for the deposition of ≥4.0 Ga detrital zircons. Precambrian Res 155:261–286

Spear FS (1988) The Gibbs method and Duhem's theorem: The quantitative relationships among P, T, chemical potential, phase composition and reaction progress in igneous and metamorphic systems. Contrib Mineral Petrol 99:249–256

Spear FS (1993) Metamorphic phase equilibria and pressure–temperature–time-paths. Mineral Soc Am 799 pp

Stern RA, Bleeker W (1998) Age of the world's oldest rocks refined using Canada's SHRIMP the Acasta gneiss complex Northwest Territories Canada. Geosci Canada 25:27–31

Stevenson DJ (1983) The nature of the earth prior to the oldest known rock record—The Hadean earth. *In:* Earth's Earliest Biosphere: Its Origin and Evolution. Princeton University Press, p. 32–40

Tabata H, Yamauchi K, Maruyama S, Liou JG (1998) Tracing the extent of a UHP metamorphic terrane: Mineral-inclusion study of zircons in gneisses from the Dabie Shan. *In:* When Continents Collide: Geodynamics and Geochemistry of Ultrahigh-Pressure Rocks. Springer Netherlands, p. 261–273

Tarduno JA, Cottrell RD, Davis WJ, Nimmo F, Bono RK (2015) A Hadean to Paleoarchean geodynamo recorded by single zircon crystals. Science 349:521–524

Taylor SR, McLennan SM (1985) The Continental Crust: Its Composition and Evolution. Oxford: Blackwell. 312 pp

Taylor HP, Sheppard SMF (1986) Igneous rocks. I. Processes of isotopic fractionation and isotope systematics. Rev Mineral 16:227–71

Tera F, Papanastassiou DA, Wasserburg G (1974) Isotopic evidence for a terminal lunar cataclysm. Earth Planet Sci Lett 22:1–21

Thern ER, Nelson DR (2012) Detrital zircon age structure within ca. 3 Ga metasedimentary rocks Yilgarn Craton: Elucidation of Hadean source terranes by principal component analysis. Precambrian Res 214:28–43

Thompson W (1863) On the secular colling of the Earth. Phil Mag 25:1–14

Tingle TN, Green HW, Finnerty AA (1988) Experiments and observations bearing on the solubility and diffusivity of carbon in olivine. J Geophys Res 93:15289–15304

Trail D, Mojzsis SJ, Harrison TM, Schmitt AK, Watson EB, Young ED (2007) Constraints on Hadean zircon protoliths from oxygen isotopes, REEs and Ti-thermometry. Geochem Geophys Geosyst 8:Q06014

Trail D, Bindeman IN, Watson EB, Schmitt AK (2009) Experimental calibration of oxygen isotope fractionation between quartz and zircon. Geochim Cosmochim Acta 73:7110–7126

Trail D, Watson EB, Tailby ND (2011a) The oxidation state of Hadean magmas and implications for early Earth's atmosphere. Nature 480:79–82

Trail D, Thomas JB, Watson EB (2011b) The incorporation of hydroxyl into zircon. Am Mineral 96:60–67

Trail D, Cherniak DJ, Watson EB, Harrison TM, Weiss BP, Szumila I (2016) Li zoning in zircon as a potential geospeedometer and peak temperature indicator. Contrib Mineral Petrol 171:1–15

Trail D, Tailby, N, Wang Y, Harrison TM, Boehnke P (2017) Al in zircon as evidence for peraluminous melts and recycling of pelites from the Hadean to modern times. Geochem Geophys Geosystem. In Press

Turner G, Harrison TM, Holland G, Mojzsis SJ, Gilmour J (2004) Xenon from extinct ^{244}Pu in ancient terrestrial zircons. Science 306:89–91

Turner G, Busfield A, Crowther SA, Harrison TM, Mojzsis SJ, Gilmour J (2007) Pu–Xe, U–Xe, U–Pb chronology and isotope systematics of ancient zircons from Western Australia. Earth Planet Sci Lett 261:491–499

Ushikubo T, Kita NT, Cavosie AJ, Wilde SA, Rudnick RL, Valley JW (2008) Lithium in Jack Hills zircons: Evidence for extensive weathering of Earth's earliest crust. Earth Planet Sci Lett 272:666–676

Valley JW, Chiarenzelli JR, McLelland JM (1994) Oxygen isotope geochemistry of zircon. Earth Planet Sci Lett 126:187–206

Valley JW, Kinny PD, Schulze DJ, Spicuzza MJ (1998) Zircon megacrysts from kimberlite: oxygen isotope variability among mantle melts. Contrib Mineral Petrol 133:1–11

Valley JW, Peck WH, King EM, Wilde SA (2002) A cool early Earth. Geology 30:351–354

Valley JW, Cavosie AJ, Fu B, Peck WH, Wilde SA (2006) Comment on "Heterogeneous Hadean Hafnium: Evidence of continental crust at 4.4 to 4.5 Ga". Science 312:1139a

Valley JW, Cavosie AJ, Ushikubo T, Reinhard DA, Lawrence DF, Larson DJ, Spicuzza MJ (2014) Hadean age for a post-magma-ocean zircon confirmed by atom-probe tomography. Nat Geosci 7:219–223

Wang H, Chen L, Sun Y, Liu X, Xu X, Chen J, Zhang H, Diwu C (2007) ~4.1 Ga xenocrystal zircon from Ordovician volcanic rocks in western part of North Qinling Orogenic Belt. Chin Sci Bull 52:3002–3010

Ward PD, Brownlee D (2000) Rare Earth: Why Complex Life is Uncommon in the Universe. Copernicus Books New York

Warren PH, Wasson JT (1979) The origin of KREEP. Rev Geophys 17:73–88

Watson EB, Cherniak DJ (1997) Oxygen diffusion in zircon. Earth Planet Sci Lett 148:527–544

Watson EB, Harrison TM (2005) Zircon thermometer reveals minimum melting conditions on earliest Earth. Science 308:841–844

Watson EB, Harrison TM (1983) Zircon saturation revisited: temperature and composition effects in a variety of crustal magma types. Earth Planet Sci Lett 64:295–304

Watson EB, Wark DA, Thomas JB (2006) Crystallization thermometers for zircon and rutile. Contrib Mineral Petrol 151:413–433

Weiss BP, Maloof AC, Tailby N, Ramezani J, Fu RR, Hanus V, Trail D, Watson EB, Harrison TM, Bowring SA, Kirschvink JL, Swanson-Hysell NL, Coe RS (2015) Pervasive remagnetization of detrital zircon host rocks in the Jack Hills Western Australia and implications for records of the early geodynamo. Earth Planet Sci Lett 430:115–128

Wendt I, Carl C (1991) The statistical distribution of the mean squared weighted deviation. Chem Geol (Isotope Geosci Sect) 86:275–285

White RW, Powell RW, Holland TJB (2001) Calculation of partial melting equilibria in the system $Na_2O–CaO–K_2O–FeO–MgO–Al_2O_3–SiO_2–H_2O$ (NCKFMASH). J Metamorph Geol 19:139–153

Wielicki MM, Harrison TM, Schmitt AK (2012) Geochemical signatures and magmatic stability of terrestrial impact produced zircon. Earth Planet Sci Lett 321:20–31

Wielicki MM, Harrison TM, Schmitt AK (2016) Reply to Kenny et al. "Differentiated impact melt sheets may be a potential source of Hadean detrital zircon". Geology 44:e398

Wilde SA, Valley JW, Peck WH, Graham CM (2001) Evidence form detrital zircons for the existence of continental crust and oceans 4.4 Ga ago. Nature 409:175–178

Williams IS (2007) Old diamonds and the upper crust. Nature 448:880–881

Wyche S (2007) Evidence of Pre-3100 Ma Crust in the Youanmi and South West Terranes, and Eastern Goldfields Superterrane, of the Yilgarn Craton. Dev Precambrian Geol 15:113–123

Wyche S, Nelson DR, Riganti A (2004) 4350–3130 Ma detrital zircons in the Southern Cross Granite Greenstone Terrane Western Australia: implications for the early evolution of the Yilgarn Craton. Aust J Earth Sci 51:31–45

Xing G, Wang X, Wan Y, Chen Z, Yang J, Jitajima K, Ushikubo T, Gopon P (2014) Diversity in early crustal evolution: 4100 Ma zircons in the Cathaysia Block of southern China. China Sci Rep 4:51–43

Zeng Y, Zhu Y, Liu J (2001) Carbonaceous material in S-type Xihuashan granite. Geochem J 35:145–153

Reviews in Mineralogy & Geochemistry
Vol. 83 pp. 365–418, 2017
Copyright © Mineralogical Society of America

12

Petrochronology Based on REE-Minerals: Monazite, Allanite, Xenotime, Apatite

Martin Engi

Department of Geological Sciences
University of Bern
Baltzerstrasse 3
3012 Bern
Switzerland

engi@geo.unibe.ch

INTRODUCTION AND SCOPE

REE-minerals

Monazite, xenotime, and allanite are REE[1]-minerals *sensu stricto* because lanthanides (La...Lu) and yttrium are critical constituents in them. Apatite does not require REE, but because it contains substantial REE in many rocks, it is included in this review. All four minerals also host unusually high radionuclide concentrations, notably Th and U, forming the basis of their utility as geochronometers.

This quartet of accessory minerals is playing an increasingly important role in petrochronology because they provide ways to link robust spot ages to petrogenetic (*P–T*) conditions so can lend petrogenetic context to chronology based on other minerals. Part I of this review assembles the basic requisites prior to integrative petrochronologic analysis. Individual characteristics of the four REE-minerals are addressed first, i.e., their crystal chemistry and stability relations. Thermobarometers and trace element geochemistry used for tracing petrogenesis are discussed next, and finally their chronology is summarized. Part II presents case studies to highlight the specific strengths of REE-minerals used to resolve the dynamics of a broad range of processes, from diagenetic to magmatic conditions. Finally, a brief section at the end outlines a few of the current challenges and promising perspectives for future work.

To introduce the four REE-minerals in style, let us recall the origins of their names. The three phosphates have well respected Greek grandparents, and allanite has solid Scottish roots, yet of all four of them show idiosyncracies in etymology or type material.

Apatite had long puzzled naturalists, as it shows great chemical and physical variability and can resemble other minerals. Once properly identified, Abraham Gottlieb Werner named it apatite. His reasoning referred to the Greek root ἀπατὰω and giving the precise Latin translation: *decipio*. Taken literally, both mean "I deceive" or "I mislead", which sounds like an apt confession from this mineral for having fooled humans so long. It is perhaps appropriate that the original material identified probably was fluorapatite.

Monazite derives from μονάζειν = to be solitary. This name probably refers to monazite's sparse occurrence in early discovered locations as well as its tendency to form isolated crystals rather than aggregates. I have been unable to trace this fully, but the now abandoned synonyms *eremite* (from ερημία = solitude) and *kryptolite* (= hidden rock) would affirm this interpretation.

---

[1] The rare earth elements (REE) *sensu stricto* comprise the lanthanides (La... Lu), plus Sc and Y

1529-6466/17/0083-0012$10.00 (print)
1943-2666/17/0083-0012$10.00 (online)   http://dx.doi.org/10.2138/rmg.2017.83.12

Xenotime again has Hellenic roots, but with a twist, as its original name was *kenotime*. An early copy-paste error led to xenotime. Both names contain the stem τιμή = honor. But κευός means "vain", and the vain honor refers to the false initial claim by Jöns Jacob Berzelius that xenotime was a new element, which he then correctly recognized to be a mineral (e.g., Berzelius 1824) but under a different name. The name xenotime, introduced later (Beudant 1832), must be considered a misnomer, for ξένος means foreign or odd, making little sense, except that it is odd to honor a mere spelling mistake. The first reported kenotime was from a pegmatite in Hidra, SW Norway.

Allanite is named for Thomas Allen (Thomson 1810), an Edinburgh mineralogist who first singled out the mineral from a load of materials he had acquired, but which had previously been pilfered by English warships (Secher and Johnsen 2008). The original collector was Karl Ludwig Giesecke, and the type locality is Aluk, SE Greenland (Ibler 2010; p.86–87).

CRYSTAL CHEMISTRY AND CONSEQUENCES

Mineral structures and compositions relevant to petrochronology are outlined here. Emphasis is placed on clarifying substitution mechanisms, since this is required to appreciate how these minerals incorporate and fractionate REE, U, Th, and other elements used in chronology and petrology. Which exchange vectors operate is essential also to formulate chemical equilibria that constrain conditions of growth or interaction with other minerals, melt or hydrous fluid. Among the REE three groups are often distinguished, i.e., light, middle, and heavy REE, where LREE: La…Nd, MREE: Sm, Eu, Gd, and HREE: Tb…Lu. Y has an ionic radius within the range of HREE and is usually included with these.

More complete and comprehensive mineralogical accounts include the following: Huminicki and Hawthorne (2002) should be consulted for phosphate structure in general; Ni et al. (1995) for monazite and xenotime, Clavier et al. (2011) for monazite, and Hetherington et al. (2008) for xenotime. For apatite refer to White et al. (2005), Hughes and Rakovan (2002), Pan and Fleet (2002) or, for an elegant overview, Hughes and Rakovan (2015). The crystal chemistry of allanite is briefly summarized in Armbruster et al. (2006) and with more background in Gieré and Sorensen (2004).

Monazite and xenotime

Crystal structure, chemical substitutions. Monazite and xenotime are large-ion orthophosphates $(REE)[PO_4]$ with related crystal structures; subtle differences between these help to explain their substantially different REE-fractionation (Ni et al. 1995): The large trivalent ion in monazite is coordinated to nine O-ions that form an irregular polyhedron $(REE)O_9$; in xenotime, Y (plus other $HREE^{3+}$) is coordinated to eight O-ions that form a regular dodecahedron $(Y)O_8$ (Fig.1). By sharing edges, these polyhedra form chains in the *b*-direction in both crystal structures. Along *c* these chains are linked by $[PO_4]$ tetrahedra to form a (100) sheet, and these layers are then stacked along *a* by sharing edges with the REE-polyhedra. While monazite is monoclinic and favors LREE, its dimorph xenotime-(REE) is tetragonal (part of the zircon group) and prefers HREE.

Large-ion phosphates show extensive solubility with related Th-phosphates and -silicates, such as cheralite $CaTh(PO_4)_2$ and huttonite $ThSiO_4$. Y (or some HREE or Gd) is a required constituent for xenotime; for monazite it is Ce (or La), which can be substituted by other LREE and Sm. Apart from exchanges involving only REE^{3+} and Y^{3+}, two main substitutions involve coupled charge balance:

$$2\,REE^{3+} \leftrightarrow Th^{4+} + Ca^{2+} \tag{1}$$

$$REE^{3+} + P^{5+} \leftrightarrow Th^{4+} + Si^{4+} \tag{2}$$

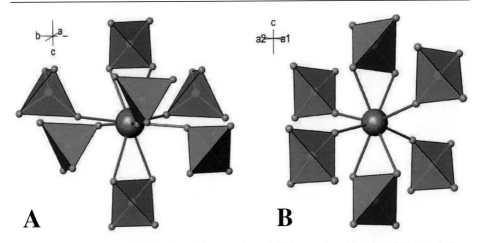

Figure 1. Local structure of REE-sites (A) in monazite and (B) in xenotime. Phosphate (PO_4) tetrahedra shown in green. The large trivalent ion (red sphere) in monazite occupies an irregular $(REE)O_9$ polyhedron that preferentially accomodates the larger, light to middle REE and Th^{4+}; in xenotime the regular $(Y,REE)O_8$ dodecahedron provides a suitable site for Y and the smaller, heavy REE.

In both cases, U^{4+} may stand for Th^{4+}; most monazites have Th/U > 1, but xenotime usually shows Th/U < 1. Data compiled by Spear and Pyle (2002) indicate that the cheralite exchange (Eqn. 1) dominates for monazite, but the huttonite exchange (Eqn. 2) is important as well (Table 1). Experimental data (Stepanov et al. 2012) indicate that huttonite dominates over cheralite at high pressure. In xenotime the chemical correlation plots are less clear; Zr^{4+} may substitute for Th^{4+}, so Zr should be included in xenotime analyses. Other minor substitutions are heterovalent as well, e.g., those involving halogens (mostly F), sulfur (in monazite), As (Janots et al. 2011), and Sr for REE (Krenn and Finger 2004).

As reviewed by Harrison et al. (2002) and Williams et al. (2007), experiments in binary systems of $LaPO_4$ with cheralite, $Ca_{1/2}U_{1/2}PO_4$, and huttonite show complete or extensive solid solutions (at 780 °C), but natural compositions cover only parts of the range (e.g., Förster and Harlov 1999). Where the two phosphates are cogenetic, different exchange vectors may dominate, e.g., cheralite in monazite, but huttonite in xenotime (Fig. 7 in Regis et al. 2012).

Compared to xenotime, natural monazite shows considerably more variability in the heterovalent substitutions; this may be due to the flexibility of the larger and less regular polyhedron. The difference in coordination may also account for much of the observed REE-partitioning: Larger LREE ions (and Th^{4+}) find a "happy home" in the larger 9-O polyhedra in monazite, while Y^{3+} and the smaller HREE clearly prefer the smaller 8-O polyhedra in xenotime. In addition, some LREE show a tetrad effect, which seems due to orbital preference (of their unpaired 4f-electrons) for the irregular REE-site in monazite (Duc-Tin and Keppler 2015). The magnitude of these REE-preferences is visible in the distribution coefficients (Fig. 2), and these data also indicate that U^{4+} fractionates but weakly between xenotime and monazite. In general, the actinide ratio (Th/U) is higher for monazite than for xenotime.

Table 1. Main compositional endmembers of REE-phosphates.

Endmember	Name in monazite	Name in xenotime	Notes
$(REE)PO_4$	monazite-(LREE)	xenotime-(LREE)	LREE: Ce > La ≈ Nd > Sm
$Ca_{1/2}Th_{1/2}PO_4$	cheralite		formerly brabantite
$ThSiO_4$	huttonite	thorite	
$USiO_4$	coffinite		

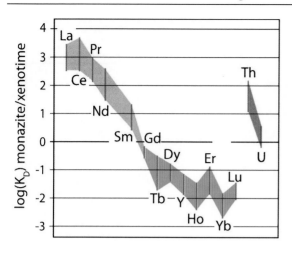

Figure 2. Distribution coefficient (concentration ratio $K_D = C_{mon}/C_{xen}$) between monazite and xenotime for REE, U, and Th. Trends for LREE to MREE highlighted in orange, for MREE to HREE in green, for actinides in blue; based on data from Franz et al. (1996).

In the REE-budget of many rocks, monazite plays a major role for the LREE, as does xenotime for the HREE. Monazite is the most common of the radioactive minerals (Overstreet 1967) and the main host of Th and U in many rocks.

Fortunately for geochronology, both monazite and xenotime do not favor incorporation of lead in their structure, although Pb^{2+} readily fits into monazite. Typically, common lead contents are low (often at ppm-level: Parrish 1990), although for precise isotopic data a common lead correction can be significant. In chemical U–Th–Pb dating (Williams and Jercinovic 2017; this volume), such correction is not possible, and its effect often would be within the analytical uncertainty due to electron probe counting statistics. It should be noted, however, that elevated common lead contents in monazite are known, e.g., in high pressure environments (Holder et al. 2015), in rocks altered by interaction with F-rich hydrous fluid (Didier et al. 2013), and occasionally in pegmatites (Kohn and Vervoort 2008); chemical dating of monazite is not advisable for such samples.

Sector zoning. Crystal-chemical controls of monazite can influence its uptake of REE. Cressey et al. (1999) documented differential incorporation of ions due to structural differences between specific growth surfaces. In that study, monazite from a carbonatite was sectorially enriched in La in {011} but depleted in {$\bar{1}$01} and {100} zones, whereas Nd showed an inverse preference; the effect is insignificant for Ce. Fractionation factors are modest in this case (~1.5), but caution is needed when concentration differences are observed, and such chemical heterogeneity in monazite is used to assess reaction relations or local REE-mobility. Stepanov et al. (2012) experimentally confirmed sector zoning in monazite from granitic melt (at 1000 °C, 1 GPa, 16 wt% H_2O); fractionation for the LREE was weak, but much more pronounced for Sm, Gd, Y, and notably also for Th. The data indicate that different growth sectors show substitutions of LREE for huttonite and for MREE (or HREE).

However, sector enrichment is an equilibrium feature of mineral surfaces, and close spatial proximity in a natural sample does not guarantee coeval growth. For instance, in a suite of lower greenschist to middle amphibolite facies metapelites, Janots et al. (2008) found substantially different La/Nd ratios in two populations of monazite (and similar fractionation factors as reported by Cressey et al. 1999), but local textures (Fig.10 in Janots et al. 2008) indicate sequential growth rims, not simultaneous sector fractionation. Sector zoning appears to be particularly common at very high metamorphic grade, notable in the granulite facies (DeWolf et al. 1993; Bingen and van Breemen 1998; Zhu and O'Nions 1999), while magmatic monazite can show oscillatory zoning (Montel 1993).

Apatite

Apatite, Earth's most abundant phosphate, has an extraordinarily flexible crystal structure. It incorporates a wide range of minor and trace elements, including some most interesting ones for geochemistry and chronology, e.g., S, Sr, U, Th, and REE. Furthermore, anion substitution in apatite is common; in most rocks apatite shows $F > OH \gg Cl$, but F/OH ratios vary. Apatite usually shows hexagonal symmetry, but anion ordering causes structural changes in $Ca_5[PO_4]_3(F,OH,Cl)$ (Hughes and Rakovan 2002).

In terms of total REE-content, natural apatite is not comparable to monazite and xenotime, but it can be significant in the LREE-budget of rocks, especially in carbonatites and mafic rocks (e.g., Y + La + Ce to 22,000 ppm: Cruft 1966; LREE > 15,000 ppm: Finger et al. 1998), in migmatites (LREE ~ 4,750 ppm: Bea 1996). REE concentrations appear to increase from amphibolite to granulite grade (LREE + MREE > 10,000 ppm: Bingen et al. 1996).

Crystal-chemically, REE do find a "happy home" in the M2 site of apatite, which is coordinated to six O-ions plus one monovalent anion (Pan and Fleet 2002). U^{4+} also favors the M2 site in most apatites (Luo et al. 2009), but Th contents are very low. The site occupancy ratio ($^{M2}REE/^{M1}REE$) changes with ionic radius: MREE such as Nd show no site preference, but HREE prefer the M1 site, which is coordinated to nine oxygen ions; M1 is smaller than M2, so ionic size is the probable cause of the site preference (Wood and Blundy 1997).

Substitution mechanisms involving REE in natural apatite involve charge compensation (Pan and Fleet 2002); the main exchanges proposed are:

$$REE^{3+} + Na^+ \leftrightarrow 2\,Ca^{2+} \tag{3}$$

$$REE^{3+} + SiO_4^{4-} \leftrightarrow Ca^{2+} + PO_4^{3-} \tag{4}$$

Such heterovalent exchange mechanisms typically lead to much slower diffusion than homovalent exchanges; this was indeed found in experiments for REE in fluorapatite (Cherniak 2000). Because coupled exchanges depend on the concentration of other mono- to pentavalent cations on different sites in the crystal lattice, REE-diffusivity in apatite must be sensitive to composition.

Sector zoning. Apatite shows sector preference for certain REE (Rakovan and Reeder 1996); in this case the MREE and HREE are most affected, and fractionation is much more pronounced than in monazite: One study (Rakovan et al. 1997) observed nearly an order of magnitude difference for Nd between {0001} and {10$\bar{1}$0} zones. On petrographic grounds, the sectors analyzed were judged to represent a single growth phase. Paired with substantially different Sm/Nd ratios in the two sectors, the compositional spread from analyses of different sectors of one crystal was sufficient to define an isochron age of 43.8 ± 4.7 Ma (2σ).

Allanite

Crystal structure, chemical substitutions. Gieré and Sorensen (2004) gave an excellent entry into the world of allanite; Armbruster et al. (2006) clarified the nomenclature for the epidote-group and the non-trivial task of site-assignment, and they nominally ended the debate of what should be called allanite. Following Ercit (2002), allanite-(Ce), allanite-(La) etc. are now recognised as distinct mineral species. Here, the term allanite refers to what is officially the allanite subgroup.

Minor amounts of REE are often present in epidote group minerals, but in allanite LREE and Fe^{2+} are structurally essential constituents. The structural formula $A_2M_3[SiO_4][Si_2O_7](O)(OH)$ describes the entire epidote group and allows the main substitutions to be understood. Two groups of sites are identified, labeled A and M: A involves A(1) and A(2), which are 9- and 11-fold coordinated, respectively; M comprises three octahedra, of which M(1) and M(2) are essentially occupied by Al. With the main constituent of A(1) being Ca (plus minor Mn^{2+}), most substitutions occur on A(2) and M(3) (Fig. 3): In natural allanite, A(2) is mostly occupied by REE^{3+}, with M(3) containing Fe^{2+} and minor Mg. By contrast, most natural epidotes (formally members of the clinozoisite subgroup) have A(2) principally occupied by Ca and M(3) by trivalent ions (Fe^{3+}, Al, Mn^{3+}). Allanite frequently shows chemical zoning or overgrowths, demonstrating solid solution with epidotes. The main exchange is heterovalent and can be written in site-specific notation as

$$^{A2}(REE)^{3+} + {}^{M3}M^{2+} \rightarrow {}^{A2}Ca^{2+} + {}^{M3}M^{3+} \tag{5}$$

Data compiled by Gieré and Sorensen (2004) show chemical analyses covering the complete range possible due to this exchange. In many cases the contents in REE^{3+} and M^{2+} are higher than will fit into the A2 and M3 sites, respectively, indicating that other sites can be involved as well. In addition to 2–3 wt% REE_2O_3, allanite shows tetravalent substitutions, notably to accommodate Th (typically 1,000–10,000 ppm) and U (mostly < 2,000 ppm):

$$^{A2}(Th^{4+},U^{4+}) + 2\,{}^{M3}M^{2+} \rightarrow {}^{A2}Ca^{2+} + 2\,{}^{M3}M^{3+} \tag{6}$$

Unless Fe^{2+}/Fe^{3+} is analyzed by Mössbauer spectrometry (Fehr and Heuss-Assbichler 1997), the ferrous fraction must be estimated when normalizing chemical analyses for epidote group minerals from electron probe data (and LA-ICP-MS[2] data for trace elements). Analyses are normalized to 12.5 oxygen atoms on an anhydrous basis, as explained by Gieré and Sorensen (2004; their Figs. 5 and 6), based on Petrík et al. (1995).

As already noted by Goldschmidt and Thomassen (1924), allanite prefers LREE (La–Nd), similar to monazite, but in marked contrast to zircon and garnet, two commonly coexisting phases that tend to be HREE-enriched. Partitioning patterns—often shown in spider diagrams, normalized to chondrite or some suitable reference mineral—can thus be used for paragenetic indications (Hermann 2002). For allanite, the absolute amounts of the REE, the slope and curvature in the diagram, the presence or absence of anomalies (Eu, Ce?) all reflect competition between allanite and other REE-bearing minerals as the allanite grew.

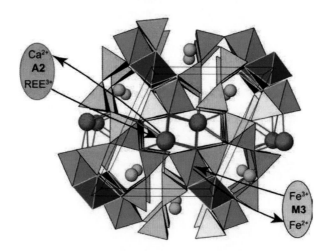

Figure 3. Structural elements of allanite shown as combined coordination polyhedra with specific ions to emphasize characteristic allanite substitutions relative to basic epidote structure. Large red spheres with sticks mark the 11-fold **A2** site for the $REE^{3+} \Leftrightarrow Ca^{2+}$ exchange; charge balance is via the $Fe^{2+} \Leftrightarrow Fe^{3+}$ on the **M3** (distorted) octahedral site, shown in bright green. Isolated SiO_4 tetrahedra in yellow, Si_2O_7 groups in turquoise. Pink spheres show 9-fold coordinated Ca in large cavities.

---

[2] Laser ablation inductively coupled plasma mass spectrometry

Sector zoning

Sharply discrepant compositions are often found in epidote group crystals, and their significance is not always clear. Apart from magmatic and metamorphic overgrowths, a possible miscibility gap has been debated, but this option was dismissed by Banno and Yoshizawa (1993) in favor of sector growth.

Radiation damage

Emission of α-particles in the decay chains of ^{238}U, ^{235}U, and ^{232}Th cause ionization and inflict radiation damage on any mineral with appreciable U or Th contents. Spontaneous ^{238}U fission is several million times rarer than α-decay, but causes much more severe structural damage to a host crystal. To indicate the extent of damage, optical effects in thin section (color, reduced anisotropy) and Raman spectra (McFarlane 2016) are useful. Minerals respond differently to the effects of both types of damage: In zircon and allanite with high actinide contents, structural damage accumulates over time, disrupting their crystalline integrity and eventually leading to a metamict (largely amorphous) state. By contrast, REE-phosphates do not become metamict, even though they can accommodate up to 30 wt% $ThO_2 + UO_2$ (van Emden et al. 1997); the damage inflicted to their structure appears to heal. Thermal annealing occurs at low temperatures, i.e., at ca. 100 °C in apatite and ca. 300 °C in monazite (Gleadow et al. 2002). While the mechanisms of such restoration are not well understood, they certainly have positive effects: Monazite, xenotime, and apatite remain structurally intact, and are less likely to lose their radiogenic Pb (Seydoux-Guillaume et al. 2004) and interact with fluid than heavily damaged allanite or zircon. Annealing can occur in allanite as well (Karioris et al. 1981), but it is more protracted. Structurally damaged or metamict grains are quite common in actinide-rich varieties, especially in Paleozoic and older rocks. But caution is indicated—even where old grains rich in U + Th now appear structurally intact, radiation-damaged grains may have healed after a period of open system behaviour. Such a scenario may explain why ages from non-metamict allanite occasionally appear implausibly young, such as for the Pacoima Canyon pegmatite (California), for which reportedly non-metamict allanite was dated to 1006 ± 37 Ma (zircon: Weber 1990; allanite: Gieré and Sorensen 2004, p. 466–472), whereas zircon is 1191 ± 4 Ma (SIMS Th–Pb, Catlos et al. 2000).

Disturbed Th–U–Pb dates in radiation-damaged allanite may result from Pb-loss or actinide remobilization (TIMS U–Pb, Barth et al. 1994). In this and other studies, discordances in ages (from TIMS and LA-ICP-MS) are larger between $^{208}Pb/^{232}Th$ and Pb/U than between $^{206}Pb/^{238}U$ and $^{207}Pb/^{235}U$. This implies unequal chemical mobility of Th and U in allanite. Barth et al. (1994) discovered minute grains of thorite and uraninite along fluid pathways (cracks) in two metamict samples. This indicates an indirect effect of radiation damage, and such secondary phases would account for the decoupling between U- and Th-systems. Beyond a critical cumulative irradiation dose, hydrothermal fluid attack appears to be much more effective. According to Smye et al. (2014), a critical dose sufficient to effect isotopic disturbance is an order of magnitude lower in allanite than in zircon. This susceptibility to damage and alteration, and the generally much higher concentration of Th + U, help explain why allanite ages > 300 Ma often show more effects of radiation damage than zircon.

Diffusion and closure temperature

Several aspects of petrochronology must consider diffusive mobility, notably

- when constraining petrogenetic conditions, applying thermobarometers etc., based on local equilibria in sample domains (Lanari and Engi 2017);

- when considering the effects of chemical re-equilibration upon subsequent changes in *P–T* (pro- and/or retrograde), whether by pure exchange- or combined net–transfer- and exchange-reactions (Kohn and Penniston-Dorland 2017);

- when analyzing geochronological data (e.g., normal / inverse discordance) to assess full/partial isotopic equilibration during mineral growth, and subsequent loss of parent or daughter isotopes (Wasserburg 1963).

The essential theoretical aspects of diffusion, experimental constraints, select analytical data from natural examples, and insight gained from petrochronological studies are presented and evaluated in this volume (Kohn and Penniston-Dorland 2017). Here, a brief account is given only to assess the current state of knowledge about isotopic closure temperatures (T_c) relevant to U–Pb, Th–Pb, and Pb–Pb chronometry. While the link between T_c and diffusivity (D) is well defined (Dodson 1973, 1986; Ganguly and Tirone 1999; Kohn and Penniston-Dorland 2017), minerals differ in their capacity to self-heal damage due to fission and α-decay, e.g., allanite can become metamict with time. Insight gained from work on natural samples occasionally is in conflict with results from experimental diffusion studies, and structural damage certainly increases ionic mobility. While this topic has been addressed in numerous studies, we are far from understanding how closure temperatures change in minerals undergoing (partial) metamictization and thus affect what ages are recorded.

For monazite, Parrish (1990) estimated $T_c = 725 \pm 25\,°C$, a value still commonly quoted, but T_c values of 800–850 °C have subsequently been proposed, based on preserved age relations in high-temperature samples (Spear and Parrish 1996; Bingen and van Breemen 1998; Kamber et al. 1998). Experimental diffusion data initially indicated $T_c \sim 500\,°C$ for 10 μm monazite (Smith and Giletti 1997), but subsequent work (Cherniak et al. 2004) found $T_c > 900\,°C$. Pb diffusion in xenotime was found to be even slower than in monazite, and T_c must be similarly high (Hawkins and Bowring 1997; Cherniak 2006, 2010).

No experimental diffusion data are available for Pb in allanite; closure temperatures have been estimated based on dated high-temperature samples. A minimum T_c of 750 °C is required to account for the retentivity of allanite (Oberli et al. 2004) crystallized in Bergell tonalite during magmatic differentiation (see Case Studies below, Fig. 21). Heaman and Parrish (1991) proposed $650 \pm 25\,°C$, but also noted that allanite U–Pb closes at higher temperature than titanite. T_c for titanite has recently been revised upward to near 800 °C (Kohn 2017), and I am not aware of any natural datasets that would indicate a lower T_c for allanite. Note, however, that allanite may seem young because of retrograde growth (Finger et al. 1998; Wing et al. 2003), well below its T_c.

At current understanding, the conclusion is that lead loss by diffusion from monazite, xenotime, and allanite is probably negligible up to temperatures of >750–800 °C over 10 Ma for grains >20 μm. Such conditions certainly encompass "normal" granulite samples, i.e., U–Th–Pb chronometry applied to high-temperature growth zones should yield formation ages. Recent studies have interpreted monazite data even from UHT terrains (>900 °C) as formation ages (e.g., Suzuki et al. 2006; Korhonen et al. 2013), but rigorous corroboration using diffusion profiles from such samples would be desirable. The shape of concentration profiles across different growth zones should allow diffusion models to be tested (e.g., reasonable $D \cdot \Delta t^3$) and may potentially constrain the T–t history (e.g., Watson and Harrison 1984). However, it is important to remember that fluid- or deformation-induced retrogression can be far more important than purely diffusional processes (Teufel and Heinrich 1997; Seydoux-Guillaume et al. 2002b; Harlov et al. 2011; Shazia et al. 2015).

---

[3] since the diffusion length is $2(D \cdot t)^{1/2}$

GEOTHERMOBAROMETRY

Monazite and xenotime thermometry

A welcome effect of the pronounced Y-HREE fractionation is the miscibility gap between xenotime and monazite, which is quite asymmetric in YPO_4–$(REE)PO_4$. The Y-content of monazite serves as a geothermometer, with only modest pressure-sensitivity (Gratz and Heinrich 1997, 1998; Heinrich et al. 1997; Andrehs and Heinrich 1998). Calibration data are summarized and discussed in Spear and Pyle (2002, Fig.24); they emphasize effects of additional components (Th, U, Ca, Si), but so far only the effect of Th on the miscibility gap has been calibrated (Seydoux-Guillaume et al. 2002a).

Figure 4 compares calibrations based on experiments. From these data, Mogilevsky (2007) developed a generalized regular solution model that can handle other components as well. For thermometry, this formalism (as presented in Mogilevsky's Equations 4–7) demands a numerical solution, and so far it seems that no applications have appeared in geology. Spear and Pyle (2010) used the same experimental data to construct a thermodynamic model based on estimated endmember properties, and approximate temperatures can be obtained from these. This approach is promising, but so far restricted to a very small compositional space; a wider basis of phase equilibrium data is sorely needed to allow a more generalized thermodynamic calibration and then applications thereof.

Apart from such technical issues, applying monazite–xenotime thermometry to rocks has its limits. Notably, xenotime should have been present when monazite grew, else the method yields minimum temperature values only. Since garnet (YAG component) is a major competitor for Y, xenotime is consumed during prograde garnet growth and commonly disappears from the assemblage; it may reappear where garnet is consumed, at higher grade or on decompression or retrogression (e.g., Hallett and Spear 2015). For these reasons, it can be tricky to ascertain that a specific monazite zone grew while saturated in xenotime, and in fact they rarely occur together in equilibrium (e.g., Pyle and Spear 1999). In zoned monazite, the Y-content is a good guide: annuli with high Y are good candidates for thermometry, whereas an abrupt decrease in Y may indicate growth in a xenotime-absent assemblage. Geochemical zoning and textural relations can convincingly discriminate such populations (e.g., Foster et al. 2004; Daniel and Pyle 2006, p. 108–113). In rocks where monazite and xenotime are both present in local assemblages, they often show zoning, and it can be unclear how to assess which compositional pairs were in equilibrium at any stage of the evolution; checking if partition coefficients (notably Gd/Y, Dy/Y: Pyle et al. 2001; perhaps also Th/U) show equilibrium values is preferable to blind trust.

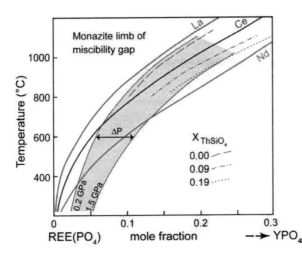

Figure 4. Monazite limb of the miscibility gap to xenotime, comparing calibrations based on experiments. Two isobars (0.2 and 1.5 GPa) from Gratz and Heinrich (1997) show sensitivity to pressure; three isopleths from Seydoux-Guillaume et al. (2002a) display effect of the Th-content of monazite; note that ~20% Th-endmember shows an effect comparable to ~1 GPa pressure increase. Curves labeled Ce, La, Ce, Nd show the dependence on the respective REE from models calibrated by Mogilevsky (2007) using experimental data.

Two further thermometers make use of the Y-contents of garnet in equilibrium with either xenotime or monazite (Pyle and Spear 2000; Pyle et al. 2001), but do not depend on both phosphates being present. The empirical garnet–xenotime thermometer is useful but over a small temperature range (~450–560 °C). As long as xenotime buffers the Y-activity of an assemblage, only the Y-content of garnet (grt) is needed for thermometry (Pyle and Spear 2000):

$$\ln\left(Y/ppm\right)_{grt} = \frac{16031\left(\pm862\right)}{\left(T/K\right)} - 13.25\left(\pm1.12\right) \tag{7}$$

The garnet–monazite thermometer extends this limited temperature range; it is most sensitive up to ~620 °C. At higher temperature the Y-concentration in garnet is low, approaching the detection limit of EMPA[4] data, hence LA-ICP-MS analyses are preferred. In addition, quartz, plagioclase and apatite must coexist, and the thermometer depends on H_2O, albeit weakly (Pyle et al. 2001):

$$T/K = \frac{\left[-1.45\times\left(P/bar\right) + 10^3\times447.8\left(\pm32.1\right)\right]}{\left[567\left(\pm40\right) - R\ln\left(K_{eq}\right)\right]} \tag{8}$$

requiring analyses and solution models for garnet (grt, Grs: $Ca_3Al_2Si_3O_{12}$, YAG: $Y_3Al_5O_{12}$), plagioclase (pl, An: $CaAl_2Si_2O_8$), monazite (mnz), and apatite (ap, OH-ap: $Ca_5[PO_4]_3(OH)$) to obtain

$$K_{eq} = \frac{\left(a_{grs,grt}^{5/4}\, a_{an,pl}^{5/4}\, a_{YPO_4,mnz}^3\, f_{H_2O}^{1/2}\right)}{\left(a_{YAG,grt}\, a_{OH-ap,ap}\right)} \tag{9}$$

Provided the mineral analyses are accurate[5], the calibration usually yields temperature estimates to within ±30 °C of independent thermometers even if a pure hydrous fluid is assumed. However, to be petrochronologically useful, the thermometer should be restricted to monazite–garnet pairs that grew simultaneously, and to ascertain this, petrographic scrutiny is essential (Pyle et al. 2001).

Monazite–melt thermobarometry

Solubility data for monazite in granitic melt were used to calibrate a thermometer for magmas (Montel 1993; Plank et al. 2009); it relies on low-pressure experiments (0.2 GPa, Montel 1986; Rapp and Watson 1986). A similar thermometric equation, but with a pressure correction, was proposed by Stepanov et al. (2012), incorporating their new data to 5 GPa. All these thermometers assume that monazite dissolves to dominantly associated $LREE(PO_4)$ species. However, recent data (Duc-Tin and Keppler 2015) indicate that at least for some magma compositions partial dissociation in the melt occurs at low pressure, and at pressures of 3 GPa (Skora and Blundy 2012) dissociated $LREE^{3+}$ and $(PO_4)^{3-}$ ions are dominant in hydrous silicic melts. This behavior parallels other hydrous fluids at high pressure (Manning 2004). As pointed out by Skora and Blundy (2012), thermometry such as advocated by Plank et al. (2009) would be seriously (>100°) in error at pressures typical of subduction systems. Stepanov et al. (2012) developed the following equation describing monazite solubility in peraluminous and metaluminous granitic melts:

$$\ln\left(C_{LREE}/ppm\right) = 16.16\left(\pm0.30\right) + 0.23\left(\pm0.07\right)\times\left(H_2O/wt.\%\right)^{1/2} + \ln\left(X_{LREE}\right)^{mnz}$$
$$-\frac{11494\left(\pm410\right)}{\left(T/K\right)} - \frac{1.94\left(\pm0.40\right)\times\left(P/GPa\right)}{\left(T/K\right)} \tag{10}$$

---

[4] EPMA: Electron probe microanalyzer
[5] As OH-apatite is obtained by difference, F in apatite requires analytical attention

(CLREE:=ΣLa...Sm in ppm; XLREE:=LREE/[LREE+Y+Th+U]). This linear fit reproduces a large dataset (750–1400 °C, 0.2–5 GPa, 1–20 wt% H₂O and 0.82–1.36 ASI 6), but includes neither the experimental data for melts with (CaO+FeO+MgO)> 3 wt% nor for peraluminous and phosphate-rich melts (Duc-Tin and Keppler 2015). Figure 5 shows that the strong temperature-dependence vanishes for highly aluminous melts (ASI > 1.0).

TRACE ELEMENT GEOCHEMISTRY AND PETROGENESIS

Magmatic and partial melting range

Phosphates. Monazite grains crystallized from melt typically show chemical zoning, and this can be very useful to single out specific growth zones for spot dating. So far, most emphasis has been placed on Th- and Y-zoning (Pyle and Spear 2003; Kohn et al. 2005; Rubatto et al. 2013). This is addressed below, but recent experimental work has clarified the interplay of these elements with the REE, and thus zoning in the LREE and MREE, even in individual REE, should be deciphered in detail. Such information is likely to add useful control on the dissolution and growth history and elucidate the control phosphate solubility plays in fractionation, e.g., of Sm over Nd (Zeng et al. 2005) or LREE showing tetrad effects for monazite (Duc-Tin and Keppler 2015).

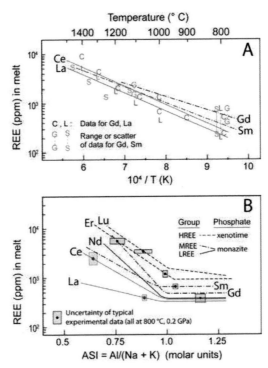

Figure 5. Experimental solubility data for REE-phosphates in granitic melt by Duc-Tin and Keppler (2015) and earlier studies. (A) Monazite solubility is more strongly temperature-dependent for the LREE than for MREE (and HREE); size of symbols shows uncertainty in data. (B) The effect of the aluminum saturation index (ASI) on REE solubility. Select trends (based on data for 800 °C, 0.2 GPa, ~0.1 wt% P₂O₅) show solubility generally increasing from LREE (La–Nd) to MREE (Sm–Gd) to HREE (Tb–Lu); differences are not significant for peraluminous melts (ASI > ~1.0), but for peralkaline melts solubility strongly increases. Steep slopes for Nd and Gd indicate tetrad effect.

---

6 ASI := Al/(Na+K+2Ca)

REE. Data on monazite solubility (Montel 1986; Rapp and Watson 1986; Rapp et al. 1987; Montel 1993; Duc-Tin 2007; Stepanov et al. 2012; Duc-Tin and Keppler 2015) show a dependence on temperature, pressure and magma composition, notably the H_2O- and phosphate-content, and the aluminum saturation index (ASI := molar $Al/(Na + K + 2Ca)$). As explained above, the pressure dependence of the solubility has been attributed to changes in complexation in the melt (Skora and Blundy 2012), but recent data for monazite and xenotime (Duc-Tin and Keppler 2015) indicate that partial dissociation also occurs at low pressures— more notably in peralkaline, less strongly in peraluminous melt (Fig. 5)—a fact that had gone unnoticed in previous studies. In the pressure-composition range of partial dissociation ($0 < y < 1$), orthophosphate (oph: monazite and xenotime) dissolution should be written as:

$$\text{REE}\left(\text{PO}_4\right)_{\text{oph}} \rightarrow \left(1-y\right)\left[\text{REE}\left(\text{PO}_4\right)^0\right]_{\text{melt}} + y\left[\text{REE}^{3+} + \left(\text{PO}_4\right)^{3-}\right]_{\text{melt}} \qquad (11)$$

The P–T–X conditions of partial dissociation are not yet fully explored, but solubility is known to increase with H_2O-content and alkalinity (for ASI < 1.0), whilst it decreases with actinide (Th + U) concentration and pressure. Solubility of monazite is most temperature-dependent for LREE, less so for MREE (Stepanov et al. 2012; Duc-Tin and Keppler 2015). With increasing dissociation (higher y-values), orthophosphate solubility increasingly depends on the phosphate contents of the melt, which can vary widely—by at least two orders of magnitude for mafic to felsic magmas, and from 0.1 to 1.5 wt% in granitic magmas alone (Pichavant et al. 1992; Breiter et al. 2008). Because REE and P_2O_5 are both incompatible, their concentration depends on the degree of melting: At high melt fractions P_2O_5 concentrations are low, whereas LREE are high; during crystallization they move in the opposite sense. Apatite solubility is low in metaluminous melts ((ASI = 1), but increases with the ASI, owing to $AlPO_4$ complexes (e.g., Wolf and London 1994), and it is also elevated at low ASI, owing to alkali phosphate complexing (Ellison and Hess 1988). In melts derived from apatite-saturated metasediments, phosphate contents may initially be buffered by apatite solubility, but since monazite is considerably less soluble in melt (and hydrous fluid) than apatite, monazite (and xenotime) largely control the REE- (and Y-) content of melts and vice versa. LREE-rich apatite dissolves incongruently (Antignano and Manning 2008; Tropper et al. 2011), so leaves behind a LREE-enriched monazite. Leucosomes in migmatites may well preserve pre-anatectic monazite and xenotime (Rapp and Watson 1986), or apatite in very hydrous melts (Zeng et al. 2005).

Recent work (Duc-Tin and Keppler 2015) shows substantial differences in solubility among pure LREE-phosphates; even neighboring LREE can differ by a factor of two (Fig. 5). This opens the possibility to refine our understanding and interpretation of zoned monazite by analyzing the patterns for individual REE or groups thereof, particularly as solubility is strongly temperature-dependent for LREE but less so for heavier REE.

Current understanding of the solubility of orthophosphates (Fig. 5), especially in granitic systems and their REE evolution, allows some specific conclusions relevant to monazite petrochronology. Attempts to model the behavior of monazite (and zircon) in the partial melting range (Kelsey et al. 2008; Yakymchuk and Brown 2014) have combined computed phase diagrams that show the stable phase relations, computed from thermodynamic models, with the solubility equations. Yakymchuk and Brown used the updated solubility equation (Eqn. 10) and also included open-system scenarios (with intermittent melt loss). The results of both studies predict that partial melting of metapelitic and metapsammitic protoliths will, depending on the exact P–T path, lead to dissolution of (most or all) monazite. Noting that melt-isopleths typically have positive P–T slope, decompression is expected to lead to dissolution, and monazite growth is expected only upon cooling. In ascending magmas, the oldest monazite age most likely dates cooling after decompression, magma emplacement or injection. This is certainly the most likely moment on a clockwise P–T–t path, say in a collisional setting, where initial decompression may be attended by heating: Most monazite is expected to crystallize sometime *after* cooling from T_{max}, and it seems most unlikely that one could use monazite to date the stage after P_{max} but before T_{max} is reached (Kohn et al. 2005). Things are more complicated if magmatic allanite is involved, i.e., in more calcic compositions (e.g., granodiorite and tonalite). At late-magmatic stages, as partial melts become enriched in volatiles, monazite resorption is likely[7].

In addition, dissolution of phosphates can be kinetically limited, especially in dry melts, where diffusion of LREE and phosphate species away from the dissolving solids is very slow (Harrison and Watson 1984).

Th, Y, and U. These elements partition into monazite, much like the LREE, but to decreasing degrees, i.e., Th most, U least. During crystallization, thorium is dominantly fractionated into monazite, and Rayleigh-like fractionation occurs, except in more calcic magmas, where allanite may interfere. Th/U ratios in monazite of course depend on Th/U in the magma, but also on the modal abundance of coexisting phases competing for actinides, notably allanite—the other main Th-sink—and zircon, xenotime, and titanite, all of which favor U over Th. In migmatites and generally in granitoid magmas, fragments or corroded relics of pre-melting monazite (from one or more prograde generations, discussed below) may survive even long periods of high temperature and extensive melting (Copeland et al. 1988; Harrison et al. 1995; Kohn et al. 2005; Dumond et al. 2015). Their eventual dissolution may explain Th-rich rims in magmatic monazite or allanite. In the absence of such complications, the earliest magmatic monazites typically show the highest Th-contents; subsequent growth zones are Rayleigh-distilled. Such patterns, reported primarily from anatectic, often haplogranitic magmas, have been put to use in deciphering and dating pre- and post-anatectic stages (Pyle and Spear 2003; Kohn et al. 2005; Corrie and Kohn 2011). Xenotime saturation, i.e., growth from melt or resorption in it, is critically dependent on and inversely correlated with garnet and monazite, the main competitors for Y. Competition for actinides during magmatic crystallization is reflected in apatite, which shows Th/U ratios that vary inversely with the modal presence of monazite (and possibly allanite?) during crystallization (Cochrane et al. 2014). Where apatite occurs as magmatic inclusions in zircon or titanite, its composition carries useful petrogenetic information (Bruand et al. 2016).

Allanite

REE-zoning in igneous allanite has long been recognized to reflect fractional crystallization (Levinson 1966), most commonly from a melt of high water contents (Johnston and Wyllie 1988; Beard et al. 2006). Experimental data on allanite solubility (Kessel et al. 2005; Klimm et al. 2008; Hermann and Rubatto 2009) cover a range of melt compositions and pressures. Where allanite forms in magma, it largely controls the LREE contents of the melt simply because it contains LREE at weight-% level. For Th, U, and Y, as noted above, competition from coexisting phases (zircon, monazite, xenotime, titanite or rutile) may intervene. As summarized by Smye et al. (2014) REE-fractionation in allanite depends on magma composition: The most LREE-enriched allanites are found in Ca-rich types (e.g., carbonatite); La/Nd and Ce/Nd are high for diorite and granodiorite, intermediate for granite, and lowest for syenite. Allanites in rhyolitic obsidian often show La/Yb 250–2300 and beyond, but only 3 to ~100 for the glass (Mahood and Hildreth 1983). Typically, chondrite-normalized REE-patterns are steep for early-formed allanite, but depletion of LREE then leads to nearly flat patterns, and the overall core-to-rim zoning in allanite may end with REE-enriched (or even REE-poor) peripheral epidote (e.g., Oberli et al. 2004). REE-contents (normalized to chondrite) in >1700 analyses compiled by Gieré and Sorensen (2004) show typical ranges for LREE vs. MREE (e.g. La: $10^4–2·10^5$, but Sm: $4·10^3–6·10^4$ only). Allanite preference for LREE is exceeded (in absolute concentration) only by its appetite for Th, i.e., allanite fractionates Th/La relative to the melt (Mahood and Hildreth 1983; Klimm et al. 2008; Hermann and Rubatto 2009).

---

[7] However, contrary to previous belief, fluorine in melt was shown to have no effect on solubility of monazite and xenotime, indicating no fluoride complexing of REE (Duc-Tin and Keppler 2015).

The actinide ratio Th/U is often used as a diagnostic means to distinguish magmatic from metamorphic allanite growth zones: Typical magmatic values are Th/U > 100(Gregory et al. 2007), metamorphic ones are < 50 (Gregory et al. 2009), except under very oxidizing conditions, where U^{4+} may lead to Th/U ≫ 100 (Janots and Rubatto 2014). Trends during magmatic differentiation depend on competing minerals; for example, Th/U increases in allanite, whereas it decreases in coexisting titanite. Fractional crystallization of allanite can produce strong zoning in major and minor elements. Figure 6 shows typical compositional data, crystal morphology, and zoning in igneous allanite from a granodiorite. In a tonalite studied by Oberli et al. (2004) allanite shows a striking magmatic evolution, from dark brown core to faintly colored rim, in wt%: REE_2O_3 17.6 → 1.9, CaO 13.7 → 23.6, Fe_2O_3 5.3 → 11.2, ThO_2 1.4 → < 0.1. In mafic rocks at high pressure allanite plays a leading role in sequestering REE and actinides. Data for eclogites show that allanite stores 90–99% of a rock's budget in LREE, Th, and U, > 60% of its Sm and still about a third of its Eu (Hermann 2002). Allanite finds its main competitor for LREE + Th in the composition range where monazite coexists. This is discussed below jointly with the evolution of the phosphates, since REE-fractionation between these and allanite is petrochronologically significant.

Subsolidus petrogenesis

Apatite, monazite, and xenotime are known to occur from diagenetic conditions (e.g., Kingsbury et al. 1993; Evans and Zalasiewicz 1996; Knudsen and Gunter 2002; Rasmussen et al. 2002; Vallini et al. 2002; Rasmussen 2005) all the way up to upper amphibolite, granulite, and eclogite grades (e.g., Black et al. 1984; Bingen et al. 1996; Zhu and O'Nions 1999; Rubatto et al. 2001; Erickson et al. 2015; Gasser et al. 2015; Tucker et al. 2015).

Diagenetic apatite, monazite and xenotime have proven datable (Evans et al. 2002; Rasmussen 2005; Davis et al. 2008), and ages can be very helpful in understanding diagenetic processes involving phosphates. Remarkably, diagenetic xenotime has proven to preserve its formation age (Vallini et al. 2002), resisting later penetrative deformation and a greenschist facies overprint. This is most welcome because textural relations are complex (Fig. 7), so present a challenge to infer specific irreversible reactions and link them to *P–T* conditions. In metaclastic sequences, accessory minerals at anchizonal conditions typically include mostly authigenic, diagenetic and/or detrital monazite, apatite, and xenotime, in addition to detrital zircon, but detrital cores show monazite or xenotime overgrowths (Fig. 7 A, F) that form in the lower greenschist (Wing et al. 2003; Rasmussen and Muhlig 2007; Janots et al. 2008, 2011; Allaz et al. 2013) and low-temperature blueschist facies (Janots et al. 2006). Low-grade apatite tends to be high in Th and U, and it is the main REE reservoir in deep sea mud (Elderfield and Pagett 1986; Kato et al. 2011; Kon et al. 2014). Relative to their (mostly clastic) precursors, low-grade metamorphic monazite is typically low in Th and U, depleted in HREE, but enriched in LREE. In many instances, apatite, thorite, and xenotime are visible as satellites or intergrown small grains, indicating local reactions and spatially limited redistribution of REE and actinides over some 50–100 μm. Reaction volumes reach much larger dimensions (mm to cm) in high-strain zones or where ample fluid interacted with the rocks (Janots et al. 2012; Seydoux-Guillaume et al. 2012), even at low temperatures. Most growth zones are small (< 10 μm), however, and *in situ* chronometry in very low-grade metamorphic monazite (< 350 °C) is probably best done by EMPA (Fig. 7 G, H).

Above the lower greenschist facies, grain coarsening and replacement reactions typically enhance petrochronological options for REE minerals. Grain to grain heterogeneity in trace element contents and isotopes of detrital phosphates, which can hinder correlations, tends to disappear due to such processes above ~500 °C (Hammerli 2014; Hammerli et al. 2014, 2016).

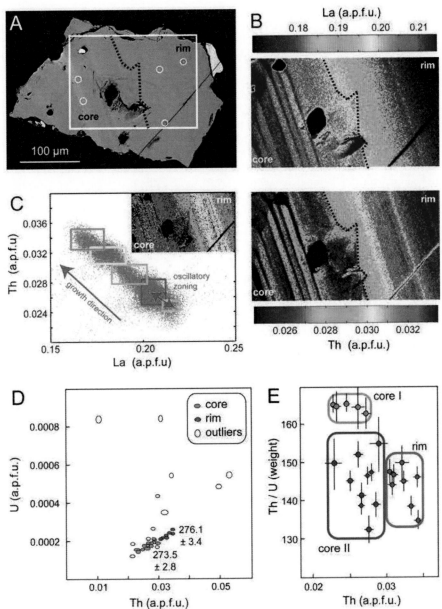

Figure 6. Magmatic allanite from Cima d'Asta pluton (CAP), a commonly used standard reference material (Burn 2016). (A) BSE image of allanite fragment shows weak oscillatory zoning in core; inclusions are apatite, zircon, and quartz; circles mark spots of LA-analysis. (B) X-ray maps (La, Th) of central part (white frame in A) show chemical zoning due to fractionation from melt during growth; rhythmic banding reflects inverse correlation of La and Th (a.p.f.u.: atoms per formula unit). From core to rim, elements with Th-affinity increase ($1.1 \, wt\% \, ThO_2$ in core, 1.6% in rim), those with La-affinity decrease in concentrations. (C) Based on the La–Th diagram five magmatic growth zones were identified. (D and E) U–Th diagrams show core to rim evolution and identify two core zones (I, II) based on their Th/U ratio and Th-contents. Their Th/Pb ages are within error, based on LA-ICP-MS data, whilst ages can be resolved between core I and rim (D). All data from Burn (2016). Ages overlap with previous TIMS and LA-ICP-MS analyses (Barth et al. 1994; Gregory et al. 2007), though cores and rim were not separately treated in these (compare Figure 12).

Monazite and allanite

Precursors to metamorphic monazite and xenotime include, in addition to their own ancestors, various low-temperature Al-phosphates, carbonates, and phyllosilicates; textures commonly indicate dissolution-reprecipitation processes (Fig. 7G,H), but few studies have worked out phase relations (e.g., Janots et al. 2006; Spear 2010) or attempted local mass balance for REE and actinides—a topic in need of study. In the greenschist or low-temperature blueschist facies, monazite in clastic metasediments breaks down to form allanite + apatite. Various reactions (Table 2) have been suggested on petrographic grounds, and *P–T* conditions for the first appearance of allanite clearly depend on rock composition. The limiting reactions of REE-phases have not been studied by experimental reversals, so remain inadequately known. For a small part of the relevant composition space experimental stability relations are available (Janots et al. 2007; Hermann and Spandler 2008) and can be compared (Fig. 8) to those computed from thermodynamic data, which were calibrated using natural assemblage data as well (Spear 2010). Effects of the bulk rock composition are evident when comparing Figure 8A and 8B: allanite is stabilized by an increase in CaO content, and substitutions such as Fe–Mg and La–Nd have a major effect on its stability relative to monazite. These thermodynamic models are a promising start, but at this point none of them adequately reproduce the reaction sequences documented in metamorphic rocks.

A next generation of monazite is known to grow at conditions around the garnet or staurolite isograd in rocks that previously contained allanite (e.g., Smith and Barreiro 1990; Catlos et al. 2001; Spear and Pyle 2002; Wing et al. 2003; Corrie and Kohn 2008; Janots et al. 2008), but it can also grow at lower grades in true (very low-Ca) metapelites and -psammites that did not contain allanite (e.g., Kingsbury et al. 1993). Allanite can persist to higher temperatures in Ca-rich rock types (Ca/Na > 1) and at high-pressure conditions (Janots et al. 2008; Kim et al. 2009; Radulescu et al. 2009), but phase relations remain poorly understood. Allanite grown in the eclogite facies can be Sr-rich (Rubatto et al. 2008; Cenki-Tok et al. 2011) due to feldspar reacting to sodic pyroxene. Typical prograde reaction textures are shown in Figure 9 from amphibolite facies metapelites and in Figure 10 from eclogite facies micaschists.

Monazite formation in the amphibolite facies has been documented from many rock types and pressure regimes, in some cases reflecting a single growth stage (e.g., Kohn and Malloy 2004), in others showing polyphase growth (Rubatto et al. 2001; Pyle and Spear 2003; Foster et al. 2004; Kohn et al. 2005; Finger and Krenn 2007). Grains that formed early in a growth sequence tend to sequester Th and Y, zones added later thus are more depleted, but since precursors and reaction mechanisms can change, patterns are not always so simple. Unless other accessories intervene, REE- and Y-patterns of monazite often reflect strong partitioning with coexisting major assemblage minerals, notably garnet and feldspar, and monazite REE-patterns essentially reflect the modally abundant REE-competitors. In granulites, for instance, monazite that formed after garnet can become very depleted in HREE (Foster et al. 2000; Holder et al. 2015), a trend noted in apatite as well (Bea and Montero 1999).

Upon partial melting, monazite tends to dissolve and may or may not completely disappear; it typically grows during crystallization of leucosome in migmatites (Williams 2001; Pyle and Spear 2003; Kohn et al. 2005; Yakymchuk and Brown 2014).

Since allanite and monazite can form at several stages, and thus samples commonly contain several (over)growth zones, chronometry is of course tempting wherever these are of sufficient size for *in situ* analysis. Constraining the physical conditions of growth stages is thus a central issue, and in the absence of thermodynamic models for the REE-minerals involved, textural relations (e.g., core–rim) and local assemblages of coexisting minerals currently are our best guides. Detailed observation of inclusions and zoned relics of characteristic phases (e.g., garnet, mica, corroded accessories) may indicate a relative

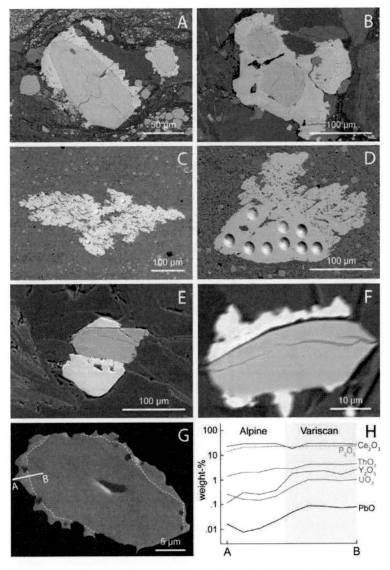

Figure 7. BSE images of low-temperature xenotime and monazite. (A, B) Authigenic/diagenetic xenotime overgrowths on detrital zircon grains and rounded detrital zircon completely engulfed in xenotime; SHRIMP dates of two growth rims yielded 1693 ± 4 and 1645 ± 3 Ma (Vallini et al. 2002, 2005). Samples are from the Mount Barren Group, Western Australia. (C, D) Greenschist-grade monazite (light) in a black shale with abundant pyrite euhedra in matrix; SHRIMP analysis (oval pits) yielded ages of 2416 ± 12 Ma for monazite, and 2430 ± 19 Ma for xenotime. Samples are from the Neoarchaean Roy Hill Shale Member, Jeerinah Group, Pilbara Craton, Western Australia (Rasmussen et al. 2005). (E) Detrital zircon with euhedral xenotime overgrowths, quartz muscovite schist, Palaeoproterozoic Mount Barren Group (Rasmussen et al. 2011). (F) Xenotime nucleating on zircon fragment in a metapelitic blueschist (350 °C, 1 GPa) from Rif, Morocco (Janots et al. 2006). Texturally similar outgrowths were reported by Rasmussen (2005, Fig. 8) from Carboniferous Lower Coal Measures (UK). (G,H) Variscan monazite core (~320 Ma) with thin Alpine rim (< 30 Ma) grown in metamarl at chloritoid grade (440 °C, 0.6 GPa). EPMA data across profile A–B show rim is low in Th, Y, U, and radiogenic Pb. Sample from Garvera, Central Swiss Alps. Data from (Janots et al. 2008). Photo credits: Birger Rasmussen (A–E) and Emilie Janots (F,G).

Table 2A. Proposed prograde reactions involving monazite, allanite, apatite, and xenotime.

Label	Reactants	Products	Reference
M0	flo ± xen	mnz + ...	Janots et al. (2006)
A1	flo + mnz + carb	aln + syn + ...	Janots et al. (2006)
A2	mnz + mus + carb + qz	aln + bio + ap	Wing et al. (2003)
A3	mnz + chl + pl + cc + qz	aln + ap + ...	Wing et al. (2003)
A4	mnz + chl + cc ± hem	aln + xen + ap + ctd	Janots et al. (2008)
A5	mnz + chl + an ± hem	aln + xen + ap + ctd	Janots et al. (2008)
A6	mnz + ...	aln + xen?	Smith and Barreiro (1990)
A7	mnz + pl + ...	aln + ep + tho + ap	Finger et al. (1998)
A8	mnz + chl + ...	aln + ctd	Radulescu et al. (2009)
A9	mnz + bt + ...	aln + F-ap + qtz + ...	(Yi and Cho 2009)
M1	aln + apa + mu + AS + qz	mnz + bio + pl + H_2O	Wing et al. (2003); Ferry (2000)
M2	aln + apa +	mnz + ...	Yang and Pattison (2006)
M3	aln + apa	mnz + pl + mag?	Tomkins and Pattison (2007)
M4	aln + apa + AFM1	mnz + an + AFM2	Janots et al. (2008)
M5	gar + apa + mu	mnz + sill + bio	Pyle and Spear (2003)
M6	gar + chl + mu	mnz + st + bio + pl	Kohn and Malloy (2004)

Table 2B. Pressure–temperature conditions estimated for allanite- and monazite-forming reactions

Label	Temperature	Pressure	Thermobarometry	Isograd	Rock type
M0	350–420 °C	1.0–1.2 GPa	ctd–chl thermometry		metapelite
A1	300–350 °C	0.3–1.0 GPa	ctd–chl thermometry		metapelite
A2/3	~400 °C	0.35 GPa	Ferry (1994)	bio–in	metapelite
A4/5	420–450 °C		TWQ	ctd–in	metapelite/marl
A6	500–550 °C	0.28–0.34 GPa	gar–bio	bio–in	metapelite
A7	550–600 °C	0.4–0.5 GPa	regional?		metagranitoid
A8	540–600 °C	2.0 GPa	phase diagrams		micaschist
A9	500–600 °C		phase relations	(vein)	leucogneiss
M1	~500 °C	0.35 GPa	Ferry (1994)	and/ky–in	metapelite
M2	480–520 °C	0.20–0.44 GPa	Helms and Labotka (1991)	~gar–in	metapelite
M3	530–580 °C	0.30 GPa	Pattison and Vogl (2005)	~cord–in	metapelite
M4	530–580 °C	0.8–0.9 GPa	TWQ		metapelite
M5	580–605 °C	0.3–0.4 GPa	gar–mnz thermometry		metapelite
M6	~600 °C	0.6–0.8 GPa		stau–in	metapelite

Minerals. AFM: Al-Fe-Mg-silicates (bio, mus, gar), aln: allanite, an: anorthite, and: andalusite, ap: apatite, AS: aluminosilicate, bio: biotite, carb: carbonate, cc: calcite, chl: chlorite, cord: cordierite, ctd: chloritoid, ep: epidote, flo: florencite, gar: garnet, hem: hematite, ky: kyanite, mag: magnetite, mnz: monazite, mu: muscovite, pl: plagioclase, qz: quartz, st: staurolite, syn: synchisite, tho: thorite.

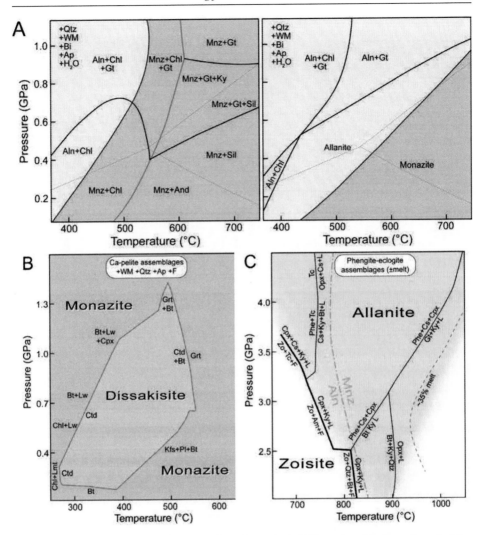

Figure 8. Models for the stability fields of allanite and monazite in *P–T* space based on thermodynamic data (A,B) and experimental phase equilibria (C). Mineral abbreviations: Aln = allanite, Am = amphibole, And = andalusite, Ap = apatite, Bt = biotite, Cpx = clinopyroxene, Chl = chlorite, Cs = coesite, Ctd = chloritoid, Dis: dissakisite, F: hydrous fluid, Gt = garnet, Kfs = K-feldspar, Ky = kyanite, L: melt (64–67 wt% SiO$_2$), Lmt = laumontite, Lw = lawsonite, Mnz: monazite, Phe = phengite, Qtz = quartz, Opx = orthopyroxene, Sil = sillimanite, Tc = talc, WM = white K-mica, Zo = zoisite. (A): Phase relations calculated for two compositions typical of clastic metasediments (both 16.57 wt% Al$_2$O$_3$; left: average pelite, 2.17 wt% CaO; right: slightly marly, 4.34 wt% CaO) using a model for Ce-allanite and Ce-monazite (Spear 2010). The predicted reaction sequence is consistent with those observed in several terrains above 500 °C, but no low-temperature monazite stability field (below 400–450 °C, Janots et al. 2008) is predicted. (B) Calculated medium-pressure stability field of dissakisite (La–Mg allanite—cliozoisite solution) (+ apatite) against various assemblages comprising La-monazite in calcic pelite model composition based on calorimetric data (Janots et al. (2007). (B) High-pressure field of Mg-allanite in phengite-eclogite model system (KNCMASH + trace elements), based on experimentally determined phase relations. Allanite first appears in experiments at temperatures just below the zoisite-out reaction. Disappearance of allanite at high temperatures is by gradual dissolution into melt, not linked to the major phase reactions; after Hermann (2002). Blue dot–dashed boundary is for a *pelitic* model system; this limit of allanite versus monazite was inferred by Hermann and Rubatto (2009) from experiments by Hermann and Spandler (2008).

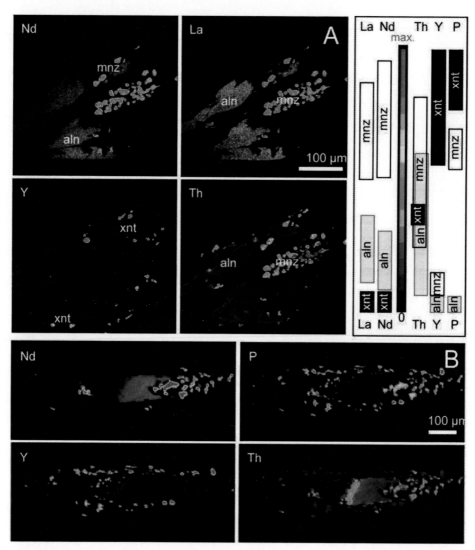

Figure 9. X-ray maps of REE-minerals at middle amphibolite facies, northern Lepontine belt (Central Alps, Switzerland). Textures indicate partial breakdown of allanite (+apatite) to monazite+xenotime. No apatite appears in the assemblage, and its consumption probably terminated the reaction (at ~560 °C, 0.8 GPa; Janots et al. 2008). Color bars (top right) indicate concentrations of elements in maps, with cold colors indicating low contents, red being maximum; typical ranges are shown for main REE-minerals. (A) La-rich allanite cores are preserved, and splay monazite aggregates interwoven with tiny xenotime are probably pseudomorphic after allanite. (B) Similar mineral textures, but preserved allanite cores are rich in Nd and Th (note heterogeneity). Both textures in same sample attest to very limited mobility of REE and actinides even during mineral reactions involving fluid. Heterogeneity in and among reaction domains poses problems in U–Th–Pb chronology (e.g., common Pb). BSE images courtesy of Daniele Regis.

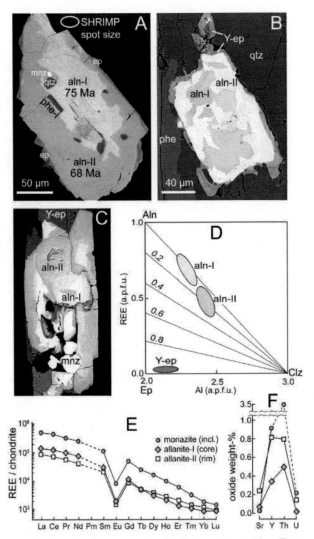

Figure 10. Allanite in eclogite facies phengite quartzite (Sesia Zone, Western Alps). Textures in BSE images show 3 HP-stages (I-III). (A) Corroded core of allanite-I (75 Ma) includes Phe-I (Si: 3.45 a.p.f.u.[1], 540 °C, 1.8 GPa), aln-II overgrowth (68 Ma) contains Phe-II (Si: 3.20 apfu, <520 °C, <1.5 GPa), and thin Y-epidote rim (60–65 Ma) is in equilibrium with matrix assemblage (550 °C, ~1.8 GPa); Ages are from SHRIMP Th/Pb isotopic analyses. Data indicate pressure-cycling (yoyo tectonics) in a subduction channel (Rubatto et al. 2011). (B, C) Samples from the same unit display similar complex resorption and replacement textures in allanite core, with Y-epidote rim. Relic monazite can be preserved inside allanite aggregates. (Regis et al. 2012). (D, E) Compositions of growth zones visible in allanite from (C). Successive generations show decrease in REE contents and increase in ferric fraction (Fe^{3+}/Fe_{total}: labels along radial lines in diagram) (after Petrík et al. 1995). Clz = clinozoisite, Ep = epidote, Aln = allanite. (E, F) Minor and trace element compositions in overgrowth sequence shown in (C). REE patterns and the sequestration of the critical minor elements Sr, Y, Th and U indicate that monazite was the main precursor of allanite. While allanite grew at eclogite facies conditions, in the absence of feldspar, REE patterns of aln-I and aln-II reflect LREE enriched monazite composition and preserve its Eu-anomaly. Note strong Th/U fractionation. All data from Regis et al. (2014). Chondrite normalization based on Sun and McDonough (1989). BSE images courtesy of Daniele Regis.

---

[1] atoms per formula unit $KAl_{2-x}Mg_x[Al_{1-x}Si_{3+x}O_{10}](OH)_2$

paragenetic sequence, and their geochemical characterization often allows estimates of *P–T* conditions (Regis et al. 2014; Mottram et al. 2015). However, textures within high-grade monazite (and less commonly allanite) tend to be complex, often patchy, and age relations are not always clear. Microstructural relations of allanite and monazite (and their inclusions) to metamorphic fabrics can be invaluable to relate *P–T–t* relations to deformation (Manzotti et al. 2012; Regis et al. 2012; Goswami-Banerjee and Robyr 2015; Wawrzenitz et al. 2015); considerable progress is being made relating monazite and allanite growth to deformation events (e.g., Ayers et al. 2002; Williams and Jercinovic 2002; Berman et al. 2005; Dahl et al. 2005; Simmat and Raith 2008; Kirkland et al. 2009; Cenki-Tok et al. 2011, 2014; Kelly et al. 2012; Wawrzenitz et al. 2012; Didier et al. 2013, 2014). Subsolidus retrograde growth of monazite (± xenotime) at the expense of allanite or apatite + garnet has been documented in several instances; it is usually related to localized or pervasive fluid-influx, and such monazite has proven datable (Bollinger and Janots 2006; Tobgay et al. 2012). From similar environments, but also in retrograde corona formation from high-grade rocks, monazite breakdown to allanite + apatite has commonly been reported (Broska et al. 2005; Krenn et al. 2012; Ondrejka et al. 2016) and occasionally dated (Yi and Cho 2009).

Xenotime. Rasmussen et al. (2011) studied a metaclastic sequence (Mount Barren Group) in SW Australia and found xenotime closely associated with detrital zircon, commonly as syntaxial outgrowths. Up to ~450 °C, detrital and diagenetic xenotime is partly preserved, but wholly metamorphic grains largely replace these at higher temperatures. Dissolution–reprecipitation reactions produced compositionally distinct rims forming at greenschist and amphibolite facies conditions. As *in situ* U–Pb chronology of xenotime yielded at least four discrete age groups (2.0 to 1.2 Ga), a discontinuous sequence of reactions is likely to be responsible for the growth of metamorphic xenotime, at the expense of detrital and diagenetic precursors. A few specific reactions have been formulated so far (Table 2). Metamorphic xenotime shows distinctly lower Th and U contents, in some instances Th-rich annuli, and tiny U- and Th-rich product grains (e.g., Figs. 3 and 8 in Rasmussen et al. 2011). Similar phase relations were reported by Allaz et al. (2013) and, in eclogite facies metaclastics, by Regis et al. (2012). The latter study showed zircon to be one of the reactant precursors and identified thorite and aeschynite among the satellite products. In more calcic domains of the same samples, monazite and xenotime broke down to produce allanite. Phase relations (Fig. 11) calculated for metapelites from a thermodynamic basis (Spear and Pyle 2010) that includes (preliminary) data for the REE-phosphates, show approximate stability limits of xenotime in the subsolidus range. To distinguish different xenotime generations, REE are useful: compared to magmatic xenotime, which is typically HREE-enriched, at subsolidus conditions enrichment increases from LREE to MREE, and patterns from MREE to HREE are either flat or decrease.

Monazite–apatite–xenotime textural relations. Mutual inclusions and overgrowth relations are often observed among REE-phosphates (and allanite) and can be of great utility in relating ages to (partial) replacement reactions and metasomatic effects, e.g., related to ore formation. For instance, inclusions of monazite and/or xenotime commonly occur in apatite, and using chemical fingerprinting (halogens, Sr, Pb), it is possible to track fluid–rock interaction (Harlov and Förster 2003; Harlov et al. 2005; Broska et al. 2014; Shazia et al. 2015; Jonsson et al. 2016).

Effects of hydrous fluids

Apatite, monazite, and xenotime are commonly found in fluid-rich environments, both at low and high temperature (Harlov et al. 2008; Hetherington and Harlov 2008). The solubility of phosphate minerals in fluids is strongly composition-dependent, and substantial evidence from experimental studies (Harlov et al. 2005, 2007, 2011; Budzyń et al. 2011; Seydoux-Guillaume et al. 2012; Grand'Homme et al. 2016b) indicates that dissolution-reprecipitation processes are particularly effective when alkaline hydrous fluids are involved; in many cases F-apatite is a characteristic product phase, and xenotime as well as monazite are involved.

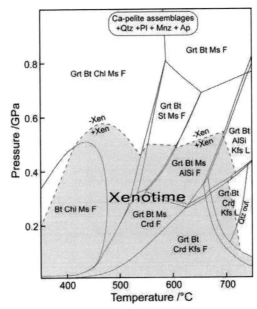

Figure 11. Phase relations calculated for a model pelite composition in $K_2O–Na_2O–CaO–MnO–MgO–FeO–Al_2O_3–SiO_2–H_2O–Y_2O_3–Ce_2O_3–P_2O_5$–Fluid from Spear and Pyle (2010); the model is substantially founded on natural assemblages from a specific locality in Maine, includes data for pure xenotime and binary Ce-Y-monazite, but does not consider allanite. Stability limits for xenotime depend on bulk Y-contents; a realistic field is shown. Mineral abbreviations as in Figure 8, plus AlSi = kyanite or sillimanite, Crd = cordierite, St = staurolite.

For chronologic efforts, partial replacement by such processes can be a challenge (Williams et al. 2011), but interaction with pervasive fluid may be a most effective way to completely reset clocks in phosphate minerals (Harlov and Hetherington 2010; Harlov et al. 2011). These studies have spawned efforts to date hydrothermal systems (Williams et al. 2011; Shazia et al. 2015), and indeed several studies have thus investigated Alpine clefts (Janots et al. 2012; Grand'Homme et al. 2016a) and altered shear zones (Berger et al. 2013).

Particular geochemical characteristics of REE-minerals, e.g., the tetrad effect (Bau 1996) have long been attributed to effects of fluids, but Duc-Tin and Keppler (2015) showed convincingly that this effect can be explained by the orbital structure of particular LREE being particularly favorable for the irregular REE-coordination polyhedra in monazite and xenotime (Fig. 1). Their REE partitioning behavior—relative to melt, hydrous fluids or minerals with regular REE polyhedra—is thus different than expected solely from ionic radius considerations (Fig. 5). As a consequence, and contrary to previous work (Bau 1996), tetrad patterns may not directly indicate the effect of fluids but of crystal-structure dependent fractionation.

CHRONOLOGIC SYSTEMS

Owing to the elevated actinide contents of the REE-minerals, chronometry has been based on the $^{208}Pb/^{232}Th$, $^{206}Pb/^{238}U$ and $^{207}Pb/^{235}U$ (or $^{207}Pb/^{206}Pb$) ratios far more often than on alternative isotopic systems. As discussed above, the closure temperatures for these systems are not precisely known, but are certainly high enough for chronometry in many applications at subsolidus environments and, with caution, into the partial melting range. For apatite, U–Pb closure is estimated at lower temperature (350–550 °C; Cochrane et al. 2014; Chew and Spikings

2015), suitable only for dating crystallization conditions of upper crustal and blueschist facies samples. Lu–Hf (T_c 675–750 °C) is an alternative for higher temperatures, but has been rarely applied so far (Barfod et al. 2003; Larsson and Söderlund 2005). And Sm–Nd dating has been applied to apatite based on intersectoral differences in partitioning (Rakovan et al. 1997).

Spatial resolution *versus* age resolution

Compared to ID-TIMS analysis, all *in situ* techniques have inferior age resolution and precision, but many studies focus on answering questions that demand high spatial resolution. For many samples and problems, finding grains of sufficient homogeneity and size for TIMS dating ranks somewhere between challenging and impossible. In other cases, the structural context can be preserved at least in part, with effort, by microdrilling (Kohn et al. 2005; Smye et al. 2011) or directly cutting monazite or allanite from a polished section. In this way, monazite has been successfully isolated and then analyzed with optimal precision using ID-TIMS (Hawkins and Bowring 1997; Baldwin et al. 2006). Allanite can also be separated into several age groups or magmatic generations based on the color and density of mineral fragments; precise ID-TIMS analysis of these then give a range of ages (Oberli et al. 2004). These indicate the timescale of plutonic evolution, though the specific significance of each age is not clear because no textural link to geochemical tracers is possible. So, precision is excellent in ID-TIMS studies, but context is sacrificed, and thus petrogenetically critical information for each fraction analyzed is missing.

Alternatives include a range of micro-analytical methods, some of which allow *in situ* dating of growth zones in polished sections. EPMA age mapping of monazite is a powerful way to resolve first-order geological problems, especially in polymetamorphic rocks with a complex evolution (Williams et al. 1999; Hermann 2002; Buick et al. 2010; Dumond et al. 2015; Johnson et al. 2015; Tucker et al. 2015) and in rocks that experienced extensive interaction with fluid (Shazia et al. 2015; Grand'Homme et al. 2016a; Kirkland et al. 2016). Allanite maps have similar potential, especially in high-pressure rocks and migmatites (Gregory et al. 2012; Regis et al. 2014), but have been much less utilized so far, since EPMA dating is not possible, and SIMS or laser-based analysis has lower spatial resolution and demands considerably more analytical effort.

U–Th–Pb

Monazite and xenotime. Monazite often contains several wt% Th, several thousand ppm U, and usually negligible initial lead (Montel et al. 1996). These high actinide concentrations result in high levels of radiogenic lead, except in very young samples. In xenotime Th and U contents are lower, but both monazite and xenotime are proven U–Th–Pb chronometers. Monazite and allanite normally contain both Th and U in concentrations sufficient to combine their decay systems for chronometry; for xenotime only U–Pb is possible. In both phosphates, REE contents are also high (e.g., 10^4–10^5 ppm Sm and Nd), and as tracer isotopes they add important geochemical constraints, especially combined with chronometry (Nemchin et al. 2013; Whitehouse et al. 2014).

The commonly high Th- and U-contents allow chemical U–Th–Pb dating of monazite and xenotime by EPMA (Suzuki and Adachi 1991; Montel et al. 1996; Williams and Jercinovic 2002; Pyle et al. 2005b; Williams et al. 2007; Hetherington et al. 2008; Spear et al. 2009). The power of the approach, its limits, technical aspects, and some applications are not reviewed here, as they are specifically addressed in this volume (Williams and Jercinovic 2017); similarly, isotopic dating by LA-ICP-MS and SIMS are covered separately (Kylander-Clark 2017; Schmitt and Vazquez 2017). Two developments are particularly suited to link ages with their geochemical and textural context: LASS ICP-MS (laser ablation split stream: Kylander-Clark et al. 2013; Goudie et al. 2014) and Scanning Ion Imaging using SIMS (Harrison et al. 2002).

In situ isotopic analysis by LA-ICP-MS or ion probe / SHRIMP involves particular concerns for each instrument, notably regarding standardization, fractionation, matrix correction, and realistic error propagation. The magnitude of the effects varies for each type of mineral; in a general way, monazite and xenotime pose fewer problems in isotopic analysis than does allanite, but successful analytical protocols for all of them have been required. In particular, the question of matrix-matched[8] standardization has been repeatedly raised (Kohn and Vervoort 2008), especially for allanite (Catlos et al. 2000), because of its large chemical variability. Since matrix-matching is more critical in ion probe analysis than in laser ablation analysis (Gregory et al. 2007; Darling et al. 2012), LA-ICP-MS may be the preferred method, given the paucity of suitable allanite standards (Smye et al. 2014). Fortunately, it now appears that the concerns over matrix-matching and standards in allanite analysis can be largely avoidable with carefully optimized LA-ICP-MS conditions (McFarlane 2016). However, matrix-matching remains essential in ion probe analysis (e.g., Fletcher et al. 2004).

Apatite

Apatite contains several thousand ppm REE and has typically high Lu/Hf ratios, making Lu–Hf dating possible (e.g., Barfod et al. 2005), with high closure temperatures similar to U–Pb in monazite (e.g., McFarlane and McCulloch 2007). Apatite also has relatively high U-contents, most useful in U–Pb thermochronometry (Chamberlain and Bowring 2001; Chew et al. 2011; Chew and Donelick 2012), often in conjunction with (U–Th)/He and/or fission track dating. These applications are beyond the scope of this review; for a short introduction consult Cochrane et al. (2014). Since metamorphic samples often contain more than one generation of apatite, U–Pb or Th–Pb isotopic analysis should be done by *in situ* methods allowing sufficient spatial resolution. Ion probe methods (Sano et al. 1999) or, more commonly, LA-ICP-MS (Willigers et al. 2002; Chew et al. 2011) has been used and is sufficiently precise to date low-U apatites, such as from mafic dykes (Pochon et al. 2016). Sr and Nd isotopes can be analyzed *in situ*, offering new possibilities of petrogenetic utility (Hammerli et al. 2014)

Allanite

Allanite usually contains 1,000–16,000 ppm Th and typically about a tenth that amount in U. Th/U > 2 is common, but shows substantial variation and may give genetic indications. Allanite has shown its potential as a robust Th/U–Pb chronometer, but several challenges remain, especially its high initial lead contents. Unlike monazite and xenotime, which commonly contain nearly 100% radiogenic lead, allanite (and apatite) have an affinity for Pb, and isotopic data used for dating must be corrected for this component. This is straightforward for ID-TIMS data using ^{204}Pb where mass interferences are eliminated. In SHRIMP analysis, energy filtering effectively reduces isobaric interference (Rubatto et al. 2001) on mass 204, but spot analyses obtained by LA-ICP-MS combine ^{204}Pb and ^{204}Hg. Several methods are in use for estimating the composition of the initial lead component (Pb_{ini}), which includes inherited radiogenic and non-radiogenic lead; some are based on models, others on analysis of the sample being dated. As petrochronology typically aims to resolve local growth stages, and relics are the rule, not the exception, overall isotopic equilibration cannot be assumed, and methods relying on a single type of common lead (Wendt 1984) are not applicable in general[9]. If the fraction of non-radiogenic Pb is high—and in allanite (Gabudianu Radulescu et al. 2009a; Gregory et al. 2012) and apatite (Cochrane et al. 2014) it often is in the range 20–70%, but may be > 90%—the common lead correction affects ages and their uncertainty very substantially. In addition, minerals with high Th/U contain excess ^{206}Pb produced from excess ^{230}Th (an intermediate daughter nuclide in the ^{238}U chain) incorporated during growth, for which a correction is needed (Schärer 1984; Barth et al. 1994). These complications enlarge age uncertainties and make it difficult to check for possible Pb-loss, which may occur at temperatures above ~750–800 °C (Kamber et al. 1998).

---

[8] Both structurally and chemically

[9] This warning does not apply where the analytical uncertainty in the common Pb measurement exceeds the effect of potential variations in its composition.

Initial lead. A complete discussion of the issues and controversies surrounding common Pb is beyond the scope of this review, but data presented in the literature are commonly based on premises that have since been questioned. As these can affect ages substantially, one should understand the basic problem involved in estimating the composition (isotopic ratios) of the initial lead, know the approaches used to solve them, and consider at least some of the perils and implications. This is true for all chronometers containing high levels of non-radiogenic lead, notably Ca-rich minerals, where Pb can replace Ca. The specific situation for allanite is presented in Gregory et al. (2007, 2012). In essence, for both the Th–Pb and the U–Pb system, two approaches are used: A single-spot correction or an isochron method. The single-spot correction assumes a model Pb composition that does not change during allanite growth, and whose isotopic ratio is predicted by a global lead evolution model (Cumming and Richards 1975; Stacey and Kramers 1975). Even though local dissolution of accessory precursors are a likely staple on which allanite fed (Romer and Rötzler 2011), single spot correction has been widely used to treat allanite dates. An alternative practice is to determine Pb_{ini} composition from internal heterogeneity within an allanite crystal. This approach assumes a single growth stage with a common age but variable chemistry. Allanite commonly shows variable U/Th and common lead, so the isochron correction usually uses a Tera-Wasserburg diagram approach (plotting $^{207}Pb/^{206}Pb$ vs. $^{238}U/^{206}Pb$). Data showing variably non-radiogenic Pb (Fig. 12) are used to obtain the age (2007; Gregory et al. 2012). The common practice to determine common lead composition by analyzing both U-rich and coexisting U-poor minerals (e.g., feldspar) assumes isotopic equilibration in the same compositional domain as the mineral chronometer; within the context of an assumed single age, the isochron method allows this assumption to be tested. As pointed out by Burn (2016), it is advisable to combine these correction methods and use them to check the quality of data by first processing them in an uncorrected Tera-Wasserburg and a $^{206}Pb_c$ normalized isochron diagram (Fig. 12). Provided a dataset from different parts of a sample define a linear array in a Tera-Wasserburg diagram, the intercept with the y-axis defines the $^{207}Pb/^{206}Pb$ initial lead composition. Each diagram returns an estimate of the age and the initial lead composition of allanite (or any other mineral with high common lead contents, such as apatite; Chew et al. 2011). The concordance of these two age estimates is used to evaluate the quality of the data and estimate the age uncertainty.

During growth, both allanite and apatite may incorporate extraneous Pb; if growth is from a melt, isotopic equilibrium is often assumed, perhaps unjustly, as indicated by pre-melting relics in leucosomes from migmatites (Berger et al. 2009). At sub-solidus condition, metamorphic growth may occur in local domains or from isotopically inhomogeneous precursors (Lanari and Engi 2017), and local lead isotope ratios may be quite different from those of the bulk rock, let alone of a terrestrial evolution model (Stacey and Kramers 1975). Inheritance is of major concern in polymetamorphic basement rocks (Romer and Siegesmund 2003; Romer and Xiao 2005; Romer and Rötzler 2011), but since many metamorphic rocks contain relic accessory minerals, local variability should always be tested because local digestion of such relics may cause major deviations from bulk rock U/Pb, Th/Pb and $^{207}Pb/^{206}Pb$ (e.g., Wawrzenitz et al. 2015). Inheritance of this kind is evident in datasets as excessive scatter in the U/Pb or Th/Pb ages, i.e., scatter beyond the purely analytical uncertainty (~0.1% for ID-TIMS, 2–3% for LA-ICP-MS or ion probe data). Major discrepancies between measured and model lead compositions have repeatedly been found in metamorphic allanite; e.g., Gabudianu Radulescu et al. (2009b) obtained a concordant U–Pb age 37.4 ± 0.9 Ma ($^{208}Pb/^{232}Th$: 38.3 ± 1.0 Ma) for allanite from micaschist (eclogite facies), in which the regressed $^{207}Pb/^{206}Pb$ was 0.498 ± 0.005. According to Pb evolution models $^{207}Pb/^{206}Pb$ would be 0.838 and imply apparent ages >200 Ma!

As emphasized by Janots and Rubatto (2014), for young samples (with low amounts of radiogenic lead) containing allanite with high Th/U and high initial Pb fraction $^{206}Pb_c$-normalized Th–Pb isochron dating will yield the most reliable age and also provides an independent control on the isotopic composition of initial Pb.

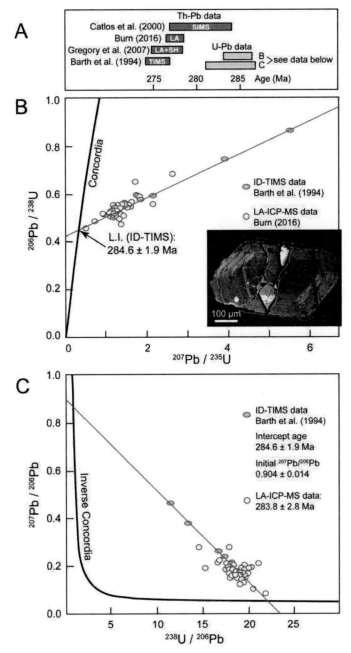

Figure 12. Allanite ages for CAP, a frequently used standard reference material from Permian granodiorite (Cima d'Asta Pluton, NE Italy). 6 datasets obtained by various techniques (ID-TIMS, SHRIMP, SIMS, LA-ICP-MS) are compared here. (A) As all datasets show, Th–Pb ages (weighted mean: 275.5 ± 1.4 Ma) are lower than U–Pb ages, indicating that ^{206}Pb is partly thorogenic (from ^{230}Th); LA + SH for LA-ICP-MS + SHRIMP. (B, C) Concordia and Tera-Wasserburg (inverse Concordia) plots. TIMS and LA-ICP-MS data show agreement within error. Symbols of ID-TIMS data are larger than analytical uncertainty; the reverse is true for LA-ICP-MS data, and gives rise to apparent excess scatter. Common lead contents are 5–46%. Inset in (B) shows BSE image of CAP allanite illustrating oscillatory zoning typical of magmatic growth. Compare with Figure 6.

CASE STUDIES

Selection, purpose

With the wealth of studies—hundreds on monazite alone—it is quite impossible to do justice to so many and diverse results. Any selection of exemplary studies is necessarily subjective—for me, looking for balance seemed futile. Rather my goal is to focus on the understanding gained by combining petrogenetic analysis and chronology based on REE-minerals. Examples were chosen, in addition to those mentioned in the first part of this review, because they either helped resolve a major scientific question—some at mineral grain scale, some at tectonic or geodynamic scales—or they shed new light into one of the classic terrains. Some studies used particularly novel thinking, others high-tech approaches or smart combinations of methods.

Single REE mineral species

Isolated inclusions. Dating isolated inclusions in porphyroblasts can be attractive for several reasons: The host may have protected inclusions from high-strain or high-temperature stages, thus preserving a trustworthy upper age limit to the host; If the host is petrogenetically useful, such as garnet, staurolite, or cordierite, inclusion ages may also be linked to *P–T* conditions. If the same phase can be dated in other structural contexts, not just from the same geological unit, but in the same sample, the age difference may give an indication of duration or rates, or it may be used to test currently held beliefs about closure. This approach has been applied to xenotime by Kamber et al. (1998) and to monazite by Montel et al. (2000). In addition, chemical gradients between inclusions and host mineral may lend themselves to diffusion chronology (Spear et al. 2012; Kohn and Penniston-Dorland 2017), especially if diffusion profiles for more than one element of unequal diffusion rates are analyzed.

Kamber et al. (1998) dated zircon, monazite, and xenotime in late Archean granulites from the Limpopo Belt (Zimbabwe) by SIMS (Fig.13). Xenotime was found in two structural contexts—as inclusions (together with zircon and monazite) in fresh cordierite, and interstitially in the matrix. Interstitial xenotime yielded $^{207/206}$Pb ages ~60 Ma younger $(2551 \pm 5\,\text{Ma})$ than xenotime inclusions $(2612 \pm 8\,\text{Ma})$, which were close to ages of monazite inclusions $(2600 \pm 5\,\text{Ma})$ and interstitial matrix monazite $(2602 \pm 17\,\text{Ma})$. Cordierite had formed by biotite dehydration-melting at ~850 °C, 0.7 GPa and included the assemblage of accessory minerals: Tiny zircon and large (150–350 μm), partly corroded monazite and xenotime were unzoned but strongly enriched in LREE and HREE, respectively. During subsequent cooling, trapped inclusions were armored from hydrothermal alteration that produced interstitial xenotime $(2551 \pm 5\,\text{Ma})$, skarns $(2544 \pm 39\,\text{Ma})$, as well as tiny zircon $(2554 \pm 27\,\text{Ma})$. Since no zoning could be detected in the large xenotime inclusion, Kamber et al. proposed that the closure temperature for xenotime had to be > 800 °C, which is 100–150° higher than previously believed. However, it would seem that the survival of older xenotime and monazite inside cordierite mere shows that retention of Pb in the phosphates was surprisingly high (probably due to low solubility or low diffusivity of Pb in cordierite).

Montel et al. (2000) used an analogous approach to constrain possible closure temperature for monazite based on age systematics in kinzigites from Beni Bousera (Morocco). Only monazite grains (< 20 μm size) included in garnet could be dated $(284 \pm 27\,\text{Ma})$ by EMPA, whereas larger interstitial grains (20–70 μm across) had Pb contents below the EMPA detection limit. Incomplete chemical and mechanical "shielding" was proposed for monazite inclusions near cracks in garnet, which show ages some 150 Ma less than those fully enclosed in garnet. Because well shielded monazite inclusions retained the age, and temperatures >850 °C were determined for the hottest samples investigated, Montel et al. (2000) proposed that the closure temperature for monazite might be higher than previously accepted. However, as in the previous study by Kamber et al. (1998), the observation that monazite shows good Pb-retention when the phosphate is included in a refractory host such as garnet, does not indicate by itself a high closure temperature for monazite. Such observations can be used, however, to interpret the age of the inclusion as a minimum formation age.

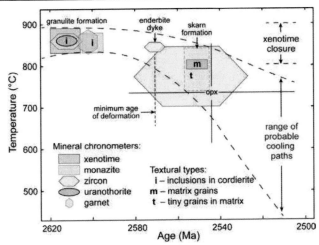

Figure 13. Mineral chronometry in the Limpopo (Zimbabwe) granulite belt based on SIMS analysis of inclusions and matrix minerals (Kamber et al. 1998). The robustness to resetting of several mineral chronometers can be assessed in response to thermal, deformational, and chemical triggers. Such HT-*t*errains cool slowly, but constraining their *T–t* paths is challenging, more because of uncertainty in thermometry (e.g., Blenkinsop and Frei 1997) than in age. Direct mineral thermometry (e.g., Y-in-monazite) would help, provided the same intragrain domains are dated.

Diversity of monazite

Following some initial studies (e.g., Suzuki et al. 1991), provenance analysis in clastic sedimentary basins has started to focus more on monazite, in addition to zircon (Hietpas et al. 2010; Moecher et al. 2011). Both minerals are resilient to transport and weathering, thus common in heavy mineral fractions; both commonly contain identifiable growth zones that are likely to retain formation ages. But zircon is extremely refractory, whereas monazite is more responsive to low- and medium-grade metamorphism and fluid stimulus, hence can preserve a more complete record of tectonic activity in hinterland source regions.

Hietpas et al. (2011) compared zircon and monazite age spectra from sandstones in the Appalachian foreland basin and found monazite to retain $^{232}Th–^{208}Pb$ age spectra that reflect all of the major Paleozoic orogenic phases known from the Appalachian orogen. By constrast, 90% of the zircon recorded none of the three major tectonic pulses, but instead retained Meso-Proterozoic or older ages (Fig. 14). The youngest detrital monazites analyzed by Hietpas et al. indicate a maximum sedimentation age some 550 Ma younger than detrital zircons. Clearly, the two mineral systems can provide complementary information in provenance studies, and arguably monazite should take on a larger role than zircon in constraining depositional ages.

Inheritance is common for zircon and monazite, and sometimes for xenotime, so textural criteria must be used to identify overgrowths (Reimink et al. 2016). To avoid meaningless mixed dates and convolution in age spectra, spatially separated ages should be acquired. EPMA analysis may be suitable for analysis of growth zones in monazite and xenotime, but SIMS or LA-ICP-MS is required for zircon.

Apatite: testing old ideas in new light

Many petrochronological studies assume that recorded growth zoning is essentially preserved, i.e., that recrystallisation did not occur and that diffusive alteration of chemistry or ages during the subsequent *T–t* evolution was negligible. But at high metamorphic grades and in the magmatic temperature range, initial concentration gradients can be diffusionally broadened, especially in apatite. While diffusion may modify or destroy the prograde record, it opens the possibility to investigate thermal timescales by studying diffusion profiles for suitable elements and minerals (e.g., Spear et al. 2012; Kohn and Penniston-Dorland 2017).

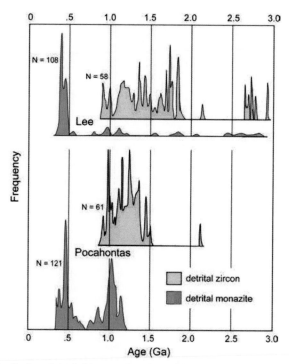

Figure 14. Age distribution of detrital heavy minerals in Middle Carboniferous to Permian sandstones from two basins (Lee, Pocahontas) in the Appalachian foreland, central eastern USA. BSE images were used to avoid overgrowth zones visible in some monazites from Pacahontas samples. Six basins were analyzed, and illustrative data for only two of them are shown here. Given the different preservation characteristics of monazite and zircon, provenance analysis is more representative if these minerals are used in combination: Whereas zircon (U/Pb, LA-ICP-MS) shows only Grenvillian (~1250–950 Ma) and older ages, monazite (Th/Pb, SIMS) retains predominantly Mesoproterozoic and Paleozoic ages, including Taconian (470–440 Ma) and Acadian (420–380 Ma) relics. Data from (Hietpas et al. 2011),

This idea was developed for Sr in apatite by Ague and Baxter (2007), who used forward modeling to constrain the duration of metamorphism in the Dalradian of Scotland, a classic metamorphic sequence (Barrow 1912). Ague and Baxter analyzed Sr-concentrations across chemically inhomogeneous apatite grains and compared these data to profiles modeled by solving the 1-D diffusion equation

$$\text{REE}(\text{PO}_4)_{\text{oph}} \rightarrow (1-y)\Big[\text{REE}(\text{PO}_4)^0\Big]_{\text{melt}} + y\Big[\text{REE}^{3+} + (\text{PO}_4)^{3-}\Big]_{\text{melt}} \qquad (11)$$

where c is concentration, t is time, D is diffusivity, and x is distance. A step function was assumed for the initial C_{Sr} profile. Then, using the temperature-dependent diffusion coefficient D_{Sr} (Cherniak and Ryerson 1993) and the maximum temperature documented (±30°) for each sample, the time required to produce the observed profile was determined. This turned out to be surprisingly short, only ~250 ka (Fig. 15). Analysis of the Sr-profiles in apatite was corroborated by analogous data for Fe, Mg, Ca, and Mn profiles in garnet, and time scales again were similarly short. The cycle of regional Barrovian metamorphism is thought to have lasted at least two orders of magnitude longer, and indeed previous studies in the Dalradian had proposed considerably longer time scales. Ague and Baxter (2007) instead explain this brief thermal pulse by advective heating due to magma or aqueous fluids, immediately followed by rapid exhumation.

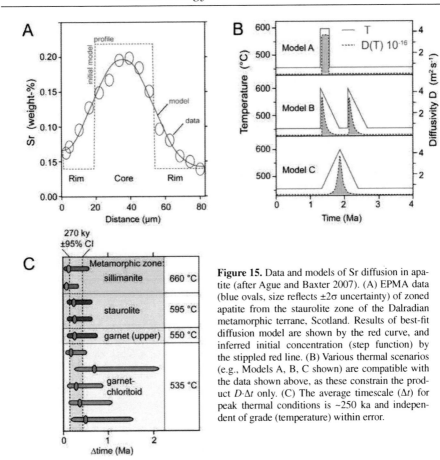

Figure 15. Data and models of Sr diffusion in apatite (after Ague and Baxter 2007). (A) EPMA data (blue ovals, size reflects ±2σ uncertainty) of zoned apatite from the staurolite zone of the Dalradian metamorphic terrane, Scotland. Results of best-fit diffusion model are shown by the red curve, and inferred initial concentration (step function) by the stippled red line. (B) Various thermal scenarios (e.g., Models A, B, C shown) are compatible with the data shown above, as these constrain the product $D·Δt$ only. (C) The average timescale ($Δt$) for peak thermal conditions is ~250 ka and independent of grade (temperature) within error.

Modeling of Pb-diffusion in 3.15 Ga old apatite (and titanite) was used by Schoene and Bowring (2007) to obtain T–t paths for the Kapvaal craton (South Africa). ID-TIMS $^{207}Pb/^{206}Pb$ ages of single grains were lower for smaller grains; numerical modeling indicates a compatible T–t evolution for all of them. In the absence of mineral chemical data, e.g., on the U-distribution within grains, it is impossible to test some of the assumptions made. For this reason, these data are not well suited to inform current debates regarding realistic diffusion coefficients for apatite and titanite (Kohn 2017; Kohn and Penniston-Dorland 2017). For example, larger grains might have begun growing earlier than smaller grains, or contain earlier-formed cores, leading to age dispersion independent of diffusion effects.

Monadic monazite

Polyphase growth (of wisdom) in New England. A metamorphic suite in the Chesham Pond Nappe (SW New Hampshire, USA) yields monazite–xenotime temperatures between 470–740 °C (Pyle and Spear 2003). As is typical in sample suites with multiple monazite generations, a perfect separation of stages is not possible; scatter and some overlap in the Y-contents $(X_{YPO_4})_{mnz}$ occur within a single sample and certainly among samples, even if closely spaced (Fig. 16). Combined uncertainties (from calibration and analytical scatter in populations) are given as ±50° (2σ). Pyle et al. (2005a) added chemical and SIMS isotopic ages for four distinct monazite generations: an early detrital one plus three produced by metamorphic reactions. Prograde evolution covers an age range of ~20 Ma.

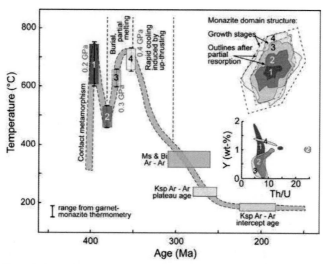

Figure 16. *P–T–t* evolution in the Chesham Pond nappe (New Hampshire, northeastern USA) based on monazite petrochronology. Pyle et al. (2005a) distinguished four monazite domains based on textural and chemical criteria. Overgrowth geometry, partial resorption, and Th/U vs. Y in monazite shown in insets. SIMS [207]Pb/[206]Pb ages (2 samples) and EPMA chemical U–Th–Pb ages (3 samples) for each domain were linked to *P–T* conditions using monazite–garnet thermometry (Pyle and Spear 2003) and metapelite phase equilibria. An early sharp thermal spike reflects contact metamorphism. Subsequent tectonically driven *P–T* evolution involved a pressure increase and fast heating (14.4±4.0 °C/Ma) towards the thermal peak (domains 2–3), growth of monazite from partial melt upon initial cooling from the thermal peak (domain 4), and rapid cooling to ~350 °C to reach Ar closure in muscovite and biotite (at -8.5±1.6 °C/Ma). All uncertainties and closure temperatures for [40]Ar/[39]Ar ages from Pyle et al. (2005a).

Such monazite ages are difficult to use for direct correlations amongst tectonic units, let alone an entire orogenic belt. Detailed analysis of individual samples with tight links to the tectono-metamorphic evolution for each of them is more promising.

Canadian Cordillera. Foster et al. (2004) documented several *P–T–t* points from a single sample containing sufficiently large monazite growth zones. For each such zone, *in situ* LA-ICP-MS ages are tied to *P–T* data (from Y in monazite thermometry and/or phase equilibrium modeling of the associated assemblage; Foster et al. 2002). X-ray maps are useful to guide trace element analysis, and relating the Y-contents in garnet and monazite is of particular importance to understand the metamorphic evolution (Pyle and Spear 1999). In two samples from the Monashee complex (Canadian Cordillera), 3–4 monazite growth zones were identified and, using textural criteria, their temporal relations to poikiloblastic kyanite and garnet were carefully established (Foster et al. 2002). U–Pb ages and thermobarometry for each monazite zone (Fig. 17) could thus be tightly linked to the overall petrogenetic evolution deduced from detailed petrography and phase diagram modeling. The resulting *P–T–t* paths show 2–3 phases of prograde monazite growth over a period of 20 Ma; the data indicate an average heating rate of 2.4±1.2 °/Ma and constrain the onset of rapid exhumation.

This study emphasized the integrated approach used. Though based on only two strategically chosen samples from the Omineca Belt, the results quantified a temporal framework of the tectonic evolution in this collisional orogen.

Accretion. Based on detailed and systematic mapping aimed to decipher the tectonic accretion of the Yukon-Tanana Terrane to western North America, Berman et al. (2007) used petrochronology on six samples in the Stewart River geology. Structurally well defined metamorphic samples from this complex, polydeformed terrane were studied to link

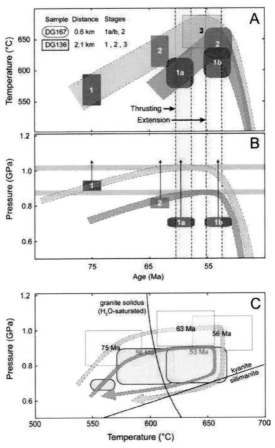

Figure 17. Petrochronological data for two samples studied by Foster et al. (2004) from the Canadian Cordillera. Samples were collected from 600 m (DG167) and 2.1 km (DG167) below the Monashee décollement, which separates two amphibolite facies continental units. Up to four generations of monazite were distinguished, 2–3 could be dated by LA-ICP-MS U–Pb analysis in each rock. Garnet–monazite, xenotime–matrix monazite, and garnet–biotite thermometry was combined with pseudosection analysis to infer *P–T* paths. Although temperature is well constrained, pressure is not. (A, B) *T–t* and *P–T* data and trends; vertical bands indicate independent constraints for the minimum age of extension and the end of thrusting, respectively. (C) *P–T* paths are schematic, and uncertainties are larger than shown. Asymmetric uncertainties in pressure indicated by position of red age labels.

in situ SHRIMP U–Pb monazite ages and a few TIMS U–Pb titanite ages to changing *P–T* conditions. Robust metamorphic constraints permit a temporal framework to be assigned to the multiple tectonic episodes (Fig. 18): Late Devonian arc formation (365–357 Ma), Mississippian magmatism and deformation—first above the subducting east-vergent proto-Pacific slab (355–340 Ma), then above a W-vergent subduction system during the Mid-Permian (260–253 Ma) involving eclogite formation and arc magmatism—followed by contact metamorphism from supra-subduction plutons (200–190 Ma), and accretion by thrusting onto North America (185 Ma). Metamorphic constraints were derived by combining forward and inverse thermodynamic models, and $^{206}Pb/^{238}U$ chronometry was used for ages. This study again demonstrates that detailed analysis of a select few samples can greatly enhance the understanding of the tectonic implications, provided the necessary groundwork was previously laid to select appropriate samples for petrochronology.

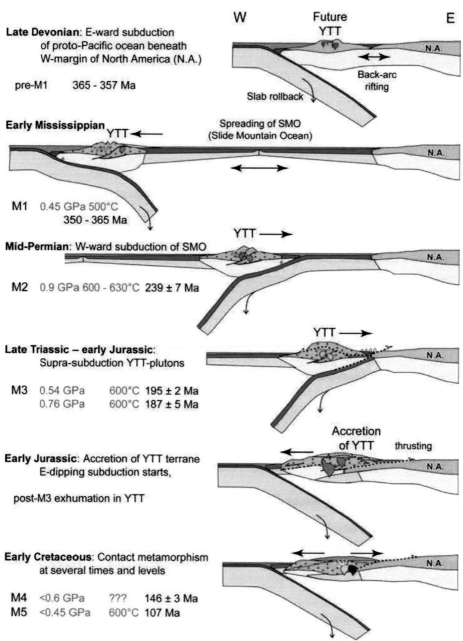

W   Future   **E**
YTT

Late Devonian: E-ward subduction
of proto-Pacific ocean beneath
W-margin of North America (N.A.)

pre-M1   365 - 357 Ma

Back-arc
rifting

Slab rollback

Early Mississippian   Spreading of SMO
YTT ←   (Slide Mountain Ocean)

M1   0.45 GPa 500°C
350 - 365 Ma

Mid-Permian: W-ward subduction of SMO   YTT →

M2   0.9 GPa 600 - 630°C 239 ± 7 Ma

YTT →

Late Triassic – early Jurassic:
Supra-subduction YTT-plutons

M3   0.54 GPa   600°C 195 ± 2 Ma
0.76 GPa   600°C 187 ± 5 Ma

Accretion
of YTT   thrusting

Early Jurassic: Accretion of YTT terrane
E-dipping subduction starts,

post-M3 exhumation in YTT

Early Cretaceous: Contact metamorphism
at several times and levels

M4   <0.6 GPa   ???   146 ± 3 Ma
M5   <0.45 GPa   600°C 107 Ma

Figure 18. Schematic cross sections (from Devonian to Cretaceous) depicting the evolution of the Yukon-Tanana Terrane (YTT) analyzed by Berman et al. (2007) in the Stewart River area, Yukon, Canada. The polydeformational and polymetamorphic evolution of the YTT spans 250 Ma, as unraveled by structural work, TWQ-thermobarometry, U–Pb dating of titanite by TIMS, and U–Pb dating of chemically distinct monazite domains by SHRIMP. Five metamorphic stages (M1–M5) found by Berman et al. in YTT samples are listed with *P–T* conditions and ages. This study demonstrated that petrochronology with detailed analysis of structurally well controlled samples can uniquely illuminate continental margin tectonics.

Himalayan tectonics. The capacity of monazite to record several sequential stages of growth and retain formation ages up to high temperature has been put to good use in many studies. Some of these have documented quite complex *P–T–D–t*-paths and helped elucidate large scale tectonics. This is notably the case in the Himalaya, where petrochronology has been combined with detailed structural studies to allow current models of collisional orogeny to be critically tested (Kohn 2014a). Specific temperature–time points—commonly from monazite—and their corresponding paths are essential, but their value is greatly enhanced if it is possible to link them to the polyphase structural evolution.

In central Nepal, four main thrust sheets were studied in detail in two main sections (Kohn et al. 2004; Corrie and Kohn 2011). The approach developed was first to identify monazite grains in thin section and produce EPMA maps of chemical zoning; select grains were then drilled out, mounted, polished, and analyzed by SIMS. X-ray maps (Th, U, Y, Si) were used to preselect spots for dating. In zoned monazite, low-Y zones (at higher grade also low-Th) were identified as late prograde monazite; high-Y rims grew around these (in the higher tectonic units), and these rims indicate growth from melt upon cooling. Xenotime was found in few samples and appears to be mostly retrograde; allanite was essentially limited to samples of the Lesser Himalayan sequence. The $^{232}Th–^{208}Pb$ ages indicate similar patterns within each tectonic unit and correlate with equally systematic *P–T* data (mostly garnet–biotite thermometry plus barometry based on Ca-contents of garnet): Within each unit, the growth age of late prograde (pre-anatectic, low-Y) monazite decreases structurally downward. Age-discontinuities were discovered between thrust sheets, and cooling in the hangingwall temporally was found to coincide with heating in the footwall (Fig. 19). These patterns indicate in-sequence thrusting of structurally higher over lower units, i.e., the unit being thrust on top started transferring heat to the one below.

In the Annapurna region, peak *P–T* conditions increase structurally upwards, from the lowest to the highest parts of the Greater Himalayan Sequence, with ~525 °C, 0.8 GPa in the lowest unit; 750 °C, 1.2 GPa in the middle units, and up to 775 °C, 1.3 GPa in the topmost tectonic unit. A similar sequence was also documented further east, in the Langtang section (Kohn 2008), starting in the lower Lesser Himalayan sequence and going up to the middle Greater Himalayan sequence. Here, migmatites in the top unit were crystallizing while the unit in their footwall was still heating. *P–T–t* paths for each unit benefit from the tight links among mineral chemistry, textures, and ages, again showing the need to identify the petrogenetic context of polyphase monazite growth; avoiding mixed dates (or at least weeding them out based on well separated ages) is crucial.

These data (and similar ones from western Nepal, Montomoli et al. 2013) appear to have resolved the long-standing question of what caused the inverted metamorphic sequence in the Himalaya, at least in central Nepal: Protracted thrusting produced sigmoidal isotherms[10]; a series of such thrusts was activated in sequence, each of them a gently inclined splay of the Main Himalayan Thrust, thus distributing the lithosphere scale strain (see also Harrison et al. 1995; Catlos et al. 2001). Stacking of these thrust sheets involved several relatively short-distance thrust periods, producing a duplex structure, which then may have (partially) collapsed by orogenic flattening. Discussions on tectonic mechanisms remain controversial, but the monazite datasets from Nepal samples certainly favor critical wedge models, even if more recent work concludes that convergence was accommodated through both wedge taper and lateral mid-crustal flow processes (Larson et al. 2013).

Combining strengths: petrochronology from monazite plus allanite

In several geological settings prograde reaction sequences involving monazite and allanite have been analyzed, providing detailed insight into reaction mechanisms (e.g., Rubatto et al. 2001; Wing et al. 2003; Janots et al. 2006). In favorable cases the age and conditions of formation of both REE-phases could be determined, with implications on the duration and rates of metamorphic processes. These studies also helped to correct long-held views of how mineral ages relate to *P–T–t* paths, i.e what monazite ages do (or do not) mean.

---

[10] the term "folded isotherms" is misleading

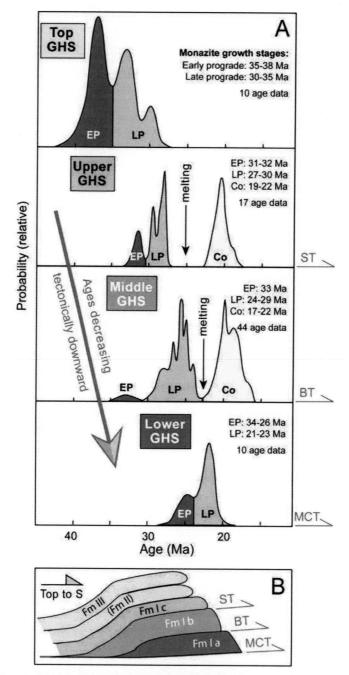

Figure 19. Monazite-based petrochronology (Corrie and Kohn 2011) in four thrust sheets of the Greater Himalayan Sequence (GHS) along Modi Khola valley, Annapurna region (Nepal). (A) Monazite data cover three segments of Formation I (Fm I). Two stages of prograde monazite growth are observed in all units, with a stage of post-anatectic growth in the central two units. Ages of each generation decrease structurally downward. Cooling was probably due to thrust emplacement (Kohn et al. 2004). This study documents that good structural control combined with detailed and petrologically solid monazite chronometry can yield tectonic (as well as thermal) rates and thus allows tectonic models to be quantitatively evaluated.

Central Alps. In metaclastic sequences of the Lepontine Alps (Switzerland) low-grade pelitic and marly rocks show evidence of first metamorphic monazite (high in La, low in Th and Y) grown at the expense of detrital monazite (dated at ~300–320 Ma) below 400 °C, but these could not be dated (diameter usually <5 μm and too young for EPMA). Along the classic metamorphic field gradient examined near Passo Lucomagno (Janots et al. 2008, 2009), at 440–450 °C (~0.4 GPa) monazite was a reactant in one of the first reactions that produced chloritoid, apatite, and allanite (zoned in REE, Y and Th). In the biotite zone (and up to 610 °C), only allanite but neither monazite nor xenotime remained in metapelites. In marly rocks between 530–500 °C (0.5–0.9 GPa), allanite retains its zoning, but is rimmed by epidote in the kyanite zone. Where staurolite formed, allanite started to break down to monazite + xenotime clusters (Fig. 20), but persisted up to 580 °C. The compositions of this second generation of metamorphic monazite reflect those of their local REE-precursors (e.g., MREE-rich allanite cores produced MREE-rich monazite, LREE-rich allanite rims produced corresponding LREE-monazite; this likely reflects a sequential breakdown); Th-rich monazite rims and xenotime satellites always ended up at the periphery of clusters. In samples of low Ca/Na-ratio only monazite + xenotime remained above 585 °C, but at high Ca/Na Y-rich epidote armored allanite, and breakdown to monazite did not occur.

In situ SHRIMP ($^{208}Pb/^{232}Th$) ages of coexisting allanite and monazite gave 31.5 ± 1.3 and 18.0 ± 0.3 Ma in one sample (and in a second one 29.2 ± 1.0 and 19.1 ± 0.3 Ma), constraining the time interval of prograde heating from 430–450 °C to 560–580 °C, implying a heating rate of 8–15 °C/m.y. (Fig. 20). It is uncertain whether monazite ages in the higher-grade parts of the Lepontine metamorphic belt (up to sillimanite grade, ~700 °C) grew by similar mechanisms and *P–T* conditions. Given the diversity of rock types and local deformation styles, other growth modes are fairly likely, though further *in situ* dating has confirmed the observed pattern (Boston et al. 2014). Many precise monazite TIMS ages (Köppel and Grünenfelder 1975, 1978; Köppel et al. 1981) indicate a gradual increase in age from ~19 to ~23 Ma across the belt, with abruptly more diverse ages in the Southern Steep belt, where monazite is ~27 Ma, and migmatites contain allanite and zircon that range between 32 and 20 Ma (Rubatto et al. 2009; Gregory et al. 2012). Yet, as in other classic orogens, many highly precise ages remain of uncertain petrogenetic and tectonic significance. The previously propagated notion that monazite ages simply document T_{max} conditions is almost as unlikely as their interpretation as cooling ages (to 450 °C, e.g., Steck et al. 2013).

Monazite at extreme conditions

Partial melting range. Prograde zoning has been found well preserved in at least some of the monazites from various migmatites (Pyle and Spear 2003; Kohn et al. 2005), and thermodynamic models (Kelsey et al. 2008; Spear and Pyle 2010) have proven very useful to analyze and interpret complexly zoned monazites, especially in settings involving partial melting (Hallett and Spear 2015). However, dissolution of monazite in melt or extended periods at high temperature reduce the record retained. Some datasets show that chemical and isotopic age zoning respond differently to temperature. Monazite at UHT has not yet been widely studied.

In migmatites (600–775 °C, 0.63–1.0 GPa) from polycyclic basement in northern Norway (Kalak Nappe Complex, Gasser et al. 2015), chemically homogeneous monazite gave concordant SIMS U–Pb ages that span >200 Ma. Despite careful structural control plus additional zircon, titanite, and rutile ages, this study could not resolve whether the 200 Ma time span reflects protracted monazite growth or partial U–Pb resetting. Ages obtained (Fig. 21) from a majority of monazite grains (698 ± 11 Ma) are within error of zircon (702 ± 5 Ma), and both reflect crystallization from partial melt, otherwise similar monazite grains cluster along Concordia between ~600 and ~800 Ma. As rutile is texturally coeval, its Zr-in-rutile temperatures (550–630 °C) clearly must be reset, and indeed its age is Caledonian, mostly 440 Ma, within error of the titanite that overgrew rutile at 440–430 Ma. In this case monazite

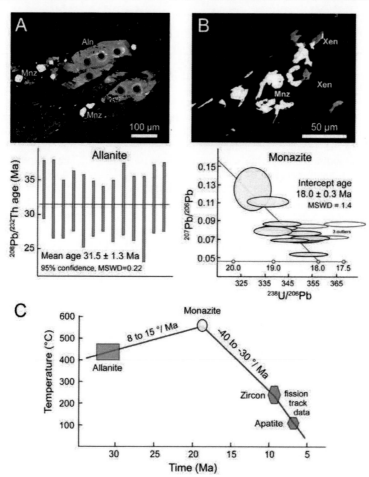

Figure 20. Ages and *P–T* data for medium-pressure amphibolite facies metapelites from the northern Lepontine belt, Central Alps, Switzerland. (A) BSE image shows metamorphic allanite and monazite. Dark pits in allanite are from SHRIMP analysis. Th/Pb ages (31.5 ± 1.3 Ma) date prograde allanite growth at chloritoid-in isograd (440 °C). (B) BSE photo of shows monazite and xenotime formed by allanite breakdown at staurolite–kyanite grade (560 °C, 0.8 GPa). 10 (out of 13) spots were used to obtain a SHRIMP U/Pb age of 18.0 ± 0.3 Ma; Tera-Wasserburg plot shows 5–12% common lead in monazite (inherited from allanite). (C) These ages, combined with silicate-based thermobarometry (Todd and Engi 1997; Janots et al. 2008) and fission track data (Janots et al. 2009), establish a *P–T–t* path for this classic orogen as well as a moderate heating rate. The high cooling rate mostly reflects tectonic unroofing.

evidently retained no chemical evidence of prograde growth. Gasser et al. were unsure whether post-migmatic cooling involved substantial decompression (from ~0.9 GPa) prior to the onset of the Caledonian cycle. But the study concluded that the samples they studied spent some 200 Ma at temperatures above 650 °C, so whatever chemical zoning monazite (in melanosome) initially had, must have been erased. These migmatites were not overprinted by the otherwise pervasive Caledonian S2 fabrics (peaking at 600 ± 30 °C, 1.13 ± 0.13 GPa), but rutile ages are partly concordant (552—432 Ma), and Zr-in-rutile yields temperatures consistent with S2, and titanites yield concordant TIMS ages of 440—429 Ma. The *P–T* evolution between the two stages is not constrained, and schematic paths such as shown (Fig. 21a, b) are possible. However, it is clear that monazite in migmatites formed over some 200 Ma at temperatures >650 °C (Gasser et al. 2015), following anatexis in the sillimanite- and kyanite-fields.

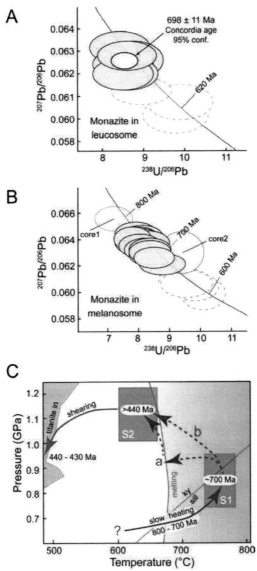

Figure 21. Ages and *P–T* path for metapelitic migmatites from the Caledonian Kalak Nappes, (northern Norway), that show pre-Caledonian (S1) fabrics. (A) Monazite in leucosome is chemically homogeneous, and a Tera-Wasserburg plot of SIMS data shows 10 concordant U–Pb ages between 710 ± 9 and 609 ± 12 Ma. The oldest 7 grains cluster at 698 ± 11 Ma (within error of zircon: 702 ± 5 Ma),whereas 3 younger monazites (stippled outline) scatter along Tera-Wasserburg. (B) Monazites from melanosome are mostly homogeneous (rare cores differ in Th and REE). Excepting two core analyses, SIMS data are concordant within 2σ, and range from 786 ± 12 to 594 ± 18 Ma. (C) Possible *P–T–t* path. Pseudosection modeling for S1-assemblages indicates peak conditions of $760 \pm 15\,°C$ and $0.93 \pm .05$ GPa at ~700 Ma for the S1 stage, and $600 \pm 30\,°C$ and 1.13 ± 0.13 GPa at ≥ 440Ma for the S2 (Caledonian) stage. Monazite growth in leucosomes entirely postdates the S1 stage, whereas in melanosomes, a few older cores remain. In melanosomes, chemical heterogeneity in monazite was largely erased along this protracted high-*T* path, but the large spread of concordant U–Pb dates probably represents reliable formation ages. Data from Gasser et al. (2015).

UHT conditions. In the Eastern Ghats (India) UHT terrain, slow cooling from peak conditions >960 °C to ~900 °C and 0.97 to 0.75 GPa (determined from phase equilibrium modeling, Korhonen et al. 2011) is reflected in monazite mean $^{207}Pb/^{206}Pb$ SHRIMP ages[11] (Korhonen et al. 2013) between 970 to 930 Ma. In this context, a spread of ages (1130 and 970 Ma) from monazites with very diverse internal textures and Y-distributions are interpreted to recall prograde stages. Not surprisingly, the data indicate partial Pb-loss (Fig. 22), though no simple core-to-rim age zoning is evident, and most of the analyses of zircon and monazite are reported as concordant (defined as <10% discordance). The *P–T–t* data indicate a counter-clockwise path, with certainly >50 Ma (and possibly 200 Ma) at UHT conditions. Discrete, concordant populations of monazite are common (more so than zircon overgrowths on zircon cores) and could be analyzed *in situ* and in grain fractions; they are clearly offset against

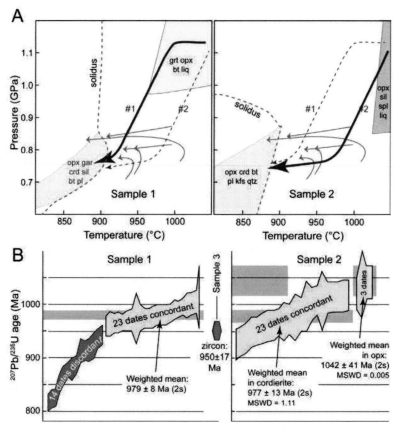

Figure 22. Petrochronology of UHT rocks (Mg-Al-rich granulites) from Sunki locality, Eastern Ghats Province, India. These rocks record cooling from >960 °C, ~1 GPa to the solidus at ~900 °C, 0.75 GPa during slow exumation (Korhonen et al. 2011). (A) *P–T* paths inferred for two samples of different composition. Initial petrographic and pseudosection analysis indicated rather steep (rapid) initial decompression (thick black paths), but later analysis inferred lower early pressures and nearly isobaric paths (thin red paths). Inclusion relations and coronal successions involving major silicates (opx, cord) and accessory minerals allowed monazite ages to be tied to mineral reactions. (B) Temperature–time paths for same samples. EPMA X-ray mapping and SHRIMP dating show slow cooling (1–2 °C/Ma) from ~1000 °C at 1000–1050 Ma to 850–900 °C at 950 Ma, when zircon in a third sample grew. Many data are discordant data in Sample 1, few in Sample 2. Weighted mean ages are shown as 2σ (grey bands), but envelopes of individual data are 1σ.

[11] except where excess ^{206}Pb was noted, in which case $^{207}Pb/^{235}U$ ages were preferred

late Neoproterozoic overprinting. Given monazite grain sizes of typically 80–200 µm, it is remarkable that only a small proportion of the data appear disturbed and had to be considered as outliers. The *P–T* conditions and duration of the UHT-evolution appear reliably defined on the basis of coherent age groups; monazite and zircon grown from (trapped) melt show overlapping ages, though some monazite ages may exceed those of zircon. In Sample 2, two stages of decompression were separately datable: monazite in opx is ~60 Ma older than in cord. The relatively large proportion of concordant dates suggests protracted monazite growth (Sample 1: 1014–959 Ma, Sample 2: 1043–922 Ma) due to multiple reactions during initial cooling. Younger discordant ages in Sample 1 were interpreted to reflect Pb loss, but monazites from Sample 2 do not show this.These data suggest that monazite chemistry and ages are resistant to diffusional resetting at temperatures of ca. 950 °C (Korhonen et al. 2013).

UHP conditions. Hacker et al. (2015) investigated the response of monazite and its host rocks to metamorphism during subduction to UHP conditions and subsequent return to the surface. A large sample suite from the Western Gneiss Region (Norway) was analyzed by LASS to obtain simultaneous spot analyses of trace-elements, U–Pb ages, and Th–U chemistry. This instrument makes an inherent spatial link between spot ages and their petrological context, so to interpret the significance of individual spots, a few well established principles were used: Spot ages from monazite showing low HREE concentrations were taken to reflect the presence of garnet (Rubatto 2002); elevated Eu and Sr indicated growth in the eclogite facies, where plagioclase had broken down to pyroxene and garnet (Finger and Krenn 2007; Holder et al. 2015); Th and U concentrations indicated where/when fluid was available (Hoskin and Schaltegger 2003) that may have facilitated monazite recrystallization (Seydoux-Guillaume et al. 2002a). Hacker et al. (2015) found that relatively few monazite ages reflect growth at UHP-conditions; in coarse-grained rocks some monazites first survived amphibolite- to granulite-facies metamorphism and partial melting, followed by UHP-subduction—with their original ages intact.

Allanite: a hot finale

Allanites from migmatites and intrusive or volcanic rocks have sparked several studies in which the control available from strong zoning in major and trace elements and isotopes was combined with retentive U–Th–Pb chronometry on allanite, zircon, and titanite.

Melt evolution and disequilibrium isotope systematics. Oberli et al. (2004) studied allanite from a single sample taken in a massive feeder dyke of the syn-orogenic Bergell tonalite (Swiss Alps) intrusion. Conventional ID-TIMS dating on mineral separates was used. Detailed documentation by EPMA and SEM are perfectly compatible with magmatic fractionation; hornblende thermobarometry indicates 940 °C and ~0.8 GPa, allanite crystallization may have started below the zircon saturation temperature (>730 °C). Fractionation is recorded in strongly zoned allanite (Fig. 23): Core compositions show REE_2O_3 ~17.6 wt%, Fe^{2+}/Fe_{total} ~0.6; both are near zero at the rim; Th/U rapidly decreases from ~170 to ~1. Apparent ages based on $^{206}Pb/^{238}U$ in allanite were 1.8 to 6.3 Ma older than $^{208}Pb/^{232}Th$ ages due to excess ^{206}Pb, reflecting initial isotopic $^{230}Th/^{238}U$ disequilibrium, and allanite's ravenous appetite for Th. The sequence of ages obtained from (in part air-abraded) fractions of growth-zoned zircon (33.0–32.0 Ma), density-separated allanite (32.0–28.0 Ma), and titanite (29.85 Ma) document growth in a differentiating melt and are related to finite intervals of melt fractionation during emplacement and *in situ* crystallization. One core-rich zircon fraction (dated at 44.7 Ma) showed evidence of inheritance, whereas six others indicate a differentiation trend over ~1 m.y. When corrected for initial isotopic disequilibrium in $^{230}Th/^{238}U$, these six data (and titanite) were concordant, whereas the upper intercept of the zircon core was at 1070 Ma.

The benefit of very precise TIMS-data was paired with painstakingly careful characterization of mineral chemical and isotopic composition. While *in situ* dating might have tightened the links between microchemical and age information, it is doubtful whether the more uncertain isotopic ratios in LA-ICP-MS or SIMS data could discern such fine age detail in a system

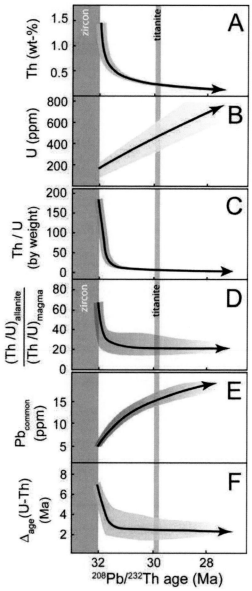

Figure 23. Systematic chemistry of allanite vs. time for 9 density-separated allanite fragments from the Bergell tonalite, Iorio Pass, Swiss/Italian Central Alps) that crystallized during magmatic differentiation and that show complex chemical zoning (Oberli et al. (2004). Painstaking U–Pb and Th–Pb dating by TIMS allowed chemical differentiation to be linked to a crystallization interval spanning 4 Ma. (A) Th is sequestered early, markedly depleting the melt in Th. (B) U increases in the melt, but its high concentration implies that crystallization of major silicates (which generally exclude U) enrich the melt faster than allanite can deplete it. (C–D) Simultaneous fractionation of Th and U, and changes to allanite and melt compositions, rapidly diminish the allanite Th–U ratio and partitioning of Th and U between allanite and melt. (E) Common lead content increases as allanite grows, demanding correction in dating. (F) ^{206}Pb/^{238}U are systematically higher than ^{208}Pb/^{232}Th ages ages because of disequilibrium in ^{230}Th/^{238}U. A decrease in Th/U as allanite crystallizes causes the disparity between ages [Δ_{age}(U–Th)] to be most pronounced (>6 Ma) for the oldest allanite, least in youngest (<2 Ma). Preservation of this initial disequilibrium signature shows that the Th–Pb data reflect growth ages during magma differentiation (between 750 and 650 °C), and that diffusive Pb-resetting was not effective.

evolving so rapidly. It is significant that the observed age differences among grains relate to magmatic differentiation, not to Pb-diffusion. Allanite thus has a sufficiently high closure temperature (>750 °C) to warrant chronometry in acid to intermediate magmatic systems.

Late Quaternary volcanoes. In very young magmatic systems, isotopic disequilibrium in $^{238}U/^{230}Th$ (i.e., the U-series age) can be used to estimate residence times (e.g., Vazquez and Reid 2004); a beautiful example utilized allanite crystals in Mono Lake magma. Cox et al. (2012) determined the expected secular equilibrium value of ^{230}Th (from the Th/U ratio analyzed in host glass) and calculated the time needed to reach the ^{230}Th measured in allanite (Fig. 24). From the difference between the U-series age of 66.0 ± 3.3 ka and the eruption age of 38.7 ± 1.2 ka (determined independently by (U–Th)/He dating), they calculated a mean residence time for these allanites of 27 ± 4 ka. This precision rivals determinations using zircon or baddelyite (Schaltegger and Davies 2017).

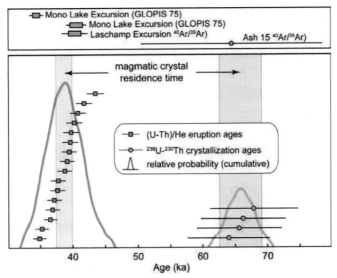

Figure 24. U-series ages from allanite and (U–Th)/He ages of eruption from ash layer 15, Mono Lake, eastern Sierra Nevada, California, USA. The U-series date (66.0 ± 3.3 ka, 2σ) was determined using ICP-MS (Cox et al. 2012) and reflects the average crystallization age in the magma. The (U–Th)/He date (38.7 ± 1.2 ka, 2σ) is the (helium) eruption age. These data yield a 27 ± 4 ka average residence time of allanite in the rhyolitic magma. Shown for comparison are $^{40}Ar/^{39}Ar$ data for Ash 15 and ages for two prominent magnetic excursions (Mono Lake and Laschamp).

FUTURE DIRECTIONS

• Further analytical developments will undoubtedly allow dating ever smaller volumes at higher precision. Of equal importance is sharpening and wielding tools that provide textural and microchemical context to these ages: X-ray maps from EPMA need to be combined with trace element maps from laser ablation methods, prior to spot dating.

• Thermodynamic models are needed for the main REE-minerals to make reliable prediction of the reaction relations involving major and accessory minerals and to enhance thermobarometry based on their composition. The lead taken by Janots et al. (2007), Spear (2010), and Spear and Pyle (2010) should be expanded, and for this purpose experimentally determined phase equilibria—ideally reaction reversals—are required that involve REE-phases and fully documented natural assemblages from well known *P–T* conditions.

• Pressure estimates need improving to link chronologic data, especially based on monazite and allanite. Since allanite frequently contains quartz inclusions, Raman micropiezometry (Kohn 2014b) on these might be attempted, though shear modulus data for epidote (Hacker et al. 2003) are but approximately valid for allanite. Radiation softening may also prove problematica. Elastic properties for the REE-phosphates were compiled by Mogilevsky (2007).

ACKNOWLEDGMENTS

I thank Thomas Armbruster for producing images of crystal structures. Emilie Janots, Rob Berman, Marco Burn, Pierre Lanari, Birger Rasmussen, and Daniele Regis donated original images. I am grateful for this and to Emilie Janots and Andrew Smye for their very helpful reviews. Thanks go to Matt Kohn for his thoughtful advice, editorial rigor and his patience, and to Christine for doing everything else while I was preparing this. The Swiss National Science foundation has provided funding for work presented here (project 200020_126946).

REFERENCES

Ague JJ, Baxter EF (2007) Brief thermal pulses during mountain building recorded by Sr diffusion in apatite and multicomponent diffusion in garnet. Earth Planet Sci Lett 261:500–516

Allaz J, Selleck B, Williams ML, Jercinovic MJ (2013) Microprobe analysis and dating of monazite from the Potsdam Formation, New York: A progressive record of chemical reaction and fluid interaction. Am Mineral 98:1106–1119

Andrehs G, Heinrich W (1998) Experimental determination of REE distributions between monazite and xenotime: potential for temperature-calibrated geochronology. Chem Geol 149:83–96

Antignano A, Manning CE (2008) Fluorapatite solubility in H_2O and H_2O–NaCl at 700 to 900 °C and 0.7 to 2.0 GPa. Chem Geol 251:112–119

Armbruster T, Bonazzi P, Akasaka M, Bermanec V, Chopin C, Gieré R, Heuss-Assbichler S, Liebscher A, Menchetti S, Yuanming PA, Pasero M (2006) Recommended nomenclature of epidote-group minerals. Eur J Mineral 18:551–567

Ayers JC, Dunkle S, Gao S, Miller CF (2002) Constraints on timing of peak and retrograde metamorphism in the Dabie Shan Ultrahigh-Pressure Metamorphic Belt, east-central China, using U–Th–Pb dating of zircon and monazite. Chem Geol 186:315–331, doi:10.1016/S0009-2541(02)00008-6

Baldwin JA, Bowring SA, Williams ML, Mahan KH (2006) Geochronological constraints on the evolution of high-pressure felsic granulites from an integrated electron microprobe and ID-TIMS geochemical study. Lithos 88:173–200

Banno S, Yoshizawa H (1993) Sector-zoning of epidote in the Sanbagawa schists and the question of an epidote miscibility gap. Mineral Mag 57:739–743

Barfod GH, Otero O, Albarède F (2003) Phosphate Lu–Hf geochronology. Chem Geol 200:241–253, doi:10.1016/S0009-2541(03)00202-X

Barfod GH, Krogstad EJ, Frei R, Albarède F (2005) Lu–Hf and PbSL geochronology of apatites from Proterozoic terranes: A first look at Lu-Hf isotopic closure in metamorphic apatite. Geochim Cosmochim Acta 69:1847–1859

Barrow G (1912) On the geology of Lower Dee-side and the Southern Highland border. Proc Geol Assoc 1912:1–17

Barth S, Oberli F, Meier M (1994) Th–Pb versus U–Pb isotope systematics in allanite from co-genetic rhyolite and granodiorite: implications for geochronology. Earth Planet Sci Lett 124:149–159

Bau M (1996) Controls on the fractionation of isovalent trace elements in magmatic and aqueous systems: Evidence from Y/Ho, Zr/Hf, and lanthanide tetrad effect. Contrib Mineral Petrol 123:323–333

Bea F (1996) Residence of REE, Y, Th, and U in granites and crustal protoliths; implications for the chemistry of crustal melts. J Petrol 37:521–552

Bea F, Montero P (1999) Behavior of accessory phases and redistribution of Zr, REE, Y, Th, and U during metamorphism and partial melting of metapelites in the lower crust: an example from the Kinzigite Formation of Ivrea-Verbano, NW Italy. Geochim Cosmochim Acta 63:1133–1153

Beard JS, Sorensen SS, Gieré R (2006) REE zoning in allanite related to changing partition coefficients during crystallisation; implications for REE behaviour in an epidote-bearing tonalite. Mineral Mag 70:419–435

Berger A, Rosenberg C, Schaltegger U (2009) Stability and isotopic dating of monazite and allanite in partially molten rocks: examples from the Central Alps. Swiss J Geosci 102:15–29, doi:10.1007/s00015-009-1310-8

Berger A, Gnos E, Janots E, Whitehouse M, Soom M, Frei R, Waight TE (2013) Dating brittle tectonic movements with cleft monazite: Fluid–rock interaction and formation of REE minerals. Tectonics 32:1176–1189, doi:10.1002/tect.20071

Berman RG, Sanborn-Barrie M, Stern RA, Carson CJ (2005) Tectonometamorphism at ca. 2.35 and 1.85 Ga in the Rae Domain, western Churchill Province, Nunavut, Canada: Insights from structural, metamorphic and in situ geochronological analysis of the southwestern Committee Bay Belt. Can Mineral 43:937–971

Berman RG, Ryan JJ, Gordey SP, Villeneuve M (2007) Permian to Cretaceous polymetamorphic evolution of the Stewart River region, Yukon-Tanana terrane, Yukon, Canada: *P–T* evolution linked with in situ SHRIMP monazite geochronology. J Metamorph Geol 25:803–827, doi:10.1111/j.1525–1314.2007.00729.x

Berzelius JJ (1824) Undersökning af några Mineralier; 1. Phosphorsyrad Ytterjord. Kungliga Svenska Vetenskaps-Akademiens handlingar, Stockholm:334–358

Beudant F-S (1832) Traité élémentaire de Minéralogie. Verdière

Bingen B, van Breemen O (1998) U–Pb monazite ages in amphibolite-to granulite-facies orthogneiss reflect hydrous mineral breakdown reactions: Sveconorwegian Province of SW Norway. Contrib Mineral Petrol 132:336–353

Bingen B, Demaiffe D, Hertogen J (1996) Redistribution of rare earth elements, thorium, and uranium over accessory minerals in the course of amphibolite to granulite facies metamorphism: the role of apatite and monazite in orthogneisses from southwestern Norway. Geochim Cosmochim Acta 60:1341–1354

Black LP, Fitzgerald JD, Harley SL (1984) Pb isotopic composition, colour, and microstructure of monazites from a polymetamorphic rock in Antarctica. Contrib Mineral Petrol 85:141–148

Blenkinsop TG, Frei R (1997) Archean and Proterozoic mineralization and tectonics at the Renco Mine (northern marginal zone, Limpopo Belt, Zimbabwe); reply. Econ Geol 92:747–748

Bollinger L, Janots E (2006) Evidence for Mio-Pliocene retrograde monazite in the Lesser Himalaya, far western Nepal. Eur J Mineral 18:289–297

Boston K (2014) Investigation of accessory allanite, monazite and rutile of the Barrovian sequence of the Central Alps (Switzerland). PhD, Australian National University, Canberra

Breiter K, Förster H-J, Skoda R (2008) Extreme P-, Bi-, Nb-, Sc-, U and F-rich zircon from fractionated perphosphorous granites: the peraluminous Podlesi granite system, Czech Republica. Lithos 88:15–34

Broska I, Williams CT, Janák M, Nagy G (2005) Alteration and breakdown of xenotime-(Y) and monazite-(Ce) in granitic rocks of the Western Carpathians, Slovakia. Lithos 82:71–83

Broska I, Krogh Ravna EJ, Vojtko P, Janák M, Konečný P, Pentrák M, Bačík P, Luptáková J, Kullerud K (2014) Oriented inclusions in apatite in a post-UHP fluid-mediated regime (Tromsø Nappe, Norway). Eur J Mineral 26:623–634, doi:10.1127/0935–1221/2014/0026–2396

Bruand E, Storey C, Fowler M (2016) An apatite for progress: Inclusions in zircon and titanite constrain petrogenesis and provenance. Geology 44:91–94, doi:10.1130/G37301.1

Budzyń B, Harlov DE, Williams ML, Jercinovic MJ (2011) Experimental determination of stability relations between monazite, fluorapatite, allanite, and REE-epidote as a function of pressure, temperature, and fluid composition. Am Mineral 96:1547–1567

Buick IS, Clark C, Rubatto D, Hermann J, Pandit M, Hand M (2010) Constraints on the Proterozoic evolution of the Aravalli–Delhi Orogenic belt (NW India) from monazite geochronology and mineral trace element geochemistry. Lithos 120:511–528, doi:10.1016/j.lithos.2010.09.011

Burn M (2016) LA-ICP-QMS Th–U/Pb allanite dating: methods and applications. PhD, University of Bern

Catlos EJ, Sorensen SS, Harrison TM (2000) Th–Pb ion microprobe dating of allanite. Am Mineral 85:633–648

Catlos EJ, Harrison TM, Kohn MJ, Grove M, Ryerson FJ, Manning C, Upreti BN (2001) Geochronologic and thermobarometric constraints on the evolution of the Main Central Thrust, central Nepal Himalaya. J Geophys Res B, Solid Earth Planets 106:16177–16204

Cenki-Tok B, Oliot E, Rubatto D, Berger A, Engi M, Janots E, Thomsen TB, Manzotti P, Regis D, Spandler C, Robyr M (2011) Preservation of Permian allanite within an Alpine eclogite facies shear zone at Mt Mucrone, Italy: Mechanical and chemical behaviour of allanite during mylonitization. Lithos 125:40–50, doi:10.1016/j.lithos.2011.01.005

Cenki-Tok B, Darling JR, Rolland Y, Dhuime B, Storey CD (2014) Direct dating of mid-crustal shear zones with synkinematic allanite: new in situ U–Th–Pb geochronological approaches applied to the Mont Blanc massif. Terra Nova 26:29–37

Chamberlain KR, Bowring SA (2001) Apatite–feldspar U–Pb thermochronometer: a reliable, mid-range (~ 450° C), diffusion-controlled system. Chem Geol 172:173–200

Cherniak D (2000) Rare earth element diffusion in apatite. Geochim Cosmochim Acta 64:3871–3885

Cherniak DJ (2006) Pb and rare earth element diffusion in xenotime. Lithos 88:1–14, doi:10.1016/j.lithos.2005.08.002

Cherniak DJ (2010) Diffusion in accessory minerals: zircon, titanite, apatite, monazite and xenotime. Rev Mineral Geochem 72:827–869, doi:10.2138/rmg.2010.72.18

Cherniak DJ, Ryerson FJ (1993) A study of strontium diffusion in apatite using Rutherford backscattering spectroscopy and ion implantation. Geochim Cosmochim Acta 57:4653–4662

Cherniak DJ, Watson EB, Grove M, Harrison TM (2004) Pb diffusion in monazite: a combined RBS/SIMS study 1. Geochim Cosmochim Acta 68:829–840, doi:10.1016/j.gca.2003.07.012

Chew DM, Donelick RA (2012) Combined apatite fission-track and U–Pb dating by LA-ICP-MS and its application in apatite provenance analysis. Quantitative Mineralogy and Microanalysis of Sediments and Sedimentary Rocks: Mineral Assoc Canada, Short Course 42:219–247

Chew DM, Spikings RA (2015) Geochronology and thermochronology using apatite: time and temperature, lower crust to surface. Elements 11:189–194, doi:10.2113/gselements.11.3.189

Chew DM, Sylvester PJ, Tubrett MN (2011) U–Pb and Th–Pb dating of apatite by LA-ICPMS. Chem Geol 280:200–216

Clavier N, Podor R, Dacheux N (2011) Crystal chemistry of the monazite structure. J Eur Ceram Soc 31:941–976

Cochrane R, Spikings RA, Chew D, Wotzlaw J-F, Chiaradia M, Tyrrell S, Schaltegger U, Van der Lelij R (2014) High temperature (> 350 °C) thermochronology and mechanisms of Pb loss in apatite. Geochim Cosmochim Acta 127:39–56

Copeland P, Parrish RR, Harrison TM (1988) Identification of inherited radiogenic Pb in monazite and its implications for U–Pb systematics. Nature 333:760–763

Corrie SL, Kohn MJ (2008) Trace-element distributions in silicates during prograde metamorphic reactions: implications for monazite formation. J Metamorph Geol 26:451–464

Corrie SL, Kohn MJ (2011) Metamorphic history of the central Himalaya, Annapurna region, Nepal, and implications for tectonic models. Geol Soc Am Bull 123:1863–1879, doi:10.1130/B30376.1

Cox SE, Farley KA, Hemming SR (2012) Insights into the age of the Mono Lake Excursion and magmatic crystal residence time from (U–Th)/He and ^{230}Th dating of volcanic allanite. Earth Planet Sci Lett 319–320:178–184, doi:10.1016/j.epsl.2011.12.025

Cressey G, Wall F, Cressey BA (1999) Differential REE uptake by sector growth of monazite. Mineral Mag 63:813–813, doi:10.1180/002646199548952

Cruft EF (1966) Minor elements in igneous and metamorphic apatite. Geochim Cosmochim Acta 30:375–398

Cumming G, Richards J (1975) Ore lead isotope ratios in a continuously changing Earth. Earth Planet Sci Lett 28:155–171

Dahl PS, Terry MP, Jercinovic MJ, Williams ML, Hamilton MA, Foland KA, Clement SM, Friberg LM (2005) Electron probe (Ultrachron) microchronometry of metamorphic monazite: Unraveling the timing of polyphase thermotectonism in the easternmost Wyoming Craton (Black Hills, South Dakota). Am Mineral 90:1712–1728

Daniel CG, Pyle JM (2006) Monazite–xenotime thermochronometry and Al$_2$SiO$_5$ reaction textures in the Picuris Range, Northern New Mexico, USA: New evidence for a 1450–1400 Ma orogenic event. J Petrol 47:97–118, doi:10.1093/petrology/egi069

Darling J, Storey C, Engi M (2012) Allanite U–Th–Pb geochronology by laser ablation ICP-MS. Chem Geol 292:103–115

Davis W, Rainbird R, Gall Q, Jefferson C (2008) In situ U–Pb dating of diagenetic apatite and xenotime: Paleofluid flow history within the Thelon, Athabasca and Hornby Bay basins. Geochim Cosmochim Acta 72:A203

DeWolf CP, Belshaw N, O'Nions RK (1993) A metamorphic history from micron-scale ^{207}Pb/^{206}Pb chronometry of Archean monazite. Earth Planet Sci Lett 120:207–220, doi:10.1016/0012-821X(93)90240-A

Didier A, Bosse V, Boulvais P, Bouloton J, Paquette J-L, Montel J-M, Devidal J-L (2013) Disturbance versus preservation of U–Th–Pb ages in monazite during fluid–rock interaction: textural, chemical and isotopic in situ study in microgranites (Velay Dome, France). Contrib Mineral Petrol 165:1051–1072, doi:10.1007/s00410-012-0847-0

Didier A, Bosse V, Cherneva Z, Gautier P, Georgieva M, Paquette J-L, Gerdjikov I (2014) Syn-deformation fluid-assisted growth of monazite during renewed high-grade metamorphism in metapelites of the Central Rhodope (Bulgaria, Greece). Chem Geol 381:206–222, doi:10.1016/j.chemgeo.2014.05.020

Dodson MH (1973) Closure temperature in cooling geochronological and petrological systems. Contrib Mineral Petrol 40:259–274

Dodson M (1986) Closure profiles in cooling systems. Mater Sci Forum 7:145–154

Duc-Tin Q (2007) Experimental studies on the behaviour of rare earth elements and tin in granitic systems. PhD, University of Tuebingen

Duc-Tin Q, Keppler H (2015) Monazite and xenotime solubility in granitic melts and the origin of the lanthanide tetrad effect. Contrib Mineral Petrol 169:1–26, doi:10.1007/s00410-014-1100-9

Dumond G, Goncalves P, Williams ML, Jercinovic MJ (2015) Monazite as a monitor of melting, garnet growth and feldspar recrystallization in continental lower crust. J Metamorph Geol 33:735–762, doi:10.1111/jmg.12150

Elderfield H, Pagett R (1986) Rare earth elements in ichthyoliths: variations with redox conditions and depositional environment. Sci Total Environ 49:175–197

Ellison A, Hess P (1988) Peraluminous and peralkaline effects upon "monazite" solubility in high-silica liquids. Eos, Trans Am Geophys Union 69:498

Ercit TS (2002) The mess that is allanite. Can Mineral 40:1411–1419

Erickson TM, Pearce MA, Taylor RJM, Timms NE, Clark C, Reddy SM, Buick IS (2015) Deformed monazite yields high-temperature tectonic ages. Geology 43:383–386

Evans J, Zalasiewicz J (1996) U–Pb, Pb–Pb and Sm–Nd dating of authigenic monazite: implications for the diagenetic evolution of the Welsh Basin. Earth Planet Sci Lett 144:421–433

Evans JA, Zalasiewicz JA, Fletcher I, Rasmussen B, Pearce NJG (2002) Dating diagenetic monazite in mudrocks: constraining the oil window? J Geol Soc 159:619–622

Fehr KT, Heuss-Assbichler S (1997) Intracrystalline equilibria and immiscibility along the join clinozoisite–epidote: An experimental and ^{57}Fe Mössbauer study. Neues Jahrb Mineral Abh 172:43–67

Ferry JM (1994) Overview of the petrologic record of fluid flow during regional metamorphism in northern New England. Am J Sci 294:905–988

Ferry JM (2000) Patterns of mineral occurrence in metamorphic rocks. Am Mineral 85:1573–1588

Finger F, Krenn E (2007) Three metamorphic monazite generations in a high-pressure rock from the Bohemian Massif and the potentially important role of apatite in stimulating polyphase monazite growth along a *PT* loop. Lithos 95:103–115

Finger F, Broska I, Roberts MP, Schermaier A (1998) Replacement of primary monazite by apatite–allanite–epidote coronas in an amphibolite facies granite gneiss from the eastern Alps. Amer Mineral 83:248–258

Fletcher IR, McNaughton NJ, Aleinikoff JA, Rasmussen B, Kamo SL (2004) Improved calibration procedures and new standards for U–Pb and Th–Pb dating of Phanerozoic xenotime by ion microprobe. Chem Geol 209:295–314, doi:10.1016/j.chemgeo.2004.06.015

Förster HJ, Harlov DE (1999) Monazite-(Ce)–huttonite solid solutions in granulite-facies metabasites from the Ivrea-Verbano Zone, Italy. Min Mag 63:587–594

Foster G, Kinny P, Vance D, Prince C, Harris N (2000) The significance of monazite U–Th–Pb age data in metamorphic assemblages; a combined study of monazite and garnet chronometry. Earth Planet Sci Lett 181:327–340

Foster G, Gibson HD, Parrish R, Horstwood M, Fraser J, Tindle A (2002) Textural, chemical and isotopic insights into the nature and behaviour of metamorphic monazite. Chem Geol 191:183–207, doi:10.1016/S0009-2541(02)00156–0

Foster G, Parrish RR, Horstwood MSA, Chenery S, Pyle J, Gibson HD (2004) The generation of prograde *P–T–t* points and paths; a textural, compositional, and chronological study of metamorphic monazite. Earth Planet Sci Lett 228:125–142, doi:10.1016/j.epsl.2004.09.024

Franz G, Andrehs G, Rhede D (1996) Crystal chemistry of monazite and xenotime from Saxothuringian-Moldanubian metapelites, NE Bavaria, Germany. Eur J Mineral:1097–1118

Gabudianu Radulescu I, Rubatto D, Gregory C, Compagnoni R (2009a) The age of HP metamorphism in the Gran Paradiso Massif, Western Alps: A petrological and geochronological study of "silvery micaschists". Lithos 110:95–108, doi:10.1016/j.lithos.2008.12.008

Gabudianu Radulescu I, Rubatto D, Gregory C, Compagnoni R (2009b) The age of metamorphism in the Gran Paradiso Massif, Western Alps: A petrological and geochronological study of "silvery micaschists". Lithos 110:95–108, doi:10.1016/j.lithos.2008.12.008

Ganguly J, Tirone M (1999) Diffusion closure temperature and age of a mineral with arbitrary extent of diffusion: theoretical formulation and applications. Earth Planet Sci Lett 170:131–140

Gasser D, Jeřábek P, Faber C, Stünitz H, Menegon L, Corfu F, Erambert M, Whitehouse M (2015) Behaviour of geochronometers and timing of metamorphic reactions during deformation at lower crustal conditions: phase equilibrium modelling and U–Pb dating of zircon, monazite, rutile and titanite from the Kalak Nappe Complex, northern Norway. J Metamorph Geol 33:513–534

Gieré R, Sorensen SS (2004) Allanite and other REE-rich epidote-group minerals. Rev Mineral Geochem 56:431–493

Gleadow AJ, Belton DX, Kohn BP, Brown RW (2002) Fission track dating of phosphate minerals and the thermochronology of apatite. Rev Mineral Geochem 48:579–630

Goldschmidt VM, Thomassen L (1924) Geochemische Verteilungsgesetze der Elemente. III. Röntgenspektrographische Untersuchungen über die Verteilung der Seltenen Erdmetalle in Mineralen. Videnskapsselskapets Skrifter I Mathematisk-Naturvidenskabelig Klasse 5:1–58

Goswami-Banerjee S, Robyr M (2015) Pressure and temperature conditions for crystallization of metamorphic allanite and monazite in metapelites: a case study from the Miyar Valley (high Himalayan Crystalline of Zanskar, NW India). J Metamorph Geol 33:535–556

Goudie DJ, Fisher CM, Hanchar JM, Crowley JL, Ayers JC (2014) Simultaneous in situ determination of U–Pb and Sm–Nd isotopes in monazite by laser ablation split-stream ICP-MS. Geochem Geophys Geosystem 15:2575–2600

Grand'Homme A, Janots E, Bosse V, Seydoux-Guillaume A, Guedes RDA (2016a) Interpretation of U–Th–Pb in-situ ages of hydrothermal monazite-(Ce) and xenotime-(Y): evidence from a large-scale regional study in clefts from the western Alps. Mineral Petrol:1–21

Grand'Homme A, Janots E, Seydoux-Guillaume A-M, Guillaume D, Bosse V, Magnin V (2016b) Partial resetting of the U–Th–Pb systems in experimentally altered monazite: Nanoscale evidence of incomplete replacement. Geology 44:431–434

Gratz R, Heinrich W (1997) Monazite–xenotime thermobarometry: experimental calibration of the miscibility gap in the system $CePO_4$–YPO_4. Am Mineral 82:772–780

Gratz R, Heinrich W (1998) Monazite–xenotime thermometry. III. Experimental calibration of the partitioning of gadolinium between monazite and xenotime. Eur J Mineral 10:579–588

Gregory CJ, McFarlane CRM, Hermann J, Rubatto D (2009) Tracing the evolution of calc-alkaline magmas: In-situ Sm–Nd isotope studies of accessory minerals in the Bergell Igneous Complex, Italy. Chem Geol 260:73–86, doi:10.1016/j.chemgeo.2008.12.003

Gregory CJ, Rubatto D, Allen CM, Williams IS, Hermann J, Ireland T (2007) Allanite micro-geochronology: A LA-ICP-MS and SHRIMP U–Th–Pb study. Chem Geol 15:162–182

Gregory C, Rubatto D, Hermann J, Berger A, Engi M (2012) Allanite behaviour during incipient melting in the southern Central Alps. Geochim Cosmochim Acta 84:433–458, doi:10.1016/j.gca.2012.01.020

Hacker BR, Abers GA, Peacock SM (2003) Subduction factory 1. Theoretical mineralogy, densities, seismic wave speeds, and H_2O contents. J Geophys Res 108(B1):2029, doi:10.1029/2001JB001127

Hacker BR, Kylander-Clark ARC, Holder R, Andersen TB, Peterman EM, Walsh EO, Munnikhuis JK (2015) Monazite response to ultrahigh-pressure subduction from U–Pb dating by laser ablation split stream. Chem Geol 409:28–41, doi:10.1016/j.chemgeo.2015.05.008

Hallett BW, Spear FS (2015) Monazite, zircon, and garnet growth in migmatitic pelites as a record of metamorphism and partial melting in the East Humboldt Range, Nevada. Am Mineral 100:951–972

Hammerli J (2014) Using microanalysis of minerals to track geochemical processes during metamorphism: examples from the Mary Kathleen fold belt, Queensland, and the Eastern Mt. Lofty Ranges, South Australia. PhD, James Cook University

Hammerli J, Kemp A, Spandler C (2014) Neodymium isotope equilibration during crustal metamorphism revealed by in situ microanalysis of REE-rich accessory minerals. Earth Planet Sci Lett 392:133–142

Hammerli J, Spandler C, Oliver NHS (2016) Element redistribution and mobility during upper crustal metamorphism of metasedimentary rocks: an example from the eastern Mount Lofty Ranges, South Australia. Contrib Mineral Petrol 171:36, doi:10.1007/s00410-016-1239-7

Harlov DE, Förster HJ (2003) Fluid-induced nucleation of (Y + REE)-phosphate minerals within apatite: Nature and experiment. Part II. Fluorapatite. Am Mineral 88:1209–1229

Harlov DE, Hetherington CJ (2010) Partial high-grade alteration of monazite using alkali-bearing fluids: Experiment and nature. Am Mineral 95:1105–1108

Harlov DE, Wirth R, Förster H-J (2005) An experimental study of dissolution–reprecipitation in fluorapatite: fluid infiltration and the formation of monazite. Contrib Mineral Petrol 150:268–286

Harlov DE, Wirth R, Hetherington CJ (2007) The relative stability of monazite and huttonite at 300–900 °C and 200–1000 MPa: metasomatism and the propagation of metastable mineral phases. Am Mineral 92:1652–1664

Harlov DE, Procházka V, Förster H-J, Matejka D (2008) Origin of monazite–xenotime–zircon–fluorapatite assemblages in the peraluminous Melechov granite massif, Czech Republic. Mineral Petrol 94:9–26

Harlov DE, Wirth R, Hetherington CJ (2011) Fluid-mediated partial alteration in monazite: the role of coupled dissolution–reprecipitation in element redistribution and mass transfer. Contrib Mineral Petrol 162:329–348

Harrison TM, Watson EB (1984) The behavior of apatite during crustal anatexis: equilibrium and kinetic considerations. Geochim Cosmochim Acta 48:1467–1477

Harrison TM, McKeegan KD, LeFort P (1995) Detection of inherited monazite in the Manaslu leucogranite by $^{208}Pb/^{232}Th$ ion microprobe dating: Crystallization age and tectonic implications. Earth Planet Sci Lett 133:271–282

Harrison TM, Catlos EJ, Montel J-M (2002) U–Th–Pb dating of phosphate minerals. Rev Mineral Geochem 48:524–558, doi:10.2138/rmg.2002.48.14

Hawkins DP, Bowring SA (1997) U–Pb systematics of monazite and xenotime: case studies from the Paleoproterozoic of the Grand Canyon, Arizona. Contrib Mineral Petrol 127:87–103

Heaman LM, Parrish RR (eds) (1991) U–Pb geochronology of accessory minerals. Mineral Assoc Canada, Toronto

Heinrich W, Andrehs G, Franz G (1997) Monazite–xenotime miscibility gap thermometry. I. An empirical calibration. J Metamorph Geol 15:3–16

Helms TS, Labotka TC (1991) Petrogenesis of Early Proterozoic pelitic schists of the southern Black Hills, South Dakota: constraints on regional low-pressure metamorphism. Geol Soc Am Bull 103:1324–1334

Hermann J (2002) Allanite: thorium and light rare earth element carrier in subducted crust. Chem Geol 192:289–306

Hermann J, Spandler CJ (2008) Sediment melts at sub-arc depths: an experimental study. J Petrol 49:717–740

Hermann J, Rubatto D (2009) Accessory phase control on the trace element signature of sediment melts in subduction zones. Chem Geol 265

Hetherington CJ, Harlov DE (2008) Partial metasomatic alteration of xenotime and monazite from a granitic pegmatite, Hidra anorthosite massif, southwestern Norway: Dissolution–reprecipitation and the subsequent formation of thorite and uraninite inclusions. Am Mineral 93:806–820

Hetherington CJ, Jercinovic MJ, Williams ML, Mahan K (2008) Understanding geologic processes with xenotime: Composition, chronology, and a protocol for electron probe microanalysis. Chem Geol 254:133–147

Hietpas J, Samson S, Moecher D, Schmitt AK (2010) Recovering tectonic events from the sedimentary record: Detrital monazite plays in high fidelity. Geology 38:167–170

Hietpas J, Samson S, Moecher D (2011) A direct comparison of the ages of detrital monazite versus detrital zircon in Appalachian foreland basin sandstones: Searching for the record of Phanerozoic orogenic events. Earth Planet Sci Lett 310:488–497

Holder RM, Hacker BR, Kylander-Clark ARC, Cottle JM (2015) Monazite trace-element and isotopic signatures of (ultra)high-pressure metamorphism: Examples from the Western Gneiss Region, Norway. Chem Geol 409:99–111, doi:10.1016/j.chemgeo.2015.04.021

Hoskin PW, Schaltegger U (2003) The composition of zircon and igneous and metamorphic petrogenesis. Rev Mineral Geochem 53:27–62

Hughes JM, Rakovan J (2002) The crystal structure of apatite, $Ca_5(PO_4)_3(F,OH,Cl)$. Rev Mineral Geochem 48:1–12

Hughes JM, Rakovan JF (2015) Structurally robust, chemically diverse: apatite and apatite supergroup minerals. Elements 11:165–170

Huminicki DM, Hawthorne FC (2002) The crystal chemistry of the phosphate minerals. Rev Mineral Geochem 48:123–253

Ibler G (2010) Karl Ludwig Giesecke (1761–1833): das Leben und Wirken eines frühen europäischen Gelehrten; Protokoll eines merkwürdigen Lebensweges. Mitt Österr Miner Ges 156:37–114

Janots E, Rubatto D (2014) U–Th–Pb dating of collision in the external Alpine domains (Urseren zone, Switzerland) using low temperature allanite and monazite. Lithos 184:155–166

Janots E, Negro F, Brunet F, Goffé B, Engi M, Bouybaouène ML (2006) Evolution of the REE mineralogy in HP–LT metapelites of the Sebtide complex, Rif, Morocco: monazite stability and geochronology. Lithos 87:214–234

Janots E, Brunet F, Goffé B, Poinssot C, Murchard M, Cemic L (2007) Thermochemistry of monazite-(La) and dissakisite-(La): implications for monazite and allanite stability in metapelites. Contrib Mineral Petrol 154:1–14, doi:10.1007/s00410-006-0176-2

Janots E, Engi M, Berger A, Allaz J, Schwarz J-O, Spandler C (2008) Prograde metamorphic sequence of REE-minerals in pelitic rocks of the Central Alps: implications for allanite–monazite–xenotime phase relations from 250 to 610 °C. J Metamorph Geol 26:509–526, doi:10.1111/j.1525–1314.2008.00774.x

Janots E, Engi M, Rubatto D, Berger A, Gregory C (2009) Metamorphic rates in collisional orogeny from in situ allanite and monazite dating. Geology 37:11–14, doi:10.1130/G25192A.1

Janots E, Berger A, Engi M (2011) Physico-chemical control on the REE-mineralogy in chloritoid-grade metasediments from a single outcrop (Central Alps, Switzerland). Lithos 121:1–11, doi:10.1016/j.lithos.2010.08.023

Janots E, Berger A, Gnos E, Whitehouse M, Lewin E, Pettke T (2012) Constraints on fluid evolution during metamorphism from U–Th–Pb systematics in Alpine hydrothermal monazite. Chem Geol 326–327:61–71, doi:10.1016/j.chemgeo.2012.07.014

Johnston AD, Wyllie PJ (1988) Constraints on the origin of Archean tronhjemites based on phase relationships of Nuk gneiss with H_2O at 15 kbar. Contrib Mineral Petrol 100:35–46

Johnson TE, Clark C, Taylor RJM, Santosh M, Collins AS (2015) Prograde and retrograde growth of monazite in migmatites: An example from the Nagercoil Block, southern India. Geosci Frontiers 6:373–387, doi:10.1016/j.gsf.2014.12.003

Jonsson E, Harlov DE, Majka J, Högdahl K, Persson-Nilsson K (2016) Fluorapatite–monazite–allanite relations in the Grängesberg apatite–iron oxide ore district, Bergslagen, Sweden. Am Mineral 101:1769–1782

Kamber B, Frei R, Gibb A (1998) Pitfalls and new approaches in granulite chronometry: an example from the Limpopo Belt, Zimbabwe. Precambrian Res 91:269–285

Karioris FG, Gowda K, Cartz L (1981) Heavy ion bombardment on monoclinic $ThSiO_4$, ThO_2, and monazite. Radiation Eff Lett 58:1–3

Kato Y, Fujinaga K, Nakamura K, Takaya Y, Kitamura K, Ohta J, Toda R, Nakashima T, Iwamori H (2011) Deep-sea mud in the Pacific Ocean as a potential resource for rare-earth elements. Nat Geosci 4:535–539

Kelly NM, Harley SL, Möller A (2012) Complexity in the behavior and recrystallization of monazite during high-T metamorphism and fluid infiltration. Chem Geol 322–323:192–208, doi:10.1016/j.chemgeo.2012.07.001

Kelsey DE, Clark C, Hand M (2008) Thermobarometric modeling of zircon and monazite growth in melt-bearing systems: examples using model metapelitic and metapsammitic granulites. J Metamorph Geol 26:199–212

Kessel R, Schmidt MW, Ulmer P, Pettke T (2005) Trace element signature of subduction-zone fluids, melts and supercritical liquids at 120–180 km depth. Nature 437:724–727

Kim Y, Yi K, Cho M (2009) Parageneses and Th–U distributions among allanite, monazite, and xenotime in Barroviantype metapelites, Imjingang belt, central Korea. Am Mineral 94:430–438

Kingsbury JA, Miller CF, Wooden JL, Harrison TM (1993) Monazite paragenesis and U–Pb systematics in rocks of the eastern Mojave Desert, California, U.S.A.: implications for thermochronometry. Chem Geol 110:147–167

Kirkland CL, Whitehouse MJ, Slagstad T (2009) Fluid-assisted zircon and monazite growth within a shear zone: a case study from Finnmark, Arctic Norway. Contrib Mineral Petrol 158:637–657, doi:10.1007/s00410-009-0401-x

Kirkland CL, Erickson TM, Johnson TE, Danišík M, Evans NJ, Bourdet J, McDonald BJ (2016) Discriminating prolonged, episodic or disturbed monazite age spectra: An example from the Kalak Nappe Complex, Arctic Norway. Chem Geol 424:96–110, doi:10.1016/j.chemgeo.2016.01.009

Klimm K, Blundy JD, Green TH (2008) Trace element partitioning and accessory phase saturation during H_2O-saturated melting of basalt with implications for subduction zone chemical fluxes. J Petrol 49:523–553

Knudsen AC, Gunter ME (2002) Sedimentary phosphorites—an example: Phosphorite formation, southeastern Idaho, U.S.A. Rev Mineral Geochem 48:363–390

Kohn MJ (2008) PTt data from central Nepal support critical taper and repudiate large-scale channel flow of the Greater Himalayan Sequence. Geol Soc Am Bull 120:259–273

Kohn MJ (2014a) Himalayan metamorphism and its tectonic implications. Ann Rev Earth Planet Sci 42:381–419

Kohn MJ (2014b) "Thermoba-Raman-try": Calibration of spectroscopic barometers and thermometers for mineral inclusions. Earth Planet Sci Lett 388:187–196

Kohn MJ (2017) Titanite petrochronology. Rev Mineral Geochem 83:419–441

Kohn MJ, Malloy MA (2004) Formation of monazite via prograde metamorphic reactions among common silicates: Implications for age determinations. Geochim Cosmochim Acta 68:101–113

Kohn MJ, Penniston–Dorland SC (2017) Diffusion: Obstacles and opportunities in petrochronology. Rev Mineral Geochem 83:103–152

Kohn MJ, Wieland MS, Parkinson CD, Upreti BN (2004) Miocene faulting at plate tectonic velocity in the Himalaya of central Nepal. Earth Planet Sci Lett 228:299–310

Kohn MJ, Wieland MS, Parkinson CD, Upreti BN (2005) Five generations of monazite in Langtang gneisses: implications for chronology of the Himalayan metamorphic core. J Metamorph Geol 23:399–406, doi:10.1111/j.1525–1314.2005.00584.x

Kohn MJ, Vervoort JD (2008) U–Th–Pb dating of monazite by single-collector ICP-MS: pitfalls and potential. Geochem Geophys Geosystem 9, doi:10.1029/2007GC001899

Kon Y, Hoshino M, Sanematsu K, Morita S, Tsunematsu M, Okamoto N, Yano N, Tanaka M, Takagi T (2014) geochemical characteristics of apatite in heavy REE-rich deep-sea mud from Minami–Torishima area, Southeastern Japan. Resour Geol 64:47–57

Köppel V, Grünenfelder M (1975) Concordant U–Pb ages of monazite and xenotime from the Central Alps and the timing of high temperature metamorphism, a preliminary report. Schweiz Mineral Petrog Mitt 55:129–132

Köppel V, Grünenfelder M (1978) The significance of monazite U–Pb ages; examples from the Lepontine area of the Swiss Alps. US Geol Survey, Open File Report 78–701:226–227

Köppel V, Günthert A, Grünenfelder M (1981) Patterns of U–Pb zircon and monazite ages in polymetamorphic units of the Swiss Central Alps. Schweiz Mineral Petrog Mitt 61:97–120

Korhonen F, Saw A, Clark C, Brown M, Bhattacharya S (2011) New constraints on UHT metamorphism in the Eastern Ghats Province through the application of phase equilibria modelling and in situ geochronology. Gondwana Res 20:764–781

Korhonen F, Clark C, Brown M, Bhattacharya S, Taylor R (2013) How long-lived is ultrahigh temperature (UHT) metamorphism? Constraints from zircon and monazite geochronology in the Eastern Ghats orogenic belt, India. Precambrian Res 234:322–350

Krenn E, Finger F (2004) Metamorphic formation of Sr-apatite and Sr-bearing monazite in a high-pressure rock from the Bohemian Massif. Am Mineral 89:1323–1329

Krenn E, Harlov DE, Finger F, Wunder B (2012) LREE-redistribution among fluorapatite, monazite, and allanite at high pressures and temperatures. Am Mineral 97:1881–1890, doi:10.2138/am.2012.4005

Kylander–Clark ARC (2017) Petrochronology by laser–ablation inductively coupled plasma mass spectrometry. Rev Mineral Geochem 83:183–198

Kylander-Clark ARC, Hacker BR, Cottle JM (2013) Laser-ablation split-stream ICP petrochronology. Chem Geol 345:99–112, doi:10.1016/j.chemgeo.2013.02.019

Lanari P, Engi M (2017) Local bulk composition effects on metamorphic mineral assemblages. Rev Mineral Geochem 83:55–102

Larson KP, Gervais F, Kellett DA (2013) A *P–T–t–D* discontinuity in east-central Nepal: Implications for the evolution of the Himalayan mid-crust. Lithos 179:275–292, doi:10.1016/j.lithos.2013.08.012

Larsson D, Söderlund U (2005) Lu–Hf apatite geochronology of mafic cumulates: An example from a Fe–Ti mineralization at Smålands Taberg, southern Sweden. Chem Geol 224:201–211

Levinson AA (1966) A system of nomenclature for rare-earth minerals. Am Mineral 51:152–158

Luo Y, Hughes JM, Rakovan J, Pan Y (2009) Site preference of U and Th in Cl, F, and Sr apatites. Am Mineral 94:345–351

Mahood G, Hildreth W (1983) Large partition coefficients for trace elements in high-silica rhyolites. Geochim Cosmochim Acta 47:11–30

Manning CE (2004) The chemistry of subduction zone fluids. Earth Planet Sci Lett 223:1–16

Manzotti P, Rubatto D, Darling J, Zucali M, Cenki-Tok B, Engi M (2012) From Permo-Triassic lithospheric thinning to Jurassic rifting at the Adriatic margin: Petrological and geochronological record in Valtournenche (Western Italian Alps). Lithos 146–147:276–292, doi:0.1016/j.lithos.2012.05.007

McFarlane CR (2016) Allanite U–Pb geochronology by 193nm LA ICP-MS using NIST610 glass for external calibration. Chem Geol 438:91–102

McFarlane CRM, McCulloch MT (2007) Coupling of in-situ Sm–Nd systematics and U–Pb dating of monazite and allanite with applications to crustal evolution studies. Chem Geol 245:45–60, doi:10.1016/j.chemgeo.2007.07.020

Moecher D, Hietpas J, Samson S, Chakraborty S (2011) Insights into southern Appalachian tectonics from ages of detrital monazite and zircon in modern alluvium. Geosphere 7:494–512

Mogilevsky P (2007) On the miscibility gap in monazite–xenotime systems. Phys Chem Minerals 34:201–214, doi:10.1007/s00269-006-0139-1

Montel J-M (1986) Experimental determination of the solubility of Ce-monazite in SiO_2–Al_2O_3–K_2O–Na_2O melts at 800°C, 2 kbar, under H_2O-saturated conditions. Geology 14:659–662

Montel J-M (1993) A model for monazite/melt equilibrium and application to the generation of granitic magmas. Chem Geol 110:127–146

Montel J-M, Foret S, Veschambre M, Nicollet C, Provost A (1996) Electron microprobe dating of monazite. Chem Geol 131:37–53

Montel J-M, Kornprobst J, Vielzeuf D (2000) Preservation of old U–Th–Pb ages in shielded monazite: example from the Beni Bousera Hercynian kinzigites (Marocco). J Metamorph Geol 18:335–342

Montomoli C, Iaccarino S, Carosi R, Langone A, Visonà D (2013) Tectonometamorphic discontinuities within the Greater Himalayan Sequence in Western Nepal (Central Himalaya): Insights on the exhumation of crystalline rocks. Tectonophysics 608:1349–1370

Mottram CM, Parrish RR, Regis D, Warren CJ, Argles TW, Harris NBW, Roberts NMW (2015) Using U–Th–Pb petrochronology to determine rates of ductile thrusting: Time windows into the Main Central Thrust, Sikkim Himalaya. Tectonics 34:1355–1374, doi:10.1002/2014TC003743

Nemchin AA, Horstwood MS, Whitehouse MJ (2013) High-spatial-resolution geochronology. Elements 9:31–37

Ni Y, Hughes JM, Mariano AN (1995) Crystal chemistry of monazite and xenotime structures. Am Mineral 80:21–26

Oberli F, Meier M, Berger A, Rosenberg CL, Gieré R (2004) U–Th–Pb and ^{230}Th/^{238}U disequilibrium isotope systematics: precise accessory mineral chronology and melt evolution tracing in the Alpine Bergell intrusion. Geochim Cosmochim Acta 68:2543–2560

Ondrejka M, Putiš M, Uher P, Schmiedt I, Pukančík L, Konečný P (2016) Fluid-driven destabilization of REE-bearing accessory minerals in the granitic orthogneisses of North Veporic basement (Western Carpathians, Slovakia). Mineral Petrol, doi:10.1007/s00710-016-0432-8

Overstreet WC (1967) The Geologic Occurrence of Monazite. USGS Professional Paper 530, p. 327

Pan Y, Fleet ME (2002) Composition of the apatite group minerals: Substitution mechanisms and controlling factors. Rev Mineral Geochem 48:13–49

Parrish RR (1990) U–Pb dating of monazite and its application to geological problems. Can J Earth Sci 27:1431–1450

Pattison DR, Vogl JJ (2005) Contrasting sequences of metapelitic mineral-assemblages in the aureole of the tilted Nelson Batholith, British Columbia: Implications for phase equilibria and pressure determination in andalusite–sillimanite-type settings. Can Mineral 43:51–88

Petrík I, Broska I, Lipka J, Siman P (1995) Granitoid allanite-(Ce): substitution relations, redox conditions and REE distributions (on an example of I-type granitoids, Western Carpathians, Slovakia). Geologica Carpathica 46:79–94

Pichavant M, Montel J-M, Richard LR (1992) Apatite solubility in peraluminous liquids: experimental data and an extension of the Harrison–Watson model. Geochim Cosmochim Acta 56:3855–3861

Plank T, Cooper LB, Manning CE (2009) Emerging geothermometers for estimating slab surface temperatures. Nat Geosci 2:611–615

Pochon A, Poujol M, Gloaguen E, Branquet Y, Cagnard F, Gumiaux C, Gapais D (2016) U–Pb LA-ICP-MS dating of apatite in mafic rocks: Evidence for a major magmatic event at the Devonian-Carboniferous boundary in the Armorican Massif (France). Am Mineral 101:2430–2442

Pyle JM, Spear FS (1999) Yttrium zoning in garnet: coupling of major and accessory phases during metamorphic reactions. Geol Mater Res 1:1–49

Pyle JM, Spear FS (2000) An empirical garnet (YAG)–xenotime thermometer. Contrib Mineral Petrol 138:51–58

Pyle JM, Spear FS (2003) Four generations of accessory-phase growth in low-pressure migmatites from SW New Hampshire. Am Mineral 88:338–351

Pyle JM, Spear FS, Rudnick RL, McDonough WF (2001) Monazite–xenotime–garnet equilibrium in metapelites and a new monazite–garnet thermometer. J Petrol 42:2083–2107

Pyle JM, Spear FS, Cheney JT, Layne G (2005a) Monazite ages in the Chesham Pond Nappe, SW New Hampshire, U.S.A.: Implications for assembly of central New England thrust sheets. Am Mineral 90:592–606

Pyle JM, Spear FS, Wark DA, Daniel CG, Storm LC (2005b) Contributions to precision and accuracy of monazite microprobe ages. Am Mineral 90:547–577

Radulescu IG, Rubatto D, Gregory C, Compagnoni R (2009) The age of HP metamorphism in the Gran Paradiso Massif, Western Alps: a petrological and geochronological study of "silvery micaschists". Lithos 110:95–108

Rakovan J, Reeder RJ (1996) Intracrystalline rare earth element distributions in apatite: Surface structural influences on incorporation during growth. Geochim Cosmochim Acta 60:4435–4445

Rakovan J, McDaniel DK, Reeder RJ (1997) Use of surface-controlled REE sectoral zoning in apatite from Llallagua, Bolivia, to determine a single-crystal Sm–Nd age. Earth Planet Sci Lett 146:329–336

Rapp RP, Watson EB (1986) Monazite solubility and dissolution kinetics: implications for the thorium and light rare earth chemistry of felsic magmas. Contrib Mineral Petrol 94:304–316, doi:10.1007/bf00371439

Rapp RP, Ryerson F, Miller CF (1987) Experimental evidence bearing on the stability of monazite during crustal anaatexis. Geophys Res Lett 14:307–310

Rasmussen B (2005) Radiometric dating of sedimentary rocks: the application of diagenetic xenotime geochronology. Earth-Sci Rev 68:197–243

Rasmussen B, Muhlig JR (2007) Monazite begets monazite: evidence for dissolution of detrital monazite and reprecipitation of syntectonic monazite during low-grade regional metamorphism. Contrib Mineral Petrol 154:675–689

Rasmussen B, Bengtson S, Fletcher IR, McNaughton NJ (2002) Discoidal impressions and trace-like fossils more than 1200 million years old. Science 296:1112–1115

Rasmussen B, Fletcher IR, Sheppard S (2005) Isotopic dating of the migration of a low-grade metamorphic front during orogenesis. Geology 33:773-776 doi:10.1130/G21666.1

Rasmussen B, Fletcher IR, Muhling JR (2011) Response of xenotime to prograde metamorphism. Contrib Mineral Petrol 162:1259–1277

Regis D, Cenki-Tok B, Darling J, Engi M (2012) Redistribution of REE, Y, Th, and U at high pressure: Allanite-forming reactions in impure meta-quartzites (Sesia Zone, Western Italian Alps). Am Mineral 97:315–328

Regis D, Rubatto D, Darling J, Cenki-Tok B, Zucali M, Engi M (2014) Multiple metamorphic stages within an eclogite-facies terrane (Sesia Zone, Western Alps) revealed by Th–U–Pb petrochronology. J Petrol 55:1429–1456, doi:10.1093/petrology/egu029

Reimink JR, Davies JHFL, Waldron JWF, Rojas X (2016) Dealing with discordance: a novel approach for analysing U–Pb detrital zircon datasets. J Geol Soc 173:577–585, doi:10.1144/jgs2015-114

Romer RL, Siegesmund S (2003) Why allanite may swindle about its true age. Contrib Mineral Petrol 146:297–307, doi:10.1007/s00410-003-0494-6

Romer RL, Xiao Y (2005) Initial Pb-Sr(-Nd) isotopic heterogeneity in a single allanite-epidote crystal: implications of reaction history for the dating of minerals with low parent-to-daughter ratios. Contrib Mineral Petrol 148:662–674

Romer RL, Rötzler J (2011) The role of element distribution for the isotopic dating of metamorphic minerals. Eur J Mineral 23:17–33

Rubatto D (2002) Zircon trace element geochemistry: partitioning with garnet and the link between U–Pb ages and metamorphism. Geology 184:123–138

Rubatto D, Williams IS, Buick IS (2001) Zircon and monazite response to prograde metamorphism in the Reynolds Range, central Australia. Contrib Mineral Petrol 140:458–468

Rubatto D, Müntener O, Barnhoorn A, Gregory C (2008) Dissolution–reprecipitation of zircon at low-temperature, high-pressure conditions (Lanzo Massif, Italy). Am Mineral 93:1519–1529

Rubatto D, Hermann J, Berger A, Engi M (2009) Protracted fluid-present melting during Barrovian metamorphism in the Central Alps. Contr Mineral Petrol 158:703–722, doi:10.1007/s00410-009-0406-5

Rubatto D, Regis D, Hermann J, Boston K, Engi M, Beltrando M, McAlpine S (2011) Yo-Yo subduction recorded by accessory minerals (Sesia Zone, Western Alps) Nat Geosci 4:338–342, doi:10.1038/ngeo1124

Rubatto D, Chakraborty S, Dasgupta S (2013) Timescales of crustal melting in the Higher Himalayan Crystallines (Sikkim, Eastern Himalaya) inferred from trace element-constrained monazite and zircon chronology. Contrib Mineral Petrol 165:349–372, doi:10.1007/s00410-012-0812-y

Sano Y, Oyama T, Terada K, Hidaka H (1999) Ion microprobe dating of apatite. Chem Geol 153:249–258

Schaltegger U, Davies JHFL (2017) Petrochronology of zircon and baddeleyite in igneous rocks: Reconstructing magmatic processes at high temporal resolution. Rev Mineral Geochem 83:297–328

Schärer U (1984) The effect of initial ^{230}Th disequilibrium on young U Pb ages: The Makalu case, Himalaya. Earth Planet Sci Lett 67:191–204

Schmitt AK, Vazquez JA (2017) Secondary ionization mass spectrometry analysis in petrochronology. Rev Mineral Geochem 83:199–230

Schoene B, Bowring SA (2007) Determining accurate temperature–time paths from U–Pb thermochronology: An example from the Kaapvaal craton, southern Africa. Geochim Cosmochim Acta 71:165–185, doi:10.1016/j.gca.2006.08.029

Secher K, Johnsen O (2008) Minerals in Greenland. Geology and Ore, GEUS 12:1–12

Seydoux-Guillaume A-M, Wirth R, Heinrich W, Montel J-M (2002a) Experimental determination of Thorium partitioning between monazite and xenotime using analytical electron microscopy and X-ray diffraction Rietveld analysis. Eur J Mineral 14:869–878

Seydoux-Guillaume AM, Paquette JL, Wiedenbeck M, Montel J-M, Heinrich W (2002b) Experimental resetting of the U–Th–Pb systems in monazite. Chem Geol 191:165–181, doi:10.1016/S0009-2541(02)00155-9

Seydoux-Guillaume AM, Wirth R, Deutsch A, Scharer U (2004) Microstructure of 24–1928 Ma concordant monazites; implications for geochronology and nuclear waste deposits. Geochim Cosmochim Acta 68:2517–2527, doi:10.1016/j.gca.2003.10.042

Seydoux-Guillaume A-M, Montel J-M, Bingen B, Bosse V, de Parseval P, Paquette J-L, Janots E, Wirth R (2012) Low-temperature alteration of monazite: Fluid mediated coupled dissolution–precipitation, irradiation damage, and disturbance of the U–Pb and Th–Pb chronometers. Chem Geol 330–331:140–158, doi:10.1016/j.chemgeo.2012.07.031

Shazia J, Harlov D, Suzuki K, Kim S, Girish-Kumar M, Hayasaka Y, Ishwar-Kumar C, Windley B, Sajeev K (2015) Linking monazite geochronology with fluid infiltration and metamorphic histories: Nature and experiment. Lithos 236:1–15

Simmat R, Raith MM (2008) U–Th–Pb monazite geochronometry of the Eastern Ghats Belt, India: Timing and spatial disposition of poly-metamorphism. Precambrian Res 162:16–39, doi:10.1016/j.precamres.2007.07.016

Skora S, Blundy J (2012) Monazite solubility in hydrous silicic melts at high pressure conditions relevant to subduction zone metamorphism. Earth Planet Sci Lett 321–322:104–114, doi:10.1016/j.epsl.2012.01.002

Smith HA, Barreiro B (1990) Monazite U–Pb dating of staurolite grade metamorphism in pelitic schists. Contrib Mineral Petrol 105:602–615, doi:10.1007/bf00302498

Smith HA, Giletti BJ (1997) Lead diffusion in monazite. Geochim Cosmochim Acta 61:1047–1055, doi:10.1016/S0016-7037(96)00396-1

Smye AJ, Bickle MJ, Holland TJ, Parrish RR, Condon DJ (2011) Rapid formation and exhumation of the youngest Alpine eclogites: a thermal conundrum to Barrovian metamorphism. Earth Planet Sci Lett 306:193–204

Smye AJ, Roberts NMW, Condon DJ, Horstwood MSA, Parrish RR (2014) Characterising the U–Th–Pb systematics of allanite by ID and LA-ICPMS: Implications for geochronology. Geochim Cosmochim Acta 135:1–28

Spear FS (2010) Monazite–allanite phase relations in metapelites. Chem Geol, doi:10.1016/j.chemgeo.2010.10.004

Spear FS, Parrish RR (1996) Petrology and cooling rates of the Valhalla complex, British Columbia, Canada. J Petrol 37:733–765

Spear FS, Pyle JM (2002) Apatite, monazite, and xenotime in metamorphic rocks. Rev Mineral Geochem 48:293–335

Spear FS, Pyle JM (2010) Theoretical modeling of monazite growth in a low-Ca metapelite. Chem Geol 273:111–119

Spear FS, Pyle JM, Cherniak D (2009) Limitations of chemical dating of monazite. Chem Geol 266:218–230, doi:10.1016/j.chemgeo.2009.06.007

Spear FS, Ashley KT, Webb LE, Thomas JB (2012) Ti diffusion in quartz inclusions: implications for metamorphic time scales. Contrib Mineral Petrol 164:977–986

Stacey JS, Kramers JD (1975) Approximation of terrestrial lead isotope evolution by a 2-stage model. Earth Planet Sci Lett 26:207–221

Steck A, Della Torre F, Keller F, Pfeifer H-R, Hunziker J, Masson H (2013) Tectonics of the Lepontine Alps: ductile thrusting and folding in the deepest tectonic levels of the Central Alps. Swiss J Geosci 106:427–450, doi:10.1007/s00015-013-0135-7

Stepanov AS, Hermann J, Rubatto D, Rapp RP (2012) Experimental study of monazite/melt partitioning with implications for the REE, Th and U geochemistry of crustal rocks. Chem Geol 300–301:200–220, doi:10.1016/j.chemgeo.2012.01.007

Sun S-s, McDonough WF (1989) Chemical and isotopic systematics of oceanic basalts: implications for mantle composition and processes. Geol Soc, London, Spec Publ 42:313–345

Suzuki K, Adachi M (1991) Precambrian provenance and Silurian metamorphism of the Tsubonosawa paragneiss in the South Kitakami terrane, Northeast Japan, revealed by the chemical Th–U-total Pb isochron ages of monazite, zircon and xenotime. Geochem J 25:357–376

Suzuki K, Adachi M, Tanaka T (1991) Middle Precambrian provenance of Jurassic sandstone in the Mino Terrane, central Japan: Th–U-total Pb evidence from an electron microprobe monazite study. Sediment Geol 75:141–147

Suzuki S, Arima M, Williams IS, Shiraishi K, Kagami H (2006) Thermal history of UHT metamorphism in the Napier Complex, East Antarctica: Insights from zircon, monazite, and garnet ages. J Geol 114:65–84, doi:10.1086/498100

Teufel S, Heinrich W (1997) Partial resetting of the U–Pb isotope system in monazite through hydrothermal experiments: An SEM and U–Pb isotope study. Chem Geol 137:273–281

Thomson T (1810) Experiments on allanite, a new mineral from Greenland. Trans R Soc Edinburgh: Earth Sci 8:371–386

Tobgay T, McQuarrie N, Long S, Kohn MJ, Corrie SL (2012) The age and rate of displacement along the Main Central Thrust in the western Bhutan Himalaya. Earth Planet Sci Lett 319–320:146–158, doi:10.1016/j.epsl.2011.12.005

Todd CS, Engi M (1997) Metamorphic field gradients in the Central Alps. J Metamorph Geol 15:513–530

Tomkins HS, Pattison DRM (2007) Accessory phase petrogenesis in relation to major phase assemblages in pelites from the Nelson contact aureole, southern British Columbia. J Metamorph Geol 25:401–421

Tropper P, Manning CE, Harlov DE (2011) Solubility of $CePO_4$ monazite and YPO_4 xenotime in H_2O and H_2O–NaCl at 800 °C and 1 GPa: Implications for REE and Y transport during high-grade metamorphism. Chem Geol 282:58–66, doi:10.1016/j.chemgeo.2011.01.009

Tucker NM, Hand M, Kelsey DE, Dutch RA (2015) A duality of timescales: Short-lived ultrahigh temperature metamorphism preserving a long-lived monazite growth history in the Grenvillian Musgrave–Albany–Fraser Orogen. Precambrian Research 264:204–234, doi:10.1016/j.precamres.2015.04.015

Vallini DA, Rasmussen B, Krapež B, Fletcher IR, McNaughton NJ (2002) Obtaining diagenetic ages from metamorphosed sedimentary rocks: U–Pb dating of unusually coarse xenotime cement in phosphatic sandstone. Geology 30:1083–1086

Vallini DA, Rasmussen B, Krapež B, Fletcher IR, Mcnaughton NJ (2005) Microtextures, geochemistry and geochronology of authigenic xenotime: constraining the cementation history of a Palaeoproterozoic metasedimentary sequence. Sedimentology 52:101–122

van Emden B, Thornber MR, Graham J, Lincoln FJ (1997) The incorporation of actinides in monazite and xenotime from placer deposits in Western Australia. Can Mineral 35:95–104

Vazquez JA, Reid MR (2004) Probing the accumulation history of the voluminous Toba magma. Science 305:991–994, doi:10.1126/science.1096994

Wasserburg G (1963) Diffusion processes in lead–uranium systems. J Geophys Res 68:4823–4846

Watson E, Harrison T (1984) Accessory minerals and the geochemical evolution of crustal magmatic systems: a summary and prospectus of experimental approaches. Phys Earth Planet Inter 35:19–30

Wawrzenitz N, Krohe A, Rhede D, Romer RL (2012) Dating rock deformation with monazite: The impact of dissolution precipitation creep. Lithos 134:52–74

Wawrzenitz N, Krohe A, Baziotis I, Mposkos E, Kylander-Clark AR, Romer RL (2015) LASS U–Th–Pb monazite and rutile geochronology of felsic high-pressure granulites (Rhodope, N Greece): Effects of fluid, deformation and metamorphic reactions in local subsystems. Lithos 232:266–285

Weber WJ (1990) Radiation-induced defects and amorphization in zircon. J Mater Res 5:2687–2697

Wendt I (1984) A three-dimensional U–Pb discordia plane to evaluate samples with common lead of unknown isotopic composition. Chem Geol 46:1–12

White T, Ferraris C, Kim J, Madhavi S (2005) Apatite–an adaptive framework structure. Rev Mineral Geochem 57:307–401

Whitehouse MJ, Kumar GR, Rimša A (2014) Behaviour of radiogenic Pb in zircon during ultrahigh-temperature metamorphism: an ion imaging and ion tomography case study from the Kerala Khondalite Belt, southern India. Contrib Mineral Petrol 168:1–18

Williams I (2001) Response of detrital zircon and monazite, and their U–Pb isotopic systems, to regional metamorphism and host–rock partial melting, Cooma Complex, southeastern Australia. Aust J Earth Sci 48:557–580

Williams ML, Jercinovic MJ (2002) Microprobe monazite geochronology: putting absolute time into microstructural analysis. J Struct Geol 24:1013–1028

Williams ML, Jercinovic MJ, Mahan KH, Dumond G (2017) Electron microprobe petrochronology. Rev Mineral Geochem 83:153–182

Williams ML, Jercinovic MJ (2012) Tectonic interpretation of metamorphic tectonites: integrating compositional mapping, microstructural analysis and in situ monazite dating. J Metamorph Geol 30:739–752, doi:10.1111/j.1525-1314.2012.00995.x

Williams ML, Jercinovic MJ, Terry MP (1999) Age mapping and dating of monazite on the electron microprobe: Deconvoluting multistage tectonic histories. Geology 27:1023–1026

Williams ML, Jercinovic MJ, Hetherington CJ (2007) Microprobe monazite geochronology: Understanding geologic processes by integrating composition and chronology. Ann Rev Earth Planet Sci 35:137–175, doi:10.1146/annurev.earth.35.031306.140228

Williams ML, Jercinovic MJ, Harlov DE, Budzyń B, Hetherington CJ (2011) Resetting monazite ages during fluid-related alteration. Chem Geol 283:218–225, doi:10.1016/j.chemgeo.2011.01.019

Willigers BJA, Baker JA, Krogstad EJ, Peate DW (2002) Precise and accurate in situ Pb–Pb dating of apatite, monazite, and sphene by laser ablation multiple-collector ICP-MS. Geochim Cosmochim Acta 66:1051–1066

Wing BA, Ferry JM, Harrison TM (2003) Prograde destruction and formation of monazite and allanite during contact and regional metamorphism of pelites: petrology and geochronology. Contrib Mineral Petrol 145:228–250

Wolf MB, London D (1994) Apatite dissolution into peraluminous haplogranitic melts: an experimental study of solubilities and mechanisms. Geochim Cosmochim Acta 58:4127–4145

Wood BJ, Blundy J (1997) A predictive model for rare earth element partitioning between clinopyroxene and anhydrous silicate melt. Contrib Mineral Petrol 129:161–181

Yakymchuk C, Brown M (2014) Behaviour of zircon and monazite during crustal melting. J Geol Soc 171:465–479

Yang P, Pattison D (2006) Genesis of monazite and Y zoning in garnet from the Black Hills, South Dakota. Lithos 88:233–253, doi:10.1016/j.lithos.2005.08.012

Yi K, Cho M (2009) SHRIMP geochronology and reaction texture of monazite from a retrogressive transitional layer, Hwacheon Granulite Complex, Korea. Geosci J 13:293, doi:10.1007/s12303-009-0028-y

Zeng L, Asimow PD, Saleeby JB (2005) Coupling of anatectic reactions and dissolution of accessory phases and the Sr and Nd isotope systematics of anatectic melts from a metasedimentary source. Geochim Cosmochim Acta 69:3671–3682

Zhu XK, O'Nions RK (1999) Zonation of monazite in metamorphic rocks and its implications for high temperature thermochronology: a case study from the Lewisian terrain. Earth Planet Sci Lett 171:209–220

Reviews in Mineralogy & Geochemistry
Vol. 83 pp. 419–441, 2017
Copyright © Mineralogical Society of America

13

Titanite Petrochronology

Matthew J. Kohn

Department of Geosciences
Boise State University
Boise, ID 83725
USA

mattkohn@boisestate.edu

INTRODUCTION AND SCOPE

Titanite ($CaTiSiO_5$) is a common mineral in calc-silicates, metamorphosed igneous rocks, and calc-alkaline plutons. The mineral was first named by Martin Klaproth in 1795 for its high content of the element titanium, which had been discovered only a few years prior, and named by Klaproth for the Titans of Greek mythology. The alternate name sphene was proposed by Rene Haüy in 1801 for the mineral's characteristic wedge-shape (sphenos in Greek means "wedge"), but in 1982 the IMA recommended that the name titanite be used in technical writing. The name sphene is still used in the gem industry, and retains a loyal following among some mineralogists.

Titanite's unusual crystal structure—including a 7-fold decahedral site—preferentially takes up numerous geochemically interesting elements, especially U, which enhances its geochronologic utility, but also other high field-strength elements like Zr, and the rare-earth elements (REE). It is one of a handful of major Ti-bearing phases that occur in almost every rock either as a silicate (titanite), as a pure Ti-oxide (rutile or anatase) or as a Fe–Ti oxide (ilmenite or magnetite). Although usually present as a minor or accessory mineral, titanite differs from many other accessory minerals in that its main chemical constituents participate in reactions with other major minerals. Significant substitution of Al and OH enhances this reactivity. Thus, although the stability and reactivity of accessory minerals such as monazite, zircon, etc. are also tied to major mineral reactions (Pyle and Spear 1999; Ferry 2000; Wing et al. 2003; Kohn and Malloy 2004; Tomkins and Pattison 2007; Spear 2010; Kohn et al. 2015, etc.), titanite's connection is much more direct and forms the basis of quantitative thermometry and barometry. More generally, titanite has served a key role in understanding igneous, metamorphic and ore-forming processes (Kerrich and Cassidy 1994; Frost et al. 2000) and is even used to constrain the depositional ages of sedimentary rocks through chronologic analysis of bacterial pseudomorphs (Banerjee et al. 2007; Calderon et al. 2013).

This review updates the outstanding comprehensive work of Frost et al. (2000), who reviewed the crystal chemistry, phase relations and chronologic utility of titanite. I specifically address the fundamental crystal structure, including crystal chemical idiosyncrasies and thermobarometric equilibria; the stability fields and reactions responsible for forming igneous and metamorphic titanite; the U–Pb and Sm–Nd chronologic systems, including potential diffusional biases; petrochronologic case studies in metamorphic and igneous systems; and recommendations for future refinements. While titanite may not be the most ubiquitous of minerals, it stands as a singularly useful petrochronometer in certain common rock types (especially calc-silicates) where other minerals such as zircon or monazite are either absent or less reactive.

1529-6466/17/0083-0013$05.00 (print)
1943-2666/17/0083-0013$05.00 (online)

http://dx.doi.org/10.2138/rmg.2017.83.13

CRYSTAL CHEMISTRY OF TITANITE

Crystal structure and chemical substitutions

Titanite contains three structural sites—a tetrahedral site with Si, an octahedral site with Ti, and an unusual 7-fold decahedral site with Ca (Fig. 1). The 7 oxygens of the decahedral site are arranged in a distorted ring of 5 oxygens, with one oxygen above and one oxygen below. The octahedral sites are arranged in chains parallel to the a-axis and are crosslinked through the tetrahedral and decahedral sites (Fig. 1). In most titanite crystals, Si essentially fills the tetrahedral site, but significant chemical substitution can occur in the octahedral and decahedral sites. The most important of these substitutions include Sr, REE, Pb and U (and possibly Na) for Ca, and Al, Fe^{2+}, Fe^{3+}, Zr, Nb and Ta for Ti. Charge balance among unlike cations is achieved in two different ways. Many proposed substitutions follow typical crystal-chemical principles of coupled cation substitutions on one or more sites (Fig. 1; Ribbe 1980; Paterson and Stephens 1992; Smith et al. 2009), for example:

$$Nb^{5+} + Al^{3+} = 2Ti^{4+} \tag{1}$$

$$REE^{3+} + Al^{3+} = Ca^{2+} + Ti^{4+} \tag{2}$$

One oxygen in the titanite structure, however, is slightly underbonded (Ribbe 1980), and consequently an unusual substitution involves replacement of this O1 oxygen with either OH or F (Fig. 1), with charge balance normally achieved through substitution of Al or Fe^{3+} for Ti:

$$(Al, Fe)^{3+} + (OH, F)^- = Ti^{4+} + O^{2-} \tag{3a}$$

Partial operation of Reaction 3a induces high F content in highly aluminous titanite (Fig. 2; Franz and Spear 1985). Alternatively, the hydroxyl substitution can be written as:

$$(Al, Fe)^{3+} + H^+ = Ti^{4+} \tag{3b}$$

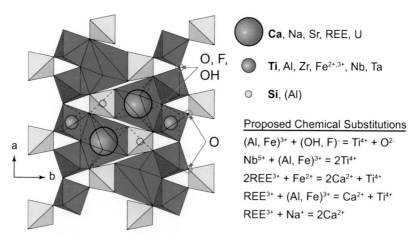

O, F, OH

O

Ca, Na, Sr, REE, U

Ti, Al, Zr, $Fe^{2+,3+}$, Nb, Ta

Si, (Al)

Proposed Chemical Substitutions

$(Al, Fe)^{3+} + (OH, F)^- = Ti^{4+} + O^{2-}$

$Nb^{5+} + (Al, Fe)^{3+} = 2Ti^{4+}$

$2REE^{3+} + Fe^{2+} = 2Ca^{2+} + Ti^{4+}$

$REE^{3+} + (Al, Fe)^{3+} = Ca^{2+} + Ti^{4+}$

$REE^{3+} + Na^+ = 2Ca^{2+}$

Figure 1. Crystal structure and chemistry of titanite projected along *c*-axis showing tetrahedral (Si), octahedral (Ti), and decahedral (Ca) sites. Geochemically important cation substitutions are listed. Oxygen at the O1 site is shared between octahedra and can exhibit major F and OH substitutions. Other O sites, e.g., between octahedra and tetrahedra, are occupied solely by O. Crystal structure image based on http://www.uwgb.edu/dutchs/petrology/Titanite Structure.HTM

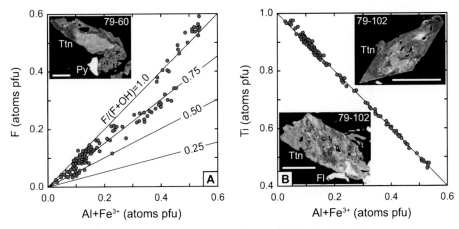

Figure 2. Compositional correlations and substitution of $Al + Fe^{3+}$ in titanite (from Franz and Spear 1985). Backscattered electron images of titanite from two marbles studied by Franz and Spear (1985) show zoning between low Al-concentration (bright areas) and high Al-concentration (dark areas). Py = pyrite; Fl = fluorite. Scale bars are $100\,\mu m$. (A) Fluorine broadly correlates with $Al + Fe^{3+}$, but additional substitution of OH is likely. (B) Substitution of $Al + Fe^{3+}$ (and $F + OH$) occurs primarily for Ti^{4+} (and O^{2-}).

where H^+ protonates the underbonded O. Comparable protonation–deprotonation substitutions are also observed in hydrous minerals such as tourmaline and amphibole. Complete operation of Reaction 3b leads to the endmember vuagnatite [$CaAl(OH|SiO_4)$], a mineral with a different structure from titanite. Difficulty in measuring OH contents in minerals means that the magnitude of the hydroxyl substitution is rarely determined in natural crystals, and because the OH content of titanite is variable, a unique anion charge cannot be assigned. Thus, most titanite compositions are normalized on a three-cation basis.

Thermometry and barometry

Thermometry and barometry may seem distinct from crystal chemistry, but require a clear understanding of chemical substitutions, so are discussed here. Key advances over the last decade now allow direct calculation of the temperature (T) and pressure (P) of crystallization using titanite chemistry in equilibrium with other common minerals. The Zr-in-titanite thermometer (Hayden et al. 2008; Fig. 3) relies on direct substitution of Zr^{4+} for Ti^{4+} according to the equilibrium:

$$CaTiSiO_5 \text{ (titanite)} + ZrSiO_4 \text{ (zircon)} =$$
$$CaZrSiO_5 \text{ (Zr-titanite)} + TiO_2 \text{ (rutile)} + SiO_2 \text{ (quartz)} \tag{4}$$

One major advantage to this equilibrium is that constituents are either sufficiently pure (titanite, zircon, rutile and quartz) that Raoult's law applies, or sufficiently dilute (Zr-titanite) that Henry's law applies, mitigating concerns about component activities except for unusually aluminous titanite (Tropper and Manning 2008). The Zr-in-titanite calibration covers a wide range of temperatures and pressures (Fig. 3A), and demonstrates a moderate P dependence (Fig. 3B). Indeed, analytical errors propagate to only $\sim\pm2\,°C$ for laser-ablation ICP-MS analyses, so the largest sources of uncertainty commonly lie in the uncertainties in P and in the activity of rutile [$a(Rt)$] in rutile-absent rocks. Pressures can be determined either in coexisting rocks or through titanite-specific equilibria (Kapp et al. 2009). However, titanite and rutile rarely coexist or exhibit equilibrium textures, so usually $a(Rt) \leq 1.0$. A nominal $a(Rt)$ of 0.75–0.85 may be assumed based on mineral equilibria applied across a wide range of metamorphic rocks (Kapp et al. 2009, Chambers and Kohn 2012), or the activity of rutile can be directly determined in quartz-bearing rocks from the chemistry of coexisting amphibole, muscovite, or biotite (Chambers and Kohn 2012), as long as compositions are within the range of calibration data.

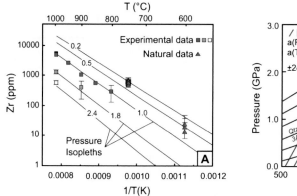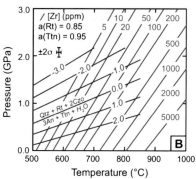

Figure 3. Basis for titanite thermobarometers. (A) Experimental and natural data used to calibrate the Zr-in-titanite thermometer (Hayden et al. 2008). Isopleths show calibration at different pressures (in GPa). (B) Pressure–temperature diagram, contoured for Zr concentrations [Zr, ppm] for the Zr-in-titanite thermometer, and log(K) for the TZARS barometer (Kapp et al. 2009). Contours for Zr-in-titanite assume saturation with quartz and zircon, but typical reduced activities of rutile [a(Rt)] and titanite [a(Ttn)] of 0.85 and 0.95 respectively. Error bar for T represents propagated analytical errors for LA-ICP-MS analysis ($\leq \pm 2\,°C$), and ± 0.1 uncertainty in the activity of rutile (c. $\pm 7\,°C$). Error bar for P represents propagated uncertainties in activities. Qtz = quartz, Rt = rutile, Czo = clinozoisite, An = anorthite, Ttn = titanite.

Pressures of titanite crystallization, particularly in calc-silicates, may be estimated from the TZARS equilibrium (Fig. 3B; Kapp et al. 2009):

$$CaTiSiO_5 \text{ (titanite)} + 3\,CaAl_2Si_2O_8 \text{ (anorthite)} + H_2O =$$
$$2\,Ca_2Al_3Si_3O_{12}(OH) \text{ ([clino]zoisite)} + TiO_2 \text{ (rutile)} + SiO_2 \text{ (quartz)} \tag{5}$$

A specific calibration has not been published, so instead automated calculations are made using an internally consistent thermodynamic database such as Thermocalc (Holland and Powell 2011). Although petrologists commonly avoid estimating P or T using equilibria that involve volatile species, this reaction is relatively insensitive to $a(H_2O)$, and many calc-silicate assemblages are restricted to $X_{H_2O} \geq 0.8$ (Kapp et al. 2009; see later discussion). Although the reaction has a moderate T-dependence, the largest source of uncertainty lies in estimating activities. These include not only the activity for rutile, as discussed above, but also the activity of zoisite or clinozoisite (most rocks contain epidote), and anorthite (plagioclase can exhibit a wide range of compositions). Still, errors less than ~$\pm 0.1\,GPa$ are possible.

Sector zoning

Titanite exhibits a strong propensity to develop sector zoning, which manifests as crystallographically controlled differences in compositions (Paterson et al. 1989). Sector zoning is ascribed to differences in chemical partitioning among distinct crystal faces (Hollister 1970), which is preserved where crystal growth rates were rapid compared to intracrystalline diffusion (Watson and Liang 1995). Cation diffusivities in titanite are generally quite low (see later discussion), so even relatively slow crystal growth can lead to sector zoning. Among all possible crystal faces, the {111} crystal form in titanite is most commonly developed, but other crystal faces [{102}, {001}, etc.] can also form. These different faces partition trace elements quite differently, for example non-{111} sectors have higher Zr and U contents by as much as a factor of ~3, while Pb and REE contents can be higher by a factor of ~2 (Fig. 4A, B; Hayden et al. 2008; Bauer 2015; Walters 2016). Exactly which sectors are exposed in a thin section or grain mount depends on the angle at which the plane of the section intersects the crystal (Fig. 4C; Paterson and Stephens 1992). Because the

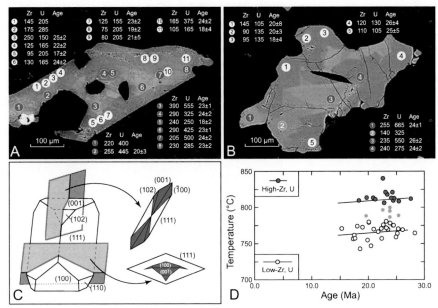

Figure 4. Titanite textures and chemistry illustrating compositional impact of sector zoning. Darkest symbols indicate highest-Zr sectors, intermediate symbols indicate moderate-Zr sectors or mixed analyses, and white symbols indicate low-Zr sectors. (A–B) Backscattered electron images of metamorphic titanite grains from a Himalayan calc-silicate gneiss (sample KN14–51a), showing dark gray, {111} zones with low Zr and U contents flanking light gray core zones with high Zr and U contents. Data from Walters (2016). (C) Typical crystal form of titanite showing dominant {111} faces with minor faces along other directions. Expected patterns of sector zoning (darker vs. lighter shading) are illustrated for transverse vs. longitudinal sections. Modified from Paterson and Stephens (1992). (D) Titanite temperature–time data for KN14–51a define parallel trends with a ~50 °C temperature offset, reflecting high- vs. low-Zr data collected in different sectors. Data from Walters (2016).

{111} faces normally dominate, however, the other sectors are most typically found in titanite cores or along an elongate medial domain that is flanked by {111} domains (Fig. 4A–C).

Not every titanite crystal exhibits sector zoning, but the intersectoral compositional differences do affect petrogenetic applications of titanite geochemistry. The Zr-in-titanite thermometer was calibrated for {111} sector chemistry (Hayden et al. 2008), so use of the unusually high Zr content of other sectors will result in calculated temperatures that are too high (e.g., by 40–80 °C; Fig. 4D; Hayden et al. 2008). Ironically, the non-{111} sectors also have the highest U/Pb, so provide superior age resolution. But until a systematic calibration of Zr partitioning among different sectors is established (beyond a nominal "factor of 3"), these higher-quality ages cannot be linked unequivocally to precise Zr-in-titanite temperatures.

IGNEOUS TITANITE

Petrogenesis

In igneous rocks, titanite occurs most commonly as a late-stage mineral in felsic calc-alkaline plutons. Its stability is strongly influenced by oxygen fugacity (f_{O_2}) and water fugacity (f_{H_2O}). In f_{O_2}–T space, two key reactions limit titanite stability (Fig. 5A; Wones 1989; Frost and Lindsley 1992; Xirouchakis and Lindsley 1998; Frost et al. 2000):

Kohn

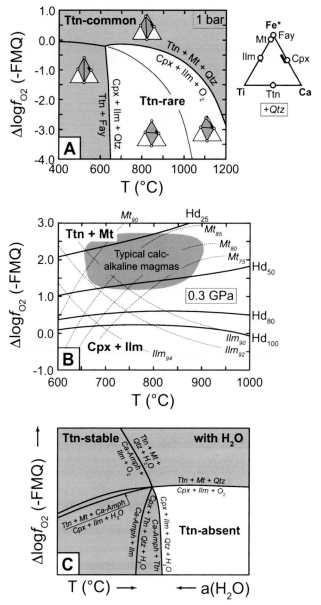

Figure 5. Phase equilibrium constraints on igneous titanite stability. (A) f_{O_2} relative to fayalite–magnetite–quartz equilibrium vs. T. Cpx = clinopyroxene, Fay = fayalite, Ilm = ilmenite, Mt = magnetite, Qtz = quartz, Ttn = titanite, Usp = ulvöspinel. The composition diagrams are projected from quartz and combine Fe^{2+} and Fe^{3+}. Ilmenite and magnetite are assumed to contain small amounts of hematite and ulvöspinel. Small black bar spans typical igneous rock compositions. Modified from Xirouchakis et al. (2001a). (B) f_{O_2} relative to fayalite–magnetite–quartz equilibrium vs. temperature, with detail of Cpx + Ilm = Ttn + Mt equilibrium contoured for the mole fraction of hedenbergite (X_{Hd}). Thin lines illustrate contours of ilmenite (X_{Ilm}) and magnetite (X_{Mt}) for coexisting oxides. Modified from Frost et al. (2000) and Xirouchakis et al. (2001b) (C) f_{O_2} relative to fayalite–magnetite–quartz equilibrium vs. temperature and water fugacity in H_2O-bearing systems. Modified from Xirouchakis et al. (2001b). See also Harlov et al. (2006).

$$3\,CaFeSi_2O_6 \text{ (clinopyroxene)} + 3\,FeTiO_3 \text{ (ilmenite)} + O_2 =$$
$$3\,CaTiSiO_5 \text{ (titanite)} + 2\,Fe_3O_4 \text{(magnetite)} + 3\,SiO_2 \text{ (quartz)} \tag{6}$$

$$CaFeSi_2O_6 \text{ (clinopyroxene)} + FeTiO_3 \text{ (ilmenite)} =$$
$$CaTiSiO_5 \text{ (titanite)} + Fe_2SiO_4 \text{ (olivine)} \tag{7}$$

High-temperatures and low f_{O_2} stabilize ilmenite rather than titanite, so with decreasing T or increasing f_{O_2}, titanite may form via Reactions like (6) and (7). Projecting from quartz into $CaO–TiO_2–(FeO + Fe_2O_3)$ space shows that these reactions reflect crossing tie-line relations (tie-line flip) that should discontinuously produce titanite as a late-stage major titanian phase after earlier-formed ilmenite. Felsic calc-alkaline magmas crystallize at relatively high f_{O_2} and at moderate- to low-temperatures, in the stability field of titanite + magnetite (Fig. 5B). In particular, evolution along ilmenite-buffered equilibria towards higher Δf_{O_2} during cooling is expected drive formation of titanite (Frost and Lindsley 1992; Fig. 5B). These relationships explain both the occurrence of titanite + magnetite + quartz assemblages in many felsic rocks (Wones 1989) and the occurrence of titanite within latest-stage distillates of igneous bodies like the Skaergaard layered mafic intrusion (Xirouchakis et al. 2001b)

High water activity can also stabilize titanite (Fig. 5C), as exemplified by the reaction:

$$7\,CaFeSi_2O_6 \text{ (clinopyroxene)} + 3\,FeTiO_3 \text{ (ilmenite)} + 5\,SiO_2 \text{ (quartz)} + 2\,H_2O =$$
$$2\,Ca_2Fe_5Si_8O_{22}(OH)_2 \text{ (Ca-amphibole)} + 3\,CaTiSiO_5 \text{ (titanite)} \tag{8}$$

which transforms clinopyroxene + ilmenite-bearing assemblages into amphibole + titanite-bearing assemblages (Fig. 5C; Frost et al. 2000). Crystallization of anhydrous minerals during cooling increases the activity of water, ultimately stabilizing titanite at the expense of ilmenite. Especially in the context of high f_{O_2} calc-alkaline magmas, titanite is expected to form with amphibole, late in the crystallization sequence (Xirouchakis et al. 2001b). Relatively high Ti solubility in amphibole tends to increase the stability of titanite-bearing assemblages towards higher temperatures and lower f_{H_2O}, whereas moderate Ti solubility in clinopyroxene increases the stability of ilmenite-bearing assemblages to slightly higher f_{O_2}.

Trace element geochemistry

Trace element geochemistry of igneous titanite has focused primarily on REE. Several experimental studies confirm titanite's preference for the middle REE (Fig. 6A; Green and Pearson 1986; Tiepolo et al. 2002; Prowatke and Klemme 2005). Fitted to lattice strain quasi-parabolic models (Blundy and Wood 1994), partition coefficients reach maxima at about the ionic radius of Dy through Sm, notably lower than the ionic radius of 7-fold coordinated Ca (c. 1.06 Å, comparable to Pr). Thus Ca may not be perfectly sized for the decahedral site, rather it is simply the most abundant divalent cation with approximately the correct ionic radius. Some dependence of partitioning on melt composition is observed, and the discrimination of middle REE from light and heavy REE becomes more pronounced with greater melt polymerization, i.e., the parabolas become tighter (Prowatke and Klemme 2005). Natural titanite–whole rock composition ratios invariably exceed experimental partition coefficients, and do not always conform to simple parabolic models (Fig. 6B). This discrepancy arises because titanite generally forms late. Many incompatible elements like REE are more concentrated in the residual melt, whereas crystallization of other minerals with different REE preferences (e.g., light REE in apatite, heavy REE in zircon, etc.) affects availability of REE and consequently the REE patterns. Thus, titanite geochemistry is expected to reflect the later stage processes of melt crystallization.

Igneous titanite crystals can exhibit sector zoning, and non-{111} sectors consistently show higher REE and Zr concentrations than {111} sectors (Fig. 6C). Extensive data from the Half Dome granodiorite (Bauer 2015) suggest differences of a factor of ~2 for REE, and a

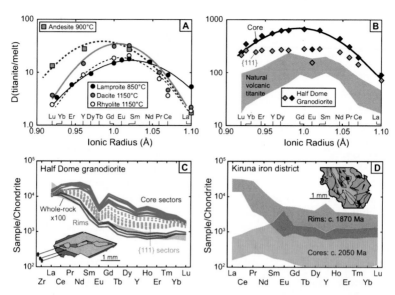

Figure 6. Trace element patterns in titanite. (A) Experimental titanite–melt partitioning data (D = mineral/melt partition coefficient) vs. ionic radius, fit with lattice strain models. Minimum strain (maximum D) typically occurs for MREE (Gd, Eu and Sm). Data and models from Green and Pearson (1986), Tiepolo et al. (2002) and Prowatke and Klemme (2005). (B) Partitioning data for igneous titanite vs. whole-rock or glass, showing generally flatter trends and larger D-values than expected from experiments (note change in vertical scale). Low-temperatures and fractional crystallization likely impact composition of melt from which titanite crystallized. Low Eu likely reflects prior plagioclase fractionation. Core vs. {111} sectors discriminate more differently for MREE than for LREE and HREE. Summarized from Green and Pearson (1986), Solgadi and Sawyer (2008), and Bauer (2015). (C) Zirconium and REE data for igneous titanite and whole rocks from the Half Dome granodiorite. Core sectors (blue) have higher MREE and Zr contents and larger negative Eu anomalies than {111} sectors (orange). Different zones are distinguishable through back-scattered electron images (inset schematic). Data from Bauer (2015). (D) REE data for titanite from the Kiruna iron district, Sweden. Patchy zoning in back-scattered electron images (inset schematic) distinguishes cores vs. rims, which have different ages, REE concentrations, and REE patterns. Data from Smith et al. (2009).

~25% enrichment in Zr for non-{111} sectors compared to {111}. Interestingly, the depth of the Eu anomaly for non-{111} sectors exceeds that of {111} sectors by ~1/3, suggesting that {111} sectors may take up more Eu^{2+}. Differences in Eu^{2+} uptake could be tested by measuring inter-sectoral differences in other divalent cations, such as Sr, Pb, and Ba.

In ore deposits, titanite commonly exhibits complex chemical zoning, and REE patterns may be used to identify fluid sources, whether magmatic or hydrothermal (Fig. 6D; e.g., Smith et al. 2009). In general, light REE enrichment indicates a magmatic origin for ore-forming fluids. For example, the REE patterns for titanite rims at Kiruna compare favorably with trends for titanite from the Half Dome granodiorite, suggesting that ore-forming fluids at Kiruna were derived from associated magmas. Other REE patterns, such as flat profiles in titanite cores, are ascribed to other processes, such as initial hydrothermal alteration, either during or soon after deposition of the original volcanic sequence (Smith et al. 2009). In some deposits, however, host-rock chemistry appears to buffer REE patterns (Chelle-Michou et al. 2015).

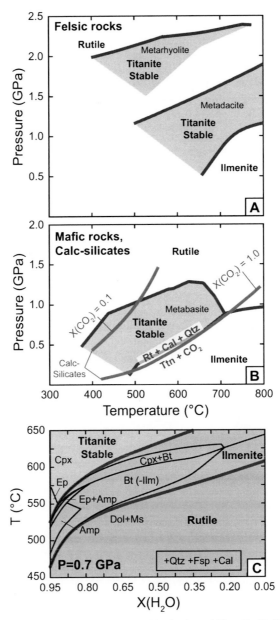

Figure 7. Phase equilibrium constraints on metamorphic titanite stability. (A, B) Composition-specific mineral assemblage diagrams for felsic through mafic bulk compositions (*metarhyolite*: Menold et al. 2009; *metadacite*: Spencer et al. 2013; *metabasite*: Carty et al. 2012; Kohn et al. 2015). Curves (thick red lines) from Frost et al. (2000) show a limiting reaction in calc-silicates that stabilizes titanite at the expense of rutile for two different $X(CO_2)$ values. (C) T–X diagram at $P = 0.7$ GPa for a typical calc-silicate composition (whole-rock data from Cottle et al. 2011).

METAMORPHIC TITANITE

Petrogenesis

The pressure–temperature (*P–T*) stability field of titanite has now been calculated for specific rocks spanning a wide range of bulk compositions (Fig. 7). For metaigneous rocks ranging in composition from metabasite to metarhyolite, the titanite stability field exhibits a wedge-like (arguably "sphenoidal") geometry (Fig. 7A, B). High Ca and Ti in metamafic rocks tend to stabilize titanite at lower *P–T* conditions than for compositionally intermediate and felsic rocks. Among titanian phases, rutile is stable at the highest pressures and lowest temperatures, titanite at intermediate *P–T* conditions, and ilmenite at the lowest pressures and highest temperatures. The common occurrence of titanite rims on rutile (e.g., Lucassen et al. 2010a) attests to *P–T* paths that evolve with increasing *T* or decreasing *P*. In contrast, the transition from titanite to rutile recorded by inclusions in garnet or, rarely, rutile overgrowths on matrix titanite, reflects prograde paths that involve increasing *P* or decreasing *T*. Examples include the Cordillera Darwin metamorphic complex, southern Chile (Kohn et al. 1993), and Franciscan complex blueschists and eclogites exposed at Jenner, California (Krogh et al. 1994). Thus, titanite in metaigneous rocks is expected to record aspects of the earlier and later stages of metamorphism, but not peak conditions.

In calc-silicate rocks, titanite becomes stable only at the highest temperatures and $X(H_2O)$ conditions (Fig. 7C). Thus, unlike in mineral assemblage diagrams for metaigneous rocks, rutile and titanite flank an intermediate stability region occupied by ilmenite. Such models further demonstrate that titanite and epidote coexist only at high $X(H_2O)$, which reduces uncertainties in *P* calculated using the TZARS barometer (Kapp et al. 2009; Fig. 3B). In contrast to its behavior in metaigneous rocks, titanite in calc-silicates may be hoped to record processes occurring near the peak of metamorphism. Note, however, that the F–OH substitution (see Fig. 2) can strongly influence the *P–T*-conditions of formation, and the irregular zoning patterns in these samples point to variable compositions of coexisting fluid on small spatial or temporal scales.

CHRONOLOGIC SYSTEMS

U–Pb

Tilton and Grünenfelder (1968) first recognized the chronologic utility of titanite, owing to a high U content that can reach hundreds of ppm (e.g., Fig. 4). Its affinity for U notwithstanding, not all rocks contain much U, and unlike other high-U minerals such as zircon and monazite, titanite also has a strong affinity for Pb. Consequently, most titanite U–Pb measurements require correction for substantial common Pb before an age can be calculated. Although historically titanite data were collected using ID-TIMS with a minimum of interferences on isotopic masses, many data are now collected *in situ* with LA-ICP-MS (Kylander-Clark 2017, this volume), where high [204]Hg backgrounds compromise measurement of [204]Pb. Because of this interference, corrections are now made in reference to inverse isochrons (Tera-Wasserburg diagrams), regressing less radiogenic and more radiogenic measurements to derive an age (Fig. 8A).

An unusual concern in dating titanite is that different grains or domains may take up different common Pb compositions (Essex and Gromet 2000). Rather than sampling a homogeneous whole-rock Pb value, titanite and other high-Pb minerals have been proposed to inherit a Pb composition that reflects local, isotopically heterogeneous, reactant phases (Romer and Rötzler 2003, 2011; Romer and Xiao 2005). Variations in common lead [207]Pb/[206]Pb may reach ~2% (data from Essex and Gromet 2000, Romer and Xiao 2005), which is at least 20 times larger than analytical precisions for ID-TIMS analysis (Schoene and Baxter 2017, this volume). Low U/Pb matrix minerals include feldspars, epidote, mica, pyroxene, amphibole and tourmaline (e.g., see Kohn and Corrie 2011), and in principle permit variations in common Pb composition to be characterized. Large common Pb corrections mean that highly precise

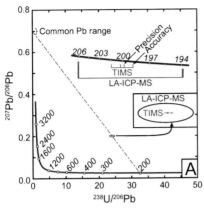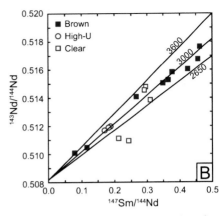

Figure 8. (A) Hypothetical example of U–Pb data for titanite, illustrating form of inverse isochron for a 200 Ma titanite analysis with moderate common Pb. Insets show relative sizes of error ellipses for ID-TIMS (0.1%) vs. LA-ICP-MS (2–3%) and importance of common Pb correction in assigning age uncertainties. Inner vs. outer brackets show age uncertainty when common Pb composition is uncertain by only 0.1% vs. 2% respectively. (B) Example of Sm–Nd dating of titanite illustrating apparent age variation. Data scatter reflects overprinting events at c. 3600 and c. 2650 Ma, as well as variation in initial ^{143}Nd/^{144}Nd. Modified from Amelin (2009).

(e.g., c. 0.1%) titanite ages are not always possible and that each titanite domain must be evaluated independently for U/Pb and potential impact of common Pb variability. For example, consider a hypothetical 200 Ma titanite grain with ^{207}Pb/^{206}Pb = 0.2 and ^{238}U/^{206}Pb = 24.4, with common Pb ^{207}Pb/^{206}Pb = 0.700. Assuming ID-TIMS measurement errors of ±0.1%, an inverse isochron would yield an age of 200.13 ± 0.22 Ma (0.11% error; Fig. 8A). If the ^{207}Pb/^{206}Pb value is subject to 2% uncertainty, however, the error increases to 1.3 Ma (0.65%; Fig. 8A).

Polygenetic or polymetamorphic rocks may be more susceptible to common Pb compositional variation, if the source of this variability indeed lies in reactant phases (Romer and Rötzler 2011). For example, a granite that is subsequently metamorphosed could pose serious chronologic problems if common Pb in two different titanite crystals is derived from low-U feldspar (virtually no radiogenic Pb) vs. high U accessory minerals (mostly radiogenic Pb). In contrast, simple crystallization processes in a magma operating on sub-Ma time scales should not generate large variations in Pb isotope ratios, as long as recharge or assimilation does not alter ^{207}Pb/^{206}Pb spatially or temporally. Clearly, the details of the geologic setting define the level of concern. A small uncertainty in the common Pb composition may be warranted in many rocks, but that assumption must be independently rationalized or demonstrated with measurements on low-U minerals.

For two reasons, use of single collector LA-ICP-MS or ion microprobe is less sensitive to assumptions about common Pb composition. First, common Pb composition and its variation can be measured directly on low-U minerals. Second, analytical uncertainties already encompass likely variation in ^{207}Pb/^{206}P. For example, a nominal ~2% variation in ^{207}Pb/^{206}Pb falls within the 2–4% analytical uncertainty of single-collector LA-ICP-MS and ion probe, so is essentially already accounted for. Moreover, the isotopic measurements on titanite, not common Pb, typically control the age uncertainty. For example, consider again a 200 Ma titanite grain with ^{207}Pb/^{206}Pb = 0.2 and ^{238}U/^{206}Pb = 24.4, but with measurement errors of 2% and 3%, respectively. Assuming common Pb ^{207}Pb/^{206}Pb = 0.700 ± 0.014 (2% error) implies an age of 200.1 ± 6.2 Ma. Reducing the uncertainty in common Pb composition from 2% to 0.1% decreases the age uncertainty by only 0.1 Ma, to ±6.1 Ma (Fig. 8A). Even doubling the error on common Pb composition to 4% increases the age uncertainty by only 0.4 Ma to ±6.6 Ma.

Sm–Nd

In principle, high REE contents, and a general preference for middle REE (Fig. 6) permits Sm–Nd dating of titanite as an alternative to U–Pb dating. For accurate dating, a $^{147}Sm/^{144}Nd$ ratio ≥0.2 is desirable, and although many igneous grains have ratios <0.2, some do reach values of c. 0.5. In his study of the Archean Itsaq Gneiss Complex, western Greenland (near Isua), Amelin (2009) compared U–Pb and Sm–Nd ages for the same separated titanite grains and multi-grain splits. Data for Sm–Nd have $^{147}Sm/^{144}Nd$ between ~0.1 and ~0.5, and scatter broadly (Fig. 8B). Much of the scatter represents mixing of ages between initial and overprinting metamorphic events at ~3600 and ~2650 Ma (Fig. 8B). Some data, however, suggest that initial $^{143}Nd/^{144}Nd$ was not constant for each grain. Because Sm–Nd ages are strongly dependent on initial $^{143}Nd/^{144}Nd$, future dating will require special efforts to ensure the initial ratio is determined accurately.

Diffusional biases

Most chronologic studies of titanite find younger U–Pb ages than for coexisting zircon, and this age difference is typically ascribed to continuous diffusional loss of Pb from titanite during cooling or to diffusional resetting during later events. A classic titanite dataset from the Western Gneiss Region, Norway illustrates discordant titanite data distributed between an orthogneiss protolith age of 1657 Ma and an overprinting metamorphic age of 395 Ma (Fig. 9A; Tucker et al. 2004). Tucker et al. (2004) clearly favored diffusive resetting to explain this discordancy, but they also note that simple mixing of protolith and metamorphic grains and domains would yield an identical distribution of data. The ubiquity of complex zoning in titanite grains (e.g., Figs. 2, 4, 6) recommends characterization of zoning patterns using back-scattered electron imaging, X-ray mapping, or trace element analysis prior to any dating attempts. Because the dated titanite separates were not characterized chemically prior to analysis, the source of discordancy (diffusive Pb loss vs. mixing) remains ambiguous.

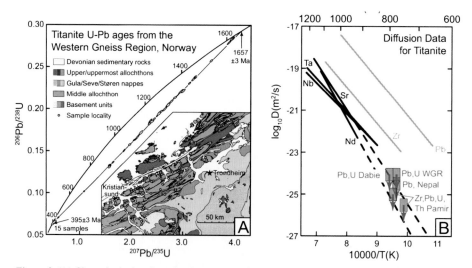

Figure 9. (A) Chronologic data from titanite separates from rocks in the Western Gneiss Region, Norway, showing a well-defined chord between 1657 ± 3 and 395 ± 3 Ma. Inset shows regional geology and location of samples. From Tucker et al. (2004). (B) Diffusivities (D=diffusion coefficient) of various elements in titanite, and constraints on D's from natural samples. Boxes with downward-pointing arrows indicate maximum limits on D for Zr, Pb, U and Th derived from natural samples. Thick gray lines show experimental data for Zr and Pb, which differ from natural constraints. Thick black lines show experimental data for other cations. Modified from Cherniak (2006, 2015); natural data from Kohn and Corrie (2011), Gao et al. (2012), Spencer et al. (2013), and Stearns et al. (2016).

The longstanding debate about Pb diffusion rates in titanite has led to a wide range of proposed closure temperatures (T_c) based on various natural settings: 450–500 °C (Mattinson 1978), 600±25 °C (Heaman and Parrish 1991; Spear and Parrish 1996), 500–670 °C (Mezger et al. 1991), >680±20 °C (Scott and St-Onge 1995), >650 °C (Pidgeon et al. 1996), >712 °C (Zhang and Schärer 1996), >775 °C (Kohn and Corrie 2011), ≥825 °C (Gao et al. 2012). Arguably, experimentally-determined Pb diffusion rates (Cherniak 1993) have influenced discussion most strongly. These relatively high diffusivities (thick gray line, Fig. 9B) imply T_c of c. 600 °C for 100 µm diameter grains cooling at 10 °C/Ma. In her subsequent study of Sr diffusion, however, Cherniak (1995) commented on the surprising inconsistency between Sr and Pb diffusion rates; their experimental diffusivities differ by ~4 orders of magnitude even though both cations are divalent, substitute at the same crystallographic site, and have nearly indistinguishable ionic radii. The difficulty in evaluating diffusive resetting or estimating T_c using natural whole- or multi-grain data as a field-check to experiments is that titanite can grow over a large range of temperatures (Kerrich and Cassidy 1994; Frost et al. 2000), and should not be expected to preserve the same age as, say, a zircon or monazite (e.g., Corfu 1996; Verts et al. 1996). So the fact that a titanite crystal records a younger age than zircon or monazite provides little direct information regarding Pb diffusivities and T_c—titanite might be quite retentive of Pb but simply grow later. For example, Lucassen and Becchio (2003) found that titanite crystallized over a time span of c. 50 Ma in different rocks from the exhumed roots of an arc.

Recent petrochronologic studies (Kohn and Corrie 2011; Gao et al. 2012; Spencer et al. 2013; Stearns et al. 2016) place relatively precise limits on D (uncertainties of a factor of 2–4) and suggest slow diffusion for Pb and other cations. Estimates of D require solving the diffusion equation, and all such solutions involve the combined parameter $D \cdot \Delta t$, where Δt is the duration at maximum T. Thus, to constrain D in titanite, the duration of peak metamorphism must be known well. The four highlighted studies are unusual in providing relatively tight constraints for Δt, as well as for T. For the studies of Gao et al. (2012; Dabie-Sulu) and Spencer et al. (2013; Western Gneiss Region), maximum D is estimated from the preservation of relict pre-metamorphic titanite cores, c. 50–100 µm in diameter, inside metamorphic titanite grains. Given protolith ages, relict cores could have lost, at most, ~50% of their Pb during subsequent metamorphism, which implies $D \cdot \Delta t / a^2 = 0.14$, where a is grain radius (Crank 1975, his Eqn. 6.19); for 90% loss, $D \cdot \Delta t / a^2 = 0.30$, so even a large uncertainty in the maximum percent loss translates into a small error in the limit for D. For Dabie-Sulu rocks, the duration at maximum temperature for the titanite core-overgrowth couple is estimated at c. 5–15 Ma, based on zircon petrochronology that demonstrates protracted heating (e.g., Liou et al. 2012). For Western Gneiss Region rocks, titanite petrochronology requires a minimum duration of 15 Ma at a temperature of c. 780 °C (Spencer et al. 2013; Kohn 2015). This latter limit is a very robust minimum (D is a robust maximum) because titanite rims post-date maximum pressures, and titanite cores could have lost Pb during prograde and maximum pressure metamorphism prior to titanite rim overgrowth. Temperatures are determined both from Zr-in-titanite and regional thermometry. Uranium shows consistent differences in core vs. overgrowth concentrations, implying broadly similar resistance to resetting.

For the studies of Kohn and Corrie (2011, Nepal) and Stearns et al. (2016, Pamir), D is estimated by solving the diffusion equation for a fixed boundary condition and identifying the maximum magnitude of diffusive loss with depth in a crystal:

$$\text{erf}^{-1}\left(\frac{C(x) - C_r}{C_0 - C_r} \right) = \left(\frac{x}{\sqrt{4 D \Delta t}} \right) \tag{9}$$

where $C(x)$, C_r and C_0 represent concentrations at position x, at the rim, and initially in the titanite interior respectively. Nepal titanite shows no resolvable age resetting at depths within 5 μm of the crystal surface for durations of 15 Ma at peak temperature, as determined from other titanite grains from the same rock. Some grains show c. 10 Ma age differences over distances of 15 μm, and in general there is close correspondence between U–Pb age and Zr-in-titanite temperature, irrespective of position within each grain. Pamir titanite shows ~4 μm age gradients produced over a maximum duration of 3–6 Ma. Similarly steep concentration gradients are observed in U, Th, and Zr. Preserved sector zoning in titanite, despite protracted high temperatures, further supports slow diffusivities for Zr and U (Fig. 4, Walters 2016).

Solving the corresponding diffusion equations for maximum D indicates Pb diffusivities at least 4 orders of magnitude lower than experiments, but similar to experimental constraints on Sr. Thus, multiple natural datasets reconcile the curious experimental mismatch between Sr and Pb diffusivities. Lead actually appears to diffuse at about the same rate as Sr, as would be expected from ionic charge and radius considerations. These data imply that Pb diffusion and chronologic resetting are ineffective at temperatures below c. 800 °C. In support of this view, at least one study that originally interpreted younger titanite ages as reflecting diffusive Pb loss (Spear and Parrish 1996) has been reinterpreted using independent data. In light of monazite petrochronology, the "young" titanite ages in the Valhalla Complex (Spear and Parrish 1996) overlap the peak of metamorphism (Spear 2004; c. 820 °C), implying that titanite can preserve peak metamorphic ages even when temperatures exceed 800 °C. Lucassen and Becchio (2003) also argued that the c. 50 Ma age differences observed among titanite separates from different rocks reflected deformation-enhanced recrystallization or growth associated with thermal pulses, not with differential diffusional resetting. Natural intracrystalline gradients for Zr, U and Th that are as steep or steeper than Pb (Stearns et al. 2016) imply that these elements diffuse at least as slowly as Pb, if not more slowly. Although diffusivities for U and Th have not been measured experimentally, results for Zr differ from experimental data (Cherniak 2006) by about 2 orders of magnitude. Natural data imply that Zr-in-titanite temperatures can be preserved on spatial scales of c. 10 μm, even at temperatures of c. 800 °C.

EXAMPLES

Temperature–time histories from single rocks

Data from high-grade calc-silicate gneisses in the central Nepal Himalaya illustrate the potential of combined titanite geochronology and thermometry to illuminate temperature–time ($T–t$) histories (Kohn and Corrie 2011). These rocks consist of relatively simple high-grade assemblages of hornblende + clinopyroxene + quartz + plagioclase + biotite + calcite + titanite + tourmaline + apatite + zircon. Analyses of low-U hornblende, clinopyroxene, biotite and tourmaline provide measures of $^{207}Pb/^{206}Pb$ to anchor inverse isochrons, whereas *in situ* spot analyses within thin sections (Fig. 10A) and depth profiles of separated titanite grains (inset, Fig. 10A,B) provide constraints on T and age, via Zr-in-titanite thermometry and U–Pb geochronology. The depth profile data (Fig. 10B) are novel in providing T and age constraints with μm spatial resolution, and demonstrate preservation of systematic changes in T and age over length scales ≤15 μm. Indeed, it was the failure of these data to conform to the predictions of experimentally estimated diffusivities that led to the recognition that the diffusivity of Pb in titanite must be quite slow (Fig. 9; Kohn and Corrie 2011). Relatively large age uncertainties (typically ~5%, but up to ~10%) reflect low Pb contents for such young grains and high proportions of common Pb, but are still sufficient to resolve $T–t$ histories with acceptable precision. It is only the high U content of these calc-silicates that permits a $T–t$ history to be resolved, and similar attempts with metabasites elsewhere in the Himalaya have failed (unpubl. data).

The $T–t$ history reveals slow heating from ~700 to ~775 °C between ~38 Ma and ~23 Ma, with slight cooling to c. 760 °C by ~21 Ma (Fig. 10C). A second, structurally lower rock yields a similar, albeit higher-T history. No titanite ages younger than ~21 Ma are observed,

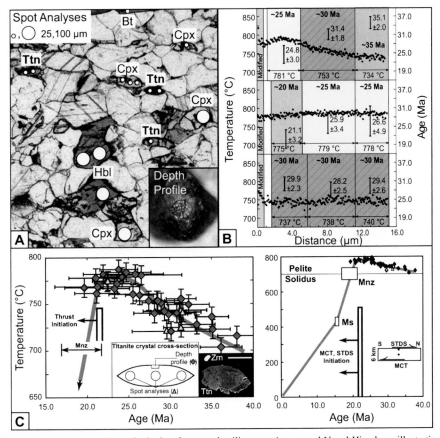

Figure 10. Titanite petrochronologic data from a calc-silicate gneiss, central Nepal Himalaya, illustrating how temperature–time curves are developed for a single rock. All data from Kohn and Corrie (2011). Titanite grains imaged in Fig. 4 reflect similar structural levels and metamorphic conditions. (A) Photomicrograph showing locations of laser ablation analyses. Bt = biotite; Cpx = clinopyroxene; Hbl = hornblende; Ttn = titanite. Other matrix minerals include zircon, quartz, calcite and feldspar. Inset shows separated titanite that was analyzed in depth-profile mode. (B) Depth profile data for 3 titanite grains, showing different zoning in Zr (temperature) and U–Pb ages. Similar background shading indicates comparable temperature and age domains. (C) Temperature time history derived from titanite and regional data showing slow heating followed by rapid cooling as bounding structures initiated above and below the samples. Mnz = monazite Th–Pb ages; Ms = muscovite $^{40}Ar/^{39}Ar$ ages; MCT = Main Central Thrust; STDS = South Tibetan Detachment System. Insets illustrate types of analysis, typical zoning observed in backscattered electron images (scale bar is 100 μm; Zrn = zircon), and structural position of sample relative to MCT and STDS. A second sample, c. 3 km structurally below, shows a similar *T–t* history.

and all temperatures are consistently above muscovite dehydration-melting. Such protracted slow heating at such high temperatures poses an interesting problem for understanding orogenesis because models of lower crustal flow (e.g., Beaumont et al. 2001) predict either that extensional shearing should nucleate at the top of a weakened crustal channel or that thrust-sense shearing should preferentially seek out weakened layers. The South Tibetan Detachment System did not form until much later (c. 21–22 Ma; see summary of Sachan et al. 2010), challenging applicability of the channel flow model. The growing recognition of cryptic metamorphic discontinuities associated with intra-GHS thrusts (first identified using monazite petrochronology by Kohn et al. 2004; cf. Montomoli et al. 2013), however, may explain the titanite data if other thrusts were active above the level of these samples.

Late-stage monazite growth during melt crystallization constrains cooling below ~700 °C to 17–21 Ma (Corrie and Kohn 2011; Fig. 10C). In combination with regional muscovite ages of c. 15 Ma (Martin et al. 2015), these data show a profound change in T–t history coinciding with regional initiation of the structurally lower Main Central Thrust and the structurally higher South Tibetan Detachment System. Cooling likely resulted from two contributing processes. Transfer of rocks to the hanging wall of underlying thrusts (e.g., the Main Central Thrust) and transport through a lateral thermal gradient would have terminated slow heating and dramatically increased cooling rates (e.g., Kohn et al. 2004; Kohn 2008). Simultaneously, initiation of extensional denudation along the South Tibetan Detachment System may have cooled these rocks if the shear zone propagated proximally to these samples. In conjunction with regional petrochronologic observations (Corrie and Kohn 2011), titanite T–t data generally support in-sequence thrust models of orogenesis modified by extensional processes, but are harder to explain by lower crustal flow.

Temperature–time histories from multiple rocks

Chronologic resolution of ±3 to 10% provides age uncertainties of only a few million years for young orogens (Alps, Himalaya, etc.), but these uncertainties expand to tens of Ma for older orogens. In these instances, a different approach is needed, specifically pooling all analyses of titanite grains and domains from a rock to deduce its average T–t point. Because different rocks experience different degrees of titanite growth and recrystallization during metamorphism (e.g., Lucassen and Becchio 2003; Spencer et al. 2013), a robust regional T–t history may be constructed.

For the ultrahigh-pressure rocks of the Scandinavian Caledonides, Spencer et al. (2013) developed T–t points on a rock-by-rock basis, typically collecting Zr and U–Pb data on c. 30 spots distributed across one or more titanite grains in each sample (inset, Fig. 11A). Titanite grains analyzed included grains within host gneisses as well as grains from leucocratic segregations. Excluding a few instances of scattered or bimodal data, each rock defines a precise inverse isochron age and T (Fig. 11A). Combining T–t points across multiple rocks defines a regional T–t path (Fig. 11B; Kohn et al. 2015; Kohn 2016). Titanite is not commonly stable at UHP conditions (Fig. 7), so must form during exhumation below ~1.8 GPa (inset, Fig. 11B), via reaction of rutile plus a calcic phase. Thus, T–t points reflect a late-stage exhumation and cooling history. These data show slow cooling between ~405 and ~390 Ma, followed by rapid cooling to muscovite closure at ~385 Ma.

The composite titanite T–t data yield several important insights. First, they demonstrate hot exhumation, markedly unlike T–t paths from typical subduction zones that should exhibit significant cooling during exhumation (e.g., Gerya et al. 2002). Clearly, mechanisms of post-UHP exhumation can differ from wedge dynamics expected in subduction zones (e.g., Cloos 1982; Gerya et al. 2002). Second, protolith U–Pb ages are at least partially preserved in the cores of some titanite grains. Preservation of Proterozoic ages and disparate U contents through the entire UHP cycle, including residence for at least 10–15 Ma at temperatures of at least 775 °C, provides strict constraints on Pb and U diffusivities (Fig. 9B). Third, temperatures above typical partial melting reactions were sustained for c. 15 Ma, and ages overlap the timing of post-tectonic pegmatite intrusion (Fig. 11B). Thus the rocks experienced protracted residence at mid/deep crustal levels, where titanite crystallized from partial melts, then reacted with other minerals or recrystallized in response to deformation (Spencer et al. 2013). Last, numerous estimates of peak UHP conditions have been proposed based on Sm–Nd ages of garnet and U–Pb ages of zircon (see summary of Kohn et al. 2015). Yet, temperatures were generally too high to preserve peak metamorphic Sm–Nd ages (e.g., see Burton et al. 1995, Ganguly et al. 1998), and zircon is expected to record retrograde, not prograde or peak ages (Kohn et al. 2015). Thus, these data unequivocally constrain exhumation to have initiated prior to ~405 Ma. In conjunction with estimates of peak UHP metamorphic ages of c. 415 to 407 Ma (e.g., Terry et al. 2000, Kylander-Clark et al. 2009), the titanite data imply exhumation rates on the order of 1–3 cm/yr.

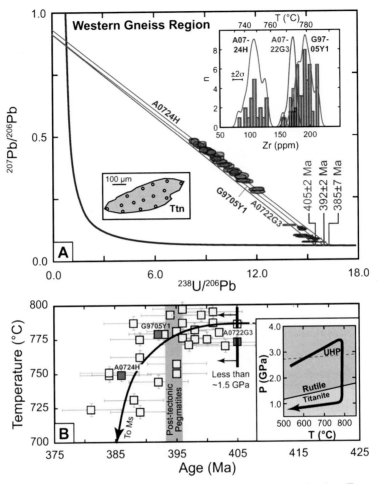

Figure 11. Titanite data from the Western Gneiss Region, Norway, illustrating how *T–t* curves are developed from multiple rocks. All data from Spencer et al. (2013). (A) U–Pb data generally suggest indistinguishable ages but variable amounts of common Pb among spots; Zr contents (upper inset) commonly exhibit heterogeneity. Lower inset illustrates analytical style, with 10 rim and 5 core analyses. Modified from Spencer et al. (2013). (B) Composite *T–t* history, showing slow cooling between ~405 and ~395 Ma, and rapid cooling afterwards. Error bars include a typical 1–2% calibration error. Inset shows general *P–T* stability of titanite. Ages must reflect post-UHP conditions. Modified from Kohn et al. (2015).

Pressure–time histories

If the pressures and ages of formation for different titanite crystals or domains can be determined, a pressure-time history can be derived, permitting burial or exhumation rates to be inferred. In their study of metamorphic titanite from two different calc-silicate nodules from the Dora Maira massif, western Italian Alps, Rubatto and Hermann (2001) identified compositionally distinctive domains in titanite whose ages could be linked to an overall *P–T* history (Fig. 12). One sample was minimally overprinted, and titanite grains contained cores inherited from a previous event and well-developed rims with highly sodic omphacite inclusions (Jd_{38}). Solving the thermodynamic expression of the reaction:

$$6\,TiO_2\,(\text{rutile}) + Ca_3Al_2Si_3O_{12}\,(\text{grossular}) + 3\,CaMgSi_2O_6\,(\text{diopside}) \tag{10}$$
$$= Mg_3Al_2Si_3O_{12}\,(\text{pyrope}) + 6\,CaTiSiO_5\,(\text{titanite})$$

for observed mineral compositions ties the titanite rims and their inclusions to UHP conditions of c. 3.5 GPa at ~35 Ma. In the second sample, relict high-pressure cores with moderately sodic omphacite inclusions (Jd_{15}) were overgrown by rims that were texturally equilibrated with late-stage amphibolite-facies mineral assemblages, including low-Na pyroxene (Jd_{06}). Solving the thermodynamic expression for the albite–jadeite–quartz reaction, in conjunction with regional thermobarometry, Rubatto and Hermann (2001) calculated much lower pressures at c. 33 Ma (c. 1 GPa) and c. 32 Ma (c. 0.5 GPa; Fig. 12). Altogether the petrologic and chronologic data indicate initial exhumation from UHP conditions at rates ≥ 3 cm/yr, slowing progressively to ~1.5 and 0.5 cm/yr as rocks reached typical crustal levels.

As in other terranes, these titanite grains preserve inherited core ages, despite peak metamorphic temperatures of c. 750 °C, perhaps suggesting slower Pb diffusion than expected from experiments (Cherniak 1993). We do not know, however, how long the Dora Maira rocks resided at maximum temperatures, so calculations of Pb diffusivity are not possible.

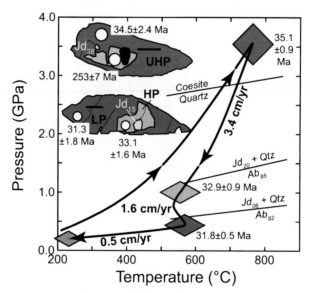

Figure 12. Titanite data and *P–T*–time history from Dora Maira massif calc-silicates, western Italian Alps. Insets show titanite textures evident in back-scattered electron images and locations of several analytical spots. Several ages were averaged to derive the pooled age listed for each stage on the *P–T* path. Titanite contains inherited cores and omphacite inclusions with compositions distinctive of ultrahigh-pressure (Jd_{38}) and later high-pressure (Jd_{15}) conditions. UHP = ultrahigh-pressure, HP = high pressure, LP = low pressure; Jd = jadeite (component in clinopyroxene), Qtz = quartz, Ab = albite. Scale bars are 50 μm. Constraint at lowest *P–T* conditions is derived from fission track age of 29.9 ± 1.4 Ma on zircon. All errors are ±2σ. From Rubatto and Hermann (2001).

Igneous processes

Despite numerous ages collected for igneous titanite, relatively little direct petrochronologic work has attempted to link ages to chemistry, temperature, etc. One major hurdle is that petrochronologic investigations inherently require more than one age. The typically short timescales of igneous processes dramatically complicate this endeavor, because rapid, spatially resolute analytical techniques like LA-ICP-MS and SIMS generally lack the precision to resolve small age differences (Kylander-Clark 2017; Schmitt and Vazquez 2017, both this volume). One ongoing study (Schmitz and Crowley 2014) examines magma chamber dynamics for the Fish Canyon Tuff through the analysis of relatively large (c. 0.5 mm-diameter) titanite grains. Crystals were extracted, doubly polished to thicknesses of ~300 μm, imaged using back-scattered electrons, and analyzed for trace elements using LA-ICP-MS. This approach discriminated two generations of titanite growth. Cores are REE-rich and U + Sr-poor with large Eu anomalies, whereas rims are REE-poor and U + Sr-rich with small Eu anomalies. Reaction zones with intermediate compositions separate cores and rims. The trace element data are interpreted to reflect rejuvenation of the magma chamber driven by underplated mafic magmas. A mafic flux reacted with original titanite grains, which were then overgrown by titanite rims with distinct trace element compositions. Crucially, ID-TIMS dating of different titanite grains suggests older ages for cores (28.4 to 29.0 Ma) than for rims (c. 28.2 Ma). The latter age is chemically linked to magma rejuvenation and is indistinguishable from the age of eruption (Kuiper et al. 2008; Wotzlaw et al. 2013).

These data support previous zircon petrochronologic interpretations of the Fish Canyon Tuff magma chamber (Wotzlaw et al. 2013) that inferred c. 200 ka of initial crystallization to produce a 75–80% crystalline mush, followed by c. 200 ka of remelting associated with underplating of andesitic magmas and infiltration of hot, hydrous fluids, culminating in eruption at 28.2 Ma. Titanite cores must have grown during the initial, nearly complete crystallization of the magma chamber, partially dissolved during fluid influx, then regrown soon prior to eruption.

FUTURE DIRECTIONS

Progress on the following topics would help promote more accurate petrochronologic interpretations of titanite:

Diffusivities

Profound differences distinguish diffusion rates determined experimentally (Cherniak 1993, 2006) vs. estimates from natural composition and age gradients (Fig. 9). A more concerted program of analysis is needed, both experimentally and based on natural samples, to explain these discrepancies. Even if natural constraints on cation diffusion rates in titanite are broadly correct, higher precision observations are needed to refine estimates of D.

Chemical domain structure

Sector-zoning aside, many metamorphic titanite crystals show patchy zoning (e.g., Figs. 2, 4, 6, 12). What causes this behavior? Does titanite dissolve and reprecipitate continuously and randomly, or do textural and local bulk compositional factors, especially heterogeneities in fluid composition and availability, influence how titanite grows or dissolves (e.g., Lucassen et al. 2010b)? Answers to these questions would help target grains more efficiently for petrochronologic analysis, and support thermodynamic and trace element modeling efforts.

Models

Major element solution models for titanite are still lacking (although see Tropper and Manning 2008), and Al- F- and OH-components are not incorporated into comprehensive thermodynamic datasets. While the basic reactions stabilizing titanite are known (Figs. 5, 7), a refined understanding of titanite stability fields and expected compositional variability of titanite across $P–T$ space would enhance the petrologic utility of titanite compositions. Simultaneously, trace element mass balance is difficult to model, particularly where different sectors partition trace elements differently, and titanite apparently undergoes chaotic dissolution-reprecipitation. Both processes can potentially bias trace element mass balance and trace element patterns (e.g., Amelin 2009), and should be characterized and quantified.

Rutile activity

Many equilibria, including Zr-in-titanite thermometry (Fig. 3), assume a(Rt). Although estimates of a(Rt) are possible in many quartz-bearing metamorphic rocks (Chambers and Kohn 2012), application to igneous rocks, which contain dissimilar compositions to the calibration dataset, is lacking. A more comprehensive assessment of a(Rt) across rock types and P–T conditions is needed to fully realize the potential of equilibria involving titanian minerals.

Common Pb

Ultimately, as the precision of isotope measurements improves, the variability in common Pb ^{207}Pb/^{206}Pb taken up by different titanite crystals and domains will limit age resolution (Fig. 8A). Techniques and case studies are needed not only to identify the range of ^{207}Pb/^{206}Pb variation but also to determine the cause of variations in common Pb among titanite grains or within domains of single grains. If local mineral reactions control common Pb (e.g., Romer and Rötzler 2011), methods will need to be developed to identify how other minerals release Pb and how titanite acquires it.

ACKNOWLEDGMENTS

This research was funded by NSF grants EAR-1321897, -1419865, and -1545903. Thanks to Gerhard Franz and Jesse Walters for providing images and data used in Figs. 2 and 4, Gerhard Franz and John Schumacher for detailed reviews, and Martin Engi for editorial handling.

REFERENCES

Amelin Y (2009) Sm–Nd and U–Pb systematics of single titanite grains. Chem Geol 261:52–60, doi:10.1016/j.chemgeo.2009.01.014

Banerjee NR, Simonetti A, Furnes H, Muehlenbachs K, Staudigel H, Heaman L, Van Kranendonk MJ (2007) Direct dating of Archean microbial ichnofossils. Geology 35:487–490, doi:10.1130/g23534a.1

Bauer JE (2015) Complex zoning patterns and rare earth element varitions across titanite crystals from the Half Dome Granodiorite, central Sierra Nevada, California. M. S. University of North Carolina at Chapel Hill

Beaumont C, Jamieson RA, Nguyen MH, Lee B (2001) Himalayan tectonics explained by extrusion of a low-viscosity crustal channel coupled to focused surface denudation. Nature 414:738–742

Blundy J, Wood B (1994) Prediction of crystal–melt partition coefficients from elastic moduli. Nature 372:452–454

Burton KW, Kohn MJ, Cohen AS, O'Nions RK (1995) The relative diffusion of Pb, Nd, Sr and O in garnet. Earth Planet Sci Lett 133:199–211

Calderon M, Prades CF, Herve F, Avendano V, Fanning CM, Massonne HJ, Theye T, Simonetti A (2013) Petrological vestiges of the Late Jurassic-Early Cretaceous transition from rift to back-arc basin in southernmost Chile: New age and geochemical data from the Capitan Aracena, Carlos III, and Tortuga ophiolitic complexes. Geochem J 47:201–217

Carty JP, Connelly JN, Hudson NFC, Gale JFW (2012) Constraints on the timing of deformation, magmatism and metamorphism in the Dalradian of NE Scotland. Scottish J Geol 48:103–117, doi:10.1144/sjg2012-407

Chambers JA, Kohn MJ (2012) Titanium in muscovite, biotite, and hornblende: Modeling, thermometry and rutile activities in metapelites and amphibolites. Am Mineral 97:543–555

Chelle-Michou C, Chiaradia M, Selby D, Ovtcharova M, Spikings RA (2015) High-resolution geochronology of the Coroccohuayco porphyry-skarn deposit, Peru: A rapid product of the Incaic Orogeny. Econ Geol 110:423–443

Cherniak DJ (1993) Lead diffusion in titanite and preliminary results on the effects of radiation damage on Pb transport. Chem Geol 110:177–194

Cherniak DJ (1995) Sr and Nd diffusion in titanite. Chem Geol 125:219–232

Cherniak DJ (2006) Zr diffusion in titanite. Contrib Mineral Petrol 152:639–647

Cherniak DJ (2015) Nb and Ta diffusion in titanite. Chem Geol 413:44–50, doi:10.1016/j.chemgeo.2015.08.010

Cloos M (1982) Flow melanges – numerical modeling and geologic constraints on their origin in the Franciscan subduction complex, California. Geol Soc Am Bull 93:330–345, doi:10.1130/0016–7606(1982)93<330:fmnmag>2.0.co;2

Corfu F (1996) Multistage zircon and titanite growth and inheritance in an Archean gneiss complex, Winnipeg River Subprovince, Ontario. Earth Planet Sci Lett 141:175–186

Corrie SL, Kohn MJ (2011) Metamorphic history of the central Himalaya, Annapurna region, Nepal, and implications for tectonic models. Geol Soc Am Bull 123:1863–1879, doi:10.1130/b30376.1

Cottle JM, Waters DJ, Riley D, Beyssac O, Jessup MJ (2011) Metamorphic history of the South Tibetan Detachment System, Mt. Everest region, revealed by RSCM thermometry and phase equilibria modelling. J Metamorph Geol 29:561–582, doi:10.1111/j.1525–1314.2011.00930.x

Crank J (1975) The Mathematics of Diffusion. Oxford University Press, London

Essex RM, Gromet LP (2000) U–Pb dating of prograde and retrograde titanite growth during the Scandian orogeny. Geology 28:419–422

Ferry JM (2000) Patterns of mineral occurrence in metamorphic rocks. Am Mineral 85:1573–1588

Franz G, Spear FS (1985) Aluminous titanite from the Eclogite Zone, south central Tauern Window, Austria. Chem Geol 50:33–46

Frost BR, Lindsley DH (1992) Equilibria among Fe-Ti oxides, pyroxenes, olivine, and quartz: Part II. Application. Am Mineral 77:1004–1020

Frost BR, Chamberlain KR, Schumacher JC (2000) Sphene (titanite): phase relations and role as a geochronometer. Chem Geol 172:131–148

Ganguly J, Tirone M, Hervig RL (1998) Diffusion kinetics of samarium and neodymium in garnet, and a method for determining cooling rates of rocks. Science 281:805–807

Gao X-Y, Zheng Y-F, Chen Y-X, Guo J (2012) Geochemical and U–Pb age constraints on the occurrence of polygenetic titanites in UHP metagranite in the Dabie orogen. Lithos 136:93–108, doi:10.1016/j.lithos.2011.03.020

Gerya TV, Stöckhert B, Perchuk AL (2002) Exhumation of high-pressure metamorphic rocks in a subduction channel: A numerical simulation. Tectonics 21:doi:10.1029/2002TC001406

Green TH, Pearson NJ (1986) Rare-earth element partitioning between sphene and coexisting silicate liquid at high pressure and temperature. Chem Geol 55:105–119, doi:10.1016/0009–2541(86)90131–2

Harlov D, Tropper P, Seifert W, Nijland T, Forster H-J (2006) Formation of Al-rich titanite (CaTiSiO$_4$O–CaAlSiO$_4$OH) reaction rims on ilmenite in metamorphic rocks as a function of f_{H_2O} and f_{O_2}. Lithos 88:72–84, doi:10.1016/j.lithos.2005.08.005

Hayden LA, Watson EB, Wark DA (2008) A thermobarometer for sphene (titanite). Contrib Mineral Petrol 155:529–540

Heaman L, Parrish R (1991) U–Pb geochronology of accessory minerals. *In*: Applications of radiogenic isotope systems to problems in geology. Short Course Handbook. Vol 19. Heaman L, Ludden JN, (eds). Mineral Assoc Canada, Toronto, Canada, p 59–102

Holland TJB, Powell R (2011) An improved and extended internally consistent thermodynamic dataset for phases of petrological interest, involving a new equation of state for solids. J Metamorph Geol 29:333–383

Hollister LS (1970) Origin, mechanism, and consequences of compositional sector-zoning in staurolite. Am Mineral 55:742–766

Kapp P, Manning CE, Tropper P (2009) Phase-equilibrium constraints on titanite and rutile activities in mafic epidote amphibolites and geobarometry using titanite–rutile equilibria. J Metamorph Geol 27:509–521, doi:10.1111/j.1525–1314.2009.00836.x

Kerrich R, Cassidy KF (1994) Temporal relationships of lode gold mineralization to accretion, magmatism, metamorphism and deformation—Archean to present: A review. Ore Geol Rev 9:263–310, doi:10.1016/0169–1368(94)90001–9

Kohn MJ (2008) *P–T–t* data from central Nepal support critical taper and repudiate large-scale channel flow of the Greater Himalayan Sequence. Geol Soc Am Bull 120:259–273, doi:10.1130/b26252.1

Kohn MJ (2016) Metamorphic chronology—a tool for all ages: Past achievements and future prospects. Am Mineral 101:25–42

Kohn MJ, Corrie SL (2011) Preserved Zr-temperatures and U–Pb ages in high-grade metamorphic titanite: evidence for a static hot channel in the Himalayan orogen. Earth Planet Sci Lett 311:136–143

Kohn MJ, Malloy MA (2004) Formation of monazite via prograde metamorphic reactions among common silicates: Implications for age determinations. Geochim Cosmochim Acta 68:101–113, doi:10.1016/s0016–7037(03)00258–8

Kohn MJ, Spear FS, Dalziel IWD (1993) Metamorphic *P–T* paths from Cordillera Darwin, a core complex in Tierra del Fuego, Chile. J Petrol 34:519–542

Kohn MJ, Wieland MS, Parkinson CD, Upreti BN (2004) Miocene faulting at plate tectonic velocity in the Himalaya of central Nepal. Earth Planet Sci Lett 228:299–310, doi:10.1016/j.epsl.2004.10.007

Kohn MJ, Corrie SL, Markley C (2015) The fall and rise of metamorphic zircon. Am Mineral 100:897–908

Krogh EJ, Oh CW, Liou JG (1994) Polyphase and anticlockwise *P–T* evolution for Franciscan eclogites and blueschists from Jenner, California, USA. J Metamorph Geol 12:121–134

Kuiper KF, Deino A, Hilgen FJ, Krijgsman W, Renne PR, Wijbrans JR (2008) Synchronizing rock clocks of Earth history. Science 320:500–504, doi:10.1126/science.1154339

Kylander-Clark ARC, Hacker BR, Johnson CM, Beard BL, Mahlen NJ (2009) Slow subduction of a thick ultrahigh-pressure terrane. Tectonics 28:doi:10.1029/2007TC002251

Kylander–Clark ARC (2017) Petrochronology by laser–ablation inductively coupled plasma mass spectrometry. Rev Mineral Geochem 83:183–198

Lucassen F, Becchio R (2003) Timing of high-grade metamorphism: Early Palaeozoic U–Pb formation ages of titanite indicate long-standing high-T conditions at the western margin of Gondwana (Argentina, 26–29°S). J Metamorph Geol 21:649–662

Lucassen F, Dulski P, Abart R, Franz G, Rhede D, Romer RL (2010a) Redistribution of HFSE elements during rutile replacement by titanite. Contrib Mineral Petrol 160:279–295, doi:10.1007/s00410–009-0477–3

Lucassen F, Franz G, Rhede D, Wirth R (2010b) Ti–Al zoning of experimentally grown titanite in the system CaO–Al_2O_3–TiO_2–SiO_2–NaCl–H_2O–(F): Evidence for small-scale fluid heterogeneity. Am Mineral 95:1365–1378, doi:10.2138/am.2010.3518

Martin AJ, Copeland P, Benowitz JA (2015) Muscovite $^{40}Ar/^{39}Ar$ ages help reveal the Neogene tectonic evolution of the southern Annapurna Range, central Nepal. Geol Soc, London, Spec Publ 412:199–220

Mattinson JM (1978) Age, origin, and thermal histories of some plutonic rocks from the Salinian Block of California. Contrib Mineral Petrol 67:233–245, doi:10.1007/bf00381451

Menold CA, Manning CE, Yin A, Tropper P, Chen XH, Wang XF (2009) Metamorphic evolution, mineral chemistry and thermobarometry of orthogneiss hosting ultrahigh-pressure eclogites in the North Qaidam metamorphic belt, Western China. J Asian Earth Sci 35:273–284, doi:10.1016/j.jseaes.2008.12.008

Mezger K, Rawnsley C, Bohlen S, Hanson G (1991) U–Pb garnet, sphene, monazite and rutile ages: Implications for the duration of high grade metamorphism and cooling histories, Adirondack Mts., New York. J Geol 99:415–428

Montomoli C, Iaccarino S, Carosi R, Langone A, Visona D (2013) Tectonometamorphic discontinuities within the Greater Himalayan Sequence in Western Nepal (Central Himalaya): Insights on the exhumation of crystalline rocks. Tectonophys 608:1349–1370

Montomoli C, Carosi R, Iaccarino S (2015) Tectonometamorphic discontinuities in the Greater Himalayan Sequence: a local or a regional feature? Geol Soc, London, Spec Publ 412:25–41

Paterson BA, Stephens WE (1992) Kinetically induced compositional zoning in titanite: implications for accessory-phase/melt partitioning of trace elements. Contrib Mineral Petrol 109:373–385, doi:10.1007/bf00283325

Paterson BA, Stephens WE, Herd DA (1989) Zoning in granitoid accessory minerals as revealed by backscattered electron imagery. Mineral Mag 53:55–61

Pidgeon RT, Bosch D, Bruguier O (1996) Inherited zircon and titanite U–Pb systems in an archaean syenite from southwestern Australia: Implications for U–Pb stability of titanite. Earth Planet Sci Lett 141:187–198, doi:10.1016/0012–821x(96)00068–4

Prowatke S, Klemme S (2005) Effect of melt composition on the partitioning of trace elements between titanite and silicate melt. Geochim Cosmochim Acta 69:695–709, doi:10.1016/j.gca.2004.06.037

Pyle JM, Spear FS (1999) Yttrium zoning in garnet: coupling of major and accessory phases during metamorphic reactions. Geol Mat Res 1:1–49

Ribbe PH (1980) Titanite. Rev Mineral 5:137–154

Romer RL, Rötzler J (2003) Effect of metamorphic reaction history on the U–Pb dating of titanite. Geol Soc Special Publ 220:147–158

Romer RL, Rötzler J (2011) The role of element distribution for the isotopic dating of metamorphic minerals. Eur J Mineral 23:17–33

Romer RL, Xiao Y (2005) Initial Pb–Sr(–Nd) isotopic heterogeneity in a single allanite–epidote crystal: implications of reaction history for the dating of minerals with low parent-to-daughter ratios. Contrib Mineral Petrol 148:662–674

Rubatto D, Hermann J (2001) Exhumation as fast as subduction? Geology 29:3–6

Sachan HK, Kohn MJ, Saxena A, Corrie SL (2010) The Malari leucogranite, Garhwal Himalaya, northern India: Chemistry, age, and tectonic implications. Geol Soc Am Bull 122:1865–1876, doi:10.1130/b30153.1

Schmitz MD, Crowley JL (2014) Titanite petrochronology in the Fish Canyon Tuff. EOS–Trans Am Geophys Union 94:V34A-08

Scott DJ, St-Onge MR (1995) Constraints on Pb closure temperature in titanite based on rocks from the Ungava orogen, Canada: Implications for U–Pb geochronology and $P–T–t$ path determinations. Geology 23:1123–1126

Schmitt AK, Vazquez JA (2017) Secondary ionization mass spectrometry analysis in petrochronology. Rev Mineral Geochem 83:199–230

Schoene B, Baxter EF (2017) Petrochronology and TIMS. Rev Mineral Geochem 83:231–260

Smith MP, Storey CD, Jeffries TE, Ryan C (2009) In Situ U–Pb and trace element analysis of accessory minerals in the Kiruna District, Norrbotten, Sweden: New Constraints on the timing and origin of mineralization. J Petrol 50:2063–2094, doi:10.1093/petrology/egp069

Solgadi F, Sawyer EW (2008) Formation of igneous layering in granodiorite by gravity flow: a field, microstructure and geochemical study of the Tuolumne Intrusive Suite at Sawmill Canyon, California. J Petrol 49:2009–2042, doi:10.1093/petrology/egn056

Spear FS (2004) Fast cooling and exhumation of the Valhalla metamorphic core complex, southeastern British Columbia. Int Geol Rev 46:193–209

Spear FS (2010) Monazite–allanite phase relations in metapelites. Chem Geol 279:55–62

Spear FS, Parrish RR (1996) Petrology and cooling rates of the Valhalla complex, British Columbia, Canada. J Petrol 37:733–765

Spencer KJ, Hacker BR, Kylander-Clark ARC, Andersen TB, Cottle JM, Stearns MA, Poletti JE, Seward GGE (2013) Campaign-style titanite U–Pb dating by laser-ablation ICP: Implications for crustal flow, phase transformations and titanite closure. Chem Geol 341:84–101

Stearns MA, Cottle JM, Hacker BR, Kylander-Clark ARC (2016) Extracting thermal histories from the near-rim zoning in titanite using coupled U–Pb and trace-element depth profiles by single-shot laser-ablation split stream (SS-LASS) ICP-MS. Chem Geol 422:13–24, doi:10.1016/j.chemgeo.2015.12.011

Terry MP, Robinson P, Hamilton MA, Jercinovic MJ (2000) Monazite geochronology of UHP and HP metamorphism, deformation, and exhumation, Nordøyane, Western Gneiss Region, Norway. Am Mineral 85:1651–1664

Tiepolo M, Oberti R, Vannucci R (2002) Trace-element incorporation in titanite: constraints from experimentally determined solid/liquid partition coefficients. Chem Geol 191:105–119, doi:10.1016/s0009–2541(02)00151–1

Tilton GR, Grünenfelder MH (1968) Sphene: uranium–lead ages. Science 159:1458–1460, doi:10.1126/science.159.3822.1458

Tomkins HS, Pattison DRM (2007) Accessory phase petrogenesis in relation to major phase assemblages in pelites from the Nelson contact aureole, southern British Columbia. J Metamorph Geol 25:401–421

Tropper P, Manning CE (2008) The current status of titanite–rutile thermobarometry in ultrahigh-pressure metamorphic rocks: The influence of titanite activity models on phase equilibrium calculations. Chem Geol 254:123–132, doi:10.1016/j.chemgeo.2008.03.010

Tucker RD, Robinson P, Solli A, Gee DG, Thorsnes T, Krogh TE, Nordgulen O, Bickford ME (2004) Thrusting and extension in the Scandian hinterland, Norway: New U–Pb ages and tectonostratigraphic evidence. Am J Sci 304:477–532, doi:10.2475/ajs.304.6.477

Verts LA, Chamberlain KR, Frost CD (1996) U–Pb sphene dating of metamorphism: The importance of sphene growth in the contact aureole of the Red Mountain pluton, Laramie Mountains, Wyoming. Contrib Mineral Petrol 125:186–199, doi:10.1007/s004100050215

Walters JB (2016) Protracted thrusting followed by late rapid cooling of the Greater Himalayan Sequence, Annapurna Himalaya, central Nepal: Insights from titanite petrochronology. MS Boise State University

Watson EB, Liang Y (1995) A simple model for sector zoning in slowly grown crystals; implications for growth rate and lattice diffusion, with emphasis on accessory minerals in crustal rocks. Am Mineral 80:1179–1187

Wing BN, Ferry JM, Harrison TM (2003) Prograde destruction and formation of monazite and allanite during contact and regional metamorphism of pelites: petrology and geochronology. Contrib Mineral Petrol 145:228–250

Wones DR (1989) Significance of the assemblage titanite + magnetite + quartz in granitic rocks. Am Mineral 74:744–749

Wotzlaw J-F, Schaltegger U, Frick DA, Dungan MA, Gerdes A, Guenther D (2013) Tracking the evolution of large-volume silicic magma reservoirs from assembly to supereruption. Geology 41:867–870, doi:10.1130/g34366.1

Xirouchakis D, Lindsley DH (1998) Equilibria among titanite, hedenbergite, fayalite, quartz, ilmenite, and magnetite: Experiments and internally consistent thermodynamic data for titanite. Am Mineral 83:712–725

Xirouchakis D, Lindsley DH, Andersen DJ (2001a) Assemblages with titanite ($CaTiOSiO_4$), Ca-Mg-Fe olivine and pyroxenes, Fe-Mg-Ti oxides, and quartz: Part I. Theory. Am Mineral 86:247–253

Xirouchakis D, Lindsley DH, Frost BR (2001b) Assemblages with titanite ($CaTiOSiO_4$), Ca-Mg-Fe olivine and pyroxenes, Fe-Mg-Ti oxides, and quartz: Part II. Application. Am Mineral 86:254–264

Zhang LS, Schärer U (1996) Inherited Pb components in magmatic titanite and their consequence for the interpretation of U–Pb ages. Earth Planet Sci Lett 138:57–65, doi:10.1016/0012-821x(95)00237-7

Reviews in Mineralogy & Geochemistry
Vol. 83 pp. 443–467, 2017
Copyright © Mineralogical Society of America

Petrology and Geochronology of Rutile

Thomas Zack

Department of Earth Sciences
University of Gothenburg
PO Box 460
41430 Gothenburg
Sweden

zack@gvc.gu.se

Ellen Kooijman

Department of Geosciences
Swedish Museum of Natural History
Box 50007
104 05 Stockholm
Sweden

ellen.kooijman@nrm.se

INTRODUCTION AND SCOPE

Rutile (TiO_2) is an important accessory mineral that, when present, offers a rich source of information about the rock units in which it is incorporated. It occurs in a variety of specific microstructural settings, contains significant amounts of several trace elements and is one of the classical minerals used for U–Pb age determination. Here, we focus on information obtainable from rutile in its original textural context. We do not present an exhaustive review on detrital rutile in clastic sediments, but note that an understanding of the petrochronology of rutile in its source rocks will aid interpretation of data obtained from detrital rutile. For further information on the important role of rutile in provenance studies, the reader is referred to previous reviews (e.g., Zack et al. 2004b; Meinhold 2010; Triebold et al. 2012). Coarse rutile is the only stable TiO_2 polymorph under all crustal and upper mantle conditions, with the exception of certain hydrothermal environments (Smith et al. 2009). As such, we will focus on rutile rather than the polymorphs brookite, anatase and ultrahigh-pressure modifications.

In this chapter, we first review rutile occurrences, trace element geochemistry, and U–Pb geochronology individually to illustrate the insights that can be gained from microstructures, chemistry and ages. Then, in the spirit of petrochronology, we show the interpretational power of combining these approaches, using the Ivrea Zone (Italy) as a case study. Finally, we suggest some areas of future research that would improve petrochronologic research using rutile.

RUTILE OCCURRENCE

Rutile is a characteristic mineral in moderate- to high pressure metapelitic rocks, in high pressure metamorphosed mafic rocks, and in sedimentary rocks (e.g., Force 1980; Frost 1991; Zack et al. 2004b; Triebold et al. 2012). Rutile also occurs rarely in magmatic rocks, e.g., anorthosites, as well as in some hydrothermal systems. Coarse-grained rutile is notably absent in many fresh quartzofeldspathic gneisses, peridotites, carbonates (with the exception of some calc-silicates, see Ferry 2000), low-grade metamorphic rocks, volcanic rocks, and fresh granites

1529-6466/17/0083-0014$05.00 (print)
1943-2666/17/0083-0014$05.00 (online)

(with the exception of porphyry copper deposits and certain highly differentiated granites that straddle the line to pegmatites or greisens) (Force 1980; Zack et al. 2004b; Carruzzo et al. 2006).

The presence of rutile depends on its stability in relation to other major Ti phases such as titanite, ilmenite, biotite and to a lesser degree amphibole. Bulk rock compositions with high Ca and Fe^{2+} contents stabilize titanite and ilmenite, respectively, over rutile, and in many metapelitic rocks, rutile occurs only if biotite is absent, in low abundance or starting to break down. Chemical reactions can be written to describe the stability of rutile in more quantitative ways. In mafic systems, Frost (1991) highlights the reaction:

$$\text{Anorthite} + \text{Quartz} + 2\,\text{Ilmenite} = \text{Garnet}\ (Gr_1Alm_2) + 2\,\text{Rutile}\ (\text{GRIPS}) \tag{1}$$

as a major reaction bounding rutile stability. This reaction was calibrated experimentally by Bohlen and Liotta (1986). It is strongly pressure dependent, with rutile being the high pressure phase. Several other reactions with ilmenite or titanite (Manning and Bohlen 1991) are feasible, all more or less pressure dependent. In an experimental study on typical N-MORB, rutile is found only above 1.2 GPa (Ernst and Liu 1998; see Fig. 1). Equilibrium phase diagrams of oxidized N-MORB in the system NCKFMASHTO show rutile stable above 1.0 GPa (Diener and Powell 2012; see Fig. 1).

In high-Al metapelites, Frost (1991) emphasizes the reaction:

$$3\,\text{Ilmenite} + Al_2SiO_5 + \text{Quartz} = \text{Almandine} + 3\,\text{Rutile}\ (\text{GRAIL}) \tag{2}$$

which was experimentally calibrated by Bohlen et al. (1983). Bohlen et al. (1983) also list examples of presumably coexisting rutile and ilmenite occurring between 0.5 and 0.8 GPa in metapelitic assemblages (see Fig. 1). In several medium-pressure metamorphic areas, e.g., the Ivrea Zone, Italy, (Zingg 1980; see below), rutile is reported in metapelitic assemblages but is largely absent in metamorphosed mafic rocks. This is consistent with the relationship shown in Figure 1, where rutile appears to be stabilized at lower pressure in metapelitic systems compared to metabasites, which can be partly explained by higher Fe contents of the latter.

In metamorphic terranes that were metamorphosed at $P \leq 0.5$ GPa, rutile is typically very rare and occurs only in restricted lithologies. For example, in amphibolite- to granulite-facies rocks of Namaqualand (South Africa), a classic low-pressure granulite-facies terrane, ilmenite is the dominant Ti-phase in felsic gneisses and garnet–cordierite metapelitic assemblages over an area of more than 10,000 km² (e.g., Waters 1986a; Willner 1995). Rutile has been reported as the dominant Ti-phase only in rare MgAl-rich rocks (Waters 1986a; Moore and Waters 1990) and sillimanite-rich rocks (Willner 1995). The protoliths of both rock types are assumed to be deeply weathered Mg-enriched soil horizons where Ca, Na and/or Fe have been leached out, whereas Al and by inference Ti have been passively enriched (Moore and Waters 1990), leaving a source composition suitable for abundant rutile over a wide *P–T* range. Examples of similar rutile distribution are found in MgAl-rich rocks from several low-pressure terranes such as the Reynolds Range, Australia (Vry and Baker 2006). Pure quartzites are equally suitable for rutile preservation/formation in low pressure metamorphic areas as they can be almost devoid of Fe and Ca, whereas Ti together with Zr can be relatively enriched as part of the original heavy mineral suite in sandstones (Hubert 1962). A detailed description of rutile in quartzites from the Reynolds Range is found in Rösel et al. (2014; see also Fig. 2I).

A locally important occurrence of rutile is certain metasomatic rock types, most notably formed by Na-metasomatism. Engvik et al. (2014) describe widespread albitisation in several areas of Scandinavia associated with large-scale fluid fluxes during high-grade metamorphism. Here, most elements were leached out of the rock, leaving almost pure albite rocks with a sometimes substantial proportion of refractory minerals, such as rutile and apatite, of economic value. The cm-sized rutile mineral standard R10 from the Bamble Sector, Norway (Luvizotto et al. 2009a) formed in such an environment.

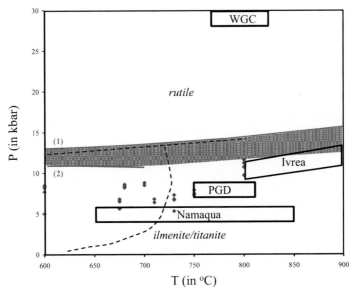

Figure 1. Pressure–temperature diagram showing stability fields for rutile and ilmenite/titanite depending on host rock composition. Transition field in metapelitic rock compositions is marked in light grey. Diamonds are *P–T* conditions for metapelites with coexisting rutile and ilmenite, from Bohlen et al. (1983). Transition field in metabasic rock compositions is marked in dark grey. Solid line (1) is the transition from titanite/ilmenite to rutile in experiments from Ernst and Liu (1998). Solid line (2) is the transition from titanite to rutile in pseudosection models from Diener and Powell (2012). Stippled lines mark the approximate fields for amphibolite-facies (lower left), granulite-facies (lower right) and eclogite facies (upper field) for reference. Solid black outlines mark the *P–T* conditions for metamorphic areas mentioned in the text: Namaqua, after Waters (1986b); Pikwitonei Granulite Domain (PGD), after Mezger et al. (1990); Ivrea, only rutile-bearing units after Redler et al. (2012); Western Gneiss Complex (WGC), after Carswell et al. (2003).

MICROSTRUCTURES OF RUTILE IN METAMORPHIC ROCKS

Microstructural relationships between accessory minerals and surrounding phases in thin section have been shown to be of fundamental importance in several studies (e.g., Engi 2017; Lanari and Engi 2017; Williams et al. 2017). Also, careful investigation of internal chemical textures, like compositional zonation, is indispensable for accessory phases, as illustrated particularly for zircon (e.g., Corfu et al. 2003). Unfortunately, only very limited use has been made so far of these layers of information to better understand processes related to rutile. The benefits of carefully describing microstructural relationships of rutile are best demonstrated when combined with Zr-in-rutile thermometry (e.g., Zack et al. 2004a; Luvizotto and Zack 2009; Kooijman et al. 2012; Ewing et al. 2013; Pape et al. 2016; Pauly et al. 2016). In this context, it is unfortunate that many studies do not report occurrences of rutile, but rather the presences of "oxides" or "opaque phases" (although rutile is not opaque) or simply ignore the occurrence of mineral phases beyond the interest of the study.

The microstructural relationships of rutile and neighboring phases can be seen best either in reflected light microscopy or in backscattered electron (BSE) imaging (see Fig. 2). If rutile is light-colored (often corresponding to low Fe-contents), thin section images in transmitted light microscopy may also be useful, although it is noted that using infrared light microscopy dramatically enhances visibility of internal features within rutile (Cabral et al. 2015). This section focuses on metamorphic processes, because rutile is found most commonly in metamorphic rocks, and because textural information is particularly important to unravel complex histories recorded by these rocks.

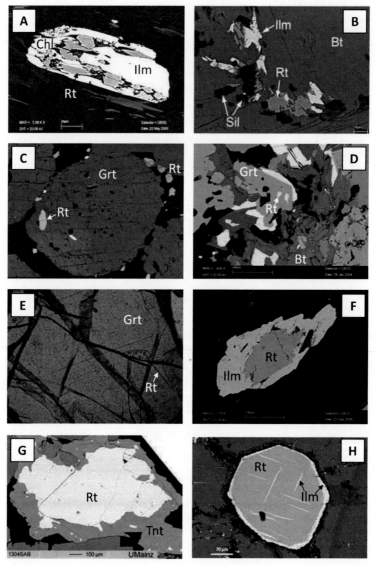

Figure 2. A–C: *prograde formation of rutile.* A) Rutile and chlorite replacing ilmenite (Erzgebirge, Germany; from Luvizotto et al. 2009b); B) Minute rutile (and ilmenite) aggregates growing around biotite during initial stages of biotite breakdown (Ivrea Zone, Italy). [Used by permission of Elsevier Limited, from Luvizotto and Zack (2009) *Chemical Geology*, Vol 261, Fig. 5b, p. 307]; C) Coarse rutile in garnet rims and in matrix from advanced to terminal biotite breakdown (Ivrea Zone, Italy). [Used by permission of Elsevier Limited, from Luvizotto and Zack (2009) *Chemical Geology*, Vol 261, Fig. 6, p. 307]. D–E: *retrograde formation of rutile.* D) Rutile crystallizing out of melt together with euhedral garnet and biotite (Itaucu Complex, Brazil; unpublished); E) Rutile exsolution needles in garnet (xenolith of mafic granulite from Sisimiut aillikite; described in Smit et al. 2016, photo courtesy of M. Smit). F–G: *replacement of rutile.* F) Detrital rutile replaced by ilmenite during prograde metamorphism (Erzgebirge, Germany; from Luvizotto et al. 2009b); G) Rutile replaced by titanite during retrograde metamorphism (garnet amphibolite from Catalina Island; photo courtesy of A. Cruz-Uribe). H–I: *Exsolution lamellae.* H) Ilmenite exsolution lamellae in rutile, with beginning corona texture (crustal xenolith from the Pamir, Tajikistan, unpublished). *Mineral abbreviations:* Bt–biotite, Chl–chlorite, Coe–coesite, Grt–garnet, Ilm–ilmenite, rt–rutile, Sil–sillimanite, Tnt–titanite, Tur–tourmaline, Zrc–zircon.

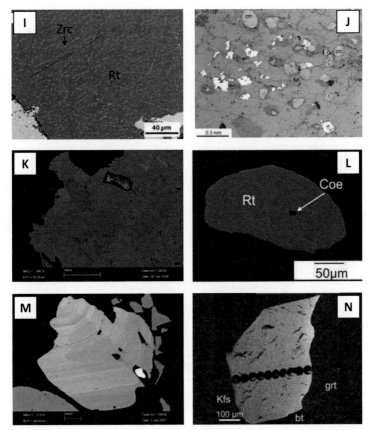

Figure 2 (cont'd). H–I: *Exsolution lamellae* I) Zircon exsolution lamellae in rutile (from UHT septa in Ivrea Zone; from Ewing et al. 2013). J–N) *other features*. J) Irregularly recrystallized rutile in an amphibolite-facies heavy mineral layer in quartzite next to unaffected round detrital zircon and tourmaline crystals (Reynolds Range, Australia; From PhD study of D. Rösel 2014); K) beginning granoblastic–polygonal recrystallization at the edge of a cm-sized rutile crystal with abundant fluid inclusions and subgrain boundaries (in chlorite-rich blackwall zone; Tiburon Peninsula, California; unpublished); L) pristine coesite inclusion in rutile (Dora Maira, Italy; from Hart et al. 2016); M) Relatively rare visible chemical zonation, here oscillatory zonation of Nb and Ta-rich (light) and Nb and Ta-poor (dark) layers (Adelaide Orogen, Australia; described in Crowhurst et al. 2002; unpublished picture); N) Zr WDS mapping of a rutile crystal illustrating asymmetric zoning ranging from ca 700 to 1350 ppm, here higher Zr content (brighter) on the right (Pikwitonei Granulite Domain, Canada). [Used by permission of John Wiley and Sons, from Kooijman, Smit, Mezger and Berndt (2012) *Journal of Metamorphic Geology*, Vol 30, Fig. 5a, p. 405]. All figures are BSE images, except E (transmitted light microphotograph), J (reflected light microphotograph) and N (Zr WDS map).
Mineral abbreviations: Bt–biotite, Chl–chlorite, Coe–coesite, Grt–garnet, Ilm–ilmenite, rt–rutile, Sil–sillimanite, Tnt–titanite, Tur–tourmaline, Zrc–zircon.

Different generations of rutile growth can be tied to different stages of a metamorphic *P–T* path. During prograde metamorphism, rutile growth has been attributed to breakdown of ilmenite, titanite and biotite. For ilmenite breakdown, pseudomorphs of rutile and chlorite intergrowth are reported at the former site of ilmenite (Fig. 2A; Luvizotto et al. 2009b). For the beginning of biotite breakdown, aggregates of small rutile are found around biotite flakes (Fig. 2B; Luvizotto and Zack 2009). At more advanced stages of biotite breakdown, large rutile is found in garnet rims and in the matrix, and biotite relicts are preserved only in garnet cores (Fig. 2C; Luvizotto and Zack 2009). Rutile growth on the retrograde path can be observed in UHT granulites from the Anapolis-Itaucu Complex, Brazil. Here, rutile is found together with and included in small euhedral garnets in a biotite-rich melanosome (Fig. 2D) that can clearly be related to peritectic garnet growth out of a crystallizing melt during early stages of exhumation (see reaction 16 in Moraes et al. 2002). Rutile commonly exsolves from other Ti-bearing phases during cooling, including quartz, biotite (often as oriented needles called sagenite), pyroxenes and garnet (see Fig. 2E).

Microstructures tied to rutile replacement are widespread and well-documented. During prograde metamorphism, pre-metamorphic detrital rutile in impure Fe-rich quartzites can be (partially) replaced by ilmenite (Fig. 2F). Similar coronas can also form during retrograde metamorphism via the GRIPS and GRAIL reactions in the down-pressure direction. As shown in Figure 2G, titanite coronas can also form around rutile, in places exhibiting spectacular resorption. The latter reaction facilitates reaction rate determination from clearly visible Nb back-diffusion into remaining rutile (Lucassen et al. 2010; Cruz-Uribe et al. 2014) although the applicability of this method has been questioned (Kohn and Penniston-Dorland 2017).

Chemical and textural modifications of rutile both during prograde and retrograde metamorphism are critical for understanding the temporal and thermal evolution of rutile-bearing rocks. Most common are sub-micron thin oriented exsolution lamellae of Fe-oxides (ilmenite, hematite, magnetite) within rutile, sometimes forming larger aggregates around rutile (Fig. 2H). Although many probably formed during cooling, changes in oxygen fugacity may also play a role in some cases (e.g., Dill et al. 2007). For Zr-in-rutile thermometry the exsolution of baddelyite and zircon (Fig. 2I) is of great interest, and has been documented in several cases (see below). Poorly documented is the potential of rutile to recrystallize. A spectacular example has been described by Rösel (2014) in a preserved heavy mineral layer within amphibolite-facies quartzites from the Reynolds Range. Here, the only abundant detrital accessory phases are zircon, tourmaline and rutile, typical for a highly mature sediment source (ZTR index of Hubert 1962). Pristine rounded detrital grains of tourmaline and zircon are preserved, whereas former detrital rutile is found only as recrystallized rutile aggregates with irregular grain boundaries (Fig. 2J). These textures show at least qualitatively that rutile has a greater tendency to recrystallize than tourmaline and zircon in this type of environment. It appears that the influence of deformation is not the dominant factor. For example, a partly recrystallized cm-sized rutile crystal is found embedded in weak chlorite-dominated matrix in a blackwall surrounding a blueschist body within the serpentinite mélange of Tiburon, California. The partly preserved early rutile, showing signs of subgrain boundaries, has been recrystallized to a perfect example of a fine-grained granoblastic-polygonal aggregate (Fig. 2K; unpublished). These internal structures are only vaguely visible in high-contrast SEM images, and more refined techniques such as recently calibrated EBSD imaging of rutile (Taylor et al. 2012) are recommended.

Finally, we emphasize the vast amount of information that may be stored within single rutile crystals with respect to microstructures. In the past, inclusions in rutile have been completely ignored. Only recently a large number of phases recording HP and UHP metamorphic conditions have been documented using careful BSE imaging and EDS and Raman identification (Hart et al. 2016), including coesite inclusions (Fig. 2L). In infrared light microscopy, rutile is largely transparent enabling detailed investigation of rutile inclusions in 3D. This has been utilized by Cabral et al. (2015) to investigate fluid inclusions in rutile related to a hydrothermal gold deposit. Internal chemical zonation is of fundamental importance in understanding accessory minerals (see e.g., Corfu et al. 2003; Engi 2017; Kohn 2017; Lanari and Engi 2017; Rubatto 2017). Unfortunately, in contrast to zircon, rutile is not sensitive to cathodoluminescence (CL) imaging. Also, BSE imaging is generally not able to resolve chemical contrasts, as rutile is always close to end-member composition. Only in extreme examples where wt% differences of heavy elements such as Nb, Ta, W and Sb occur is BSE imaging able to resolve these (e.g., Smith and Perseil 1997; Rice et al. 1998; Fig. 2M). In the current absence of a simple imaging technique, chemical heterogeneities within rutile crystals mostly have been made visible by time-consuming WDS mapping, as shown for a Zr mapping in Figure 2N.

ZIRCONIUM-IN-RUTILE THERMOMETRY
AND RUTILE–QUARTZ OXYGEN ISOTOPE THERMOMETRY

Rutile provides the opportunity to obtain precise temperature information for a large variety of rocks. There are several reasons for making such an optimistic statement, in particular: (1) two independent geothermometers have been experimentally calibrated that are applicable for moderate to high temperatures (Zr-in-rutile) as well as for moderate to low temperatures (rutile–quartz oxygen isotope exchange) and (2) rutile occurs in many mineral assemblages that are otherwise inappropriate for temperature estimations (e.g., hydrothermal quartz veins, quartzites). The Zr incorporation in rutile is based on the simple relationship:

$$ZrO_2 \text{ (in rutile)} + SiO_2 \text{ (quartz/coesite)} = ZrSiO_4 \text{ (zircon)} \tag{3}$$

where the two other phases (quartz/coesite and zircon) represent saturating near-endmember phases when present. It is important that rutile coexists with quartz/coesite and zircon in order to apply Zr-in-rutile thermometry as outlined below. Alternatively, the activities of SiO_2 and $ZrSiO_4$ need to be constrained (Zack et al. 2004a). Rutile Zr concentrations are remarkably temperature-sensitive over the range of naturally occurring values from a few 10s to 10000 ppm (Zack et al. 2004a). This relationship has been experimentally calibrated by Watson et al. (2006) at 1 GPa and a significant pressure effect has been calibrated by Tomkins et al. (2007). Both experimental data sets are incorporated in the equation:

$$T\left(^\circ C\right) = \frac{83.9 + 4.10P}{0.1428 - R\ln\left(Zr \text{ in ppm}\right)} - 273 \tag{4}$$

(Tomkins et al. 2007; see Fig. 3), which currently defines the best calibration in the α-quartz stability field. Other expressions apply to the β-quartz and coesite stability fields. Here, P is pressure given in GPa, and R is the gas constant ($0.0083144 \text{ kJ K}^{-1}$). This equation technically is calibrated for α-quartz, but is less than 5 °C divergent from the calibration for β-quartz (Tomkins et al. 2007). Temperature determinations are remarkably consistent with other geothermometers at medium temperature (ca. 500–750 °C), especially in eclogites (e.g., Zack and Luvizotto 2006; Miller et al. 2007). However, both at higher and lower temperatures, results are more difficult to interpret. In HT and UHT granulites, rutile may show a large variation in Zr concentration, even within one thin section (e.g., Kooijman et al. 2012). Zirconium concentrations are typically homogeneous within the core of a given grain, but strongly different between grains. In some cases, highest Zr concentrations were observed for rutile grains included within garnet and orthopyroxene (Zack et al. 2004a); in other cases such systematics do not occur (e.g., Luvizotto and Zack 2009; Ewing et al. 2013). In one case, increasing Zr contents of rutile inclusions in garnet cores to garnet rims are consistent with a prograde heating path in UHT granulites from Antarctica (Pauly et al. 2016). In some cases, pronounced intra-grain zoning, including asymmetric zoning, has been observed (Kooijman et al. 2012). Additionally, decreasing Zr concentrations occur towards grain boundary contacts to zircon (Kooijman et al. 2012; Pape et al. 2016). Different mechanisms have been proposed to explain these features. Most of these models involve a combination of Zr diffusion within rutile as experimentally calibrated by Cherniak et al. (2007; see Fig. 4), and grain boundary diffusion of Zr. The latter is difficult to quantify and is possibly strongly dependent on the presence and nature of a grain-boundary fluid (Luvizotto and Zack 2009) or on the attachment/detachment of ions on buffering phases (Kohn et al. 2016). In consequence, the intra-grain differences in Zr concentration, typically up to values of 1000 ppm Zr at the contact to zircon, have been interpreted to correspond to a closure temperature for Zr diffusion in rutile, corresponding to ca. 700 °C for the given Zr concentration (Luvizotto and Zack 2009; Kooijman et al. 2012). The large inter-grain variation can be explained through different scenarios that are not mutually exclusive: (1) rutile formed at different times during *P–T* histories (Kooijman et al. 2012; Pape et al. 2016; Pauly et al. 2016); (2) differences in the rates of grain-boundary diffusion

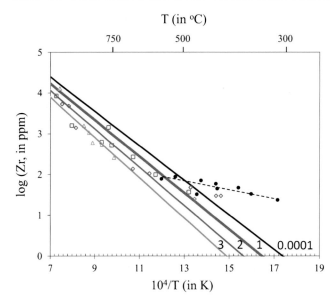

Figure 3. Zr content in rutile as a function of temperature. Solid lines are calculated from experiments of Tomkins et al. (2007; Eqn. 4). The influence of pressure is shown by different lines with pressure in GPa. Open symbols represent results for 24 metamorphic areas from studies of Zack et al. (2004a), Watson et al. (2006) and Zack and Luvizotto (2006), with squares from ca 1 GPa, diamonds from ca 2 GPa and triangles from ca 3 GPa. Black solid circles are results for rutilated quartz veins from low pressures (Shulaker et al. 2015) that seem to indicate a different temperature dependence at low temperatures (stippled line).

(Luvizotto and Zack 2009; drawing on the concept of a fluid diffusion-controlled regime of Dohmen and Chakraborty 2003; see also Kohn et al. 2016), (3) distance of a given rutile crystal to the nearest zircon (Luvizotto and Zack 2009) and (4) potential recrystallization of some rutile crystals (Ewing et al. 2013). A robust distinction between the exact mechanisms in different environments has not yet been laid out. Nevertheless, there is general consensus that temperatures derived from the maximum Zr content in a given rutile population is a good indication for the minimum peak metamorphic temperature of a sample (e.g., Luvizotto and Zack 2009; Kooijman et al. 2012; Ewing et al. 2013; Pape et al. 2016).

A common property of rutile from HT and UHT granulites is the occurrence of baddelyite and/or zircon exsolution lamellae (Fig. 2I). These have been interpreted to be produced during cooling caused by the decreasing Zr solubility of rutile at lower temperature (Kooijman et al. 2012; Ewing et al. 2013). Formation of zircon lamellae within rutile is possible by transporting Si from the surrounding matrix, for example at sufficient Si diffusion rates within rutile, speculated to be a rate-limiting step (Taylor-Jones and Powell 2015). However, the presence of fluids along fractures in rutile may enable sufficient Si supply (Pape et al., 2016). Rutile may contain sufficient Si to form zircon without requiring Si in-diffusion (Kohn et al. 2016), although it is currently unconstrained how much Si typically occurs in natural rutile. In the absence of sufficient Si supply, baddelyite exsolution can occur (Kooijman et al. 2012). The formation of zircon and baddeleyite may depend on factors other than chemical transport such as nucleation (Kohn et al. 2016; Kohn and Penniston-Dorland 2017). Reintegrating the Zr content in rutile before exsolution has been shown to give good estimates of minimum peak metamorphic temperature of a sample (Ewing et al. 2013; Pape et al. 2016; Smit et al. 2016) although no consensus exists for the ideal quantification method. For UHT samples from the Ivrea Zone (see below), significantly lower temperatures were calculated when reintegration was performed using a 60-μm defocused beam EPMA measurement (940–1060 °C; Pape et al. 2016) and a defocused spot LA-ICP-MS measurement (1000–1020 °C; Ewing et al. 2013) compared to reintegration based on grayscale BSE imaging (1085–1194 °C; Pape et al. 2016). In any case, Zr-in-rutile temperatures often provide robust estimates on peak metamorphic conditions for UHT granulites (e.g., Zack et al. 2004a; Harley 2008; Ewing et al. 2013) where traditional thermobarometers (e.g., Fe–Mg exchange) are re-equilibrated.

Different challenges performing Zr-in-rutile thermometry exist at low temperature ($<500\,°C$): (1) natural conditions are well outside the calibrated range ($650–1300\,°C$; Watson et al. 2006; Tomkins et al. 2007); (2) intra-grain Zr diffusion in rutile becomes very slow (Cherniak et al., 2007); (3) expected Zr concentrations become increasingly challenging to measure precisely by some techniques such as electron probe micro analysis ($<30\,ppm$ below $500\,°C$); and (4) Zr solubility in any grain boundary fluid will also be substantially reduced, making it increasingly difficult for equilibrium to be attained among rutile, quartz and zircon grains that are not in direct contact. Unfortunately, the latter parameter has not been experimentally quantified.

An alternative that avoids these problems is provided by the rutile–quartz oxygen isotope thermometer. Experimentally calibrated down to $300\,°C$ (Matthews et al. 1979), it has the advantage that oxygen diffusion in rutile is ca four orders of magnitude faster than Zr in rutile (see Fig. 4) and touching grain boundaries are easy to find. An attractive bonus is that the rutile–quartz oxygen isotope thermometer is, next to magnetite–quartz, the second most temperature-sensitive oxygen isotope-exchange thermometer (e.g., Matthews and Schliestedt 1984), potentially distinguishing temperature differences to better than $±50\,°C$ (Shulaker et al. 2015). In quartz-rich and rutile-poor lithologies (in particular quartzites) effective closure temperatures are defined by O diffusion in rutile, which is rarely below $600\,°C$ even for slow cooling and small grain sizes (see Valley 2001). Traditionally, the rutile–quartz oxygen isotope thermometer has been used successfully for more than 30 years (see examples in Matthews et al. 1979 and Agrinier 1991). Surprisingly, it is only recently that it has received renewed attention. Shulaker et al. (2015) calibrated oxygen isotope measurements of rutile by HR-SIMS, and applied the new protocol to 'rutilated' quartz from alpine clefts from a range of different metamorphic settings. The oxygen isotope exchange temperatures derived from rutile–quartz pairs span from 310 to $540\,°C$, consistent with the range of regional metamorphic conditions in that area. Interestingly, Zr-in-rutile temperatures correlate strongly with oxygen isotope temperature, but do not follow the temperatures expected from the Tomkins et al. (2007) calibration (see Fig. 3). An explanation cannot be given so far, but a misfit of the Zr-in-rutile thermometer with other temperature estimates is apparent in several other low-T metamorphic and hydrothermal examples (Cabral et al. 2015). Shulaker et al. (2015) caution against the use of the Zr-in-rutile thermometry at low-T environments and recommend using the rutile–quartz oxygen isotope thermometer instead.

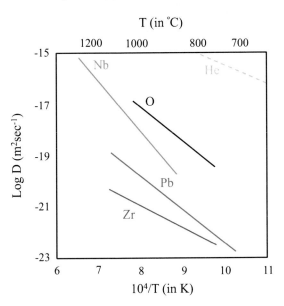

Figure 4. Arrhenius diagram illustrating seven orders of magnitude different diffusion systematics in rutile. Solid lines mark the temperature range of experiments for each system; stippled line for He is extrapolated from lower temperature range. Data source: Zr-Cherniak et al. (2007); Pb-Cherniak (2000); Nb-Marschall et al. (2013); O-Moore et al. (1998); He-Cherniak and Watson (2011).

NIOBIUM AND Cr DISTRIBUTION AS SOURCE ROCK INDICATORS

Niobium and Cr in detrital rutile have been used extensively to distinguish different source rocks in provenance studies. The concept is based on the observation that the bulk Nb and Ti contents in many rocks are predominantly concentrated in rutile when rutile is stable; consequently, Nb concentrations in rutile reflect Nb/Ti ratios in host rock, which vary widely (see Zack et al. 2002, 2004b). Several attempts have been made to distinguish, geochemically, rutile that originates from metapelites and metabasites, which are the main source lithologies for detrital rutile (e.g., Triebold et al. 2012). All such attempts are entirely empirical, have a pragmatic approach and are not meant to be based on first principles. Rather these two rock types reflect endmembers on a compositional continuum from mafic to felsic and aluminous rocks. Furthermore, Nb and Ti are not always controlled only by rutile, and Nb/Ti fractionation can occur between rutile and biotite (Luvizotto and Zack 2009) and rutile and titanite (Lucassen et al. 2010; Cruz-Uribe et al. 2014).

Despite all those shortcomings, discrimination diagrams based on Nb and Cr for a large suite of detrital rutile grains provide a statistical perspective on the distribution of other trace elements derived from different source rocks. We have selected four studies on trace element distribution of detrital rutile that report Nb and Cr (in addition to several other trace elements, including U): Jurassic to recent sediments from the North Sea (Morton and Chenery 2009), Paleozoic sediments from Germany (Rösel 2014) and Turkey (Okay et al. 2011) as well as Mesoproterozoic greenschist facies quartzites from Australia (Rösel et al. 2014). Altogether 1170 analyses are considered (excluding only 55 rutiles with Ce and/or Y with > 10 ppm from the North Sea data set, potentially mixed analyses including phases like zircon, monazite, etc). About 19% of detrital rutile grains fall in the metabasite field and 81% in the metapelite field based on the latest discrimination diagram (Fig. 5; Triebold et al. 2012).

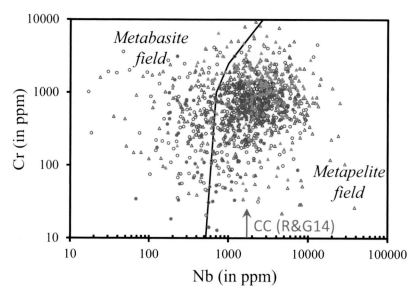

Figure 5. Log Cr versus log Nb concentrations for 1170 detrital rutile analyses. Symbols correspond to four different studies covering a wide range of depositional age and geographic position: open triangles-North Sea sediments, Jurassic to Recent (Morton and Chenery 2009, *n* = 463); solid circles—Carboniferous sediments from Turkey (Okay et al. 2011, *n* = 105); open circles—Ordovician sediments from Germany (Rösel 2014, *n* = 426); solid triangles—Mesoproterozoic sediments from Reynolds Range, Australia (Rösel et al. 2014, *n* = 186). Metabasite versus metapelite fields from Triebold et al. (2012). Arrow represents Nb content of a hypothetical rutile from average continental crust (CC) from Rudnick and Gao (2014).

This compilation of more than 1100 detrital rutile grains illustrates how a clear distinction between those two rock types is not possible (see for example Meyer et al. 2011, Kooijman et al. 2012). Overall, there are not two separate fields: only one main cluster exists within a diffuse cloud of possible Nb and Cr concentrations. Still, it is interesting that the median of all Nb values of 1790 ppm closely follows the expected hypothetical Nb content in rutile of 1875 ppm calculated from average continental crust (0.64 wt% TiO_2 and 12 ppm Nb; Rudnick and Gao 2014), assuming all Ti and Nb are concentrated in rutile (see Fig. 5). On the one hand it supports our assumption that detrital rutile chosen for this compilation is representative of normal crustal sections. On the other hand it hints to the exciting prospect that large numbers of detrital rutile of different ages may provide direct insight into the Nb/Ti fractionation in the crust throughout Earth's history. The oldest rutile grains analysed for these trace elements (from the Mesoproterozoic Reynolds Range) have the highest median Nb/Ti ratio, which nevertheless is based only on a small population and may not be representative.

URANIUM–LEAD GEOCHRONOLOGY

Uranium, Th and common Pb distribution in rutile

To retrieve U–Pb age information from rutile, three chemical components need to be considered: U, Th and Pb. Unlike zircon, the U contents of some rutile crystals are too low (<0.1 ppm, see Zack et al. 2011) to be datable. Single spot $^{206}Pb/^{238}U$ 1σ age errors of <6% for the lowest-U rutile are reported down to 1.0 ppm (Li et al. 2011; Topuz et al. 2013). On the other extreme, U contents of >100 ppm have been reported from several locations (e.g., Mezger et al. 1989). A predictive model for targeting high-U rutile has not been presented yet, and several factors (e.g., whole rock U content, coexisting accessory phases, temperature, oxygen fugacity) seem to be important parameters. In the absence of an encompassing predictive model, we exploit the detrital rutile data set presented above to survey U–Th–Pb systematics of rutile in general.

Figure 6A shows that for detrital rutile classified as "metapelitic", only 27% contains <4 ppm U. In contrast, for "metabasic" rutile 60% has U contents of <4 ppm (Fig. 6B). This confirms the common knowledge that metapelitic rocks are more likely to contain datable rutile than coexisting metabasic rocks. In the absence of choice, it is nevertheless possible to date rutile in some metabasic rocks (e.g., tectonic sliver of blueschist or eclogite). However, it requires a thorough search for high-U rutile (e.g., Li et al. 2011). Additionally, rutile with >100 ppm U is rare, but it can be discovered through a systematic search of a large pool of lithologies.

An advantageous property of most rutile is the absence of significant amounts of Th. This property, and its corresponding lack of ^{208}Pb ingrowth, facilitates common Pb correction via ^{208}Pb measurement (Zack et al. 2011). Measurement of ^{208}Pb instead of ^{204}Pb is favorable, because it is ca. 40 times more abundant and is not compromised by ^{204}Hg interference—a common problem in laser ablation ICP-MS (Kylander-Clark 2017). As can be seen in Figure 6C, the majority of analyzed detrital rutile (52%) has Th contents <0.5 ppm. This can be explained by the large size of Th^{4+} compared to the much smaller Ti^{4+} (Zack et al. 2002). However, not all rutile is low in Th, and exceptions with Th contents of up to 35 ppm are reported from diagenetically grown rutile from Oman (Dunkl and von Eynatten 2009). This is also visible in the detrital data set in Figure 6C, where 3% of all analyzed grains had >10 ppm Th. Such cases compromise the common ^{208}Pb correction, although we are not able to fully evaluate the data set used, and part of the 3% may be based on mixed analyses. Nevertheless, we strongly recommend monitoring Th contents when using the common ^{208}Pb correction method during U–Pb rutile dating.

It is a widespread misconception that rutile can be rich in common Pb, as most rutile has common Pb contents <0.2 ppm (Zack et al 2011). In many cases it is the low U content that leads to an unfavorably high ratio of common to radiogenic Pb. To a certain degree this can and should be corrected by monitoring $^{206}Pb/^{208}Pb$ (Zack et al. 2011). Down to $^{206}Pb/^{208}Pb$ values of ca 10, the common Pb correction does not lead to a severe loss in age precision.

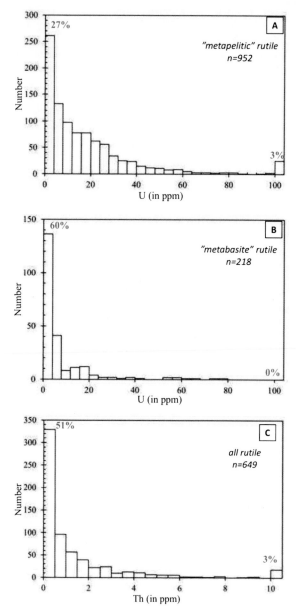

Figure 6. Histograms for U and Th concentrations of same detrital rutile grains as listed in Figure 5A)
Uranium concentrations for "metapelitic" rutile; B) Uranium concentrations for "metabasic" rutile; C)
Thorium concentrations for all rutile where data are available.

Analytics

Uranium-Lead geochronology of rutile has been applied for decades to constrain
metamorphic histories (e.g., Ludwig and Cooper 1984; Schärer et al. 1986; Corfu and Muir
1989; Mezger et al. 1989), to date hydrothermal alteration processes (e.g., Richards et al. 1988),
and eruption of xenoliths (Davis et al. 1997). These pioneering studies applied isotope-dilution

thermal ionization mass spectrometry (ID-TIMS) analysis of grain separates. This method has the advantage of high precision, but has the potential complication of providing inaccurate results for grains having more than one age component or grains having U/Pb-bearing mineral inclusions such as zircon. More recently, high spatial resolution techniques have been developed for rutile geochronology including laser ablation inductively coupled plasma mass spectrometry (LA-ICP-MS; e.g., Vry and Baker 2006; Kooijman et al. 2010; Zack et al. 2011; Kylander-Clark 2017) and secondary ion mass spectrometry (SIMS; e.g., Clark et al. 2000; Li et al. 2011; Taylor et al. 2012; Schmitt and Zack 2012; Ewing et al. 2015; Schmitt and Vasquez 2017). These in situ techniques enable accurate dating of heterogeneous rutile and are largely complementary; LA-ICP-MS allows rapid analysis of large populations required for detrital provenance studies, whereas SIMS has higher spatial resolution and is less destructive. Both techniques are not without challenges; depth profiling by LA-ICP-MS is at a stage where only integrating several profiles yields a level of precision where age zoning becomes resolvable (Smye and Stockli 2011), whereas U–Pb SIMS analysis may be compromised by strong crystal orientation effects (Li et al. 2011; Taylor et al. 2012; Schmitt and Zack 2012). Nevertheless, in situ rutile dating has been successfully applied in an increasing number of studies covering most of the geological time-scale, from as old as 3.3 Ga (Upadhyay et al. 2014) to as young as only 9 Ma (Smit et al. 2014). Further improvement of in situ techniques will be key to advancing our ability to resolve zoning in heterogeneous grains and enable dating of samples with low radiogenic Pb contents.

The data quality of in situ techniques such as LA-ICP-MS and SIMS largely depends on the availability of suitable matrix-matched calibration standards. Such standards should: 1) be homogeneous in age; 2) have a high U concentration, 3) contain little or no common Pb, and 4) be commonly available. The number of natural rutile standards has significantly increased in the last decade with various groups developing material for large-scale distribution (Luvizotto et al. 2009a; Zack et al. 2011; Bracciali et al. 2013). Currently available standards cover a range of ages from 489.5 Ma (R19; Zack et al. 2011) to 2625 Ma (WHQ; Taylor et al. 2012) with the most commonly used standards being R10 and R10b (1090 Ma; Luvizotto et al. 2009a; Schmitt and Zack 2012). Bracciali et al. (2013) show that these standards can sometimes be heterogeneous or high in common Pb, which may be a problem if this is not corrected for. They present two new potential standards of ~1.8 Ga (Sugluk-4 and PCA-S207), which show good reproducibility (1–2% RSD). Although such precision is only slightly larger than the long-term reproducibility of most zircon standards, there is an ongoing community effort to characterize more abundant and more homogeneous reference materials (e.g., E. Axelsson, pers. comm.).

Uranium–lead systematics in rutile

A reliable interpretation of U–Pb ages of any mineral requires knowledge of the closure temperature (T_c) for Pb diffusion. The mathematical expression to determine T_c is defined by Dodson (1973) as:

$$T_c = \frac{E_a / R}{\ln\left[\frac{\left(ART_c^2 D_0 / a^2\right)}{E_a \frac{dT}{dt}}\right]} \tag{5}$$

in which E_a is the activation energy, R is the gas constant, A is a geometric factor, D_0 is the frequency factor, i.e., the diffusion coefficient at infinite temperature, a is the effective diffusion radius, and dT/dt is the cooling rate. This expression gives a weighted average T_c for a whole grain. Important assumptions of the model are: (1) the matrix around the crystal acts as a homogeneous infinite reservoir for the diffusing species, (2) the concentration of the species at the onset of cooling was homogeneous, (3) no overgrowth, resorption or recrystallization occurred after the onset of cooling, (4) there was sufficient diffusive loss during cooling to modify the concentration of the species in the core of the grain (i.e., cooling was slow), and (5) temperature is inversely proportional to time during cooling. The first estimate of T_c for Pb in rutile was made by Mezger et al. (1989), who applied ID-TIMS analysis to rutile grains of 90–210 μm of the Proterozoic Adirondacks and compared the resulting ages by assuming closure temperatures of other dated minerals in the same terrane. Their estimates were between 380–420 °C, one of the lowest closure temperatures for Pb in accessory minerals. Vry and Baker (2006) recalculated this value to 500–540 °C using higher T_c estimates for Pb in monazite, Pb in titanite, Ar in hornblende, and Ar in biotite in the same geological setting. The revised T_c estimate is more consistent with the diffusion coefficients determined in experiments by Cherniak (2000; $E_a = 250 \pm 12$ kJ mol^{-1} and $D_0 = 3.92 \times 10^{-10}$ for synthetic rutile, $E_a = 242 \pm 10$ kJ mol^{-1} and $D_0 = 1.55 \times 10^{-10}$ for natural rutile; Fig. 4). These coefficients predict a whole-grain T_c of ~620 °C for a spherical rutile of 200 μm at an average cooling rate of 1 °C/Ma.

The diffusion rates for Pb in rutile imply that diffusive age gradients will develop in rutile during cooling from lower crustal conditions. The description of T_c as a function of the position within the grain $[T_c(x)]$ was mathematically defined before it was viable to demonstrate analytically (Dodson 1986). When age gradients are detectable, in principle it is possible to obtain a continuous thermal history by forward modelling T–t histories that can match the observed age gradient (Grove and Harrison 1999; Harrison et al. 2005). With the advancements of analytical techniques it has become possible to resolve U/Pb zoning or age profiles in rutile (Kooijman et al. 2010, 2015; Smye and Stockli 2014). These studies present

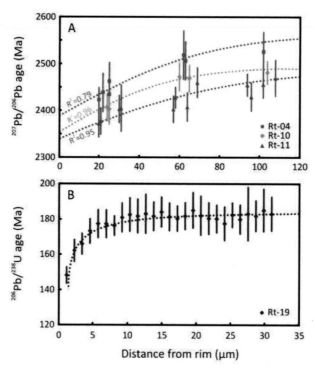

Figure 7. Age profiles in rutile grains. A) ^{207}Pb/^{206}Pb age profiles for three rutile grains from sample 462d from the Pikwitonei Granulite Domain, Canada. Dashed lines represent error function best fit to the data and R^2 values indicate goodness of fit (Kooijman et al. 2010). B) error-weighted ^{206}Pb/^{238}U age depth-profiles calculated for sample IVZR19 ($n = 10$) from the Ivrea Zone, Southern Alps, Italy. Dashed line represents power law best fit to the data (Smye and Stockli 2014).

resolvable diffusion profiles in very different geological settings (Fig. 7). As such, each of these profiles is of a different length scale and amplitude, but all adhere closely to Fickian diffusion behavior allowing modeling of cooling rates.

Extremely slow cooling of the Archaean lower crust in the Pikwitonei Granulite Domain, Canada resulted in age profiles in rutile that show a core-to-rim age range of up to 200 Ma in grains of up to 270 μm (Fig. 7; Kooijman et al. 2010). Their study involved error-function based diffusion modelling of Pb in rutile using the diffusion coefficients of Cherniak (2000). They show that the T_c as a function of position in the grain ranged from 640 °C in the core to 490 °C in the rim of grains. The latter temperature is currently the lowest T_c estimate from an empirical study. Although the spatial resolution of the LA-ICP-MS analyses was slightly limiting, the cooling rate modeling results fitted well with previous estimates of average cooling rate based on comparison with other thermochronometers (Mezger et al. 1989). The time-resolved model provided additional constraints by enabling construction of a continuous cooling history from 2.2 °C/Ma at 640 °C to 0.4 °C/Ma at 490 °C (Kooijman et al. 2010). The study by Smye and Stockli (2014) demonstrates age closure profiles in rutile at a higher resolution and over a much smaller scale of ~20 μm using LA-ICP-MS depth profiling. Their study targeted rutile grains from the Ivrea Zone where modeling of the age profiles constrained rapid cooling from 180 Ma (Fig. 7; see below). High-resolution depth profiling permitted the identification of a transient thermal event that so far remained undetected with traditional spot analysis (Smye and Stockli 2014). In the well-studied UHP zone of the Western Gneiss Complex, Norway, Kooijman et al. (2015) resolved Pb diffusion profiles in mm-size rutile grains. The diffusion length in these grains is 400 μm, which could be resolved well by LA-ICP-MS using a typical 30-μm spot size. Such diffusion profiles allow for reconstructing a significant part of the cooling history of a metamorphic terrane (Dodson 1986; Kooijman et al 2010). These studies show that U–Pb age profiles in rutile can occur, and be resolved, on different scales, depending on the geological setting. This is in contrast to Villa (2016) who stated that "diffusion profiles of isotope ratios in minerals used for geochronology are absent". Current analytical techniques allow for precise determination of such profiles in rutile and, theoretically, permit forward modelling of the profiles as a function of *T–t* to obtain a continuous thermal history for temperature intervals that are inaccessible to conventional thermochronology (Smye and Stockli 2014; Kooijman et al. 2015).

Finally, it is important to note that the effective diffusion radius needs to be accurately defined before invoking volume-diffusion controlled cooling ages. One of the main issues that could pose a problem for interpreting U–Pb rutile ages is the common presence of exsolution lamellae (Fig. 2I) and/or twinning planes. These features may reduce the effective diffusion radius and hence the effective T_c of Pb at any spot in the grain if 1) they act as effective pathways for Pb movement, or 2) if Pb partitions into the lamellae. If either one of these occurs, the effective T_c of rutile will be controlled by the distance between the lamellae, and be significantly different from the whole-grain T_c. If the lamellae form at temperatures above T_c, the U–Pb systematics will be affected. To date, there are no studies that have quantified these potential effects. A further complication is that exsolution lamellae a priori do not have to form above T_c, but can be a late feature.

Cooling vs formation ages

The intermediate closure temperature of Pb in rutile implies that it is possible to find crystallization ages if rutile grains are large enough or if cooling is very fast. Diffusion-induced resetting is as yet unresolvable in low-grade metamorphic rocks such as blueschists and low-temperature eclogites (diffusion length scale < 1 μm) and formation ages can be found in hydrothermal deposits, where, e.g., U–Pb ages for rutile and monazite are identical (Shulaker et al. 2015). In higher-grade metamorphic rocks, the preservation of a formation age in the core depends on a number of factors including peak-metamorphic temperature, cooling rate, and grain size. Figure 8 shows the relation between these variables based on a diffusion

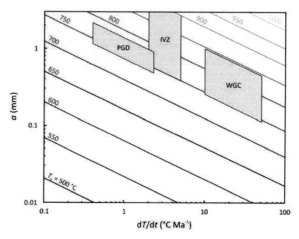

Figure 8. Relationship between effective diffusion length (a), average cooling rate (dT/dt) and peak metamorphic temperature for Pb in rutile based on diffusion modeling (Smit et al. 2013) using the diffusion coefficients for natural rutile (Cherniak 2000). Shaded areas show peak metamorphic temperatures and average cooling rates for the Pikwitonei Granulite Domain (PGD) in Canada (Mezger et al. 1989; Kooijman et al. 2010), the Ivrea Zone (IVZ) in the Southern Alps, Italy, and the Western Gneiss Complex (WGC) in Norway (Carswell et al. 2003).

modeling approach by Smit et al. (2013) using the diffusion coefficients for natural rutile (Cherniak 2000) for a range of geologically relevant conditions. Assumptions for this model are the same as described earlier for the definitions of Dodson (1973, 1986).

The isotherms in Figure 8 indicate that higher temperatures and slower cooling will require larger grains to preserve crystallization ages in grain cores. To illustrate the potential applicability of this model, metamorphic terranes discussed in the text have been marked (Fig. 8). In the case of the Pikwitonei Granulite Domain, which has peak T estimates of 750–800 °C and cooling rates of 1–2 °C/Ma the minimum diffusion radius would be 0.5–1.5 mm. This means that a grain size larger than 1–3 mm is required to find a crystallization age in the core. Although diffusion profiles have been demonstrated in rutile from the Pikwitonei Granulite Domain (Kooijman et al. 2010), the largest grains were 270 μm implying that the core ages must be far removed from the original crystallization age. The Western Gneiss Complex in Norway underwent slightly higher peak T conditions at 750–850 °C, which was followed by much faster cooling at rates of 10–20 °C/Ma (e.g., Carswell et al. 2003; Kylander-Clark et al. 2008). This implies that a diffusion length on the order of 0.2–1.0 mm could be expected in rutile grains from this terrane (Fig. 8). Kooijman et al. (2015) resolved age profiles in rutile from this area with a diffusion length of 400 μm in mm-sized grains, demonstrating that cooling in this terrane was fast enough to allow for crystallization ages to be preserved in the cores of rutile grains.

In many cases in metamorphic systems rutile will record only cooling ages, but this still yields a wealth of useful information about a different part of the rock's history, for example cooling and exhumation. This makes rutile ages very much complementary to zircon and monazite ages in the same rock. Even in metamorphic rock samples where crystallization ages still can be found, parts of rutile grains would be expected to record cooling ages, which can also be exploited. In this regard, rutile is not likely to record pre-metamorphic information, i.e., multiple generations, because rutile is not stable under most low- to medium-grade metamorphic conditions and will either break down completely, or fully recrystallise (Zack et al. 2004b). It therefore exhibits strikingly different behavior compared to zircon or monazite, which commonly preserve multiple growth zones.

Comparison with U–Pb titanite ages

There is a significant body of literature where cooling paths of high grade metamorphic terranes are derived using ages from several metamorphic phases and isotope systems. In most cases, rutile ages are younger than U–Pb ages from titanite (e.g., Mezger et al. 1998; Möller et al. 2000; Flowers et al. 2006), and similar to Ar–Ar ages from hornblende and Rb–Sr ages from muscovite (e.g., Vry and Baker 2006). We will not discuss complexities that may arise with Ar–Ar and Rb–Sr systems or with recrystallization behavior of micas (see e.g., Villa 1998), but rather focus on a comparison of U–Pb rutile ages with U–Pb titanite ages. Rutile and titanite are among a handful of phases where experimental diffusion rates are available that can be directly related to thermochronology (Cherniak 1993, 2000). One of the challenges in thermochronology is that experimental diffusion rates for Pb in both phases are broadly similar, which contrasts with natural data.

One possibility is that the experimentally derived diffusion parameters for Pb in rutile may be wrong, as advocated in several studies. For example, Schmitz and Bowring (2003) suggest that experiments for rutile may be compromised by not considering the effect of water and defect distribution in the crystal lattice and therefore prefer to adhere to the original T_c estimate of ~400 °C by Mezger et al (1989). It has to be stated that this may be applicable to titanite experiments as well, which Schmitz and Bowring do not question. To this end, several studies show that titanite can record pre-metamorphic/magmatic U–Pb ages in areas heated up beyond the closure temperature predicted (e.g., Pidgeon et al. 1996; Frost et al. 2000; Rubatto and Hermann 2001; Spencer et al. 2013). Kohn and Penniston-Dorland (2017) therefore argue that Pb diffusion in titanite must be much slower in nature than indicated experimentally.

Another possibility for the conflict between experiment and some studies on natural samples can be fundamentally addressed by questioning the validity of volume diffusion for the U–Pb system. Villa (1998, 2016) argues that in the presence of fluids, processes like recrystallization, resorption and overgrowth can be much faster than volume diffusion and should dominate. While we have shown above that clear examples for volume diffusion-controlled age zonation profiles exist for rutile, the same is currently missing in the literature for titanite, with Kohn and Corrie (2011) giving examples of U–Pb age and Zr-in-titanite temperature relationships attributed to growth zoning and Stearns et al. (2016) documenting U–Pb age and trace element variations due to recrystallization.

CASE STUDY: THE IVREA ZONE

The Ivrea Zone (also Ivrea-Verbano Zone) in the Southern Alps, Italy is arguably the area best-studied in terms of rutile occurrence, geochemistry and U–Pb age behavior. It is a classical area that exposes a near-complete section through all levels of the lower crust (Zingg 1980, 1983; Handy et al. 1999; Zingg et al. 1990) with clear connections to contemporaneous upper crustal sections, like the metamorphic Serie de Laghi (also called Strona-Ceneri zone; Zingg 1983) and sedimentary and volcanic cover rocks, e.g., in the Monte Nudo Basin, the Lombardian Basin and the Canavese Zone (e.g., Beltrando et al. 2015). The geologic importance of this area in combination with a wide geographic occurrence of assemblages containing large rutile grains (up to 1 mm in diameter) makes the Ivrea Zone a testbed for new methodologies applicable to rutile. In the following we focus on the unique advantages that rutile petrochronology affords in unraveling the geologic history of the Ivrea Zone.

The geologic history of the Ivrea Zone spans more than 300 Ma, including regional metamorphism up to granulite-facies during the Carboniferous, massive Permian magma intrusions, Jurassic rifting and finally exhumation during the Alpine orogeny (Handy et al. 1999). The first clearly resolvable episode is the regional metamorphism at around 316 ± 3 Ma

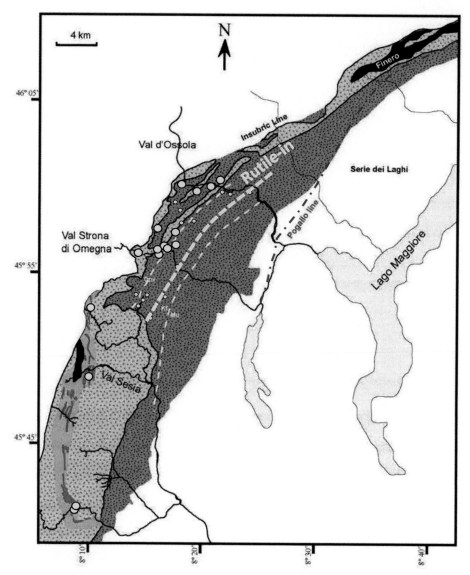

Figure 9. Geologic map of the Ivrea Zone, modified after Ewing et al. (2015). Dark stippled unit- mostly amphibolite- and granulite-facies metapelites (called Kinzigite Formation); light stippled unit- Mafic Complex; light unstippled unit- Paragneiss-bearing belt with so-called metapelite septa (dark unstippled); black unit- mantle peridotite; white stippled lines are isograds (Mu–muscovite out; Kfs–K-feldspar in; Opx–orthopyroxene in) from Zingg (1980). Large circles are coarse-grained rutile occurrences in metapelites described in Luvizotto and Zack (2009), and Ewing et al. (2013). Small circles are additional rutile occurrences mapped in Zingg (1980). The rutile-in isograd is inferred after Zingg (1980). [Used by permission of Springer, from Ewing, Rubatto, Beltrando and Hermann (2015) *Contributions to Mineralogy and Petrology*, Vol 169, Fig. 1, p. 44]

(Ewing et al. 2013), which affected the entire Ivrea Zone, but to various peak *P–T* conditions. It is best recorded in Val Strona di Omegna and Val d'Ossola (see Fig. 9), where a general increase in metamorphic grade is visible from ca 650 °C / 0.5 GPa in the eastern, mid-crustal section to >900 °C / 1.1 GPa in the western, lowermost crustal section (Redler et al. 2012).

P–T estimates are based on pseudosections that include a Ti-bearing biotite model, which is crucial as Ti stabilizes biotite to higher temperatures compared to models without Ti (see also Tajčmanová et al. 2009). It is this area where one of the few published cases where a regional rutile-in isograd has been mapped (Zingg 1980), observable in metapelitic assemblages. The dominance of ilmenite as the major Ti-phase in metabasites (e.g., Kunz et al. 2014) broadly corresponds to different minimum pressures required to stabilize rutile in those different lithologies (Fig. 1). The rutile-in isograd is attributed to biotite breakdown, because the products (garnet, K-feldspar and sillimanite) and pressures exceed the GRAIL equilibrium for these rocks (Zingg 1980; Luvizotto and Zack 2009). The increasing metamorphic gradient in the rutile-bearing assemblages in Val Strona di Omegna and Val d'Ossola is also recorded in rutile in the form of an increase in maximum Zr contents in a given rutile population. Here, the maximum Zr contents of rutile range from 2500 ppm Zr close to the rutile-in isograd, to 5000 ppm Zr in the highest grade rocks (Luvizotto and Zack 2009). The calculated temperatures of 850 °C (at 0.8 GPa) and 950 °C (at 1.1 GPa), respectively, are in good agreement with Redler et al. (2012), and are strong evidence that previous temperature estimates of this part of the Ivrea Zone based on Fe–Mg exchange thermometers (e.g., a maximum of 860 °C by Henk et al. 1997 for the highest grade section) are prone to diffusional resetting.

In Permian times (288 ± 4 Ma; Peressini et al. 2007), a massive layered intrusion, the Mafic Complex (Fig. 9), ca. 10 km in thickness intruded the lower crust (e.g., Quick et al. 2003). Although no rutile is recorded in a wide range of intrusive rock types from norites up to granodiorite (gabbro is the main rock type), spectacular rutile occurrences are nevertheless found in metapelitic slivers (so-called septa; Sinigoi et al. 1995) ingested by mafic intrusive units (Quick et al. 2003; Ewing et al. 2013). In these ≤ 100 m thick septa, highly restitic lithologies contain biotite-free, garnet- and sillimanite-rich gneisses with abundant rutile (Ewing et al. 2013), sometimes with additional corundum (Quick et al. 2003; Pape et al. 2016). Having obviously been exposed to contact metamorphism by a gabbroic magma, those rutile grains record some of the highest published Zr contents (up to 10000 ppm Zr, corresponding to temperatures of ca. 1000 °C; Ewing et al. 2013). Other rutile grains show textbook-example exsolution lamellae of zircon and in one case baddeleyite formed during cooling (Ewing et al. 2013; Pape et al. 2016). Reintegrating the expelled Zr in rutile, Ewing et al. (2013) and Pape et al. (2016) concluded that peak metamorphic temperatures were even higher than 1000 °C. Interestingly, Ti-in-zircon does not record such high temperatures (only up to 870 °C), likely because it crystallized from in situ anatectic melts during cooling (Ewing et al. 2013).

Between intrusion of the Mafic Complex and the Jurassic rifting, it is uncertain what magmatic and/or tectonic events affected the entire Ivrea Zone. In the northern end, gabbros around the prominent mafic-ultramafic Finero body formed at 232 ± 3 Ma (Zanetti et al. 2013). Whether sporadic zircon ages between 260 and 200 Ma throughout the entire Ivrea Zone (Vavra et al. 1999; Ewing et al. 2013) are attributed to accompanying magmatic activities or due to local fluid influx, are a matter of debate (see discussion in Ewing et al. 2013 and references therein). In any case, a geochemical or age imprint on rutile has not yet been ascribed to these events.

Jurassic rifting is recorded in many parts of the Alps and beyond (e.g., Handy et al. 1999), leading to the opening of the Alpine Tethys Ocean. In the Ivrea Zone, the most prominent feature recording this event is the Pogallo line (Fig. 9), an amphibolite-facies shear zone partly juxtaposing the lower crustal Ivrea Zone against upper crustal Serie dei Laghi units. Movement along the Pogallo line has been dated at 182.0 ± 1.6 Ma (Mulch et al. 2002). It appears that rutile may be the phase that provides the most coherent picture of these thermal events during this time span. Several studies record U–Pb rutile ages between 160 and 200 Ma throughout the Ivrea Zone (Zack et al. 2011; Smye and Stockli 2014; Ewing et al. 2015). However, what makes rutile such a key phase here is the view that rutile crystallized before the Jurassic (Ewing et al. 2015). In particular, Zr contents of > 4000 ppm in rutile can originate only at temperatures acquired during Variscan granulite-facies conditions and especially during the intrusion of the Permian Mafic Complex for rutile in the septae. Therefore, U/Pb rutile ages have to be interpreted in the framework of Pb loss. Complications from recrystallization and/or overgrowth can be excluded (Ewing et al. 2015).

As can be expected and as is shown in Figure 8, slow cooling granulite-facies areas like the rutile-bearing parts of the Ivrea Zone do not preserve formation ages in rutile even in the largest recorded grains of 1 mm. Therefore, widespread rutile core ages of 180 Ma (Zack et al. 2011; Smye and Stockli 2014; Ewing et al. 2015) correspond to temperature conditions of ca. 600 °C. However, Smye and Stockli (2014) were able to detect age zoning profiles in the outermost 15 microns of euhedral rutile crystals (Fig. 7), which they cannot interpret with a monotonous cooling path, but rather with rapid cooling at ca. 180 Ma. In combination with Zr diffusion profiles, the ages and cooling paths can be interpreted to originate after a short-lived transient temperature spike up to ≥ 700 °C. Those results show the advantage of diffusion-controlled age zoning profiles, where instead of T–t points from bulk mineral results, continuous sections of a T–t evolution can be determined. In comparison with extensive previous age determinations of multiple dating systems in the same section of the Ivrea Zone (Siegesmund et al. 2008), a rift-related reheating event is apparently visible only from the new U–Pb rutile data set. We stress that such a Jurassic reheating event has also been proposed previously (see e.g., Fig. 13 in Siegesmund et al. 2008), but it became evident only using rutile petrochronology.

The Jurassic U/Pb rutile ages of ca 180 Ma are observable in all rutile-bearing sections in the Ivrea Zone (Zack et al. 2011; Smye and Stockli 2014; Ewing et al. 2015). Ewing et al. (2015) also describe a second age population of ca 160 Ma using SIMS, not found in nearby locations studied by Zack et al. (2011) and Smye and Stockli (2014) who used LA-ICP-MS, and ascribe this to a second reheating event at this time. Future studies should examine whether one or several reheating events affected the Ivrea Zone and how this is recorded in the entire rutile-bearing section, spanning a paleodepth of almost 4 km.

CONCLUDING REMARKS AND RECOMMENDATIONS

Rutile is an ideal recorder of Earth's (in particular metamorphic) processes due to the fact that it provides several layers of information regarding temperature and time. Although petrologic information such as microstructures and trace element composition can be interpreted independently from geochronologic information, it is the combination of those two fields in the spirit of petrochronology, where progress will be made. This is particularly relevant when considering U–Pb thermochronology of rutile. Knowledge of the T_c of Pb in rutile is crucially important to accurately interpret the age. Caution is warranted when the following textural features occur, because these may result in a higher or lower apparent T_c than would be predicted by grain size:

• Crystals with resorbed grain shapes (e.g., Fig. 2G),

• Multicrystal aggregates (e.g., Figs. 2J and 2K) and

• Grains with twinning planes and/or exsolution lamellae (Figs. 22H and 2I).

In high-grade metamorphic rocks, specifically of granulite-facies grade, rutile is potentially the most suitable mineral for studying the influence of intra- and intergrain diffusion during cooling on thermochronological interpretation. If maximum temperatures and cooling rates are known, it can be estimated what minimum grain sizes of rutile crystals are required to resolve age zoning profiles consistent with Fickian diffusion behavior (Fig. 8). Following Ewing et al. (2015), we recommend an active search for microstructural and chemical criteria that demonstrate the absence of modification of crystal cores during cooling such as:

• Evidence for older textural relationships, e.g., contact with older minerals like zircon or garnet (e.g., Fig. 2C) or original rounded grain shape of detritus in quartzites

• Preservation of inclusions (Fig. 2L)

• Chemical composition pointing towards an older history (e.g., Zr contents of > 1000 ppm)

The same criteria also apply for the search of U–Pb formation ages in rutile from lower grade metamorphic areas, specifically in blueschists and eclogites, except that zoning profiles may not be resolvable.

The element Zr, used for Zr-in-rutile thermometry, diffuses much more slowly than Pb in rutile (Fig. 4) and, hence, can be linked to crystallization or recrystallization of rutile. Depending on rock type and metamorphic history Zr may record prograde, peak or retrograde conditions, and thus should be commonly decoupled from U–Pb chronology. Only when formation ages are found in rutile can they be directly tied with temperature information from Zr contents or, at temperatures below ca 500 °C, to O isotopes.

ACKNOWLEDGMENTS

We would like to express our gratitude towards Tanya Ewing and Julie Vry for their very extensive and constructive reviews. Additional comments and suggestions by Matt Kohn are gratefully acknowledged. Pierre Lanari is thanked for his editorial guidance and patience. Tanya Ewing is also thanked for providing an editable version of Figure 9. We are grateful to Alicia Cruz-Uribe, George Luvizotto, Delia Rösel and Matthijs Smit for providing (partly unpublished) microphotographs for this article. Matthijs Smit is also thanked for discussions about this review and help with some figures. They are also thanked, along with Jasper Berndt, John Cottle, Helen Degeling, Istvan Dunkl, Tanya Ewing, Andreas Kronz, Andrew Kylander-Clark, Craig Manning, Horst Marschall, Klaus Mezger, Andreas Möller, Renato Moraes, Patty O'Brien, Roger Powell, Axel Schmitt, Andrew Smye, Danny Stockli, Silke Triebold and Hilmar von Eynatten for stimulating discussions regarding Rutilology throughout the years. TZ acknowledges funding from Svenska Vetenskapsradet (VR E0582701) and previous rutile-related grants from the German Science Foundation.

REFERENCES

Agrinier P (1991) The natural calibration of $^{18}O/^{16}O$ geothermometers—application to the quartz–rutile mineral pair. Chem Geol 91:49–64

Beltrando M, Stockli DF, Decarlis A, Manatschal G (2015) A crustal-scale view at rift localization along the fossil Adriatic margin of the Alpine Tethys preserved in NW Italy. Tectonics 34:1927–1951

Bohlen SR, Liotta JJ (1986) A barometer for garnet amphibolites and garnet granulites. J Petrol 27:1025–1034

Bohlen SR, Wall VJ, Boettcher AL (1983) Experimental investigations and geological applications of equilibria in the system FeO–TiO_2–Al_2O_3–SiO_2–H_2O. Am Mineral 68:1049–1058

Bracciali L, Parrish RR, Horstwood MSA, Condon DJ, Najman Y (2013) U–Pb LA-(MC)-ICP-MS dating of rutile: New reference materials and applications to sedimentary provenance. Chem Geol 347:82–101

Cabral AR, Rios FJ, de Oliveira LAR, de Abreu FR, Lehmann B, Zack T, Laufek F (2015) Fluid-inclusion microthermometry and the Zr-in-rutile thermometer for hydrothermal rutile. Int J Earth Sci 104:513–519

Carruzzo S, Clarke DB, Pelrine KM, MacDonald MA (2006) Texture, composition, and origin of rutile in the South Mountain Batholith, Nova Scotia. Can Mineral 44:715–730

Carswell DA, Brueckner HK, Cuthbert SJ, Mehta K, O'Brien PJ (2003) The timing of stabilisation and the exhumation rate for ultra-high pressure rocks in the Western Gneiss Region of Norway. J Met Geol 21:601–612

Cherniak DJ (1993) Lead diffusion in titanite and preliminary results on the effects of radiation damage on Pb transport. Chem Geol 110:177–194

Cherniak DJ (2000) Pb diffusion in rutile. Contrib Mineral Petrol 139:198–207

Cherniak DJ, Watson EB (2011) Helium diffusion in rutile and titanite, and consideration of the origin and implications of diffusional anisotropy. Chem Geol 288:149–161

Cherniak DJ, Manchester J, Watson EB (2007) Zr and Hf diffusion in rutile. Earth Planet Sci Lett 261:267–279

Clark DJ, Hensen BJ, Kinny PD (2000) Geochronological constraints for a two-stage history of the Albany-Fraser Orogen, Western Australia. Precam Res 102:155–183

Corfu F, Muir TL (1989) The Hemlo-Heron Bay greenstone belt and Hemlo Au-Mo deposit, Superior Province, Ontario, Canada; 2. Timing of metamorphism, alteration and Au mineralization from titanite, rutile, and monazite U–Pb geochronology. Chem Geol 79:201–223

Corfu F, Hanchar JM, Hoskin PWO, Kinny P (2003) Atlas of zircon textures. Rev Mineral Geochem 53:469–500

Cruz-Uribe AM, Feineman MD, Zack T, Barth M (2014) Metamorphic reaction rates at similar to 650–800 °C from diffusion of niobium in rutile. Geochim Cosmochim Acta 130:63–77

Davis WJ (1997) U–Pb zircon and rutile ages from granulite xenoliths in the Slave province: evidence for mafic magmatism in the lower crust coincident with Proterozoic dike swarms. Geology 25:343–346

Diener JFA, Powell R (2012) Revised activity-composition models for clinopyroxene and amphibole. J Met Geol 30:131–142

Dill HG, Melcher F, Füßl, Weber B (2007) The origin of rutile–ilmenite aggregates ("nigrine") in alluvial–fluvial placers of the Hagendorf pegmatite province, NE Bavaria, Germany. Mineral Petrol 89:133–158

Dodson MH (1973) Closure temperature in cooling geochronological and petrological systems. Contrib Mineral Petrol 40:343–346

Dodson MH (1986) Closure profiles in cooling systems. Mat Sci Forum 7:145–154

Dohmen R, Chakraborty S (2003) Mechanism and kinetics of element and isotopic exchange mediated by a fluid phase. Am Mineral 88:1251–1270

Dunkl I, von Eynatten H (2009) Anchizonal-hydrothermal growth and (U–Th)/He dating of rutile crystals in the sediments of Hawasina window, Oman. Geochim Cosmochim Acta 73:A314.

Engi M (2017) Petrochronology based on REE–minerals: monazite, allanite, xenotime, apatite. Rev Mineral Geochem 83:365–418

Engvik A, Ihlen PM, Austrheim H (2014) Characterisation of Na-metasomatism in the Sveconorwegian Bamble Sector of South Norway. Geosci Front 5:659–672

Ernst WG, Liu J (1998) Experimental phase-equilibrium study of Al- and Ti-contents of calcic amphibole in MORB - A semiquantitative thermobarometer. Am Mineral 83:952–969

Ewing TA, Hermann J, Rubatto D (2013) The robustness of the Zr-in-rutile and Ti-in-zircon thermometers during high-temperature metamorphism (Ivrea-Verbano Zone, northern Italy). Contrib Mineral Petrol 165:757–779

Ewing TA, Rubatto D, Beltrando M, Hermann J (2015) Constraints on the thermal evolution of the Adriatic margin during Jurassic continental break-up: U–Pb dating of rutile from the Ivrea-Verbano Zone, Italy. Contrib Mineral Petrol 169:44. Doi:10.1007/s00410–015-1135–6

Ferry JM (2000) Patterns of mineral occurrence in metamorphic rocks. Am Mineral 85:1573–1588

Flowers RM, Mahan KH, Bowring SA, Williams ML, Pringle MS, Hodges KV (2006) Multistage exhumation and juxtaposition of lower continental crust in the western Canadian Shield: Linking high-resolution U–Pb and ^{40}Ar/^{39}Ar thermochronometry with pressure–temperature–deformation paths. Tectonics 25:TC4003. Doi:10.1029/2005TC001912

Force ER (1980) The provenance of rutile. J Sediment Petrol 50:485–488

Frost BR (1991) Stability of oxide minerals in metamorphic rocks. Rev Mineral 25:469–488

Frost BR, Chamberlain KR, Schumacher JC (2000) Sphene (titanite): phase relations and role as a geochronometer. Chem Geol 172:131–148

Grove M, Harrison TM (1999) Monazite Th–Pb age depth profiling. Geology 27:487–490

Handy MR, Franz L, Heller F, Janott B, Zurbriggen R (1999) Multistage accretion and exhumation of the continental crust (Ivrea crustal section, Italy and Switzerland). Tectonics 18:1154–1177

Harley SL (2008) Refining the *P–T* records of UHT crustal metamorphism. J Metamorph Geol 26:125–154

Harrison TM, Grove M, Lovera OM, Zeitler PK (2005) Continuous thermal histories from inversion of closure profiles. Rev Mineral Geochem 58:389–409

Hart E, Storey C, Bruand E, Schertl H-P, Alexander BD (2016) Mineral inclusions in rutile: A novel recorder of HP-UHP metamorphism. Earth Planet Sci Lett 446:137–148

Henk A, Franz L, Teufel S, Oncken O (1997) Magmatic underplating, extension and crustal equilibration: insights from a cross-section through the Ivrea Zone and Strona–Ceneri Zone, Northern Italy. J Geol 105:367–377

Hubert JF (1962) A zircon–tourmaline–rutile maturity index and the interdependence of the composition of heavy mineral assemblages with the gross composition and texture of sandstone. J Sediment Petrol 32:440–450

Kohn MJ (2017) Titanite petrochronology. Rev Mineral Geochem 83:419–441

Kohn MJ, Penniston–Dorland SC (2017) Diffusion: Obstacles and opportunities in petrochronology. Rev Mineral Geochem 83:103–152

Kohn MJ, Corrie SL (2011) Preserved Zr-temperatures and U–Pb ages in high-grade metamorphic titanite: Evidence for a static hot channel in the Himalayan orogen. Earth Planet Sci Lett 311:136–143

Kohn MJ, Penniston-Dorland SC, Ferreira JCS (2016) Implications of near-rim compositional zoning in rutile for geothermometry, geospeedometry, and trace element equilibration. Contrib Mineral Petrol 171:78, doi 10.1007/s00410–016-1285–1

Kooijman E, Mezger K, Berndt J (2010) Constraints on the U–Pb systematics of metamorphic rutile from in situ LA-ICP-MS analysis. Earth Planet Sci Lett 293:321–330

Kooijman E, Smit MA, Mezger K, Berndt J (2012) Trace element systematics in granulite facies rutile: implications for Zr geothermometry and provenance studies. J Metamorph Geol 30:397–412

Kooijman E, Hacker BR, Smit MA, Kylander-Clark ARC (2015) Rutile thermochronology constrains time-resolved cooling histories in orogenic belts. Goldschmidt Abstr:1657

Kunz BE, Johnson TE, White RW, Redler C (2014) Partial melting of metabasic rocks in Val Strona di Omegna, Ivrea Zone, northern Italy. Lithos 190:1–12

Kylander–Clark ARC (2017) Petrochronology by laser–ablation inductively coupled plasma mass spectrometry. Rev Mineral Geochem 83:183–198

Kylander-Clark ARC, Hacker BR, Mattinson JM (2008) Slow exhumation of UHP terranes: Titanite and rutile ages of the Western Gneiss Region, Norway. Earth Planet Sci Lett 272:531–540

Lanari P, Engi M (2017) Local bulk composition effects on metamorphic mineral assemblages. Rev Mineral Geochem 83:55–102

Li Q-L, Lin W, Su W, Li X-H, Shi Y-H, Liu Y, Tang G-Q (2011) SIMS U–Pb rutile age of low-temperature eclogites from southwestern Chinese Tianshan, NW China. Lithos 122:76–86

Lucassen F, Dulski P, Abart R, Franz G, Rhede D, Romer RL (2010) Redistribution of HFSE elements during rutile replacement by titanite. Contrib Mineral Petrol 160:279–295

Ludwig KR, Cooper JA (1984) Geochronology of Precambrian granites and associated U–Ti–Th mineralization, northern Olary province, South Australia. Contrib Mineral Petrol 86:298–308

Luvizotto GL, Zack T (2009) Nb and Zr behavior in rutile during high-grade metamorphism and retrogression: An example from the Ivrea–Verbano Zone. Chem Geol 261:303–317

Luvizotto GL, Zack T, Meyer HP, Ludwig T, Triebold S, Kronz A, Muenker C, Stockli DF, Prowatke S, Klemme S, Jacob DE, von Eynatten H (2009a) Rutile crystals as potential trace element and isotope mineral standards for microanalysis. Chem Geol 261:346–369

Luvizotto GL, Zack T, Triebold S, von Eynatten H (2009b) Rutile occurrence and trace element behavior in medium-grade metasedimentary rocks: example from the Erzgebirge, Germany. Mineral Petrol 97:233–249

Manning CE, Bohlen SR (1991) The reaction titanite+kyanite=anorthite+rutile and titanite–rutile barometry in eclogites. Contrib Mineral Petrol 109:1–9

Marschall HR, Dohmen R, Ludwig T (2013) Diffusion-induced fractionation of niobium and tantalum during continental crust formation. Earth Planet Sci Lett 375:361–371

Matthews A (1994) Oxygen-isotope geothermometers for metamorphic rocks. J Met Geol 12:211–219

Matthews A, Schliestedt M (1984) Evolution of the blueschist and greenschist facies rocks of Sifnos, Cyclades, Greece- a stable isotope study of subduction-related metamorphism. Contrib Mineral Petrol 88:150–163

Matthews A, Beckinsale RD, Durham JJ (1979) Oxygen isotope fractionation between rutile and water and geothermometry of metamorphic eclogites. Mineral Mag 43:405–413

Meinhold G (2010) Rutile and its applications in earth sciences. Earth-Sci Rev 102:1–28

Meyer M, John T, Brandt S, Klemd R (2011) Trace element composition of rutile and the application of Zr-in-rutile thermometry to UHT metamorphism (Epupa Complex, NW Namibia). Lithos 126:388–401

Mezger K, Hanson GN, Bohlen SR (1989) High-precision U–Pb ages of metamorphic rutile- application to the cooling history of high-grade terranes. Earth Planet Sci Lett 96:106–118

Mezger K, Bohlen SR, Hanson GN (1990) Metamorphic history of the Archean Pikwitonei granulite domain and the Cross Lake Subprovince, Superior Province, Manitoba, Canada. J Petrol 31:483–517

Miller C, Zanetti A, Thoeni M, Konzett J (2007) Eclogitisation of gabbroic rocks: Redistribution of trace elements and Zr in rutile thermometry in an Eo-Alpine subduction zone (Eastern Alps). Chem Geol 239:96–123

Möller A, Mezger K, Schenk V (2000) U–Pb dating of metamorphic minerals: Pan-African metamorphism and prolonged slow cooling of high pressure granulites in Tanzania, East Africa. Precambrian Res 104:123–146

Moore JM, Waters DJ (1990) Geochemistry and origin of cordierite–orthoamphibole/orthopyroxene–phlogopite rocks from Namaqualand, South Africa. Chem Geol 85:77–100

Moore DK, Cherniak DJ, Watson EB (1998) Oxygen diffusion in rutile from 750 to 1000 °C and 0.1 to 1000 MPa. Am Mineral 83:700–711

Moraes R, Brown M, Fuck RA, Camargo MA, Lima TM (2002) Characterization and *P–T* evolution of melt-bearing ultrahigh-temperature granulites: An example from the Anapolis-Itaucu Complex of the Brasilia Fold Belt, Brazil. J Petrol 43:1673–1705

Morton AC, Chenery SP (2009) Detrital rutile geochemistry and thermometry as guides to provenance of Jurassic–Paleocene sandstones of the Norwegian Sea. J Sediment Res 79:540–553

Mulch A, Cosca MA, Handy MR (2002) In-situ UV-laser ^{40}Ar/^{39}Ar geochronology of a micaceous mylonite: an example of defect-enhanced argon loss. Contrib Mineral Petrol 142:738–752

Okay N, Zack T, Okay AI, Barth M (2011) Sinistral transport along the Trans-European Suture Zone: detrital zircon–rutile geochronology and sandstone petrography from the Carboniferous flysch of the Pontides. Geol Mag 148:380–403

Pape J, Mezger K, Robyr M (2016) A systematic evaluation of the Zr-in-rutile thermometer in ultra-high temperature (UHT) rocks. Contrib Mineral Petrol 171:44, doi:10.1007/s00410–016-1254–8

Pauly J, Marschall HR, Meyer HP, Chatterjee N, Monteleone B (2016) Prolonged Ediacaran-Cambrian metamorphic history and short-lived high-pressure granulite-facies metamorphism in the H.U. Sverdrupfjella, Dronning Maud Land (East Antarctica): evidence for continental collision during Gondwana assembly. J Petrol 57:185–228

Peressini G, Quick JE, Sinigoi S, Hofmann AW, Fanning M (2007) Duration of a large mafic intrusion and heat transfer in the lower crust: A SHRIMP U–Pb zircon study in the Ivrea-Verbano Zone (Western Alps, Italy). J Petrol 48:1185–1218

Pidgeon RT, Bosch D, Bruguier O (1996) Inherited zircon and titanite U–Pb systems in an Archaen sysenite from southwestern Australia: implications for U–Pb stability of titanite. Earth Planet Sci Lett 141:187–198

Quick JE, Sinigoi S, Snoke AW, Kalakay TJ, Mayer A, Peressini G (2003) Geological Map of the Southern Ivrea-Verbano Zone, Northwestern Italy. Geologic Investigations Series Map I-2776 and booklet (22 p) US Geological Survey

Redler C, Johnson TE, White RW, Kunz BE (2012) Phase equilibrium constraints on a deep crustal metamorphic field gradient: metapelitic rocks from the Ivrea Zone (NW Italy). J Metamorph Geol 30:235–254

Rice CM, Darke KE, Still JW (1998) Tungsten-bearing from the Kori Kollo gold mine, Bolivia. Mineral Mag 62:421–429

Richards JP, Krogh TE, Spooner ETC (1988) Fluid inclusions characteristics and U–Pb rutile age of late hydrothermal alteration veining at the Suoshi stratiform copper deposit, central African copper belt. Econ Geol 83:118–139

Rösel D (2014) U–Pb dating of detrital rutile: implications for sedimentary provenance and Pb diffusion in rutile. PhD study, Universität Mainz, Germany, 196 p

Rösel D, Zack T, Boger SD (2014) LA-ICP-MS U–Pb dating of detrital rutile and zircon from the Reynolds Range: A window into the Palaeoproterozoic tectonosedimentary evolution of the North Australian Craton. Precambrian Res 255:381–400

Rubatto D (2017) Zircon: The metamorphic mineral. Rev Mineral Geochem 83:261–29

Rubatto D, Hermann J (2001) Exhumation as fast as subduction? Geology 29:3–6

Rudnick RL, Gao S (2014) Composition of the continental crust. Treatise in Geochemistry (2nd Edition) *In:* Holland H, Turekian K (ed) Elsevier, Vol. 4, p. 1–51

Schärer U, Krogh TE, Gower CF (1986) Age and evolution of the Grenville Province in eastern Labrador from U–Pb systematics of accessory minerals. Contrib Mineral Petrol 94:438–451

Schmitt AK, Vazquez JA (2017) Secondary ionization mass spectrometry analysis in petrochronology. Rev Mineral Geochem 83:199–230

Schmitt AK, Zack T (2012) High-sensitivity U–Pb rutile dating by secondary ion mass spectrometry (SIMS) with an O_2^+ primary beam. Chem Geol 332:65–73

Schmitz MD, Bowring SA (2003) Constraints on the thermal evolution of continental lithosphere from U–Pb accessory mineral thermochronometry of lower crustal xenoliths, southern Africa. Contrib Mineral Petrol 144:592–618

Shulaker DZ, Schmitt AK, Zack T, Bindeman I (2015) In-situ oxygen isotope and trace element geothermometry of rutilated quartz from Alpine fissures. Am Mineral 100:915–925

Siegesmund S, Layer P, Dunkl I, Vollbrecht A, Steenken A, Wemmer K, Ahrendt H (2008) Exhumation and deformation history of the lower crustal section of the Valstrona di Omegna in the Ivrea Zone, southern Alps. *In:* Tectonic Aspects of the Alpine–Dinaride–Carpathian System. Vol 298. Siegesmund S, Fugenschuh B, Froitzheim N (eds). Geol Soc Spec Publ 298:45–68

Sinigoi S, Quick JE, Mayer A, Demarchi G (1995) Density-controlled assimilation of underplated crust, Ivrea-Verbano zone, Italy. Earth Planet Sci Lett 129:183–191

Smit MA, Scherer EE, Mezger K (2013) Peak metamorphic temperatures from cation diffusion zoning in garnet. J Metamorph Geol 31:339–358

Smit MA, Ratschbacher L, Kooijman E, Stearns MA (2014) Early evolution of the Pamir deep crust from Lu–Hf and U–Pb geochronology and garnet thermometry. Geology 42:1047–1050

Smit MA, Waight TE, Nielsen TFD (2016) Millennia of magmatism recorded in crustal xenoliths from alkaline provinces in Southwest Greenland. Earth Planet Sci Lett 451: 241–250

Smith DC, Perseil EA (1997) Sb-rich rutile in the manganese concentrations at St. Marcel-Praborna, Aosta valley, Italy: Petrology and crystal-chemistry. Mineral Mag 61:655–669

Smith SJ, Stevens R, Liu S, Li G, Navrotsky A, Boerio-Goates J, Woodfield BF (2009) Heat capacities and thermodynamic functions of TiO_2 anatase and rutile: Analysis of phase stability. Am Mineral 94:236–243

Smye AJ, Stockli DF (2014) Rutile U–Pb age depth profiling: A continuous record of lithospheric thermal evolution. Earth Planet Sci Lett 408:171–182

Spencer KJ, Hacker BR, Kylander-Clark ARC, Andersen TB, Cottle JM, Stearn MA, Poletti JE, Seward GGE (2013) Campaign-style titanite U–Pb dating by laser ablation ICP: Implications for crustal flow, phase transformation and titanite closure. Chem Geol 341:84–101

Stearns MA, Cottle JM, Kylander-Clark AR, Hacker BR (2016) Extracting thermal histories from the near-rim zoning in titanite using coupled U–Pb and trace element depth profiles by single shot-split stream laser ablation (SS-LASS) ICP-MS: Chem Geol 422:13–24

Tajčmanová L, Connolly JAD, Cesare B (2009) A thermodynamic model for titanium and ferric iron solution in biotite. J metamorphic Geol 27:153–165

Taylor R, Clark C, Reddy SM (2012) The effect of grain orientation on secondary ion mass spectrometry (SIMS) analysis of rutile. Chem Geol 300:81–87

Taylor-Jones K, Powell R (2015) Interpreting zirconium-in-rutile thermometric results. J Metamorph Geol 33:115–122

Tomkins HS, Powell R, Ellis DJ (2007) The pressure dependence of the zirconium-in-rutile thermometer. J Metamorph Geol 25:703–713

Topuz G, Göcmengil G, Rolland Y, Celik ÖF, Zack T, Schmitt AK (2013) Jurassic accretionary complex and ophiolite from northeast Turkey: No evidence for Cimmerian continental ribbon. Geology 41:255–258

Triebold S, von Eynatten H, Zack T (2012) A recipe for the use of rutile in sedimentary provenance analysis. Sediment Geol 282:268–275

Upadhyay D, Chattopadhyay S, Kooijman E, Mezger K, Berndt J (2014) Magmatic and metamorphic history of Paleoarchean tonalite–trondhjemite–granodiorite (TTG) suite from the Singhbhum craton, eastern India. Precambrian Res 252:180–190

Valley JW (2001) Stable isotope thermometry at high temperature. Rev Mineral Geochem 43:365–413

Vavra G, Schmid R, Gebauer D (1999) Internal morphology, habit and U–Th–Pb microanalysis of amphibolite-to-granulite facies zircons: geochronology of the Ivrea Zone (Southern Alps). Contrib Mineral Petrol 134:380–404

Villa IM (1998) Isotopic closure. Terra Nova 10:42–47

Villa IM (2016) Diffusion in mineral geochronometers: present and absent. Chem Geol 420:1–10

Vry JK, Baker JA (2006) LA-MC-ICPMS Pb–Pb dating of rutile from slowly cooled granulites: Confirmation of the high closure temperature for Pb diffusion in rutile. Geochim Cosmochim Acta 70:1807–1820

Waters DJ (1986a) Metamorphic history of sapphirine-bearing and related magnesian gneisses from Namaqualand, South Africa. J Petrol 27:541–565

Waters DJ (1986b) Metamorphic zonation and thermal history of pelitic gneisses from western Namaqualand, South Africa. Trans Geol Soc S Afr 89:97–102

Watson EB, Wark DA, Thomas JB (2006) Crystallization thermometers for zircon and rutile. Contrib Mineral Petrol 151:413–433

Williams ML, Jercinovic MJ, Mahan KH, Dumond G (2017) Electron microprobe petrochronology. Rev Mineral Geochem 83:153–182

Willner AP (1995) Pressure–temperature evolution of a low-pressure amphibolite-facies terrane in central Bushmanland (Namaqua Mobile Belt, South Africa). Commun Geol Surv Namibia 10:5–20

Zack T, Luvizotto GL (2006) Application of rutile thermometry to eclogites. Mineral Petrol 88:69–85

Zack T, Kronz A, Foley SF, Rivers T (2002) Trace element abundances in rutiles from eclogites and associated garnet mica schists. Chem Geol 184:97–122

Zack T, Moraes R, Kronz A (2004a) Temperature dependence of Zr in rutile: empirical calibration of a rutile thermometer. Contrib Mineral Petrol 148:471–488

Zack T, von Eynatten H, Kronz A (2004b) Rutile geochemistry and its potential use in quantitative provenance studies. Sed Geol 171:37–58

Zack T, Stockli DF, Luvizotto GL, Barth MG, Belousova E, Wolfe MR, Hinton RW (2011) In situ U–Pb rutile dating by LA-ICP-MS: [208]Pb correction and prospects for geological applications. Contrib Mineral Petrol 162:515–530

Zanetti A, Mazzucchelli M, Sinigoi S, Giovanardi T, Peressini G, Fanning M (2013) SHRIMP U–Pb Zircon Triassic Intrusion Age of the Finero Mafic Complex (Ivrea-Verbano Zone, Western Alps) and its Geodynamic Implications. J Petrol 54:2235–2265

Zingg A (1980) Regional metamorphism in the Ivrea zone (Southern Alps, N-Italy): field and microscopic investigations. Schweiz Mineral Petrograph Mitteil 60:153–179

Zingg A (1983) The Ivrea and Strona-Ceneri zones (Southern Alps, Ticino and N-Italy—a review. Schweiz Mineral Petrograph Mitteil 63:361–392

Zingg A, Handy MR, Hunziker JC, Schmid SM (1990) Tectonometamorphic history of the Ivrea Zone and its relationship to the crustal evolution of the Southern Alps. Tectonophysics 182:169–192

Reviews in Mineralogy & Geochemistry
Vol. 83 pp. 469–533, 2017
Copyright © Mineralogical Society of America

Garnet: A Rock-Forming Mineral Petrochronometer

E.F. Baxter

Department of Earth and Environmental Sciences
Boston College
Chestnut Hill, Massachusetts 02467
USA

baxteret@bc.edu

M.J. Caddick

Department of Geosciences
Virginia Tech
Blacksburg, Virginia 24060
USA

caddick@vt.edu

B. Dragovic

Department of Geosciences
Virginia Tech
Blacksburg, Virginia 24060
USA

dragovic@vt.edu

INTRODUCTION

Garnet could be the ultimate petrochronometer. Not only can you date it directly (with an accuracy and precision that may surprise some), but it is also a common rock-forming and porphyroblast-forming mineral, with wide ranging—yet thermodynamically well understood—solid solution that provides direct and quantitative petrologic context. While accessory phase petrochronology is based largely upon establishing links to the growth or breakdown of key rock-forming pressure–temperature–composition (P–T–X) indicators (e.g., Rubatto 2002; Williams et al. 2007), garnet *is* one of those key indicator minerals. Garnet occurs in a great variety of rock types (see Baxter et al. 2013) and is frequently zoned (texturally, chemically) meaning that it contains more than just a snapshot of metamorphic conditions, but rather a semi-continuous history of evolving tectonometamorphic conditions during its often prolonged growth. In this way, garnet and its growth zonation have been likened to dendrochronology: garnet as the tree rings of evolving tectonometamorphic conditions (e.g., Pollington and Baxter 2010).

In some ways, the dream of 'petrochronology' all started with garnet (Fig. 1). When Atherton and Edmunds (1965) or Hollister (1966) recognized the chemical zonation in garnet, when Rosenfeld (1968) noted the spiral 'snowball' of inclusions in rotated garnet, or when Tracy et al. (1976) drew the first 2-D map of garnet chemical zonation, illuminating those 'tree-rings' for the first time, they could only imagine what is now a reality decades later—direct zoned garnet geochronology of those concentric rings of growth. Geoscientists soon thereafter attempted the first garnet geochronology (van Breemen and Hawkesworth 1980), though several factors severely limited the development and wider-spread use of

1529-6466/17/0083-0015$5.00 (print)
1943-2666/17/0083-0015$5.00 (online)

http://dx.doi.org/10.2138/rmg.2017.83.15

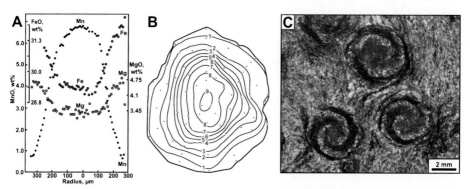

Figure 1. Important observations in the development of garnet petrochronology. **A.** Chemical zoning in a 1-D traverse across garnet from the Kwoiek area of British Columbia (redrawn from Hollister 1966). The 'bell-shaped' Mn zoning profile led to the now classic interpretation of garnet growth adhering to a Rayleigh fractionation model. **B.** 2-D zoning of Mn in garnet from central Massachusetts (redrawn from Tracy et al 1976, numbers equate to atomic percent Mn and dots indicate position of microprobe analyses). **C.** Spiral 'snowball' inclusions in rotated garnet described by Rosenfeld (1968); image shown is from thin section with first-order red filter provided by John Rosenfeld.

garnet geochronology from that point. These factors included 1) contamination of garnet by micro-mineral inclusions, 2) analytical limitations of small sample size, 3) the requirement of anchoring a garnet age analysis with another point on an isochron, and 4) the significant time and effort required for age determination. Even today, whether via MC-ICPMS or TIMS (e.g., Schoene and Baxter 2017, this volume), garnet geochronology requires weeks of time-consuming sample preparation. So, while petrologists boldly forged ahead in the use and development of garnet as probably the premier mineral recorder of evolving metamorphic and tectonic processes during the 1970's, 80's and 90's, garnet geochronology was mostly (though not completely) supplanted by the excitement, relative ease and undoubted utility of accessory phase geochronology (as reviewed by Engi 2017, this volume; Rubatto 2017, this volume). Then, because the petrology and thermodynamics of accessory phases had not been as well studied, the task of 'petrochronology' was to bring petrologic context to the age information accessible in phases such as zircon and monazite. Often, this endeavor came back to linking the growth of accessory phases with a key rock-forming mineral—garnet. The last 20 years have now seen the advancement of direct garnet geochronology driven by several factors, including 1) the addition of Lu–Hf to Sm–Nd as viable systems to date garnet (e.g., Duchene et al. 1997; Scherer et al. 2000), 2) robust methods to eliminate the effects of contaminating inclusions (e.g., Amato et al. 1999; Baxter et al. 2002; Anczkiewicz and Thirlwall 2003), 3) improved analytical techniques to reduce sample size limitations (e.g., Harvey and Baxter 2009; Bast et al. 2015), and 4) microsampling methods whereby those individual tree-rings can be sampled at higher and higher spatial resolution (e.g., Stowell et al. 2001; Ducea et al. 2003; Pollington and Baxter 2010, 2011). What this makes possible today is the introduction—or re-introduction—of garnet into the cadre of modern 'petrochronometers'.

Our purpose in this chapter is threefold. We begin with a review of the 'petro-' of garnet, followed by the '-chrono-' of garnet, setting us up for the modern possibilities—many yet to be fully explored—of garnet 'petrochronology'. First, we re-acquaint the reader with the myriad ways in which garnet has been used directly to reconstruct past tectonometamorphic conditions. Ranging from foundational thermodynamic *P–T* modeling and textural analysis to recent advances in inclusion barometry and trace element zonation, our aim is to illuminate for all readers the remarkable scope and potential for garnet as a recorder of tectonometamorphic context. It is not garnet per se that we are interested in, but rather the conditions and processes that it records. Second, we bring the reader up to date on recent advances in direct chronology of garnet via Sm–Nd and Lu–Hf systems that have made it the robust and precise chronometer that it is. Our hope is to dispel the misconceptions that continue to limit the use and credence of garnet geochronology, and clearly lay out the wide-ranging opportunities for direct garnet geochronology, appropriately framed within the challenges and limitations which still exist for this (as with any!) geochronometer. This section on the 'chrono' of garnet also discusses diffusion geospeedometry of garnet, which provides complementary information on heating/cooling durations and rates. Finally, we bring the 'petro' and 'chrono' together and present several examples from the past decade of true garnet petrochronology wherein these methods are integrated in increasingly innovative ways. Figure 2 serves as both an outline of our chapter, and as our definition of garnet petrochronology: the effort to integrate any aspect(s) of garnet-based petrology (on the left of Fig. 2) with any aspect(s) of garnet geochronology, and in so doing, to gain insights about evolving tectonometamorphic conditions that complement what may be extracted from other methodologies. Perhaps the most important thing to emphasize is that garnet petrochronology is really still in its infancy. Many challenges have been overcome, and it is time now for creative petrologists to come back to garnet and try new ways of integrating the 'petro' and the 'chrono' (by choosing from the left and the right of Fig. 2) to make true advances in our understanding of petrologic and tectonic evolution of the crust-mantle system. Never before has the time been so ripe, and we hope the reader is inspired to take these next steps in creatively deploying garnet petrochronology in the future.

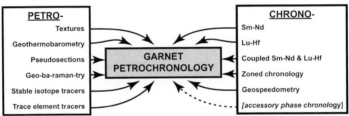

Figure 2. Outline of the chapter and definition of garnet petrochronology. Any "petro-" of garnet integrated with any "chrono" of garnet makes for garnet petrochronology. Accessory phase geochronology is included with a dashed line to note its integrative value, as valuable efforts have been made to link it to garnet.

PETRO- OF GARNET

Said the professor to their student: "*please try to collect some samples with garnet in them; there's not nearly as much we can do without it*". Although this statement is apocryphal, the sentiment will be familiar to many readers. Metamorphic petrology has been tied to the search for and study of garnet-bearing samples since soon after Barrow defined the 'garnet zone' in the eastern Scottish Highlands (Barrow 1893, 1912). Advancements in our ability to characterize the compositions and textures of metamorphic minerals, and in thermochemical models to describe the conditions of their evolution have commonly focused on garnet-bearing samples. In this section we review some of the reasons why garnet has been so central to the endeavor of reconstructing tectonometamorphic processes and conditions. We give an overview of some of the mechanisms that have been developed to infer pressure, temperature, deformation, mineral reaction or fluid–rock interaction, highlighting some of the recent and current developments that will aid this. Much has already been written about most of this, and we have thus made no attempt to be absolutely comprehensive, instead directing the reader to several of the other excellent reviews where each provide more detail about each technique. But by bringing all of this content together, our goal is to provide a comprehensive taste of what garnet has previously been used to quantify and constrain, with the expectation that these methods will increasingly be coupled with each other and with direct dating, as described later in the 'Chrono- of garnet' section of this chapter. We will begin with the most easily observable (often at the hand sample scale) textural information locked in garnet, before reviewing its role in geothermobarometry and in tracing evolving petrologic reactions and open system fluid flow processes.

Textures of garnet—provider of tectonic context

Garnet's rigid porphyroblastic growth endows it with a propensity to retain and record its growth history in chemical as well as textural manifestations. This section provides several brief examples of how textural observations can provide valuable petrologic and tectonic context for garnet's growth span.

Garnet as a deformation and strain monitor. Garnet crystals are frequently riddled with cracks and inclusions of other minerals that can reveal a reaction history and the co-genetic evolution of tectonic stresses. For example Angiboust et al. (2011) interpreted cracks in UHP eclogite garnets as reflecting seismic brecciation during subduction to ~80 km depth in the Western Alps. Shear stress can also deform the matrix and cause garnet to rotate (Spry 1963; i.e., 'snowball garnets'; Rosenfeld 1968) or record changing metamorphic fabrics during porphyroblast growth (i.e., Ramsay 1962; Bell 1985). Controversy still exists between these two end-member interpretations (i.e., Johnson 1993; Ikeda et al. 2002; Stallard et al. 2002) but in any case, these common and often vivid patterns of inclusions inside garnets (e.g., Fig. 1c) clearly record rock deformation that may reflect localized structural—(e.g., Robyr et al. 2009) or regional tectonic-scale (e.g., Aerden et al. 2013; Sayab et al. 2016) processes. Beyond documentation of inclusion patterns in thin section, valuable textural information can be gleaned from additional methods of observation. Robyr et al. (2009) employed EBSD (electron backscattered diffraction) analysis to show that crystallographic orientations within single garnet porphyroblasts remained constant while the garnet was rotating. Then, those crystallographic orientations shifted to reflect surrounding phyllosilicate foliation when garnet grew during subsequent non-rotational regimes. Aerden et al. (2013) relate different textural generations of inclusions within garnets, each separated by a FIA (fold intersection axis) to major shifts in tectonic convergence vectors during the assembly and deformation of the Iberian Peninsula. These FIA are carefully mapped through analysis of radial sets of vertical thin sections and would not otherwise be readily apparent in a simple 2D thin section. Sayab et al. (2016) use similar methods to document tectonic convergence vectors in the Himalayan Orogen. Among the very first applications of true garnet petrochronology, Christensen et al. (1989) and Vance and Onions (1992) dated the cores and rims of such rotated snowball garnets to place constraints on the tectonic shear strain rate. Cases in which the identity or composition of the included minerals change from core to rim of the host garnet reveal additional, powerful information, as detailed below (e.g., St-Onge 1987).

Garnet resorption textures. As systems evolve and *P–T–X* changes, it is common for garnet to become unstable and start to break down by consumption from its rim inward (see Lanari et al. 2017 for additional examples). This resorption is driven by changes in *P–T–X* occurring during prograde metamorphism (e.g., Florence and Spear 1993), during retrograde metamorphism, due to chemical change related to influx of an external fluid, and sometimes due to polymetamorphism (when old garnet is introduced into a new orogenic cycle). Such resorbed surfaces often exhibit irregular morphologies, such as embayments, which often contrast with sharp original crystal growth faces. On the one hand, once a crystal (or its outermost portion) has been resorbed, there is little direct 'garnet petrochronology' that can be done to study that resorption event. However, if a resorption surface can be identified (texturally or chemically), the relative timing and conditions surrounding the resorption event may still be constrained. For example, if an element liberated during garnet resorption cannot be readily incorporated into product phases, a small amount of garnet might re-precipitate, preferentially incorporating that element, or the element itself might be reincorporated into any residual garnet (e.g., see discussion below relating to Y and HREE incorporation into resorbed crystal rims). In general, such elements would be those that tend to be strongly partitioned into garnet over other matrix phases including Mn, Y, Lu and other HREE (e.g., Kohn and Spear 2000; Kelly et al. 2011; Gatewood et al. 2015). In some cases, secondary minerals may form at a garnet resorption surface driven by the sudden and proximal influx of those particular elements. Xenotime, for example, has been observed decorating the resorbed surfaces of garnets where localized flux of Y (from the resorbed garnet) promotes growth of xenotime (e.g., Gatewood et al. 2015). In this case, there is a useful textural link between resorbed garnet surface and neocrystallized accessory phase xenotime.

Textural evidence for polymetamorphism. We use the term 'polymetamorphic garnet' to describe crystals that grew during more than one tectonic 'event' separated by a significant hiatus that can be resolved texturally, chemically, or temporally with existing methods. Several examples of polymetamorphic garnet have been recognized in the Alpine-Himalayan system that grew during multiple orogenic cycles, separated by millions to hundreds of millions of years. Argles et al. (1999) dated fractions of garnet crystals, identifying that only crystal rims record Tertiary metamorphism, with crystal cores hundreds of millions of years older. Gaidies et al. (2006), Herwartz et al. (2011), Robyr et al. (2014) and (Lanari et al. 2017) recognized, garnet zonation textures in which Alpine overgrowths are vividly separated by prominent microstructural and chemical discontinuities from older Variscan cores. Manzotti and Ballevre (2013) recognized chemically heterogeneous detrital garnet cores—derived from multiple sources—that were overgrown by a homogeneous Alpine garnet rim. All of these examples serve to illustrate some of the textural and chemical means by which polymetamorphic garnet may be recognized. Of course, only by adding direct geochronology to each of these growth generations may we establish the absolute chronology and length of the hiatus between phases of garnet growth.

Porphyroblast nucleation and growth models. Metamorphic porphyroblastic minerals comprise vivid records of progressive nucleation and growth kinetics. Treated individually, each porphyroblast may record aspects of the rock's overall history colored by local chemical-textural features. Taken as a whole, porphyroblast crystal size distributions, their relative spatial disposition, and chemical zonation reveal information about the nucleation and growth process rock wide (e.g., Kretz 1973; Carlson et al. 1995; Chernoff and Carlson 1997; Meth and Carlson 2005). Several forward models have been developed to predict and interpret garnet CSD's and spatial dispositions within porphyroblastic rocks (Galwey and Jones 1966; Kretz 1966, 1974; Cashman and Ferry 1988; Carlson 1989, 2011; Spear and Daniel 2001; Gaidies et al. 2008b; Gaidies et al. 2011; Schwarz et al. 2011; Ketcham and Carlson 2012; Kelly et al. 2013a,b). These models offer testable predictions about the rate limiting processes for crystal growth and progressive metamorphism, including important implications for the attainment of local and rock-wide chemical equilibrium. Existing models make justified assumptions and interpretations about whether the rate-limiting step is breakdown of parent minerals (e.g., Schwarz et al. 2011), surface energetics of garnet (e.g., Gaidies et al. 2011), or transport of the least mobile nutrient (often aluminum) through the intergranular medium (e.g., Ketcham and Carlson 2012). The importance of Ostwald ripening (favoring growth of larger porphyroblasts at the expense of smaller porphyroblasts) has been debated (e.g., Carlson 1999), and some studies have cited heating rate as the key limitation. Each of these published cases shows persuasively that numerical simulations of one or several processes can successfully model natural crystal occurrences, suggesting that different rate-limiting processes may dominate under different circumstances. Furthermore, EBSD analysis has revealed that garnets porphyroblasts may often (perhaps >20% of the time) include more than one primary nucleation site and, thus, more than one fundamental crystal lattice orientation. These early forming crystal clusters may ultimately coalesce, separated by high-angle grain boundaries, to form a macroscopically visible porphyroblast (e.g., Daniel and Spear 1998; Spiess et al. 2001; Whitney et al. 2008; Whitney and Seaton 2010).

Garnet as a pressure and temperature sensor

One of the most important, and familiar, aspects of the 'petro-' in petrochronology involves deciphering metamorphic pressure and temperature to infer rocks' journeys through the crust and mantle. Garnet has arguably been the single most useful mineral for the estimation of evolving metamorphic *P–T* conditions for over forty years[1]. It is especially useful because of 1) its occurrence in diverse lithologies and *P–T* conditions, and 2) its relatively simple and well understood chemistry, which is dominated by divalent Fe, Mg, Ca, Mn, and trivalent Al, Fe cations (e.g., Grew et al. 2013) but is also sensitive to certain trace elements like Y. Because garnet is a major rock-forming mineral, its growth history is directly linked to broader rock wide petrologic evolution; so we must also develop the use of broader descriptions of the entire mineral assemblage and provide context for the growth of garnet and its changing chemistry. Here we review some of the techniques that have been developed to utilize garnet's chemical and mechanical properties to constrain *P–T*.

Compositional thermometry—Major elements. The advent of the electron microprobe in the 1950s sparked a revolution in metamorphic petrology as systematic characterization revealed that the compositions of co-existing mineral phases are often strongly correlated with metamorphic grade (e.g., Ramberg 1952; Kretz 1959). Garnet composition almost immediately became a focus, leading to the discovery that crystals commonly exhibit zoning from core to rim (Atherton and Edmunds 1965; Evans 1966; Evans and Guidotti 1966; Harte and Henley 1966; Hollister 1966; Atherton 1968) and in complex patterns around inclusions and upon contact with specific matrix phases (Tracy et al. 1976; Thompson et al. 1977; Tracy 1982; Figures 1 and 3). Relationships between mineral composition and metamorphic temperature suggested a geothermometer based upon the partitioning of Fe and Mg between garnet and biotite (e.g., Frost 1962; Kretz 1964; Saxena 1968, 1969; Thompson 1976b; Goldman and Albee 1977). This results primarily from garnet's large distorted cubic site favoring Fe rather than Mg at low temperature, under which conditions biotite's octahedral sites prefer Mg. Indeed, garnet typically contains far lower Mg content than most coexisting phases, with the ratio Mg/Mg+Fe(X_{Mg}) decreasing in the order cordierite > chlorite > biotite > chloritoid > staurolite > garnet (Albee 1965, 1972; Hensen 1971). The preference of each phase for Mg or Fe is reduced upon heating, thus imparting a temperature dependence of Fe–Mg partitioning between most co-existing phases (the volumetric effect of Mg–Fe exchange is comparatively minor). Garnet plays a particularly important role here because its low Mg content at low temperature establishes a much larger ΔX_{Mg} between it and a second equilibrated mineral (e.g., orthopyroxene or biotite) than a pairing between most other phase combinations would permit, explaining why garnet often occurs as a key mineral in exchange thermometer calibrations. The garnet side of these equilibria is shown in Figure 4, which demonstrates the relatively extensive Mg and Fe variation expected under possible crustal conditions, assuming a fixed bulk rock composition. After the classic experimental calibration by Ferry and Spear (1978), numerous refinements to the garnet–biotite exchange thermometer (Perchuk and Lavrent'eva 1983; Ganguly and Saxena 1984; Indares and Martignole 1985; Kleeman and Reinhardt 1994; Berman and Aranovich 1996; Gessmann et al. 1997; Holdaway et al. 1997; Mukhopadhyay et al. 1997; Holdaway 2000) have made this one of the most often used geothermometers.

Temperature dependent Fe–Mg equilibria between garnet and many other phases have also been calibrated as thermometers, all based on similar principles to the garnet–biotite example outlined above. It is beyond the scope of this review to discuss all of these in detail, but the reader is directed to Essene (1982, 1989) and Spear (1993) for more information.

---

[1] For example, Spear (1993) lists 37 geothermobarometers on page 517–519 of his book; 25 of these include garnet, far more than any other mineral.

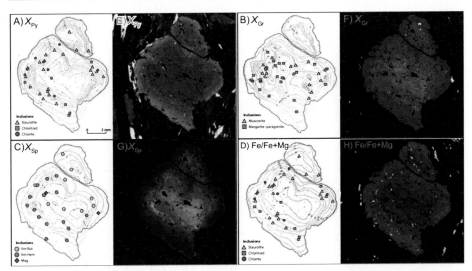

Figure 3. Garnet zoning, then and now. Panels A-D are some of the first 2-D garnet element zoning maps, hand-contoured from point analyses by R.J. Tracy as part of his graduate studies and published as Thompson et al (1977). These images of a crystal from the Gassetts Schist, Vermont, first hinted at the complexity of zoning in natural garnet crystals and were used to advance a model of progressive crystal growth during evolving *P* and *T*. The very first 2-D images had appeared a year earlier (Tracy et al 1976). Panels E-H are element maps of the same crystal, collected ~ 35 years later by the same R.J. Tracy in a ~12 hour run with an automated electron microprobe (Tracy et al 2012). The general zoning patterns were extremely well documented by the previous work, but modern techniques yield a far more detailed picture, particularly highlighting fine-scale zoning in the crystal core of the Ca map that is still difficult to construct a satisfactory growth model for. Computerized automation has revolutionized many aspects of metamorphic petrology, paving the way for the petrochronology discussed in this volume.

Compositional barometry. While low ΔV exchange equilibria were developed into mineral thermometers, high ΔV equilibria were explored as potential barometers, again with an early emphasis on garnet-bearing assemblages. Observation of and experimentation on the garnet–plagioclase system (Kretz 1959; Ghent 1976; Goldsmith 1980; Newton and Haselton 1981) led to calibration of a mineral barometer that uses the end member 'GASP' (garnet–aluminosilicate–silicate–plagioclase) reference reaction:

$$Ca_3Al_2Si_3O_{12} + 2Al_2SiO_5 + SiO_2 = 3CaAl_2Si_2O_8$$

This is pressure sensitive because of the large molar volume difference of reactants and products, and has been calibrated and recalibrated multiple times as new experimental data and activity–composition (*a–X*) models for garnet and plagioclase have become available (e.g., Ghent 1976, 1977; Newton and Haselton 1981; Koziol and Newton 1988; McKenna and Hodges 1988; Berman 1990; Holdaway 2001; Caddick and Thompson 2008). The relative performance of most of these calibrations was compared by Wu and Cheng (2006), but taken as a group the GASP calibrations have become probably the most widely used chemical geobarometer for metamorphic rocks, due in part to the large range of *P–T* conditions over which garnet and plagioclase can both co-exist with a suitable buffering assemblage. Uncertainties on temperature of equilibration, end-member thermodynamic properties (particularly enthalpy and entropy), mineral solution behavior, and measurements of garnet and plagioclase composition, likely combine to yield GASP uncertainties reaching or exceeding ±2 kbars in many cases (see discussions by, Hodges and McKenna (1987), McKenna and Hodges (1988), Kohn and Spear (1991) and Holdaway (2001) for more information).

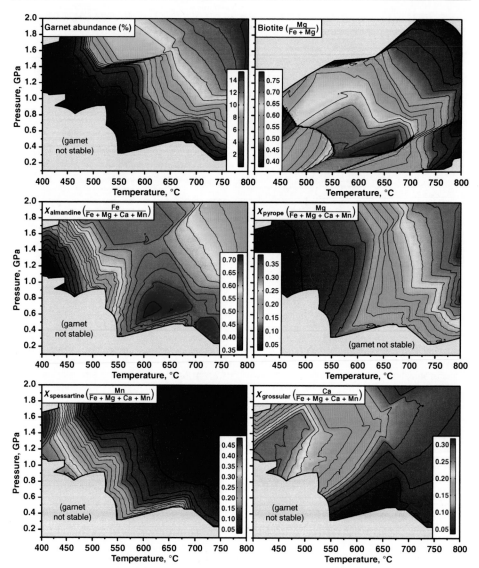

Figure 4. Mineral abundances and compositions in a pelitic rock at crustal conditions as a function of *P* and *T* for the 'average pelitic bulk-rock' composition from Caddick and Thompson (2008). Calculated using Perple_X (Connolly 2005) with the THERMOCALC ds5.5 dataset (updated from Holland and Powell 1998), and activity models described in Caddick and Thompson (2008), except for garnet (White et al 2005), amphibole and omphacite (both Diener and Powell 2012). The underlying calculation from which contours were extracted considers many additional phases that are not shown, with biotite and garnet in most regions of the diagrams co-existing with 5 or 6 additional phases (see annotated pseudosection for this rock, which appeared as Figure 1 in Caddick and Thompson (2008), noting that that diagram was calculated with earlier thermodynamic data and is thus not identical).

Many additional mineral systems have been suggested and calibrated for inferring pressure and temperature, with a significant proportion of these relying partly on the properties of garnet, e.g., garnet–plagioclase–biotite–muscovite, garnet–muscovite–plagioclase–quartz, garnet–biotite–plagioclase–quartz, garnet–ilmenite–rutile–kyanite–quartz, garnet–ilmenite–rutile–

plagioclase–quartz, garnet–clinopyroxene–plagioclase–quartz, garnet–orthopyroxene, garnet–clinopyroxene–phengite (Bohlen and Essene 1980; Ghent and Stout 1981; Bohlen et al. 1983; Hodges and Crowley 1985; Bohlen and Liotta 1986; Moecher et al. 1988; Hoisch 1990, 1991; Waters and Martin 1993; Pattison et al. 2003; Wu et al. 2004; Wu and Zhao 2006). Again, more details about many of these were provided by Essene (1989) and by Spear (1993).

Compositional thermometry—Trace elements. While the distribution of various trace elements in garnet can be used to determine reaction sequence, the applicability of geochronometers, and fluid/melt history (see discussions below), trace element (particularly yttrium) concentrations of coexisting accessory phases and garnet have also been utilized as thermometers in pelitic lithologies.

Pyle and Spear (2000) developed an empirical trace element thermometer for the temperature range of 450–550 °C, by study of regional metamorphic rocks from New England. This method utilizes the yttrium content of garnet in xenotime-bearing pelites, assuming that the presence of xenotime buffers the rock in Y. Temperature and pressure estimates for calibration purposes, along with *P–T* paths, were determined by traditional thermobarometry and differential thermodynamics, respectively. Pyle et al. (2001) expanded this approach to the assessment of equilibrium partitioning of coexisting monazite–xenotime–garnet, developing a garnet–monazite thermometer. Critical in employing this thermometer is 1) accurate compositional information about all phases involved in the reaction (Foster et al. 2004; Hallett and Spear 2015) and 2) that monazite was a stable phase during garnet growth, i.e., that garnet and monazite were in equilibrium. Inclusion of homogeneous monazite in homogeneous or continuously zoned garnet may be considered a good candidate for such equilibration, though zoning in that monazite and the chances of its isolation from the reactive part of the rock warrant particular care, as further discussed by Lanari and Engi (2017, this volume). Uncertainties in the other parameters for calculation likely amount to a combined uncertainty of approximately ±25 °C (Pyle et al. 2001). Pyle et al. (2001) note that garnet and monazite maintain compositional equilibrium even as garnet fractionation dramatically changes the Y budget in the reactive bulk composition, allowing for a consistent element partitioning during the rock's prograde history.

Thermobarometry of zoned garnet. If mineral composition is to be used to infer pressure and temperature but natural phases such as garnet are commonly chemically zoned (e.g., Figs. 1 and 3), where should one make a microprobe analysis, where should analyses of coexisting phases be made, and how can results be put into the context of an evolving *P–T* history? These questions have been addressed in detail (e.g., Essene 1989; Kohn and Spear 2000), and extreme caution should be taken before averaging data whose variability may be geologically meaningful. For example, the apparent repeatability of garnet zoning patterns suggests that this zoning records changes in pressure and temperature during crystal growth, leading to early models of garnet zoning due to sequential mineral reactions (e.g., Hollister 1966; Loomis 1975; Thompson 1976a; Tracy et al. 1976; Tracy 1982), which in turn led to direct use of garnet zoning to infer *P–T* paths during prograde metamorphism. Development of a differential thermodynamic approach, commonly referred to as the Gibbs method, allowed derivation of *P–T* paths for rocks, based upon garnet crystal zoning that is assumed to have been established during growth and an estimate of the *P–T* point at which garnet first began to grow (Spear and Selverstone 1983; Spear et al. 1984). Zoned garnet porphyroblasts containing complex suites of mineral inclusions present an additional opportunity, with thermobarometry on the hosts and their inclusions yielding complex information that may be interpreted as a simple *P–T* progression within a single orogenic event (e.g St-Onge 1987) or as overprinted conditions associated with multiple events or tectonic processes (e.g., Dorfler et al. 2014). As such, the most appropriate course for thermobarometry generally begins with careful consideration of mineral texture and mineral zoning (e.g., Essene 1989; Kohn and Spear 2000) with garnet often at the heart of this endeavor.

Forward models of phase equilibria. A powerful use of equilibrium thermodynamics in the understanding of metamorphic processes involves the calculation and interpretation of phase equilibria, and much modern petrochronology involves the coupling of geochronological methods with phase equilibria constraints. Prediction of the proportion and composition of each major phase in a rock, and how these would change as functions of pressure, temperature and system composition (in so-called 'pseudosections', 'isochemical phase diagrams' or 'mineral assemblage diagrams' such as Fig. 4) has had enormous influence on the petrologic community. This has been reviewed recently by Spear et al. (2016) and in this volume by Yakymchuk et al. (2017, this volume), and here we focus on the primary value from a garnet petrochronology standpoint: revealing how garnet growth or dissolution would be expected along any *P–T* path, how this might modify the residual rock composition (discussed also in detail by Lanari and Engi 2017, this volume), how garnet chemistry would evolve during growth, and how this is associated with changes in the co-existing assemblage, including the production or consumption of other minerals, fluids, or melt.

Most studies of garnet in crustal metamorphism are concerned with pelitic, greywacke, basaltic or carbonate lithologies, in which case reasonable approximations can be made by considering a thermodynamic system comprised of Na_2O, K_2O, CaO, FeO, MgO, Al_2O_3, SiO_2, H_2O, $\pm MnO$, $\pm TiO_2$, $\pm Fe_2O_3$, $\pm CO_2$. Incorporation of all or most of these components permits description of garnet as $(Fe^{2+},Mg,Ca,Mn)_3(Al,Fe^{3+})_2Si_3O_{12}$ and calculation of phase equilibria involving most *major* phases that this garnet is likely to co-exist with. The thermodynamics of MnO-bearing mineral end-members are generally less well-constrained than others, but MnO plays such a significant role in stabilizing garnet at low *P–T* conditions and on the modal proportion of garnet at higher *P–T* conditions (Symmes and Ferry 1991; Mahar et al. 1997; Tinkham et al. 2001; Caddick and Thompson 2008; White et al. 2014b) that its inclusion is often warranted. A plethora of relevant mineral, fluid and melt models have been described, with most recent contributions associated with various updates of the 'THERMOCALC' end-member dataset (Holland and Powell 1985, 1990, 1998, 2011). Even models for compositionally well-constrained minerals such as garnet have been reassessed numerous times (inlcuding but not limited to Wood and Banno 1973; Engi and Wersin 1987; Powell and Holland 1988; Hackler and Wood 1989; Koziol and Newton 1989; Berman 1990; Ganguly et al. 1996; Mukhopadhyay et al. 1997; White et al. 2000, 2005, 2014a,b; Stixrude and Lithgow-Bertelloni 2005; Malaspina et al. 2009). Seemingly minor updates to solid solution models can lead to significant changes in calculated phase equilibria unless end-member thermodynamic properties are modified accordingly (see discussion in White et al. 2014a). Furthermore, the relative stabilities and coexisting compositions of complex mineral solutions may respond inappropriately if applied beyond the limits (in composition, pressure or temperature) of their original calibration, though it can often be difficult to ascertain whether this is the case. It is thus informative to make direct comparison between $K_{Mg-Fe}^{Gar-Bio}$ derived from the mineral compositions predicted by phase equilibria calculations and $K_{Mg-Fe}^{Gar-Bio}$ derived from experimental calibrations such as Ferry and Spear (1978) across a range of *P–T* conditions and for a range of bulk-rock compositions (and thus additional buffering minerals in the calculated equilibria). Such comparisons, (e.g., Figure 6 of Caddick and Thompson 2008) highlight that, although experimentally determined partitioning is well recovered at many *P–T* conditions, it is poorly fit elsewhere, and reveal some unexpected consequences of utilizing complex activity-compositions models for numerous phases.

Comparison between calculated and natural garnet composition is one of the primary constraints on segments of *P–T* paths inferred from pseudosection calculations (Konrad-Schmolke et al. 2006; Gaidies et al. 2008a; Chambers et al. 2009; Cutts et al. 2010; Vorhies and Ague 2011; Hallett and Spear 2014a; Mottram et al. 2015). Many factors can lead to calculated compositions that fail to intersect at *P–T* conditions consistent with other observations, including inappropriate assumptions of bulk composition or short length-scale compositional heterogeneity (e.g., Tinkham and Ghent 2005; Kelsey and Hand 2015; Guevara and Caddick 2016; Palin et al. 2016; Lanari and Engi 2017, this volume), the potential for crystal nucleation and growth at conditions removed from sample-wide thermodynamic equilibrium (e.g., Kretz 1973; Waters and Lovegrove 2002; Gaidies et al. 2011; Pattison et al. 2011; Kelly et al. 2013a,b, b; Carlson et al. 2015b) and the possibility of diffusive modification of crystal composition following growth (discussed in more detail below). Several recent approaches have focused more explicitly on garnet, developing computer codes that compare measured natural crystal zoning with thermodynamic predictions to automatically search for the most appropriate $P–T$ path experienced, generally also considering modification of the bulk-rock composition upon garnet growth (e.g., Moynihan and Pattison 2013; Vrijmoed and Hacker 2014; Lanari et al. 2017). To some extent these are natural successors to the pioneering Gibbs method (Spear and Selverstone 1983; Spear et al. 1991; Menard and Spear 1993) and to calculations that forward model the chemical zoning established along prescribed $P–T$ paths (e.g., Florence and Spear 1991; Gaidies et al. 2008a,b; Konrad-Schmolke et al. 2008; Caddick et al. 2010).

Given the relative ease of calculating phase equilibria, what is their value? As emphasized recently (e.g., Pattison 2015; Spear et al. 2015, 2016), the ability to calculate diagrams may have exceeded our capability to assess their predictions. For example, if a pseudosection predicts a 5 °C, 0.5 kbar window within which a garnetiferous assemblage is stable in a relatively dry lithology, it may be unlikely that this assemblage will actually be found at these $P–T$ conditions in nature due to (i) uncertainties in the thermodynamic data, and (ii) possible kinetic effects that may hinder sample-scale equilibration on the timescale that the rock passes through the $P–T$ conditions of the field (Carlson et al. 2015b). This can be closely associated with the concept that the volume of any rock in mutual thermodynamic equilibrium (i.e., the equilibrium length-scale) probably changes throughout metamorphism as a function of P, T and fluid availability (e.g., Stüwe 1997; Guiraud et al. 2001; Carlson 2002). In fact, the equilibrium length-scale may be different for each element depending on the relative diffusive transport rate (in turn a function of elemental partitioning/solubility into the intergranular transporting medium; Baxter and DePaolo 2002b; Carlson 2002). The choice of appropriate bulk composition for a pseudosection calculation is thus often not trivial and this composition may have to change as subsequently refractory phases such as garnet (e.g., Spear et al. 1991; Marmo et al. 2002; Tinkham and Ghent 2005; Caddick et al. 2007; Konrad-Schmolke et al. 2008), hydrous fluid (e.g., Guiraud et al. 2001) or melt (e.g., Tajčmanová et al. 2007; Yakymchuk and Brown 2014; Guevara and Caddick 2016) are produced and fractionated from the reactive system. Lanari and Engi (2017, this volume) review bulk composition in more detail, demonstrating that use of an XRF-based rock composition to retrieve $P–T$ conditions of zoned garnet growth, without modification of that composition to account for components sequestered by garnet, can produce significantly erroneous estimates if more than ~2 vol% garnet has been produced. Eventually the length-scales of equilibration may become small enough to establish domainal textures such as coronae, in which substantial chemical potential (μ) gradients are preserved over geological timescales and can best be interpreted through $\mu–\mu$ equilibria (e.g., White et al. 2008; Štípská et al. 2010; Baldwin et al. 2015). The retention of chemical zoning in garnet is one obvious indicator that μ gradients are commonly maintained in metamorphic rocks.

Despite these various complications and limitations, equilibrium thermodynamic models are of immense importance to garnet petrochronology and to metamorphic petrology more broadly, as testified by the fact that the combined citations of the main papers describing the THERMOCALC, THERIAK-DOMINO and Perple_X programs (e.g., de Capitani and Brown 1987; Powell and Holland 1988; Connolly 1990; Powell et al. 1998; Connolly 2005, 2009; de Capitani and Petrakakis 2010) and the THERMOCALC dataset (e.g., Holland and Powell 1990, 1998, 2011) significantly exceed 10,000 at the time of writing. Recent successes with respect to application within garnet petrochronology are given at the end of this chapter, but we would like here to highlight that the potential power of pseudosections in garnet petrochronology (albeit without the title) was elegantly demonstrated ~25 years ago when Vance and Holland (1993) coupled garnet compositions from the Gassetts Schist rocks that had previously been element mapped by Thompson et al. (1977; e.g., Fig. 3) with pseudosection constraints on the *P–T* path of garnet growth and isotopic dating of that garnet. This early adoption of garnet petrochronology led to estimations of heating and decompression rate, and to inferences about tectonic and thermal mechanisms for this metamorphism.

The loss of information. Compositional thermobarometry and comparison of pseudosections with natural mineral assemblages rely on the fundamental assumption that rocks faithfully record information associated with equilibration at a set of *P–T* conditions, often assumed to be peak *T*. However, a substantial body of empirical (e.g., Anderson and Olimpio 1977; Woodsworth 1977; Yardley 1977), experimental (e.g., Elphick et al. 1981; Loomis et al. 1985; Chakraborty and Ganguly 1991; Ganguly et al. 1998a; Vielzeuf and Saúl 2011) and modeled (e.g., Carlson 2006; Chu and Ague 2015) evidence suggests that major element zoning in garnet will relax at high temperature through volume diffusion. Furthermore, detailed observations indicate that exchange and net-transfer reactions can operate on the rims of garnet crystals after peak metamorphic conditions and that these can substantially modify the apparent *P–T* conditions recovered from mineral thermometry (e.g., Tracy 1982; Spear and Florence 1992; Kohn and Spear 2000; Pattison et al. 2003; Spear 2004). An illustrative example comes from a transect through the western Himalaya, where oxygen isotope thermometry and phase equilibria constraints (Vannay et al. 1999) reveal far higher temperatures than previously inferred for the same samples through major element garnet-based thermobarometry (Vannay and Grasemann 1998). The implication is that garnet–biotite compositions were reset by Mg–Fe exchange during cooling from peak temperatures of ~750 °C to ~600 °C, where exchange and diffusion became sufficiently sluggish to inhibit further substantial modification (effectively representing a closure temperature).

Diffusional resetting of metamorphic systems is covered in depth in this volume (Kohn and Penniston-Dorland 2017, this volume) and later in this chapter we discuss how diffusional *partial* resetting can be used to constrain heating and cooling timescales (i.e., geospeedometry). But here, we must consider the conditions at which compositional information in garnet is altered due to volume diffusion, thus compromising geothermobarometry. This is particularly pertinent to petrochronology because garnet is such an important constituent of many thermobarometers and major element diffusion within the garnet lattice is typically considered to be slower than in many other major silicate minerals in metamorphic rocks (Brady and Cherniak 2010). Numerous studies have thus sought to quantify the extent to which garnet zoning established during prograde growth will be modified at subsequent stages of metamorphism, with several trying to generalize results for a range of conditions (Florence and Spear 1991; Gaidies et al. 2008a; Caddick et al. 2010). Figure 5 is an example that shows the extent to which garnet prograde zoning will be preserved as a pelitic rock heats and is buried along the indicated *P–T* path, exploring a wide range T_{max}, heating rate and eventual crystal sizes. It indicates, for example, that if a crystal nucleates at thermodynamically-constrained 'garnet-in' and grows spherically to 1 mm diameter during ca. 6 Ma of heating to T_{Max} of 580 °C (e.g., by following path '*b*' in Fig. 5, equating to ~25 °C/Ma), its core will effectively preserve initial X_{Gar}^{Mg} and X_{Gar}^{Ca} contents but X_{Gar}^{Fe} (not shown) and X_{Gar}^{Mn} (not shown) will have been modified by more than 1 mole fraction unit due to volume diffusion subsequent to growth. Crystals achieving a 5 mm diameter along the same prograde path will retain their initial compositions at T_{max}. If heating continues at the same rate to 750 °C, the core composition of crystals exceeding 1 cm diameter will have been partially modified and crystals smaller than 500 μm may preserve little or no major element zoning. Slower heating rates increase the crystal sizes for which crystal core compositions are lost (e.g., paths '*c*' and '*f*' in Fig. 5), while fast heating tend to preserve growth information (e.g., path '*a*').

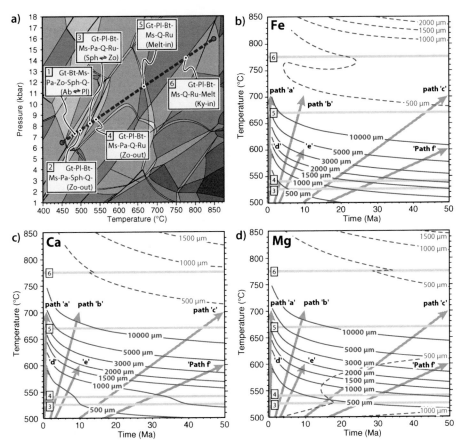

Figure 5. Garnet opening to resetting upon heating, modified from Caddick et al (2010). Phase stabilities in a pelitic rock were calculated along the burial and heating path shown in panel A, which also highlights significant equilibria encountered. Equilibrium garnet compositions at successive points along that path were used to establish growth zoning, and diffusional relaxation of this zoning was calculated between each successive point. Along this *P–T* path, a rock encounters reactions in the same order regardless of heating rate, which only controls the time available for diffusion at each temperature step. This crystal growth and diffusion model was run for ~ 13,000 variations of heating rate, maximum temperature and crystal size at that T_{Max}. Panels B–D summarize results of these calculations, contoured to show the T_{Max}–heating-duration–crystal-diameter relations for which garnet crystals first 'open' sufficiently for diffusion to noticeably modify the crystal core composition. Note that results concern prograde metamorphism only, simply showing which combination of parameters will retain initial garnet crystal core growth compositions upon reaching T_{Max} and which will be modified by diffusion (solid curves, labeled for crystal diameter). Horizontal gray lines numbered 3-6 indicate the assemblage phase boundaries traversed in the model *P–T* path shown in A. Dashed curves show the conditions for which diffusion becomes sufficiently effective to completely eradicate zoning in garnet. See Caddick et al (2010) for additional details.

A consequence is that crystals may appear to preserve prograde zoning *patterns*, such as bell-shaped Mn profiles, up to granulite-facies conditions but the absolute composition at all interior points within the crystal may have been substantially modified from their initial growth composition, rendering thermobarometry misleading. These resetting estimates are somewhat faster than previous estimates (e.g., Florence and Spear 1991), partly because of the use of different diffusivity data. Discretization of garnet porphyroblasts into sub-grains that may not be obvious optically will further reduce timescales of diffusional homogenization, sometimes substantially so (e.g., Konrad-Schmolke et al. 2007). This loss of compositional information remains both a challenge and an opportunity for garnet petrochronology, reducing the sensitivity of P–T estimates but providing additional methods for inferring duration, as summarized in the geospeedometry section, below.

The strength of garnet: Geo-ba-Raman-try[2]. Previously discussed methods of thermobarometry rely on phase compositions that are generally assumed to have equilibrated near metamorphic peak temperature and/or pressure and to have experienced limited subsequent modification. These assumptions can be difficult to assess in natural samples, so alternative methods that make fewer (or at least different) assumptions are valuable. One of the most promising of these relies on a very different property of garnet: its resistance to elastic deformation.

Minerals trapped as inclusions during metamorphic growth of porphyroblasts generally experience approximately lithostatic pressure at the time of entrapment. Rocks expand (decompress) upon exhumation and undergo phase transformations accordingly, but mineral inclusions can retain a substantial proportion of the pressure at which they were trapped if the host phase can resist the deformation required to permit expansion of the inclusion, and if pressure is not 'lost' through development of cracks in the host. Identification of deformation around inclusions in high pressure diamond (Sorby and Butler 1869; Sutton 1921) first led to the concept of thermobarometry by study of stress in minerals hosting inclusions (Rosenfeld and Chase 1961). Focus turned to garnet porphyroblasts (Rosenfeld and Chase 1961; Rosenfeld 1969) because phases with low thermal expansivity and high bulk and shear moduli make good natural pressure vessels: their inclusions, which would naturally expand upon exhumation, are likely to retain a substantial fraction of their entrapment pressure (Zhang 1998; Izraeli et al. 1999; Guiraud and Powell 2006). A classic example comes from preservation of coesite and palisade quartz inclusions in garnet porphyroblasts, which led to the first identification of continental rocks that had been subducted to >90 km (Chopin 1984) based on the inference that coesite once existed as the stable SiO_2 polymorph but was pervasively inverted to quartz upon exhumation unless trapped as inclusions within garnet.

Substantial literature suggests that residual pressure should also be maintained in rocks lacking the obvious palisades inclusion textures, leading to studies that recovered ultra-high pressure conditions by measuring pressure-sensitive laser Raman peak positions of coesite and olivine inclusions in diamond and garnet (Izraeli et al. 1999; Parkinson and Katayama 1999; Sobolev et al. 2000). Correlations between pressure and the laser Raman spectra of several phases have been developed, due partly to the need for calibrants in high-pressure experiments (e.g., Hemley 1987; Schmidt and Ziemann 2000; Schmidt et al. 2013). Residual pressure (i.e., current pressure felt by a mineral inclusion, which may be substantially different to the pressure felt by crystals in other textural settings) can thus now be determined for these phases *in situ* with Raman techniques. However, a mineral inclusion with pressure sensitive Raman characteristics only makes an effective thermo-barometer in natural samples if its elastic properties contrast sufficiently with its hosting phase (otherwise the initial pressure felt by the inclusion may be lost as it deforms its host). A detailed survey of candidate mineral pairs (Kohn 2014) revealed several promising combinations, with quartz-in-garnet among the most sensitive barometers because of the relative compressibility of quartz and rigidity of garnet. Additionally, zircon inclusions in garnet might make a very sensitive thermometer (Kohn 2014), with uncertainties of quartz-in-garnet and zircon-in-garnet thermobarometry potentially as small as several hundreds of bars and tens of degrees, respectively.

The quartz-in-garnet laser Raman barometer was applied to quartz–eclogite, epidote–amphibolite and amphibolite facies metamorphic samples by Enami et al. (2007), who used a simple numerical model (Van der Molen 1981) to infer original metamorphic pressure from

---

[2] Kudos to editor Kohn for coining the term Thermoba-Raman-try (Kohn 2014).

measured residual pressure and reveal the high pressure history of samples that otherwise preserve little evidence of eclogite-facies equilibration. The need for more sophisticated modeling approaches has since been identified (Angel et al. 2014; Ashley et al. 2014a; Kohn 2014; Kouketsu et al. 2014) and available software now offers automated calculation with several choices of elastic model and simple correction of elastic properties for garnet composition (Ashley et al. 2014b). In an important validation of this methodology, application to experimentally derived quartz inclusions in garnet yielded "an entrapment pressure at 800 °C of 19.880 kbar—essentially identical to the experimental pressure of 20 kbar" (Spear et al. 2014).

Application of Raman barometry (or ba-Raman-try) to high pressure lithologies has retrieved pressures of quartz or coesite trapping (e.g., Korsakov et al. 2010; Zhukov and Korsakov 2015), analysis of more complex suites of inclusions has revealed *P–T* paths of subduction zone garnet growth (Ashley et al. 2014a), and comparison with thermodynamic modeling has indicated the likely extent of overstepping the garnet isograd in subduction zone and Barrovian-sequence samples (Spear et al. 2014; Castro and Spear 2016). However, application to high temperature metamorphic rocks remains challenging, with Sato et al. (2009) describing Raman shifts which imply that the interface between quartz inclusion and garnet host can be subjected to tensile stress upon cooling and that the quartz inclusions thus preserve negative residual pressure, as discussed further by Kouketsu et al. (2014). Ashley et al. (2015) detailed an extreme case in which this effective under-pressuring led to polymorphic transformation of quartz inclusions to cristobalite. A further challenge for high temperature metamorphic rocks is that modeling currently relies on an assumption that only elastic deformation acted on the host and inclusion, raising substantial problems for rocks that may have experienced significant plastic deformation. Indeed, given that chemical diffusion in garnet could either establish pressure gradients after crystal growth (Baumgartner et al. 2010) or could act as a mechanism to reduce pressure variations, garnet crystals experiencing substantial intra-crystalline diffusion after inclusion of quartz probably make unreliable pressure vessels without additional consideration and modeling. However, for the many cases in which diffusive re-equilibration is minimal, thermoba-Raman-try is an exciting, and complementary, method of inferring pressure–temperature evolution during garnet growth.

An exciting future development involves the combined application of multiple systems. In particular, given that quartz-in-garnet and zircon-in-garnet respectively act a barometer and a thermometer (Kohn 2014), we note that their combined use would be compelling. This would raise the possibility of extremely detailed petrochronology in samples containing garnet (whose *P–T* and duration of growth can be determined chemically and with zoned isotope geochronology) with inclusions of both zircon (providing additional *P–T* information with Raman and trace element thermometry and additional time constraints with U–Pb geochronology) and quartz (providing sensitive constraint on evolution of *P* during garnet porphyroblast growth). This represents a very powerful suite of techniques for deciphering rates of change of pressure and temperature in the mid crust, and is clearly an important potential avenue for further development.

Garnet as a tracer of reaction pathways and fluid–rock interaction

Thus far, we have focused on the use of garnet as a monitor of the *P–T* evolution of the systems wherein it grows. In the following section we move from the '*P–T*' to the '*X*'— or composition—of the system during garnet growth. As system chemistry changes, often reflecting the role of fluids or the local production or consumption of key phases, garnet crystals can record aspects of that change especially in their trace element and isotopic zonation.

Stable Isotopes in Garnet. Oxygen is the most abundant element in the Earth's crust and a main component of the fluids that transport heat and mass during crustal metamorphism. Oxygen isotopes in whole rocks and minerals have been used to characterize the composition, relative timing, and sources of these metamorphic fluids (e.g., Bickle and Baker 1990; Baumgartner and Valley 2001). In cases where the fluid history is complex, distinct garnet growth zones may be associated with evolving fluid compositions and sources. Additionally, the oxygen isotopic values of co-existing mineral pairs have been utilized as a thermometer due to the measurable and temperature-dependent fractionation between different minerals (e.g., Chacko et al. 2001; Valley 2001; Valley et al. 2003). In this section, we review the use of oxygen isotopes in garnet as a marker of open system fluid flow, local reaction history, and as a crustal thermometer.

Historically, a principal objective of stable oxygen isotope studies was to determine peak temperatures by utilizing the temperature-dependent fractionation of oxygen isotopes between coexisting mineral pairs, determined either experimentally or empirically (Taylor and Sheppard 1986; Richter et al. 1988; Clayton et al. 1989; Zheng 1991, 1993; Kohn and Valley 1998; Chacko et al. 2001; Valley et al. 2003). The principles of equilibrium fractionation factors (Chacko et al., 2001) and stable isotope thermometry (Valley 2001) have been reviewed in great detail in a previous volume of this series, so only more recent contributions, specifically in relation to its application with garnet ($\Delta^{18}O$ mineral–garnet thermometry), will be discussed briefly here. Quartz–garnet oxygen isotope pairs have been used most often to constrain metamorphic temperatures (Rumble and Yui 1998; Peck and Valley 2004; Moscati and Johnson 2014). Peck and Valley (2004) utilize garnet–quartz pairs for thermometry of quartzites from the southern Adirondacks, NY. Conceptually, if most of the oxygen in the quartzites is in the quartz, any effect of oxygen isotope exchange will be recorded in the $\delta^{18}O$ of the minor phase, in this case garnet. Refractory accessory minerals like garnet are closed to oxygen diffusion during growth (Valley 2001), and as quartz in a quartzite acts as an infinite reservoir for oxygen diffusion, this mineral pair can be retentive of the peak temperature isotope fractionation, even if the rock is slowly cooled (Kohn and Valley 1998). Metamorphic temperatures of ~700–800 °C were calculated from the quartzites from the southern Adirondacks, with no grain size dependence, suggestive of slow oxygen diffusion in garnet and closure temperatures of at least 730 °C (Peck and Valley 2004).

The analysis of oxygen isotopes in garnet continues to benefit from increased accuracy and precision related to the improvement of *in-situ* analytical techniques (Vielzeuf et al. 2005; Kita et al. 2009; Page et al. 2010), and also from texturally constrained samples (see Electronic Appendix for a discussion on the historical challenges related to analysis of $\delta^{18}O$ in garnet). When quartz is fully armored by a mineral with slow diffusion of oxygen, such as garnet, then any diffusive exchange of oxygen between quartz and host during cooling is minimal. Hence, garnet–quartz mineral pairs as determined by *in-situ* techniques can offer a detailed temperature and fluid ($\delta^{18}O$) history (Strickland et al. 2011). As Russell et al. (2013) note, if combined with constraints on the timing (via garnet geochronology) and *P–T* conditions (via phase equilibria modeling, or even Raman barometry of quartz in garnet) of garnet growth, then a robust *P–T–t–f* metamorphic history is attainable.

Without metasomatism at elevated pressures and temperatures, metamorphic rocks will tend to inherit the bulk rock oxygen isotopic composition of the protolith (Fig. 6). This can be recorded in the $\delta^{18}O$ of the individual minerals, though an accurate knowledge of the equilibrium isotope fractionation factors among coexisting phases is required in order to properly interpret these measured isotope compositions. Garnet has one of the lowest $\delta^{18}O$ of all minerals, lower than that of bulk rock $\delta^{18}O$, and thus increases with increasing temperature (Kohn 1993) . In some lithologies (e.g., metabasites), garnet $\delta^{18}O$ is close to the whole rock and can encode the $\delta^{18}O$ of the whole rock (Putlitz et al. 2000). Any deviation in garnet $\delta^{18}O$ from initial values can be produced by processes including a) change in temperature inducing changing equilibration fractionation factors (see above), b) changing mineral assemblages and modal proportions during progressive metamorphism (Kohn 1993; Young and Rumble 1993), and c) open system change in the reactive $\delta^{18}O$ composition of the bulk rock from the

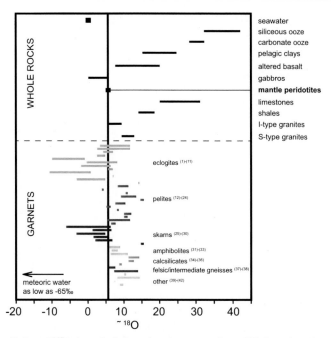

Figure 6. Compilation of δ^{18}O data of whole rock and garnet analyses. Whole rock analyses and average mantle value (δ^{18}O = 5.80‰) from Eiler (2001). References to garnet data include: [(1)(31)]Errico et al (2013), [(2)]Martin et al (2014), [(3)(33)]Page et al (2014), [(4)]Rubatto and Angiboust (2015), [(5)]Rumble and Yui (1998), [(6)]Russell et al (2013), [(7)]Zheng et al (1998), [(8)]Burton et al (1995), [(9)]Chen et al (2014), [(10)(12)]Gordon et al (2012), [(11)(24)]Masago et al (2003), [(13)]Kohn and Valley (1994), [(14)]Kohn et al (1993), [(15)]Kohn et al (1997), [(16)]Lancaster et al (2009), [(17)(32)(34)]Martin et al (2006), [(18)]Martin et al (2011), [(19)]Raimondo et al (2012), [(20)] Skelton et al (2002), [(21)]van Haren et al (1996), [(22)]Young and Rumble (1993), [(23)(30)(36)(39)]Ferry et al (2014), [(25)]Crowe et al (2001), [(26)]Clechenko and Valley (2003), [(27)]D'Errico et al (2012), [(28)]Jamtveit and Hervig (1994), [(29)]Page et al (2010), [(35)]Sobolev et al (2011), [(37)]Gauthiez-Putallaz et al (2016), [(38)]Vielzeuf et al (2005), [(39)]Peck and Valley (2004), [(41)]Moscati and Johnson (2014), [(42)]Abart (1995).

interaction of externally-derived fluids (Kohn and Valley 1994). Because oxygen diffusion in garnet at crustal conditions is sluggish (Coghlan 1990; Burton et al. 1995; Vielzeuf et al. 2005), any change in garnet δ^{18}O during its growth can be preserved in its concentric zonation (Kohn et al. 1993; Skelton et al. 2002), in one case even after a regional granulite facies overprint (Clechenko and Valley 2003). These effects on garnet δ^{18}O are further discussed below.

Kohn (1993) developed a model showing the effect of net transfer reactions during prograde heating for systems that are closed to externally-derived fluids. It was predicted that the fractional crystallization of (low δ^{18}O) garnet during metamorphism would result in an increase in the δ^{18}O of garnet from core to rim. However, this effect was calculated to be minor (~1‰ or less over 150 °C of heating) and independent of bulk composition (Kohn 1993). This was further tested in natural samples from regionally metamorphosed rocks from Chile (Kohn et al. 1993), where the difference between garnet cores and rims from metapelites and amphibolites was 0.5‰ and < 0.1‰, respectively. These observations suggest that garnet growth occurred in a system closed to infiltration of fluids out of δ^{18}O equilibrium with the bulk rock, and are similar to observations from regionally metamorphosed amphibolites and schists from Vermont (Kohn and Valley 1994) and New Hampshire, U.S.A (Kohn et al. 1997). Comparably, the fractionation of δ^{18}O due to (internally-derived) devolatilization of a meta-basalt was shown to be small, with garnet rims having a δ^{18}O less than 1‰ compared to garnet cores (Valley 1986; Kohn et al. 1993).

The mechanism most likely to result in large changes in garnet $\delta^{18}O$ from core to rim is open system fluid infiltration. A large amount of work on $\delta^{18}O$ in garnet has focused on its ability to elucidate the source of externally derived fluids in various metamorphic settings. The infiltration of these fluids, often over substantial thermal and chemical gradients, can cause significant zoning of $\delta^{18}O$ in garnet, offering a relative time marker for fluid flow (Skelton et al. 2002), especially when the protolith composition differs significantly from mantle-like compositions (~5–6‰; Fig. 6), or when the fluids are derived from devolatilization of high $\delta^{18}O$ sediments (Errico et al. 2013), interaction with low $\delta^{18}O$ meteoric water (Russell et al. 2013; Fig. 6), or rehydration in a mid-crustal shear zone (Raimondo et al. 2012).

Russell et al. (2013) examined the $\delta^{18}O$ record of garnets from a series of orogenic eclogites and found that these crystals can display a range of $\delta^{18}O$ that falls well outside that expected for mantle-derived protoliths. While the intracrystalline variation of $\delta^{18}O$ in these garnets is small (increase of ~1–2‰ between core and rim), the absolute $\delta^{18}O$ values record the origin of these eclogites, distinguishing between those derived from altered, oceanic upper crust (Trescolmen, Alps; 7.7 to 8.3‰) and high-temperature infiltration of meteoric water into mafic intrusions buried *in-situ* with subaerial, continental crust (Western Gneiss Region, Norway; −1.2 to −0.2‰). Indeed, extremely negative $\delta^{18}O$ values in eclogitic garnet from the Dabie (−10‰; Zheng et al. 2006) and Kokchetav (−3.9‰; Masago et al. 2003) terranes have been interpreted as evidence of extreme infiltration of meteoric water, possibly derived from a rift environment involving low $\delta^{18}O$ glacial melt water. Rumble and Yui (1998) analyzed eclogitic garnets from the Qinglongshan ridge in the Sulu terrane, and found bulk garnet $\delta^{18}O$ values as low as −11.1‰ (Fig. 6), also attributing such negative values as reflecting alteration of the protolith by meteoric waters at high paleo-latitudes and/or –altitudes prior to subduction. Other low garnet $\delta^{18}O$ values are observed in several studies from hydrothermal skarns where variable fluid sources in open systems are recorded in zoned garnet (Jamtveit and Hervig 1994; Crowe et al. 2001; Clechenko and Valley 2003; D'Errico et al. 2012). D'Errico et al. (2012) observed a large range in intragrain $\delta^{18}O$ values (−4 to 4‰) from a hydrothermal system in the Sierra Nevada batholith, attributing early, low $\delta^{18}O$ values to meteoric water input, with higher values resulting from variable mixing of magmatic and metamorphic fluids.

Some examples offer instances where garnet $\delta^{18}O$ records changing fluid compositions (e.g., Kohn and Valley 1994; Martin et al. 2011; Errico et al. 2013; Page et al. 2014; Rubatto and Angiboust 2015). Many of these cases of strongly zoned garnet $\delta^{18}O$ are recorded in eclogites. Rubatto and Angiboust (2015) observed eclogitic garnet cores from the Monviso ophiolite in the Western Alps with $\delta^{18}O$ values of 0.2–2.0‰, recording ocean floor metasomatism of the basaltic protolith. High-pressure metasomatism is recorded in the garnet rims, with $\delta^{18}O$ values of 3.5–6.0‰. While these rocks are found in a shear zone adjacent to down-going lithospheric mantle, Monviso serpentinites ($\delta^{18}O$ of 3.0–3.6‰) were not found to be the source of the metasomatizing fluid; high $\delta^{18}O$ fluids from dehydrating metasediments were instead suggested to be the primary source for fluid influx at depth.

A final example of garnet $\delta^{18}O$ recording multiple metasomatic events comes from two separate findings from the well-studied Franciscan Complex (Errico et al. 2013; Page et al. 2014). Garnet crystals from eclogite and amphibolite blocks record contrasting fluid flow histories (Errico et al. 2013). Amphibolites, which are inferred to record an early subduction initiation history (Anczkiewicz et al. 2004), are metasomatized by an early influx of sediment-derived fluids, recorded in their garnet rims (~8‰; Errico et al. 2013; see Fig. 6). Eclogitic blocks have high $\delta^{18}O$ garnet cores (~6–11‰), interpreted to record extreme ocean floor alteration of the protolith, with fluid-mediated exchange with the overlying mantle wedge at depth recorded in low $\delta^{18}O$ garnet rims (3–5‰; Errico et al. 2013). Page et al. (2014) suggest that their observed shifts in garnet $\delta^{18}O$ (2–3‰) are evidence of limited fluid flow during burial where garnet rims are formed during metasomatism at high fluid/rock ratios during exhumation/re-burial, with possible interaction with highly oxidized fluids.

In all of the cases shown above, analysis of garnet $\delta^{18}O$ has elucidated the source and relative timing of fluid–rock interaction in a variety of tectonic settings. Garnet $\delta^{18}O$, combined with a detailed *P–T* and temporal record, has great potential to provide an important marker for metamorphic fluid flow, especially in hydrothermal and subduction zone systems where the role of exotic external fluids is most observable. Combining $\delta^{18}O$ with studies of fluid inclusions in garnet (where preserved) can provide an even more detailed fluid history of burial/prograde (primary fluid inclusions) and exhumation/retrogression (secondary fluid inclusions), coupling these techniques to elucidate the source and composition of fluids.

Increasingly, stable isotopes other than oxygen are utilized in metamorphic systems to study the nature and scale of fluid flow, particularly in subduction zone systems, and relying (mainly) on the interpretation of whole rock isotopic analyses. Stable isotopes of lithium, calcium, and magnesium are ideal for studying the effects of diffusion due to their large relative mass differences, and are likely to be particularly useful in natural systems where strong chemical potential gradients exist. If garnet growth occurs in an environment with high fluid flux or large chemical gradients, then large diffusion-driven mass fractionation may be recorded in growing porphyroblasts, and *in-situ* or microsampled stable isotope analyses of garnet can produce a detailed record of such fluid and mass transfer. The subduction interface, where the overlying mantle wedge comes into contact with subducting metasediments and metabasaltic rocks, is an ideal locale for such studies, given the strong chemical contrasts in several chemical components between these lithologies. Studies that have focused on the length-scales and timescales of subduction zone fluid flow have utilized isotope systems including calcium (John et al. 2012), lithium (Zack et al. 2003; Penniston-Dorland et al. 2010) and magnesium (Pogge von Strandmann et al. 2015). Bulk isotopic analyses on garnet separates have been limited to just a few studies (e.g., on magnesium; Pogge von Strandmann et al. 2015). Notably also, Bebout et al. (2015) developed a technique to measure the variation in $\delta^{7}Li$ in garnet using SIMS, in order to determine whether the lithium isotopic record could help elucidate the nature of fluid flow during subduction zone metamorphism of metasediments from Lago Di Cignana, Western Alps. Finally, Cr (Wang et al. 2016) and Fe (Williams et al. 2009) isotopes have been utilized for studying evolving redox conditions (using Cr and Fe) and the effects of partial melting processes (using Fe).

As these isotopes are primarily limited to measurement by solution ICP-MS or TIMS, bulk analyses of garnet have been the primary focus of earlier studies. A promising new direction may involve the use of SIMS for *in-situ* analysis, where applicable. However, detailed microdrilling from texturally or compositionally constrained garnet growth zones, in tandem with various stable isotope analyses (oxygen and otherwise), has the potential to link these geochemical tracers to *P–T-t* constraints for various metamorphic (and igneous) processes, thus expanding the incorporation of various stable isotope systems to the garnet petrochronologist's toolbox.

Trace elements in garnet. Garnet is often zoned in trace elements, with cores that are enriched in Y and HREE (Hickmott et al. 1987; Chernoff and Carlson 1999; Pyle and Spear 1999; Otamendi et al. 2002). These highly compatible elements tend to resist re-equilibration at elevated temperatures owing to their low inter- and intragranular diffusivities (Lanzirotti 1995; Spear and Kohn 1996; Otamendi et al. 2002; Carlson 2012; Bloch et al. 2015). Thus, trace element zoning can be preserved at upper amphibolite and even granulite facies, in which major element zoning is often erased (Fig. 5), providing a useful tool for deciphering high-temperature metamorphic processes (Spear and Kohn 1996; Hermann and Rubatto 2003). Early garnet trace element studies required secondary ion mass spectrometry (Hickmott et al. 1987) or synchrotron (Lanzirotti 1995) techniques. However, LA-ICPMS is increasingly now used for trace element analyses at adequate spatial resolution and sensitivity for many applications.

The analyzed distribution of trace elements, specifically REEs, can be utilized for interpretation of garnet geochronology (e.g., zonation of radioactive parent elements; see later section), changes in garnet growth rate, and changes in reaction history, specifically with relation to accessory phases. Often, trace element zonation is sensitive to different specific mineral reactions or matrix processes than the major elements, thus offering complementary information. Insofar as trace element zonation relates to dateable accessory phases, trace elements in garnet represent an often crucial correlative part of accessory phase petrochronology. Garnet can be vividly zoned in REE's (see expanded discussion below), as well as elements such as Zr (Anczkiewicz et al. 2007), Cr (Spear and Kohn 1996); Yang and Rivers (2001); (Martin 2009; Angiboust et al. 2014), P (Spear and Kohn 1996; Chernoff and Carlson 1999; Kawakami and Hokada 2010; Hallett and Spear 2015; Jedlicka et al. 2015; Ague and Axler 2016), and As (Jamtveit et al. 1993). Here, we present a summary of the controlling mechanisms for partitioning of trace elements in garnet.

Hickmott et al. (1987), Hickmott and Spear (1992), and Hickmott and Shimizu (1990) were among the earliest contributions to interpret remnant trace element zoning in garnet (in metapelites from the Tauern Window and Massachusetts), finding controls on trace element concentrations related to elemental fractionation during progressive metamorphism, fluid–rock interaction, and the breakdown of trace element-rich minerals. Numerous contributions have built upon these earlier studies in attempting to decipher metamorphic processes and rock reaction histories from garnet trace element zoning, with additional examples cited below.

Carlson (2012), Moore et al. (2013), Cahalan et al. (2014) and Carlson et al. (2014) have sought to determine how various trace elements are structurally incorporated into natural garnet, and to calibrate the rates and mechanism of diffusion of these elements in garnet, utilizing samples collected from the aureole of the Makhavinekh Lake Pluton, Labrador. Using lattice dynamic calculations, Carlson et al. (2014) found that the incorporation of trace elements into garnet is likely dominated by menzerite- (for REE) and alkali-type (for Li and Na) coupled substitutions at a low energetic cost. This energetic cost will (i) decrease as the host garnet unit-cell increases, (ii) decrease as temperature increases or pressure decreases, and (iii) decrease substantially with contraction of the ionic radius across the lanthanide series. These observations, as well as diffusivities calculated through the use of stranded diffusion profiles in the Labrador garnets (Carlson 2012) have been shown to have profound implications for the interpretation of Lu–Hf garnet ages (Kelly et al. 2011; see also later sections), the preservation of matrix trace element distributions during porphyroblast crystallization (or overprint zoning; see examples below), element mobility, and intergranular solubility (Carlson et al. 2015a). Finally, Cahalan et al. (2014) found that the diffusion of Li in garnet is strongly governed by coupled substitution with slowly diffusing REE, and thus Li zoning may be retained during metamorphism (even at elevated temperatures) and utilized as a monitor of fluid– (and/or melt–) rock interaction.

Simple Rayleigh fractionation between a growing garnet crystal and the rock matrix would result in the incorporation of Y + HREE into garnet cores, with smoothly decreasing 'bell-shaped' profiles of these elements towards garnet crystal rims due to their relatively high garnet/matrix partition coefficients (e.g., Hollister 1966; Otamendi et al. 2002; Lapen et al. 2003; Anczkiewicz et al. 2007; Kohn 2009; Fig. 7). In contrast, MREE and LREE generally display a slight increase in abundance towards garnet crystal rims, or show no zonation at all. Garnet REE profiles can, however, deviate from these simple profiles, as observed in a number of settings. This may be due to a number of factors including 1) diffusion-limited REE uptake during prograde metamorphism, 2) resorption (or recrystallization), 3) syn-growth breakdown of a REE-bearing accessory phase, 4) breakdown of REE-bearing major phases, 5) change in the kinetics of garnet growth, 6) overprint zoning, and 7) infiltration of trace element-rich fluid. Let us treat these each in turn.

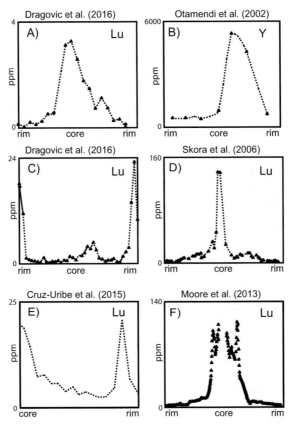

Figure 7. Garnet trace element zonation from various studies. A. Example of Rayleigh fractionation; modified from Dragovic et al (2016). B. Rayleigh fractionation example; modified from Otamendi et al (2002). C. Lu enriched in garnet rim, possibly from breakdown of accessory phases; modified from Dragovic et al (2016). D. Diffusion-limited REE uptake example; modified from Skora et al (2006). E. Core to rim profile showing Lu annulus from breakdown of accessory phase; modified from Cruz-Uribe et al (2015). F. Example showing intermediate peaks in Lu, possibly from change in local garnet-forming reaction; modified from Moore et al (2013).

Diffusion-limited REE uptake. Diffusion of trace elements to a growing garnet crystal can be relatively slow compared to the crystal's growth rate (Skora et al. 2006; Moore et al. 2013). An early nucleating garnet crystal will sequester highly compatible elements ($K_D > 1$), forming a high concentration central peak in zoning profiles (Fig. 7). As a result of the slow intergranular diffusion, a steep chemical potential 'depletion halo' will form around the early-growing crystal (supply is slower than uptake). Increased temperature during progressive metamorphism will increase intergranular mobility, relaxing the chemical potential gradient of REE around the growing garnet crystal and possibly permitting a secondary REE peak in the growing garnet. REE abundance in the crystal core is dependent on the K_D (relative to garnet precursor phases) and matrix concentration of each element, with HREE exhibiting the strongest fractionation into garnet. The location and magnitude of the secondary peak will depend on the intergranular diffusivity of the element and its dependence on temperature, with MREE exhibiting a rimward secondary peak compared to the HREE, owing to their lower intergranular diffusion rate (and higher ionic radii). For LREE, this will manifest itself as a low abundance in crystal cores, potentially increasing at the crystal rim.

Resorption (/recrystallization). Annular maxima in trace element abundances may result from the partial consumption of garnet during an orogenic cycle (i.e., during prograde heating/ burial or retrograde dissolution). At elevated temperatures, Y + HREE would be liberated from the resorbed garnet rim. Elements that have a strong affinity for garnet will preferentially re-partition into the remnant crystal rim, and subsequently diffuse inward at a rate controlled partly by temperature (Lanzirotti 1995; Pyle and Spear 1999). Kelly et al. (2011), utilized Lu–Hf garnet data from Makhavinekh Lake Pluton, Labrador, to suggest that the strong preferential reincorporation of Lu relative to Hf results in false apparent Lu–Hf garnet ages. Using their model, higher degrees of garnet resorption and Lu retention would result in younger apparent ages and stranded diffusion profiles of Lu at remnant crystal rims. Gatewood et al. (2015) make a similar observation based on Sm-enriched rims likely formed during resorption and leading to young apparent Sm–Nd rim ages.

Interface-coupled dissolution–reprecipitation (ICDR) can also result in unique trace element zoning. Ague and Axler (2016) present garnet Na, P, and Ti distribution maps from a high-pressure felsic granulite from the Saxon Granulite Massif, showing sharply defined cross-cutting chemical domains that likely record ICDR during retrograde fluid–rock interaction. While major element zonation is absent at such temperatures (>900 °C) due to intracrystalline diffusion, retention of zonation of these trace elements highlights their far lower diffusivities at these conditions (Ague and Axler 2016).

Breakdown of an accessory phase. The presence of accessory phases can buffer trace element activities, thereby exerting strong controls on their partitioning into garnet. Several studies have linked REE zoning in garnet with the growth and breakdown of phases such as monazite, allanite, epidote, apatite, or xenotime (Hickmott and Shimizu 1990; Hickmott and Spear 1992; Chernoff and Carlson 1999; Pyle and Spear 1999; Yang and Rivers 2002; Corrie and Kohn 2008; Stowell et al. 2010; Gieré et al. 2011; Cruz-Uribe et al. 2015; Dragovic et al. 2016; Engi 2017, this volume; Fig. 7). Pyle and Spear (1999) showed that in xenotime-bearing, garnet-zone assemblages, the early presence of xenotime would buffer Y and result in high-Y garnet cores. Continued growth of garnet would deplete the matrix in Y, eventually resulting in loss of xenotime and a subsequent decrease in Y towards garnet rims. Yang and Rivers (2002) also suggested that enrichment of certain REE in garnet can be associated with the breakdown of particular minerals: enrichment in LREE associated with the breakdown of allanite, in MREE with epidote, and in HREE with xenotime or zircon. Using trace element zoning patterns in both garnet and tourmaline, Gieré et al. (2011) modeled the metamorphic evolution of rocks from the Central Alps, relating this evolution to the growth and breakdown of accessory and major phases. The presence of subhedral annuli enriched in Ca, Sr, Y, and HREE was associated with the breakdown of allanite, while internal zones of these annuli were associated with garnet growth coincident with the breakdown of detrital monazite. Hallett and Spear (2014a) suggest that P zoning in garnets from the Humboldt Range, NV, results from breakdown of a phosphate phase such as apatite, or melting reactions involving the breakdown of plagioclase. Finally, the discontinuous breakdown of Y-rich mineral phases such as epidote and allanite was invoked to explain Y annuli in garnets from metapelitic rocks of the Black Hills, South Dakota (Yang and Pattison 2006).

Breakdown of major rock-forming phases. Konrad-Schmolke et al. (2008) used a path dependent thermodynamic forward model and published trace element partition coefficients to model major and trace element distribution in garnet, comparing modeled elemental distributions to those observed in UHP garnets from the Western Gneiss Region, Norway. The modeled trace element distributions result from a sequence of garnet-producing mineral breakdown reactions (specifically the breakdown of chlorite, epidote, and amphibole) during subduction, combined with progressive fractional modification of the effective bulk composition upon garnet growth. Along an inferred subduction *P–T* path, changes in the calculated abundance of major phases and in trace element partitioning between those phases and garnet generated synthetic distributions of major and trace elements in garnet. HREE zoning patterns are largely predicted by fractionation into the garnet core by a Rayleigh-type process, with possible perturbations towards the garnet rim for some effective bulk compositions resulting from breakdown of epidote and amphibole. The liberation of MREE and LREE into garnet results from epidote- and amphibole-consuming reactions, and is expressed as intermediate peaks in a growing garnet crystal. These peaks are shown to form progressively further from the crystal core for decreasing atomic numbers (or garnet/matrix partition coefficient), as is also described in Zermatt-Saas Fee garnets (Skora et al. 2006).

Konrad-Schmolke et al. (2008) also relate trace element distributions in garnet to possible interpretations of garnet geochronological methods. Strong partitioning of radioactive parent elements such as Lu and Sm into a growing garnet crystal will have significant effects on the interpretation of a garnet age (see below). The models from Konrad-Schmolke et al. (2008) show that most of the Lu partitioned into garnet during early growth comes from the breakdown of chlorite. Therefore, Lu–Hf in garnet from a subducting lithology could date chlorite breakdown reactions, with the Sm–Nd system possibly tracking the growth of garnet via the breakdown of epidote or amphibole and thus the amphibolite/blueschist to eclogite facies transition.

Changes in the kinetics of garnet growth. For elements that strongly partition into a growing garnet crystal (i.e., Y + HREE), a decrease in the garnet growth rate can lead to an increase in the concentration of that particular element in the garnet's outermost surface (Hickmott and Spear 1992). The combined presence of these elemental annuli in a zone of relatively inclusion-free garnet has been partly attributed to such decreases in garnet growth rate (Lanzirotti 1995; Yang and Rivers 2001; Yang and Pattison 2006).

Overprint zoning. Pre-existing matrix heterogeneities or former accessory phases can be retained as overprint zoning in the chemistry of garnet that grows in these locations (Menard and Spear 1996; Spear and Kohn 1996; Yang and Rivers 2001, 2002; Kohn 2004; Vielzeuf et al. 2005; Martin 2009; Carlson et al. 2015a). Martin (2009) shows sub-millimeter scale Y and Cr zoning in garnet from central Nepal that defines an internal foliation, spiraling from the rim to the core of crystals, similar to patterns often portrayed by inclusion trails. These garnets lack well-defined inclusion patterns, so only the trace element zoning in garnet offers a diagnostic tool for determining the relative timing of deformation, possibly even helping to determine the sense of rotation of a porphyroblast relative to the matrix foliation.

Infiltration of trace element-rich fluid. Infiltration of an externally-derived fluid whose trace element concentrations are out of equilibrium with the mineral assemblage may result in distinct trace element zoning patterns (Jamtveit and Hervig 1994; Stowell et al. 1996; Moore et al. 2013; Angiboust et al. 2014; Hallett and Spear 2014b). Invoking the 'exotic fluid' interpretation has often proved challenging because examples of patchy or oscillatory zoning can often be equally well explained by pre-existing heterogeneities (see above). This is especially true with respect to REE zoning in garnet, as corroborating evidence from fluid inclusions or coincident zoning in major elements is usually missing. Moore et al. (2013) explain REE zoning in rims of Franciscan blueschist garnets as related to fluid infiltration, because zoning in these elements is also coincident with annuli in Ca and Mn, presumably also enriched in the metasomatizing fluid. Moore et al. (2013) note that such zonation can act as a time marker because rock-wide infiltration would result in similar trace element zoning for all garnet crystals growing during the interval influenced by the fluid flow event.

A vivid example of garnet trace element zonation involving the infiltration of exotic fluids comes from the fossil subduction zone of the Western Alps, where Angiboust et al. (2014) observe patchy and oscillatory Cr zoning in garnets from shear zone eclogites (Fig. 8). The complex Cr zoning, coupled with both enrichments of Mg, Ni, Cr, and LILEs and with boron isotopic signatures in metasomatic minerals, was interpreted to reflect the large-scale episodic infiltration of serpentinite-derived fluids (via antigorite breakdown) at ca. 80 km depth at the slab–mantle interface (Angiboust et al. 2014). While this appears to contrast with garnet oxygen isotope interpretations presented earlier (sediment-derived fluid source), Monviso eclogites experienced multiple phases of metasomatism, and as the Tethyan seafloor (slow-spreading ridge) dominantly comprised serpentinized mantle and pelites, mafic blocks were likely to have interacted with multiple sources of fluid (Angiboust, pers. comm.).

Ultimately, linking the concentrations and zoning of trace elements in garnet to the evolution of the mineral assemblage and the timing of garnet growth can be a powerful petrochronologic tool, especially when comparing the competing roles of fluid infiltration, intergranular element mobility, and crystal growth rate.

Concluding this review of the ways in which garnet can record petrologic or tectonic processes and conditions, we turn now to the topic of direct garnet chronology. As we shift gears, we encourage the reader to keep in mind the "petro-" of garnet, and realize that direct garnet chronology allows the petrochronologist to date any of these conditions and processes, and in the best case, constrain their rate or duration (e.g., Fig. 2). The "petro-" of garnet provides the motivation for garnet petrochronology; otherwise we would just be dating a mineral.

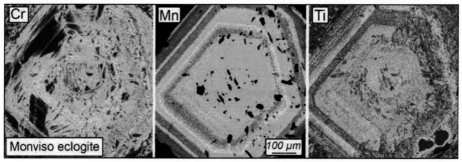

Figure 8. Cr, Mn, and Ti zoning from a garnet from the Monviso ophiolite of the Western Alps, modified from Angiboust et al (2014). Note patchy zoning in the core and oscillatory-zoned rims. Trace element annuli attributed to influx of externally derived fluid from antigorite breakdown of associated mantle wedge serpentinites.

CHRONO- OF GARNET

Even today, many in the geoscience community are unaware, or unconvinced, that the growth of garnet can be dated directly, precisely, and accurately with the Sm–Nd or Lu–Hf isotope systems. Much of that skepticism stems from the early history of garnet geochronology in the 1980's and 1990's before certain modern methods were developed. Since then, direct garnet geochronology has evolved to approach the hopes of would-be petrochronologists of decades past. At the same time, garnets have also become one of most often utilized minerals for 'geospeedometry' (Lasaga 1983; e.g., Chakraborty and Ganguly 1991) whereby the durations of tectonic and metamorphic events (generally events related to heating and cooling) can be accurately reconstructed by modeling stranded diffusion profiles within garnet. Valuable advances have also been made in linking the growth (and chronology) of accessory phases such as monazite and zircon to the growth or breakdown of garnet via textural and chemical means (e.g., Engi 2017, this volume; Lanari and Engi 2017, this volume; Rubatto 2017, this volume; Williams et al. 2017, this volume); we view the integration of accessory phase petrochronology with true garnet petrochronology as one of the more exciting future applications which may be inspired by this full RiMG volume. Indeed, the decision to approach time-consuming garnet geochronology might be established and motivated by preliminary accessory phase petrochronology, thermodynamic modeling, and thin section petrography. While we leave it to other authors to describe advances in linking accessory phase dating to garnet, in this section, we will review the past, present, and future of direct garnet geochronology and garnet geospeedometry. Together, these two methods provide the 'chrono' in garnet petrochronology including absolute ages, absolute durations, and relative rates of thermal processes at conditions wherein garnets exist.

Garnet geochronology

van Breemen and Hawkesworth (1980) and Griffin and Brueckner (1980) first recognized that garnet's relatively high Sm/Nd ratio (as compared to most other common minerals) could be exploited for geochronology. Garnet growth is dated with the isochron method by pairing it with the hosting 'whole rock' and/or other mineral separates with which it grew (and subsequently evolved isotopically) in an isochron diagram (for a review of garnet isochron basics, see Baxter and Scherer 2013). Further exploration of Sm–Nd garnet geochronology continued in a few labs through the 1980's until 1989 when the first attempts to date garnet via U–Pb (Mezger et al. 1989) and Rb–Sr (Christensen et al. 1989) were published. Despite the novelty of these studies, concerns surrounding the problem of contaminating mineral inclusions in garnet, matrix heterogeneity, and open system mobility of daughter isotopes in the rock system pushed garnet geochronology via U–Pb and Rb–Sr out of favor (for example, see Zhou and Hensen 1995; DeWolf et al. 1996; Romer and Xiao 2005; Sousa et al. 2013). In a positive twist, it was in part the recognition that such U–Pb analyses of 'garnet' were dominated by accessory mineral inclusions that led to the advancement of accessory phase petrochronology with textural and chemical links to garnet growth (e.g., Hermann and Rubatto 2003; Pyle and Spear 2003; Wing et al. 2003; Foster et al. 2004). Sm–Nd garnet geochronology survived this period, but not without lingering skepticism stunting its further development, especially given the time and analytical effort required to overcome the challenges of contaminating inclusions and low Nd concentrations. In 1997, the first Lu–Hf garnet geochronology was published by Duchene et al. (1997). Since that time, significant advances have been made for both Lu–Hf and Sm–Nd garnet geochronology, both in terms of sample preparation, isotopic analysis, and data interpretation. With modern methods, garnet geochronology via Lu–Hf and Sm–Nd can yield accurate and precise ages of garnet directly from the garnet itself. In this section of the chapter, we will review some of the key advances, and remaining caveats, of which any user or interpreter of garnet geochronology should be aware.

Garnet isochron age precision. The precision of an isochron age (Fig. 9) depends on how well the slope of the isochron is constrained. Age precision (i.e., slope precision) depends on three main factors: 1) the analytical precision of the isotopic datapoints, 2) the spread in parent/daughter between lowest and highest points in the isochron, and 3) the scatter in the isochron. In general, the slope of the isochron is constrained most precisely when the analytical precision of each point is best, the spread in parent/daughter ratio is largest, and the scatter is smallest. Let us first address the importance of analytical precision and isochron spread in theoretical two-point isochrons. Then in the next section, we will explore the effect of adding additional points to an isochron and their scatter therefrom.

Modern mass spectrometry (TIMS or MC-ICP-MS for Sm–Nd, MC-ICP-MS for Lu–Hf) permits <10 ppm 2RSD analytical precision for $^{143}Nd/^{144}Nd$ and $^{176}Hf/^{177}Hf$ when sample size is unlimited (~100 ng of Hf or Nd). TIMS is the optimal tool for Nd analysis given higher net ion efficiency (especially as sample size decreases), whereas MC-ICP-MS is the optimal tool for Hf for the converse reason. Oftentimes, especially in applications requiring smaller amounts of garnet to be extracted and analyzed (such as zoned garnet chronology,

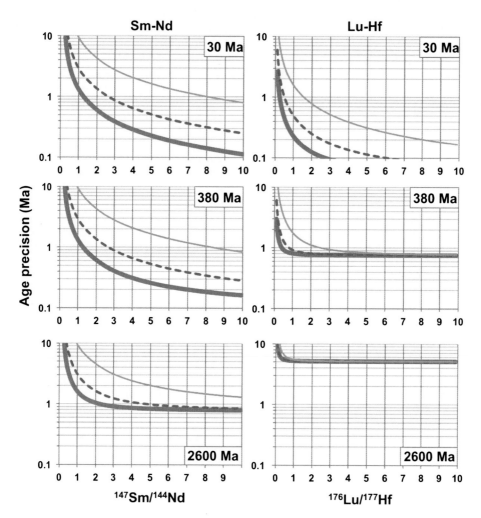

Figure 9. Garnet–matrix two-point isochron age precision as a function of parent/daughter isotope ratio for garnets of three different ages: Alpine (30 Ma), Acadian (380 Ma), Archean (2.6 Ga). In all plots, the bold lower line is for 10ppm (2σ) analytical precision on the garnet daughter isotope analysis, the dashed middle line is for 30ppm, and the thin upper line is for 100ppm. 10ppm analytical precision is achievable for $^{176}Hf/^{177}Hf$ when load size is unlimited (10s to 100s of ng Hf run on MC-ICPMS) and for $^{143}Nd/^{144}Nd$ (run as NdO^+ on TIMS) when load size is >4ng. As sample size decreases, as is usually the case for small amounts of clean garnet with low Hf and Nd concentration, analytical precision worsens (see text for discussion). $^{147}Sm/^{144}Nd$ analytical precision is 0.03% and $^{176}Lu/^{177}Hf$ is 0.2% indicative of optimal performance on TIMS and MC-ICPMS, respectively. Matrix $^{147}Sm/^{144}Nd$ is 0.15 whereas matrix $^{176}Lu/^{177}Hf$ is 0.02, each typical values for crustal rocks. If additional points are included in the isochron, the age precision can change depending on the MSWD (see text for discussion).

see below), the amount of Nd and Hf is much less than desired for optimal analysis. This sample size limitation has been one of the major factors limiting garnet geochronology. Figure 10 shows a compilation of garnet data, cleaned as well as possible with modern methods (see discussion below). Daughter element concentrations in clean garnet are generally less than 0.5 ppm and often below 0.1 ppm for Nd and Hf, meaning that tens of

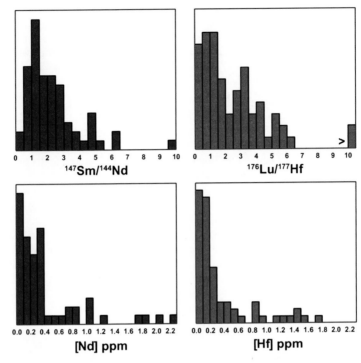

Figure 10. Histograms of reported garnet Sm–Nd and Lu–Hf data after cleansing efforts to remove inclusions using the methods of Pollington and Baxter (2011) for Sm–Nd and Lagos et al (2007) for Lu–Hf. $^{147}Sm/^{144}Nd$ and $^{176}Lu/^{177}Hf$ data are taken from Baxter and Scherer (2013), [Nd ppm] and [Lu ppm] concentration data correspond to the parent/daughter data and are compiled from the Baxter Lab and Scherer Lab (Erik Scherer, pers. comm.). Some $^{176}Lu/^{177}Hf$ data are off scale reaching as high as 50. Most well cleansed garnet has $^{147}Sm/^{144}Nd$ and $^{176}Lu/^{177}Hf$ > 1.0, and [Nd ppm] and [Hf ppm] < 0.4.

milligrams of clean garnet may still only provide a few nanograms of Nd or Hf for analysis. Note that in estimating required sample size, one must consider the sample loss that occurs during mineral separation and garnet cleansing (see below) which can often exceed 80% of the raw garnet mass (e.g., Pollington and Baxter 2011). As analytical methods continue to improve, we can push to smaller and smaller garnet sample sizes and higher resolution petrochronology. Fortunately, recent and continuing advances in mass spectrometry permit improved analytical precision even as sample size decreases (Harvey and Baxter 2009; Schoene and Baxter 2017, this volume; Baxter and Scherer 2013; see Bast et al. 2015). For example, sub-nanogram loads of pure Nd standard solution analyzed as NdO^+ via TIMS can still yield external precision < 30 ppm 2RSD, though experience shows that sub-nanogram samples run through column chemistry typically return internal precisions 2–5 times worse (Schoene and Baxter 2017, this volume). Small sample analysis of Hf via MC-ICP-MS remains a limitation in most labs, though use of a desolvating nebulizer (Aridus), refined cone geometry, and 10^{12} ohm resistors, permits sub-nanogram loads of Hf from dissolved samples run through column chemistry and analyzed via MC-ICP-MS to yield 50–180 ppm 2RSE internal precision (Bast et al. 2015). Daughter isotope precision plays a larger role in controlling age precision for Paleozoic and younger samples and when parent/daughter ratio is lower, whereas parent/daughter isotope precision plays a larger role for Proterozoic and older samples and when parent/daughter ratio is higher (see Baxter and Scherer 2013). In general, $^{147}Sm/^{144}Nd$ ratios measured by isotope dilution (ID)TIMS can yield external precision (i.e., by assessing reproducibility of mixed standard solutions, or homogeneous dissolved natural rock samples) better than 0.1% with modern methods. $^{176}Lu/^{177}Hf$ analysis via ID-MC-ICPMS is less precise—more typically 0.1 to 1.0%—because making precise mass bias corrections and spike subtraction is more challenging given that there are only two isotopes of Lu (though MC-ICP-MS permits addition of a secondary element, such as Erbium, for purposes of mass bias correction; e.g., Bast et al. 2015).

Most clean garnet yields parent/daughter greater than 1.0 for both isotopic systems (Fig. 10), perhaps slightly more often for $^{147}Sm/^{144}Nd$ than for $^{176}Lu/^{177}Hf$. It is also interesting to note that the very highest parent/daughter ratios (as high as 50; Lagos et al. 2007) are from $^{176}Lu/^{177}Hf$. These observations are an indication that $^{176}Lu/^{177}Hf$ tends to be much more strongly zoned from core (highest Lu/Hf) to rim (lowest Lu/Hf) owing to strong Rayleigh fractionation of parent Lu (over daughter Hf) into garnet, than $^{147}Sm/^{144}Nd$ which tends to have more uniform zonation (for example, see Fig. 7, or modeling of Kohn (2009), or LA-ICP-MS data from Anczkiewicz et al. (2007) and Anczkiewicz et al. (2012). Overall, as shown in Figure 9, theoretical two-point isochrons between clean garnet with parent/daughter > 1.0 and a whole rock or matrix as the lower second point on the isochron can generally yield age precision between ±1.0 and ±0.1 million years (Baxter and Scherer 2013) assuming optimal analytical precision. If the garnet has parent/daughter «1.0, age precision (and sometimes accuracy, see below) degrades quickly. As sample size decreases and analytical precision predictably worsens, poorer age precision also results, though ±1–5 million years age precision is often still achievable in most cases.

Multi-point isochrons and the MSWD. If more than two points populate an isochron (which is always desirable), a statistical opportunity exists to evaluate that scatter and include it in the uncertainty of the isochron's slope, and the age. The most geochronologically relevant statistical measure of this scatter is the 'mean square of weighted deviates' or MSWD (e.g., Wendt and Carl 1991). It compares the scatter expected given the reported analytical uncertainty to the actual scatter of data from an isochron. When the observed scatter matches the statistically predicted scatter given analytical uncertainty, the MSWD is near 1.0; in this case additional points to the isochron can lead to better precision than that shown in Figure 9. Much higher MSWD will in turn lead to poorer (higher) age uncertainty and means that there is real geologic scatter from a single isochron; that is, some of the points populating the isochron fail one of the fundamental isochron assumptions. High MSWD isn't necessarily bad news though. If the high MSWD is due to multiple garnet analyses that scatter off the isochron, this may be an indication of resolvable age variation in the garnets being analyzed. Since it has been shown that garnet from the same rock (e.g., Skora et al. 2008; Herwartz et al. 2011) or the same crystal (e.g., Pollington and Baxter 2010; Dragovic et al. 2015) can span growth ages of many millions of years or even tens to hundreds of millions of years, such scatter and high MSWD is a useful and expected indicator of resolvable age zonation (also see Kohn 2009) that might encourage further textural or zoned garnet geochronologic analysis. In this regard, a range of individual two-point garnet–matrix ages is of greater meaning than a lumped average growth age produced by a multi-point isochron with high MSWD (e.g., Pollington and Baxter 2010; and see Fig. 16 discussed in the section on Zoned Garnet Geochronolgy below). If high MSWD is due to scatter from multiple whole rock, matrix, or mineral analyses (including poorly cleansed garnet) on the low side of the isochron, that could mean that one or more of those points doesn't belong and should be removed (this is discussed in a later section). Or, it may reflect real heterogeneity in the local rock matrix that, unfortunately, must be included in the age and its uncertainty (also discussed in a later section). MSWD less than 1.0 generally means that analytical uncertainties have been overestimated. Overestimation—sometimes referred to as 'conservative' estimation—of analytical uncertainty can also change a dataset that would yield a very high MSWD into a dataset that yields a lower MSWD, unintentionally masking what may in fact be important geological scatter. The savvy geochronologist, or interpreter of geochronologic data, will be careful to look for such high 'conservative' estimates of analytical uncertainty in reported isochron data. It is therefore crucial that estimates of analytical uncertainty are accurate, not 'conservative', as both over and underestimates can have deleterious effects on age interpretation masking what could be important information about the system.

The garnet point on the isochron and the problem of inclusion contamination. Garnet geochronology with Lu–Hf or Sm–Nd works because garnet uniquely fractionates parent from daughter creating an unusually high parent/daughter ratio. The garnet always represents the high point (or points, if one makes multiple measurements of the garnet) on the isochron. However, essentially all garnets contain micro-inclusions of other minerals that may be older (if inherited) or younger (if accessed by cracks) than the crystallization age of the garnet itself. The problem of inclusions continues to be the greatest challenge for successful garnet geochronology; if inclusions in garnet are not sufficiently removed, resulting 'garnet' ages can be imprecise, and worse, grossly inaccurate. Unfortunately, the published literature is full of such examples and the reader is invited to critically evaluate the literature themselves in light of the perspective and data we offer in this section. At the same time, the notion that we cannot overcome the challenge of inclusions remains the greatest misconception that continues to limit the credibility and broader use of garnet geochronology. In fact, numerous methodologies have been developed that provide solutions to the inclusion problem in almost all cases, leading to clean garnet and robust, precise, and accurate garnet ages.

Why inclusions are a problem. The problem of contaminating inclusions was recognized in the very first paper on garnet geochronology in 1980 (van Breemen and Hawkesworth 1980). Micro-inclusions can be inherited from significantly older episodes, and some inclusions can be reset or precipitated after garnet growth during retrograde cracking and fluid influx. Therefore, in the worst case, a 'garnet' analysis that still contains abundant inclusions can lead to grossly *inaccurate* apparent ages that can be younger or older than the true garnet age (e.g., Fig. 11). In general, inherited (i.e., older) inclusions that have a parent/daughter isotope ratio lower than the host rock (or younger inclusions that have a parent/daughter ratio greater than the host rock) will pull the contaminated garnet down off the true isochron to create falsely young ages. The converse is also true. Of course, the more contaminated the 'garnet' analysis is, the lower its apparent parent/daughter, the greater the potential inaccuracy (Fig. 12). Even if included phases are in age equilibrium with the garnet (i.e., they lie perfectly on the garnet–matrix isochron) their presence in the 'garnet' analysis will pull the 'garnet' point down along the isochron, reducing the spread along the isochron, thus worsening the age precision (Fig. 12). The reader is referred to a great number of published studies that define and describe the problem of inclusion contamination for garnet geochronology (e.g., Zhou and Hensen 1995; Scherer et al. 2000; Prince et al. 2001; Thoni 2002; Baxter and Scherer 2013).

How to solve the problem of inclusions. In the past 20 years numerous methods have been developed to eliminate most inclusions from most garnets thereby solving the inclusion problem. Any attempt to clean a garnet separate begins and ends with careful handpicking of finely crushed separates. However, good handpicking alone is usually inadequate to fully alleviate the inclusion problem (e.g., Thoni 2002). Micro-inclusions that may not be visible to the naked eye can escape even the most diligent handpicking. So, the most successful methods to cleanse garnets of their micro-inclusions all involve a 'leaching' or 'partial dissolution' procedure in various strong acids either to dissolve away problem inclusions in discarded solution leaving pure garnet for analysis, or, to dissolve away pure garnet in the analyzed solution leaving problem inclusions behind in a solid residue. The former has been successfully employed for Sm–Nd geochronology (e.g., Zhou and Hensen 1995; DeWolf et

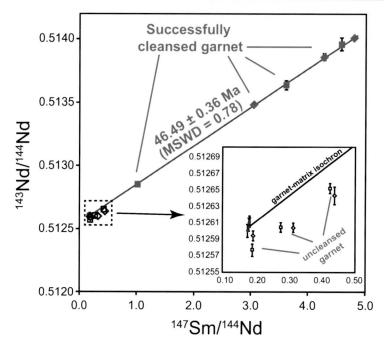

Figure 11. The problem of inclusions, and the success of partial dissolution to solve it. Sm–Nd isochron data are from a mafic blueschist (modified from Dragovic et al 2012). The lower inset shows uncleansed garnet (open symbols) dominated by inclusions prior to partial dissolution. Any attempt to create an age by pairing an uncleansed garnet with the matrix (filled circles) would lead to falsely young—even negative—ages. Main figure shows the same garnets after proper partial dissolution cleansing (filled diamonds and squares) and resultant accurate and precise isochron age.

al. 1996; Amato et al. 1999; Scherer et al. 2000; Baxter et al. 2002; Thoni 2002; Anczkiewicz and Thirlwall 2003; Pollington and Baxter 2011) where the most insidious inclusions are REE-rich minerals like monazite or clinozoisite which will dissolve in acid more readily than their garnet host. The latter has been successfully employed for Lu–Hf geochronology (e.g., Scherer et al. 2000; Lagos et al. 2007) where the most insidious inclusion is Hf-rich zircon, which is extremely resistant even to hydrofluoric acid at typical hotplate temperatures. This dichotomy does mean it is difficult to design a single cleansing method optimized for both Lu–Hf and Sm–Nd geochronology on the same sample aliquot. This, along with the analytical differences described earlier, in turn explains the dearth of studies where both Lu–Hf and Sm–Nd geochronology on the same samples yield optimal quality results from both systems. Some of the variables to consider when testing a partial dissolution method on a given garnet sample include: 1) acids used (e.g., Anczkiewicz et al. (2004) use sulfuric acid which attacks phosphates well, but does little to silicate inclusions like epidote if they are a factor, whereas HF partial dissolution of Amato et al. (1999) removes silicates well, but dissolves much of the garnet too requiring a delicate balance), 2) duration (anywhere from 15 to 180 min in HF have proven useful in different cases; e.g., Baxter and Scherer (2013)), 3) grain size (anywhere between 250 and 100 mesh size is recommended; finer grain size may lead to problems with garnet reactive surface area being too high; Pollington and Baxter 2011).

The term 'partial dissolution' cleansing is preferred over 'leaching' as the latter syntax could be interpreted to mean that something is actually leaching out of the garnet lattice itself. This leads to questions about whether the 'leaching' procedure is preferentially 'leaching' out

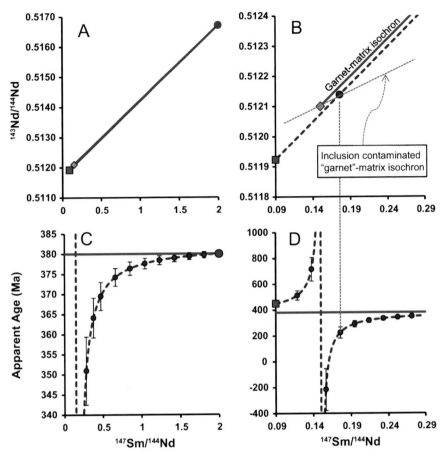

Figure 12. Effects of contamination by inclusions. These theoretical diagrams shown are for the Sm–Nd system but also qualitatively apply to the Lu–Hf system. Clean garnet (circle) has high $^{147}Sm/^{144}Nd$ ratio. Diamond is the matrix. Square is a low $^{147}Sm/^{144}Nd$ inherited inclusion (e.g., monazite for Sm–Nd or zircon for Lu–Hf) yielding an age of 450 Ma when paired with the host matrix. A and B show the true garnet–matrix isochron (solid line, with an age of 380 Ma) and a garnet–inclusion mixing array (bold dashed line). C and D show the apparent "garnet"-matrix age for "garnet" data points (circles) with varying amounts of inclusion contamination along the garnet–inclusion mixing line. Note that heavily contaminated "garnet" plotting near the matrix $^{147}Sm/^{144}Nd$ can yield imprecise and grossly inaccurate ages that are older or younger than the true garnet–matrix age, including even negative ages. B and D show a single example of a heavily contaminated "garnet" data point paired with the matrix to create a falsely young apparent isochron age with shallower slope (fine dashed) than the true garnet–matrix age. Two-point isochron age error bars shown are determined assuming the analytical parameters outlined in Figure 9.

parent vs. daughter (or vice versa). Here, it is important to note that none of the aforementioned studies have shown any evidence that the partial dissolution methods being employed create any fractionation or change of the garnet chemistry itself. When garnets are well cleaned they give remarkably consistent age results that would be impossible if some fractionation via 'leaching' was occurring (see Fig. 11). Differential 'leaching' of parent and daughter is also extremely unlikely because the mechanism for 'leaching' would be solid state diffusion through the garnet lattice; diffusivities at hot-plate temperatures (~100–200 °C) are too slow to allow for any significant diffusion of REE and even slower Hf. Alpha-decay induced lattice damage that may enhance diffusive loss in U/Th-rich minerals like metamict zircon is not an issue in garnet given the orders of magnitude smaller alpha flux in garnet due to lower concentrations of the alpha producing elements (e.g., Sm), the much slower decay rate of Sm, and single alpha decay for ^{147}Sm to stable ^{143}Nd. Should any fractionation effects be found (and no evidence yet exists to suggest there should be) it would be more likely to affect the Lu–Hf rather than the Sm–Nd system given the stronger chemical difference between the REE Lu and the HFSE Hf (as opposed to the REE's Sm and Nd). In summary, there is no evidence to suggest that partial dissolution cleansing of inclusions is affecting the remaining *clean* garnet's Sm–Nd and Lu–Hf chemistry at all.

How to know the garnet is clean (or clean enough). Because contaminating inclusions have much lower parent/daughter ratios, and much higher daughter element concentrations than their garnet hosts, well-cleaned garnet generally will exhibit high parent/daughter ratio and low daughter element concentration. Figures 10 and 13 provide two different compilations of 'garnet' parent/daughter ratio and daughter element concentration data to illustrate this. Figure 13 depicts the full range of published garnet data from a partial compilation of the literature. No screening was used in this compilation except to include only data reported as 'garnet' by the authors. Note the remarkable range in the data spanning many orders of magnitude, and the conspicuous negative relationship between parent/daughter ratio and daughter element concentration. In general, higher ^{147}Sm/^{144}Nd and lower Nd concentration indicates a cleaner garnet. But where do we draw the line between clean, clean enough, and dirty garnet? Rather than single out specific studies as good or bad, we present in Figure 10 similar compilations of published and unpublished Sm–Nd and Lu–Hf garnet data from over 100 different garnet-bearing rocks analyzed in the Boston University Lab (data from Ethan Baxter) and the Muenster Lab (personal communication from Erik Scherer) as they represent a subset of data prepared in a consistent manner, reflecting modern practices for Sm–Nd (by Baxter and colleagues using partial dissolution methods described by Pollington and Baxter 2011) and Lu–Hf (by Scherer and colleagues using partial dissolution methods described by Lagos et al. 2007). By no means does this mean that every sample in these plots is a perfectly cleansed garnet (that is surely not the case). But it does give an indication of the range of measured garnet compositions that typically results from best practices in these two particular labs that is instructive for sake of comparison. For Sm–Nd, 90% of the data indicate ^{147}Sm/^{144}Nd > 1.0 and [Nd] ppm < 0.4. For Lu–Hf, 90% of the data indicate ^{176}Lu/^{177}Hf > 0.5 and [Hf] ppm < 0.4. If the garnet doesn't exceed these cutoffs after the first attempt to cleanse it, we suggest additional attempts. Baxter and Scherer (2013) recommend that garnet is clean enough to avoid significant effects of inclusion contamination when it has parent/daughter ratio > 1.0, generally coinciding with daughter element concentration < 1.0. These of course are arbitrary cutoffs but are meant to provide some guidance in establishing the confidence (i.e., accuracy) of a given garnet age. Figure 12 also shows an example of the progressively diminished effect of inclusion contamination as a garnet is cleansed.

A good way to evaluate whether a garnet separate is clean is by comparison to in situ LA-ICP-MS or SIMS analysis (e.g., Prince et al. 2001; Anczkiewicz et al. 2007; Stowell et al. 2010; Stowell et al. 2014; Gatewood et al. 2015; Dragovic et al. 2016). Ideally, any garnet would first be lasered *in situ* to establish a baseline for clean parent/daughter and daughter element concentration in the garnet. Careful screening of laser data (considering multiple elements) is crucial to eliminate even slightly contaminated spots from the data (i.e., contamination by non-garnet mineral inclusions), because garnet Nd and Hf concentrations in clean garnet are so low (< 1 ppm, sometimes < 0.1 ppm). Still, LA-ICP-MS analysis can provide a valuable comparison to full garnet analysis of cleansed samples to see if partial dissolution was successful. Figure 14 shows an example from large garnets from Townshend Dam, Vermont from Gatewood et al. (2015). Note first the raw dataset for [Nd] ppm showing abundant spikes of high Nd (over 100 ppm!) within the garnet vividly depicting the significant presence of inclusions. After screening away most of the inclusion-contaminated points,

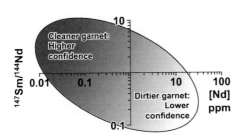

Figure 13. ^{147}Sm/^{144}Nd vs. [Nd] ppm for "garnet" data. The oval outlines a compilation of hundreds of published "garnet" data in the literature, a great many of which had not been properly cleansed of inclusions. Compare to the data from cleansed and uncleansed garnet shown in Figures 10, 11. Garnets with ^{147}Sm/^{144}Nd >1 and [Nd] <1 ppm is more likely to be clean and will return an accurate garnet age (darker shading). Garnets with very low ^{147}Sm/^{144}Nd and/or very high [Nd] ppm (lighter shading) constitute a large amount of the published data, these are almost certainly contaminated by inclusions; use of such analyses for a garnet isochron age will lead to less precise and possibly inaccurate results (see Figure 12).

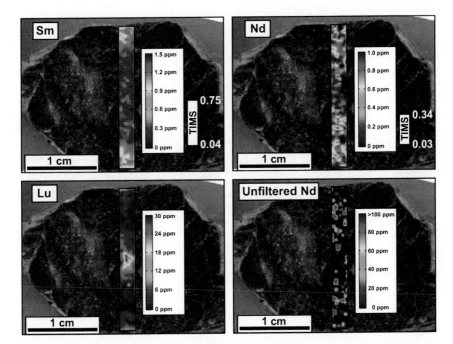

Figure 14. LA-ICPMS maps of Sm, Nd, Lu [ppm] in garnet from Townshend Dam, Vermont (modified from Gatewood et al 2015). Note Sm is slightly zoned with enrichments at the rim, whereas Lu is more strongly zoned with enrichments at the core. Nd is unzoned, except for numerous high spots reflecting Nd-rich inclusions. Bottom right panel shows raw unfiltered Nd data where the highest analyses show [Nd] in excess of 100ppm. Top right panel shows [Nd] data after carefully filtering out the most egregious contamination, revealing the underlying mostly clean garnet with concentrations <1ppm. White bars and values show the range of ID-TIMS analyses of the same Vermont garnets after proper cleansing reported by Gatewood et al (2015), showing good agreement with the laser data.

the refined LA-ICP-MS data show low [Nd] from 0.03 ppm to 0.5 ppm. In this study, these values are a close match for the cleaned bulk garnet data analyzed via TIMS, indicating success in cleansing away these inclusions and recovering true clean garnet.

Another vivid example of the success of partial dissolution cleansing, and the great importance of doing so, is shown in Figure 11 from Dragovic et al. (2012). Here, note where the 'garnet' data plot when they were intentionally analyzed without partial dissolution cleansing.

These data yield low $^{147}Sm/^{144}Nd < 0.45$ and high [Nd] ppm between 0.9 and 2.8 ppm along a crude mixing trend with an inherited inclusion population with low $^{147}Sm/^{144}Nd$ (akin to the theoretical example shown in Fig. 12b). Any of these 'garnets', when paired with the matrix, would give falsely young ages; a few would even give negative ages. But consider these very same garnet samples after experiencing a proper partial dissolution. Now, these data give high $^{147}Sm/^{144}Nd$ between 1 and 5 with [Nd] between 0.03 ppm and 0.07 ppm and plot together on a remarkably tight isochron indicating the time of garnet growth in this mafic blueschist.

The second point on the isochron. It takes a slope to calculate an age, a line to calculate a slope, and two points to make a line. Three, four, or five points lend further statistical credence to a linear isochron relationship but without at least a second point, isochron geochronology is a non-starter. In this regard, the second point on the isochron is just as important and powerful as the garnet point on the isochron in determining age precision and accuracy. The second point on the isochron must be an accurate and precise measurement of the rock matrix reservoir with which the garnet initially grew in isotopic equilibrium (e.g., $^{143}Nd/^{144}Nd$ or $^{176}Hf/^{177}Hf$). In the very simplest scenario (rarely achieved in real systems), an isotopically homogeneous parent rock or melt simultaneously crystallizes all its minerals, including garnet, such that the entire rock is in initial isotopic equilibrium and all minerals remain closed systems from that point until the present day. In this case, one could measure anything else in the rock—the entire whole rock itself, just the matrix, or any subset of the various other minerals in the rock—as a representation of that initial rock reservoir and pair it with the pure garnet to form a two point isochron. Additional garnets (with high Sm/Nd or Lu/Hf) or additional whole rock, matrix, or mineral analyses (with low Sm/Nd of Lu/Hf) would further populate the isochron and a happy multi-point isochron with a perfect MSWD of 1.0 would result. In reality, many rocks (or protoliths) are not perfectly isotopically homogeneous at the onset of garnet growth, the whole rock and matrix may differ due to parent/daughter Rayleigh fractionation during garnet growth, the matrix may experience open system change of parent/daughter or daughter isotope composition after/during garnet growth, and different mineral phases may grow or close to isotopic exchange with the rock reservoir at different times than the garnet. Any of these factors can conspire to create an inaccurate or imprecise garnet isochron age if such data that don't belong are included on an isochron. Stated differently, more points on an isochron are only a good idea if those points belong on the isochron. The good news is that most of these effects are very minor to negligible most of the time. Let us explore these each in turn.

Initial matrix heterogeneity. If the rock reservoir is initially heterogeneous for whatever reason, how can we know what the garnet actually grew in equilibrium with? Igneous protoliths are probably least susceptible to this issue given that they are fairly homogenous when they first form. However, ancient layered sedimentary protoliths where each layer may include inherited minerals of varying provenance can present problems in this regard. For isotopic systems like U–Pb or Rb/Sr where the range of parent/daughter ratio among phases can be several orders of magnitude, this can create enormous heterogeneities (via differential radiogenic ingrowth) the effects of which can be severely problematic for garnet (or any) isochron geochronology in those systems (e.g., Romer and Xiao 2005). Fortunately, with the exception of garnet, most common minerals (and rocks) have a limited range of Lu/Hf and Sm/Nd ratios such that the magnitude of matrix heterogeneities is much smaller. Still, matrix heterogeneity can be significant given the high precision and accuracy often desired (and achievable) with modern methods.

Let us consider a layered sedimentary protolith. The layers have varying $^{147}Sm/^{144}Nd$ (or $^{176}Lu/^{177}Hf$) which has led to varying $^{143}Nd/^{144}Nd$ (or $^{176}Hf/^{177}Hf$). Let us now permit garnet to grow instantaneously and with uniform distribution in that layered rock. [In reality, the different layers may have different enough major element chemistry so as to preferentially crystallize garnet more on certain layers than in others, but let us ignore that for the present discussion]. At any given location in the layered rock system, the new garnet crystallizes with an average composition reflective of the equilibrium lengthscale (*Le*) for that element in the system (see Baxter and DePaolo (2002a,b), and DePaolo and Getty (1996) for further discussion). The equilibrium lengthscale for a given element, *Le*, is dependent on the effective diffusivity (*D**) of the element within the intergranular transporting medium (ITM; Baxter and DePaolo 2002b), and the local reaction/exchange rate (*R*) for that element between the matrix minerals and the ITM: $Le = (D^*/R)^{1/2}$. The effective diffusivity (*D**) includes both the diffusivity in the ITM, and the partitioning of that element between the solid minerals and the ITM. Elements (like Sr) that are strongly partitioned (i.e., soluble) into the fluid filled ITM have high *D** and smear out and average $^{87}Sr/^{86}Sr$ over a large equilibrium lengthscale. Elements like Nd and Hf are very weakly partitioned (i.e., insoluble) into the fluid filled ITM of most crustal fluids and thus have relatively low *D**; thus $^{143}Nd/^{144}Nd$ and $^{176}Hf/^{177}Hf$ of the ITM (and any garnet crystallizing from it) will closely match the local bulk matrix solid. In practice, this means you wouldn't want to pair a garnet from one layer with bulk matrix from another layer. But, if you crushed up the entire rock and sampled a representative average garnet separate and matrix from the entire volume, the heterogeneities average out and the problem goes away. However, now consider the growth of a very large garnet porphyroblast large enough to grow over several compositional layers. A single crystal like this would be attractive for high-resolution microsampling and zoned geochronology. In this scenario, each concentric growth ring, or even different portions of the same growth ring, of the garnet may inherit a different starting $^{143}Nd/^{144}Nd$ (or $^{176}Hf/^{177}Hf$) depending on the layer(s) it is in. When focusing on a single crystal, or portions thereof, it is no longer advisable to use a large averaged whole rock to anchor the isochron in a heterogeneous protolith. Instead, one must carefully analyze and evaluate the extent of porphyroblast scale matrix heterogeneity via multiple measurements and include them all on an isochron. This is an important statistical acknowledgement of the uncertainty about the second point on the isochron that will result in larger age errors and larger MSWD. Gatewood et al. (2015) shows a vivid example of the effects of such cm scale matrix heterogeneity on zoned garnet geochronology from a layered meta-sedimentary rock. On the contrary, Pollington and Baxter (2010) saw no such effect when conducting zoned garnet chronology in a homogeneous rock matrix, most likely inherited from an igneous protolith. Overall, it is always advisable to evaluate the extent of matrix heterogeneity of any rock, try to avoid major heterogeneities and lithologic contacts if possible, and match multiple garnets (or bulk garnet separates) and average matrix from the same rock volume to average out heterogeneities.

Whole rock vs. matrix. We define a 'whole rock' as the entire rock volume including the garnet itself. We define the 'matrix' as the entire rock volume excepting the garnet. As a garnet with much higher Sm/Nd and Lu/Hf than the original whole rock grows, it will alter the Sm/Nd and Lu/Hf of the remaining matrix via simple Raleigh fractionation. Strictly speaking, the core of the garnet (the initial fraction of garnet to grow) should be paired with the 'whole rock', whereas the outermost rim of the garnet (the final bit to grow) should be paired with the matrix. In general it is advisable to measure both matrix and whole rock from the same rock averaged volume to establish the significance of that difference for your particular rock. For Sm–Nd, the general finding is that matrix and whole rock are very rarely different enough to make a significant difference to the age. This is largely due to the fact that the concentration of Nd and Sm in clean garnet is usually one or more orders of magnitude smaller than the concentration in the matrix. It would require a lot of garnet to create a significant difference in parent/daughter. The Lu–Hf system is more susceptible to this process as Lu concentrations in garnet can be much higher (up to an order of magnitude or more) than the whole rock especially at the onset of garnet growth, whereas Hf concentrations are very low. This could lead to more significant Rayleigh fractionation effects leaving the residual matrix with even lower Lu/Hf after garnet growth. Fortunately, for either system, if the time duration between growth of core and growth of rim is short, negligible radiogenic daughter in-growth (and rotation of the isochron beyond horizontal) will have occurred. In this case, shifting Sm/Nd or Lu/Hf of the matrix to different values along a near horizontal line near the time of garnet growth will have negligible effect on the final isochron age determination. Finally, because most garnets also include (and effectively sequester) other minerals the net change to the host rock composition is offset and minimized. Rocks with low Nd or Hf concentrations (< 10 ppm) and a large timespan between growth of garnet core and rim are therefore more susceptible to this easily accounted for effect.

Open system change of matrix. If the matrix experienced any kind of open system exchange, or loss, or gain of Sm, Nd or Lu, Hf there is potential to skew the garnet–matrix isochron age relationship. Mechanisms may include partial melting where a melt is lost from the system, the injection of a melt into the system, or the passage of fluids through the system. Most common crustal fluids have low solubilities of REE and Hf so only in extreme cases of focused fluid flow should we observe major Sm/Nd or Lu/Hf loss or gains. Important examples of such open system mobility of REE have been found in crustal fluids (Zack and John 2007; Ague 2011). Loss of an internally derived partial melt would leave the solid residue with a higher Sm/Nd or Lu/Hf ratio. But as long as the internally derived melt would be in Nd and Hf isotopic equilibrium with the bulk solid there would be no isotopic fractionation in the melt depleted residue. This should generally be the case, unless incongruent melting of a very old protolith (where significant mineral scale Nd or Hf isotopic differences had already evolved) created a melt with different isotopic composition (e.g., Zeng et al. 2005) and that melt was extracted quickly, before isotopic equilibrium could be restored via diffusion; in this case the resulting effect on age should be evaluated. Fortunately for the Sm–Nd and Lu–Hf systems (as opposed to U–Pb or Rb–Sr), such radiogenic differences are relatively small given the narrow range of Sm/Nd and Lu/Hf that exist among most common minerals (e.g., Romer and Xiao 2005). Introduction of an external melt or fluid into the system has the potential to be much more problematic, regardless of when the metasomatic event occurred with respect to garnet growth. External melts or fluids could bring completely different Nd or Hf isotopic chemistry that could mix with the original rock matrix pulling it vertically up or down well off the garnet–matrix isochron. Changes in Sm/Nd or Lu/Hf may also occur which could add to the effect if the metasomatic event happened a long time after garnet growth (when the isochron had already rotated well past horizontal). In general, the garnet geochronologist should avoid samples bearing evidence of open system metasomatism: in a vein, next to pegmatite, near a lithologic contact. Or, great care should be taken to evaluate the isotopic extent of that metasomatism to try and recreate the original rock matrix as well as possible. In some cases, the garnet inclusions may serve as a guide towards reconstructing the original rock matrix, but as we have seen, these minerals inclusions (if they are inherited from earlier events) do not always accurately reflect the garnet's original isotopic growth environment.

Non-garnet mineral growth and/or closure. The very first garnet geochronology papers chose minerals (like clinopyroxene in an eclogite, for example; Griffin and Brueckner 1980) rather than a whole rock to pair with the garnet in two-point isochrons. Some papers continue to add other minerals along with a whole rock or matrix to anchor the garnet isochron. Most of the time, these mineral data plot very near the whole rock or matrix (i.e., with similarly low parent/ daughter) and have little effect on the age. But sometimes these individual minerals plot slightly off the isochron, leading to higher MSWD if included, or different absolute ages if employed instead of the whole rock. Which point is a better representation of the rock reservoir with which the garnet grew in isotopic equilibrium? On the one hand, the clinopyroxene and garnet in an eclogite both reflect growth at eclogite facies conditions (whereas the whole rock may have other phases that grew before or since eclogite facies) and thus one could argue they are a good match if the goal is to date eclogite facies conditions. However, it is almost certainly the case that clinopyroxene has a different effective 'closure' time (due to processes like diffusion or matrix recrystallization) than the garnet, so if the clinopyroxene has a different $^{147}Sm/^{144}Nd$ than the matrix it will evolve off the garnet–matrix isochron and should not be included. The magnitude of this effect depends on the difference in closure/growth age of clinopyroxene vs. garnet, and on the difference in $^{147}Sm/^{144}Nd$ between clinopyroxene and the matrix. But, if the matrix itself is made up of all these minerals, how then can we argue that the matrix is any better a choice? The reason is that the matrix itself has remained a closed system and represents an appropriate rock-averaged Sm–Nd and Lu–Hf composition from which the garnet first grew. Resistant accessory phases (like monazite or zircon) are the exception in that they can retain their inherited isotopic signatures and generally do not participate as reactants in garnet growth.

The second point(s) on the isochron cannot be overlooked. At a minimum, it is worth measuring a representative whole rock and matrix. Pure non-garnet mineral separates may not be good choices for the isochron, especially when their parent/daughter ratio differs greatly from the matrix and/or their growth or closure time differs greatly from that of garnet. Avoid rocks with evidence for open system exchange, of significant layering or heterogeneity. If heterogeneity exists, it must be evaluated and included in the isochron to acknowledge that uncertainty (e.g., Gatewood et al. 2015). While many of these issues have the potential to completely ruin U–Pb or Rb–Sr garnet geochronology, they are rarely a major problem for Sm–Nd and Lu–Hf geochronology if reasonable care is taken to evaluate them.

Why Do Lu–Hf and Sm–Nd Ages Differ? The Complementarity of Lu–Hf and Sm–Nd Garnet Geochronology. One of the powerful opportunities of garnet geochronology is the theoretical ability to date the same garnet (or garnet-bearing rock) with both Lu–Hf and Sm–Nd. As discussed above, this hasn't been done as often as we would like given the different instruments required for optimal analysis (i.e., TIMS for Sm–Nd and MC-ICPMS for Lu–Hf) and because the most popular methods employed for removing inclusions are opposite for Lu–Hf and Sm–Nd. Thus, most labs are optimally equipped for one or the other, but rarely for both. Still, a growing number of labs have attempted to date the same garnet-bearing rocks with both methods (e.g., Lapen et al. 2003; Kylander-Clark et al. 2007; Cheng et al. 2008, 2016; Skora et al. 2009; Anczkiewicz et al. 2012; Smit et al. 2013) leading to interesting observations that, when fully understood can lead to valuable insights about the rock's history. Both systems are well suited for garnet geochronology, though their different pros and cons may make one more suitable for a given sample suite or application, and thus powerfully complementary when used in concert.

A notable observation made in many (though not all) of such combined studies is that Lu–Hf garnet ages tend to be older than Sm–Nd ages from the same rock. Why is this the case? The possible answers can be grouped into three categories, one (or more) of which may apply in any given situation:

1. Different parent isotope zonation for Lu–Hf and Sm–Nd (e.g., Skora et al. 2009)

2. Different 'closure temperature' for Lu–Hf and Sm–Nd (e.g., Yakymchuk et al. 2015)

3. Something is wrong with the Sm–Nd age or the Lu–Hf age

Let us address each scenario in turn as each provides an opportunity to further explore some of the important differences and complementarity of Lu–Hf and Sm–Nd garnet geochronology.

Different parent isotope zonation for Lu–Hf and Sm–Nd. As discussed in the section on trace elements above, Lu is very strongly partitioned into garnet during its growth as compared to other common matrix minerals. On the contrary, Sm, Nd, and Hf are all weakly partitioned into garnet as compared to other matrix minerals (though Sm not as weakly as Nd, hence the reason that garnet can still have high Sm/Nd *ratios* despite relatively low Sm and very low Nd *concentrations*). Thus, garnets tend to be strongly zoned in Lu (with highest Lu concentrations in the early grown core as much as ~100 times higher than rims with low Lu; Figs. 7 and 14) whereas Sm, Nd, and Hf tend to be unzoned or with slight enrichments toward the rim (e.g., Lapen et al. 2003; Cheng et al. 2008; Skora et al. 2008; Peterman et al. 2009; Anczkiewicz et al. 2012; Smit et al. 2013; Gatewood et al. 2015; Fig. 15). Because the age information contained within a garnet crystal is based on the concentration of the parent element (Lu or Sm) this chemical zonation means that Lu–Hf ages will generally be skewed towards the age of core growth (where more of the Lu resides) whereas Sm–Nd ages will be skewed towards the age of rim growth (where more of the Sm resides). It is also important to note that while these general trends for Lu and Sm zonation are common, more complex zonation in Lu and/ or Sm can exist in certain garnets ((see trace element section, above). This underscores the value of analyzing the core-to-rim Lu and Sm zonation of the actual samples in question as they have implications for interpreting the different 'bulk' Lu–Hf and Sm–Nd garnet ages. Lapen et al. (2003), Skora et al. (2009), and Kohn (2009) show examples of how measured Lu and Sm zonation can be used to model bulk ages by identifying where/when during garnet growth the parent element and mass averaged age would plot (e.g., Fig. 15). Such modeling requires certain assumptions including growth symmetry (e.g., perfect spherical shells), growth rate (constant vs. episodic), but are generally very useful in explaining age differences and, in the best case, placing constraints on total growth duration (e.g., Skora et al. 2009). For garnets that never experience temperatures above ~650 °C (above which diffusive mobilization may become significant; see below) the difference in Lu vs. Sm zonation is the dominant reason why Lu–Hf ages are different (generally older) than Sm–Nd ages. Both systems date primary growth, but they reflect different stages of that prolonged growth duration.

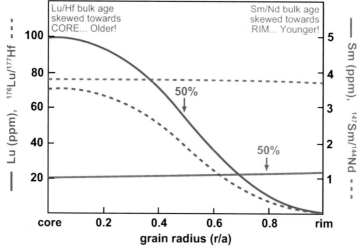

Figure 15. Conceptual zonation diagram of Lu, Hf, Sm, Nd resulting from simple Rayleigh fractionation during growth (modified from Kohn 2009). "*r/a*" on the *x*-axis indicates relative radial location from garnet core to rim of radius "*a*". "50%" shows the radial location where exactly half of the parent isotope is in core-ward and rim-ward garnet portions. Lu is strongly zoned, varying by two orders of magnitude from core to rim, whereas Sm is only slightly zoned. Note spherical symmetry involved in this model. This zonation is the primary reason that Lu–Hf bulk grain ages tend to be older than Sm–Nd bulk grain ages. Because $^{147}Sm/^{144}Nd$ is nearly unzoned, the Sm–Nd system offers a better chance at precise zoned garnet chronology from core-to-rim, whereas Lu–Hf offers a better chance to precisely date the earliest growth of garnet core.

Different 'closure temperature' for Lu–Hf and Sm–Nd. Because most garnets grow below their 'closure temperature' for Sm–Nd and Lu–Hf, and never experience higher temperatures, diffusive 'closure' is generally *not* at play when interpreting garnet geochronologic data, including the difference between Sm–Nd and Lu–Hf ages. However, for garnets that did experience temperatures in excess of ~650°C during or after their growth, the issue of diffusive 'closure' (i.e., Dodson 1973) or perhaps more appropriately diffusive 're-opening' (i.e., Caddick et al. 2010; Watson and Cherniak 2013) may come into play. Numerous papers have attempted to place constraints on the 'closure temperature' of garnet for both systems but, as pointed out and modeled by Ganguly and Tirone (1999) the classic Dodson formalism for closure is rarely appropriate for garnet because it is rarely completely reset before dropping through a closure temperature interval. As Watson and Cherniak (2013) describe, the matter is more often one of diffusive 'opening' whereby a garnet is heated for some period of time to temperatures high enough that diffusive resetting may begin. For example, Baxter et al. (2002) used the diffusion data of Ganguly et al. (1998b) to show that only the outermost few 10s of microns of a garnet crystal would be affected by age resetting during heating up to ~660°C (over a span of 12 million years), resulting in a negligible degree of bulk age resetting (depending on the grain size of course). Since then, numerous studies—both experimental and empirical—have sought to constrain the diffusivity of the REE (i.e., Sm, Nd, Lu: e.g., Carlson (2012), Tirone et al. (2005), Van Orman et al. (2002) and Hf: Bloch et al. (2015)). While all REE (including Sm, Nd, and Lu) appear to diffuse at the same rate within existing uncertainties, some overall discrepancy exists amongst existing studies of REE diffusion. Ganguly et al. (1998b) and Tirone et al. (2005) return diffusivities for the REE about an order of magnitude higher than those of Van Orman et al. (2002), Carlson (2012), and Bloch et al. (2015). Bloch et al. (2015) suggest that the difference may reflect a concentration dependence on REE diffusivity (i.e., lower concentrations diffuse via a faster diffusion mechanism) such that the faster Tirone et al. (2005) data are more appropriate for geochronologic applications. In any case, the data of Tirone et al. (2005) represent the fastest and thus provide a lower constraint on the temperatures at which REE in garnet become mobile for garnets, which of course itself is also dependent on grain size and cooling rate. Rather than quote blanket 'closure temperatures', interpreters of garnet geochronologic data should explore the partial diffusive resetting of ages by means of simple analytical or numerical modeling for the grain size and specific heating/cooling history that may have existed in the given system. Ganguly and Tirone (2001) and Watson and Cherniak (2013) provide useful analytical formalisms that may be used in general cases. Baxter et al. (2002), Korhonen et al. (2012), and Dragovic et al. (2016) are examples where simple case specific modeling was employed to constrain the likely extent of resetting. Baxter and Scherer (2013) model the range of temperatures required to reset a bulk garnet age just 5% (i.e., just beginning to 're-open') and 95% (i.e., essentially completely reset) for given grain size, temperature, and dwell time at that temperature. The reader is encouraged to consult these papers and model the diffusive resetting of Nd or Hf age information in their specific case.

The recent experimental data of Bloch et al. (2015) confirmed previous empirical inferences (e.g., Scherer et al. 2000; Kohn 2009; Anczkiewicz et al. 2012; Smit et al. 2013) that the diffusivity of Hf is considerably lower (by an order of magnitude, or more) than that of all the REE, including Lu. Without question, Hf diffusivity is much slower than Nd diffusivity; thus the Sm–Nd system is more susceptible to thermal diffusive resetting than the Lu–Hf system. That is, the 'closure temperature' of Hf is significantly higher than that of Nd. Bloch and Ganguly (2015) present diffusion modeling to show that the time required to fully reset Hf isotopic composition of garnet is 10–1000 times longer (at a given temperature) than the time required to fully reset Nd isotopes. Thus, for garnets experiencing temperatures above ~650°C, the difference in Hf vs. Nd diffusivity should result in generally younger Sm–Nd ages (which have been partially reset) vs. Lu–Hf ages. Yakymchuk et al. (2015) shows a good example of this in a high temperature (~850°C) migmatite where other possibilities (Lu vs. Sm zonation) can be ruled out.

Something wrong with the Lu–Hf or the Sm–Nd Age? As discussed above, there are numerous other reasons why a given Sm–Nd or Lu–Hf age might simply be flawed, in which case comparisons to good Lu–Hf or Sm–Nd data are misguided. Perhaps most notorious is the effect of inclusions, especially as most samples are not cleaned with methods optimized for both Sm–Nd and Lu–Hf (see above). While this chapter prefers not to enter the business of identifying 'bad' published data, we will point out that both good and bad certainly do exist in the literature. As previously discussed, Sm–Nd or Lu–Hf garnet data where the parent/daughter ratio is < 1.0 and/or the daughter element concentration is > 0.5 ppm should be scrutinized carefully as these are the hallmarks of inclusion contamination that can create significant age inaccuracies (see Figs. 12, 13, 14). While high Nd or Hf concentration and low $^{147}Sm/^{144}Nd$ or $^{176}Lu/^{177}Hf$ do not necessarily prove age-distorting contamination, experience shows that such data usually indicate an inclusion problem that should be addressed or evaluated in some way (Smit et al. 2013) before the resulting age is accepted.

An additional factor unique to the Lu–Hf system is the possibility of diffusive decoupling of parent Lu from daughter Hf. Because Lu (and all REEs) diffusive at a much faster rate than Hf (see above discussion and Bloch et al. 2015), it is theoretically possible that Lu zonation may smooth out within a garnet, or that Lu may diffuse into or out of a garnet as *P–T* changes (only if garnet–matrix Lu partitioning changes) subsequent to garnet growth, all while the Hf isotopic composition remains locked in. While the effects of such Lu mobility on resulting age can vary depending on the situation, most often this would result in a counterclockwise rotation of the garnet–matrix isochron and falsely old ages. As with all closure-related arguments for garnet geochronology, it is important to note again that this process is only possible when garnet is heated above the 'closure temperature' for Lu (> ~650 °C) for a significant period of time, and most enhanced for smaller garnets, and the highest temperatures. This argument rarely applies to the majority of greenschist and amphibolite facies garnets that never experience such high temperatures. When this process is at play, Lu–Hf ages can be compromised, thus rendering meaningful comparisons to good Sm–Nd data impossible. Papers by Kohn (2009), Anczkiewicz et al. (2012), Bloch et al. (2015), Bloch and Ganguly (2015), and (Kohn and Penniston-Dorland 2017, this volume) discuss and develop this potentially confounding problem for Lu–Hf geochronology while Anczkiewicz et al. (2012) shows compelling evidence for its occurrence in a natural setting.

Zoned garnet geochronology. The vast majority of published garnet geochronology, and the entire discussion in this chapter up to this point, involves what we call 'bulk garnet' geochronology. As long as no unintended fractionation of the garnet occurs (for example, due to magnetic separation which may select against garnet portions with differing amount of magnetic inclusions; e.g., Lapen et al. 2003), a 'bulk age' simply represents the parent isotope weighted (based on parent element zonation) average age of the garnet (Fig. 15). For a garnet that grew very rapidly (i.e., whose growth duration is within the age precision), such a bulk age is a valuable and precise measure of the growth event. However, if the garnet growth duration exceeds the attainable age precision (as may often be the case) the 'bulk age' is of limited petrochronologic utility, representing only an average age within an unresolved timespan of garnet growth. Insofar as multi-point bulk garnet isochrons might display scatter amongst clean garnet data with high parent/daughter, the bulk age precision (poorer) and MSWD (higher) represent a valuable indication that the garnet may have resolvable age zoning. Consider for example, the dataset in Figure 16 from Pollington and Baxter (2010). All of these garnet data are from the same rock. All of these garnet data are clean, passing any test of inclusion contamination. Matched with the matrix data, we can calculate a multi-point age of 25.5 Ma with a disappointing precision of ±5.3 Ma and eyebrow-raising MSWD of 270. Here is a reminder that poor MSWD, especially when caused by scatter of clean garnet at high parent/daughter, is not necessarily (nor even often) an indication that something has gone wrong. Rather it is a statistical invitation to explore the hypothesis of resolvable age zonation through means of zoned garnet geochronology.

In fact, the data shown in Figure 16 are from different zones of the same single garnet crystal, sampled as concentric rings from core to rim. With this context, the dataset suddenly makes sense. Instead of a single ill-advised multi-point isochron, what we really have are twelve individual two-point isochrons that reveal a 7.5 million year growth duration (Pollington and Baxter 2010). With the textural context of each garnet growth zone, we find that multiple two-point isochrons are better, and more appropriate, than a single multi-point isochron.

Zoned garnet geochronology is not new. In fact, the first attempts to extract and date different garnet growth zones spanned work between 1988–1999 (Cohen et al. 1988; Christensen et al. 1989, 1994; Vance and Onions 1990, 1992; Burton and Onions 1991; Mezger et al. 1992; Getty et al. 1993; Vance and Harris 1999). These papers used Rb–Sr and/or Sm–Nd to date two or three concentric growth zones in single 1–5 cm garnet crystals, separated via some combination of sawing, crushing, and plucking. Stowell et al. (2001) pioneered a core-drilling procedure to extract different garnet zones for Sm–Nd geochronology in a number of settings including contact metamorphism in Alaska, granulite facies migmatites in Fiordland, New Zealand, and granulites from the Cascades of Washington USA. Solva et al. (2003) separately analyzed core and rim of magmatic pegmatite garnet constraining Permian magmatism in the Alps. Ducea et al. (2003) were the first to employ a microdrilling apparatus (the MicroMill™) to extract garnet from three concentric growth zones from high temperature garnets in two Western North American sites. They modeled their data to constrain cooling rates from peak magmatic arc temperatures. In their study, the MicroMill™ was used to directly extract garnet in powdered form from individual drilled pits in core and rim. Pollington and Baxter (2010) used the MicroMill™ for chemically contoured sampling of concentric zones in a large 6 cm garnet from the Austrian Alps (Fig. 16). The spatial and age resolution of the Pollington and Baxter (2010) study permitted not just a constraint on total growth duration, but also the recognition of two significant pulses in the rate of garnet growth, correlated with chemical and textural features indicative of evolving thermodynamic and tectonic conditions, that would have otherwise have been missed. Pollington and Baxter (2010, 2011) used the Micromill™ to drill out concentric trenches, between which solid garnet annuli were left for collection, crushing, partial dissolution cleansing, and Sm–Nd TIMS analysis. They found that garnet powders derived from the drill trenches could not be cleaned of inclusions due to the extremely fine grain size; thus, the material from drilled trenches was discarded. Since then, three other studies (Dragovic et al. 2012, 2015; Gatewood et al. 2015) have employed the microdrilling methodology outlined in Pollington and Baxter (2011) extracting between 3 and 10 annuli per garnet crystal in blueschist facies rocks from Sifnos, Greece (Dragovic et al. 2012, 2015) and regional metamorphic schists from Townshend Dam, Vermont (Gatewood et al. 2015; the same rocks first dated by Christensen et al. 1989). Most of the studies cited in this paragraph show that < 1.0 Myr Sm–Nd age precision is achievable even as sample size decreases with more highly resolved sampling and smaller sample sizes.

Until recently, Lu–Hf zoned geochronology had not been attempted, in large part due to sample size limitations given that most labs still require 10's–100's ng Hf for MC-ICPMS analysis, but this is changing. Herwartz et al. (2011) separated cores and rims of strongly color zoned garnet from the Alps by crushing and handpicking, which were subsequently dated via Lu–Hf. This first study of garnet age zonation with Lu–Hf revealed an old inherited Variscan (~333 Ma) core surrounded by much younger Alpine (~38 Ma) rims, which these authors interpreted as evidence of two distinct cycles of subduction to mantle depths separated by three hundred million years. Nesheim et al. (2012) separated and grouped cores and rims of multiple garnet crystals in a single Mesoproterozoic sample from Idaho, USA, revealing mixed garnet ages bracketed by two major garnet growth events at ~1330 and ~1080 Ma. Anczkiewicz et al. (2014) extracted and dated three growth zones, based on textural and chemical patterns, from a large 3 cm garnet crystal from the Himalaya (see further discussion below). Schmidt et al. (2015) used saw cuts to sample up to 4–6 solid growth zones in four different 3–5 cm garnets for Lu–Hf geochronology. Rather than the surrounding matrix (which could not be sampled in this study), Schmidt et al. (2015) anchored each interior garnet analysis with the Lu-depleted

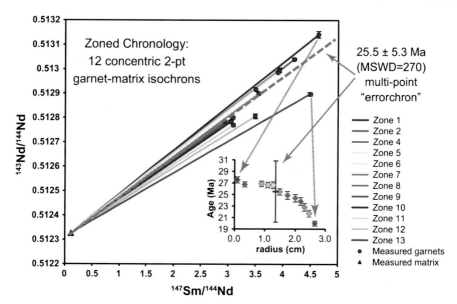

Figure 16. Zoned garnet geochronology for a single porphyroblast from Stillup Tal, Austria (modified from Pollington and Baxter 2010). Each garnet "zone" represents a concentric growth zone from core (zone 1) to rim (zone 13). When each is paired with the matrix, a pattern of systematically younging two-point isochron ages is seen from core to rim, spanning about 7.5 million years as shown (inset). Bold dashed line is the ill-advised multi-point 'errorchron' resulting from combining all garnet data and the matrix.

outermost rim in two-point 'garnet only' isochrons. The resulting age data suggest a total span of garnet growth of at least 12 million years. Cheng et al. (2016) presented Lu–Hf analysis of 12 micro-sawed garnet zones (not concentric) within two single crystals from the Huwan shear zone, China. Resulting two-point isochrons display ages ranging from 400 to 264 Ma which are interpreted to reflect mixing of at least two distinct episodes of garnet growth spanning that time.

Zoned garnet geochronology, despite its first applications almost 30 years ago, remains in its infancy, but its potential is great. Methodologies have now been developed to extract well-defined chemically contoured growth zones of garnet from single crystals of ~5 mm of greater diameter, clean those garnet portions of inclusions, analyze their isotopic composition even as sample size diminishes, and extract accurate and precise Sm–Nd, or Lu–Hf ages on each zone. Pollington and Baxter (2011) review the potential and current limitations for zoned garnet geochronology by Sm–Nd. As analytical approaches continue to improve for Sm–Nd and Lu–Hf smaller, more highly resolved records will become increasingly feasible. This is the exciting future that awaits garnet petrochronology.

Geospeedometry with garnet

Given the time and expense required to obtain high precision zoned garnet geochronology data, additional methods of inferring approximate metamorphic duration are desirable, particularly for understanding events on timescales too short to resolve isotopically. Although diffusion was discussed previously as an obstacle to thermobarometry, chemical profiles that retain stranded and incompletely flattened chemical (or more precisely, chemical potential) gradients provide an opportunity to infer maximum metamorphic duration. This obstacle/opportunity duality often referred to as 'geospeedometry' (c.f. Lasaga 1983) is discussed in detail by Kohn and Penniston-Dorland (2017, this volume) and reviewed briefly here because garnet has proven to be a particularly useful phase for geospeedometric study of tectono-metamorphism (e.g., Chakraborty and Ganguly 1991; Perchuk et al. 1999; Dachs and Proyer

2002; Fernando et al. 2003; Faryad and Chakraborty 2005; Storm and Spear 2005; Ague and Baxter 2007; Galli et al. 2011; Hallett and Spear 2011; Viete et al. 2011; Spear 2014).

Fick's first and second laws relate flux, J, and chemical concentration, C, of component i to distance, x, and time, t:

$$J_i = -D_i \frac{\partial C_i}{\partial x}$$

$$\frac{\partial C_i}{\partial t} = D_i \frac{\partial^2 C_i}{\partial x^2}$$

where D_i is the diffusivity. Numerous experimental and modeling studies have ascertained Arrhenius relationships between D and T for various major (e.g., Elphick et al. 1981; Loomis et al. 1985; Chakraborty and Ganguly 1991; Ganguly et al. 1998a; Carlson 2006; Ganguly 2010; Vielzeuf and Saúl 2011; Chu and Ague 2015) and trace elements (e.g., Tirone et al. 2005; Carlson 2012; Bloch et al. 2015) in garnet. The primary fitting parameters in most cases are a pre-exponential constant and an activation energy term, although many formulations consider activation volume and oxygen-fugacity terms, and some (e.g., Carlson 2006; Chu and Ague 2015) explicitly account for changes in diffusion coefficient across the range of possible garnet compositional space (e.g., stating that Fe would diffuse at a different rate through a grossular-dominated crystal than through an almandine-dominated crystal).

Geospeedometry (Lasaga 1983) typically uses Ficks' second law cast into an appropriate geometry (e.g., Crank 1975) to model how an imposed initial chemical zoning profile would relax with time, with the aim of finding the duration that best-fits an analyzed profile (e.g., in a microprobe traverse through a garnet crystal). For diffusion of trace elements, D is essentially independent of chemical composition (though see Bloch et al. 2015 who posit a concentration dependence on REE diffusivity) and diffuses only in response to its own chemical potential gradients, so that calculation is straightforward with analytical (e.g., Crank 1975) or numerical techniques. Major element diffusion through the garnet lattice is less straightforward, with mass and charge balance constraints requiring neutrality at each point throughout the crystal and mandating coupling of diffusion. In essence one can think of this as follows: Fe cannot freely flow left through a 1-D model of a garnet crystal unless one or several other components are flowing right to compensate and maintain charge and volume. The four main divalent cations in garnet are thus often considered simultaneously, by generation of a 3×3 element diffusion matrix that couples tracer diffusivities of each component with their relative compositions. The fourth element is generally considered a dependent component (Lasaga 1979):

$$D_{ij} = D_i^* \delta_{ij} - \left[\frac{D_i^* z_i z_j C_i}{\sum_{k=1}^{n} z_k^2 C_k D_k^*} \right] \left[D_j^* - D_n^* \right]$$

where D^* is a tracer diffusivity, z is the charge of the component of interest, n is the dependent component, and δ_{ij} is the Kronecker delta. For the example of Ca as the dependent component, Ficks' second law thus expands to:

$$
\begin{bmatrix} \dfrac{\partial c_{Fe}}{\partial t} \\[2ex] \dfrac{\partial c_{Mg}}{\partial t} \\[2ex] \dfrac{\partial c_{Mn}}{\partial t} \end{bmatrix}
=
\begin{bmatrix} \dfrac{\partial D_{FeFe}}{\partial x} & \dfrac{\partial D_{FeMg}}{\partial x} & \dfrac{\partial D_{FeMn}}{\partial x} \\[2ex] \dfrac{\partial D_{MgFe}}{\partial x} & \dfrac{\partial D_{MgMg}}{\partial x} & \dfrac{\partial D_{MgMn}}{\partial x} \\[2ex] \dfrac{\partial D_{MnFe}}{\partial x} & \dfrac{\partial D_{MnMg}}{\partial x} & \dfrac{\partial D_{MnMn}}{\partial x} \end{bmatrix}
\begin{bmatrix} \dfrac{\partial c_{Fe}}{\partial x} \\[2ex] \dfrac{\partial c_{Mg}}{\partial x} \\[2ex] \dfrac{\partial c_{Mn}}{\partial x} \end{bmatrix}
+
\begin{bmatrix} D_{FeFe} & D_{FeMg} & D_{FeMn} \\[1ex] D_{MgFe} & D_{MgMg} & D_{MgMn} \\[1ex] D_{MnFe} & D_{MnMg} & D_{MnMn} \end{bmatrix}
\begin{bmatrix} \dfrac{\partial^2 c_{Fe}}{\partial x^2} \\[2ex] \dfrac{\partial^2 c_{Mg}}{\partial x^2} \\[2ex] \dfrac{\partial^2 c_{Mn}}{\partial x^2} \end{bmatrix}
$$

which can be solved simply with a finite difference approach, as described in more detail in Spear's MSA monograph (1993) and in Ganguly's EMU Notes in Mineralogy (2002) and Reviews in Mineralogy and Geochemistry (2010) chapters.

Although the numerical solution is straightforward, constraining the initial, pre-diffusion profile is often difficult. Typically, initial profiles are either drawn subjectively, based upon the remnant preserved zoning and a petrologically and geometrically reasonable estimate of what the 'undiffused' profile must have looked like (e.g., Dachs and Proyer 2002; Ague and Baxter 2007), or are assumed to be insignificantly zoned at peak temperature and develop zoning on rims due to changing equilibrium conditions during exhumation and cooling (e.g., O'Brien and Vrána 1995; Ganguly et al. 2000; Storm and Spear 2005). Kohn and Penniston-Dorland (2017, this volume) review such retrograde resetting in detail. A final constraint on initial profile comes from incorporation of information from equilibrium thermodynamic calculations to generate model profiles that would form during growth along a given P–T path (e.g., Florence and Spear 1991, 1993; Gaidies et al. 2008b; Caddick et al. 2010; Galli et al. 2011).

In all cases, the timescales retrieved by diffusion speedometry are only as good as (i) the initial profile modeled, (ii) the diffusivity data applied, (iii) the appropriate boundary conditions being set on the model, (iv) knowledge of the P–T and f_{O_2} conditions that the sample experienced. Garnet-based speedometry is probably most useful for defining maximum timescales of metamorphic heating that are consistent with retention of observed chemical zoning, thus yielding minimum permissible rates. Most diffusion occurs at or near to T_{max} and a characteristic T of diffusion can be quantified for a complex P–T history accordingly (e.g., Chakraborty and Ganguly 1991). This means that speedometry is generally most sensitive to the duration spent near peak temperature, with more subtle information required to infer lower temperature metamorphic rates. Methodologies are also generally incapable of distinguishing single long 'events' from multiple, briefer pulses, typically only revealing the total integrated $D(T) \cdot t$ (e.g., Ague and Baxter 2007; Kohn and Penniston-Dorland 2017, this volume). Thus without additional textural or compositional (thermodynamic) constraints there can be significant non-uniqueness in results, with strong, negative correlation between apparent peak temperature and apparent metamorphic duration. An example of this (from Galli et al. 2011) is described in the following section, which draws together aspects of both the 'petro' and 'chrono' of garnet into true petrochronology.

EXAMPLES OF PETRO-CHRONOLOGY OF GARNET

Figure 2 of this chapter provided a list of techniques to extract the 'petro-' of garnet and a list of techniques to extract the 'chrono' of garnet. The first two sections of the chapter have detailed those methodologies. Now, the stage is set for their integration into true garnet petrochronology. Below, we have chosen five examples from the literature of the past decade that show the great potential of garnet petrochronology. Other good examples exist (most of them have already been cited elsewhere in this chapter) and many more are in progress. Our hope is that, through these examples and our template in Figure 2, the reader will appreciate how to conceive of a garnet petrologic application, appreciate the unique and complementary value of the information extracted therefrom, and independently decide that it is worthy of the time and effort required.

Petrochronology of garnet: High pressure crustal metamorphism

Cheng et al. (2013; 2016) and Cheng and Cao (2015) present an impressive garnet petrochronologic dataset from eclogite samples of the Huwan shear zone in the northwestern boundary of the Hong-an orogen (Western Dabie). This suite of papers represents an excellent example of the power of garnet petrochronology via: 1) combined Lu–Hf and Sm–Nd geochronology, 2) zoned garnet geochronology, 3) P–T constraints on garnet growth via thermodynamic modeling, and 4) comparison to zircon ages from the same sample suites. Cheng et al. (2013) presented combined Lu–Hf and Sm–Nd geochronology from the same handpicked eclogite garnet aliquots. Multi-point isochrons (Fig. 17) show remarkable consistency yielding 260.4 ± 2.0 Ma ($n = 10$; MSWD = 1.4) for Sm–Nd and 260.0 ± 1.0 ($n = 9$; MSWD = 1.0) for Lu–Hf. The tight agreement of each isochron data point, and between Sm–Nd and Lu–Hf ages are indicative of successful geochronology. The agreement between Lu–Hf and Sm–Nd ages itself is noteworthy, contradicting the oft-cited generalization that Lu–Hf ages are older than Sm–Nd ages. Here, according to the authors, handpicking efforts appear to have preferentially removed the Lu-rich garnet cores (that would normally skew a bulk crystal Lu–Hf age toward the early stages of garnet growth); thus both Lu–Hf and Sm–Nd ages date the same rim portion of the garnet. Eclogite facies mineral inclusions and *P–T* estimates from garnet rim chemistry pin the 260 Ma age to peak eclogite facies conditions. However the story, and the innovative garnet petrochronology, does not end there. The 260 Ma age of eclogite facies metamorphism is not reflected in zircon age data from the same area, which instead cluster mostly around 310 Ma. Thus, the garnet records an entire eclogite facies event that is apparently missed by the zircons. Subsequent papers by Cheng and Cao (2015) and Cheng et al. (2016) presented zoned garnet chronology on other eclogites from the field area to test whether garnet cores might in fact record an age more consistent with the zircon data. In the 2015 paper, garnet cores yielded Lu–Hf ages of 296.7 ± 3.8 Ma and rims gave both Lu–Hf and Sm–Nd ages of ~255 Ma. The Lu–Hf core age likely represents a mix between ~310 Ma eclogitic garnet and ~255 Ma eclogitic garnet, reflecting 40 million years of prolonged subduction zone metamorphic conditions between 1.9 and 2.4 GPa at 500 °C to 575 °C, as constrained by intersecting core and rim isopleths in pseudosections. Finally, the 2016 paper revealed an even broader scatter of Lu–Hf data from multiple microsawed zones in a large garnet megacryst spanning 400 to 250 Ma; these ages matched age spectra of newly acquired zircon data. The data presented in this series of papers shows a range of ages because each garnet bearing sample, and each zone of individual garnet crystals, reflects and records a different stage in the complex evolution of this terrane. Only through direct garnet petrochronology were these authors able to recognize at least two distinct (ca. 400 Ma and ~310–255 Ma), and prolonged, episodes of subduction zone metamorphism, thus motivating further study on convergent margin geodynamics.

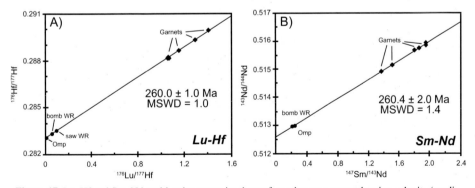

Figure 17. Lu–Hf and Sm–Nd multi-point garnet isochrons from the same garnet-bearing eclogite (modified from Cheng et al 2013). Note the perfect agreement of the two ages. When Lu–Hf and Sm–Nd are used to date the exact same age generation of garnet (in this case the rim) the two chronometers agree. Later work by Cheng et al (2016) found that garnet cores from other eclogites in this field area indeed contain an older generation of garnet. All garnet growth generations were linked to *P–T* conditions via thermodynamic modeling of garnet chemical isopleths, revealing multiple subduction episodes. "Bomb-WR" indicates sample in which the whole rock was fully dissolved in a Parr bomb.

Petrochronology of garnet: Geospeedometry and the timescales of granulite facies metamorphism

A primary motivation for deciphering metamorphic *P–T* paths is that knowledge of *P–T* evolution can help to reveal processes that control the evolution of plate margins and influence heating and cooling of Earth's crust. Garnet petrochronology helped to address this in the case of granulite facies rocks from the Gruf Complex, eastern Central Alps. Assemblages in these rocks have long been thought to record substantially higher temperatures than in many surrounding lithologies (e.g., Droop and Bucher-Nurminen 1984) and they do not correlate easily with simple tectonic models of Alpine evolution. In a study of Gruf granulites and charnockites, Galli et al. (2011) constrained reaction sequences through careful textural analysis before estimating *P–T* conditions by (i) combining experimental constraints to determine the position of key reactions, (ii) mineral thermobarometry, (iii) construction of appropriate pseudosections for peak metamorphic conditions, and (iv) multiphase mineral thermobarometry to constrain the equilibration *P–T* of late-stage coronae and symplectites. Results suggest peak temperatures in excess of 900 °C at ~9 kbar, with subsequent symplectite formation at closer to 720–740 °C and ~7 kbar (Fig. 18a). Intriguingly, garnet crystals retain remnant 'prograde' chemical zoning (dashed curves in Fig. 18d), raising the question of how such high temperatures were achieved, how briefly they must have been maintained for, and what tectonic mechanisms were responsible. Zircons from Gruf charnockites contain Permian (~290–260 Ma) cores overgrown with Tertiary (~34–29 Ma) rims, with orthopyroxene inclusions implying that core ages record or post-date granulite facies metamorphism (Galli et al. 2012). Galli et al. (2011) thus described a two stage model, with ~900 °C Permian metamorphism overprinted by symplectite formation during Alpine orogenesis (Fig. 18b-c). A coupled garnet growth and diffusion model was constructed for an assumed *P–T* path that experiences both events, using an appropriate bulk-rock composition and thermodynamic constraints to establish growth zoning during prograde metamorphism. Results were inconsistent with observed garnet zoning unless the initial event was rapid, experiencing ultra-high temperature conditions for less than ~1 Myrs (Fig. 18d). Typical Alpine metamorphic timescales of several tens of millions of years result in complete loss of prograde zoning if peak *T* is ~900 °C, though a rapid early 900 °C event overprinted by tens of millions of years of Alpine metamorphism reaching more typical peak temperatures for the region (~720 °C) yielded relatively good fits to measured garnet compositions. In this petrochronology study, Galli et al. (2011) thus concluded that granulite facies temperatures were only sustained for a very brief period and were probably unrelated to Alpine collisional tectonics. Rocks then resided in the mid crust for ~250 Myrs, before experiencing exhumation and a second phase of metamorphism at conditions and rates compatible with interpretations from elsewhere in the region, possibly due to mechanical interaction with melts that now form an adjacent pluton (Galli et al. 2013).

Petrochronology of garnet: Timescales of lower crustal melting

Stowell et al. (2010, 2014) provide excellent examples of the use of petrochronology in understanding the timescales and conditions of high-temperature metamorphism, partial melting, and the generation of high Sr/Y magmas in the lower arc crust of Fiordland, NZ. Stowell et al. (2010) combined Sm–Nd garnet geochronology on drilled cores and rims, trace element zoning in garnet, phase equilibria modeling and U–Pb zircon geochronology to provide an integrated history of a short duration, high-pressure melting event recorded by the Pembroke Granulite. Garnet core and rim ages were indistinguishable within each sample (e.g., core age of 123.5 ± 2.1 Ma; rim age of 122.5 ± 2.1 Ma for a dioritic gneiss), attesting to the short duration of garnet growth. The duration of garnet growth and subsequent partial melting event

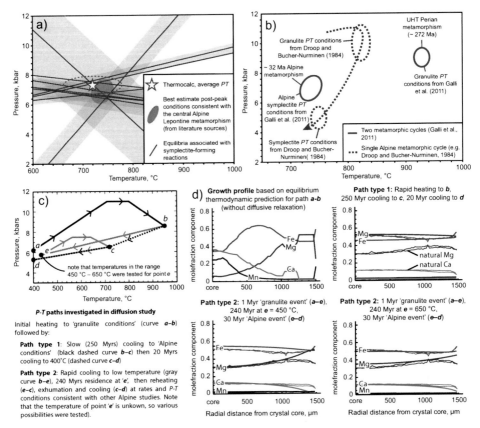

Figure 18. Evolution of the Gruf complex modified from Galli et al (2011). A. Calculated locations of observed reactions in coronae and symplectites. B. *P–T* estimate for corona formation in the context of the peak *P–T* experienced, with age constraints where available. C. *P–T* path types for which model results are shown in subsequent panels. E. Garnet growth and diffusion model results showing: core-to-rim zoning in the absence of intra-crystalline diffusion in garnet; 'type 1' path experiencing slow cooling from peak temperature; 'type 2' path experiencing rapid UHT conditions followed by a long duration at 450 °C, then a 30 Ma 'Alpine metamorphic event'; 'type 2 path' but with intermediate residence at 650 °C. In each case, the solid curves are model results and the dashed curves show natural crystal zoning. See Galli et al (2011) for more details.

is constrained by comparing bulk garnet and zircon ages from a gabbroic gneiss showing no evidence of melting (garnet age of 126.1 ± 2.0 Ma) to bulk garnet and zircon ages from a dioritic gneiss that shows melting evidence (a garnet reaction zone and leucosome; peritectic garnet age of 122.6 ± 2.0 Ma). Well-preserved HREE zoning in the dioritic gneiss was attributed to the continued consumption of accessory phases during garnet growth. Phase equilibria modeling of the gabbroic gneiss indicates peak metamorphic conditions at 1.1–1.4 GPa and 680–815 °C, with garnet growth occurring during a >0.5 GPa increase. Intrusion of the Western Fiordland Orthogneiss (125–115 Ma) is synchronous with the granulite-facies event, indicating that voluminous arc magmatism may have triggered garnet growth and partial melting.

The combined geochronologic and thermodynamic approach allowed for the interpretation that thickening of arc crust resulted in high-pressure metamorphism and that granulite-facies garnet growth lasted 3–7 Ma, with subsequent partial melting occurring shortly thereafter, perhaps as little as 3 Ma later.

Stowell et al. (2014) dated an additional nine granulite-facies rocks from around the Malaspina pluton using Sm–Nd garnet geochronology, and obtained ages ranging from 115.6 ± 2.6 Ma to 110.6 ± 2.0 Ma, roughly 10 Ma later than that from the Pembroke Granulite (at conditions of ~920°C and 1.4–1.5 GPa), suggesting that high-pressure metamorphism and partial melting was diachronous over a > 3000 km² area of the mid- to lower crust. Stowell et al. (2014) suggest this may result from pulsed underplating of magma in the lower crust or 'drip-style' delamination of the lowermost crust. High-precision garnet geochronology allowed resolution of the diachronous nature of high-pressure metamorphism and partial melting, and when combined with constraints on the *P–T* conditions (and *P–T* path) during garnet growth, allows for interpretation of the regional tectonic framework upon loading, partial melting, and extensional collapse of the Fiordland arc crust.

Petrochronology of garnet: Subduction zone dehydration

Garnet growth typically results from the breakdown of hydrous phases during subduction zone metamorphism, and can thus be a powerful proxy for subduction zone devolatilization (Baxter and Caddick 2013). Dragovic et al. (2012, 2015) utilized zoned Sm–Nd garnet geochronology coupled with phase equilibria modeling in order to constrain the duration and rate of devolatilization during subduction of mafic and felsic lithologies, respectively, from the Cycladic Blueschist Unit of Sifnos, Greece. These papers combined some of the techniques described in previous sections (i.e., zoned geochronology and pseudosections), but also highlighted and resolved some of the challenges associated with Sm–Nd isotopic analyses of very small sample sizes. Dragovic et al. (2012) microdrilled three chemically distinct, concentric zones from each of two separate garnet crystals and dated them using Sm–Nd. All six garnet fractions could be placed onto a single isochron (see Fig. 11), indicating to a first order that garnet growth was very brief. When separated into three isochrons, the duration of garnet growth was calculated to be *0.04 Ma*, with a 2σ *maximum* duration of ~*1 Ma*. Dragovic et al. (2015) took this *much* further, microdrilling and dating ten growth zones from each of two garnet crystal which revealed three distinct pulses of crystal growth, with the final pulse growing a majority of the garnet and occurring in < 0.8 Ma (Fig. 19a).

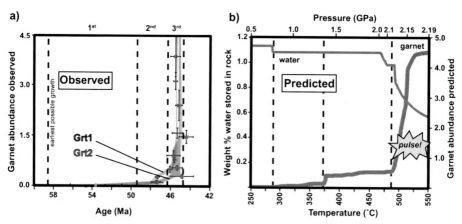

Figure 19. Devolatilization and garnet growth history recorded in zoned garnet porphyroblasts from Sifnos, Greece (modified from Dragovic et al. 2015). A. Volume abundance of garnet over time showing pulses of garnet growth in two garnet crystals from nineteen microdrilled growth zones. The final pulse lasted < 0.8 Ma. B. Associated fluid release as modeled using a path dependent thermodynamic forward model. The aforementioned final pulse released > 0.5 wt.% fluid from the bulk rock.

In Dragovic et al. (2012), the *P–T* conditions for the interval of garnet growth were calculated with separate pseudosections for whole rock (for initiation of growth) and 'garnet depleted rock' (for the termination of growth) bulk compositions. Results imply initiation of crystal core growth at ~2.0 GPa and ~460 °C and termination of rim growth ~2.2 GPa and ~560 °C. In Dragovic et al. (2015), a path dependent thermodynamic forward model was employed in which phase equilibria were calculated at 0.5 °C intervals along a prescribed *P–T* path, with stable garnet and fluid fractionated from the effective bulk composition at every interval. In both contributions, the change in bulk rock water content during the < 1 Ma garnet growth interval was calculated at ~0.3–0.5 wt%. The path dependent thermodynamic model was especially useful here, simultaneously 1) discerning the likely subduction *P–T* path, 2) modeling observed growth pulses of garnet, and 3) predicting the evolution of fluid release (Fig. 19b).

Both studies thus suggested a pulse of fluid release and heating (~100 °C/Ma) spanning just hundreds of thousands of years. The high spatial resolution attainable by zoned garnet geochronology, the ability to measure smaller samples sizes (thinner growth zones), and the coupling with thermodynamic constraints gave the ability to distinguish whether garnet growth, and thus metamorphic devolatilization, was temporally continuous or focused into pulses of mineral growth and fluid production.

Petrochronology of garnet: Collisional tectonics and inverted metamorphic gradients

An improved understanding of the evolution of the well-known Himalayan inverted metamorphic gradient has been a goal of tectonic studies for several decades. The Lesser Himalayan sequence in Sikkim exposes an unusually complete section through the inverted Barrovian sequence, and a remarkable group of studies have repeatedly analyzed a suite of samples with various techniques to build a detailed picture of the evolution of these rocks. For example, sample 24/99 is an aluminous garnet schist from the Lesser Himalaya, first described by Dasgupta et al. (2009) and representing the low temperature base of the inverted metamorphic sequence. Dasgupta et al. (2009) subjected 24/99 to element mapping and quantitative microprobe analysis, using garnet–biotite thermometry and two methods of multiphase thermobarometry to infer that it reached peak conditions of ~525 °C, 5 kbar. Pseudosection constraints suggested a somewhat higher peak temperature but confirmed texturally-derived interpretations that garnet grew at the expense of chlorite and chloritoid.

Garnets from all of the Barrovian zones were dated with Lu–Hf at very high precision in order to constrain the temporal evolution of the sequence. For example, garnet in sample 24/99 was dated using Lu–Hf at 10.6 ± 0.2 Ma, suggesting that it experienced the youngest garnet growth in the sampled transect (Anczkiewicz et al. 2014). In the same study, zoned Lu–Hf geochronology on a kyanite grade sample implied garnet core growth at 13.7 ± 0.2 Ma and rim growth at 9.9 ± 3.8 Ma (Anczkiewicz et al. 2014). Taken together, data constrain the duration of prograde metamorphism (using first garnet growth as a marker) across the Lesser Himalayan sequence, showing that ages decrease towards structurally lower levels (and lower metamorphic grade) and suggesting that the garnet isograd swept through the entire sequence from garnet-grade to sillimanite-grade in ~6 Myrs.

The size, shape and distribution of every garnet crystal in a several cubic centimeter block of sample 24/99 was then mapped in 3-D (Gaidies et al. 2015). Analysis of results suggested clustering and no spatial ordering of garnet, providing little evidence for diffusion-controlled crystallization and suggesting that garnet nucleation and growth could have been controlled by crystal interface processes (Gaidies et al. 2015). Thin sections were cut through the geometric cores of the largest garnet crystals and were carefully characterized to define inclusion mineralogy and fabrics, and to quantify major element zoning profiles. Thermodynamic models helped to refine *P–T* paths, and garnet diffusion speedometry gave constraints on both the heating and cooling rates, though the extent of diffusion was limited in 24/99 because it did not reach a particularly high maximum temperature.

This entire remarkable dataset is shown to be consistent with models of a coherent Lesser Himalayan block being buried during collision (Anczkiewicz et al. 2014; Chakraborty et al. 2016). All rocks reached their respective peak metamorphic conditions simultaneously but with those samples that eventually reach highest grade passing through garnet-in reactions first and thus recording earliest garnet growth (Anczkiewicz et al. 2014). This truly represents an integrated garnet petrochronology application.

OUTLOOK

Both the 'petro' and the 'chrono' of garnet have evolved significantly over the past several decades, and that evolution will continue as analytical methodologies improve and as the creativity of the scientific applications grows. We have touched on many aspects of that evolution here, showcasing a few examples of the current state of the art garnet petrochronology. Once, we were barely able to see the 'tree rings' of garnet growth. If the concentric growth rings in a garnet crystal can be likened to the pages of a history book (as suggested in Baxter and Scherer 2013), while we first only got to read the title, and then progressed to reading the title of each chapter, we are now able to read the record with the resolution of several pages. New petrologic techniques (thermodynamic, isotopic, textural, mechanical) are like learning new vocabulary that allows us to extract greater meaning and context from every chapter and page of that garnet chronology. What will the future hold? How soon and with what new analytical breakthroughs will we be able to read every single word of that garnet tree ring history book? Garnet petrochronology is not a panacea; its limitations and challenges must also be understood to fully harness its potential and it is quite possible that we will never be fully satisfied with the resolution of the story preserved in metamorphic rocks, or our ability to read that story. But as Lu–Hf and Sm–Nd geochronology become increasingly precise and accessible, and our understanding of the processes that embed chemical and isotopic signatures in crystals improve, garnet will surely become an ever more important member of the cadre of modern petrochronometers discussed in this RiMG volume. So although we have dwelled mostly here on previous examples and current methods, above all we encourage the reader to consider the *potential* of garnet petrochronology in deciphering the evolution of crustal processes.

ACKNOWLEDGMENTS

EFB gratefully acknowledges NSF award 1250497/1561882. MJC gratefully acknowledges NSF awards EAR–1250470 and EAR–1447568 (to MJC and BD). We thank E.E. Scherer for providing data in Figure 10 as well as numerous discussions on garnet geochronology. We thank Robert Anczkiewicz, David Pattison, and Pierre Lanari for valuable reviews, Matthew Kohn for his feedback, and Martin Engi for his patient, detailed and constructive editorial handling. We thank R.J. Tracy for providing the materials for Figure 3, for patiently emphasizing the importance of 'chemo-petrography', and for guiding our collection and interpretation of countless 'chemo-petrographic' images. We thank John Rosenfeld for providing his 'favorite rotated garnet photo' from an original print for Figure 1c.

REFERENCES

Abart R (1995) Phase-equilibrium and stable-isotope constraints on the formation of metasomatic garnet–vesuvianite veins (SW Adamello, N Italy). Contrib Mineral Petrol 122:116–133
Aerden D, Bell TH, Puga E, Sayab M, Lozano JA, de Federico AD (2013) Multi-stage mountain building vs. relative plate motions in the Betic Cordillera deduced from integrated microstructural and petrological analysis of porphyroblast inclusion trails. Tectonophysics 587:188–206

Ague JJ (2011) Extreme channelization of fluid and the problem of element mobility during Barrovian metamorphism. Am Mineral 96:333–352

Ague JJ, Axler JA (2016) Interface coupled dissolution–reprecipitation in garnet from subducted granulites and ultrahigh-pressure rocks revealed by phosphorous, sodium, and titanium zonation. Am Mineral 101:1696–1699

Ague JJ, Baxter EF (2007) Brief thermal pulses during mountain building recorded by Sr diffusion in apatite and multicomponent diffusion in garnet. Earth Planet Sci Lett 261:500–516

Albee AL (1965) Phase equilibria in three assemblages of kyanite-zone pelitic schists, Lincoln Mountain Quadrangle, central Vermont. J Petrol 6:246–301

Albee AL (1972) Metamorphism of pelitic schists: reaction relations of chloritoid and staurolite. Geol Soc Am Bull 83:3249–3268

Amato JM, Johnson CM, Baumgartner LP, Beard BL (1999) Rapid exhumation of the Zermatt–Saas ophiolite deduced from high-precision Sm–Nd and Rb–Sr geochronology. Earth Planet Sci Lett 171:425–438

Anczkiewicz R, Thirlwall MF (2003) Improving precision of Sm–Nd garnet dating by H_2SO_4 leaching: a simple solution to the phosphate inclusion problem. Geol Soc London Spec Pub 220:83–91

Anczkiewicz R, Platt JP, Thirlwall MF, Wakabayashi J (2004) Franciscan subduction off to a slow start: evidence from high-precision Lu–Hf garnet ages on high grade-blocks. Earth Planet Sci Lett 225:147–161

Anczkiewicz R, Szczepanski J, Mazur S, Storey C, Crowley Q, Villa IM, Thirlwall ME, Jeffries TE (2007) Lu–Hf geochronology and trace element distribution in garnet: Implications for uplift and exhumation of ultra-high pressure granulites in the Sudetes, SW Poland. Lithos 95:363–380

Anczkiewicz R, Thirlwall M, Alard O, Rogers NW, Clark C (2012) Diffusional homogenization of light REE in garnet from the Day Nui Con Voi Massif in N-Vietnam: Implications for Sm–Nd geochronology and timing of metamorphism in the Red River shear zone. Chem Geol 318:16–30

Anczkiewicz R, Chakraborty S, Dasgupta S, Mukhopadhyay D, Kołtonik K (2014) Timing, duration and inversion of prograde Barrovian metamorphism constrained by high resolution Lu–Hf garnet dating: A case study from the Sikkim Himalaya, NE India. Earth Planet Sci Lett 407:70–81

Anderson DE, Olimpio JC (1977) Progressive homogenization of metamorphic garnets, south Molnar, Scotland: Evidence for volume diffusion. Can Mineral 15:205–216

Angel RJ, Mazzucchelli ML, Alvaro M, Nimis P, Nestola F (2014) Geobarometry from host–inclusion systems: The role of elastic relaxation. Am Mineral 99:2146–2149

Angiboust S, Agard P, Raimbourg H, Yamato P, Huet B (2011) Subduction interface processes recorded by eclogite-facies shear zones (Monviso, W. Alps). Lithos 127:222–238

Angiboust S, Pettke T, De Hoog JCM, Caron B, Oncken O (2014) Channelized fluid flow and eclogite-facies metasomatism along the subduction shear zone. J Petrol 55:883–916

Argles TW, Prince CI, Foster GL, Vance D (1999) New garnets for old? Cautionary tales from young mountain belts. Earth Planet Sci Lett 172:301–309

Ashley KT, Caddick MJ, Steele-MacInnis M, Bodnar RJ, Dragovic B (2014a) Geothermobarometric history of subduction recorded by quartz inclusions in garnet. Geochem Geophys Geosystem 15:350–360

Ashley KT, Steele-MacInnis M, Caddick MJ (2014b) QuIB Calc: A MATLAB® script for geobarometry based on elastic modeling of quartz inclusions in garnet. Comp and Geosci 60:155–157

Ashley KT, Darling RS, Bodnar RJ, Law RD (2015) Significance of "stretched" mineral inclusions for reconstructing *P*–*T* exhumation history. Contrib Mineral Petrol 169:1–9

Atherton MP (1968) The variation of garnet, biotite and chlorite composition in medium grade pelitic rocks from the Dalradian, Scotland, with particular reference to the zonation in garnet. Contrib Mineral Petrol 18:347–371

Atherton MP, Edmunds WM (1965) An electron microprobe study of some zoned garnets from metamorphic rocks. Earth Planet Sci Lett 1:185–193

Baldwin JA, Powell R, White RW, Štípská P (2015) Using calculated chemical potential relationships to account for replacement of kyanite by symplectite in high pressure granulites. J Metamorph Geol 33:311–330

Barrow G (1893) On an intrusion of muscovite–biotite gneiss in the east Highlands of Scotland, and its accompanying metamorphism. Quart J Geol Soc London 49:330–358

Barrow G (1912) On the geology of lower Deeside and the southern Highland border. Proc Geol Assoc 23:268–284

Bast R, Scherer EE, Sprung P, Fischer-Godde M, Stracke A, Mezger K (2015) A rapid and efficient ion-exchange chromatography for Lu–Hf, Sm–Nd, and Rb–Sr geochronology and the routine isotope analysis of sub-ng amounts of Hf by MC-ICP-MS. J Anal At Spectrom 30:2323–2333

Baumgartner LP, Valley JW (2001) Stable isotope transport and contact metamorphic fluid flow. Rev Mineral Geochem 43:415–467

Baumgartner LP, Floess D, Podladchikov Y, Foster CT (2010) Pressure gradients in garnets induced by diffusion relaxation of major element zoning. Geol Soc Denver Ann Meeting

Baxter EF, DePaolo D (2002a) Field measurement of high temperature bulk reaction rates II: Interpretation of results from a field site near Simplon Pass, Switzerland. Am J Sci 302:465–516

Baxter EF, DePaolo DJ (2002b) Field measurement of high temperature bulk reaction rates I: Theory and technique. Am J Sci 302:442–464

Baxter EF, Caddick MJ (2013) Garnet growth as a proxy for progressive subduction zone dehydration. Geology 41:643–646

Baxter EF, Scherer EE (2013) Garnet Geochronology: Timekeeper of Tectonometamorphic Processes. Elements 9:433–438

Baxter EF, Ague JJ, Depaolo DJ (2002) Prograde temperature–time evolution in the Barrovian type-locality constrained by Sm/Nd garnet ages from Glen Clova, Scotland. J Geol Soc London 159:71–82

Baxter EF, Caddick MJ, Ague JJ (2013) Garnet: common mineral, uncommonly useful. Elements 9:415–419

Bebout GE, Tsujimori T, Ota T, Shimaki Y, Kunihiro T, Carlson WD, Nakamura E (2015) Lithium in UHP metasedimentary garnets from Lago Di Cignana, Italy: Coupled substitutions and the role of accessory phases. Geol Soc Am Ann Meeting

Bell TH (1985) Deformation partitioning and porphyroblast rotation in metamorphic rocks: a radical reinterpretation. J Metamorph Geol 3:109–118

Berman RG (1990) Mixing properties of Ca–Mg–Fe–Mn garnets. Am Mineral 75:328–344

Berman RG, Aranovich LY (1996) Optimized standard state and solution properties of minerals. I. Model calibration for olivine, orthopyroxene, cordierite, garnet, and ilmenite in the system FeO–MgO–CaO–Al$_2$O$_3$–TiO$_2$–SiO$_2$. Contrib Mineral Petrol 126:1–24

Bickle MJ, Baker J (1990) Advective diffusive transport of isotopic fronts—an example from Naxos, Greece. Earth Planet Sci Lett 97:78–93

Bloch E, Ganguly J (2015) ^{176}Lu–^{176}Hf geochronology of garnet II: numerical simulations of the development of garnet–whole-rock ^{176}Lu–^{176}Hf isochrons and a new method for constraining the thermal history of metamorphic rocks. Contrib Mineral Petrol 169:1–16

Bloch E, Ganguly J, Hervig R, Cheng W (2015) ^{176}Lu–^{176}Hf geochronology of garnet I: experimental determination of the diffusion kinetics of Lu^{3+} and Hf^{4+} in garnet, closure temperatures and geochronological implications. Contrib Mineral Petrol 169:1–18

Bohlen SR, Essene EJ (1980) Evaluation of coexisting garnet–biotite, garnet–clinopyroxene and other Mg–Fe exchange thermometers in Adirondack granulites. Geol Soc Am Bull 91:685–719

Bohlen SR, Liotta JJ (1986) A barometer for garnet amphibolites and garnet granulites. J Petrol 27:1025–1034

Bohlen SR, Wall VJ, Boettcher AL (1983) Experimental investigations and gelogical applications of equilibria in the system FeO–TiO$_2$–Al$_2$O$_3$–SiO$_2$–H$_2$O. Am Mineral 68:1049–1058

Brady JB, Cherniak DJ (2010) Diffusion in minerals: an overview of published experimental diffusion data. Rev Mineral Geochem 72:899–920

Burton KW, Onions RK (1991) High-resolution garnet chronometry and the rates of metamorphic processes. Earth Planet Sci Lett 107:649–671

Burton KW, Kohn MJ, Cohen AS, Onions RK (1995) The relative diffusion of Pb, Nd, Sr and O in garnet. Earth Planet Sci Lett 133:199–211

Caddick MJ, Thompson AB (2008) Quantifying the tectono-metamorphic evolution of pelitic rocks from a wide range of tectonic settings: Mineral compositions in equilibrium. Contrib Mineral Petrol 156:177–195

Caddick MJ, Bickle MJ, Harris NBW, Holland TJB, Horstwood MSA, Ahmad T (2007) Burial and exhumation history of a Lesser Himalayan schist: Recording the formation of an inverted metamorphic sequence in NW India. Earth Planet Sci Lett 264:375–390

Caddick MJ, Konopásek J, Thompson AB (2010) Preservation of garnet growth zoning and the duration of prograde metamorphism. J Petrol 51:2327–2347

Cahalan RC, Kelly ED, Carlson WD (2014) Rates of Li diffusion in garnet: Coupled transport of Li and Y + REEs. Am Mineral 99:1676–1682

Carlson WD (1989) The significance of intergranular diffusion to the mechanisms and kinetics of porphyroblast crystallization. Contrib Mineral Petrol 103:1–24

Carlson WD (1999) The case against Ostwald ripening of porphyroblasts. Can Mineral 37:403–414

Carlson WD (2002) Scales of disequilibrium and rates of equilibration during metamorphism. Am Mineral 87:185–204

Carlson WD (2006) Rates of Fe, Mg, Mn and Ca diffusion in garnet. Am Mineral 91:1–11

Carlson WD (2011) Porphyroblast crystallization: linking processes, kinetics, and microstructures. Int Geol Rev 53:406–445

Carlson WD (2012) Rates and mechanism of Y REE, and Cr diffusion in garnet. Am Mineral 97:1598–1618

Carlson WD, Denison C, Ketcham RA (1995) Controls on the nucleation and growth of porphyroblasts: Kinetics from natural textures and numerical models. Geol J 30:207–225

Carlson WD, Gale JD, Wright K (2014) Incorporation of Y and REEs in aluminosilicate garnet: Energetics from atomistic simulation. Am Mineral 99:1022–1034

Carlson WD, Hixon JD, Garber JM, Bodnar RJ (2015a) Controls on metamorphic equilibration: the importance of intergranular solubilities mediated by fluid composition. J Metamorph Geol 33:123–146

Carlson WD, Pattison DR, Caddick MJ (2015b) Beyond the equilibrium paradigm: How consideration of kinetics enhances metamorphic interpretation. Am Mineral 100:1659–1667

Cashman KV, Ferry JM (1988) Crystal size distribution (CSD) in rocks and the kinetics and dynamics of crystallization. 3. Metamorphic crystallization. Contrib Mineral Petrol 99:401–415

Castro AE, Spear FS (2016) Reaction overstepping and re-evaluation of peak *P–T* conditions of the blueschist unit Sifnos, Greece: implications for the Cyclades subduction zone. Int Geol Rev:1–15

Chacko T, Cole DR, Horita J (2001) Equilibrium oxygen, hydrogen and carbon isotope fractionation factors applicable to geologic systems. Rev Mineral Geochem 43:1–81

Chakraborty S, Ganguly J (1991) Compositional zoning and cation diffusion in garnets. *In:* Diffusion, Atomic Ordering, and Mass Transport. Ganguly J (ed) Springer-Verlag, New York, p 120–175

Chakraborty S, Anczkiewicz R, Gaidies F, Rubatto D, Sorcar N, Faak K, Mukhopadhyay D, Dasgupta S (2016) A review of thermal history and timescales of tectonometamorphic processes in Sikkim Himalaya (NE India) and implications for rates of metamorphic processes. J Metamorph Geol 34:785–803

Chambers J, Caddick M, Argles T, Horstwood M, Sherlock S, Harris N, Parrish R, Ahmad T (2009) Empirical constraints on extrusion mechanisms from the upper margin of an exhumed high-grade orogenic core, Sutlej valley, NW India. Tectonophysics 477:77–92

Cheng H, Cao D (2015) Protracted garnet growth in high-Peclogite: constraints from multiple geochronology and *P–T* pseudosection. J Metamorph Geol 33:613–632

Cheng H, King RL, Nakamura E, Vervoort JD, Zhou Z (2008) Coupled Lu–Hf and Sm–Nd geochronology constrains garnet growth in ultra-high-pressure eclogites from the Dabie orogen. J Metamorph Geol 26:741–758

Cheng H, Zhang C, Vervoort JD, Zhou Z (2013) New Lu–Hf and Sm–Nd geochronology constrains the subduction of oceanic crust during the Carboniferous–Permian in the Dabie orogen. J Asian Earth Sci 63:139–150

Chen YX, Zheng YF, Li L, Chen RX (2014) Fluid–rock interaction and geochemical transport during protolith emplacement and continental collision: A tale from Qinglongshan ultrahigh-pressure metamorphic rocks in the Sulu orogen. Am J Sci 314:357–399

Cheng H, Liu XC, Vervoort JD, Wilford D, Cao DD (2016) Micro-sampling Lu–Hf geochronology reveals episodic garnet growth and multiple high-*P* metamorphic events. J Metamorph Geol 34:363–377

Chernoff CB, Carlson WD (1997) Disequilibrium for Ca during growth of pelitic garnet. J Metamorph Geol 15:421–438

Chernoff CB, Carlson WD (1999) Trace element zoning as a record of chemical disequilibrium during garnet growth. Geology 27:555–558

Chopin C (1984) Coesite and pure pyrope in high-grade blueschists of the Western Alps: a first record and some consequences. Contrib Mineral Petrol 86:107–118

Christensen JN, Rosenfeld JL, Depaolo DJ (1989) Rates of tectonometamorphic processes from rubidium and strontium isotopes in garnet. Science 244:1465–1469

Christensen JN, Selverstone J, Rosenfeld JL, DePaolo DJ (1994) Correlation by Rb–Sr geochronology of garnet growth histories from different structural levels within the Tauern Window, Eastern Alps. Contrib Mineral Petrol 118:1–12

Chu X, Ague JJ (2015) Analysis of experimental data on divalent cation diffusion kinetics in aluminosilicate garnets with application to timescales of peak Barrovian metamorphism, Scotland. Contrib Mineral Petrol 170

Clayton RN, Goldsmith JR, Mayeda TK (1989) Oxygen isotope fractionation in quartz, albite, anorthite and calcite. Geochim Cosmochim Acta 53:725–733

Clechenko CC, Valley JW (2003) Oscillatory zoning in garnet from the Willsboro Wollastonite Skarn, Adirondack Mts, New York: a record of shallow hydrothermal processes preserved in a granulite facies terrane. J Metamorph Geol 21:771–784

Coghlan RAN (1990) Studies in diffusional transport: grain boundary transport of oxygen in feldspars, diffusion of oxygen, REE's in garnet. Doctoral dissertation, Brown Univ.

Cohen AS, O'Nions RK, Siegenthaler R, Griffin WL (1988) Chronology of the pressure–temperature history recorded by a granulite terrain. Contrib Mineral Petrol 98:303–311

Connolly JAD (1990) Multivariable phase diagrams: an algorithm based on generalized thermodynamics. Am J Sci 290:666–718

Connolly JAD (2005) Computation of phase equilibria by linear programming: A tool for geodynamic modeling and its application to subduction zone decarbonation. Earth Planet Sci Lett 236:524–541

Connolly JAD (2009) The geodynamic equation of state: What and how. Geochem Geophys Geosystem 10

Corrie SL, Kohn MJ (2008) Trace-element distributions in silicates during prograde metamorphic reactions: implications for monazite formation. J Metamorph Geol 26:451–464

Crank J (1975) The Mathematics of Diffusion, 2nd Edition. Clarendon Press, Oxford

Crowe DE, Riciputi LR, Bezenek S, Ignatiev A (2001) Oxygen isotope and trace element zoning in hydrothermal garnets: Windows into large-scale fluid-flow behavior. Geology 29:479–482

Cruz-Uribe AM, Hoisch TD, Wells ML, Vervoort JD, Mazdab FK (2015) Linking thermodynamic modelling, Lu–Hf geochronology and trace elements in garnet: new *P–T–t* paths from the Sevier hinterland. J Metamorph Geol 33:763–781

Cutts K, Kinny P, Strachan R, Hand M, Kelsey D, Emery M, Friend C, Leslie A (2010) Three metamorphic events recorded in a single garnet: Integrated phase modelling, in situ LA-ICPMS and SIMS geochronology from the Moine Supergroup, NW Scotland. J Metamorph Geol 28:249–267

D'Errico ME, Lackey JS, Surpless BE, Loewy SL, Wooden JL, Barnes JD, Strickland A, Valley JW (2012) A detailed record of shallow hydrothermal fluid flow in the Sierra Nevada magmatic arc from low-^{18}O skarn garnets. Geology 40:763–766

Dachs E, Proyer A (2002) Constraints on the duration of high-pressure metamorphism in the Tauern Window from diffusion modelling of discontinuous growth zones in eclogite garnet. J Metamorph Geol 20:769–780

Daniel CG, Spear FS (1998) Three-dimensional patterns of garnet nucleation and growth. Geology 26:503–506

Dasgupta S, Chakraborty S, Neogi S (2009) Petrology of an inverted Barrovian sequence of metapelites in Sikkim Himalaya, India: Constraints on the tectonics of inversion. Am J Sci 309:43–84

de Capitani C, Brown TH (1987) The computation of chemical equilibrium in complex systems containing non-ideal solutions. Geochim Cosmochim Acta 51:2639–2652

de Capitani C, Petrakakis K (2010) The computation of equilibrium assemblage diagrams with Theriak/Domino software. Am Mineral 95:1006–1016

DePaolo DJ, Getty SR (1996) Models of isotopic exchange in reactive fluid–rock systems: Implications for geochronology in metamorphic rock. Geochim Cosmochim Acta 60:3933–3947

DeWolf CP, Zeissler CJ, Halliday AN, Mezger K, Essene EJ (1996) The role of inclusions in U–Pb and Sm–Nd garnet geochronology: Stepwise dissolution experiments and trace uranium mapping by fission track analysis. Geochim Cosmochim Acta 60:121–134

Diener J, Powell R (2012) Revised activity–composition models for clinopyroxene and amphibole. J Metamorph Geol 30:131–142

Dodson MH (1973) Closure temperature in cooling geochronological and petrological systems. Contrib Mineral Petrol 40:259–274

Dorfler KM, Tracy RJ, Caddick MJ (2014) Late-stage orogenic loading revealed by contact metamorphism in the northern Appalachians, New York. J Metamorph Geol 32:113–132

Dragovic B, Samanta LM, Baxter EF, Selverstone J (2012) Using garnet to constrain the duration and rate of water-releasing metamorphic reactions during subduction: An example from Sifnos, Greece. Chem Geol 314–317:9–22

Dragovic B, Baxter EF, Caddick MJ (2015) Pulsed dehydration and garnet growth during subduction revealed by zoned garnet geochronology and thermodynamic modeling, Sifnos, Greece. Earth Planet Sci Lett 413:111–122

Dragovic B, Guevara VE, Caddick MJ, Baxter EF, Kylander-Clark AR (2016) A pulse of cryptic granulite-facies metamorphism in the Archean Wyoming Craton revealed by Sm–Nd garnet and U–Pb monazite geochronology. Precambrian Res 283:24–49

Droop GTR, Bucher-Nurminen K (1984) Reaction textures and metamorphic evolution of sapphirine-bearing granulites from the Gruf Complex, Italian Central Alps. J Petrol 25:766–803l

Ducea MN, Ganguly J, Rosenberg EJ, Patchett PJ, Cheng WJ, Isachsen C (2003) Sm–Nd dating of spatially controlled domains of garnet single crystals: a new method of high-temperature thermochronology. Earth Planet Sci Lett 213:31–42

Duchene S, BlichertToft J, Luais B, Telouk P, Lardeaux JM, Albarede F (1997) The Lu–Hf dating of garnets and the ages of the Alpine high-pressure metamorphism. Nature 387:586–589

Eiler JM (2001) Oxygen isotope variations of basaltic lavas and upper mantle rocks. Rev Mineral Geochem 43:319–364

Elphick SC, Ganguly J, Loomis TP (1981) Experimental study of Fe–Mg interdiffusion in aluminosilicate garnet. Trans Am Geophys Union 62:411

Enami M, Nishiyama T, Mouri T (2007) Laser Raman microspectrometry of metamorphic quartz: A simple method for comparison of metamorphic pressures. Am Mineral 92:1303–1315

Engi M (2017) Petrochronology based on REE–minerals: monazite, allanite, xenotime, apatite. Rev Mineral Geochem 83:365–418

Engi M, Wersin P (1987) Derivation and application of a solution model for calcic garnet. Schweiz Mineral Petrograph Mitteil 67:53–73

Errico JC, Barnes JD, Strickland A, Valley JW (2013) Oxygen isotope zoning in garnets from Franciscan eclogite blocks: evidence for rock-buffered fluid interaction in the mantle wedge. Contrib Mineral Petrol 166:1161–1176

Essene EJ (1982) Geologic thermometry and barometry. Rev Mineral Geochem 10:153–206

Essene EJ (1989) The current status of thermobarometry in metamorphic rocks. *In:* Evolution of Metamorphic Belts. Daly JS, Cliff, RA and Yardley BWD (eds) Geological Society of London, p 1–44

Evans BW (1966) Microprobe study of zoning in eclogite garnet. Geol Soc Am Spec Pap 87:54

Evans BW, Guidotti CV (1966) The sillimanite–potash feldspar isograd in Western Maine, USA. Contrib Mineral Petrol 12:25–62

Faryad SW, Chakraborty S (2005) Duration of Eo-Alpine metamorphic events obtained from multicomponent diffusion modeling of garnet: a case study from the Eastern Alps. Contrib Mineral Petrol 150:306–318

Fernando G, Hauzenberger CA, Baumgartner LP, Hofmeister W (2003) Modeling of retrograde diffusion zoning in garnet: evidence for slow cooling of granulites from the Highland Complex of Sri Lanka. Mineral Petrol 78:53–71

Ferry JM, Spear FS (1978) Experimental calibration of the partitioning of Fe and Mg between biotite and garnet. Contrib Mineral Petrol 66:113–117

Ferry JM, Kitajima K, Strickland A, Valley JW (2014) Ion microprobe survey of the grain-scale oxygen isotope geochemistry of minerals in metamorphic rocks. Geochim Cosmochim Acta 144:403–433

Florence FP, Spear FS (1991) Effects of diffusional modification of garnet growth zoning on $P–T$ path calculations. Contrib Mineral Petrol 107:487–500

Florence FP, Spear FS (1993) Influences of reaction history and chemical diffusion on $P–T$ calculations for staurolite schists from the Littleton Formation, northwestern New Hampshire. Am Mineral 78:345–359

Foster GL, Parrish RR, Horstwood MSA, Chenery S, Pyle JM, Gibson HD (2004) The generation of prograde $P–T–t$ points and paths; a textural, compositional, and chronological study of metamorphic monazite. Earth Planet Sci Lett 228:125–142

Frost MJ (1962) Metamorphic grade and iron–magnesium distribution between coexisting garnet–biotite and garnet–hornblende. Mineral Mag 99:427–438

Gaidies F, Abart R, De Capitani C, Schuster R, Connolly JAD, Reusser E (2006) Characterization of polymetamorphism in the Austroalpine basement east of the Tauern Window using garnet isopleth thermobarometry. J Metamorph Geol 24:451–475

Gaidies F, De Capitani C, Abart R (2008a) THERIA_G: a software program to numerically model prograde garnet growth. Contrib Mineral Petrol 155:657–671

Gaidies F, De Capitani C, Abart R, Schuster R (2008b) Prograde garnet growth along complex $P–T–t$ paths: results from numerical experiments on polyphase garnet from the Wölz Complex (Austroalpine basement). Contrib Mineral Petrol 155:673–688

Gaidies F, Pattison DRM, de Capitani C (2011) Toward a quantitative model of metamorphic nucleation and growth. Contrib Mineral Petrol 162:975–993

Gaidies F, Petley-Ragan A, Chakraborty S, Dasgupta S, Jones P (2015) Constraining the conditions of Barrovian metamorphism in Sikkim, India: $P–T–t$ paths of garnet crystallization in the Lesser Himalayan Belt. J Metamorph Geol 33:23–44

Galli A, Le Bayon B, Schmidt MW, Burg, J-P, Caddick MJ, Reusser E (2011) Granulites and charnockites of the Gruf Complex: evidence for Permian ultra-high temperature metamorphism in the Central Alps. Lithos 124:17–45

Galli A, Le Bayon B, Schmidt MW, Burg JP, Reusser E, Sergeev SA, Larionov A (2012) U–Pb zircon dating of the Gruf Complex: disclosing the late Variscan granulitic lower crust of Europe stranded in the Central Alps. Contrib Mineral Petrol 163:353–378

Galli A, Le Bayon B, Schmidt MW, Burg, J-P, Reusser E (2013) Tectonometamorphic history of the Gruf complex (Central Alps): exhumation of a granulite–migmatite complex with the Bergell pluton. Swiss J Geosci 106:33–62

Galwey AK, Jones KA (1966) Crystal size frequency distribution of garnets in some analysed metamorphic rocks from Mallaig Inverness Scotland. Geol Mag 103:143-and

Ganguly J (2002) Diffusion kinetics in minerals: Principles and applications to tectono- metamorphic processes. *In:* EMU Notes in Mineralogy: Energy Modelling in Minerals. Gramaccioli CMO (ed) Eur Mineral Union, p 271–309

Ganguly J (2010) Cation diffusion kinetics in aluminosilicate garnets and geological applications. Rev Mineral Geochem 72:559–601

Ganguly J, Saxena SK (1984) Mixing properties of aluminosilicate garnets: constraints from natural and experimental data, and applications to geothermo-barometry. Am Mineral 69:88–97

Ganguly J, Tirone M (1999) Diffusion closure temperature and age of a mineral with arbitrary extent of diffusion: theoretical formulation and applications. Earth Planet Sci Lett 170:131–140

Ganguly J, Tirone M (2001) Relationship between cooling rate and cooling age of a mineral: Theory and applications to meteorites. Meteorit Planet Sci 36:167–175

Ganguly J, Cheng WJ, Tirone M (1996) Thermodynamics of aluminosilicate garnet solid solution: New experimental data, an optimized model, and thermometric applications. Contrib Mineral Petrol 126:137–151

Ganguly J, Cheng W, Chakraborty S (1998a) Cation diffusion in aluminosilicate garnets: experimental determination in pyrope-almandine diffusion couples. Contrib Mineral Petrol 131:171–180

Ganguly J, Tirone M, Hervig RL (1998b) Diffusion kinetics of samarium and neodymium in garnet, and a method for determining cooling rates of rocks. Science 281:805–807

Ganguly J, Dasgupta S, Cheng W, Neogi S (2000) Exhumation history of a section of the Sikkim Himalayas, India: records in the metamorphic mineral equilibria and compositional zoning of garnet. Earth Planet Sci Lett 183:471–486

Gatewood MP, Dragovic B, Stowell HH, Baxter EF, Hirsch DM, Bloom R (2015) Evaluating chemical equilibrium in metamorphic rocks using major element and Sm–Nd isotopic age zoning in garnet, Townshend Dam, Vermont, USA. Chem Geol 401:151–168

Gauthiez-Putallaz L, Rubatto D, Hermann J (2016) Dating prograde fluid pulses during subduction by in situ U–Pb and oxygen isotope analysis. Contrib Mineral Petrol 171:1–20

Gessmann CK, Spiering B, Raith M (1997) Experimental study of the Fe–Mg exchange between garnet and biotite: Constraints on the mixing behavior and analysis of the cation-exchange mechanisms. Am Mineral 82:1225–1240

Getty SR, Selverstone J, Wernicke BP, Jacobsen SB, Aliberti E, Lux DR (1993) Sm–Nd dating of multiple garnet growth events in an arc-continent collision zone, northwestern United-States Cordillera. Contrib Mineral Petrol 115:45–57

Ghent ED (1976) Plagioclase–garnet–Al$_2$SiO$_5$–quartz: a potential geothermometer–geobarometer. Am Mineral 61:710–714

Ghent ED (1977) Applications of activity-composition relations to displacement of a solid-solid equilibrium anorthite = grossular + kyanite + quartz. *In:* Short Course in Application of Thermodynamics to Petrology and Ore Deposits. Greenwood HJ (ed) Mineral Assoc Canada, p 99–108

Ghent ED, Stout MZ (1981) Geobarometry and geothermometry of plagioclase–biotite–garnet–muscovite assemblages. Contrib Mineral Petrol 76:92–97

Gieré R, Rumble D, Gunther D, Connolly J, Caddick MJ (2011) Correlation of growth and breakdown of major and accessory minerals in metapelites from Campolungo, Central Alps. J Petrol 52:2293–2334

Goldman DS, Albee AL (1977) Correlation of Mg/Fe partitioning between garnet and biotite with O^{18}/O^{16} partitioning between quartz and magnetite. Am J Sci 277:750–761

Goldsmith JR (1980) Melting and breakdown reactions of anorthite at high pressures and temperatures. Am Mineral 65:272–284

Gordon SM, Luffi P, Hacker B, Valley J, Spicuzza M, Kozdon R, Kelemen P, Ratshbacher L, Minaev V (2012) The thermal structure of continental crust in active orogens: insight from Miocene eclogite and granulite xenoliths of the Pamir Mountains. J Metamorph Geol 30:413–434

Grew ES, Locock AJ, Mills SJ, Galuskina IO, Galuskin EV, Halenius U (2013) Nomenclature of the garnet supergroup. Am Mineral 98:785–810

Griffin WL, Brueckner HK (1980) Caledonian Sm–Nd ages and a crustal origin for Norwegian eclogites. Nature 285:319–321

Guevara VE, Caddick MJ (2016) Shooting at a moving target: phase equilibria modeling of high-temperature metamorphism. J Metamorph Geol 34:209–235

Guiraud M, Powell R (2006) *P–V–T* relationships and mineral equilibria in inclusions in minerals. Earth Planet Sci Lett 244:683–694

Guiraud M, Powell R, Rebay G (2001) H$_2$O in metamorphism and unexpected behaviour in the preservation of metamorphic mineral assemblages. J Metamorph Geol 19:445–454

Hallett BW, Spear FS (2011) Insight into the cooling history of the Valhalla complex, British Columbia. Lithos 125:809–824

Hackler RT, Wood BJ (1989) Experimental determination of Fe and Mg exchange between garnet and olivine and estimation of Fe–Mg mixing properties in garnet. Am Mineral 74:994–999

Hallett BW, Spear FS (2014a) The *P–T* history of anatectic pelites of the Northern East Humboldt Range, Nevada: Evidence for tectonic loading, decompression, and anatexis. J Petrol 55:3–36

Hallett BW, Spear FS (2015) Monazite, zircon, and garnet growth in migmatitic pelites as a record of metamorphism and partial melting in the East Humboldt Range, Nevada. Am Mineral 100:951–972

Harte B, Henley KJ (1966) Occurence of compositionally zoned almanditic garnets in regionally metamorphosed rocks. Nature 210:689–692

Harvey J, Baxter EF (2009) An improved method for TIMS high precision neodymium isotope analysis of very small aliquots (1–10ng). Chem Geol 258:251–257

Hemley RJ (1987) Pressure dependence of Raman spectra of SiO$_2$ polymorphs; α-quartz, coesite and stishovite. *In:* High-Pressure Research in Mineral Physics. Murli, HM and Syono Y (eds) Am Geophys Union, Washington DC, p 347–359

Hensen BJ (1971) Theoretical phase relations involving cordierite and garnet in the system MgO–FeO–Al$_2$O$_3$–SiO$_2$. Contrib Mineral Petrol 33:191–214

Hermann J, Rubatto D (2003) Relating zircon and monazite domains to garnet growth zones: age and duration of granulite facies metamorphism in the Val Malenco lower crust. J Metamorph Geol 21:833–852

Herwartz D, Nagel TJ, Munker C, Scherer EE, Froitzheim N (2011) Tracing two orogenic cycles in one eclogite sample by Lu–Hf garnet chronometry. Nat Geosci 4:178–183

Hickmott D, Shimizu N (1990) Trace-element zoning in garnet from the Kwoiek area, British-Columbia—disequilibrium partitioning during garnet growth. Contrib Mineral Petrol 104:619–630

Hickmott D, Spear FS (1992) Major-element and trace-element zoning in garnets from calcareous pelites in the NW Shelburne Falls quadrangle, Massachusetts—garnet growth histories in retrograded rocks. J Petrol 33:965–1005

Hickmott DD, Shimizu N, Spear FS, Selverstone J (1987) Trace-element zoning in a metamorphic garnet. Geology 15:573–576

Hodges KV, Crowley PD (1985) Error estimation and empirical geothermobarometry for pelitic systems. Am Mineral 70:702–709

Hodges KV, McKenna LW (1987) Realistic propagation of uncertainties in geologic thermobarometry. Am Mineral 72:671–680

Hoisch TD (1990) Empirical calibration of six geobarometers for the mineral assemblage quartz + muscovite + biotite + plagioclase + garnet. Contrib Mineral Petrol 104:225–234

Hoisch TD (1991) Equilibria within the mineral assemblage quartz + muscovite + biotite + garnet + plagioclase, and implications for the mixing properties of octahedrally-coordinated cations in muscovite and biotite. Contrib Mineral Petrol 108:43–54

Holdaway MJ (2000) Application of new experimental and garnet Margules data to the garnet–biotite geothermometer. Am Mineral 85:881–892

Holdaway MJ (2001) Recalibration of the GASP geobarometer in light of recent garnet and plagioclase activity models and versions of the garnet–biotite geothermometer. Am Mineral 86:1117–1129

Holdaway MJ, Mukhopadhyay B, Dyar MD, Guidotti CV, Dutrow BL (1997) Garnet–biotite geothermobarometry revised: New Margules parameters and a natural specimen data set from Maine. Am Mineral 82:582–595

Holland TJB, Powell R (1985) An internally consistent thermodynamic dataset with uncertainties and correlations: 2. Data and results. J Metamorph Geol 3:343–370

Holland TJB, Powell R (1990) An enlarged and updated internally consistent thermodynamic dataset with uncertainties and correlations—the system $K_2O–Na_2O–CaO–MgO–MnO–FeO–Fe_2O_3–Al_2O_3–TiO_2–SiO_2–C–H_2–O_2$. J Metamorph Geol 8:89–124

Holland TJB, Powell R (1998) An internally consistent thermodynamic data set for phases of petrological interest. J Metamorph Geol 16:309–343

Holland TJB, Powell R (2011) An improved and extended internally consistent thermodynamic dataset for phases of petrological interest, involving a new equation of state for solids. J Metamorph Geol 29:333–383

Hollister LS (1966) Garnet zoning: an interpretation based on the Rayleigh fractionation model. Science 154:1647–1651

Ikeda T, Shimobayashi N, Wallis SR, Tsuchiyama A (2002) Crystallographic orientation, chemical composition and three-dimensional geometry of sigmoidal garnet: evidence for rotation. J Struct Geol 24:1633–1646

Indares AD, Martignole J (1985) Biotite–garnet geothermometry in the granulite facies: the influence of Ti and Al in biotite. Am Mineral 70:272–278

Izraeli ES, Harris JW, Navon O (1999) Raman barometry of diamond formation. Earth Planet Sci Lett 173:351–360

Jamtveit B, Hervig RL (1994) Constraints on transport and kinetics in hydrothermal systems from zoned garnet crystals. Science 263:505–508

Jamtveit B, Wogelius RA, Fraser DG (1993) Zonation patterns of skarn garnets—records of hydrothermal system evolution. Geology 21:113–116

Jedlicka R, Faryad SW, Hauzenberger C (2015) Prograde Metamorphic History of UHP Granulites from the Moldanubian Zone (Bohemian Massif) Revealed by Major Element and Y + REE Zoning in Garnets. J Petrol 56:2069–2088

John T, Gussone N, Podladchikov YY, Bebout GE, Dohmen R, Halama R, Klemd R, Magna T, Seitz HM (2012) Volcanic arcs fed by rapid pulsed fluid flow through subducting slabs. Nat Geosci 5:489–492

Johnson SE (1993) Testing models for the development of spiral-shaped inclusion trails in garnet porphyroblasts—to rotate or not to rotate, that is the question. J Metamorph Geol 11:635–659

Kawakami T, Hokada T (2010) Linking *P–T* path with development of discontinuous phosphorus zoning in garnet during high-temperature metamorphism—an example from Lützow-Holm Complex, East Antarctica. J Mineral Petrol Sci 105:175–186

Kelly ED, Carlson WD, Connelly JN (2011) Implications of garnet resorption for the Lu–Hf garnet geochronometer: an example from the contact aureole of the Makhavinekh Lake Pluton, Labrador. J Metamorph Geol 29:901–916

Kelly ED, Carlson WD, Ketcham RA (2013a) Crystallization kinetics during regional metamorphism of porphyroblastic rocks. J Metamorph Geol 31:963–979

Kelly ED, Carlson WD, Ketcham RA (2013b) Magnitudes of departures from equilibrium during regional metamorphism of porphyroblastic rocks. J Metamorph Geol 31:981–1002

Kelsey DE, Hand M (2015) On ultrahigh temperature crustal metamorphism: phase equilibria, trace element thermometry, bulk composition, heat sources, timescales and tectonic settings. Geoscience Frontiers 6:311–356

Ketcham RA, Carlson WD (2012) Numerical simulation of diffusion-controlled nucleation and growth of porphyroblasts. J Metamorph Geol 30:489–512

Kita NT, Ushikubo T, Fu B, Valley JW (2009) High precision SIMS oxygen isotope analysis and the effect of sample topography. Chem Geol 264:43–57

Kleeman U, Reinhardt J (1994) Garnet–biotite thermometry revised: the effect of AlVI and Ti in biotite. Eur J Mineral 6:925–941

Kohn MJ (1993) Modeling of prograde mineral $\delta^{18}O$ changes in metamorphic systems. Contrib Mineral Petrol 113:249–261

Kohn MJ (2004) Oscillatory- and sector-zoned garnets record cyclic (?) rapid thrusting in central Nepal. Geochem Geophys Geosystem 5:9

Kohn MJ (2009) Models of garnet differential geochronology. Geochim Cosmochim Acta 73:170–182

Kohn MJ (2014) "Thermoba-Raman-try": Calibration of spectroscopic barometers and thermometers for mineral inclusions. Earth Planet Sci Lett 388:187–196

Kohn MJ, Penniston–Dorland SC (2017) Diffusion: Obstacles and opportunities in petrochronology. Rev Mineral Geochem 83:103–152

Kohn MJ, Spear FS (1991) Error propagation in barometers: 2. Application to rocks. Am Mineral 76:138–147

Kohn MJ, Spear FS (2000) Retrograde net transfer reaction insurance for pressure–temperature estimates. Geology 28:1127–1130

Kohn MJ, Valley JW (1994) Oxygen-isotope constraints on metamorphic fluid-flow, Townshend Dam, Vermont, USA. Geochim Cosmochim Acta 58:5551–5566

Kohn MJ, Valley JW (1998) Obtaining equilibrium oxygen isotope fractionations from rocks: theory and examples. Contrib Mineral Petrol 132:209–224

Kohn MJ, Valley JW, Elsenheimer D, Spicuzza MJ (1993) O isotope zoning in garnet and staurolite—evidence for closed-system mineral growth during regional metamorphism. Am Mineral 78:988–1001

Kohn MJ, Spear FS, Valley JW (1997) Dehydration-melting and fluid recycling during metamorphism: Rangeley Formation, New Hampshire, USA. J Petrol 38:1255–1277

Konrad-Schmolke M, Babist J, Handy MR, O'Brien PJ (2006) The physico-chemical properties of a subducted slab from garnet zonation patterns (Sesia Zone, Western Alps). J Petrol 47:2123–2148

Konrad-Schmolke M, O'Brien PJ, Heidelbach F (2007) Compositional re-equilibration of garnet: the importance of sub-grain boundaries. Eur J Mineral 19:431–438

Konrad-Schmolke M, O'Brien PJ, De Capitani C, Carswell DA (2008) Garnet growth at high- and ultra-high pressure conditions and the effect of element fractionation on mineral modes and composition. Lithos 103:309–332

Korsakov AV, Zhukov VP, Vandenabeele P (2010) Raman-based geobarometry of ultrahigh-pressure metamorphic rocks: applications, problems, and perspectives. Anal Bioanal Chem 397:2739–2752

Korhonen FJ, Brown M, Grove M, Siddoway CS, Baxter EF, Inglis JD (2012) Separating metamorphic events in the Fosdick migmatite–granite complex, West Antarctica. J Metamorph Geol 30:165–191

Kouketsu Y, Nishiyama T, Ikeda T, Enami M (2014) Evaluation of residual pressure in an inclusion-host system using negative frequency shift of quartz Raman spectra. Am Mineral 99:433–442

Koziol AM, Newton RC (1988) Redetermination of the anorthite breakdown reaction and improvement of the plagioclase– garnet–Al_2SiO_5–quartz geobarometer. Am Mineral 73:216–223

Koziol AM, Newton RC (1989) Grossular activity-composition relationships in ternary garnets determined by reversed displaced-equilibrium experiments. Contrib Mineral Petrol 103:423–433

Kretz R (1959) Chemical study of garnet, biotite and hornblende from gneisses of southwestern Quebec, with emphasis on distri- bution of elements in coexisting minerals. J Geol 67:371–403

Kretz R (1964) Analysis of equilibrium in garnet–biotite–sillimanite gneisses from Quebec. J Petrol 5:1–20

Kretz R (1966) Interpretation of the shape of mineral grains in metamorphic rocks. J Petrol 7:68–94

Kretz R (1973) Kinetics of the crystallization of garnet at two localities near Yellowknife. Can Mineral 12:1–20

Kretz R (1974) Some models for the rate of crystallization of garnet in metamorphic rocks. Lithos 7:123–131

Kylander-Clark ARC, Hacker BR, Johnson CM, Beard BL, Mahlen NJ, Lapen TJ (2007) Coupled Lu–Hf and Sm–Nd geochronology constrains prograde and exhumation histories of high- and ultrahigh-pressure eclogites from western Norway. Chem Geol 242:137–154

Lagos M, Scherer EE, Tomaschek F, Munker C, Keiter M, Berndt J, Ballhaus C (2007) High precision Lu–Hf geochronology of eocene eclogite-facies rocks from syros, cyclades, Greece. Chem Geol 243:16–35

Lanari P, Engi M (2017) Local bulk composition effects on mineral assemblages. Rev Mineral Geochem v. 83

Lanari P, Giuntoli F, Loury C, Burn M, Engi M (2017) An inverse modeling approach to obtain *P–T* conditions of metamorphic stages involving garnet growth and resorption. Eur J Mineral in press:doi:10.1127/ejm/2017/0029–2597

Lancaster PJ, Fu B, Page FZ, Kita NT, Bickford ME, Hill BM, McLelland JM, Valley JW (2009) Genesis of metapelitic migmatites in the Adirondack Mountains, New York. J Metamorph Geol 27:41–54

Lanzirotti A (1995) Yttrium zoning in metamorphic garnets. Geochim Cosmochim Acta 59:4105–4110

Lapen TJ, Johnson CM, Baumgartner LP, Mahlen NJ, Beard BL, Amato JM (2003) Burial rates during prograde metamorphism of an ultra-high-pressure terrane: an example from Lago di Cignana, western Alps, Italy. Earth Planet Sci Lett 215:57–72

Lasaga AC (1979) Multicomponent exchange and diffusion in silicates. Geochim Cosmochim Acta 43:455–469

Lasaga AC (1983) Geospeedometry: An extension of geothermometry. *In:* Kinetics and Equilibrium in Mineral Reactions. Saxena SK (ed) Springer-Verlag, New York, p 81–114

Loomis TP (1975) Reaction zoning of garnet. Contrib Mineral Petrol 52:285–305

Loomis TP, Ganguly J, Elphick SC (1985) Experimental determination of cation diffusivities in aluminosilicate garnets II. Multicomponent simulation and tracer diffusion coefficients. Contrib Mineral Petrol 90:45–51

Mahar E, Powell R, Holland TJB, Howell N (1997) The effect of Mn on mineral stability in metapelites. J Metamorph Geol 15:223–238

Malaspina N, Poli S, Fumagalli P (2009) The oxidation state of metasomatized mantle wedge: insights from C–O–H-bearing garnet peridotite. J Petrol 50:1533–1552

Manzotti P, Ballevre M (2013) Multistage garnet in high-pressure metasediments: Alpine overgrowths on Variscan detrital grains. Geology 41:1151–1154

Marmo BA, Clarke GL, Powell R (2002) Fractionation of bulk rock composition due to porphyroblast growth: effects on eclogite facies mineral equilibria, Pam Peninsula, New Caledonia. J Metamorph Geol 20:151–165

Martin AJ (2009) Sub-millimeter heterogeneity of yttrium and chromium during growth of semi-pelitic garnet. J Petrol 50:1713–1727

Martin L, Duchene S, Deloule E, Vanderhaeghe O (2006) The isotopic composition of zircon and garnet: A record of the metamorphic history of Naxos, Greece. Lithos 87:174–192

Martin LAJ, Ballevre M, Boulvais P, Halfpenny A, Vanderhaeghe O, Duchene S, Deloule E (2011) Garnet re-equilibration by coupled dissolution-reprecipitation: evidence from textural, major element and oxygen isotope zoning of 'cloudy' garnet. J Metamorph Geol 29:213–231

Martin LAJ, Rubatto D, Crépisson C, Hermann J, Putlitz B, Vitale-Brovarone A (2014) Garnet oxygen analysis by SHRIMP-SI: Matrix corrections and application to high-pressure metasomatic rocks from Alpine Corsica. Chem Geol 374–375:25–36

Masago H, Rumble D, Ernst WG, Parkinson CD, Maruyama S (2003) Low $\delta^{18}O$ eclogites from the Kokchetav massif, northern Kazakhstan. J Metamorph Geol 21:579–587

McKenna LW, Hodges KV (1988) Accuracy versus precision in locating reaction boundaries: Implications for the garnet–plagioclase–aluminum silicate–quartz geobarometer. Am Mineral 73:1205–1205

Menard T, Spear FS (1993) Metamorphism of calcic pelitic schists, Strafford Dome, Vermont: Compositional zoning and reaction history. J Petrol 34:977–1005

Menard T, Spear FS (1996) Interpretation of plagioclase zonation in calcic pelitic schist, south Strafford, Vermont, and the effects on thermobarometry. Can Mineral 34:133–146

Meth CE, Carlson WD (2005) Diffusion-controlled synkinematic growth of garnet from a heterogeneous precursor at Passo del Sole, Switzerland. Can Mineral 43:157–182

Mezger K, Hanson GN, Bohlen SR (1989) U–Pb systematics of garnet—dating the growth of garnet in the late Archean Pikwitonei Granulite Domain at Cauchon and Natawahunan lakes, Manitoba, Canada. Contrib Mineral Petrol 101:136–148

Mezger K, Essene EJ, Halliday AN (1992) Closure temperatures of the Sm–Nd system in metamorphic garnets. Earth Planet Sci Lett 113:397–409

Moecher DP, Essene EJ, Anovitz LM (1988) Calculation and application of clinopyroxene–garnet–plagioclase–quartz geobarometers. Contrib Mineral Petrol 100:92–106

Moore SJ, Carlson WD, Hesse, Ma (2013) Origins of yttrium and rare earth element distributions in metamorphic garnet. J Metamorph Geol 31:663–689

Moscati RJ, Johnson CA (2014) Major element and oxygen isotope geochemistry of vapour-phase garnet from the Topopah Spring Tuff at Yucca Mountain, Nevada, USA. Mineral Mag 78:1029–1041

Mottram CM, Parrish RR, Regis D, Warren CJ, Argles TW, Harris NB, Roberts NM (2015) Using U–Th–Pb petrochronology to determine rates of ductile thrusting: Time windows into the Main Central Thrust, Sikkim Himalaya. Tectonics 34:1355–1374

Moynihan DP, Pattison DRM (2013) An automated method for the calculation of *P–T* paths from garnet zoning, with application to metapelitic schist from the Kootenay Arc, British Columbia, Canada. J Metamorph Geol 31:525–548

Mukhopadhyay B, Holdaway MJ, Koziol AM (1997) A statistical model of thermodynamic mixing properties of Ca–Mg–Fe^{2+} garnets. Am Mineral 82:165–181

Nesheim TO, Vervoort JD, McClelland WC, Gilotti JA, Lang HM (2012) Mesoproterozoic syntectonic garnet within Belt Supergroup metamorphic tectonites: Evidence of Grenville-age metamorphism and deformation along northwest Laurentia. Lithos 134:91–107

Newton RC, Haselton HT (1981) Thermodynamics of the garnet–plagioclase–Al₂SiO₅–quartz geobarometer. *In:* Thermodynamics of Minerals and Melts. Newton RC, Navrotsky, A and Wood BJ (eds) Springer-Verlag, New York, p 129–145

O'Brien PJ, Vrána S (1995) Eclogites with a short-lived granulite facies overprint in the Moldanubian Zone, Czech Republic: Petrology, geochemistry and diffusion modelling of garnet zoning. Geol Rundsch 84:473–488

Otamendi JE, de la Rosa JD, Douce AEP, Castro A (2002) Rayleigh fractionation of heavy rare earths and yttrium during metamorphic garnet growth. Geology 30:159–162

Page FZ, Kita NT, Valley JW (2010) Ion microprobe analysis of oxygen isotopes in garnets of complex chemistry. Chem Geol 270:9–19

Page FZ, Essene EJ, Mukasa SB, Valley JW (2014) A garnet–zircon oxygen isotope record of subduction and exhumation fluids from the Franciscan Complex, California. J Petrol 55:103–131

Palin RM, Weller OM, Waters DJ, Dyck B (2016) Quantifying geological uncertainty in metamorphic phase equilibria modelling; a Monte Carlo assessment and implications for tectonic interpretations. Geosci Frontiers 7:591–607

Parkinson C, Katayama I (1999) Present-day ultrahigh-pressure conditions of coesite inclusions in zircon and garnet: Evidence from laser Raman microspectroscopy. Geology 27:979–982

Pattison D (2015) Challenges in phase diagram modelling, focusing on metapelites. Geological Society of America

Pattison DRM, Chacko T, Farquhar J, McFarlane CRM (2003) Temperatures of granulite-facies metamorphism: constraints from experimental phase equilibria and thermobarometry corrected for retrograde exchange. J Petrol 44:867–900

Pattison DRM, De Capitani C, Gaidies F (2011) Petrological consequences of variations in metamorphic reaction affinity. J Metamorph Geol 29:953–977

Peck W, Valley J (2004) Quartz–garnet isotope thermometry in the southern Adirondack Highlands (Grenville Province, New York). J Metamorph Geol 22:763–773

Penniston-Dorland SC, Sorensen SS, Ash RD, Khadke SV (2010) Lithium isotopes as a tracer of fluids in a subduction zone melange: Franciscan Complex, CA. Earth Planet Sci Lett 292:181–190

Perchuk LL, Lavrent'eva IV (1983) Experimental investigation of exchange equilibria in the system cordierite–garnet–biotite. *In:* Kinetics and Equilibrium in Mineral Reactions. Saxena SK (ed) Springer-Verlag, New York, p 199–239

Perchuk A, Philippot P, Erdmer P, Fialin M (1999) Rates of thermal equilibration at the onset of subduction deduced from diffusion modeling of eclogitic garnets, Yukon–Tanana terrane, Canada. Geology 27:531–534

Peterman EM, Hacker BR, Baxter EF (2009) Phase transformations of continental crust during subduction and exhumation: Western Gneiss Region, Norway. Eur J Mineral 21:1097–1118

Pogge von Strandmann PAE, Dohmen R, Marschall HR, Schumacher JC, Elliott T (2015) Extreme magnesium isotope fractionation at outcrop scale records the mechanism and rate at which reaction fronts advance. J Petrol 56:33–58

Pollington AD, Baxter EF (2010) High resolution Sm–Nd garnet geochronology reveals the uneven pace of tectonometamorphic processes. Earth Planet Sci Lett 293:63–71

Pollington AD, Baxter EF (2011) High precision microsampling and preparation of zoned garnet porphyroblasts for Sm–Nd geochronology. Chem Geol 281:270–282

Powell R, Holland TJB (1988) An internally consistent dataset with uncertainties and correlations. 3. Applications to geobarometry, worked examples and a computer-program. J Metamorph Geol 6:173–204

Powell R, Holland TJB, Worley B (1998) Calculating phase diagrams involving solid solutions via non-linear equations, with examples using THERMOCALC. J Metamorph Geol 16:577–588

Prince C, Harris N, Vance D (2001) Fluid-enhanced melting during prograde metamorphism. J Geol Soc London 158:233–241

Putlitz B, Matthews A, Valley JW (2000) Oxygen and hydrogen isotope study of high-pressure metagabbros and metabasalts (Cyclades, Greece): implications for the subduction of oceanic crust. Contrib Mineral Petrol 138:114–126

Pyle J, Spear F (1999) Yttrium zoning in garnet: Coupling of major and accessory phases during metamorphic reactions. Geol Mater Res 1:1–49

Pyle JM, Spear FS (2000) An empirical garnet (YAG)–xenotime thermometer. Contrib Mineral Petrol 138:51–58

Pyle JM, Spear FS (2003) Four generations of accessory-phase growth in low-pressure migmatites from SW New Hampshire. Am Mineral 88:338–351

Pyle JM, Spear FS, Rudnick RL, McDonough WF (2001) Monazite–xenotime–garnet equilibrium in metapelites and a new monazite- garnet thermometer. J Petrol 42:2083–2107

Raimondo T, Clark C, Hand M, Cliff J, Harris C (2012) High-resolution geochemical record of fluid–rock interaction in a mid-crustal shear zone: a comparative study of major element and oxygen isotope transport in garnet. J Metamorph Geol 30:255–280

Ramberg H (1952) Chemical bonds and distribution of cations in silicates. J Geol 60:331–355

Ramsay JG (1962) The geometry and mechanics of formation of "similar" type folds. J Geol 70:309–327

Richter R, Spiering B, Hoernes S (1988) Petrology and isotope-geochemistry of granulite-facial marbles and calcium-silicate rocks of Sri-Lanka. Fortschr Mineral 66:134–134

Robyr M, Carlson WD, Passchier C, Vonlanthen P (2009) Microstructural, chemical and textural records during growth of snowball garnet. J Metamorph Geol 27:423–437

Robyr M, Darbellay B, Baumgartner LP (2014) Matrix-dependent garnet growth in polymetamorphic rocks of the Sesia zone, Italian Alps. J Metamorph Geol 32:3–24

Romer RL, Xiao YL (2005) Initial Pb–Sr(Nd) isotopic heterogeneity in a single allanite–epidote crystal: implications of reaction history for the dating of minerals with low parent-to-daughter ratios. Contrib Mineral Petrol 148:662–674

Rosenfeld JL (1968) Garnet rotations due to the major Paleozoic deformations in southeast Vermont. *In:* Studies of Appalachian Geology: Northern and Maritime. Zen E (ed) John Wiley and Sons, Inc., New York, p 185–202

Rosenfeld JL (1969) Stress effects around quartz inclusions in almandine and the piezothermometry of coexisting aluminum silicates. Am J Sci 267:317–351

Rosenfeld JL, Chase AB (1961) Pressure and temperature of crystallization from elastic effects around solid inclusions in minerals? Am J Sci 259:519–541

Rubatto D (2002) Zircon trace element geochemistry: partitioning with garnet and the link between U–Pb ages and metamorphism. Chem Geol 184:123–138

Rubatto D (2017) Zircon: the metamorphic mineral. Rev Mineral Geochem 83

Rubatto D, Angiboust S (2015) Oxygen isotope record of oceanic and high-pressure metasomatism: a *P–T–time–fluid* path for the Monviso eclogites (Italy). Contrib Mineral Petrol 170

Rumble D, Yui TF (1998) The Qinglongshan oxygen and hydrogen isotope anomaly near Donghai in Jiangsu Province, China. Geochim Cosmochim Acta 62:3307–3321

Russell AK, Kitajima K, Strickland A, Medaris LG, Schulze DJ, Valley JW (2013) Eclogite-facies fluid infiltration: constraints from $\delta^{18}O$ zoning in garnet. Contrib Mineral Petrol 165:103–116

Sato K, Santosh M, Tsunogae T (2009) A petrologic and laser Raman spectroscopic study of sapphirine–spinel–quartz–Mg-staurolite inclusions in garnet from Kumiloothu, southern India: Implications for extreme metamorphism in a collisional orogen. J Geodyn 47:107–118

Saxena SK (1968) Distribution of elements between coexisting minerals and the nature of solid solution in garnet. Am Mineral 53:994–1014

Saxena SK (1969) Silicate solid solutions and geothermometry: 3. Distribution of Fe and Mg between coexisting garnet and biotite. Contrib Mineral Petrol 22:259–267

Sayab M, Shah SZ, Aerden D (2016) Metamorphic record of the NW Himalayan orogeny between the Indian plate-Kohistan Ladakh Arc and Asia: Revelations from foliation intersection axis (FIA) controlled *P–T–t–d* paths. Tectonophysics 671:110–126

Scherer EE, Cameron KL, Blichert-Toft J (2000) Lu–Hf garnet geochronology: Closure temperature relative to the Sm–Nd system and the effects of trace mineral inclusions. Geochim Cosmochim Acta 64:3413–3432

Schmidt C, Ziemann MA (2000) *In-situ* Raman spectroscopy of quartz: A pressure sensor for hydrothermal diamond-anvil cell experiments at elevated temperatures. Am Mineral 85:1725–1734

Schmidt C, Steele-MacInnis M, Watenphul A, Wilke M (2013) Calibration of zircon as a Raman spectroscopic pressure sensor to high temperatures and application to water-silicate melt systems. Am Mineral 98:643–650

Schmidt A, Pourteau A, Candan O, Oberhansli R (2015) Lu–Hf geochronology on cm-sized garnets using microsampling: New constraints on garnet growth rates and duration of metamorphism during continental collision (Menderes Massif, Turkey). Earth Planet Sci Lett 432:24–35

Schoene B, Baxter EF (2017) Petrochronology and TIMS. Rev Mineral Geochem 83:231–260

Schwarz JO, Engi M, Berger A (2011) Porphyroblast crystallization kinetics: the role of the nutrient production rate. J Metamorph Geol 29:497–512

Skelton A, Annersten H, Valley J (2002) $\delta^{18}O$ and yttrium zoning in garnet: time markers for fluid flow? J Metamorph Geol 20:457–466

Skora S, Baumgartner LP, Mahlen NJ, Johnson CM, Pilet S, Hellebrand E (2006) Diffusion-limited REE uptake by eclogite garnets and its consequences for Lu–Hf and Sm–Nd geochronology. Contrib Mineral Petrol 152:703–720

Skora S, Baumgartner LP, Mahlen NJ, Lapen TJ, Johnson CM, Bussy F (2008) Estimation of a maximum Lu diffusion rate in a natural eclogite garnet. Swiss J Geosci 101:637–650

Skora S, Lapen TJ, Baumgartner LP, Johnson CM, Hellebrand E, Mahlen NJ (2009) The duration of prograde garnet crystallization in the UHP eclogites at Lago di Cignana, Italy. Earth Planet Sci Lett 287:402–411

Smit MA, Scherer EE, Mezger K (2013) Lu–Hf and Sm–Nd garnet geochronology: Chronometric closure and implications for dating petrological processes. Earth Planet Sci Lett 381:222–233

Sobolev NV, Fursenko BA, Goryainov SV, Shu J, Hemley RJ, Mao A, Boyd FR (2000) Fossilized high pressure from the Earth's deep interior: the coesite-in-diamond barometer. PNAS 97:11875–11879

Sobolev NV, Schertl HP, Valley JW, Page FZ, Kita NT, Spicuzza MJ, Neuser RD, Logvinova AM (2011) Oxygen isotope variations of garnets and clinopyroxenes in a layered diamondiferous calcsilicate rock from Kokchetav Massif, Kazakhstan: a window into the geochemical nature of deeply subducted UHPM rocks. Contrib Mineral Petrol 162:1079–1092

Solva H, Thoni M, Habler G (2003) Dating a single garnet crystal with very high Sm/Nd ratios (Campo basement unit, Eastern Alps). Eur J Mineral 15:35–42

Sorby HC, Butler PJ (1869) On the structure of rubies, sapphires, diamonds, and some other minerals. Proc R Soc London 17:291–302

Sousa J, Kohn MJ, Schmitz MD, Northrup CJ, Spear FS (2013) Strontium isotope zoning in garnet: implications for metamorphic matrix equilibration, geochronology and phase equilibrium modelling. J Metamorph Geol 31:437–452

Spear FS (1993) Metamorphic phase equilibria and pressure–temperature–time paths. Mineral Soc Am, Washington, DC

Spear FS (2004) Fast cooling and exhumation of the Valhalla metamorphic core complex, southeastern British Columbia. Int Geol Rev 46:193–209

Spear FS (2014) The duration of near-peak metamorphism from diffusion modelling of garnet zoning. J Metamorph Geol 32:903–914

Spear FS, Daniel CG (2001) Diffusion control of garnet growth, Harpswell Neck, Maine, USA. J Metamorph Geol 19:179–195

Spear FS, Florence FP (1992) Thermobarometry in granulites: pitfalls and new approaches. Precambr Res 55:209–241

Spear FS, Kohn MJ (1996) Trace element zoning in garnet as a monitor of crustal melting. Geology 24:1099–1102

Spear FS, Selverstone J (1983) Quantitative *P–T* paths from zoned minerals: theory and tectonic applications. Contrib Mineral Petrol 83:348–357

Spear FS, Selverstone J, Hickmott D, Crowley PD, Hodges KV (1984) *P–T* paths from garnet zoning—a new technique for deciphering tectonic processes in crystalline terranes. Geology 12:87–90

Spear FS, Kohn MJ, Florence FP, Menard T (1991) A model for garnet and plagioclase growth in pelitic schists: Implications for thermobarometry and *P–T* path determinations. J Metamorph Geol 8:683–696

Spear FS, Thomas JB, Hallett BW (2014) Overstepping the garnet isograd: a comparison of QuiG barometry and thermodynamic modeling. Contrib Mineral Petrol 168

Spear FS, Pattison DRM, Cheney JT (2015) It's a mad, mad, mmad world. GSA Annual Meeting. Baltimore

Spear FS, Pattison DRM, Cheney JT (2016) The metamorphosis of metamorphic petrology. Geol Soc Am Spec Pap 523:31–73

Spiess R, Peruzzo L, Prior DJ, Wheeler J (2001) Development of garnet porphyroblasts by multiple nucleation, coalescence and boundary misorientation-driven rotations. J Metamorph Geol 19:269–290

Spry A (1963) The origin and significance of snowball structure in garnet. J Petrol 4:211–222

St-Onge MR (1987) Zoned poikiloblastic garnets: *P–T* paths and syn-metamorphic uplift through 30 km of structural depth, Wopmay Orogen, Canada. J Petrol 28:1–21

Stallard A, Ikei H, Masuda T (2002) Numerical simulations of spiral-shaped inclusion trails: can 3D geometry distinguish between end-member models of spiral formation? J Metamorph Geol 20:801–812

Štípská P, Powell R, White RW, Baldwin JA (2010) Using calculated chemical potential relationships to account for coronas around kyanite: an example from the Bohemian Massif. J Metamorph Geol 28:97–116

Stixrude L, Lithgow-Bertelloni C (2005) Thermodynamics of mantle minerals—I. Physical properties. Geophys J Inter 162:610–632

Storm LC, Spear FS (2005) Pressure, temperature and cooling rates of granulite facies migmatitic pelites from the southern Adirondack Highlands, New York. J Metamorph Geol 23:107–130

Stowell HH, Menard T, Ridgway CK (1996) Ca-metasomatism and chemical zonation of garnet in contact-metamorphic aureoles, Juneau Gold Belt, southeastern Alaska. Can Mineral 34:1195–1209

Stowell HH, Taylor DL, Tinkham DL, Goldberg SA, Ouderkirk KA (2001) Contact metamorphic *P–T–t* paths from Sm–Nd garnet ages, phase equilibria modelling and thermobarometry: Garnet Ledge, south-eastern Alaska, USA. J Metamorph Geol 19:645–660

Stowell H, Tulloch A, Zuluaga C, Koenig A (2010) Timing and duration of garnet granulite metamorphism in magmatic arc crust, Fiordland, New Zealand. Chem Geol 273:91–110

Stowell H, Parker KO, Gatewood M, Tulloch A, Koenig A (2014) Temporal links between pluton emplacement, garnet granulite metamorphism, partial melting and extensional collapse in the lower crust of a Cretaceous magmatic arc, Fiordland, New Zealand. J Metamorph Geol 32:151–175

Strickland A, Russell AK, Quintero R, Spicuzza MJ, Valley JW (2011) Oxygen isotope ratios of quartz inclusions in garnet and implications for mineral pair thermometry. Geol Soc Am Abstr with Programs 43:93

Stüwe K (1997) Effective bulk composition changes due to cooling: a model predicting complexities in retrograde reaction textures. Contrib Mineral Petrol 129:43–52

Sutton JR (1921) Inclusions in diamond from South Africa. Mineral Mag 19:208–210

Symmes GH, Ferry JM (1991) Evidence from mineral assemblages for infiltration of pelitic schists by aqueous fluids during metamorphism. Contrib Mineral Petrol 108:419–438

Tajčmanová L, Konopásek J, Connolly JAD (2007) Diffusion-controlled development of silica-undersaturated domains in felsic granulites of the Bohemian Massif (Variscan belt of Central Europe). Contrib Mineral Petrol 153:237–250

Taylor HP, Sheppard SMF (1986) Igneous rocks. 1. Processes of isotopic fractionation and isotope systematics. Rev Miner 16:227–271

Thompson AB (1976a) Mineral reactions in pelitic rocks: I. Prediction of P–T–X (Fe–Mg) phase relations. Am J Sci 276:401–424

Thompson AB (1976b) Mineral reactions in pelitic rocks: II. Calculation of some P–T–X (Fe–Mg) phase relations. Am J Sci 276:425–454

Thompson AB, Tracy RJ, Lyttle PT, Thompson JB (1977) Prograde reaction histories deduced from compositional zonation and mineral inclusions in garnet from the Gassetts schist, Vermont. Am J Sci 277:1152–1167

Thoni M (2002) Sm–Nd isotope systematics in garnet from different lithologies (Eastern Alps): age results, and an evaluation of potential problems for garnet Sm–Nd chronometry. Chem Geol 185:255–281

Tinkham DK, Ghent ED (2005) Estimating P–T conditions of garnet growth with isochemical phase-diagram sections and the problem of effective bulk-composition. Can Mineral 43:33–50

Tinkham DK, Zuluaga CA, Stowell HH (2001) Metapelite phase equilibria modeling in MnNCKFMASH: The effect of variable Al_2O_3 and MgO/(MgO+FeO) on mineral stability. Geol Mater Res 3:1–42

Tirone M, Ganguly J, Dohmen R, Langenhorst F, Hervig R, Becker, H-W (2005) Rare earth diffusion kinetics in garnet: Experimental studies and applications. Geochim Cosmochim Acta 69:2385–2398

Tracy RJ (1982) Compositional zoning and inclusions in metamorphic minerals. Rev Mineral 10:355–397

Tracy RJ, Robinson P, Thompson AB (1976) Garnet composition and zoning in the determination of temperature and pressure of metamorphism, central Massachusetts. Am Mineral 61:762–775

Tracy RJ, Caddick MJ, Thompson AB (2012) Garnet growth zoning in metapelites: not so simple after all, and revealing more than we once thought? GSA Annual Meeting

Valley JW (1986) Stable isotope geochemistry of metamorphic rocks. Rev Mineral 16:445–489

Valley JW (2001) Stable isotope thermometry at high temperatures. Rev Mineral Geochem 43:365–413

Valley JW, Bindeman IN, Peck WH (2003) Empirical calibration of oxygen isotope fractionation in zircon. Geochim Cosmochim Acta 67:3257–3266

van Breemen O, Hawkesworth CJ (1980) Sm–Nd isotopic study of garnets and their metamorphic host rocks. Transactions of the Royal Society of Edinburgh: Earth Sciences 71:97–102

Van der Molen I (1981) The shift of the α–β transition temperature of quartz associated with the thermal expansion of granite at high pressure. Tectonophysics 73:323–342

van Haren JLM, Ague JJ, Rye DM (1996) Oxygen isotope record of fluid infiltration and mass transfer during regional metamorphism of pelitic schist, Connecticut, USA. Geochim Cosmochim Acta 60:3487–3504

Van Orman JA, Grove TL, Shimizu N, Layne GD (2002) Rare earth element diffusion in a natural pyrope single crystal at 2.8 GPa. Contrib Mineral Petrol 142:416–424

Vance D, Harris N (1999) Timing of prograde metamorphism in the Zanskar Himalaya. Geology 27:395–398

Vance D, Holland T (1993) A detailed isotopic and petrological study of a single garnet from the Gassetts Schist, Vermont. Contrib Mineral Petrol 114:101–118

Vance D, Onions RK (1990) Isotopic chronometry of zoned garnets—growth-kinetics and metamorphic histories. Earth Planet Sci Lett 97:227–240

Vance D, Onions RK (1992) Prograde and retrograde thermal histories from the central Swiss Alps. Earth Planet Sci Lett 114:113–129

Vannay JC, Grasemann B (1998) Inverted metamorphism in the High Himalaya of Himachal Pradesh (NW India): Phase equilibria versus thermobarometry. Schweiz Mineral Petrograph Mitteil 78:107–132

Vannay JC, Sharp ZD, Grasemann B (1999) Himalayan inverted metamorphism constrained by oxygen isotope thermometry. Contrib Mineral Petrol 137:90–101

Vielzeuf D, Saúl A (2011) Uphill diffusion, zero-flux planes and transient chemical solitary waves in garnet. Contrib Mineral Petrol 161:638–702

Vielzeuf D, Champenois M, Valley JW, Brunet F, Devidal JL (2005) SIMS analyses of oxygen isotopes: Matrix effects in Fe–Mg–Ca garnets. Chem Geol 223:208–226

Viete DR, Hermann J, Lister GS, Stenhouse IR (2011) The nature and origin of the Barrovian metamorphism, Scotland: diffusion length scales in garnet and inferred thermal time scales. J Geol Soc London 168:115–131

Vorhies SH, Ague JJ (2011) Pressure–temperature evolution and thermal regimes in the Barrovian zones, Scotland. J Geol Soc London 168:1147–1166

Vrijmoed JC, Hacker BR (2014) Determining *P–T* paths from garnet zoning using a brute-force computational method. Contrib Mineral Petrol 167:1–13

Wang X, Planavsky NJ, Reinhard CT, Zou H, Ague JJ, Wu Y, Gill BC, Schwarzenbach EM, Peucker-Ehrenbrink B (2016) Chromium isotope fractionation during subduction-related metamorphism, black shale weathering, and hydrothermal alteration. Chem Geol 423:19–33

Waters D, Martin H (1993) The garnet–clinopyroxene–phengite barometer. Terra Abst 5:410–411

Waters DJ, Lovegrove DP (2002) Assessing the extent of disequilibrium and overstepping of prograde metamorphic reactions in metapelites from the Bushveld Complex aureole, South Africa. J Metamorph Geol 20:135–149

Watson EB, Cherniak DJ (2013) Simple equations for diffusion in response to heating. Chem Geol 335:93–104

Wendt I, Carl C (1991) The statistical distribution of the mean squared weighted deviation Chem Geol 86:275–285

White RW, Powell R, Holland TJB, Worley B (2000) The effect of TiO_2 and Fe_2O_3 on metapelitic assemblages at greenschist and amphibolite facies conditions: mineral equilibria calculations in the system $K_2O–FeO–MgO–Al_2O_3–SiO_2–H_2O–TiO_2–Fe_2O_3$. J Metamorph Geol 18:497–511

White RW, Pomroy NE, Powell R (2005) An in situ metatexite–diatexite transition in upper amphibolite facies rocks from Broken Hill, Australia. J Metamorph Geol 23:579–602

White RW, Powell R, Baldwin JA (2008) Calculated phase equilibria involving chemical potentials to investigate the textural evolution of metamorphic rocks. J Metamorph Geol 26:181–198

White RW, Powell R, Holland TJB, Johnson TE, Green ECR (2014a) New mineral activity–composition relations for thermodynamic calculations in metapelitic systems. J Metamorph Geol 32:261–286

White RW, Powell R, Johnson TE (2014b) The effect of Mn on mineral stability in metapelites revisited: new *a–X* relations for manganese-bearing minerals. J Metamorph Geol 32:809–828

Whitney DL, Seaton NCA (2010) Garnet polycrystals and the significance of clustered crystallization. Contrib Mineral Petrol 160:591–607

Whitney DL, Goergen ET, Ketcham RA, Kunze K (2008) Formation of garnet polycrystals during metamorphic crystallization. J Metamorph Geol 26:365–383

Williams ML, Jercinovic MJ, Hetherington CJ (2007) Microprobe monazite geochronology: understanding geologic processes by integrating composition and chronology. Ann Rev Earth Planet Sci 35:137

Williams HM, Nielsen SG, Renac C, Griffin WL, O'Reilly SY, McCammon CA, Pearson N, Viljoen F, Alt JC, Halliday AN (2009) Fractionation of oxygen and iron isotopes by partial melting processes: Implications for the interpretation of stable isotope signatures in mafic rocks. Earth Planet Sci Lett 283:156–166

Williams ML, Jercinovic MJ, Mahan KH, Dumond G (2017) Electron microprobe petrochronology. Rev Mineral Geochem 83:153–182

Wing BA, Ferry JM, Harrison TM (2003) Prograde destruction and formation of monazite and allanite during contact and regional metamorphism of pelites: petrology and geochronology. Contrib Mineral Petrol 145:228–250

Wood BJ, Banno S (1973) Garnet–orthopyroxene and orthopyroxene–clinopyroxene relationships in simple and complex systems. Contrib Mineral Petrol 42:109–124

Woodsworth GJ (1977) Homogenization of zoned garnets from pelitic schists. Can Mineral 15:230–242

Wu C-M, Zhao G (2006) Recalibration of the garnet–muscovite (GM) geothermometer and the garnet–muscovite–plagioclase–quartz (GMPQ) geobarometer for metapelitic assemblages. J Petrol 47:2357–2368

Wu C-M, Zhang J, Ren L-D (2004) Empirical garnet–biotite–plagioclase–quartz (GBPQ) geobarometry in medium- to high-grade metapelites. J Petrol 45:1907–1921

Wu C-M, Cheng B-H (2006) Valid garnet–biotite (GB) geothermometry and garnet–aluminum silicate–plagioclase–quartz (GASP) geobarometry in metapelitic rocks. Lithos 89:1–23

Yakymchuk C, Brown M (2014) Consequences of open-system melting in tectonics. J Geol Soc London

Yakymchuk C, Brown M, Clark C, Korhonen FJ, Piccoli PM, Siddoway CS, Taylor RJM, Vervoort JD (2015) Decoding polyphase migmatites using geochronology and phase equilibria modelling. J Metamorph Geol 33:203–230

Yakymchuk C, Clark C, White RW (2017) Phase relations, reaction sequences and petrochronology. Rev Mineral Geochem 83:13–53

Yang PS, Pattison D (2006) Genesis of monazite and Y zoning in garnet from the Black Hills, South Dakota. Lithos 88:233–253

Yang P, Rivers T (2001) Chromium and manganese zoning in pelitic garnet and kyanite: Spiral, overprint, and oscillatory (?) zoning patterns and the role of growth rate. J Metamorph Geol 19:455–474

Yang P, Rivers T (2002) The origin of Mn and Y annuli in garnet and the thermal dependence of P in garnet and Y in apatite in calc- pelite and pelite. Gagnon terrane, western Labrador. Geol Mater Res 4:1–35

Yardley BWD (1977) An empirical study of diffusion in garnet. Am Mineral 62:793–800

Young ED, Rumble D (1993) The origin of correlated variations in insitu $^{18}O/^{16}O$ and elemental concentrations in metamorphic garnet from southeastern Vermont, USA. Geochim Cosmochim Acta 57:2585–2597

Zack T, John T (2007) An evaluation of reactive fluid flow and trace element mobility in subducting slabs. Chem Geol 239:199–216

Zack T, Tomascak PB, Rudnick RL, Dalpe C, McDonough WF (2003) Extremely light Li in orogenic eclogites: The role of isotope fractionation during dehydration in subducted oceanic crust. Earth Planet Sci Lett 208:279–290

Zeng L, Asimow PD, Saleeby JB (2005) Coupling of anatectic reactions and dissolution of accessory phases and the Sr and Nd isotope systematics of anatectic melts from a metasedimentary source. Geochim Cosmochim Acta 69:3671–3682

Zhang Y (1998) Mechanical and phase equilibria in inclusion–host systems. Earth Planet Sci Lett 157:209–222

Zheng YF (1991) Calculation of oxygen isotope fractionation in metal-oxides. Geochim Cosmochim Acta 55:2299–2307

Zheng YF (1993) Calculation of oxygen-isotope fractionation in hydroxyl-bearing silicates. Earth Planet Sci Lett 120:247–263

Zheng Y-F, Fu B, Li Y, Xiao Y, Li S (1998) Oxygen and hydrogen isotope geochemistry of ultrahigh-pressure eclogites from the Dabie Mountains and the Sulu terrane. Earth Planet Sci Lett 155:113–129

Zheng YF, Zhao ZF, Wu YB, Zhang SB, Liu XM, Wu FY (2006) Zircon U–Pb age, Hf and O isotope constraints on protolith origin of ultrahigh-pressure eclogite and gneiss in the Dable orogen. Chem Geol 231:135–158

Zhou B, Hensen BJ (1995) Inherited Sm/Nd isotope components preserved in monazite inclusions within garnets in leucogneiss from East Antarctica and implications for closure temperature studies. Chem Geol 121:317–326

Zhukov VP, Korsakov AV (2015) Evolution of host-inclusion systems: a visco-elastic model. J Metamorph Geol 33:815–828.

Reviews in Mineralogy & Geochemistry
Vol. 83 pp. 535–575, 2017
Copyright © Mineralogical Society of America

Chronometry and Speedometry of Magmatic Processes using Chemical Diffusion in Olivine, Plagioclase and Pyroxenes

Ralf Dohmen, Kathrin Faak

Institut für Geologie, Mineralogie und Geophysik
Ruhr-Universität Bochum
Universitätsstraße 150
44801 Bochum
Germany

ralf.dohmen@rub.de

kathrin.faak@rub.de

Jon D. Blundy

School of Earth Sciences
University of Bristol
Wills Memorial Building
Bristol BS8 1RJ
United Kingdom

jon.blundy@bristol.ac.uk

INTRODUCTION

The magmatic processes that fuel volcanism, crustal growth, ore formation and discharge of volcanic gases and aerosols to the atmosphere occur across a range of timescales, from millions of years to just a few seconds. For example, the production of new oceanic crust at mid-ocean ridges is a near-continuous process that can operate in any one ocean basin on timescales of more than 100 m.y. However, the driving force for such processes is the spreading of the ocean plates that happens on a cm/yr timescale. At the other end of the spectrum, explosive volcanic eruptions involve the ascent and fragmentation of magma at velocities of the order of 100 m/s such that the journey from a magma chamber to an ash cloud may take place in a matter of minutes. In this case the driving force is the rapid expansion of magmatic gas in response to changes in pressure. At intermediate timescales magmatic processes may give rise to hydrothermal ore deposits on timescales of less than a million years for an individual deposit, while growth of giant granite batholiths may require piecemeal assembly of magma batches on timescales of a few million years. Although each of these processes has a characteristic, time-averaged timescale on which it operates, this is typically the end result of one or more natural processes that operate on much shorter timescales. For example, mid-ocean ridges do not extrude magma continuously onto the ocean floor, mineralising fluids do not discharge continuously through the shallow crust, and granitic magmas do not dribble continuously into evolving batholithic chambers. In some cases it is the long-term timescales that are important, for example the spreading rate of ocean basins, in others it is the short-term timescales that are important, for example the episodic growth of lava domes at active volcanoes. Although the long-term timescales are reasonably well known, accessing the shorter timescales is notoriously difficult, yet is vital if we are to successfully model magmatic systems. The importance of gauging the appropriate timescale for a given magmatic

1529-6466/17/0083-0016$05.00 (print)
1943-2666/17/0083-0016$05.00 (online)

http://dx.doi.org/10.2138/rmg.2017.83.16

process is that, coupled to an extensive parameter such as volume, mass or length, it enables calculations of rates or fluxes of matter and heat. As the tempo of magmatism in the widest sense is controlled ultimately by the interplay of several different processes each with their own characteristic rate, e.g., magma input, convection, degassing, cooling etc, so the determination of timescales is fundamental to the understanding of magmatism. In this chapter we review first the available petrochronometers and then focus specifically on the burgeoning field of diffusion chronometry as a means of accessing timescales for a variety of magmatic processes.

Types of chronometers

Magmatic timescales can be constrained primarily by three methods: observational, radiometric and diffusive. Observations are limited to active magmatic systems, such as volcanoes that are erupting or degassing passively. For example the growth of lava domes at Mount St. Helens (Washington) and Soufrière Hills (Montserrat) volcanoes has been documented with unprecedented detail over the course of more than two decades. Visual observations of dome characteristics (volume, height, temperature, etc.) have been supplemented by geodesy, tiltmeters, seismology and measurements of gas chemistry. The latter can also be achieved remotely, using satellites, for volcanoes that degas syn- or inter-eruptively, such as Etna (Italy). All of these observations provide precise timescales that can be used to refine physical understanding of sub-volcanic processes. For example, at Soufrière Hills observations of lava dome extrusions have been used to develop physical models of eruption periodicity that embrace conduit flow and gas exsolution (Melnik and Sparks 2005). Likewise, coupled observations of lava dome extrusion and inter-eruption gas discharge have been used to invoke mushy, sub-volcanic magmatic systems that are subject to periodic instability that leads to a decoupling of gas and magma (Christopher et al. 2015). Kilauea volcano (Hawaii) is a classic example of a well-monitored volcano where frequent, long-lasting observations have been used to formulate a detailed image of the magma plumbing system and dynamics (Tilling and Dvorak 1993; Dvorak and Dzurisin 1997).

The drawback of quantitative observational timescales is that they are limited to the modern era, notably since the advent of routine volcanic monitoring of the type that is now widespread at many active and restless volcanoes. To extend the observational timescale further back in time requires the availability of archival records, such as those found in eyewitness accounts. For example, colonial records of volcanism in Ecuador have provided insights into eruptions of Cotopaxi volcano (Pistolesi et al. 2011); Albert et al. (2016) used historical accounts of seismicity to explore the build-up to eruptions at monogenetic volcanoes. Other temporal indicators of volcanism, such as tree rings or ice-cores, can also be exploited, notably because they contain robust chronologies. Linking historic eyewitness accounts to dendrochronology and ice-core records is, perhaps, a relatively unexplored approach to volcanic timescales. Yet still, this approach is limited, at the very most, to the last several hundred years.

For longer timescales the methodology of choice has been radiometric dating, whereby the decay of naturally occurring radioisotopes to daughter isotopes has long been exploited as a geochronometer. Originally this approach was limited to radioisotopes with multi-million year half-lives, restricting applicability to long timescales. The lower limit on these timescales is constrained by the analytical precision with which the ratio of daughter to parent isotopes can be measured and the half-life of the parent. In the case of the well-established Ar–Ar dating technique, in the most favourable of circumstances (i.e., high initial potassium contents) ages down to as little as 2000 yr can be recovered (Lanphere et al. 2007). For most other long-lived radioisotopes, such as ^{87}Sr, ^{238}U, ^{235}U, ^{230}Th, ^{147}Sm, etc., timescales of less than about 1 m.y. are inaccessible. Carbon-14 dating ($t_{1/2} = 5730$ yr) affords glimpses of shorter timescales, although its application relies on knowledge of the global ^{14}C flux and the availability of carbonised material that can be directly related to volcanism, such as charred tree trunks in pyroclastic flows.

With the advent of improved mass spectrometric methods a new class of daughter isotopes became accessible; those of the short-lived uranium and thorium decay series that form en route to the stable daughter isotopes of lead. The so-called U-series dating method (e.g., Turner and Costa 2007; Cooper and Reid 2008) has proven especially valuable in constraining many magmatic timescales, not least because the daughter isotopes comprise elements with widely differing chemical affinity. For example, radon is a gas whose loss from the decay chain can provide valuable insights into degassing processes (e.g., Gauthier and Condomines 1999; Berlo et al. 2006; Kayzar et al. 2009). Other daughter isotopes are preferentially sequestered by particular minerals, thereby constraining the onset of crystallisation of those minerals in an evolving magmatic system. The uptake of radium by plagioclase is a case in point (e.g., Cooper and Reid 2003; Turner et al. 2003). The wide variety of half-lives of U-series radionuclides means that, in principle, a wealth of timescales from seconds to millennia can be recovered. However, once secular equilibrium between parent decay and daughter ingrowth is established, typically after about 5 daughter half-lives, all chronometric information is lost. Thus, the central problem of radiometric dating persists; the parent and daughter isotopes can become decoupled, such that what is dated is a time of decoupling (or cessation of secular equilibrium). A further limitation of the U-series approach is that the analytical techniques used do not lend themselves to high spatial resolution. Consequently it may be difficult to disentangle multiple timescales from zoned crystals, especially where crystal cores and rims differ greatly in age (e.g., Cooper and Reid 2003). The application of U-series methods to constrain timescales of magmatic processes is discussed in some detail by Cooper and Reid (2008). U-series dating of zircons is covered by Schaltegger and Davies (2017, this volume).

The focus of this chapter is an alternative family of geochronometers that rely on the time-dependent flux (diffusion) of elements in response to chemical potential gradients established during magmatic processes. These gradients are most readily manifest in zoned crystals of common minerals in magmatic systems, e.g., Costa et al. (2008), Kahl et al. (2011). As a crystal grows, it takes up major and trace elements in response to changes in the melt with which it is bathed. Perturbations to the melt chemistry or the physical parameters that control uptake, such as pressure, temperature and redox, result in chemical changes between one layer (zone) and the next in the growing crystal. At the instant of growth, each layer has some simple relationship to the melt from which it grew, but no such relationship to the previous layer. The relationship between each layer and the melt may be one of chemical equilibrium or, if growth is relatively rapid, kinetic factors may dominate (e.g., Watson and Liang 1995; Watson 1996). Foremost amongst these is the rate at which chemical components can be supplied diffusively to the boundary layer around the growing crystal (Albarède and Bottinga 1972) and the relative preference of trace species for the surface of growing crystals relative to their interior (Pinilla et al. 2012). There is a rich literature on the uptake of trace elements into growing crystals that is beyond the scope of this chapter. What is important is that the growth of zoned magmatic crystals sets up chemical potential gradients between successive zones. Diffusion strives to eliminate such gradients by moving chemical components from one layer to another. The rate at which this happens, the chemical diffusivity, is a complex function of the diffusion mechanism, the point defect chemistry of the crystal, the temperature at which diffusion occurs and the geometry of the interface across which diffusion takes place. The significance of these various parameters is relatively well established through experiment and theory for a range of geological substrates, both crystals and melts, and the diffusivity of many trace species has been determined experimentally. Diffusion is, ultimately, a thermally mediated process, such that it occurs most rapidly at high (magmatic) temperatures and is effectively quenched once a system cools below a characteristic closure temperature (e.g., Dodson 1986; Kohn and Penniston-Dorland 2017, this volume). Consequently, diffusion chronometry is a very effective means to constrain the timescale between the event that caused the chemical perturbation responsible for a zone to form in a crystal and the moment at which this system cooled below some closure temperature when diffusion was effectively arrested.

For explosive volcanic systems closure occurs upon eruption such that diffusion chronometry is a valuable tool for recovering the timescales of pre-eruptive processes that are responsible for modifying the chemistry of volcanic crystals. The sensitivity of crystal chemistry to both intensive parameters (P, T, f_{O_2}) and melt composition means that diffusion chronometry can be used to establish the time-lapse between a change in the configuration of the sub-volcanic system and its eventual eruption. The challenge is to establish the temperature at which the original perturbation occurred, the nature of the diffusive interface, the original chemical potential driving diffusion and the type of diffusion that is occurring. Conversely, for effusive volcanic systems diffusion of some species may continue during cooling of the lava flow. In that case, diffusion chronometry may be used to constrain cooling times or flow rates of lava flows. For example, Newcombe et al. (2014) have shown how diffusion of major and volatile (H_2O, F) elements across the walls of olivine-hosted melt inclusions may be used as a cooling rate speedometer for basaltic lava flows, while Sio et al. (2013) used Fe–Mg interdiffusion in olivine phenocrysts to determine the cooling rate of the Kilauea Iki lava lake.

In this review we will explore the use of diffusion-based geochronometers in deriving magmatic timescales by highlighting a number of recent applications. The considerable potential of diffusion chronometry in understanding magmatic systems is evidenced by the significant number of recent papers in high profile journals such as Nature and Science (Martin et al. 2008; Druitt et al. 2012; John et al. 2012; Saunders et al. 2012a; Ruprecht and Plank 2013; Cooper and Kent 2014). Here we will discuss the assumptions implicit in these and other applications, their limitations and the insights that can (and cannot) be gained from this approach. This review can be read as an update of that given by Costa et al. (2008). Therefore, we will not give a detailed introduction into the theoretical background of diffusion modelling, which is given elsewhere (e.g., Ganguly 2002; Watson and Baxter 2007; Costa et al. 2008; Costa and Morgan 2010; Zhang 2010). However, to make it easier for the reader to follow the text and the logic within this chapter we will illustrate some basics of the approach, wherever necessary, without going into the mathematical details. In addition the review of Kohn and Penniston-Dorland (2017, this volume) introduces into some of the basics of diffusion and its role for petrochronology.

Basic approach of diffusion chronometry

Diffusion chronometry (sometimes called "geospeedometry", e.g., Lasaga 1983; Kohn and Penniston-Dorland 2017, this volume) is based on the extent to which ion exchange reactions that are sensitive to a change in some environmental condition (e.g., temperature, pressure, or chemical potential) proceed during this change. In general, the approach used is forward modelling of concentration distributions (typically, element or isotope profiles measured along traverses across sections of crystals, glasses or bulk rocks). Firstly, a diffusion model needs to be defined by describing the element fluxes and geometry for the real system. The model has to be translated to partial differential equations that must be solved for the given initial conditions (the concentration distributions at some defined point in time) and boundary conditions (the element fluxes at the geometric boundaries of the system). The boundary conditions define the interaction of the system (typically a chemically zoned crystal) with the exterior, thermodynamic environment, e.g., a silicate melt. The measured distributions of concentration are then fitted by this solution, where time (or the temperature–time path) is the only unknown, provided the relevant diffusion coefficients are known for the conditions of interest. For the fundamentals of diffusion chronometry, different strategies to define appropriately the diffusion problem, and methods to solve the respective diffusion equation, we refer the reader to the detailed reviews of Ganguly (2002), Chakraborty (2008), Costa et al. (2008) and Costa and Morgan (2010), including the various textbooks on diffusion cited therein. In addition, some discussion on diffusion and diffusion coefficients can be found in Kohn and Penniston-Dorland (2017, this volume). For the purposes of this chapter we first summarize briefly the pre-requisites, advantages, and shortcomings of diffusion chronometry (see also Chakraborty 2006, 2008), before we address specific examples for olivine, orthopyroxene, and plagioclase.

Prerequisites. (1) A measurable chemical or isotope zoning that is conserved in a mineral and that has been at least partly produced by diffusion. A homogeneous crystal would allow for an estimate of the minimum time scales if it can be assumed that the crystal was originally zoned. (2) Accurate diffusion coefficients: ideally, for the experimental determination of the diffusion coefficients, their dependence on various intensive thermodynamic parameters should be quantified, which in addition to P, T, and major element composition could be oxygen fugacity, water fugacity or, for example, activity of silica in the case of silicates (e.g., Dohmen and Chakraborty 2007; Zhukova et al. 2014). The relevance of these different parameters is discussed in detail in Chakraborty (2008). Diffusion of major as well as trace elements in ionic solids involves the coupling of diffusion to other ions, which might require the application of a multicomponent diffusion equation (see also discussion on effect of crystal chemistry and substitution mechanisms in Kohn and Penniston-Dorland 2017, this volume). (3) An appropriate diffusion model, which includes geometry (e.g., simplified geometry vs. real crystal shape; 1D vs. 2D vs. 3D models), initial condition (e.g., homogenous concentration vs. step-like zoning vs. complex zoning), and boundary condition (e.g., fixed rim composition vs. variable rim composition as controlled by a temperature-dependent exchange reaction, a geothermometer, or by variable element fluxes from the environment; open system vs. closed system, Chakraborty and Ganguly 1991).

General Strengths and Advantages. The main strength of diffusion chronometry is that we can determine a time scale of the respective process (duration) that is independent of the absolute age of the crystal. Thus the same methodology can be applied to crystals in volcanic rocks from any age, i.e., Archean to the present day. Moreover, with the right choice of the element-mineral pair there is no limit on the time scale that can be investigated, from seconds up to billions of years. Short-lived processes can be determined in old crystals. For example, with diffusion modelling the cooling rate of refractory high temperature condensates (CAIs) in the solar nebula or the peak temperature of metamorphism were determined for chondrite parent bodies at the onset of our solar system (Simon et al. 2011; Schwinger et al. 2016). Further examples are the determinations of short residence times or ascent rates of crystals from historical volcanic eruptions (e.g., Druitt et al. 2012; Ruprecht and Plank 2013). In addition, petrology can be used to relate the chemical gradient in the system to a specific process (e.g., magma ascent, magma mixing, cooling), and hence the duration of this process can be determined by diffusion modelling—that is the essence of diffusion chronometry. Moreover, it is possible to access timescales that are commensurate with observational timescales of magmatic processes, as discussed above. Consequently there is potential to link timescales derived from young volcanic crystals with geophysical or geodetic observations of the volcanoes that erupted them (e.g., Kahl et al. 2011; Saunders et al. 2012a).

Shortcomings and Uncertainties. (1) The accuracy of diffusion coefficients, D, is typically within 0.2–0.5 log units for the experimental parameter range (T, P, f_{O_2}, etc.). This uncertainty becomes significantly larger when the data have to be extrapolated beyond the experimental parameter range. Furthermore, considering an isothermal process, the temperature needs to be known from thermometry, and any uncertainties in T further affect the uncertainty in D. Since the total extent of diffusion that occurred at this temperature T (generally equivalent to the loosely defined diffusion profile length) is proportional to $\sqrt{Dt}$, the uncertainty in the respective time, t, as determined by fitting the measured concentration distribution, comprises both the quality of the fit and the uncertainty in D. Assuming an almost perfect fit, an uncertainty in D of 0.5 log units translates to an uncertainty in the duration of the isothermal event by the same 0.5 log units. Even considering these relative large uncertainties in D, the order of magnitude of the time scale can be determined by diffusion modelling for a specific geologic process, whether it lasts for seconds, days, months or millions of years. It should be also noted that for trace elements (as experimentally demonstrated for H and Li diffusion in olivine and clinopyroxene, see sections below) different diffusion mechanisms and rates may be involved. In this case the operating diffusion mechanism needs to be identified for the given crystal, since the rates may differ by orders of magnitudes (see also Dohmen et al. 2016b). (2) The difference between the initial and the final (observed) concentration distribution (or, more correctly, the chemical potential distribution) of the element/isotope of interest is the measure of the total extent of diffusion, which basically contains the time information. The reconstruction of the initial profile can be ambiguous, but various possible strategies are discussed in the specific case studies below. In general, the observed concentration profile/distribution to be fitted limits the overall range of possible initial profiles (see also discussion in Chakraborty and Ganguly 1991; Costa et al. 2008). (3) As for the initial condition, the modeller has to make a decision regarding the boundary condition based on the petrologic understanding of the system. When minerals are in contact with a melt it is usually assumed that transport within the melt is fast enough (at least much faster than in the crystal) and therefore (i) the melt is treated as a homogeneous phase and (ii) local equilibrium at the crystal–melt interface is attained. These two assumptions may be incorrect for very compatible or very slow-diffusing elements (see for example the model examples of Watson and Müller (2009) or Dohmen et al. (2003) or in the case of very rapid crystal growth. It should be noted that an incorrect boundary condition can have a stronger affect on the inferred time scale than a wrong initial condition (Costa et al. 2008). Further discussions on uncertainties of diffusion chronometry are given in Costa et al. (2008) and Faak et al. (2013). (4) The obtainable timescale and its accuracy are not limited only by the relevant diffusion rate but also by the spatial resolution of the available analytical method. If the concentration gradient is very steep, the measured diffusion profile could be convolved (i.e., not adequately resolved spatially), thus a larger apparent timescale would be obtained (Ganguly et al. 1988). Considering a Gaussian distribution function for the intensity of the signal centred at the measured spot, characterized by the standard deviation σ with unit length, only for a diffusion profile length $\sqrt{Dt} \gg \sigma$ we can exclude significant contributions of the convolution effect to the obtained time scale (Ganguly et al. 1988). However, if a steep concentration gradient can be well resolved spatially it provides a better constraint for the timescale compared to a shallower gradient due to the fitting procedure, which needs to account for the precision of the individual data points along a diffusion profile (see for example Figs. 3 and 4 in Ruprecht and Plank 2013). If the gradient is too steep to be resolved by a given analytical method, it yields a constraint for a maximum permissible timescale of the process.

OLIVINE

Olivine (Ol) has been used to constrain time scales of processes related to mafic rocks (e.g., Gerlach and Grove 1982; Coogan et al. 2002, 2007; Pan and Batiza 2002; Costa and Dungan 2005; Kahl et al. 2011, 2013, 2015; Ruprecht and Plank 2013; Newcombe et al. 2014; Albert et al. 2015; Oeser et al. 2015; Faak and Gillis 2016; Hartley et al. 2016; Rae et al. 2016), where it is an ubiquitous phase. Application to more silicic volcanic rocks (andesite, dacite) can be found in Nakamura (1995), Coombs et al. (2000), Costa and Chakraborty (2004), and Martin et al. (2008), where olivine is often present in the form of xenocrysts or as part of more mafic inclusions/enclaves; and to basanite in Martí et al. (2013) and Longpré et al. (2014). We will use olivine as a "reference case" to illustrate the general approach of diffusion chronometry where, in addition, many recent developments have been made to improve the accuracy of the method, for example to distinguish diffusion zoning from growth zoning (e.g., Oeser et al. 2015; Shea et al. 2015a) and user-friendly diffusion modeling algorithms are now available for olivine (e.g., DIPRA, Girona and Costa 2013). This and the following sections will be organized in a similar way whereby the available diffusion coefficients are first discussed briefly (more details can be found for olivine in Chakraborty 2010).

Diffusion coefficients

The diffusion rates of major and minor cations occupying the metal site in olivine are relatively fast when compared to pyroxenes and garnet (e.g., Müller et al. 2013). However, the published diffusion rates of trace cations in olivine are controversial (e.g., Cherniak 2010 vs. Spandler and O'Neill 2010, see also Chakraborty 2010). This controversy might be related to the different diffusion mechanisms that are possible in olivine as demonstrated for Li and H diffusion (Dohmen et al. 2010 and Mackwell and Kohlstedt 1990, respectively). In any case, for most practical applications where the electron microprobe was used to map concentration distributions or measure element profiles, the most relevant elements are Fe, Mg, Ca, Ni and Mn for which diffusion datasets exist that have been reproduced in various studies using different experimental approaches over a large temperature range (700–1250 °C), and therefore data extrapolation is, for the most part, unnecessary (Fe–Mg interdiffusion: Chakraborty 1997; Dohmen and Chakraborty 2007; Holzapfel et al. 2007; Ca "tracer" diffusion: Coogan et al. 2005; Ni and Mn "tracer" diffusion: Petry et al. 2004). Diffusion of these cations (and other minor or trace elements occupying the M-site) is in principle a coupled diffusion process that would require a multi-component diffusion model (e.g., Lasaga 1979), but it can be shown numerically that diffusion of minor or trace elements having the same charge as the major cations (Fe^{2+} and Mg^{2+}) can be treated independently by using the so-called "tracer diffusion coefficient" as measured experimentally (see Costa et al. 2008 or Vogt et al. 2015 for numerical examples). The diffusion coefficients of Fe–Mg interdiffusion, and Mn, and Ni tracer diffusion are rather similar and respective diffusion profiles are of similar length, but the Ca tracer diffusion coefficient is about an order of magnitude smaller. Other trace elements in olivine have also been used as a diffusion chronometer in cases where their concentration zoning is large enough to be resolved by other micro-analytical methods such as laser ablation inductively coupled mass spectrometry (LA-ICP-MS, e.g., Qian et al. 2010), Fourier transform infrared spectroscopy (FT-IR) for H (e.g., Demouchy et al. 2006), or secondary ion mass spectroscopy (SIMS) for Li (e.g., Jeffcoate et al. 2007). To measure short chemical zoning profiles in olivine nm-scale analytical methods can be also applied, like Nano-SIMS, or field emission electron microprobe (FEG-probe), or time-of-flight SIMS (TOF-SIMS). Alternatively, back-scattered electron (BSE) imaging can be used to infer the major component (fayalite–forsterite) zoning in olivine (Martin et al. 2008; Hartley et al. 2016; Rae et al. 2016), for details of this approach see the section on orthopyroxene (Opx) and Morgan et al. (2004).

Determination of magma residence times using Fe–Mg, Ca, Ni, and Mn diffusion

Magma residence times by diffusion modelling of chemical zoning in olivine phenocrysts or xenocrysts have been determined for a large variety of volcanic rocks, from basanite to dacite (e.g., Costa and Chakraborty 2004; Costa and Dungan 2005; Marti et al. 2013; Albert et al. 2015). In all of these studies a straightforward diffusion chronometry approach was applied whereby the system is treated as isothermal, such that a constant diffusion coefficient can be used. The simplistic scenario commonly used is that the olivine crystal became suddenly exposed to a different chemical/magmatic environment (which could be related to a sudden change in temperature, magma mixing, decompression, magma recharge etc...), the olivine rim re-equilibrates with this new chemical environment by element exchange or an overgrowth in equilibrium with the melt is formed, then progressive internal re-equilibration of olivine with time occurs by solid-state diffusion. The system and hence the chemical zoning is eventually quenched by the volcanic eruption and the chemical zoning is therefore a measure of the time span between the change of the environment and the eruption. Thus, we do not measure the total residence time (or life time) of a given crystal, but infer instead timescales related to the dynamics of the magmatic system.

Figure 1 shows a chemical zoning map from Kahl et al. (2011), which evidently cannot be fitted by a model wherein the olivine was initially homogeneous and only the rim composition changed. Compositional plateaus that could be identified in olivine phenocrysts in basalts of Mt. Etna indicate various growth stages. Equally, a compositional profile such as shown in Figure 1 cannot be produced by fractional crystallization in a continuously cooled system. Thus, Kahl et al. (2011) assumed that chemical composition of each of these plateaus reflects a different magmatic environment. Based on a system analysis of the diverse compositions and types of zonings, three different magma environments were identified. Diffusional relaxation between these compositional plateaus contains information on the time spans between the different growth steps and the final eruption. Here two sequential diffusion-modelling steps were performed assuming that specific regions in the profile reflect different growth stages of the crystal in two respective chemical environments (concept illustrated in Figure 2). Finally, from simulation of chemical zoning in 28 crystals from the 1991–1993 SE-flank eruptions, the dynamic evolution of the plumbing system was reconstructed by incorporating information from volcano monitoring. The same approach was applied to eruptions between 2001 and 2008 of Mt. Etna (Kahl et al. 2013, 2015) where, in addition, thermodynamic phase equilibrium calculations with MELTS (Ghiorso and Sack 1995) were performed. It was confirmed that different types of olivine core compositions (plateau regions) "reflect compositionally distinct batches of magma that have resided in different sections of the plumbing system". The MELTS calculations allowed Kahl et al. (2013, 2015) to identify the intensive parameters that control the different types of olivine compositions in these magmatic environments. Water content in the melt and temperature of the magma appear to exercise the dominant control. In addition the connectivity between these different magmatic environments was established and, finally, diffusion modelling was used to obtain the time scales (ranging from days to 2 yr) of mixing between these different magma batches and the time of final eruption.

Pronounced compositional steps in olivine were also found in basaltic andesites from the 1963–1965 eruption of Irazu volcano, Costa Rica, indicating again magma mixing (Ruprecht and Plank 2013). In particular Ni shows strong variations in primitive olivine crystals with high and relatively constant forsterite contents (>90%). This variation was ascribed primarily to an olivine crystallization trend in primary melts of the mantle, as olivine controls the Ni budget in the melt. It further implies that primary melts coexist and mix in the source region. Corresponding zoning of P was used to infer the growth zoning and estimate the effect of diffusion for Ni. Diffusion modelling of the Ni zonations indicates that, between the magma

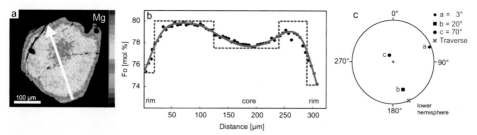

Figure 1. (a) Mg distribution map, (b) Fo concentration profile with model fit, and (c) respective profile orientation data of an olivine crystal from the 1991–1993 SE-flank eruptions products of Mt. Etna (Sicily). The dotted line in (b) is the assumed initial step profile formed during three different growth stages (core, 1st overgrowth, 2nd overgrowth) reflecting different magmatic environments. The model fit (solid line) reproduces the measured concentration profile by two sequential diffusion-modelling steps, which indicates that the 1st overgrowth was formed about 250 days before the 2nd overgrowth, which was formed 113 days before eruption. Figure modified after Kahl et al. (2011). Note the asymmetric shape of the zoning and the crystal habit in 2D, which is most likely a sectioning effect based on the simulations of Shea et al. (2015b).

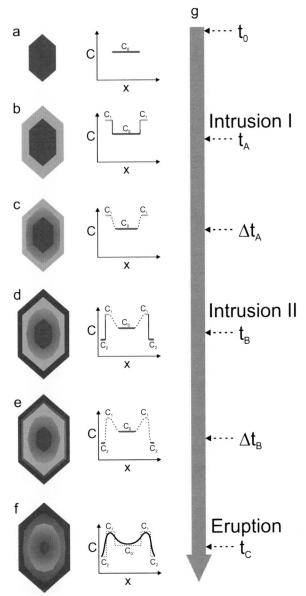

Figure 2. Illustration of the concept to reconstruct the growth history and residence times in different magmatic environments of a phenocryst by sequential diffusion modelling. (a)–(f) illustrate concentration profiles (x denotes distance) induced sequentially by magmatic events during time line (g): (a) At some time t_0 a homogeneous crystal with composition C_0 was formed; (b) After the crystal came into contact with a different magmatic environment at time t_A (e.g., a new batch of melt arrived) a new rim with composition C_1 was formed; (c) After the formation of the rim, a diffusive flux starts to smooth the initial step profile; (d) At time t_B a second magmatic event (intrusion, magma mixing) changes the local thermodynamic conditions and a second rim is formed with composition C_2; (e) Concentration gradients between C_0, C_1 and C_2 tend to homogenize the entire concentration profile by chemical diffusion. (f) The eruption cools the magma quickly and freezes the concentration profile, which is eventually a product of growth and diffusion zoning (black solid line). The black dashed line represents the concentration profile produced only by growth without diffusion (the initial profile used for diffusion modelling). The duration of diffusion at different stages before eruption, here Δt_A and Δt_B, are the quantities that can be obtained by simulating the observed concentration profiles. (Figure modified after Kahl et al. 2011).

mixing processes deep within the mantle and final eruption, a time between months and years elapsed. That implies the erupted crystals record recharge of the magma during the course of the eruption. Because the mixing process occurred in a source region at about 35 km depth, from this timescale an integrated ascent rate of the order 10–50 m per day was estimated. We discuss below how an ascent rate can be inferred more directly from H diffusion profiles in olivine.

Growth vs. diffusive zoning: stable isotopes as diffusive fingerprints

A fundamental problem of the approach described above is to identify how much of the zoning was formed simply by growth (the initial profile) and how much was modified subsequently by solid-state diffusion. Several strategies can be applied to exclude that the inferred time scale was biased due to pre-diffusional crystal growth:

First, diffusion in olivine is anisotropic (e.g., for Fe–Mg interdiffusion, $D_{//c}=6\times D_{//a}=6\times D_{//b}$ in the T range 700–1250 °C, diffusion anisotropy changes at higher temperatures, Tachibana et al. 2013); this allows to check whether the relative length of diffusion profiles in the same crystal along different directions is in accordance with the predicted relative profile lengths (e.g., Costa and Chakraborty 2004; Costa et al. 2008); crystal growth would not be expected to be similarly anisotropic. Crystallographic orientations of measured profiles are usually measured using electron back-scatter diffraction (EBSD).

Secondly, the zoning of major or trace elements that apparently have slow diffusivities could be a tracer of the original zoning type and different growth stages. For example P in olivine conserves typically complex zoning (e.g., Milman-Barris et al. 2008; Welsch et al. 2013) that in some cases indicates dendritic growth of olivine in the initial phase (Welsch et al. 2014; Shea et al. 2015a), but zoning of major elements and other divalent cations is typically smooth. Bouvet de Maisonneuve et al. (2016) compared trace element zoning (P, Ti, Sc, V, Al) vs. major and minor element zoning (Fe–Mg, Ni, Mn, Ca) in olivine and plagioclase. They concluded that trace elements like P, Ti, Sc, and Al in olivine record more and earlier events in the magmatic system than Fe–Mg, Ni, Mn, and Ca. However, it was also concluded that incorporation of these trace elements is decoupled from those of the major and minor elements, and their zoning is controlled by kinetic factors related to rapid growth rather than equilibrium uptake into olivine.

Thirdly, zoning of Fe–Mg, Ca, Ni, and Mn due simply to crystal growth is distinct from that due to diffusion (Costa et al. 2008). Consequently, if diffusion, rather than growth, is the cause of the observed zoning patterns, then simultaneous diffusion modelling of different elements should ideally give the same or similar time scale (e.g., Costa and Dungan 2005).

Fourthly, isotopic zoning provides detailed insights into the origin of chemical zoning (Sio et al. 2013; Oeser et al. 2015). Light isotopes diffuse slightly faster than heavy isotopes (the "isotope effect", e.g., Schoen 1958). Thus, in contrast to equilibrium partitioning, isotopic diffusion produces strong isotope anomalies even at magmatic temperatures (Richter et al. 1999, 2003). In the case of Fe–Mg interdiffusion in olivine, such diffusive fractionation results in negatively correlated isotope anomalies for Mg and Fe due to their opposing diffusion flux directions. Strong negative correlations of δ^{56}Fe with δ^{26}Mg isotopes were first identified by Teng et al. (2011) (see also Dauphas et al. 2010 for first indications of this effect) in analyses of olivine fragments from Hawaiian basalts (Fig. 3), a feature that can be explained by diffusive fractionation. Diffusion modelling and *in situ* measurements of Fe and Mg isotopes by microdrilling and of Fe isotopes by *in situ* measurements using LA-ICP-MS and multi collector SIMS (MC-SIMS) found a similar correlation as in Figure 3 (Sio et al. 2013), confirming this explanation. In addition Sio et al. (2013) were able to reproduce the known cooling rate for these samples by diffusion modelling of the fayalite zoning (which also confirms the experimental diffusion data). New analytical developments using a femtosecond laser ablation system allow *in situ* analysis of both Fe and Mg isotopes (Oeser et al. 2014). Chemically zoned olivine crystals from different types of basaltic rocks have been analysed using this technique by Oeser et al. (2015, Fig. 4).

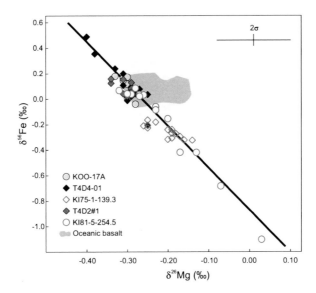

Figure 3. Mg and Fe isotope fractionations in olivine fragments from Hawaiian lavas. The values for $\delta^{56}Fe = ((^{56}Fe/^{54}Fe)_{Sample}/(^{56}Fe/^{54}Fe)_{Standard} - 1) \times 1000$ and $\delta^{26}Mg = ((^{26}Mg/^{24}Mg)_{Sample}/(^{26}Mg/^{24}Mg)_{Standard} - 1) \times 1000$ define a linear relationship with a slope of -3.3 ± 0.3 (solid line). Olivine data are reported in Table 1 of Teng et al. (2011). Oceanic basalt data are reported in Teng et al. (2008, 2010, 2011). Figure modified after Teng et al. (2011).

As in Sio et al. (2013) the clear fingerprint of diffusion on Fe and Mg isotopes was observed. In addition Oeser et al. (2015) discussed and modelled different scenarios, including growth and dissolution, to produce chemical zoning and associated isotope zoning. Based on these models it can be concluded that solid state diffusion leaves a unique fingerprint on the Fe and Mg isotopes and, for example, it can be distinguished whether the olivine zoning was formed by element exchange at the surface or whether an overgrowth was formed (Fig. 5). In most of the olivine crystals investigated by Oeser et al. (2015) the isotope variations could be reproduced by the diffusion model shown in Figure 4, indicating element exchange at the rim, potentially driven by cooling of the magma. However, the parameter β, which for two isotopes with masses M_i and M_j is used to empirically describe the isotope effect according to the relation $D_i / D_j = (M_j / M_i)^\beta$, was used as a free parameter for the fitting procedure and varied for Mg and Fe isotopes between 0.06–0.12 and 0.08–0.2, respectively, with the value for Mg systematically about a factor of two smaller. Whether this variation is related to additional processes responsible for the isotope anomalies or related to anisotropy and compositional effects should be clarified by experiments to calibrate β. As, for example, discussed by Van Orman and Krawczynski (2015) this parameter is controlled by the diffusion mechanism and should be somehow coupled for elements that interdiffuse with each other such as Fe and Mg in olivine.

Determination of cooling rates from Fe–Mg zoning in olivine

Based on the assumption that the Fe–Mg zoning was produced during cooling, Oeser et al. (2015) determined the respective cooling rates of basaltic rocks by applying an approach suggested by Ganguly (2002) (for another simplified approach to determine cooling rates from diffusion models, see also Watson and Cherniak 2015). In this model the diffusion equation is transformed and the diffusion coefficient is integrated over time (see also Lai et al. 2015 or Schwinger et al. 2016), which is a measure of the total extent of diffusion ("diffusion profile length"), and is used as a fitting parameter. The reciprocal cooling rate for an assumed initial

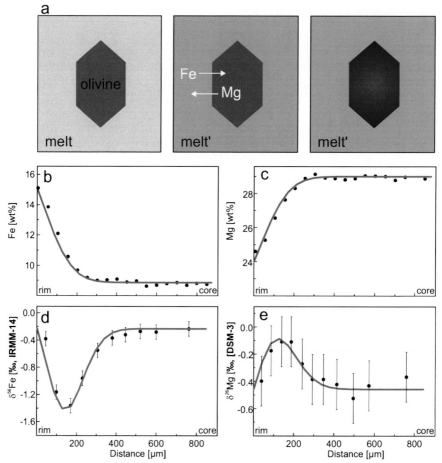

Figure 4. Simplified modeling of Fe–Mg chemical and isotopic diffusion profiles for the Massif Central olivine assuming element exchange at the crystal rim after a homogenous olivine was formed. (a) Illustration of the modelled scenario: pure Fe–Mg interdiffusion due to a compositional contrast between olivine and melt. (b)–(e) Measured data and modelled profiles for (b) Fe, (c) Mg, (d) $\delta^{56}Fe$, and (e) $\delta^{26}Mg$. Modelling results are shown as solid lines, which indicate either an effective residence time of 3.26 yr at 97% of the peak temperature of 1297 °C, or alternatively a cooling rate of 57 °C/yr^{-1} from this peak temperature. Note the negative and positive bumps for $\delta^{56}Fe$ and $\delta^{26}Mg$, respectively, whose minimum/maximum coincides with the position of the strongest curvature in the chemical profiles. The negative bump for $\delta^{56}Fe$ forms because the diffusive flux of Fe into the olivine is faster for the lighter Fe isotope (^{56}Fe) than for the heavier one (^{58}Fe). In the case of Mg, diffusion is in the opposite direction, i.e., out of the crystal, hence a positive bump forms for $\delta^{26}Mg$. Figure modified after Oeser et al. (2015).

temperature can be determined from this fitting parameter. For the different crystals and intra-plate volcanoes studied, the inferred cooling rates varied typically between 30–300 °C/yr.

Fe–Mg exchange between olivine and melt is sensitive to temperature but apparently in the basaltic rocks studied by Sio et al. (2013) and Oeser et al. (2015) this is not reflected in the profile shape (e.g., Fig. 4). Here a fixed rim composition is assumed, which during cooling did not change significantly. The interplay between the exchange reaction at the rim controlled by temperature and the resulting diffusion profile is discussed in more detail below for Ca in olivine and Mg in plagioclase.

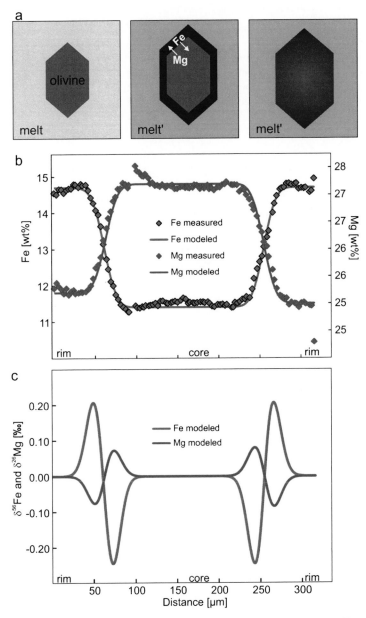

Figure 5. Simulation of Fe and Mg diffusion for a scenario (a) where an initial step profile was formed due to quick overgrowth onto a homogeneous olivine core. (b) Fe and Mg concentration profiles across an olivine phenocryst (EH_31-10-11_OL9) investigated by Longpré et al. (2014) and the respective simulation by diffusion modelling. (c) The corresponding Fe–Mg isotopic profiles across this olivine as expected from modelling the diffusion-generated chemical zoning. The Fe and Mg concentration data are from Table DR4 in Longpré et al. (2014). Note that the bulk crystal does not show any isotope fractionation due to internal diffusive redistribution of elements and their respective isotopes. Internally, the δ^{56}Fe and δ^{26}Mg values are negatively correlated because of the interdiffusion process, but each show both a negative and a positive bump in the profile. This isotopic signature is clearly different to that produced in the scenario for Figure 4. Figure modified after Oeser et al. (2015).

Determination of cooling rates from Ca zoning in olivine

The cooling rate of rocks can alternatively be determined by using the closure profiles concept of Dodson (1986) or the extension of this formulation by Ganguly and Tirone (1999) (see also Kohn and Penniston-Dorland 2017, this volume). This approach requires that the composition at the crystal rim changes significantly during cooling (approximately linear with time) and that the external environment behaves as an infinite reservoir. An example, applied by Coogan et al. (2002), is the temperature-sensitive exchange of Ca between olivine and clinopyroxene (Cpx) (Köhler and Brey 1990). The net exchange reaction may be written as:

$$Mg_2SiO_4 \ (Ol) + CaMgSi_2O_6 \ (Cpx) = CaMgSiO_4 \ (Ol) + Mg_2Si_2O_6 \ (Cpx) \tag{1}$$

For a minor variation in the major element composition of Ol and Cpx (the enstatite component in Cpx and forsterite component in Ol) according to this reaction an effective partition coefficient of Ca between Ol and Cpx, $\ln K_{p(Ca)}^{Ol\text{-}Cpx}$, can be defined. The temperature (and pressure) dependence of $\ln K_{p(Ca)}^{Ol\text{-}Cpx}$ was calibrated by Köhler and Brey (1990) (a more recent calibration considering also the effect of the forsterite content is given in Shejwalkar and Coogan 2013):

$$\ln K_{p(Ca)}^{Ol\text{-}Cpx} = \frac{-0.425 P\left[MPa\right] - 5792 - 1.25\left(T\right)}{\left(T\right)} \tag{2}$$

Equation (2) predicts that $\ln K_{p(Ca)}^{Ol\text{-}Cpx}$ decreases with decreasing temperature, leading to a diffusive flux of Ca out of olivine and into clinopyroxene during cooling. According to the exchange reaction in Equation (1) this directly implies an opposing flux of Mg. Assuming that this exchange reaction is controlled by Ca diffusion in olivine (i.e., clinopyroxene is considered to be an infinite reservoir for Ca in this case), a closure temperature, T_c, can be defined, which depends on the distance, x, from the interface (Dodson 1986; Onorato et al. 1981). At $T_c(x)$ the diffusive flux becomes too small to effectively change the composition (in this case the Ca content). Thus, the rims of an olivine crystal will be able to maintain equilibrium Ca-concentrations down to lower temperatures than the core, leading to the development of a closure profile that is convex upwards for a continuous cooling history (Fig. 6). According to Dodson (1986) this closure profile (T_c-profile) can be calculated as:

$$T_c\left(x\right) = \frac{E}{R \ln\left(\left(\frac{R T_c^2 D_0}{E s a^2}\right) + G\left(x\right)\right)} \tag{3}$$

where R = ideal gas constant, s = cooling rate around the closure temperature T_c, and a = radius of the grain. The parameters E and D_0 refer to the activation energy and pre-exponential factor in the diffusion equation. Coogan et al. (2005) provide experimentally determined diffusion coefficients for Ca in olivine (e.g., $E = 207$ kJ/mol and $D_0 = 1 \times 10^{-10}$ m^2/s for diffusion along the c-axis and $f_{O_2} = 10^{-7}$ Pa). The closure function $G(x)$ depends on the geometry of the cooling object and the position x within it. For example, using the geometry of a sphere (Dodson 1986):

$$G(x) = \gamma + 4\sum_{n=1}^{\infty} \frac{\left(-1\right)^{n+1} \sin(n\pi x) \ln\left(n\pi\right)}{n\pi x} \tag{4}$$

where $\gamma \approx 0.57721$ is Euler's constant. As T_c appears on both sides of Equation (3), the equation has to be solved iteratively.

Coogan et al. (2002, 2007) applied the above set of equations to determine cooling rates of gabbros from the Oman ophiolite using the measured Ca-concentration profiles in olivine. These profiles can be translated into closure temperatures using the thermometer given in Equation (2):

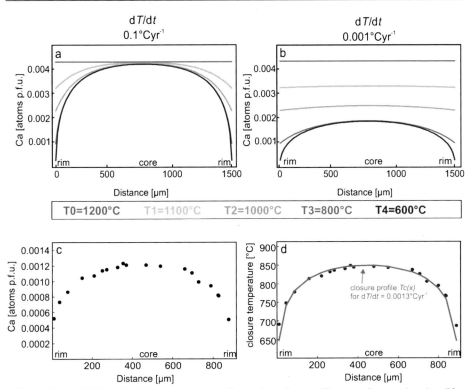

Figure 6. (a) and (b) Simulated Ca concentration profiles in olivine for two different cooling rates based on diffusive exchange between Cpx and Ol with the temperature-sensitive partition coefficient given in Equation (2). (c) A measured Ca concentration profile and corresponding closure profile fit (d) according to Equation (5).

$$T_c(x) = \frac{-0.425P[\text{MPa}] - 5792}{\ln\left(\dfrac{C_{Ca}^{Ol}(x)}{C_{Ca}^{Cpx}}\right) + 1.25} - 273 \tag{5}$$

where P = pressure, $C_{Ca}^{Ol}(x)$ = measured concentration of Ca in olivine at position x, C_{Ca}^{Cpx} = measured concentration of Ca in clinopyroxene (adjacent to the olivine grain). The assumption that Cpx behaves as an infinite reservoir directly implies that C_{Ca}^{Cpx} is effectively homogeneous. The cooling rate s in Equation (3) is iteratively changed until the best fit between the closure profile calculated from Equation (3) and the closure profile calculated from Equation (5) is attained (Fig. 6).

Coogan et al. (2007) performed numerical diffusion simulations to test whether the assumptions of the Dodson model are justified. For example, whether the environment of olivine can be treated as an infinite reservoir for Ca depends on the relative diffusion rates in the exchange couple (Ca diffusion in clinopyroxene and olivine) and their relative modal abundances. Although Coogan et al.(2007) have demonstrated that in this specific case this assumption is not exactly fulfilled, it could be shown that the final error for the derived cooling rate with the Dodson model is on the order of 0.1–0.2 log units. In addition, further errors were introduced because of a simplified geometry for the olivine (sphere) and the fact that diffusion anisotropy was not considered.

VanTongeren et al. (2008) used the same approach as Coogan et al. (2002, 2007) to obtain cooling rates from the lower oceanic crust exposed in the Wadi Khafifah section of the Oman ophiolite. However, instead of measuring and fitting full concentration profiles, these authors measured between 2 and 7 (on average 3) analyses per olivine grain and then used the highest measured Ca values from each crystal core for the calculation of a single T_c and bulk cooling rate. Following this approach, they report uncertainties up to 1.5 log units on the obtained cooling rates. Faak and Gillis (2016) show that chemical alteration as well as secondary fluorescence during electron microprobe analysis from adjacent Ca-rich phases can lead to anomalously high Ca contents, which can only be filtered out meaningfully by looking at complete zoning profiles. They demonstrate how the cooling rate extracted from averaged analyses in the core can overestimate the cooling rate by up to 2 orders of magnitude compared to that extracted from diffusion profiles over the whole grain.

Determination of cooling rates to constrain thermal structure of a crustal segment

Quantifying the rate of temperature changes of rocks during magmatic processes additionally provides insights into energy transfer processes. Different processes of heat exchange (e.g., conductive vs. convective heat transport via hydrothermal circulation) leave behind different patterns of cooling rates as a function of lateral and vertical position within a cooling body. Mapping cooling rates recorded in natural rock samples provides the possibility to determine the thermal structure and thermal evolution of a crustal segment formed by igneous processes. Coogan et al. (2002, 2007) used the Ca-in-olivine geospeedometer to determine cooling rates as a function of stratigraphic depth (i.e., cooling rate profiles demonstrated) within five different segments of the plutonic section of the oceanic crust. They demonstrated that the Ca content in olivine was systematically reset to lower temperatures for deeper gabbros, indicating systematically lower cooling rates with increasing depth (e.g., for fast-spreading mid-ocean ridges cooling rates decreased from $0.1\,°Cyr^{-1}$ at the top of the plutonic section to $0.0001\,°Cyr^{-1}$ at the Moho, Fig. 7). An issue here is that the Ca concentrations towards the olivine rim could not be reproduced by the model, which was explained by fluorescence effects of the electron microprobe measurements. A comparison between cooling rate profiles predicted from different thermal models and their associated cooling rate profiles from diffusion modelling allowed them to extract information about the dominant mechanism of heat loss in the lower oceanic crust.

Faak and Gillis (2016) derived cooling rates from olivine gabbros and troctolites formed at the fast-spreading East Pacific Rise and drilled from the Hess Deep Rift. Here, the Ca exchange between olivine and clinopyroxene as well as Mg exchange between plagioclase and clinopyroxene were used to obtain cooling rates from diffusion modelling (the details on the Mg-in-plagioclase thermometer are given below). The obtained cooling rates for the deeper level plutonics ($>2km$ below the dike/gabbro boundary) span a range from 0.005 to $0.0001\,°Cyr^{-1}$, with a mean value of $0.0011\,°Cyr^{-1}$. In combination with cooling rates obtained from shallow level gabbros from the Hess Deep Rift (Faak et al. 2014), they were able to construct a cooling rate profile over ~2km, starting from the top of the plutonic sequence (Fig. 7). According to Faak and Gillis (2016), both the shape of the cooling rate profile and the actual values are best explained by thermal models with near-conductive cooling of the lower crust and hydrothermal heat extraction at the top.

Determination of magma ascent rates using H diffusion

By analogy with the determination of cooling rates where diffusion modelling is coupled to a thermometer, for the determination of decompression or ascent rates a barometer is needed. Olivine, a nominally anhydrous mineral (NAM), contains H in trace amounts up to 1000 ppm, and experimental calibrations demonstrate that the solubility of H in olivine increases strongly with pressure in water-saturated settings (e.g., Kohlstedt et al. 1996). Thus, during isothermal or adiabatic decompression in fluid-saturated magmas a strong

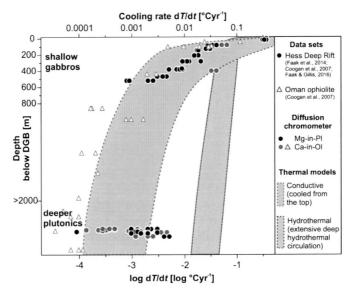

Figure 7. Cooling rates obtained from the Ca in olivine and Mg in plagioclase diffusion modelling plotted against the stratigraphic depth of gabbroic rocks from the Hess Deep Rift and the Oman ophiolite. The cooling rate profile is compared to predictions of cooling rates with depth from different thermal models (for details see Faak and Gillis 2016). Figure modified after Faak and Gillis (2016).

driving force for H diffusion out of olivine exists. Unlike most other elements, since H is the fastest diffusing element in olivine (Kohlstedt and Mackwell 1998; Demouchy and Mackwell 2006) its H content is able to respond rapidly to magma ascent (e.g., Fig. 8), as demonstrated by Demouchy et al. (2006) and Peslier and Luhr (2006) who measured H profiles in olivine from mantle xenoliths using Fourier transform infrared spectroscopy. FT-IR spectroscopy is very sensitive to O–H bonds in different materials and is routinely used to analyse the H content in NAMs as well as to infer its chemical environment. From diffusion modelling of H, a total travel time to the Earth's surface can be determined and after estimating an initial depth, an ascent rate can be calculated. With this approach, the following olivine ascent rates were inferred: ~3–9 m/s for garnet–peridotite xenoliths in alkali basalts (Fig. 8; Demouchy et al. 2006); 0.2–0.5 m/s for spinel–peridotite xenoliths in alkali basalts (Peslier and Luhr 2006); 5–37 m/s for garnet and spinel–peridotite xenoliths in kimberlites, Peslier et al. (2008); 3–12 m/s for spinel–peridotite xenoliths in (Eifel volcanoes, Denis et al. 2013); and 0.2–25.3 m/s (Hawaian volcanoes, Peslier et al. 2015). These ascent rates are broadly consistent with independent methods (see for example Fig. 7 in Peslier et al. 2015).

However, there are a number of additional complexities that need to be considered in this case:

(i) Estimation of the initial profile: Comparison of the water contents between pyroxenes and olivine indicate that the olivine cores have lost their original H content, once in equilibrium with the thermodynamic environment at greater depth (e.g., Demouchy et al. 2006; Denis et al. 2013). The H contents of pyroxene are in general not zoned, which indicates that they still preserve their initial H content. This observation appears to be consistent with the known relative diffusion rates of H in orthopyroxene and olivine. The initial H content in olivine can be then reconstructed by using the H partition coefficient between orthopyroxene and olivine, $K_{p(H)}^{Ol/Opx}$. Either the experimentally measured $K_{p(H)}^{Ol/Opx}$ were used (e.g., Peslier and Luhr 2006; Denis et al. 2013) or estimated indirectly using experimental calibrations of the water content in orthopyroxene as a function of water fugacity, which was used to estimate the water fugacity in the magma chamber at depth. In thermodynamic equilibrium this water fugacity reflects also the initial water content

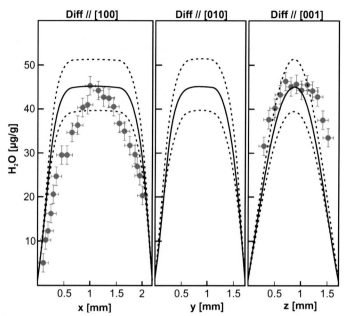

Figure 8. Water content of olivine (in ppm H_2O by weight) as a function of position parallel to each axis using calibration of Bell et al. (2003). Polarized infrared profiles are across an olivine crystal from Pali-Aike alkali basalt, Patagonia, Chile. Solid lines are not a fit but represent calculated diffusion profiles assuming the slower diffusion mechanism for H in olivine. As a function of temperature, different durations for dehydrogenation are extracted and yield a global ascent rate of 6 ± 3 m/s Dotted lines give $\pm5\%$ on ascent time. Note that the measured data indicate a stronger H loss parallel to the *a*-axis compared to the *c*-axis, which would be more consistent with the faster diffusion mechanism according to Kohlstedt and Mackwell (1998). See also the more recent model calculations of Thoraval and Demouchy (2014) for these profiles. Figure modified from Demouchy et al. (2006).

in olivine and which can be calculated in an analogous way by using experimental solubility data (Demouchy et al. 2006). In the case that the pyroxenes may have also lost their initial H content, an alternative way of estimating the initial H content in olivine could be used for fluid-saturated systems: the water fugacity can be calculated from available thermodynamic models (for example using the software VolatileCalc, Newman and Lowenstern 2002) if realistic estimates of the initial volatile contents can be made, e.g., from melt inclusions. Assuming thermodynamic equilibrium at depth, the experimental calibration of the water content in olivine as a function of water fugacity, P, T, and X_{Fe} (e.g., Zhao et al. 2004) could be used to calculate the equilibrium content of H_2O in olivine (Fig. 9). In any of these cases independent estimates of temperature and pressure are necessary to perform these calculations. If, for example, the estimate of the initial H content were too high, a slower ascent rate would be predicted.

(ii) The boundary condition: In all published models so far a fixed boundary condition was used, where it has been assumed that from the outset of magma ascent the entire H content at the rim is lost. This assumption is clearly erroneous as the H content at the rim tends to equilibrate with the local environment, which defines at depth a water fugacity that is significantly greater than 0 Pa (see for example the discussion in Peslier et al. 2008). With decreasing depth the water fugacity decreases (more strongly in fluid-saturated systems) imposing an increased driving force for H diffusion out of olivine (or any other NAMs), as illustrated in Figure 9. As a net result, by assuming a fixed boundary condition the ascent rate would be overestimated (e.g., Costa et al. 2010b).

(iii) Another consequence of the loss of H_2O in the core is that a 3D model is required since diffusion fluxes from all directions out of the core need to be taken into account. Due to the diffusion anisotropy of H, it is also necessary to measure the H concentration distribution in a cross section along two main crystallographic axes, ideally including the fastest diffusion direction to compare with the model calculations (Demouchy et al. 2006). Measurement along one just profile direction can be very misleading (as illustrated in Thoraval and Demouchy 2014). In a recent paper Ferriss et al. (2015) illustrated an approach by performing transmission FT-IR on whole crystals without cutting them to compare the measurements with 3D simulations.

(iv) Diffusion and incorporation mechanisms of atoms in crystals are strongly coupled. How H in olivine is incorporated is still debated with little agreement between research groups in the interpretation of the IR spectrum and hence the structural site of water (e.g., see the review article of Beran and Libowitzky 2006). This is chiefly related to the presence of different peaks in the IR spectrum of olivine, which vary markedly from one olivine to another, in contrast to other NAMs. The following sites have been suggested for olivine: hydroxyl groups associated with vacancies either on the metal site or the Si site; or hydroxyls coupled to trace elements like Al or Ti. Certain peaks in the spectrum have been clearly identified to be characteristic of association of a hydroxyl group with trivalent cations (Fe^{3+}, Cr^{3+}, Al^{3+}) and Ti^{4+} (Berry et al. 2005, 2007). Interestingly, the latter peaks dominate in most IR spectra of mantle olivines, whereas in most experimental high-pressure studies where large amounts of water are incorporated (> 200 ppm) these peaks are absent or very minor. Consequently, it has been established that at higher pressures (Mosenfelder et al., 2006; Withers and Hirschmann 2008) there is no variation of solubility with different trace element concentrations e.g., Al, Cr, Ti, and Fe^{3+}. These observations show that various sites for hydroxyl in olivine have to be considered and in particular for hydroxyls that are associated with other trace elements this must have also a consequence for the dehydrogenation mechanism and the corresponding rate. For example, for the "Titanium-clinohumite" defect (Walker et al. 2007),

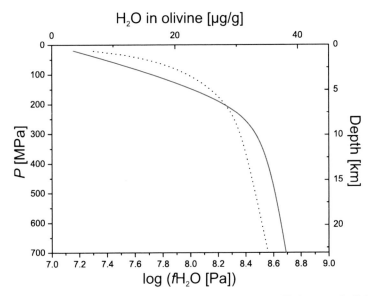

Figure 9. Model calculation to illustrate the driving force for water diffusion out of olivine during ascent. Water fugacity (dotted line) for a basaltic melt was calculated as a function of pressure using VolatileCalc (Newman and Lowenstern 2002) assuming a closed system, $T = 1200\,°C$, and initial H_2O and CO_2 contents of 5 wt% and 3000 ppm, respectively. The corresponding H_2O content in olivine in equilibrium with the melt (solid line) was calculated using the experimental calibration of Zhao et al. (2004). Note that only in the second stage during ascent, after H_2O degasses from the melt, can a strong decrease of H_2O in the olivine be expected.

dehydrogenation could form Ti-bearing precipitates and/or involve the coupled flux of Ti and H. However, diffusion of H in olivine has only been investigated previously by hydrogenation experiments (Mackwell and Kohlstedt 1990; Kohlstedt and Mackwell 1998; Demouchy and Mackwell 2006) (with one unpublished study on H–D exchange in forsterite cited by Ingrin and Blanchard 2006). Two different rates of hydrogenation have been observed, where the faster one only allows a partial equilibration of olivine (Kohlstedt and Mackwell 1998). The different rates are related to different charge-balancing mechanisms and it was suggested that the faster mechanism involves interdiffusion of electron holes and protons and the slower mechanisms a coupled diffusion of metal vacancies and two protons. Metal vacancies would then be the favoured site for the protons as suggested by Kohlstedt et al. (1996). In any case, for both proposed diffusion mechanisms, it is unclear which one controls the rate of dehydrogenation, or even if olivine will be dehydrogenated by the slow mechanism and whether once a certain level is reached the dehydrogenation accelerates. In Demouchy et al. (2006), Peslier and Luhr (2006), Denis et al. (2013), and Peslier et al. (2008, 2015) only the slow diffusion mechanism has been used to infer ascent rates since using the faster diffusion rate would indicate unrealistically fast rates. However, the two diffusion mechanisms have a different anisotropy and the measured profiles of Demouchy et al. (2006) appear more consistent with the anisotropy of the faster mechanism, whereby $D_{//a} > D_{//c}$, opposite to the slow mechanism (Fig. 8). Different mobilities of different hydrogen defects were also documented by Peslier et al. (2015) in olivine crystals from Hawaiian basalts. Here H associated with metal vacancies is apparently more mobile compared to H associated with Ti (see also Padron-Navarta and O'Neill 2014). Therefore, the main issue of this approach is the uncertain dehydrogenation rate in olivine, which strongly depends on the various possible incorporation mechanisms and structural H sites. Clearly more experiments are needed for better constraints on the derived ascent rates.

Ferriss et al. (2016) concluded that the common observation of hydrogen zonation in mantle xenolith olivines but not in clinopyroxenes implies that hydrogen diffusion is much faster in olivine than in pyroxene, which then requires the operation of the fastest diffusion mechanism quantified in olivine and diffusivities in clinopyroxene at the lower end of the measured range (see also section on Cpx below). Such high diffusivities in olivine strongly suggest that water in mantle xenoliths has at least partially equilibrated with the host magma, and that the diffusion profiles observed in mantle xenolith olivine reflect only the final stage of ascent after water begins to exsolve.

PLAGIOCLASE

Since the fundamental experiments of Bowen (1913), plagioclase (Pl) feldspar has a long tradition of being used as an archive of magmatic processes. Various zoning types for major element (albite–anorthite) zoning have been classified (normal, reversed, oscillatory, patchy, e.g., Streck 2008), which are related to different types of growth conditions and related magmatic settings. The ability of plagioclase to conserve even nm-scaled major element zoning (e.g., Ginibre et al. 2002a; Saunders et al. 2014) is due to the very sluggish coupled inter-diffusion rates of NaSi and CaAl (e.g., Grove et al. 1984). Minor and trace element zoning have been studied to infer open versus closed system processes as these elements are thought to be less dependent on the intensive thermodynamic parameters of the system (e.g., Blundy and Shimizu 1991; Singer et al. 1995; Davidson and Tepley 1997; Ginibre et al. 2002b; Berlo et al. 2007; Ruprecht and Wörner 2007; Smith et al. 2009). However, diffusivities of trace and minor elements like Li, K, Mg, and potentially Sr and Ba, are fast enough that their respective growth zoning patterns could be disturbed, depending mainly on the magma temperature and crystal residence times. An extreme case is Li diffusion, which is extremely fast in plagioclase (Giletti and Shanahan 1997) and has been used to determine the final ascent rate of an explosive eruption (Charlier et al. 2012).

The first approach to reproduce chemical zoning in plagioclase by diffusion modelling was applied by Zellmer et al. (1999) for Sr. Based on the strong dependence of Sr plagioclase–melt partition coefficients on the anorthite content (Blundy and Wood 1991) a crucial point was already realized—if the albite (Ab)-anorthite (An) content of plagioclase is zoned, and can be considered to be effectively immobile, the re-equilibration of more mobile trace elements results in a non-uniform element distribution within the plagioclase. This is related to the dependence of the chemical potential of trace element components in plagioclase with the anorthite content, which can be described by non-ideal interactions of individual trace elements with albite and anorthite (Blundy and Wood 1991; Dohmen and Blundy 2014). However, as shown by Costa et al. (2003) this variation of the chemical potential with anorthite has also consequences for the respective diffusion equation of the trace element since the diffusive flux is proportional to the chemical potential gradient, not the concentration gradient, in the more general case (e.g., Ganguly 2002). Consequently, diffusion modelling for chemically zoned plagioclase with non-ideal interactions is more complex and requires knowledge of how the chemical potential of the trace or minor element (or the respective chemical component) varies with the anorthite content.

Diffusion coefficients and specifics for trace element diffusion in plagioclase

Costa et al. (2003) derived an equation for diffusion modeling of Mg in plagioclase, where the variation of the chemical potential of the Mg-component with anorthite is considered:

$$\frac{\partial C_{Mg}}{\partial t} = \frac{\partial}{\partial x}\left(D_{Mg} \cdot \frac{\partial C_{Mg}}{\partial x} - \frac{D_{Mg} C_{Mg}}{RT} \cdot A_{Mg} \cdot \frac{\partial X_{An}}{\partial x}\right) \qquad (6)$$

where D_{Mg} stands for the diffusion coefficient of Mg in plagioclase, which may be also a function of the fraction of anorthite, X_{An}, and/or the activity of SiO_2, a_{SiO_2} (Faak et al. 2014 vs. Van Orman et al. 2014, see discussion below). The parameter A_{Mg} is a thermodynamic factor reflecting the non-ideal interaction of the Mg-component with Ab and An and was inferred from the empirical fit of Bindeman et al.(1998) for plagioclase–melt partition coefficients, as follows:

$$RT \cdot \ln K^{pl-l}_{p(Mg^{2+})} = A_{Mg} \cdot X_{An} + B_{Mg} \qquad (7)$$

Equation (6) can be also generalized for other elements showing similar linear compositional dependence of the partition coefficient as in Equation (7). In the derivation of Equation (6) by Costa et al. (2003) it was assumed that Equation (7) reflects the explicit dependence of the chemical potential of Mg on An. However, as discussed in Bindeman et al. (1998) the parameters in Equation (7) were obtained from Drake and Weill's (1975) experiments on basaltic melts where the variation of the An content was related to a simultaneous change in T and liquid composition. Thus, as demonstrated by thermodynamic modelling of Dohmen and Blundy (2014), Equation (7) also includes the implicit dependence of An on T and liquid composition. In the thermodynamic model of Dohmen and Blundy (2014) the chemical potential variation with anorthite of various trace components were inferred from the application of the lattice strain model of Blundy and Wood (1994) to experimental plagioclase–melt partitioning data from isothermal series in the diopside–albite–anorthite system. The derivation of the diffusion equation applying this thermodynamic model and the respective quasi steady state condition, which describes the partial equilibrium of trace elements in plagioclase with the surrounding melt (open system condition) or internal equilibration (closed system condition) is presented in the Electronic Appendix, including a stable numerical scheme to solve this diffusion equation (a Mathematica or C++ source code can be requested from the authors). Finally, the derived diffusion equation (Eqn. A8) is shown to be effectively equivalent to Equation (6) since the variation of $RT \ln(\gamma_i)$ with X_{An} is approximately linear.

The main implication of the thermodynamic model of Dohmen and Blundy (2014) is that the thermodynamic parameter A_i differs from the values inferred from the relations presented in Bindeman et al. (1998), as shown in Table 1. The latter values were used in various diffusion modelling studies with Sr and Mg (e.g., Costa et al. 2003; Druitt et al. 2012; Ruprecht and Cooper 2012; Cooper and Kent 2014) and are potentially incorrect according to Dohmen and Blundy (2014) and Faak et al. (2013). Different values for the parameter A_i have consequences for the diffusion profiles and the inferred time scales. In particular for Mg the sign of A_i of Dohmen and Blundy (2014) and Faak et al.(2013) compared to those inferred from Bindeman et al. (1998) is the opposite, which has implications both for the diffusion flux and the "equilibrated profile". In Figure 10 model results are shown for Mg and Sr diffusion in a synthetic, normally zoned plagioclase crystal (albitic overgrowth on anorthite-rich core and corresponding positive correlations of Mg and Sr). A particular advantage of using Sr is that it is compatible in plagioclase. Consequently the variation in Sr with X_{An} resulting from growth will have a positive slope, whereas diffusive re-equilibration will impart a negative slope (see also Cooper and Kent 2014). According to the diffusion data of Cherniak and Watson (1994) Sr diffusion should be detectable with ion probe measurements (5–10 µm spatial resolution) after about 10 yr at 890 °C or a month at 1200 °C.

For Mg the distinction between growth zoning and zoning affected by diffusion is more difficult. According to Dohmen and Blundy (2014) Mg has an A-parameter of positive sign, opposite to those of other relevant divalent cations like Sr or Ba, and opposite to those inferred by Bindeman et al. (1998). According to the lattice strain model adopted by Dohmen and Blundy (2014) the reason for the positive sign is that, unlike Sr^{2+} and Ba^{2+}, the ionic radius of Mg^{2+} is smaller than the optimum ionic radius for divalent cations, r_0^{2+}, which decreases from Ab to An. Therefore with increasing X_{An}, cations like Sr^{2+}, with a larger ionic radius than r_0^{2+}, become energetically less favourable for this site, whereas Mg^{2+} becomes more favourable. These energetics quantified by the lattice strain model are reflected in the A-parameters reported in Table 1. According to the positive A_{Mg}, at partial equilibrium (or quasi steady state) of Mg it would show a positive, roughly linear, variation with X_{An}. Unlike for Sr, the growth zoning of Mg in plagioclase depends on the co-precipitating mafic minerals that control the Mg budget in the melt. In case of the cotectic growth with Pl of mafic minerals like Ol, Opx, Cpx, the Mg in the melt tends to decrease and for normal zoning of X_{An}, Mg should decrease as well. Thus, during growth a positive correlation of Mg with X_{An} is obtained, potentially difficult to distinguish from diffusive equilibration, depending on the exact Mg vs. X_{An} slope. However, in a number of studies of plagioclase phenocrysts from a variety of volcanic rocks (Costa et al. 2003, 2010a; Moore et al. 2014) a negative correlation of Mg with X_{An} was observed, which were reproduced assuming partial equilibrium and using the negative A_{Mg} parameter of Bindeman et al. (1998) (e.g., see also Fig. 10). Alternatively, if the positive A_{Mg} of Dohmen and Blundy (2014) and Faak et al. (2013) is correct, the observed

Table 1 Values for the thermodynamic parameter A_i (kJ/mol) representing the non-ideal interaction of the trace element *i* with the major components Ab and An, as inferred from plagioclase–melt and plagioclase–clinopyroxene partitioning data.

Reference	A_{Mg}	A_{Sr}	A_{Ba}	A_{Li}	A_{K}	A_{Rb}
Dohmen and Blundy (2014) 1200 °C	15.8	−17.4	−35.1	−1.7	−8.0	−15.9
Dohmen and Blundy (2014) 900 °C	13.7	−15.1	−30.5	−2.5	−8.8	−16.7
Bindeman et al. (1998)	−26.1	−30.4	−55.0	−6.9	−25.5	−40.0
Faak et al. (2013)	16.9					

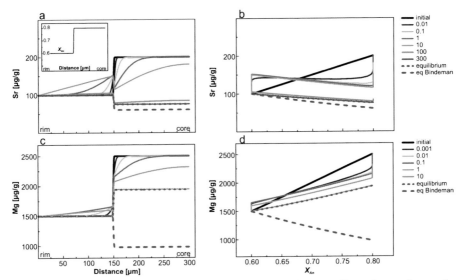

Figure 10. Results of model calculations for Sr (a) and Mg (c) diffusion profiles and respective correlations of their concentrations with X_{An} in plagioclase, (b) and (d), respectively. For the initial step growth zoning of X_{An} (see inset in panel (a)) and trace elements (synthetic crystal) a positive correlation of both, Sr and Mg, with X_{An} was assumed. Temperature was assumed to be 1200 °C. The numbers in the legend denote the time in years. For D_{Sr} and D_{Mg} we used the experimental data of Cherniak and Watson (1994) and Van Orman et al. (2014), respectively. Note the uphill diffusion flux for Sr at the interface between low and high anorthite, which is the result of the second term in the diffusion equation (Eqn. 6). It should be emphasized that such concentration gradients have been not observed so far in any natural sample, which would be an independent test of the diffusion model and would also allow to estimate the parameter A_{Sr} or A_{Mg}. Therefore, the diffusion model should be experimentally tested.

Mg concentration patterns would lie far from equilibrium and hence should reflect growth zoning, whereby Mg in plagioclase increases as An decreases. Growth zoning with such a negative Mg vs. X_{An} slope would require plagioclase-dominated crystallization, such that Mg behaves as an incompatible element in the crystallising magma. This need not mean that no mafic minerals were present or growing in the system but simply that the larger plagioclase phenocrysts that are typically analysed may reflect a stage of magma differentiation where mostly plagioclase crystallizes (or mafic minerals are resorbed). An alternative resolution to this apparent discrepancy between experimental and natural observations is if Mg occupies the T-site, rather than the M-site, in natural plagioclases. The lattice strain model described above pertains to substitution of Mg^{2+} for Ca^{2+} or Na^{1+} on the large, approximately 10-fold co-ordinated M-site. It has been proposed (e.g., Longhi 1976; Peters et al.1995; Miller et al. 2006) that in Ca-rich plagioclases Mg may also be incorporated onto the tetrahedral site, via the substituent molecule $CaMgSi_3O_8$. In that case it is the size mismatch between Mg^{2+} ions and the T-site that will control partitioning and the relationship with X_{An} may differ from that described above for M-site incorporation. Further work on the Mg environment in natural and synthetic plagioclases is required to resolve this matter.

For other incompatible trace elements, such as Ba, Li, K, and Rb, both growth and diffusion will impart negative correlations with X_{An}, making the effects of growth and diffusion difficult to deconvolve. Thus, Sr and potentially Mg (in case of reversed growth zoning) are the elements for which, within reasonable timescales, an effect of diffusive re-equilibration should be clearly observed, without knowing in detail the initial growth zoning.

According to the experimental diffusion data of Cherniak and Watson (1994) for D_{Sr}^{Pl} and Van Orman et al. (2014) for D_{Mg}^{Pl}, the concentration profiles of Sr and Mg change at different time scales (e.g., Fig. 10), whereby Mg re-equilibrates about a factor 30 faster than Sr for X_{An} between 0.6 and 0.8. Both diffusion coefficients depend strongly on the anorthite content (Fig. 11). In contrast, Faak et al. (2013) did not observe a dependence of the diffusion rate of Mg in plagioclase on X_{An} (for an investigated range of X_{An} between 0.5 and 0.8) but found a strong influence of D_{Mg}^{Pl} on the chemical potential of silica in the system. Both studies, Faak et al. (2013) and Van Orman et al. (2014), are consistent with earlier data of LaTourrette and Wasserburg (1998) measured for natural single crystals of anorthite ($X_{An}=0.95$) at 0.1 MPa between temperatures of 1200 and 1400 °C. However, based on Faak et al. (2013), D_{Mg}^{Pl} is rather similar to D_{Sr}^{Pl} for plagioclase with X_{An} around 0.4. Whether Mg diffuses on a similar timescale to Sr may be inferred from investigation of natural plagioclases.

Determination of magma residence times using Mg or Sr diffusion

Magma residence times or, to be more precise, timescales between mixing (or mingling) events using Mg and Sr diffusion modelling in plagioclase were determined by Zellmer et al. (1999), Tepley et al. (2000), Costa et al. (2003, 2010a), Druitt et al. (2012), Ruprecht and Cooper (2012), Cooper and Kent (2014), and Singer et al. (2016). In these studies typically either the Mg or the Sr zoning was modelled to extract timescales. A general strategy to reconstruct the growth zoning of trace elements was to use the anorthite growth zoning and find a correlation of the Sr or Mg contents with X_{An} during growth. For example in Cooper and Kent (2014) for plagioclases of Mt. Hood the initial Sr profiles were estimated by two separate methods: (1) using the observed general correlation between Sr and anorthite and (2) using the results of Rhyolite-MELTS modelling of the liquid line of descent to calculate Sr contents of plagioclase from the plagioclase–melt partition coefficient. From both methods only a minor difference of the slopes for Sr vs. X_{An} was obtained. This similarity suggests here that very little equilibration of Sr by diffusion occurred. As illustrated in Figure 10, the residence time was then inferred from the deviation of the measured Sr vs. X_{An} correlation to the "partial" chemical equilibrium correlation, which is negative. Cooper and Kent (2014) proposed strong temperature fluctuations in the magmatic system between 750–900 °C, making Sr effectively immobile in plagioclase throughout most of the residence time.

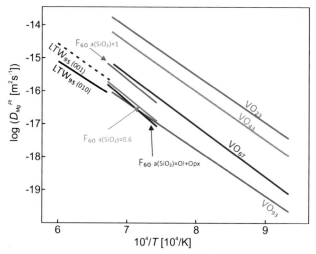

Figure 11. Arrhenius plot of measured Mg diffusion coefficients from the literature (LTW: LaTourrette and Wasserburg 1998; F: Faak et al. 2013; VO: Van Orman et al. 2014) and their effect on X_{An} and a_{SiO_2} (X_{An} and a_{SiO_2} are given as subscripts).

Thus only diffusive effects close to the peak temperature were visible. By this model, the apparent discrepancies between the residence times inferred from the short-lived U-series isotopes and those of diffusion modelling was explained. Clearly the effect of a fluctuating temperature field on diffusion should be borne in mind when interpreting timescales.

Determination of cooling rates using Mg diffusion

A widespread mineral association in mafic and intermediate igneous rocks is plagioclase–clinopyroxene. Faak et al. (2014) developed an Mg-in-plagioclase geospeedometer, based on the diffusive exchange of Mg between plagioclase and clinopyroxene during cooling. The Mg-in-plagioclase geospeedometer builds upon the diffusion model proposed by Costa et al. (2003) to account for the effect of zoned X_{An} (Eqn. 6).

According to Faak et al. (2013) the partition coefficient of Mg between plagioclase and clinopyroxene can be parameterised, as follows:

$$\ln K^{Pl\text{-}Cpx}_{p(Mg)} = 1.6 + \frac{-9219}{T} + \frac{16913 \left[Jmol^{-1} \right]}{RT} X_{An} + \ln a_{SiO_2} \tag{8}$$

where T is in Kelvin. The partition coefficient $K^{Pl\text{-}Cpx}_{p(Mg)}$ decreases with temperature, i.e., during cooling an exchange of Mg between plagioclase and clinopyroxene rims is required to maintain equilibrium, which can be described by one of the following net transfer reactions (dependent on the assumed site occupancy of Mg in plagioclase):

$$\text{Mg on T-site: } CaSiO_3(Cpx) + MgSiO_3(Cpx) + SiO_2 = CaMgSi_3O_8(Pl) \tag{9}$$

$$\text{Mg on M-site: } MgSiO_3(Cpx) + SiO_2 + Al_2O_3 = MgAl_2Si_2O_8(Pl) \tag{10}$$

Consequently, this thermometer is not based on a simple exchange reaction and, in addition to temperature, also depends on the silica activity (or more precisely the chemical potential of silica). The relevance of the chemical environment is common for trace element thermometry (e.g., effect of silica or TiO_2 activity for the Ti-in-zircon or Zr-in-rutile thermometer, e.g., Ferry and Watson 2007). Since Mg is a major element in clinopyroxene, the Mg-content (the $MgSiO_3$ component) is effectively buffered and does not change due to the Mg flux from plagioclase to clinopyroxene during cooling (cf. Ca exchange between Ol and Cpx discussed above). Thus, for a known concentration of Mg in clinopyroxene, an apparent temperature can be calculated at each point along a measured Mg-profile in plagioclase. Based on Reaction (9), the $CaMgSi_3O_8$ component dissolves into clinopyroxene and produces some extra silica that in many cases could in turn be used to produce minor amounts of $MgSiO_3$ (En) from Mg_2SiO_4 (Fo) according to reaction Fo + SiO_2 = En. Therefore, local equilibrium between plagioclase and clinopyroxene interfaces is not controlled only by lattice diffusion within these minerals but also involves processes such as grain boundary diffusion (e.g., transport of silica), which might be the rate limiting process (e.g., Dohmen and Chakraborty 2003).

The diffusion model of Faak et al. (2014) is based on the assumption that local equilibrium is achieved instantaneously at the interface between both phases, and hence the concentration of Mg at the interface of the plagioclase crystal can be calculated from $K^{Pl\text{-}Cpx}_{p(Mg)}$ at any T. According to Equations 6 and 8, the evolution of the resulting concentration profile depends on the sub-solidus cooling history, as well as on the X_{An}-profile (see Fig. 12). For that reason, the standard Dodson model (or any recent extension, e.g, Ganguly and Tirone 1999) cannot be used to obtain the closure profile and hence cooling rates (e.g., as for Ca in olivine). Hence the Mg profiles were calculated numerically by solving the appropriate diffusion equation (Eqn. 6). The temperature interval over which cooling rates can be recorded ranges from T_c at the crystal core (T_c^{core}) to T_c at the crystal rim (T_c^{rim}). The exact values for T_c^{core} and T_c^{rim}

depend on the cooling rate and on the size of the crystals (for plutonic rocks from the lower oceanic crust, this interval is between 1100–750 °C; Faak et al. 2015). However, modelling the redistribution of Mg between plagioclase and clinopyroxene during cooling requires an estimate of the initial conditions (T_{start}) for the system. Faak et al. (2014) show that for a system with peak temperatures above 1100 °C (i.e., most mafic igneous systems), a grain size of 1 mm and a cooling rate of 0.1 °Cyr⁻¹, the memory of the initial conditions would be erased by diffusion (thus one assumption of the Dodson model would be fulfilled). In these cases, a reasonable approach is to calculate an equilibrium concentration profile of Mg in plagioclase based on equilibrium partitioning with clinopyroxene as starting profile.

Two crucial parameters in the Mg-in-plagioclase geospeedometer, the diffusion coefficient D_{Mg}^{Pl} and the partition coefficient $K_{p(Mg)}^{Pl-Cpx}$, depend on the silica activity a_{SiO_2} of the system (Faak et al. 2013). Therefore, the application of this approach requires knowledge of the silica activity (as a function of temperature). In many systems of interest, a_{SiO_2} is either buffered by a given mineral assemblage or can at least be constrained to lie between two buffer reactions (e.g., Carmichael et al. 1970). For example, Faak et al. (2013, 2014) provide a method to constrain silica activity for the case of a system buffered either by pure SiO_2 (e.g., quartz) or by coexisting olivine–orthopyroxene.

The Mg-in-plagioclase geospeedometer has been tested with regard to the robustness of the calculated cooling rates and the precision is found to be better than half an order of magnitude (Faak et al. 2014, 2015). Faak et al. (2014) investigated the effect of different

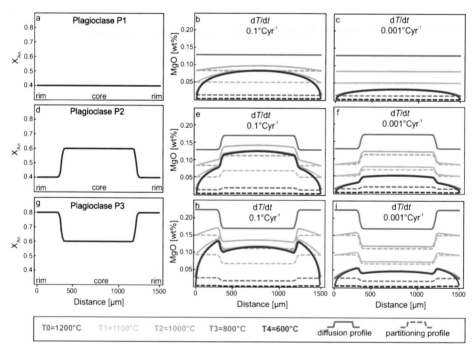

Figure 12. Schematic evolution of Mg-profiles in plagioclase during cooling (in contact with clinopyroxene) for three different assumed plagioclase crystals (length = 1500 μm, plane sheet geometry). P1 (a, b, c) has a flat X_{An}-profile, P2 (d, e, f) and P3 (g, h, i) have stepped X_{An}-profiles that are reversely zoned. Panels (b), (e) and (h) show the calculated Mg-profiles at different temperatures for a linear cooling rate of 0.001 °Cyr⁻¹. Panels (c), (f) and (i) shows the calculated Mg-profiles for the same temperatures as above, but for faster linear cooling of 0.1 °Cyr⁻¹. Dashed lines show the calculated theoretical partitioning profiles for Mg in plagioclase in equilibrium with clinopyroxene at different temperatures. (Figure modified after Faak et al. 2014).

diffusion coefficients (Faak et al. 2013 vs. Van Orman et al. 2014) on the cooling rate that would be obtained for plagioclase crystals with normally and reversely zoned X_{An} profiles. For their specific example, they show that the cooling rate obtained for a normally zoned plagioclase crystal will be by a factor of 3 faster when using an X_{An}-dependent form of D_{Mg}^{Pl} compared to a D_{Mg}^{Pl} that does not depend on X_{An}. For a reversely zoned plagioclase crystal, the obtained cooling rate will be a factor of 4 slower when an X_{An}-dependent D_{Mg}^{Pl} is used.

Additionally, Faak and Gillis (2016), in a combined approach, applied the Mg-in-plagioclase geospeedometer and the Ca-in-olivine geospeedometer (Coogan et al. 2002, 2007) on the same samples from the lower oceanic crust (Fig. 7). They demonstrate that these independent geospeedometers yield similar cooling rates, implying that the absolute values obtained from the two methods are robust.

PYROXENES

Pyroxenes have been used to constrain time scales of processes in silicic systems that lack olivine. The observed diffusion profile lengths of Fe-Mg in natural orthopyroxene are often much shorter (a few mm or less) than in olivine. This observation appears to be consistent with the measured diffusion rates in Opx and Cpx that are generally slower than in olivine (e.g, about 1–2 log units for Fe–Mg: Müller et al. 2013; Dohmen et al. 2016a), considering similar timescales for the respective magmatic processes. Therefore BSE imaging (for Fe–Mg, e.g., Morgan et al. 2004, Saunders et al. 2012a) or other higher resolution methods have been used (TOF-SIMS: Saunders et al. 2012b; NanoSIMS: Saunders et al. 2014) to resolve in detail the chemical zoning. Fast processes can be determined using Li diffusion in Cpx. For example, concentration profiles of Li in Cpx from martian meteorites (nakhlites) were modelled to determine cooling rates of the respective magmatic system (Beck et al. 2006). Here, *in situ* zoning of the Li isotopes was measured, and found to be consistent with diffusive fractionation, as was also demonstrated in other cases (e.g., Gallagher and Elliott 2009). For a review of diffusion data in pyroxenes we refer to Cherniak and Dimanov (2010) and for more recent data to Müller et al. (2013) for Cpx and Dohmen et al. (2016a) for Opx. Here only a short overview and an update of diffusion data are given.

Diffusion coefficients

Orthopyroxene. Until very recently (Dohmen et al. 2016a) no directly measured diffusion data for Fe–Mg interdiffusion in Opx were available. Therefore in the past, D_{Fe-Mg}^{Opx} for diffusion chronometry was inferred from Mg tracer diffusion measured by Schwandt et al. (1998) in Opx with 88% of enstatite component (En_{88}) at f_{O_2} of the iron–wuestite buffer, IW, (e.g., Saunders et al. 2012a) or from theoretical correlations between Fe–Mg diffusion and Fe–Mg order–disorder rates on the M1 and M2 sites (Ganguly and Tazzoli 1994; Allan et al. 2013). Based on the theoretical work of Ganguly and Tazzoli (1994) and the experimental data for order–disorder data of Stimpfl et al. (2005), Allan et al. (2013) parameterised the dependence of D_{Fe-Mg}^{Opx} on T, f_{O_2}, and $X_{Fe} = Fe/(Fe + Mg)$. However, the first direct experimental measurements of Fe–Mg interdiffusion in Opx by Dohmen et al. (2016a) demonstrate that the dependence on f_{O_2} is much smaller than estimated by Allan et al. (2013). In addition, the available data on order–disorder rates of Fe–Mg in Opx were discussed in Dohmen et al. (2016a) and it was concluded that at constant P, T, and f_{O_2}, D_{Fe-Mg}^{Opx} varies less than 0.5 log units for $0.1 < X_{Fe} < 0.5$. As a consequence, the Mg tracer diffusion coefficients of Schwandt et al. (1998) measured for Opx close to enstatite composition and at $f_{O_2} = IW$ are very similar to D_{Fe-Mg}^{Opx} with $X_{Fe} < 0.5$ at f_{O_2} close to the Ni–NiO buffer, NNO, (Fig. 13). According to Dohmen et al. (2016a), D_{Fe-Mg}^{Opx} parallel to the *c*-axis is a factor of 3.5 faster than parallel to the *a*-axis. Therefore for accurate determination of time scales the crystallographic orientation of the profile should be established.

With the exception of H (Li has so far not been measured for Opx), other elements (Cr, REE, Pb, and Ti) diffuse significantly more slowly than Fe–Mg, by at least an order of magnitude. Recent data measured after the review of Cherniak and Dimanov (2010) are those of Sano et al. (2011) for Nd and of Cherniak and Liang (2012) for Ti. As for olivine the behaviour of H diffusion in Opx is complex and depends on the relevant incorporation and diffusion mechanism (e.g., Stalder and Skogby 2003, 2007; Stalder and Behrens 2006).

Clinopyroxene. D_{Fe-Mg}^{Cpx} was measured by Müller et al. (2013) between 800–1200 °C and is slower than in Opx with a similar X_{Fe} (Müller et al. 2013), but has the same activation energy within experimental error, 320 kJ/mol and 308 kJ/mol, respectively. The D_{Fe-Mg}^{Cpx} of Müller et al. (2013) are consistent with $D_{Fe-Mg-Mn}^{Cpx}$ measured by Dimanov and Wiedenbeck (2006), although in the latter work a stronger effect of f_{O_2} could be identified (for details, see discussion in Müller et al. 2013). Ca tracer diffusion rates or inter-diffusion processes involving Ca are at least one log unit slower than Fe–Mg interdiffusion in Cpx.

Faster diffusing elements are H and Li. As in olivine and orthopyroxene, their respective diffusion rates are dependent on more than just temperature. H diffusion experiments by, for example, Sundvall et al. (2009) and Ferriss et al. (2016) have shown that diffusion rates of H vary by up to five orders of magnitude at a given temperature where the total Fe content (and respective effects on the Fe^{3+} content) and tetrahedral Al content in Cpx appear to have a dominant control (Ferriss et al. 2016).

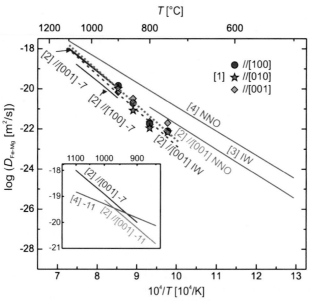

Figure 13. Measured and estimated D_{Fe-Mg} from the literature in an Arrhenius plot: [1] Schwandt et al. (1998); [2] Dohmen et al. (2016); [3] Ganguly and Tazzoli (1994); [4] Allan et al. (2013). Different symbols are used for Mg tracer diffusion data from different crystallographic directions of Schwandt et al. (1998), as indicated in the figure. The diffusional anisotropy of D_{Mg} is not well resolved with this dataset measured at f_{O_2}=IW, but can be well described by the Arrhenius relation of Dohmen et al. (2016), which has only a weak dependence on f_{O_2}. Also shown are D_{Fe-Mg} for En_{91} as calculated from the expressions of Ganguly and Tazzoli (1994) at f_{O_2}=IW [2] and from Allan et al. (2013) at f_{O_2}=NNO [3]. The calculated values of D_{Fe-Mg} at a constant oxygen fugacity are shown in the inset of this figure (log f_{O_2} [Pa]=−11 according to Allan et al. (2013) and according to Dohmen et al. (2016) at log f_{O_2} [Pa]=−11 and log f_{O_2} [Pa]=−7). The stronger f_{O_2} dependence of diffusivity postulated by Allan et al. (2013) produces a low activation energy and significant deviation from measured values at various temperatures. Figure modified from Dohmen et al. (2016).

The first measurements of Li diffusion in clinopyroxene found diffusion profiles with a simple Arrhenius relationship for f_{O_2} close to the wüstite–magnetite buffer, WM, (Coogan et al. 2005). In a more recent study by Richter et al. (2014), where f_{O_2} was varied between IW and NNO, a more complex behaviour of Li diffusion was found, analogous to that in olivine (Dohmen et al. 2010), where a model with two Li species (Li on interstitial and on metal sites) was required to simulate the unusual diffusion profiles. The two mechanisms may operate simultaneously to produce variable diffusion rates (by more than two log units) at a given temperature depending on the Li gradient as well as f_{O_2} (Richter et al. 2014, see also Tomascak et al. 2016). These conditions also impact on the observed Li isotope profiles and total extent of diffusive fractionation between ^{6}Li and ^{7}Li (Richter et al. 2014). Richter et al. (2014) modelled two Li isotope profiles from the literature and showed that only the single species (slow) mechanism operated. Furthermore they could distinguish between different boundary conditions—a sudden change (fixed boundary condition) or a continuous change at the rim of the crystal. Depending on the boundary condition, the diffusive flux into the crystal changes and hence the total extent of diffusive isotope fractionation is affected, which is an effect of mass balance. The stable isotopes therefore also leave a fingerprint of the operating diffusion mechanisms and, if measured in natural samples, help to identify the correct boundary condition and diffusion rate.

Determination of magma residence times using Fe–Mg diffusion in Opx and Cpx

In BSE images, pyroxenes in silicic volcanic rocks typically show a diversity of zoning styles, with normal, reversed, multiple, or patchy zoning. However, it is quite common that a final, outermost rim with a thickness of 10–100 μm is formed at some point before the magma erupts (e.g., Fig. 14). The compositional jump at the interface between the outermost rim and the interior of the crystal is typically very sharp and cannot be resolved using electron microprobe analysis without convolution effects (Ganguly et al. 1988; Morgan et al. 2004). Therefore, Morgan et al. (2004) inferred variations in the Fe content using the grey scale of BSE images of clinopyroxene, since the contribution to the changes in the grey scale by variations of other elements, Al or Si, are minor. For Cpx it must also be assumed that the variation of Ca was small enough to not affect the grey scale. The linear correlation of the grey scale with the $Mg\# = Mg/(Fe + Mg)$ in Opx was demonstrated by Allan et al. (2013). BSE imaging allows for a spatial resolution down to 100 nm or less (e.g., Saunders et al. 2012b) and is a very time efficient way to map Fe in pyroxene (or olivine) crystals. Using image analysis software, Morgan et al. (2004) transferred the grey scale of a BSE image into pixelated image data and residence times at magmatic temperatures were obtained from the following equation

$$C_{(x,t)} = \frac{n}{2(n+1)} = 1 - \left[0.5\,\text{erfc}\left(\frac{x}{2\sqrt{Dt}} \right) \right] \qquad (11)$$

at which C represents the compositional contrast on a scale of 0–0.5 between the junction of the observed maximum extent of diffusion at the half width x (see Fig. 14 and Appendix A in Morgan et al. 2004 for details), t is the residence time, n is the number of different grey shades in the BSE image in the zone effected by diffusion, and D is the diffusion coefficient of Fe-Mg in clinopyroxene. For the application of Equation (11), D needs to be approximately constant, an initial step profile is assumed, and the diffusion profiles must be short enough to assume an infinite one-dimensional medium as geometry (see Crank 1975 for explicit initial and boundary conditions; e.g., Fig. 14). A main advantage of the short profiles measured along a traverse at the half width of the crystal is that sectioning effects are less likely and a one-dimensional geometry for the diffusion model is often justified (Krimer and Costa 2016). A risk is that the time scale is still overestimated if the Fe–Mg zoning was not measurably modified following growth, meaning the diffusion distance is smaller than the spatial resolution of the BSE imaging. The measurement of more slowly diffusing elements, e.g., Al and Ca, provides

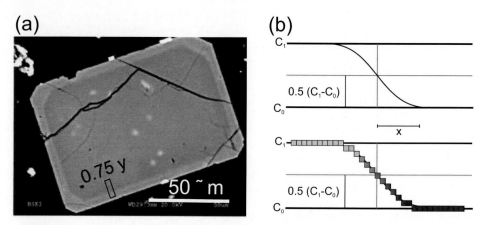

Figure 14. (a) BSE image of a chemically zoned orthopyroxene from Mt St. Helens (crystal SH127_ KS4, Saunders et al. 2012a). Note the geometrically well-defined rim with a relatively sharp interface (b) Illustration of the method of Morgan et al. (2004) to invert the grey scale zoning of the BSE image to a timescale (see Eqn. 11).

clues to the original growth zoning (Morgan et al. 2004). This has been demonstrated by Saunders et al. (2012a,b, 2014) for orthopyroxene (and Pl) by using TOF-SIMS and Nano-SIMS measurements that allowed for sub-μm resolution of slow-diffusing elements.

The appropriate diffusion coefficient depends on the magmatic temperature T, which Morgan et al. (2004) determined based on the composition of melt inclusions in clinopyroxene using the geothermometer of Cioni et al. (1999). Using the diffusion data of Dimanov and Sautter (2000) for D_{Fe-Mg}^{Cpx}, Morgan et al. (2004) determined residence times in multiple clinopyroxene crystals (following growth of the final rim) from the 1944 eruption of Vesuvius and obtained a range from 0.4 to 9 yr with a distribution skewed to young ages. They used the shape of the obtained cumulative residence curve for comparison with modelled populations with different magma chamber volumes. Their analysis reveals that a constant magma chamber volume is inadequate for Vesuvius and suggest instead a volume of 8.0×10^7 m^3 that expanded to $1.15–1.5 \times 10^8$ m^3 prior to the 1944 eruption.

Following the basic approach of Morgan et al. (2004) five recent studies have explored the potential of Fe–Mg zoning in Opx to quantify magmatic time scales (Saunders et al. 2012a, Allan et al. 2013; Chamberlain et al. 2014; Kilgour et al. 2014). Here we focus on the methodological developments for diffusion chronometry and the time scales inferred in these studies. As illustrated already for olivine by Kahl et al. (2011) a main strength of diffusion chronometry is the direct comparison of timescales of magmatic processes at depth with monitoring signals at Earth's surface. This potentially allows for the interpretation of certain surface signals (seismic, degassing, or ground deformation) in terms of a specific causative magmatic process at depth (see also Albert et al. 2016). For example from the zoning at the outer rims of Opx crystals from Mount St. Helens (e.g., Fig. 14a) Saunders et al. (2012a) established the time elapsed between the final rim growth and the eruption, which varied between days to months. The value they used for D_{Fe-Mg}^{Opx} was 4.72×10^{-21} m^2/s, as calculated at 884 °C (estimated from Fe-Ti oxides, Blundy et al. 2008) from Mg tracer diffusion of Schwandt et al. (1998). The recent experimental data for D_{Fe-Mg}^{Opx} parallel to the c-axis measured for Opx with En$_{91}$ of Dohmen et al. (2016a) predict at the same temperature for f_{O_2}=NNO+2 a value of D_{Fe-Mg}^{Opx}= 7.39×10^{-21} m^2/s, which justifies the use of Schwandt et al. (1998). The Opx in Saunders et al. (2012) were more Fe-rich, with a range of X_{Fe}=0.24–0.46. The higher Fe content might enhance the Fe–Mg diffusion rates by a factor 2–3 compared to Opx with X_{Fe} = 0.09 (see discussion in Dohmen et al. 2016a on the compositional dependence of D_{Fe-Mg}^{Opx}), but for profile orientations parallel to the *a*-axis, the diffusion rate is smaller by a factor 3.5, compensating the effect of composition. From the types of zoning and the trace element systematics, Saunders et al. (2012a) concluded that the final rim growth was formed by injection of hotter magma in the shallow magma chamber, and hence the date of magma intrusion could be determined from diffusion modelling. It was shown that this timing correlated with recorded episodes of seismicity, suggesting that swarms of small earthquakes heralded the ascent of new magma into the sub-volcanic reservoir.

In Allan et al. (2013) effects of growth zoning were identified in Opx crystals from the 25.4 ka, 530 km³ Oruanui eruption (Taupo volcano, New Zealand) for profile directions parallel to the *c*-axis, and hence only diffusion profiles roughly parallel to the *a*- or *b*-axis were considered. The overgrown rim was typically thicker for directions parallel to the *c*-axis compared to other directions and hence the growth rate was larger. This might have been responsible for a diffusion boundary layer in the melt and a respective evolution in the rim composition during growth. The measured profile shapes for Fe could not be fitted by a simple diffusion model that does not consider simultaneous growth. Four different thermometers were applied, which gave a consistent temperature of $770 \pm 30\,°C$. To relate the compositional change and textures of Opx crystals to physical conditions, the trace element composition of amphibole (Amp) crystals (mainly Zn and Mn contents) were correlated to pressure calculations using the thermobarometric formulations of Ridolfi et al. (2010). Allan et al. (2013) formulated their own parameterized D_{Fe-Mg}^{Opx} based on order–disorder rates and the model of Ganguly and Tazzoli (1994). For the relevant conditions, $T = 770\,°C$ and $f_{O_2} = NNO - 0.2$, $D_{Fe-Mg}^{Opx} = 1.28 \times 10^{-22}\,m^2/s$ according to Dohmen et al. (2016a), which is more than two log units smaller than the $D_{Fe-Mg}^{Opx} = 3.97 \times 10^{-20}\,m^2/s$ predicted from the relation in Allan et al. (2013) for Opx with Mg# = 0.5. Consequently, the magma residence times of Allan et al. (2013), varying between tens of years up to 2000 yr, would be longer (1–200 ka) if D_{Fe-Mg}^{Opx} of Dohmen et al. (2016a) had been used. Time scales of 200 ka appear to be very long and inconsistent with other estimates using U–Th model ages with zircon. One explanation could be that for $T \ll 900\,°C$ extrapolation of the expression of Dohmen et al. (2016a) underestimates D_{Fe-Mg}^{Opx} because of a change in the diffusion mechanism, as discussed in Dohmen et al. (2016a). From the kinetic data of the Fe–Mg order–disorder for $T < 900\,°C$ a smaller activation energy of Fe–Mg interdiffusion could be inferred as well as a lack of any effect of f_{O_2}, which would imply an about one log unit faster diffusion rate at the relevant temperature of $770\,°C$.

The strong discrepancy between estimated D_{Fe-Mg}^{Opx} using Allan et al. (2013) and measured D_{Fe-Mg}^{Opx} is most likely the reason for some of the inconsistencies obtained by Chamberlain et al. (2014) who compared timescales obtained from four different diffusion chronometers for multiple samples of Bishop Tuff, California: Fe–Mg in Opx, Ti in Qtz, and Sr and Ba in sanidine. Here again the relation of Allan et al. (2013) was used to estimate D_{Fe-Mg}^{Opx} but compared to the Ti in Qtz chronometer, the residence times were a factor of 10 shorter. Temperatures were derived form two-feldspar thermometry and span a range of 753–813 °C for the different samples. Based on the previous discussion and Dohmen et al. (2016a), it is likely that Chamberlain et al. (2014) used an estimate for D_{Fe-Mg}^{Opx} that is too large by at least a factor of 10. Therefore, considering the new D_{Fe-Mg}^{Opx} data, timescales inferred from Fe–Mg in Opx and Ti in Qtz might be reconciled. Timescales inferred from Sr and Ba zoning in sanidine are much longer, which according to Chamberlain et al. (2014) are probably related to only a minor diffusive modification of the Sr and Ba growth zoning.

Singer et al. (2016) determined time scales of magma recharge for the 1912 Novarupta-Katmai eruption from diffusion modelling of Mg in Opx and Pl. In this case for Pl only rough estimates for the timescale could be made, but these are in general on the order of hundreds of years. From Fe–Mg zoning in Opx significantly shorter time scales (weeks) were inferred.

SYNOPSIS/PERSPECTIVES

The great potential of reconstructing chemical and isotope zoning by diffusion modelling is that it not only allows for the determination of timescales (duration or rate) of a specific magmatic process, but also can be used in combination with petrological and observational (e.g., volcano monitoring) constraints to provide an integrated picture of a magmatic system. However, the methodology relies on robust estimates of magmatic temperatures, measurement of the concentration profiles without convolution effects, appropriate diffusion coefficients, a knowledge of chemical potential (in order to establish the equilibrium profile), and appropriate choice of initial and boundary conditions. Without careful appraisal of these factors, a timescale can always be obtained, but its accuracy is poor and its significance dubious. As illustrated above for a number of case studies, different strategies have been developed to improve the accuracy of the various input parameters of the forward modelling procedure. In addition to the quality of the individual diffusion profile fits (standard deviations), which is a first indicator of a diffusion controlled zoning, for non-cubic minerals the consistency of the timescales obtained from different elements measured along different crystallographic directions in the same crystal is one way to gauge the quality of the model and the inferred time scale. This of course requires a good experimental calibration of the diffusion anisotropy. In addition, measurement and fitting of chemical zoning in a larger number of crystals allows to identify sectioning effects (e.g., Pearce 1984), and whether 2D or 3D modelling is required to improve the accuracy of the timescale (Costa and Chakraborty 2004). For example, for Fe–Mg diffusion modelling with olivine it was shown numerically by Shea et al. (2015b) that with the appropriate choice of crystals and fitting of 20 concentration profiles with a 1D model an accuracy of 5% for the time scale can be achieved (ignoring the uncertainty of the experimentally determined diffusion coefficient and input parameters such as T, f_{O_2}, etc.). However, if the core composition remained unchanged by diffusion, and with the appropriate choice of the profile direction, sectioning effects are in general less likely.

In-situ analysis of stable isotopes for the relevant elements has been shown to provide strong constraints on the origin of the chemical zoning (growth/dissolution vs. diffusion), the initial zoning, as well as the boundary condition (e.g., Oeser et al. 2015). The simultaneous fitting of the chemical and isotope zoning reduces greatly the degrees of freedom for the boundary condition, which is often coupled to the detailed thermal history. Consequently, the thermal history can be reconstructed in greater detail compared to the case when only chemical zoning is fitted (Sio and Dauphas 2016). This technique is particularly useful for interdiffusion processes and related correlations of isotope anomalies for the interdiffusing elements. Experimental calibrations of the respective isotope effects in the minerals of interest are needed to obtain even stronger constraints. Currently, based on the spatial resolution of the available techniques and constraints related to the counting statistics, this approach requires long diffusion profiles on the order of 100 µm or more, which is rarely the case in natural samples and hard to achieve in experiments. But even bulk analysis of olivine fragments provides some hints to the nature of the chemical zoning as shown by Teng et al. (2011). In the near future it is likely that the spatial resolution of these isotope profiling techniques will improve and even shorter diffusion profiles or those of minor elements could be analysed.

Diffusion coefficients are the most fundamental input parameter for diffusion chronometry and, as discussed for Fe–Mg interdiffusion in orthopyroxene or Mg diffusion in plagioclase, controversies persist for important diffusion data. Only detailed experimental studies with the focus of identifying the diffusion mechanism and factors controlling the relevant point defect concentrations can provide robust and appropriate diffusion data sets for natural systems. Conflicting diffusion data sets that still exist, for example REE diffusion in olivine, clinopyroxene and garnet, may be explained by different diffusion mechanisms operating in the various types of experiments (e.g., Dohmen et al. 2016b). Two types of diffusion mechanisms with drastically different rates (3–4 orders of magnitude) have been identified for Li and H diffusion in olivine or pyroxene. The relevant diffusion rate can depend on the detailed trace element chemistry of the mineral and the chemical environment (boundary condition) of the crystal. Thus, in certain cases a simple Arrhenius type of equation to describe the respective *D* is not appropriate, and a more complex parameterization is required that considers in detail the effect of oxygen fugacity or trace elements present as heterovalent substitutions on the point defect chemistry of the mineral and hence its diffusion properties (e.g., Dohmen and Chakraborty 2007). In the case of an element occupying two lattice sites, the interaction of the different diffusing species and point defects (e.g., Dohmen et al. 2010; Richter et al. 2014) need to be considered in the governing diffusion equations. Again, measurement of stable isotopes provides additional constraints on the operating diffusion mechanisms (Richter et al. 2014).

In addition to Ol, Opx, Cpx, and Pl, chemical zoning in quartz (Qtz), alkali feldspar (Afs), spinel (Sp) and melts/glasses have been used to estimate timescales of magmatic processes. The general approaches used in these studies do not differ fundamentally from those discussed above and therefore we do not discuss them in detail here. For example Ba zoning in sanidine crystals from the A.D. 79 eruption of Vesuvius was modelled to determine the time elapsed between magma recharge and eruption (Morgan et al. 2006). As described for Fe–Mg in pyroxenes, the zoning of Ba was directly inferred from BSE images. In a similar study diffusion of Sr and Ba in sanidine have been used to introduce the concept of binary element diffusion modelling and determine magma residence times for sanidine crystals in the Bishop Tuff (Morgan and Blake 2006). Here again a slower (Ba) diffusing element was used to reconstruct the initial profile of a faster diffusing element (Sr) (see also Till et al. 2015). A more comprehensive reconstruction of the crystal history can be obtained, as demonstrated by Zellmer and Clavero (2006), when trace element zoning in different zones of a sanidine crystal are modelled simultaneously (analogous to the example shown in Figs. 1 and 2). In their study Sr diffusion was modelled in a sanidine crystal with multiple growth zones from Taapaca volcano, Central Andes. A simplified procedure was applied to model the oscillatory zoned Sr concentration profiles, which showed that the core was older by about 750 yr compared to the outer growth zone, indicating an average growth rate of the crystal on the order of 10^{-10} cm s^{-1}. Ti diffusion in Qtz can be used to reconstruct the thermal history of a magma as its content is strongly sensitive to temperature (Wark et al. 2007). However, complex zoning patterns of Ti in Qtz, typically inferred indirectly by cathodoluminescence (CL) intensity, could equally be the result of fast growth and related kinetic effects and therefore no equilibrium of Ti between quartz and co-existing melt can be assumed as a means to estimate temperatures (Seitz et al. 2016).

In a number of recent papers the potential of H diffusion in melt inclusions (e.g., Liu et al. 2007; Chen et al. 2013; Newcombe at al. 2014) and melt embayments (e.g., Humphreys et al. 2008; Lloyd et al. 2014; Ferguson et al. 2016) to determine magma ascent rates was explored. Compared to olivine, the diffusion rate of H appears to be better constrained but involves additional complexities related to the presence of two H species in glasses and melts, molecular water and hydroxyl ions (e.g., Wasserburg 1988; Zhang et al. 1991). As for olivine, the initial water content and the depth needed to be estimated, and a fixed boundary condition was assumed, whereby the H content is completely lost from the boundary, which is not realistic since the H content in melt is a function of pressure and related water fugacity in the system. Clearly more diffusion chronometers will be developed in the future (for example Nb and Zr diffusion in rutile as, for example, applied for metamorphic systems, e.g., see Zack and Kooijman 2017, this volume) but this is obviously dependent on the availability and accuracy of the respective diffusion data set. For example diffusion of Fe–Mg or other elements in amphibole could be useful for magmatic systems, but diffusion data are currently lacking.

Finally, it can be concluded that for diffusion chronometry we cannot apply a simple recipe to obtain reliable timescales for a generic magmatic scenario. But scientific progress in this field is fast and, in conjunction with experimental studies, modelling strategies are developed for the diffusion chronometer that is appropriate for the magmatic process of interest. Diffusion chronometry of magmatic processes has yet to reach its full potential, although rapid and concerted developments in the field suggest that this may not be the case for much longer.

ACKNOWLEDGMENTS

The authors would like to thank Martin Engi, Fidel Costa, and Philip Rupprecht for their constructive and helpful reviews that improved the manuscript. In addition we thank Sumit Chakraborty for helpful comments. JB acknowledges support from ERC Advanced Grant "CRITMAG"

REFERENCES

Albarède F Bottinga Y (1972) Kinetic disequilibrium in trace element partitioning between phenocrysts and host lava. Geochim Cosmochim Acta 36:141–156
Albert H, Costa F, Marti J (2015) Timing of magmatic processes and unrest associated with mafic historical monogenetic eruptions in Tenerife Island. J Petrol 56:1945–1965
Albert H, Costa F, Marti J (2016) Years to weeks of seismic unrest and magmatic intrusions precede monogenetic eruptions. Geology 44:211–214
Allan ASR, Morgan DJ, Wilson CJN, Millet M-A (2013) From mush to eruption in centuries: assembly of the super-sized Oruanui magma body. Contrib Mineral Petrol 166:143–164
Beck P, Chaussidon M, Barrat JA, Gillet Ph, Bohn M (2006) Diffusion induced Li isotopic fractionation during the cooling of magmatic rocks: the case of pyroxene phenocrysts from nakhlite meteorites. Geochim Cosmochim Acta 70:4813–4825
Bell DR, Rossman GR, Maldener J, Endisch D, Rauch F (2003) Hydroxide in olivine: a quantitative determination of the absolute amount and calibration of the IR spectrum. J Geophys Res 108:2105, doi:10.1029/2001JB000679
Beran A, Libowitzky E (2006) Water in natural mantle minerals II: Olivine, garnet and accessory minerals. Rev Mineral Geochem 62:169–191
Berlo K, Turner S, Blundy J, Black S, Hawkesworth C (2006) Tracing pre-eruptive magma degassing using (^{210}Pb/^{226}Ra) disequilibria in the volcanic deposits of the 1980–1986 eruption of Mount St. Helens. Earth Planet Sci Lett 249:337–349
Berlo K, Blundy J, Turner S, Hawkesworth C (2007) Textural and chemical variation in plagioclase phenocrysts from the 1980 eruptions of Mount St. Helens, USA Contrib Mineral Petrol 154:291–308
Berry A, Hermann J, O'Neill HSC, Foran GJ (2005) Fingerprinting the water site in mantle olivine. Geology 33:869–872
Berry AJ, O'Neill HStC, Hermann J, Scott DR (2007) The infrared signature of water associated with trivalent cations in olivine. Earth Planet Sci Lett 261:134–142
Bindeman IN, Davies AM, Drake MJ (1998) Ion microprobe study of plagioclase–basalt partition experiments at natural concentration levels of trace elements. Geochim Cosmochim Acta 62:1175–1193.
Blundy JD, Shimizu N (1991) Trace element evidence for plagioclase recycling in calc-alkaline magmas. Earth Planet Sci Lett 102:178–197
Blundy JD, Wood BJ (1991) Crystal-chemical controls on the partitioning of Sr and Ba between plagioclase feldspar, silicate melts and hydrothermal solutions. Geochim Cosmochim Acta 55:193–209
Blundy JD, Wood BJ (1994) Prediction of crystal–melt partition coefficients from elastic moduli. Nature 372:452–454
Blundy J, Cashman KV, Berlo K (2008) Evolving magma storage conditions beneath Mount St. Helens inferred from chemical variations in melt inclusions from the 1980–1986 and current (2004–2006) eruptions. *In:* Sherrod DR, Scott WE, Stauffer PH (Eds) A volcano rekindled: the renewed eruption of Mount St. Helens 2004–2006. US Geol Surv Prof Pap 1750:755–790
Bouvet de Maisonneuve C, Costa F, Huber C, Vonlanthen P, Bachmann O, Dungan M (2016) Do olivines faithfully record magmatic events? Contrib Mineral Petrol 171: doi 10.1007/s00410-016-1264-6
Bowen NL (1913) The melting phenomena of the plagioclase feldspars. Am J Sci 35:577–599
Carmichael ISE, Nicholls J, Smith AL (1970) Silica activity in igneous rocks. Am Mineral 55:246–263
Chakraborty S (1997) Rates and mechanisms of Fe–Mg interdiffusion in olivine at 980–1300°C J Geophys Res 102:12317–12331

Chakraborty S (2006) Diffusion modeling as a tool for constraining timescales of evolution of metamorphic rocks. Mineral Petrol 88:7–27

Chakraborty S (2008) Diffusion in solid silicates: a tool to track timescales of processes comes of age. Ann Rev Earth Planet Sci 36:153–190

Chakraborty S (2010) Diffusion coefficients in olivine, wadsleyite and ringwoodite. Rev Mineral Geochem 72:603–639

Chakraborty S, Ganguly J (1991) Compositional zoning and cation diffusion in garnets. *In:* Ganguly J (ed) Diffusion, atomic ordering and mass transport. Advances in physical geochemistry, vol 8. Springer, New York, pp 120–175

Chamberlain, KJ, Morgan, DJ, Wilson, CJN (2014) Timescales of mixing and mobilisation in the Bishop Tuff magma body: perspectives from diffusion chronometry. Contrib Mineral Petrol 168:1034 DOI 10.1007/s00410-014-1034-2

Charlier BLA, Morgan DJ, Wilson CJN, Wooden JL, Allan ASR, Baker JA (2012) Lithium concentration gradients in feldspar and quartz record the final minutes of magma ascent in an explosive supereruption. Earth Planet Sci Lett 319–320:218–227

Chen Y, Provost A, Schiano P, Cluzel N (2013) Magma ascent rate and initial water concentration inferred from diffusive water loss from olivine-hosted melt inclusions. Contrib Mineral Petrol 165:525–541

Cherniak DJ (2010) Cation diffusion in feldspars. Rev Mineral Geochem 72:691–733

Cherniak DJ, Watson EB (1994) A study of strontium diffusion in plagioclase using Rutherford backscattering spectroscopy. Geochim Cosmochim Acta 58:5179–5190

Cherniak DJ, Dimanov A (2010) Diffusion in pyroxene, mica and amphibole. Rev Mineral Geochem 72:641–690

Cherniak DJ, Liang Y (2012) Ti diffusion in natural pyroxene. Geochim Cosmochim Acta 98:31–47. doi:10.1016/j.gca.2012.09.021

Christopher TE, Blundy J, Cashman K, Cole P, Edmonds M, Smith PJ, Sparks RSJ, Stinton A (2015) Crustal-scale degassing due to magma system destabilization and magma–gas decoupling at Soufriere Hills Volcano, Montserrat. Geochem Geophys Geosystem 16:2797–2811

Cioni R, Marianelli P, Santacroce R (1999) Temperature of Vesuvius magma. Geology 27:443–446

Coogan LA, Jenkin GRT, Wilson RN (2002) Constraining the cooling rate of the lower oceanic crust: a new approach applied to the Oman ophiolite. Earth Planet Sci Lett 199:127–146

Coogan LA, Hain A, Stahl S, Chakraborty S (2005) Experimental determination of the diffusion coefficient for calcium in olivine between 900 and 1500 °C Geochim Cosmochim Acta 69:3683–3694

Coogan LA, Jenkin GRT, Wilson RN (2007) Contrasting cooling rates in the lower oceanic crust at fast- and slow-spreading ridges revealed by geospeedometry. J Petrol 48:2211–2231

Coombs ML, Eichelberger JC, Rutherford MJ (2000) Magma storage and mixing conditions for the 1953–1974 eruptions of Southwest Trident Volcano, Katmai National Park, Alaska. Contrib Mineral Petrol 140:99–118

Cooper KM, Kent AJR (2014) Rapid remobilization of magmatic crystals kept in cold storage. Nature 506:480–483

Cooper M, Reid MR (2003) Re-examination of crystal ages in recent Mount St. Helens lavas: implications for magma reservoir processes. Earth Planet Sci Lett 213:149–167

Cooper KM, Reid MR (2008) Uranium-series crystal ages. Rev Mineral Geochem 69:479–544

Costa F, Chakraborty S (2004) Decadal time gaps between mafic intrusion and silicic eruption obtained from chemical zoning patterns in olivine. Earth Planet Sci Lett 227:517–530

Costa F, Dungan M (2005) Short timescales of magmatic assimilation from diffusion modelling of multiple elements in olivine. Geology 33:837–840

Costa F, Morgan D (2010) Time constraints from chemical equilibration in magmatic crystals. *In:* Dosseto A, Turner SP, Van Orman JA (eds) Timescales of magmatic processes: from core to atmosphere. Wiley, Chichester, p. 125–159

Costa F, Chakraborty S, Dohmen R (2003) Diffusion coupling between trace and major elements and a model for calculation of magma residence times using plagioclase. Geochim Cosmochim Acta 67:2189–2200

Costa F, Dohmen R, Chakraborty S (2008) Timescales of magmatic processes from modeling the zoning patterns of crystals. Rev Mineral Geochem 69:545–594

Costa F, Coogan LA, Chakraborty S (2010a) The timescales of magma mixing and mingling involving primitive melts and melt–mush interaction at mid-ocean ridges. Contrib Mineral Petrol 159:173–194

Costa F, Dohmen R, Demouchy S (2010b) Modelling the dehydrogenation of mantle olivine with implications for the water content of the Earth's upper mantle, and ascent rates of kimberlite and alkali basaltic magmas. AGU Fall Meeting 2010, V24C-06.

Crank J (1975) The Mathematics of Diffusion, 2nd edition, 414 p. Oxford Science Publication, Oxford.

Dauphas N, Teng F-Z, Arndt NT (2010) Magnesium and iron isotopes in 2.7 Ga Alexo komatiites: Mantle signatures, no evidence for Soret diffusion, and identification of diffusive transport in zoned olivine. Geochim Cosmochim Acta 74:3274–3291

Davidson JP, Tepley FJ (1997) Recharge in volcanic systems; evidence from isotope profiles of phenocrysts. Science 275:826–829

Demouchy S, Mackwell S (2006) Mechanisms of hydrogen incorporation and diffusion in iron-bearing olivine. Phys Chem Miner 33:347–355

Demouchy S, Jacobsen SD, Gaillard F, and Stern CR (2006) Rapid magma ascent recorded by water diffusion profiles in mantle olivine. Geology 34:429–432

Denis CMM, Demouchy S, Shaw CSJ (2013) Evidence of dehydration in peridotites from Eifel volcanic field and estimates of the rate of magma ascent. J Volcanol Geotherm Res 258:85–99, doi:10.1016/j.jvolgeores.2013.04.010

Dimanov A, Sautter V (2000) "Average" interdiffusion of (Fe,Mn)–Mg in natural diopside. Eur J Mineral 12:749–760

Dimanov A, Wiedenbeck M (2006) (Fe,Mn)–Mg interdiffusion in natural diopside: effect of pO_2. Eur J Mineral 18:705–718

Dodson MH (1986) Closure profiles in cooling systems. Mater Sci Forum 7:145–154

Dohmen R, Chakraborty S, Palme H, Rammensee W (2003) The role of element solubility on the kinetics of element partitioning: in situ observations and a thermodynamic kinetic model. J Geophys. Res 108: B3 2157, doi:10.1029/2001JB000587

Dohmen R, Blundy J (2014) A predictive thermodynamic model for element partitioning between plagioclase and melt as a function of pressure, temperature and composition. Am J Sci 314:1319–1372

Dohmen R, Chakraborty S (2007) Fe–Mg diffusion in olivine II: Point defect chemistry, change of diffusion mechanisms and a model for calculation of diffusion coefficients in natural olivine. Phys Chem Mineral 34:409–430

Dohmen R, Kasemann SA, Coogan L, Chakraborty S (2010) Diffusion of Li in olivine. Part I: Experimental observations and a multi species diffusion model. Geochim Cosmochim Acta 74:274–292

Dohmen R, Ter Heege JH, Becker H-W, Chakraborty S (2016a) Fe–Mg interdiffusion in orthopyroxene. Am Mineral 101:2210–2221

Dohmen R, Marschall H, Wiedenbeck M, Polednia J, Chakraborty S (2016b) Trace element diffusion in minerals: the role of multiple diffusion mechanisms operating simultaneously. AGU, Fall meeting 2016, MR51A-2682.

Drake MJ, Weill DF (1971) Petrology of Apollo-11 sample-10071—Differentiated mini-igneous complex. Earth Planet Sci Lett 13:61–70

Druitt TH, Costa F, Deloule E, Dungan M, Scaillet B (2012) Decadal to monthly timescales of magma transfer and reservoir growth at a caldera volcano. Nature 482:77–80

Dvorak JJ, Dzurisin D (1997) Volcano geodesy: The search for magma reservoirs and the formation of eruptive vents. Rev Geophys 35: doi: 10.1029/97RG00070

Faak K, Gillis KM (2016) Slow cooling of the lowermost oceanic crust at the fast-spreading East Pacific Rise. Geology 44:115–118

Faak K, Chakraborty S, Coogan LA (2013) Mg in plagioclase: experimental calibration of a new geothermometer and diffusion coefficients. Geochim Cosmochim Acta 123:195–217

Faak K, Chakraborty S, Coogan LA (2014) A new Mg-in-plagioclase geospeedometer for the determination of cooling rates of mafic rocks. Geochim Cosmochim Acta 140:691–707

Faak K, Coogan LA, Chakraborty S (2015) Near conductive cooling rates in the upper-plutonic section of crust formed at the East Pacific Rise. Earth Planet Sci Lett 423:36–47, doi:10.1016/j.epsl.2015.04.025

Ferguson DJ, Gonnermann HM, Ruprecht P, Plank T, Hauri EH, Houghton BF, Swanson DA (2016) Magma decompression rates during explosive eruptions of Kilauea volcano, Hawaii, recorded by melt embayments. Bull Volcanol 78:71, doi: 10.1007/s00445-016- 1064-x

Ferriss E, Plank T, Walker D, Nettles M (2015) The whole-block approach to measuring hydrogen diffusivity in nominally anhydrous minerals. Am Mineral 100:837–851

Ferris E, Plank T, Walker D (2016) Site-specific hydrogen diffusion rates during clinopyroxene dehydration. Contrib Mineral Petrol 171:55 DOI 10.1007/s00410-016-1262-8

Ferry JM, Watson EB (2007) New thermodynamic models and revised calibrations for the Ti-in-zircon and Zr-in-rutile thermometers. Contrib Mineral Petrol 154:429–437

Gallagher K, Elliott T (2009) Fractionation of lithium isotopes in magmatic systems as a natural consequence of cooling. Earth Planet Sci Lett 278:286–296

Ganguly J (2002) Diffusion kinetics in minerals: Principles and applications to tectono-metamorphic processes. Eur Mineral Union Notes 4:271–309

Ganguly J, Tazzoli V (1994) Fe^{2+}–Mg interdiffusion in orthopyroxene: Retrieval from the data on intracrystalline exchange reaction. Am Mineral 79:930–937

Ganguly J, Tirone M (1999) Diffusion closure temperature and age of a mineral with arbitrary extent of diffusion; theoretical formulation and applications. Earth Planet Sci Lett 170:131–140

Ganguly J, Bhattacharya RN, Chakraborty S (1988) Convolution effect in the determination of compositional profiles and diffusion coefficients by microprobe step scans. Am Mineral 73:901–909

Gauthier PJ, Condomines M (1999) [210]Pb–[226]Ra radioactive disequilibria in recent lavas and radon degassing: inferences on the magma chamber dynamics at Stromboli and Merapi volcanoes. Earth Planet Sci Lett 172:111–126

Gerlach D, Grove T (1982) Petrology of Medicine Lake Highland volcanics: Characterization of endmembers of magma mixing. Contrib Mineral Petrol 80:147–159

Ghiorso MS, Sack RO (1995) Chemical mass transfer in magmatic processes. IV A revised and internally consistent thermodynamic model for the interpolation and extrapolation of liquid–solid equilibria in magmatic systems at elevated temperatures and pressures. Contrib Mineral Petrol 119:197–212

Giletti BJ, Shanahan TM (1997) Alkali diffusion in plagioclase feldspar. Chem Geol 139:3–20

Ginibre C, Kronz A, Wörner G (2002a) High-resolution quantitative imaging of plagioclase composition using accumulated backscattered electron images: New constraints on oscillatory zoning. Contrib Mineral Petrol 142:436–448

Ginibre C, Wörner G, Kronz A (2002b) Minor- and trace- element zoning in plagioclase: implications for magma chamber processes at Parinacota volcano, northern Chile. Contrib Mineral Petrol 143:300–315

Girona T, Costa F (2013) DIPRA: a user-friendly program to model multi-element diffusion in olivine with applications to timescales of magmatic processes. Geochem Geophys Geosyst 14:422–431, doi:10.1029/2012gC004427

Grove TL, Baker MB, Kinzler RJ (1984) Coupled CaAl–NaSi diffusion in plagioclase feldspar: experiments and applications to cooling rate speedometry. Geochim Cosmochim Acta 48:2113–2121

Hartley ME, Morgan DJ, Maclennan J, Edmonds M, Thordarson T (2016) Tracking timescales of short-term precursors to large basaltic fissure eruptions through Fe–Mg diffusion in olivine. Earth Planet Sci Lett 439:58–70

Holzapfel C, Chakraborty S, Rubie DC, Frost DJ (2007) Effect of pressure on Fe–Mg, Ni and Mn diffusion in $(Fe_xMg_{1-x})_2SiO_4$ olivine: Phys Earth Planet Inter 162:186–198

Humphreys MCS, Menand T, Blundy JD, Klimm K (2008) Magma ascent rates in explosive eruptions: Constraints from H_2O diffusion in melt inclusions. Earth Planet Sci Lett 270:25–40

Ingrin J, Blanchard M (2006) Diffusion of hydrogen in minerals. Rev Mineral Geochem 63:291–320

Jeffcoate AB, Elliott T, Kasemann SA, Ionov D, Cooper K, Brooker R (2007) Li isotope fractionation 4144 in peridotites and mafic melts. Geochim Cosmochim Acta 71:202–218

John T, Gussone N, Podladchikov YY, Bebout GE, Dohmen R, Halama R, Klemd R, Magna T, Seitz HM (2012) Volcanic arcs fed by rapid pulsed fluid flow through subducting slabs. Nat Geosci 5:489–492

Kahl M, Chakraborty S, Costa F, Pompilio M (2011) Dynamic plumbing system beneath volcanoes revealed by kinetic modeling and the connection to monitoring data: An example from Mt. Etna. Earth Planet Sci Lett 308:11–22

Kahl M, Chakraborty S, Costa F, Pompilio M, Liuzzo M, Viccaro M (2013) Compositionally zoned crystals and real-time degassing data reveal changes in magma transfer dynamics during the 2006 summit eruptive episodes of Mt. Etna. Bull Volcanol 75:1–14

Kahl M, Chakraborty S, Pompilio M, Costa F (2015) Constraints on the nature and evolution of the magma plumbing system of Mt. Etna Volcano (1991–2008) from a combined thermodynamic and kinetic modelling of the compositional record of minerals. J Petrol 56:2015–2068

Kayzar TM, Cooper KM, Reagan MK, Kent AJR (2009) Gas transport model for the magmatic system at Mount Pinatubo, Philippines: Insights from $(^{210}Pb)/(^{226}Ra)$. J Volcanol Geotherm Res 181:124–140

Kilgour GN, Saunders KE, Blundy JD, Cashman KV, Scott BJ, Miller CA (2014) Timescales of magmatic processes at Ruapehu volcano from diffusion chronometry and their comparison to monitoring data. J Volcanol Geotherm Res 288:62–75

Köhler TP, Brey GP (1990) Calcium exchange between olivine and clinopyroxene calibrated as a geothermometer for natural peridotites from 2 to 60kb with applications. Geochim Cosmochim Acta 54:2375–2388

Kohlstedt DL, Mackwell SJ (1998) Diffusion of hydrogen and intrinsic point defects in olivine. Z Phys Chem 207:147–162

Kohlstedt DL, Keppler H, Rubie DC (1996) Solubility of water in the α, β and γ phases of $(Mg,Fe)_2SiO_4$. Contrib Mineral Petrol 123:345–357

Kohn MJ, Penniston–Dorland SC (2017) Diffusion: Obstacles and opportunities in petrochronology. Rev Mineral Geochem 83:103–152

Krimer D, Costa F (2016) Evaluation of the effects of 3d diffusion, crystal geometry, and initial conditions on retrieved time-scales from Fe–Mg zoning in natural oriented orthopyroxene crystals. Geochim Cosmochim Acta 196:271–288

Lanphere M, Champion D, Melluso L, Morra V, Perrotta A, Scarpati C, Tedesco D, Calvert A (2007) $^{40}Ar/^{39}Ar$ ages of the AD 79 eruption of Vesuvius, Italy. Bull Volcanol 69:259–263

Lasaga AC (1979) Multicomponent exchange and diffusion in silicates. Geochim Cosmochim Acta 43:455–469

Lasaga AC (1983) Geospeedometry: an extension of geothermometry. *In:* Kinetics and equilibrium in mineral reactions (ed. Saxena SK). Springer-Verlag, New York. pp. 82–114

Lai Y-J, Pogge von Strandmann PAE, Dohmen R, Takazawa E, Elliott T (2015) The influence of melt infiltration onthe Li and Mg isotopic composition of the Horoman Peridotite Massif. Geochim Cosmochim Acta 164:318–332

LaTourrette T, Wasserburg GJ (1998) Mg diffusion in anorthite: implications for the formation of early solar system planetesimals. Earth Planet Sci Lett 158:91–108

Liu Y, Anderson AT, Wilson CJN (2007) Melt pockets in phenocrysts and decompression rates of silicic magmas before fragmentation. J Geophys Res, Solid Earth 112:B06204, doi: 10.1029/2006JB004500

Lloyd AS, Ruprecht P, Hauri EH, Rose W, Gonnerman HM, Plank T (2014) NanoSIMS results from olivine-hosted melt embayments: magma ascent rate during explosive basaltic eruptions. J Volc Geotherm Res 283:1–18

Longhi, J, Walker D, Hays J (1976) Fe and Mg in plagioclase. Proc Lunar Sci Conf 7:1281–1300

Longpré MA, Klügel A, Diehl A, Stix J (2014) Mixing in mantle magma reservoirs prior to and during the 2011–2012 eruption at El Hierro, Canary Islands. Geology 42:315–318

Mackwell SJ, Kohlstedt DL (1990) Diffusion of hydrogen in olivine: implications for water in the mantle. J Geophys Res 95(B4):5079, doi:10.1029/JB095iB04p05079

Marti J, Castro A, Rodriguez C, Costa F, Carrasquilla S, Pedreira R, Bolos × (2013) Correlation of magma evolution and geophysical monitoring during the 2011–2012 El Hierro (Canary Islands) Submarine Eruption. J Petrol 54:1349–1373

Martin VM, Morgan DJ, Jerram DA, Caddick MJ, Prior DJ, Davidson JP (2008) Bang! Month-scale eruption triggering at Santorini volcano. Science 321:1178

Melnik O, Sparks RSJ (2005) Controls on conduit magma flow dynamics during lava dome building eruptions. J Geophys Res 110: B02209, doi:10.1029/2004JB003183

Mierdel K, Keppler H, Smyth JR, Langenhorst F (2007) Water solubility in aluminous orthopyroxene and the origin of Earth's asthenosphere. Science 315:364–368

Milman-Barris MS, Beckett JR, Michael MB, Hofmann A, Morgan Z, Crowley MR, Vielzeuf D, Stolper E (2008) Zoning of phosphorus in igneous olivine. Contrib Mineral Petrol 155:739–765

Miller SA, Asimow PD, Burnett DS (2006) Determination of melt influence on divalent element partitioning between anorthite and CMAS melts. Geochim Cosmochim Acta 70:4258–4274

Moore A, Coogan L, Costa F, Perfit MR (2014) Primitive melt replenishment and crystal-mush disaggregation in the weeks preceding the 2005–2006 eruption 9° 50'N, EPR Earth Planet Sci Lett 403:15–26

Morgan DJ, Blake S (2006) Magmatic residence times of zoned phenocrysts: Introduction and application of the binary element diffusion modelling (BEDM) technique: Contrib Mineral Petrol 151:58–70

Morgan DJ, Blake S, Rogers NW, DeVivo B, Rolandi G, Macdonald R, Hawkesworth J (2004) Timescales of crystal residence and magma chamber volume from modeling of diffusion profiles in phenocrysts: Vesuvius 1944. Earth Planet Sci Lett 222:933–946

Morgan, DJ, Blake S, Rogers NW, DeVivo B, Rolandi G, Davidson J (2006) Magma recharge at Vesuvius in the century prior to the eruption AD 79. Geology 34:845–848

Mosenfelder JL, Deligne NI, Asimow PD, Rossman GR (2006) Hydrogen incorporation in olivine from 2–12 GPa. Am Mineral 91:285–294

Müller T, Dohmen R, Becker HW, ter Heege J, Chakraborty S (2013) Fe–Mg interdiffusion rates in clinopyroxene: experimental data and implications for Fe–Mg exchange geothermometers. Contrib Mineral Petrol 166:1563–1576

Nakamura M (1995) Residence time and crystallization history of nickeliferous olivine phenocrysts from the northern Yatsugatake volcanoes, Central Japan: Application of a growth and diffusion model in the system Mg-Fe-Ni. J Volcanol Geotherm Res 66:81–100

Newcombe ME, Fabbrizio A, Zhang Y, Ma C, Le Voyer M, Guan Y, Eiler JM, Saal AE, Stolper EM (2014) Chemical zonation in olivine-hosted melt inclusions. Contrib Mineral Petrol 168:1–26

Newman S, Lowenstern JB (2002) VolatileCalc: a silicate melt–H_2O–CO_2 solution model written in Visual Basic for Excel. Comp Geosci 28:597–604

Oeser M, Weyer S, Horn I, Schuth S (2014) High-precision Fe and Mg isotope ratios of silicate reference glasses determined in situ by femtosecond LA-MC-ICP-MS and by solution nebulisation MC-ICP-MS Geostand Geoanal Res 38:311–328

Oeser M, Dohmen R, Horn I, Schuth S, Weyer S (2015) Processes and time scales of magmatic evolution as revealed by Fe–Mg chemical and isotopic zoning in natural olivines. Geochim Cosmochim Acta 154:130–150

Onorato PIK, Hopper RW, Yinnon H, Uhlmann DR, Taylor LA, Garrison JR, Hunter R (1981) Solute partitioning under continuous cooling conditions as a cooling rate indicator. J Geophys Res 86:9511–9518

Padron-Navarta JA, Hermann J, O'Neill HStC (2014) Site-specific hydrogen diffusion rates in forsterite. Earth Planet Sci Lett 392:100–112

Pan Y, Batiza R (2002) Mid-ocean ridge magma chamber processes: Constraints from olivine zonation in lavas from the East Pacific Rise at 9°30'N and 10°30'N J Geophys Res 107:2022

Pearce TH (1984) The analysis of zoning in magmatic crystals with emphasis on olivine. Contrib Mineral Petrol 86:149–154

Peslier AH, Luhr JF (2006) Hydrogen loss from olivines in mantle xenoliths from Simcoe (USA) and Mexico: Mafic alkalic magma ascent rates and water budget of the sub-continental lithosphere. Earth Planet Sci Lett 242:302–319

Peslier AH, Woodland AB, Wolff JA (2008) Fast kimberlite ascent rates estimated from hydrogen diffusion profiles in xenolithic olivines from Southern Africa. Geochim Cosmochim Acta 72:2711–2722

Peslier AH, Bizimis M, Matney M (2015) Water disequilibrium in olivines from Hawaiian peridotites: Recent metasomatism, H diffusion and magma ascent rates. Geochim Cosmochim Acta 154:98–117

Peters MT, Shaffer EE, Burnett DS, Kim SS (1995) Magnesium and titanium partitioning between anorthite and Type B CAI liquid: Dependence on oxygen fugacity and liquid composition Geochim Cosmochim Acta 59:2785–2796

Petry C, Chakraborty S, Palme H (2004) Experimental determination of Ni diffusion coefficients in olivine and their dependence on temperature, composition, oxygen fugacity, and crystallographic orientation. Geochim Cosmochim Acta 68:4179–4188

Pinilla C, Davis SA, Scott TB, Allan, NL, Blundy JD (2012) Interfacial storage of noble gases and other trace elements in magmatic systems. Earth Planet Sci Lett 319:287–294

Pistolesi M, Rosi M, Cioni R, Cashman KV, Rossotti A, Aguilera E (2011) Physical volcanology of the post-XII century activity at Cotopaxi Volcano, Ecuador: behavior of an andesitic central volcano. Geol Soc Am Bull 123:1193–1215

Qian Q, O'Neill HSC, Hermann J (2010) Comparative diffusion coefficients of major and trace elements in olivine at 950 °C from a xenocryst included in dioritic magma. Geology 38:331–334

Rae ASP, Edmonds M, Maclennan J, Morgan D, Houghton B, Hartley ME, Sides I (2016) Time scales of magma transport and mixing at Kilauea Volcano, Hawai'i. Geology 44:463–466

Richter FM, Liang Y, Davis AM (1999) Isotope fractionation by diffusion in molten oxides. Geochim Cosmochim Acta 63:2853–2861

Richter FM, Davis AM, DePaolo DJ, Watson EB (2003) Isotope fractionation by chemical diffusion between molten basalt and rhyolite. Geochim Cosmochim Acta 67:3905–3923

Richter F, Watson B, Chaussidon M, Mendybaev R, Ruscitto D (2014) Lithium isotope fractionation by diffusion in minerals. Part 1: Pyroxenes. Geochim Cosmochim Acta 126:352–370

Ridolfi F, Renzulli A, Puerini M (2010) Stability and chemical equilibrium of amphibole in calc-alkaline magmas: an overview, new thermobarometric formulations and application to subduction-related volcanoes. Contrib Mineral Petrol 160:45–66

Ruprecht P, Cooper KM (2012) Integrating uranium-series and elemental diffusion geochronometers in mixed magmas from Volcan Quizapu, Central Chile. J Petrol 53:841–871

Ruprecht P, Plank T (2013) Feeding andesitic eruptions with a high-speed connection from the mantle. Nature 500:68–72

Ruprecht P, Wörner G (2007) Variable regimes in magma systems documented in plagioclase zoning patterns: El Misti stratovolcano and Andahua monogenetic cones. J Volcanol Geotherm Res 165:142–162

Sano J, Ganguly J, Hervig R, Dohmen R, Zhang X(2011) Neodymium diffusion in orthopyroxene: Experimental studies and applications to geological and planetary problems. Geochim Cosmochim Acta 75:4684–4698

Saunders K, Blundy J, Dohmen R, Cashman K (2012a) Linking petrology and seismology at an active volcano. Science 336:1023–1027

Saunders K, Rinnen S, Blundy J, Dohmen R, Klemme S, Arlinghaus HF (2012b) TOF-SIMS and electron microprobe investigation of zoned magmatic orthopyroxenes: first results of trace and minor element analysis with implications for diffusion modelling. Am Mineral 97:532–542

Saunders K, Buse B, Kilburn MR, Kearns S, Blundy J (2014) Nanoscale characterisation of crystal zoning. Chem Geol 364:20–32

Schaltegger U, Davies JHFL (2017) Petrochronology of zircon and baddeleyite in igneous rocks: Reconstructing magmatic processes at high temporal resolution. Rev Mineral Geochem 83:297–328

Schoen AH (1958) Correlation and the isotope effect for diffusion in crystalline solids. Phys Rev Lett 1:138–140

Schwandt CS, Cygan RT, Westrich HR (1998) Magnesium self-diffusion in orthoenstatite. Contrib Mineral Petrol 130:390–396

Schwinger S, Dohmen R, Schertl HP (2016) A combined diffusion and thermal modeling approach to determine peak temperatures of thermal metamorphism experienced by meteorites. Geochim Cosmochim Acta 191:255–276

Seitz S, Putlitz B, Baumgartner LP, Escrig S, Meibom A, Bouvier AS (2016) Short magmatic residence times of quartz phenocrysts in Patagonian rhyolites associated with Gondwana breakup. Geology 44:67–70

Shea T, Lynn KJ, Garcia MO (2015a) Cracking the olivine zoning code: Distinguishing between crystal growth and diffusion. Geology 43:935–938

Shea T, Costa F, Krimer D, Hammer JE (2015b) Accuracy of timescales retrieved from diffusion modeling in olivine: A 3D perspective. Am Mineral 100:2026–2042

Simon JI, Hutcheon ID, Simon SB, Matzel JEP, Ramon EC, Weber PK, Grossman L, DePaolo DJ (2011) Oxygen isotope variations at the margin of a CAI records circulation within the solar nebula. Science 331:1175–1178, doi: 10.1126/science.1197970

Singer BS, Dungan MA, Layne GD (1995) Textures and Sr, Ba, Mg, Fe, K, and Ti compositional profiles in volcanic plagioclase: Clues to the dynamics of calc-alkaline magma chambers. Am Mineral 80:776–798

Singer B, Costa F, Herrin JS, Hildreth W, Fierstein J (2016) The timing of compositionally zoned magma reservoirs and mafic 'priming' weeks before the 1912 Novarupta-Katmai rhyolite eruption. Earth Planet Sci Lett 451:125–137

Sio CK, Dauphas N (2016) Thermal and crystallization histories of magmatic bodies by Monte Carlo inversion of Mg–Fe isotopic profiles in olivine. Geology DOI: 10.1130/G38056.1

Sio CK, Dauphas N, Teng F-Z, Chaussidon M, Helz RT, Roskosz M (2013) Discerning crystal growth from diffusion profiles in zoned olivine by in-situ Mg-Fe isotopic analyses. Geochim Cosmochim Acta 123:302–321

Smith VC, Blundy JD, Arce JL (2009) A Temporal Record of Magma Accumulation and Evolution beneath Nevado de Toluca, Mexico, Preserved in Plagioclase Phenocrysts. J Petrol 50:405–426

Spandler C, O'Neill HStC (2010) Diffusion and partition coefficients of minor and trace elements in San Carlos olivine at 1300 °C with some geochemical implications. Contrib Mineral Petrol 159:791–818

Sundvall R, Skogby H, Stalder R (2009) Dehydration–hydration mechanisms in synthetic Fe-poor diopside. Eur J Mineral 21:17–26. doi:10.1127/0935-1221/2009/0021-1880

Stalder R, Behrens H (2006) D/H exchange in pure and Cr-doped enstatite: implications for hydrogen diffusivity. Phy Chem Mineral 33:601–611

Stalder R, Skogby H (2003) Hydrogen diffusion in natural and synthetis orthopyroxene. Phy Chem Mineral 30:12–19

Stalder R, Skogby H (2007) Dehydration mechanisms in synthetic Fe-bearing enstatite. Eur J Mineral 19:201–216

Streck MJ (2008) Mineral textures and zoning as evidence for open system processes. Rev Mineral Geochem 69:595–622

Tachibana S, Tamada S, Kawasaki H, Ozawa K, Nagahara H (2013) Interdiffusion of Mg–Fe in olivine at 1,400–1,600 °C and 1 atm total pressure. Phys Chem Minerals 40:511–519

Teng F-Z, Dauphas N, Helz RT (2008) Iron isotope fractionation during magmatic differentiation in Kilauea Iki lava lake. Science 320:1620–1622

Teng F-Z, Li W-Y, Ke S, Marty B, Dauphas N, Huang S, Wu F-Y, Pourmand A, (2010) Magnesium isotopic composition of the Earth and chondrites. Geochim Cosmochim Acta 74:4150–4166

Teng F-Z, Dauphas N, Helz R T, Gao S, Huang S (2011) Diffusion-driven magnesium and iron isotope fractionation in Hawaiian olivine. Earth Planet Sci Lett 308:317–324

Tepley FJ III, Davidson JP, Tilling RI, Arth JG (2000) Magma mixing, recharge and eruption histories recorded in plagioclase phenocrysts from El Chichón volcano, Mexico. J Petrol 41:1397–1411

Thoraval C, Demouchy S (2014) Numerical models of ionic diffusion in one and three dimensions: Application to dehydration of mantle olivine. Phys Chem Minerals 41:709–723. doi:10.1007/s00269-014-0685-x

Till CB, Vazquez JA, Boyce JW (2015) Months between rejuvenation and volcanic eruption at Yellowstone caldera, Wyoming. Geology 43:695–698

Tilling RI, Dvorak JJ (1993) Anatomy of a basaltic volcano. Nature 363:125–132

Tomascak PB, Magna T, Dohmen R (2016) Advances in Lithium Isotope Geochemistry. Springer Cham Heidelberg

Turner S, George R, Jerram DA, Carpenter N, Hawkesworth C (2003) Case studies of plagioclase growth and residence times in island arc lavas from Tonga and the Lesser Antilles, and a model to reconcile discordant age information. Earth Planet Sci Lett 214:279–294

Turner S, Costa F (2007) Measuring timescales of magmatic evolution. Elements 3:267–272

Van Orman JA, Cherniak DJ, Kita NT (2014) Magnesium diffusion in plagioclase: dependence on composition, and implications for thermal resetting of the $^{26}Al–^{26}Mg$ early solar system chronometer. Earth Planet Sci Lett 385:79–88

Van Orman JA, Krawczynski MA (2015) Theoretical constraints on the isotope effect for diffusion in minerals. Geochim Cosmochim Acta 164:365–381

VanTongeren JA, Kelemen PB, Hanghoj K (2008) Cooling rates in the lower crust of the Oman ophiolite: Ca in olivine, revisited. Earth Planet Sci Lett. 267:69–82, doi:10.1016/j.epsl.2007.11.034

Vogt K, Dohmen R, Chakraborty S (2015) Fe–Mg diffusion in spinel—New experimental data and a point defect model. Am Mineral 100:2112–2122

Walker AM, Hermann J, Berry AJ, O'Neill HSC (2007) Three water sites in upper mantle olivine and the role of titanium in the water weakening mechanism. J Geophys Res 112:B05211, doi:10.1029/2006JB004620

Wark DA, Hildreth W, Spear FS, Cherniak DJ, Watson EB (2007) Pre-eruption recharge of the Bishop magma system. Geology 35:235–238

Wasserburg GJ (1988) Diffusion of water in silicate melts. J Geol 96:363–367

Watson EB (1996) Surface enrichment and trace-element uptake during crystal growth. Geochim Cosmochim Acta 60:5013–5020

Watson EB, Baxter EF (2007) Diffusion in solid-Earth systems. Earth Planet Sci Lett 253:307–327

Watson EB, Cherniak DJ (2015) Quantitative cooling histories from stranded diffusion profiles. Contrib Mineral Petrol 169:1–14

Watson EB, Liang Y (1995) A simple model for sector zoning in slowly grown crystals: Implications for growth rate and lattice diffusion, with emphasis on accessory minerals in crustal rocks. Am Mineral 80:1179–1187

Watson EB, Müller T (2009) Non-equilibrium isotopic and elemental fractionation during diffusion-controlled crystal growth under static and dynamic conditions. Chem Geol 267:111–124

Welsch B, Faure F, Famin V, Baronnet A, Bachèlery P (2013) Dendritic crystallization: a single process for all the textures of olivine in basalts? J Petrol 54:539–574

Welsch B, Hammer JE, Hellebrand E (2014) Phosphorus zoning reveals dendritic architecture of olivine. Geology 42:867–870

Withers AC, Hirschmann MM (2008) Influence of temperature, composition, silica activity and oxygen fugacity on the H_2O storage capacity of olivine at 8 GPa. Contrib Miner Petrol 156:595–605

Zack T, Kooijman E (2017) Petrology and geochronology of rutile. Rev Mineral Geochem 83:443–467

Zellmer GF, Blake S, Vance D, Hawkesworth C, Turner S (1999) Plagioclase residence times at two island arc volcanoes (Kameni Islands, Santorini, and Soufriere, St. Vincent) determined by Sr diffusion systematics. Contrib Mineral Petrol 136:345–357

Zellmer GF, Clavero JE (2006) Using trace element correlation patterns to decipher a sanidine crystal growth chronology: An example from Taapaca volcano, Central Andes. J Volcanol Geotherm Res 156:291–301

Zhang Y (2010) Diffusion in minerals and melts: Theoretical background. Rev Mineral Geochem 72:5–59

Zhang YX, Stolper EM, Wasserburg GJ (1991) Diffusion of water in rhyolitic glasses. Geochim Cosmochim Acta 55:441–456

Zhao Y-H, Ginsberg SB, Kohlstedt DL (2004) Solubility of hydrogen in olivine: dependence on temperature and iron content. Contrib Miner Petrol 147:155–161, doi:10.1007/s00410-003-0524-4

Zhukova I, O'Neill H StC, Cambell IH, Kilburn MR (2014) The effect of silica activity on the diffusion of Ni and Co in olivine. Contrib Mineral Petrol 168:1029, DOI 10.1007/s00410-014-1029-z

RiMG Series

HISTORY OF RiMG

Volumes 1–38 were published as "*Reviews in Mineralogy*" (ISSN 0275-0279). Volumes 1-6 originally appared as "*Short Course Notes*" (no ISSN). The name was changed to "*Reviews in Mineralogy & Geochemistry*" (RiMG) (ISSN 1529-6466) starting with Volume 39. Paul Ribbe was sole editor for volumes 1–41. He was joined by Jodi Rosso as series editor for volumes 42–53 in the RiMG series submitted through the Geochemistry Society. With his retirement, Jodi Rosso became sole editor for volumes 54–79 in the RiMG series. With Jodi Rosso's move to Executive Editor of Elements magazine, Ian Swainson became Series Editor starting with volume 80.

HOW TO PUBLISH IN RiMG

RiMG volumes are based on topics that have been proposed and appoved by the MSA Council or Geochemical Society Board of Directors. If you have an idea for a future RiMG volume, or a short course accompanied by a RiMG volume, you should read the Short Course Guide which describes how to develop and propose a topic for consideration for either case. Proposals should be submitted to the Short Course Committee (http://www.minsocam.org/msa/SC/SCCommittee.html). Contributions to an appoved volume are by invitation only.

A listing of the previous volume numbers and their volume editors is below. Selecting the titles at http://www.minsocam.org/msa/RIM/index2.html gives you to a detailed description of each volume, the table of contents, and any errata or supplementary material

Previous volumes can be ordered from https://msa.minsocam.org/orders.html.

Volume 23: *Mineral–Water Interface Geochemistry*

| 1990 | ISBN 0-939950-28-6;
ISBN13 978-0-939950-28-7 | MF Hochella, Jr., AF White | i-xvi + 603 pp |

Volume 22: *The Al₂SiO₅ Polymorphs*

| 1990 | ISBN 0-939950-27-8;
ISBN13 978-0-939950-27-0 | DM Kerrick | i-xii + 406 pp |

Volume 21: *Geochemistry and Mineralogy of Rare Earth Elements*

| 1989 | ISBN 0-939950-25-1;
ISBN13 978-0-939950-25-6 | BR Lipin, GA McKay | i-x + 348 pp |

Volume 20: *Modern Powder Diffraction*

| 1989 | ISBN 0-939950-24-3;
ISBN13 978-0-939950-24-9 | DL Bish, JE Post | i-xii + 369 pp |

Volume 19: *Hydrous Phyllosilicates (exclusive of micas)*

| 1988 | ISBN 0-939950-23-5;
ISBN13 978-0-939950-23-2 | SW Bailey, | i-xiii + 725 pp |

Volume 18: *Spectroscopic Methods in Mineralogy and Geology*

| 1988 | ISBN 0-939950-22-7;
ISBN13 978-0-939950-22-5 | FC Hawthorne | i-xvi + 512 pp |

Volume 17: *Thermodynamic Modeling of Geological Materials: Minerals, Fluids and Melts*

| 1987 | ISBN 0-939950-21-9;
ISBN13 978-0-939950-21-8 | ISE Carmichael, HP Eugster | i-xiv + 499 pp |

Volume 16: *Stable Isotopes in High Temperature Geological Processes*

| 1986 | ISBN   0-939950-20-0;
ISBN13 978-0-939950-20-1 | JW Valley, HP Taylor, Jr., JR O'Neil | i-xvi + 570 pp |

Volume 15: *Mathematical Crystallography*

| 1985 | ISBN 0-939950-19-7;
ISBN13 978-0-939950-19-5 | MB Boisen, Jr., GV Gibbs | i-xii + 460 pp |

Volume 14: *Microscopic to Macroscopic*

| 1985 | ISBN 0-939950-18-9;
ISBN13 978-0-939950-18-8 | SW Kieffer, A Navrotsky | i-x + 428 pp |

Volume 13: *Micas*

| 1984 | ISBN   0-939950-17-0;
ISBN13 978-0-939950-17-1 | SW Bailey | i-xii + 584 pp |

Volume 12: *Fluid Inclusions*

| 1984 | ISBN 0-939950-16-2;
ISBN13 978-0-939950-16-4 | Edwin Roedder | i-vi + 646 pp |

Volume 11: *Carbonates: Mineralogy and Chemistry*

| 1983,
1990 | ISBN 0-939950-15-4;
ISBN13 978-0-939950-15-7 | RJ Reeder | i-xii + 399 pp |

Volume 10: *Characterization of Metamorphism through Mineral Equilibria*

| 1982 | ISBN 0-939950-12-X;
ISBN13 978-0-939950-12-6 | JM Ferry | i-xiv + 397 pp |

Volume 9B: *Amphiboles and Other Hydrous Pyriboles—Mineralogy*

| 1982 | ISBN 0-939950-11-1;
ISBN13 978-0-939950-11-9 | DR Veblen, PH Ribbe | i-x + 390 pp |

Volume 9A: *Amphiboles: Petrology and Experimental Phase Relations*

| 1981 | ISBN 0-939950-10-3;
ISBN13 978-0-939950-10-2 | DR Veblen, PH Ribbe | i-xii + 372 pp |

Volume 8: *Kinetics of Geochemical Processes*

1981   ISBN 0-939950-08-1;   AC Lasaga, RJ Kirkpatrick   i-x + 398 pp
      ISBN13 978-0-939950-08-9

Volume 7: *Pyroxenes*

1980   ISBN 0-939950-07-3;   CT Prewitt   i-x + 525 pp
      ISBN13 978-0-939950-07-2

Volume 6: *Marine Minerals*

1979   ISBN 0-939950-06-5;   RG Burns   i-x + 380 pp
      ISBN13 978-0-939950-06-5

Volume 5: *Orthosilicates*

1980   ISBN 0-939950-13-8;   RG Burns   i-xii + 450 pp
      ISBN13 978-0-939950-13-3

Volume 4: *Mineralogy and Geology of Natural Zeolites*

1977   ISBN 0-939950-04-9;   FA Mumpton   i-xii + 233 pp
      ISBN13 978-0-939950-04-1

Volume 3: *Oxide Minerals*

1976   ISBN 0-939950-03-0;   D Rumble, III   i-3 + 706 pp
      ISBN13 978-0-939950-03-4

Volume 2: *Feldspar Mineralogy*

1975   ISBN 0-939950-14-6;   PH Ribbe   i-vii + 362 pp
1983   ISBN13 978-0-939950-14-0

Volume 1: *Sufide Mineralogy*

1974   ISBN 0-939950-01-4;   PH Ribbe   i-v + 301 pp
      ISBN13 978-0-939950-01-0